Vascular Innervation and Receptor Mechanisms

Vascular Innervation and Receptor Mechanisms

New Perspectives

Edited by

Lars Edvinsson and Rolf Uddman
Department of Experimental Research
University of Lund
Malmö General Hospital
Malmö, Sweden

ACADEMIC PRESS, INC.
Harcourt Brace Jovanovich, Publishers
San Diego New York Boston London Sydney Tokyo Toronto

Cover photograph: Whole mount preparation of a cat middle cerebral artery that contains a network of varicose nerve fibers displaying substance P immunoreactivity. (Adapted from Figure 12.4; courtesy of S. Gulbenkian and J. Wharton.)

This book is printed on acid-free paper. ♾

Academic Press, Inc.
1250 Sixth Avenue, San Diego, California 92101-4311

United Kingdom Edition published by
Academic Press Limited
24–28 Oval Road, London NW1 7DX

Library of Congress Cataloging-in-Publication Data

Vascular innervation and receptor mechanisms : new perspectives /
edited by Lars Edvinsson and Rolf Uddman.
p. cm.
Includes bibliographical references and index.
ISBN 0-12-232350-5
1. Vasomotor system. 2. Chemoreceptors. 3. Blood--Circulation--Regulation. 4. Endothelins. 5. Neuropeptides. I. Edvinsson, Lars. II. Uddman, Rolf.
[DNLM: 1. Cardiovascular System--innervation. 2. Receptors, Endogenuous Substances--physiology. WG 102 V331]
QP109. V36 1993
612'. 18--dc20
DNLM/DLC
for Library of Congress 92-49263
CIP

PRINTED IN THE UNITED STATES OF AMERICA
93 94 95 96 97 98 MM 9 8 7 6 5 4 3 2 1

Contents

PART II
The Sympathetic System

PART IV
Sensory Afferent System

Part V
Clinical Application

Contributors

Numbers in parentheses indicate the pages on which the authors' contributions begin.

T. BENNETT (**41**), Department of Physiology and Pharmacology, University of Nottingham Medical School, Queen's Medical Centre, Nottingham NG7 2UH, United Kingdom.

G. BURNSTOCK (**1**), Department of Anatomy and Developmental Biology, and Centre for Neuroscience, University College London, London WC1E 6BT, United Kingdom.

JOHN CLARKE (**367**), Department of Cardiovascular Medicine, Hammersmith Hospital, London W12 0NN, United Kingdom.

T. COWEN (**25**), Department of Anatomy, Royal Free Hospital, School of Medicine, London NW3 2PF, United Kingdom.

PEDRO D'ORLÉANS-JUSTE (**281**), Department of Pharmacology, Medical School, University of Sherbrooke, Sherbrooke, JIH 5N4 Québec, Canada.

CARL DAHLÖF (**159**), GLF Medical Research Centre AB, Sociala Huset, S-411 17 Göthenburg, Sweden.

LARS EDVINSSON (**175, 263, 351, 399**), Department of Experimental Research, University of Lund, Malmö General Hospital, S-214 01 Malmö, Sweden.

ANTHONY EGLEZOS (**135**), National Research Institute of Gerontology and Geriatric Medicine, North West Hospital, Parkville, Melbourne, Victoria, Australia.

DAVID ERLINGE (**351**), Department of Internal Medicine, University Hospital of Lund, 211 85 Lund, Sweden.

S. M. GARDINER (**41**), Department of Physiology and Pharmacology, University of Nottingham Medical School, Queen's Medical Centre, Nottingham NG7 2UH, United Kingdom.

PETER J. GOADSBY (**415**), Department of Neurology, The Prince Henry Hospital, Sydney NSW 3026, Australia.

SÉRGIO GULBENKIAN (**247**), Department of Cell Biology, Gulbenkian Institute of Science, 2781 Oeiras Codex, Portugal.

ROLF HÅKANSON (**203**), Department of Pharmacology, University of Lund, Lund, Sweden.

PAVO HEDNER (**351**), Department of Internal Medicine, University Hospital of Lund, 221 85 Lund, Sweden.

THOMAS HEDNER (**175, 351**), Department of Clinical Pharmacology, Sahlgrenska University Hospital, S-413 45 Göteborg, Sweden.

ROBERT D. HELME (**135**), National Research Institute of Gerontology, and Geriatric Medicine, North West Hospital, Parkville, Melbourne, Victoria 3052, Australia.

PETER HOLZER (**299**), Department of Experimental and Clinical Pharmacology, University of Graz, A-8010 Graz, Austria.

M. HUANG (**221**), Department of Medicine, University of Calgary, Calgary, T2N 4N1 Alberta, Canada.

INGER JANSEN (**263**), Department of Experimental Research, Malmö General Hospital, S-214 01 Malmö, Sweden.

ROAR JUUL (**399**), Department of Neurosurgery, University Hospital, 7006 Trondheim, Norway.

CHARLES KENNEDY (**187**), Department of Physiology and Pharmacology, University of Strathclyde, Glasgow G1 1XW, Scotland, United Kingdom.

ZEINAB KHALIL (**135**), National Research Institute of Gerontology and Geriatric Medicine, North West Hospital, Parkville, Melbourne, Victoria, Australia.

PIA LINTON-DAHLÖF (**159**), GLF Medical Research Centre AB, Sociala Huset, S-411 17 Göthenburg, Sweden.

OVE LUNDGREN (**449**), Department of Physiology, University of Göteborg, S-413 90 Göteborg, Göteborg, Sweden.

ANDERS LUTS (**203**), Department of Medical Cell Research, University of Lund, Lund, Sweden.

TOMOH MASAKI (**87**), Department of Pharmacology, Kyoto University, Sakyo-ku, Kyoto 606, Japan.

LEIF NILSSON (**263**), Department of Surgery, University Hospital of Lund, S-221 85 Lund, Sweden.

JOHN PERNOW (**325**), Departments of Pharmacology and Cardiology, Karolinska Hospital, Karolinska Institute, S-104 01 Stockholm, Sweden.

DOMENICO REGOLI (**281**), Department of Pharmacology, Medical School, University of Sherbrooke, Sherbrooke, JIH 5N4 Québec, Canada.

O. RORSTAD[1] (**221**), Department of Medicine, University of Calgary, Calgary, T2N 4N1 Alberta, Canada.

NOURREDINE ROUISSI (**281**), Department of Pharmacology, Medical School, University of Sherbrooke, Sherbrooke, JIH 5N4 Québec, Canada.

GABOR M. RUBANYI[2] (**107**), Research Center, Schering AG, 1000 Berlin 65, Germany.

H. SHIRAHASE (**221**), Department of Medicine, University of Calgary, Calgary, T2N 4N1 Alberta, Canada.

XIANG-YING SUN (**175**), Department of Clinical Pharmacology, Sahlgrenska University Hospital, S-413 45 Göteborg, Sweden.

1. *Present address:* Department of Medicine, Calgary General Hospital, Calgary, Alberta T2E 0A1, Canada.
2. *Present address:* Vascular Biology Research, Berlex Biosciences, Alameda California 94501.

FRANK SUNDLER (**203**), Department of Medical Cell Research, University of Lund, Lund, Sweden.

THOMAS THULIN (**351**), Department of Internal Medicine, University Hospital of Lund, S-221, 85 Lund, Sweden.

ROLF UDDMAN (**203**), Department of Otorhinolaryngology, University of Lund, Malmö General Hospital, S-214 01 Malmö, Sweden.

STIG VALDEMARSSON (**351**), Department of Internal Medicine, University Hospital of Lund, S-221 85 Lund, Sweden.

JOHN WHARTON (**247**), Department of Histochemistry, Royal Postgraduate Medical School, Hammersmith Hospital, London W12 0NN, United Kingdom.

J. G. WIDDICOMBE (**425**), Department of Physiology, St. George's Hospital Medical School, London SW17 0RE, United Kingdom.

Preface

Interest in the control of vascular resistance in different parts of the circulation has increased rapidly over the past decade. Blood vessel dysfunction is responsible for many of the manifestations of a number of widespread diseases, such as hypertension, coronary heart disease, stroke, and migraine. This recognition has triggered considerable effort in basic research into vascular smooth-muscle function and its control by autonomic nerves, endothelium, and hormones. As a result, an abundance of new findings in vascular science has accumulated during the last decades; a number of symposia have covered some of the more recent developments. This volume reviews the rapidly growing area of neurobiological research, dealing with structure, function, and pathophysiology of the circulatory system. The contributors are all actively involved in the study of these areas and have all reviewed the results of their own work as well as contributions by other scientists in the field. From their efforts a picture emerges of increasing complexity in the local regulation of cardiovascular tone.

It has not been our primary goal to give a balanced presentation of all fields in the pathophysiology of the circulatory system. Rather, we have aimed at recent pertinent results in which neuropeptides may play a role. Particular attention has been paid to those fields of research that are most intensely studied today. The volume is organized to provide the basic background, including a brief account on recent research in the structural function of the endothelium. Using this background the peptidergic innervation is analyzed, followed by a section on the current knowledge on peptides in vascular control in various disorders.

We are very grateful to the contributors to this volume for the effort and care they have put into completing their chapters. This volume provides a synopsis of current trends in the studies of vascular innervation and receptor mechanisms and illustrates the relationship with endothelial, hormonal, and pathophysiological control. Many colleagues in the field, both basic scientists and clinicians, will find this effort useful in the understanding of a potentially important future arena for drug design and treatment of vascular disorders.

Lars Edvinsson
Rolf Uddman

CHAPTER 1

Introduction: Changing Face of Autonomic and Sensory Nerves in the Circulation

G. BURNSTOCK

I. Introduction

Studies on neurohumoral control of the vasculature focused for many years on the catecholamines released from sympathetic perivascular nerves and from the adrenal medulla into the bloodstream. In recent years, new and improved techniques in immunohistochemistry, electron microscopy, electrophysiology, and pharmacology have led to a wealth of discoveries that have profoundly reshaped our understanding of the perivascular nervous system (Burnstock and Griffith, 1988; Burnstock, 1990a; Mione *et al.*, 1990; Mulvany and Aalkjaer, 1990; Owman, 1990).

In addition to the classical transmitters noradrenaline (NA) and acetylcholine (ACh), other transmitters have been identified in perivascular nerves, including 5-hydroxytryptamine (5-HT), adenosine 5′-triphosphate (ATP), and a number of peptides. Neuromodulatory mechanisms have been recognized, including prejunctional inhibition or enhancement of transmitter release, postjunctional modulation of transmitter action, and the secondary involvement of locally synthesized agents, such as histamine, bradykinin, and prostaglandins. The presence of more than one transmitter substance in some nerves, termed cotransmission, is now also widely recognized (Burnstock, 1976; Burnstock, 1990b). While the concept that antidromic impulses in sensory nerve collaterals form the basis of "axon reflex" vasodilatation of skin vessels was described many years ago (Lewis, 1927), it is now recognized that this mechanism is widespread and that sensory-motor perivascular nerves play an important role in the control of many blood vessels (Maggi and Meli, 1988; Burnstock, 1990b). Subpopulations of endothelial cells have been shown to store and release a variety of substances, including ACh, substance P (SP), ATP, 5-HT, vasopressin (VP), and angiotensin II (AII), that, on release during changes in blood flow or hypoxia, act on receptors on endothelial cells and lead to the production of endothelium-derived relaxing factor (EDRF) (identified as nitric oxide) which, in turn, produces vasodilatation. Endothelium-derived contracting factors, such as endothelin, may also be released. There appears to be a resting dynamic balance

Vascular Innervation and
Receptor Mechanisms: New Perspectives

between endothelium-derived vasodilator tone and sympathetic vasoconstrictor tone, which is altered under different physiological and pathophysiological circumstances (Burnstock, 1988a; Lincoln *et al.*, 1991).

The plasticity of perivascular nerves and endothelial cells is beginning to be investigated. Changes in expression of transmitters and receptors occur not only during development and aging but also in fully mature adults under the influence of hormones and growth factors and in the nerves that remain following trauma and surgery and in a variety of disease situations (Burnstock, 1990c).

II. The Vascular Neuromuscular Junction

The vascular neuroeffector junction consists of varicose nerve fibers within a plexus at the adventitial-medial border. Transmitter is released "en passage" from varicosities to reach vascular smooth muscle cells that are in electrical continuity with each other via gap junctions (Burnstock, 1986a,b). The varicosities do not have a fixed relationship with particular smooth muscle cells, and the junctional cleft varies between about 60 nm in some arterioles and small arteries and as much as 2 μm in some large elastic arteries; muscle cells do not have postjunctional specializations. This means that the vascular neuromuscular junction differs in a significant way from "synapses" at the motor end plate in striated muscle or within ganglia, where there is a fixed relationship with both pre- and postjunctional specialization. The variable geometry of the vascular neuromuscular junction means that neuromodulation is an important feature. A neuromodulator is defined as a substance that modifies the process of neurotransmission. It may act as a prejunctional modulator by decreasing or increasing the amount of transmitter released by the nerve varicosity, or it may act as a postjunctional modulator by altering the time course or extent of action of the neurotransmitter. Neuromodulators may be circulating neurohormones, local agents, or neurotransmitter substances released from other nerves nearby or even from the same nerve varicosity.

III. Multiplicity of Transmitters; Cotransmission

For over 50 years, the only transmitters considered in perivascular nerves were NA and ACh. Since the discovery of nonadrenergic, noncholinergic transmission in the early 1960s, over 12 new chemical messengers have been identified, including monoamines, purines, amino acids, and polypeptides (see Table 1).

The concept of cotransmission, namely that nerves synthesize, store, and release more than one transmitter, was first proposed in 1976 (Burnstock, 1976), bringing together various hints from the earlier literature on adrenal medullary cells and sympathetic nerves, invertebrate neurons, and developing nerves models. This hypothesis is now generally accepted (Cuello, 1982; Osborne, 1983; Chan-Palay and Palay, 1984; Hökfelt *et al.*, 1986; Bartfai *et al.*, 1988; Burnstock, 1990b; Kupfermann, 1991).

Table 1 Established and Putative Transmitters in Perivascular Nerves

Noradrenaline (NA)
Acetylcholine (ACh)
Adenosine 5′-triphosphate (ATP)
5-Hydroxytryptamine (5-HT)
Dopamine (DA)
Enkephalin-dynorphin (ENK-DYN)
Vasoactive intestinal polypeptide (VIP)-Peptide histidine isoleucine (PHI)
Substance P (SP)
Gastrin-releasing peptide (GRP)
Somatostatin (SOM)
Neurotensin (NT)
Vasopressin-like peptide (VP)
Cholecystokinin-gastrin (CCK-GAS)
Neuropeptide Y-pancreatic polypeptide (NYP-PPP)
Galanin (GAL)
Angiotensin (ANG)
Adrenocorticotrophic hormone (ACH)
Calcitonin gene-related peptide (CGRP)

While at first sight the multiplicity of transmitters released in various computations from different perivascular nerves appears formidable and unnecessary, a pattern is now emerging that clarifies the situation. This is the idea that autonomic nerves have a "chemical coding" (i.e., individual neurones contain a particular combination of transmitter substances, have processes that project to identifiable target sites, and have defined central connections). This concept has been developed most fully for the enteric nervous system (see Furness and Costa, 1987), but it also applies to perivascular nerves.

IV. Sympathetic Perivascular Nerves

There is now a substantial body of evidence to show that NA and ATP act as cotransmitters, being released from sympathetic nerves in variable proportions depending on the tissue and species (Burnstock, 1990d; see Kennedy, Chapter 9, this volume). Most of the early and more detailed studies were made on the vas deferens, but many studies of sympathetic cotransmission, involving ATP and NA, have now also been carried out on a number of different blood vessels, including rat tail artery (Sneddon and Burnstock, 1984; Bao *et al.*, 1989), rabbit ear artery (Kennedy and Burnstock, 1985; Suzuki, 1985; Saville and Burnstock, 1988), dog basilar artery (Muramatsu and Kigoshi, 1987), mesenteric artery (Ishikawa, 1985; Von Kügelgen and Starke, 1985; Muramatsu, 1986; Ramme *et al.*, 1987; Muir and Wardle, 1988; Machalay *et al.*, 1988), rabbit pulmonary artery (Katsuragi and Su, 1982), guinea pig and rabbit saphenous artery (Cheung and Fujioka, 1986; Burnstock and Warland, 1987; Warland and Burnstock, 1987), and rabbit hepatic artery

(Brizzolara and Burnstock, 1990). Sympathetic cotransmission involving NA and ATP has also been shown in the circulation of skeletal muscle (Shimada and Stitt, 1984), cat intestine (Taylor and Parsons, 1989), kidney (Schwartz and Malik, 1989), dog skin (Flavahan and Vanhoutte, 1986), and in the pithed rat (Bulloch and McGrath, 1988; Schlicker *et al.*, 1989).

Evidence for purinergic cotransmission includes block of the prazosin-resistant component of the response to sympathetic nerve stimulation by the ATP antagonist arylazido aminoproprionyl-ATP ($ANAPP_3$) or by selective desensitization of the P_{2X}-purinoceptor by α,β-methylene ATP; release of ATP during nerve stimulation, which is prevented by tetrodotoxin, guanethidine, or by destruction of sympathetic nerves by 6-hydroxydopamine, but is unaffected by selective depletion of NA by reserpine; and mimicry of excitatory junction potentials (EJPs) by ATP but not by NA.

One of the problems raised about sympathetic cotransmission in blood vessels is that in some arteries, while the EJPs are completely blocked by ATP antagonists, the mechanical response is barely affected. The explanation for this discrepancy is partly that the purinergic component is optimal with short bursts of sympathetic nerve stimulation (1 s or less), while the traditional period of nerve stimulation in experimental *in vitro* preparations is 30 s or more, under these conditions the NA component dominates the mechanical responses (Kennedy *et al.*, 1986). This is because ATP acts via P_{2X}-purinoceptors that lead to contractions resulting from electromechanical coupling involving voltage-dependent Ca^{2+} channels, whereas NA acts via receptor-operated Ca^{2+} channels. A second reason is that the proportion of the cotransmitters NA and ATP varies considerably between different vessels. For example, ATP is the major component of sympathetic cotransmission in the rabbit saphenous artery and a significant component in mesenteric arteries, but it appears to be a relatively minor component in rabbit ear artery and rat tail artery where the relationship between EJPs and mechanical responses is more difficult to demonstrate. The purinergic component of sympathetic cotransmission is selectively affected by the dihydropyridines, nifedipine, and Bay K 8644 (Stone, 1981; Mackenzie *et al.*, 1988).

Recent studies in our laboratory have shown that, in rabbit coronary vessels, in contrast to other vessels where NA and ATP cause synergistic constriction via α_1-adrenoceptors and P_{2X}-purinoceptors, respectively, the predominant effect of ATP is vasodilatation via P_{2Y}-purinoceptors (Corr and Burnstock, 1991). Since in this vessel the predominant effect of NA is vasodilatation via β-adrenoceptors, this is consistent with the synergism that appears to be characteristic of cotransmission.

Neuropeptide Y (NPY) is also stored in and released from sympathetic nerves (Lundberg *et al.*, 1983; Lundberg *et al.*, 1984), but in the vas deferens and many vessels it has little direct postjunctional action (Stjärne *et al.*, 1986; Pernow *et al.*, 1986). However, it has potent prejunctional actions reducing the release of NA and ATP and postjunctional actions enhancing the actions of NA and ATP (Wahlestedt *et al.*, 1986; Stjärne, 1989; Saville *et al.*, 1990). NPY has direct vasoconstrictor actions in some vessels, for example, those in heart, brain, spleen, and skeletal muscle. In the spleen, NA release is favored by continuous stimulation at low frequencies, while NPY release is optimal with high-frequency intermittent bursts of stimulation (Lundberg *et al.*, 1986).

The geometry of particular sympathetic neuromuscular junctions appears to influence the type of neuromodulation (Fig. 1); with wide junctional clefts, postjunctional potentiation by NPY dominates, while narrow clefts favor prejunctional inhibition by NPY (Burnstock, 1990e).

5-HT-immunofluorescent nerves have been localized in a number of vessels (Griffith and Burnstock, 1983; Burnstock and Griffith, 1988). However, it seems that, for the most part, 5-HT is not synthesized and stored in separate nerves but is taken up, stored in, and released as a "false transmitter" from sympathetic nerves (Jackowski *et al.*, 1989).

In a study of blood vessels in guinea pig skin, a differential chemical coding has been demonstrated (i.e., while sympathetic nerves in the distributing arteries contain NPY and NA, in the smaller arteries they contain dynorphin as well as NPY and NA, while in precapillary arterioles only dynorphin and NA are present) (Gibbins and Morris, 1990).

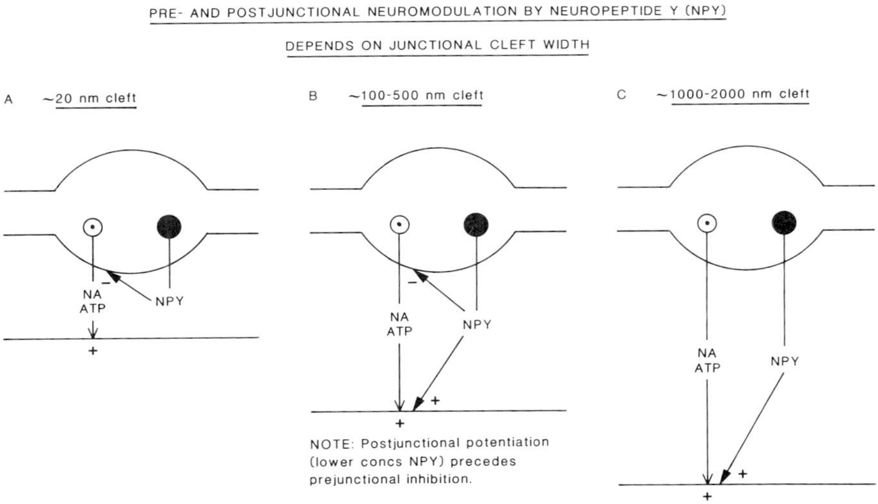

Figure 1 A diagram illustrating that pre- and/or postjunctional modulation of sympathetic cotransmission by NPY depends on the junctional cleft width. (A) Close (20 nm) anterior neuromuscular cleft as seen for example in the vas deferens, where prejunctional inhibition of transmitter release by NPY is dominant. (B) Medium-size junctional cleft (100–500 nm) characteristic of many blood vessels, where postjunctional potentiation of transmitter action occurs with low concentrations of NPY and, later, prejunctional modulation as the concentration of NPY in the cleft increases during transmission. (C) Wide (1000–2000 nm) cleft typical of large elastic arteries, where postjunctional modulation by NPY is dominant. (From Burnstock, 1990e.)

V. Parasympathetic Perivascular Nerves

In the salivary gland of the cat, vasoactive intestinal polypeptide (VIP) appears to be stored together with ACh in parasympathetic nerves, probably in separate vesicles. During low-frequency stimulation, ACh is released to increase salivary secretion from acinar cells and also to elicit some minor dilatation of blood vessels in the gland (Bloom and Edwards, 1980; Lundberg, 1981). VIP is released from the same nerves, especially at high stimulation frequencies to produce marked dilatation and, although it has no direct effect on acinar cells, it acts as a neuromodulator to enhance substantially the postjunctional effect of ACh on acinar cell secretion and increase the release of ACh from the nerve varicosities via prejunctional receptors. VIP is a potent vasodilator of many vessels, notably penile vessels; it appears to play a major role in erection, but these nerves are severely damaged in diabetic neuropathy (Crowe *et al.*, 1983). In a study of ageing of cerebral vessels of the rat, VIP-containing perivascular vasodilator nerves have been shown to *increase* markedly in old age, in contrast to NA- and 5-HT-containing excitatory nerves that decline with age (Mione *et al.*, 1988a).

VI. Sensory-Motor Perivascular Nerves

Sensory nerves have been shown to store and release SP (Furness *et al.*, 1982; Duckles and Buck, 1982; Barja *et al.*, 1983), calcitonin gene-related peptide (CGRP) (Lundberg *et al.*, 1985; Terenghi *et al.*, 1986), and ATP (Jahr and Jessel, 1983; Salt and Hill, 1983; Fyffe and Perl, 1984; Krishtal *et al.*, 1988). CGRP and SP have been shown to coexist in sensory nerve terminals in many perivascular nerves (Gibbins *et al.*, 1985; Uddman *et al.*, 1985) and, with the use of colloidal gold particles of different sizes, they have been shown to coexist in the same large granular vesicles (Gulbenkian *et al.*, 1986). By analogy with other systems, it seems likely that ATP coexists in different proportions with these two peptides, perhaps cooperating in axon reflex vasodilatation (Burnstock, 1977). Since the role of these varicose nerve fibers during the axon reflex to many organs (Maggi and Meli, 1988) is motor rather than sensory, I have termed them "sensory-motor nerves" to distinguish them from the other subpopulation of afferent fibers that have an entirely sensory role and whose terminals contain few vesicles and a predominance of mitochondria (Fig. 2) (Burnstock, 1986b).

VII. Perivascular Nerves Arising from Intramural Neurones

Little is known about the physiological roles or the pharmacology of intrinsic neurons of the heart because it is so difficult to study them *in situ*. However, a novel culture preparation from the atria of newborn guinea pigs has been developed in our laboratory to study the intrinsic innervation of the heart under conditions of unequivocal extrinsic denervation (Hassall and Burnstock, 1984, 1986). Some of these

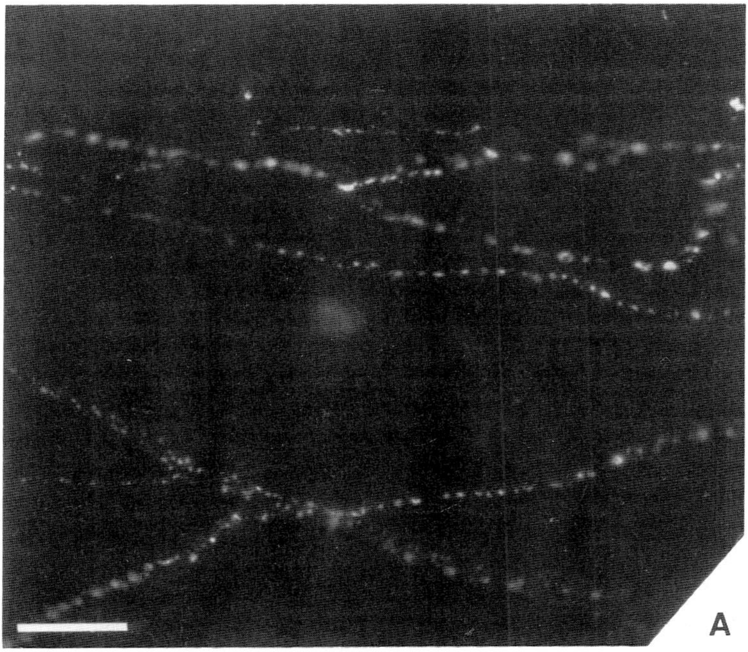

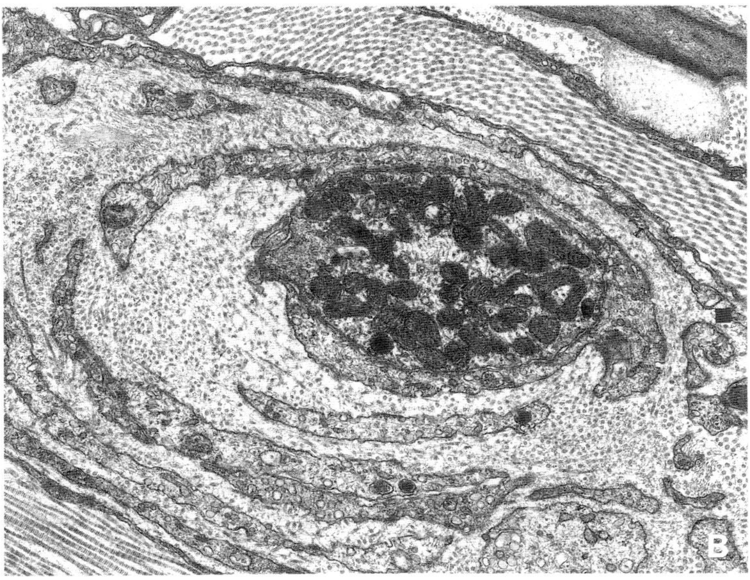

Figure 2 (A) Calcitonin gene-related peptide-like immunoreactivity in varicose, perivascular sensory-motor nerve fibers in the rabbit central ear artery. Calibration bar: 30 μm. (B) Adventitia of the external iliac artery of an adult guinea pig. An axon packed with mitochondria and a few vesicles is surrounded by cell processes and collagen fibrils, and probably represents a specialized nerve ending (possibly a sensory corpuscle). A muscle cell of the tunica media is visible at the top right corner, close to an elastic fiber. Electron micrograph ×12,000. (Courtesy of G. Gabella, unpublished.)

neurons show immunofluorescence for NPY, some for 5-HT, and some for variable mixtures of both transmitter substances. Projections of these neurons *in situ* form perivascular plexuses in small resistance coronary vessels (Corr *et al.*, 1990). Both NPY and 5-HT are potent vasoconstrictors of coronary vessels and may have synergistic actions.

Few studies have been carried out on the projections of intrinsic neurones to blood vessels in other organs, but intrinsic enteric neurones are known to supply some vessels in the gut and mesentery and it is well known that monoamine-containing neurones in the brain contribute to the innervation of some cerebral vessels (Burnstock and Griffith, 1988; Uddman and Edvinsson, 1989; Edvinsson, 1991).

VIII. Neural–Endothelial Interactions in Local Control of Vascular Tone

Since 1980, when Furchgott and Zawadzki first reported that the vasodilatation response to ACh requires the presence of an intact endothelium, the role of the endothelium in the regulation of vascular tone has attracted considerable interest (Vanhoutte and Rimele, 1983; Furchgott, 1984). Action on endothelial receptors by a number of vasoactive substances stimulates the production of EDRF or endothelium-derived constricting factors (EDCF) and/or prostaglandins. These subsequently modify vascular tone by causing contraction or relaxation of the vascular smooth muscle. Much of the current research has centred on the characterization of the endothelial factors. EDRF has been identified as nitric oxide (Ignarro *et al.*, 1986; Palmer *et al.*, 1987), while the peptide endothelin is considered one of the constricting factors (Yanagisawa *et al.*, 1988). It should be noted that there is considerable heterogeneity in the endothelium-dependent responses of mammalian blood vessels, with variations between arteries and veins and between different vascular beds.

In addition to ACh, endothelium-dependent vasodilatation has been shown to occur in response to ATP, adenosine 5′-diphosphate, arachidonic acid, SP, neurokinin A, 5-HT, bradykinin, histamine, neurotensin, VP, AII, and thrombin (Mione *et al.*, 1990). Some of the vasoactive substances that require the endothelium to produce vasodilatation are also neurotransmitters that, on release from perivascular nerves, cause vasoconstriction. Pharmacological studies have therefore investigated the receptors involved in mediating the two opposing responses. In many cases, they have demonstrated that different subtypes of the receptors to such vasoactive substances occur on the endothelium and on the vascular smooth muscle. An example of this is ATP, which acts on P_2-purinoceptors. Studies of the rank order of potency of the agonists ATP, α,β-methylene ATP, and 2-methylthio-ATP, together with the effects of antagonism by reactive blue 2 and desensitization by α,β-methylene ATP, have resulted in the subclassification of P_2-purinoceptors. Thus, P_{2X}-purinoceptors are present on the vascular smooth muscle and are acted on by ATP released as a cotransmitter from perivascular sympathetic nerves to produce vasoconstriction, while in many vessels ATP causes vasodilatation via P_{2Y}-purinoceptors on the endothelial cells (Burnstock and Kennedy, 1985; Burnstock, 1988b).

While transmitter substances released from perivascular nerves can conceivably reach endothelial cell receptors in the microvasculature (although these vessels are often sparsely innervated), it is neither likely nor desirable that the same neurotransmitter can diffuse through the media and basal lamina of a blood vessel (without degradation) to act on endothelial receptors to produce the opposite effect. In order to establish that endothelium-dependent responses in larger vessels have a role to play in the control of vascular tone in the intact organism, it is necessary to identify the source of the vasoactive substances that act on the endothelial receptors. For some substances, a readily available source is the blood. Indeed, it has been suggested that the endothelium may have a protective role without which substances released from aggregating platelets, such as 5-HT and ADP, would produce vasoconstriction. Furthermore, platelet aggregation at sites of endothelial cell damage could provide a mechanism for vasospasm in coronary vessel disease (Houston *et al.*, 1985). In the case of ACh and SP, however, circulating levels are low because of their rapid breakdown. The possibility that endothelial cells themselves may be the source of such substances was first proposed in 1985, when Parnavelas *et al.* (1985) reported that choline acetyltransferase (ChAT), the enzyme responsible for the synthesis of ACh, could be localized in endothelial cells lining capillaries and small vessels in the rat cortex. Since this time, using the same technique of immunocytochemical staining combined with electron microscopy, ChAT, SP, 5-HT, VP, and AII have all been localized in endothelial cells from a variety of blood vessels (Loesch and Burnstock, 1988; Burnstock *et al.*, 1988; Milner *et al.*, 1989; Lincoln *et al.*, 1990). In addition, SP levels have been measured in endothelium isolated from cerebral arteries and aorta (Linnik and Moskowitz, 1989). Other workers have also demonstrated that endothelial cells have the capability of synthesizing AII and histamine (Kifor and Dzau, 1987; Hollis and Rosen, 1972).

Experiments have been carried out to investigate whether physiological stimuli can cause the release of vasoactive substances from their endothelial stores. 5-HT, ATP, SP, and ACh, all of which are present in coronary endothelial cells, have been shown to be released during hypoxia of the Langendorf heart preparation from the rat (Paddle and Burnstock, 1974; Burnstock *et al.*, 1988; Milner *et al.*, 1989). Hypoxic vasodilatation has been shown to be endothelium dependent (Hopwood *et al.*, 1989). In the perfused rat hindlimb, increased flow causes the release of SP, which is unaffected by removal of SP-containing sensory nerve fibers with capsaicin; however, after removal of the endothelium by perfusion with air bubbles (Ralevic *et al.*, 1989), increased flow no longer induced the release of SP. SP has been localized by the use of electron-immunolabeling methods in the endothelial cells of the rat femoral artery (Loesch and Burnstock, 1988). SP has also been shown to be released from columns of endothelial cells grown on microcarrier beads following increased flow (Milner *et al.*, 1990a). This supports the view that the source of the SP is the endothelial cells and suggests that SP and other vasoactive substances within endothelial cells, by their release, contribute to flow-induced vasodilatation, which is known to be an endothelium-dependent response. A model depicting the interactions of perivascular nerves and endothelial factors during changes in flow and hypoxia is shown in Fig. 3.

It has been proposed that the endothelium mediates vasoconstriction via production of an EDCF in response to various chemical and physical stimuli, such as NA, thrombin, high extracellular potassium, hypoxia, and stretch (De Mey and Vanhoutte, 1983; Rubanyi and Vanhoutte, 1985; Katusic *et al.*, 1986; Katusic *et al.*, 1987a,b). Endothelial Ca^{2+} channels have been shown to be activated in response to stretch and it has been suggested that they operate as mechanotransducers (Lansman *et al.*, 1987). Thus, the role of endothelial cells is twofold, sensory and effector, such that vasoconstriction may occur independently of the action of extraneous vasoactive substances. Although the nature of the endothelium-dependent contracting factor is still uncertain and appears to be different in blood vessels of

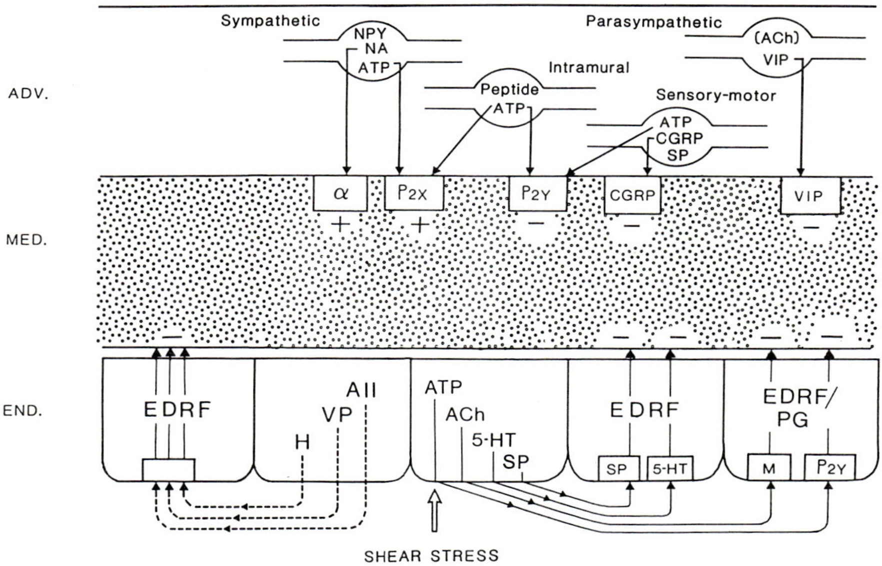

Figure 3 Regulation of vascular tone by perivascular nerves and endothelial cells. Neuropeptide Y (NPY), noradrenaline (NA), adenosine 5′-triphosphate (ATP), calcitonin gene-related peptide (CGRP), substance P (SP), and vasoactive polypeptide (VIP) can be released from nerve varicosities in the adventitia (ADV) to act on receptors in the media (MED), causing vasoconstriction or vasodilatation. ATP, acetylcholine (ACh), 5-hydroxytryptamine (5-HT), and SP, released from endothelial cells (END) by shear stress or hypoxia, act on their receptors on endothelial cells to cause a release of endothelium-dependent relaxing factor (EDRF) or prostaglandins (PG), which in turn act on the smooth muscle to cause relaxation. Angiotensin II (AII), arginine-vasopressin (VP), and histamine (H) are also contained in and may be released from subpopulations of endothelial cells. In areas denuded of endothelial cells, opposite effects may be produced by receptors on the smooth muscle cells, for example via P_{2X}- and P_{2Y}-purinoceptors and muscarinic receptors (M). (Modified from Burnstock, 1989.)

different anatomical origin, at least three different classes of endothelial vasoconstrictor substances have been recognized: (1) metabolites of arachidonic acid; (2) a polypeptide-like factor (or factors) produced by cultured endothelial cells; (3) a still unidentified diffusible factor released from anoxic/hypoxic endothelial cells (Lüscher, 1988). An endothelium-derived 21-residue vasoconstrictor peptide, *endothelin,* has been isolated from porcine aortic endothelial cells, and the complementary DNA of its precursor, pre-pro-endothelin, has been cloned and sequenced (Yanagisawa *et al.*, 1988). It has been shown to be a potent constrictor in, for example, the rabbit skin microvasculature (Brain *et al.*, 1988), in isolated human resistance vessels (Hughes *et al.*, 1988), and in rat mesenteric resistance vessels (Warner *et al.*, 1988). Endothelin and ATP, but not VIP, have been shown to be released from isolated aortic endothelial cells exposed to increased flow (Milner *et al.*, 1990b). Endothelin is also expressed in the nervous system (Giaid *et al.*, 1989). Receptors for endothelin have been localized by autoradiography on cultured rat aortic smooth muscle cells (Hirata *et al.*, 1988) and human and porcine coronary arteries (Power *et al.*, 1989).

IX. Plasticity of Perivascular Nerves and Endothelial Cells

A. *Long-Term ("Trophic") Interactions between Perivascular Nerves and Endothelial Cells*

Since perivascular nerves are separated from endothelial cells by vascular smooth muscle, the possibility of trophic interactions between perivascular nerves and endothelial cells has received little direct investigation. However, studies of the vasculature in disease and following denervation or mechanical injury do provide some indication that such interactions may exist. For example, it has been shown that 2–8 weeks after sympathetic and sensory denervation of the rabbit ear artery, endothelium-dependent relaxation responses to methacholine are significantly depressed (Mangiarua and Bevan, 1986). The reduction in response was not due to any impairment of the ability of the muscle to relax since the maximal relaxation to sodium nitroprusside (an endothelium-independent agent) was unaffected by denervation. Long-term sympathetic denervation in rabbits resulted in an increase in the sensitivity of cerebral arteries to hypercapnia, hypoxia, and 5-HT (Aubinea *et al.*, 1989). Although morphological changes in the endothelial cells were not detected under these conditions (Dimitriadou *et al.*, 1988), it is possible that alterations in the endothelial control of the cerebral vasculature after sympathetic denervation contributed to this effect. In contrast, when the endothelium of the dog coronary artery was injured mechanically without disruption of the elastic lamina, neuron-specific enolase-positive nerve fibers were increased in number at both 1 and 3 months (Taguchi *et al.*, 1986). An increased density of SP-containing nerve fibers was also observed in the dog coronary artery 3 months after mechanical injury to the endothelium. Surgical sympathectomy or long-term adrenoceptor blockade by propranolol is claimed to prevent or reduce the induction of atherosclerosis by diet

(Lichtor *et al.*, 1987). It has been proposed that NPY and NA in cerebral perivascular nerves, which increase during the development of hypertension in rats (Dhital *et al.*, 1988), are involved in protection against disruption of the blood–brain barrier and cerebral haemorrhage caused by hypertension. Sympathetic denervation before the development of hypertension results in increased incidence of stroke and increased permeability of the blood–brain barrier.

Perivascular nerve varicosities have been demonstrated in close apposition to endothelial cells in capillaries; this raises the possibility of direct trophic interactions between nerves and endothelial cells in the microvasculature. Adenosine can be formed from the extracellular breakdown of ATP released from nerves. Chronic inhibition of adenosine uptake with dipyridamole has been shown to cause proliferation of capillary endothelium and increased capillary density in skeletal muscle and heart (Hudlická, 1984). Furthermore, it has been suggested that neuropeptides may have a role in controlling neurochemical differentiation, cell proliferation, hypertrophy, and regeneration (Burnstock, 1982, 1985). Such interactions require further investigation in the context of vascular cell biology.

In a recent study in our laboratory, we have shown that stimulation *in vivo* for the perivascular nerves supplying ear artery of the rabbit results in changes in the structure and, remarkably, turn on the expression of the neuropeptides CGRP and NPY in endothelial cells in these vessels (Loesch *et al.*, 1992). The mechanisms involved in the trophic interactions between these cells, which are widely separated by a smooth muscle coat, are under investigation. However, whatever these mechanisms turn out to be, there are implications for diseases, such as hypertension, where increase in sympathetic perivascular nerve activity may lead to secondary changes in endothelial regulatory function. Thus, it appears that there are trophic interactions between endothelial cells and perivascular nerves and that changes in these interactions may be involved in the progression of some cardiovascular diseases. Clearly, the vascular smooth muscle is likely to form an integral part of these processes since trophic influences between endothelial and smooth muscle cells and between perivascular nerves and vascular smooth muscle have already been demonstrated (Dimitriadou *et al.*, 1988; Karnovsky, 1981).

B. Changes in Perivascular Nerves in Development and Aging

In a study of the changes in density of sympathetic adrenergic nerves in blood vessels of the rabbit, using image analysis quantitation, Cowen *et al.* (1982) recognized that the pattern of change with age varied considerably between different vessels. While the early stages of development of vascular innervation were similar in all the vessels studied and reached an initial peak density at about 6 weeks after birth, the density of innervation of some vessels (e.g., femoral artery) declined thereafter; other vessels (e.g., renal artery) reached peak density at 6 months and then rapidly declined while in the basilar artery, density of innervation continued to increase into old age (3 yr).

Changes in the development of peptide-containing perivascular nerves of guinea pig vessels were studied between 6 weeks *in utero* and old age and compared with changes in perivascular adrenergic nerves (Dhital *et al.*, 1986). Again, variation in the pattern of development of perivascular nerves in different vessels was demonstrated. In addition, in mesenteric and carotid arteries, whereas adrenergic nerve density reached a peak 4 weeks after birth and declined thereafter, the peptide-containing nerves (VIP, CGRP, and SP) reached a peak at birth and declined thereafter to about half maximum density in old age, raising the possibility that perivascular neuropeptides may play a trophic role in early development.

Another study from our laboratory showed that whereas there is a decrease in expression of vasoconstrictor, cerebrovascular neurotransmitters (NA and 5-HT) in aging rats, there is an increase in vasodilator neurotransmitters (VIP and CGRP) (Mione *et al.*, 1988a).

C. Changes in Perivascular Nerves following Trauma, Surgery, and Chronic Exposure to Hormones

Unilateral removal of the superior cervical ganglion results in the reinnervation of denervated cerebral vessels by sprouting nerves from the contralateral superior cervical ganglion (Kåhrström *et al.*, 1986). Other marked compensatory changes following superior cervical ganglionectomy include increased SP levels in the ipsilateral iris and ciliary body (Cole *et al.*, 1983), increased CGRP content of pial vessels (Schon *et al.*, 1985), and increased expression of NPY in nonadrenergic VIP-containing nerves in the cerebral vasculature (Gibbins and Morris, 1988). Long-term chemical sympathectomy of developing rats, induced by chronic guanethidine treatment, leads to increased brightness and density of CGRP-positive immunofluorescent nerves innervating blood vessels (Aberdeen *et al.*, 1990). In late pregnancy, sympathetic innervation of guinea pig uterine blood vessels exhibits a remarkable switch from adrenergic vasoconstrictor to cholinergic vasodilator control (Bell, 1968) although ultrastructural studies of the guinea pig uterine artery did not show any degeneration of serotonergic or peptidergic (NPY, VIP, SP, and CGRP)-containing nerves in late pregnancy (Mione *et al.*, 1988b). A recent study has shown that a 4-week treatment with estrogen, but not progesterone, leads to a marked reduction in the density and varicosity diameters of 5-HT-containing nerves supplying the rabbit basilar artery (Dhall *et al.*, 1988). In view of the possible involvement of 5-HT in the pathogenesis of headache, this finding suggests that contraceptive pills with a high estrogen content may be contraindicated in women prone to migraine attacks. The effect of crush lesions on perivascular noradrenergic nerves has shown differential rates of reinnervation in different blood vessels, suggesting the presence of characteristic levels of local neurotrophic activity (Cowen *et al.*, 1982).

D. Changes in Perivascular Nerves in Disease

The distribution of NA and NPY in nerves was compared during the early development of cerebral vessels in normotensive rats and spontaneously hypertensive rats (SHR) before and after the time when hypertension becomes apparent at about 5 weeks of age (Dhital *et al.*, 1988). Three interesting findings emerged from this study: (1) the levels of both NA and NPY were significantly higher in cerebral perivascular sympathetic nerves in SHR compared with normotensive rats; (2) in both normotensive rats and SHR, there was a discrepancy between the time course of changes in the expression of NA and NPY (i.e., the density of NA-fluorescent fibers increased rapidly between 4 and 6 weeks, while NPY-immunofluorescent nerves showed a rapid increase between 6 and 8 weeks) (Fig. 3); since NA and NPY coexist in sympathetic perivascular nerves, this shows that the expression of cotransmitters is not necessarily identical; (3) the increase in NA and NPY in SHR does not occur in the sympathetic nerve cell bodies in the superior cervical ganglion from which the cerebral perivascular nerves arise. There have been reports that purinergic mechanisms are altered in SHR. Purinergic modulation of NA release from perivascular sympathetic nerves is attenuated in SHR (Kamikawa *et al.*, 1980; Jackson, 1987), and the ATP component of sympathetic cotransmission to the tail artery is claimed to be more dominant in SHR (Vidal *et al.*, 1986; Cheung, 1989). Finally, impairment of endothelium-dependent relaxation to ATP appears to develop in parallel with hypertension (Miller *et al.*, 1987).

In 8-week streptozotocin-diabetic rats, the adrenergic innervation of the vasa vasorum in optic nerves is virtually absent, while in sciatic and vagal nerves it is significantly increased (Dhital *et al.*, 1986). Perivascular nerves in penile vessels containing the potent vasodilator VIP are seriously damaged or lost in both diabetic man and in streptozotocin-diabetic rats (Crowe *et al.*, 1983). In contrast, VIP expression is increased in the gut (Belai *et al.*, 1987). Reduction in the expression of VIP and 5-HT, but not NPY and NA, has been demonstrated in perivascular nerves supplying the cerebral blood vessels of 8-week streptozotocin-induced diabetic rats (Lagnado *et al.*, 1987). Attenuation of endothelium-dependent relaxation of aorta has been claimed in diabetic rat (Oyama *et al.*, 1986; Durante *et al.*, 1988; Kamata *et al.*, 1989).

Impairment of sympathetic and neural function has been claimed in cholesterol-fed animals (Panek *et al.*, 1985). It has also been suggested that surgical sympathectomy may be useful in controlling atherosclerosis in certain arterial beds (Lichtor *et al.*, 1987). Defective cholinergic arteriolar vasodilation has been claimed in atherosclerotic rabbits (Yamamoto *et al.*, 1988) and, in our laboratory, we have recently shown impairment of responses to perivascular nerves supplying the mesenteric, hepatic, and ear arteries of Watanabe heritable hyperlipidemic rabbits (Burnstock *et al.*, 1991).

Loss of adrenergic innervation has been reported in alcoholism (Low *et al.*, 1975), amyloidosis (Rubenstein *et al.*, 1983), orthostatic hypotension (Bannister *et al.*, 1981), and subarachnoid hemorrhage (Hara and Kobayashi, 1988). Recent evidence shows that there is also a loss of noradrenergic innervation of blood vessels supplying malignant, as compared with benign, human intracranial tumors (Crockard *et al.*, 1987).

References

Aberdeen, J., Corr, L., Milner, P., Lincoln, J., and Burnstock, G. (1990). Marked increases in calcitonin gene-related peptide-containing nerves in the developing rat following long-term sympathectomy with guanethidine. *Neuroscience 35,* 175–184.

Aubinea, P., Pearce, W., Reynier-Rebuffel, A. M., Cuevas, J., and Issertial, O. (1989). Long-term sympathetic denervation increases sensitivity of cerebral arteries to CO_2 and 5-HT *in vitro*. *J. Cereb. Blood Flow Metab. 9* (Suppl. 1), S506.
Bannister, R., Crowe, R., Eames, R., and Burnstock, G. (1981). Adrenergic innervation in autonomic failure. *Neurology 31,* 1501–1506.
Bao, J. X., Eriksson, I. E., and Stjärne, L. (1989). On pre- and/or postjunctional roles of ATP and an unknown "substance X" as sympathetic co-transmitters in rat tail artery. *Acta Physiol. Scand. 135,* 65–66.
Barja, F., Mathison, R., and Huggel, M. (1983). Substance P-containing nerve fibres in large peripheral blood vessels of the rat. *Cell Tissue Res. 229,* 441–422.
Bartfai, T., Iverfeldt, K., and Fisone, G. (1988). Regulation of the release of coexisting neurotransmitters. *Annu. Rev. Pharmacol. Toxicol. 28,* 285–310.
Belai, A., Lincoln, J., and Burnstock, G. (1987). Lack of release of vasoactive intestinal polypeptide and calcitonin gene-related peptide during electrical stimulation of enteric nerves in streptozotocin-diabetic rats. *Gastroenterology 93,* 1034–1040.
Bell, C. (1968). Dual vasoconstrictor and vasodilator innervation of the uterine arterial supply in guinea-pig. *Circ. Res. 23,* 279–289.
Bloom, S. R., and Edwards, A. V. (1980). Vasoactive intestinal polypeptide in relation to atropine resistant vasodilatation in the sub-maxillary gland of the cat. *J. Physiol. 300,* 41–53.
Brain, S. D., Tippins, J. R., and Williams, T. J. (1988). Endothelin induces potent microvascular constriction. *Br. J. Pharmacol. 95,* 1005–1007.
Brizzolara, A. L., and Burnstock, G. (1990). Evidence for noradrenergic-purinergic cotransmission in the hepatic artery of the rabbit. *Br. J. Pharmacol. 99,* 835–839.
Bulloch, J. M., and McGrath, J. C. (1988). Blockade of vasopressor and vas deferens responses by α,β-methylene ATP in the pithed rat. *Br. J. Pharmacol. 94,* 103–108.
Burnstock, G. (1976). Do some nerve cells release more than one transmitter? *Neuroscience 1,* 239–248.
Burnstock, G. (1977). Autonomic neuroeffector junctions—reflex vasodilatation of the skin. *J. Invest. Dermatol. 69,* 47–57.
Burnstock, G. (1982). Neuromuscular transmitters and trophic factors. *In* "Advanced Medicine 18." (M. Sarner, ed.), pp. 143–148. Pitman Medical, London.
Burnstock, G. (1985). Neurohumoral control of blood vessels: Some future directions. *J. Cardiovasc. Pharmacol. 7,* S137–S146.
Burnstock, G. (1986a). The changing face of autonomic neurotransmission. (The First von Euler Lecture in Physiology). *Acta Physiol. Scand. 126,* 67–91.
Burnstock, G. (1986b). Autonomic neuromuscular junctions: Current developments and future directions. *J. Anat. 146,* 1–30.
Burnstock, G. (1988a). Regulation of local blood flow by neurohumoral substances released from perivascular nerves and endothelial cells. *Acta Physiol. Scand. 133* (Suppl. 571), 53–59.
Burnstock, G. (1988b). Local purinergic regulation of blood pressure. (The First John T. Shepherd Lecture). *In* "Vasodilatation: Vascular Smooth Muscle, Peptides, Autonomic Nerves, and Endothelium." (P. M. Vanhoutte, ed.), pp. 1–14. Raven Press, New York.
Burnstock, G. (1989). Vascular control by purines with emphasis on the coronary system. *Eur. Heart J., 10* (Suppl. F), 15–21.
Burnstock, G. (1990a). Local mechanisms of blood flow control by perivascular nerves and endothelium. *J. Hypertens. 8* (Suppl. 7), S95–S106.
Burnstock, G. (1990b). Co-transmission. The Fifth Heymans Lecture—Ghent, February 17, 1990. *Arch. Int. Pharmacodyn. Ther. 304,* 7–33.
Burnstock, G. (1990c). Changes in expression of autonomic nerves in aging and disease. *J. Auton. Nerv. Syst. 30,* 525–534.
Burnstock, G. (1990d). Noradrenaline and ATP as cotransmitters in sympathetic nerves. *Neurochem. Int. 17,* 357–368.
Burnstock, G. (1990e). Overview—purinergic mechanisms. *In* "Biological Actions of Extracellular ATP. Ann. N.Y. Acad. Sci. vol. 603" (G. R. Dubyak, and J. A. Fedan, eds.), pp. 1–14. New York Academy of Sciences, New York.

Burnstock, G., and Griffith, S. G. (1988). "Nonadrenergic Innervation of Blood Vessels." Vol. I, pp. 1–149, and Vol. II pp. 1–233. CRC Press, Boca Raton, FL.

Burnstock, G., and Kennedy, C. (1985). Is there a basis for distinguishing two types of P_2-purinoceptor? *Gen. Pharmacol. 16,* 433–440.

Burnstock, G., and Warland, J. J. I. (1987). A pharmacological study of the rabbit saphenous artery *in vitro:* A vessel with a large purinergic contractile response to sympathetic nerve stimulation. *Br. J. Pharmacol. 90,* 111–120.

Burnstock, G., Lincoln, J., Fehér, E., Hopwood, A. M., Kirkpatrick, K., Milner, P., and Ralevic, V. (1988). Serotonin is localized in endothelial cells of coronary arteries and released during hypoxia: A possible new mechanism for hypoxia-induced vasodilatation of the rat heart. *Experientia 44,* 705–707.

Burnstock, G., Stewart-Lee, A. L., Brizzolara, A. L., Tomlinson, A., and Corr, L. (1991). Dual control by nerves and endothelial cells of arterial blood flow in atherosclerosis. *In* "Atherosclerotic Plagues." (R. W. Wissler, M. G. Bond, M. Mercuri, and P. Tanganelli, eds.), pp. 285–292. Plenum Press, New York.

Chan-Palay, V., and Palay, S. L. (1984). "Co-existence of Neuroactive Substances in Neurones." p. 433. Wiley, New York.

Cheung, D. W. (1989). Vascular neuroeffector mechanisms in hypertension. *Can. J. Physiol. Pharmacol. 67,* 1146–1150.

Cheung, D. W., and Fujioka, M. (1986). Inhibition of the excitatory junction potential in the guinea-pig saphenous artery by $ANAPP_3$. *Br. J. Pharmacol. 89,* 3–5.

Cole, D. F., Bloom, S. R., Burnstock, G., Butler, J. M., McGregor, G. P., Saffrey, M. J., Unger, W. G., and Zhang, S. Q. (1983). Increase in SP-like immunoreactivity in nerve fibres of rabbit iris and ciliary body one to four months following sympathetic denervation. *Exp. Eye Res. 37,* 191–197.

Corr, L., and Burnstock, G. (1991). Vasodilator response of coronary smooth muscle to the sympathetic co-transmitters noradrenaline and adenosine 5′-triphosphate. *Br. J. Pharmacol. 104,* 337–342.

Corr, L. A., Aberdeen, J. A., Milner, P., Lincoln, J., and Burnstock, G. (1990). Sympathetic and nonsympathetic neuropeptide Y-containing nerves in the rat myocardium and coronary arteries. *Circ. Res. 66,* 1602–1609.

Cowen, T., Haven, A. J., Wen-Qin, C., Gallen, D. D., Franc, F., and Burnstock, G. (1982). Development and ageing of perivascular adrenergic nerves in the rabbit. A quantitative fluorescence histochemical study using image analysis. *J. Auton. Nerv. Syst. 5,* 317–336.

Crockard, H. A., Alafaci, C., Dhital, K. K., and Burnstock, G. (1987). Noradrenergic innervation of blood vessels supplying human intracranial tumours. *J. Cereb. Blood Flow Metab. 7* (Suppl. 1), S596.

Crowe, R., Lincoln, J., Blacklay, P. F., Pryor, J. P., Lumley, J. S. P., and Burnstock, G. (1983). Vasoactive intestinal polypeptide-like immunoreactive nerves in diabetic penis. A comparison between streptozotocin-treated rats and man. *Diabetes 32,* 1075–1077.

Cuello, A. C. (1982). "Co-transmission. Proc. Symp. 50th Anniversary Meet. Brit. Pharmacol. Soc., Oxford." p. 264. Macmillan Press, London.

De Mey, J. G., and Vanhoutte, P. M. (1983). Anoxia and endothelium-dependent reactivity of the canine femoral artery. *J. Physiol. (Lond.) 335,* 65–74.

Dhall, U., Cowen, T., Haven, A. J., and Burnstock, G. (1988). Effect of oestrogen and progesterone on noradrenergic nerves and on nerves showing serotonin-like immunoreactivity in the basilar artery of the rabbit. *Brain Res. 442,* 335–339.

Dhital, K., Lincoln, J., Appenzeller, O., and Burnstock, G. (1986). Adrenergic innervation of vasa and nervi nervorum of optic, sciatic, vagus and sympathetic nerve trunks in normal and streptozotocin-diabetic rats. *Brain Res. 367,* 39–44.

Dhital, K. K., Gerli, R., Lincoln, J., Milner, P., Tanganelli, P., Weber, G., Fruschelli, C., and Burnstock, G. (1988). Increased density of perivascular nerves to the major cerebral vessels of the spontaneously hypertensive rat: Differential changes in noradrenaline and neuropeptide Y during development. *Brain Res. 444,* 33–45.

Dimitriadou, V., Aubineau, P., Taxi, J., and Seylaz, J. (1988). Ultrastructural changes in the cerebral artery wall induced by long-term sympathetic denervation. *Blood Vessels 25,* 122–143.

Duckles, S. P., and Buck, S. M. (1982). Substance P in the cerebral vasculature: Depletion by capsaicin suggests a sensory role. *Brain Res. 245,* 171–174.

Durante, W., Sen, A. K., and Sunahara, F. A. (1988). Impairment of endothelium-dependent relaxation in aortae from spontaneously diabetic rats. *Br. J. Pharmacol. 94,* 463–468.

Edvinsson, L. (1991). Innervation and effects of dilatory neuropeptides on cerebral vessels. *Blood Vessels 28,* 35–45.

Flavahan, N. A., and Vanhoutte, P. (1986). Sympathetic purinergic vasoconstriction and thermosensitivity in a canine cutaeneous vein. *J. Pharmacol. Exp. Ther. 239,* 784–789.

Furchgott, R. F. (1984). Role of endothelium in responses of vascular smooth muscle. *In* "Frontiers in Physiological Research." (D. G. Garlick and P. I. Korner, eds.), pp. 116–133. Cambridge University Press, Cambridge.

Furchgott, R. F., and Zawadski, J. V. (1980). The obligatory role of endothelial cells in the relaxation of arterial smooth muscle by acetylcholine. *Nature 288,* 373–376.

Furness, J. B., and Costa, M. (1987). "The Enteric Nervous System." Churchill Livingstone, Edinburgh.

Furness, J. B., Papka, R. E., Della, N. G., Costa, M., and Eskay, R. L. (1982). Substance P-like immunoreactivity in nerves associated with the vascular system of guinea-pigs. *Neuroscience 7,* 447–459.

Fyffe, R. E. W., and Perl, E. R. (1984). Is ATP a central synaptic mediator for certain primary afferent fibres from mammalian skin?. *Proc. Natl. Acad. Sci. U.S.A. 81,* 6890–6893.

Giaid, A., Gibson, S. J., Ibrahim, N. B. N., Legon, S., Bloom, S. R., Yanagisawa, M., Masaki, T., Varndell, I. M., and Polak, J. M. (1989). Endothelin 1, and endothelium derived peptide, is expressed in neurones of the human spinal cord and dorsal root ganglia. *Proc. Natl. Acad. Sci. U.S.A. 86,* 7634–7638.

Gibbins, I. L., and Morris, J. L. (1988). Co-existence of immunoreactivity to neuropeptide Y and vasoactive intestinal polypeptide in nonadrenergic axons innervating guinea-pig cerebral arteries after sympathectomy. *Brain Res. 444,* 402–406.

Gibbins, I. L., and Morris, J. L. (1990). Sympathetic noradrenergic neurons containing dynorphin but not neuropeptide Y innervate small cutaneous blood vessels of guinea-pigs. *J. Auton. Nerv. Syst. 29,* 137–150.

Gibbins, I. L., Furness, J. B., Costa, M., Macintyre, I., Hillyard, C. J., and Girgis, S. (1985). Co-localization of calcitonin gene-related peptide-like immunoreactivity with substance P in cutaneous, vascular and visceral sensory neurons of guinea-pigs. *Neurosci. Lett. 57,* 125–130.

Griffith, S. G., and Burnstock, G. (1983). Immunohistochemical demonstration of serotonin in nerves supplying human cerebral and mesenteric blood vessels: Some speculations about their involvement in vascular disorders. *Lancet i,* 561–562.

Gulbenkian, S., Merighi, A., Wharton, J., Varndell, I. M., and Polak, J. M. (1986). Ultrastructural evidence for the coexistence of calcitonin gene-related peptide and substance P in secretory vesicles of peripheral nerves in the guinea pig. *J. Neurocytol. 15,* 535–542.

Hara, H., and Kobayashi, S. (1988). Reduced tyrosine hydroxylase-like immunoreactivity around cerebral arteries after experimental subarachnoid haemorrhage in rats. An immunohistochemical study. *Acta Neuropathol. (Berl.) 75,* 538–540.

Hassall, C. J. S., and Burnstock, G. (1984). Neuropeptide Y-like immunoreactivity in cultured intrinsic neurones of the heart. *Neurosci. Lett. 52,* 111–115.

Hassall, C. J. S., and Burnstock, G. (1986). Intrinsic neurones and associated cells of the guinea-pig heart in culture. *Brain Res. 364,* 102–113.

Hirata, Y., Yoshimi, H., Takata, S., Watanabe, T. X., Kumagai, S., Nakajima, K., and Sakakibara, S. (1988). Cellular mechanism of action by a novel vasoconstrictor endothelin in cultured rat vascular smooth muscle cells. *Biochem. Biophys. Res. Commun. 154,* 868–875.

Hökfelt, T., Fuxe, K., and Pernow, B. (1986). "Coexistence of Neuronal Messengers: A New Principle in Chemical Transmission. Progress in Brain Research, Vol. 68." Elsevier, Amsterdam.

Hollis, T. M., and Rosen, L. A. (1972). Histidine decarboxylase activities of bovine aortic endothelium and intima-media. *Proc. Soc. Exp. Biol. Med. 141,* 978–981.

Hopwood, A. M., Lincoln, J., Kirkpatrick, K. A., and Burnstock, G. (1989). Adenosine 5′-triphosphate, adenosine and endothelium-derived relaxing factor in hypoxic vasodilatation of the heart. *Eur. J. Pharmacol. 165,* 323–326.

Houston, D. S., Shepherd, J. T., and Vanhoutte, P. M. (1985). Adenine nucleotides, serotonin and endothelium-dependent relaxations to platelets. *Am. J. Physiol. 248,* H389–H395.

Hudlická, O. (1984). Development of microcirculation: Capillary growth and adaptation. *In* "Handbook of Physiology—The Cardiovascular System, Vol. IV" (E. M. Renkin and C. C. Michel, eds.), pp. 165–216, American Physiological Society, Bethesda.

Hughes, A. D., Schachter, M., Hair, W. M., and Sever, P. S. (1988). Endothelin is a potent constrictor of isolated human resistance arteries. *Br. J. Pharmacol. Proc. 95* (Suppl.), 722P.

Ignarro, L. J., Byrns, R. E., and Wood, K. S. (1986). Pharmacological and biochemical properties of endothelium-derived relaxing factor (EDRF): Evidence that it is closely related to nitric oxide (NO) radical. *Circulation 74* (Suppl. II), 287.

Ishikawa, S. (1985). Actions of ATP and α,β-methylene ATP on neuromuscular transmission and smooth muscle membrane of the rabbit and guinea-pig mesenteric arteries. *Br. J. Pharmacol. 86,* 777–787.

Jackowski, A., Crockard, A., and Burnstock, G. (1989). 5-Hydroxytryptamine demonstrated immunohistochemically in rat cerebrovascular nerves largely represents 5-hydroxytryptamine uptake into sympathetic nerve fibres. *Neuroscience 29,* 453–462.

Jackson, E. K. (1987). Role of adenosine in noradrenergic neurotransmission in spontaneously hypertensive rats. *Am. J. Physiol. 253,* H909–H918.

Jahr, C. E., and Jessel, T. M. (1983). ATP excites a subpopulation of rat dorsal horn neurones. *Nature 304,* 730–733.

Kåhrström, J., Hardebo, J. E., Nordberg, C., and Owman, C. (1986). Experiments on cerebrovascular nerve plasticity and trophic vascular adaption in young and adult rats. *In* "Neural Regulation of Brain Circulation." (C. Owman and J. E. Hardebo, eds.), pp. 589–606. Elsevier, Amsterdam.

Kamata, K., Miyata, N., and Kasuya, Y. (1989). Impairment of endothelium-dependent relaxation and changes in levels of cyclic GMP in aorta from streptozotocin-induced diabetic rats. *Br. J. Pharmacol. 97,* 614–618.

Kamikawa, Y., Cline, W. H., and Su, C. (1980). Diminished purinergic modulation of the vascular adrenergic neurotransmission in spontaneously hypertensive rats. *Eur. J. Pharmacol. 66,* 347–353.

Karnovsky, M. J. (1981). Endothelial-vascular smooth muscle cell interaction. *Am. J. Pathol. 105,* 200–206.

Katsuragi, T., and Su, C. (1982). Augmentation by theophylline of [^{3}H]purine release from vascular adrenergic nerves: Evidence for presynaptic autoinhibition. *J. Pharmacol. Exp. Ther. 220,* 152–156.

Katusic, Z. S., Shepherd, J. T., and Vanhoutte, P. M. (1986). Endothelium-dependent contractions to stretch in canine basilar arteries. *Fed. Proc. 45,* 289.

Katusic, Z. S., Shepherd, J. T., and Vanhoutte, P. M. (1987a). Potassium-induced endothelium-dependent rhythmic activity in the canine basilar artery. *J. Cardiovasc. Pharmacol. 12,* 37–41.

Katusic, Z. S., Shepherd, J. T., and Vanhoutte, P. M. (1987b). Endothelium-dependent contractions to stretch in canine basilar artery. *Am. J. Physiol. 252,* H671–H673.

Kennedy, C., and Burnstock, G. (1985). ATP produces vasodilation via P_1 purinoceptors and vasoconstriction via P_2 purinoceptors in the isolated rabbit central ear artery. *Blood Vessels 22,* 145–155.

Kennedy, C., Saville, V. L., and Burnstock, G. (1986). The contributions of noradrenaline and ATP to the responses of the rabbit central ear artery to sympathetic nerve stimulation depend on the parameters of stimulation. *Eur. J. Pharmacol. 122,* 291–300.

Kifor, I., and Dzau, V. J. (1987). Endothelial renin-angiotensin pathway: Evidence for intracellular synthesis and secretion of angiotensin. *Circ. Res. 60,* 422–428.

Krishtal, O. A., Marchenko, S. M., and Obukhov, A. G. (1988). Cationic channels activated by extracellular ATP in rat sensory neurons. *Neuroscience 27,* 995–1000.

Kupfermann, I. (1991). Functional studies of cotransmission. *Physiol. Rev. 71,* 683–732.
Lagnado, M. L. J., Crowe, R., Lincoln, J., and Burnstock, G. (1987). Reduction of nerves containing vasoactive intestinal polypeptide and serotonin, but not neuropeptide Y and catecholamine, in cerebral blood vessels of the 8-week streptozotocin-induced diabetic rat. *Blood Vessels 24,* 169–180.
Lansman, J. B., Hallam, T. J., and Rink, T. J. (1987). Single stretch-activated ion channels in vascular endothelial cells as mechanotransducers? *Nature 325,* 811–813.
Lewis, J. (1927). "The Blood Vessels of the Human Skin and their Responses." Shaw & Sons, London.
Lichtor, T., Davis, H. R., Johns, L., Vesselinovitch, D., Wissler, R. W., and Mullan, S. (1987). The sympathetic nervous system and atherosclerosis. *J. Neurosurg. 67,* 906–914.
Lincoln, J., Loesch, A., and Burnstock, G. (1990). Localization of vasopressin, serotonin and angiotensin II in endothelial cells of the renal and mesenteric arteries of the rat. *Cell Tissue Res. 259,* 341–344.
Lincoln, J., Ralevic, V., and Burnstock, G. (1991). Neurohumoral substances and the endothelium. *In* "Cardiovascular Significance of Endothelium-Derived Vasoactive Factors." (G. M. Rubanyi, ed.), pp. 83–110. Futura, Mount Kisco, NY.
Linnik, M. D., and Moskowitz, M. A. (1989). Identification of immunoreactive substance P in human and other mammalian endothelial cells. *Peptides 10,* 957–962.
Loesch, A., and Burnstock, G. (1988). Ultrastructural localisation of serotonin and substance P in vascular endothelial cells of rat femoral and mesenteric arteries. *Anat. Embryol. 178,* 137–142.
Loesch, A., Maynard, K. I., and Burnstock, G. (1992). CGRP- and NPY-like immunoreactivity in endothelial cells after long-term stimulation of perivascular nerves. *Neuroscience 48,* 723–736.
Low, P. A., Walsh, J. C., Huang, C. Y., and McLeod, J. G. (1975). The sympathetic nervous system in alcoholic neuropathy. A clinical and pathological study. *Brain 98,* 357–364.
Lundberg, J. M. (1981). Evidence for coexistence of vasoactive intestinal polypeptide (VIP) and acetylcholine in neurons of cat exocrine glands. Morphological, biochemical and functional studies. *Acta Physiol. Scand. 112* (Suppl. 496), 1–57.
Lundberg, J. M., Terenius, L., Hökfelt, T., and Goldstein, M. (1983). High levels of neuropeptide Y in peripheral noradrenergic neurons in various mammals including man. *Neurosci. Lett. 42,* 167–172.
Lundberg, J. M., Ånggård, A., Theodorsson-Norheim, E., and Pernow, J. (1984). Guanethidine-sensitive release of neuropeptide Y-like immunoreactivity in the cat spleen by sympathetic nerve stimulation. *Neurosci. Lett. 52,* 175–180.
Lundberg, J. M., Franco-Cereceda, A., Hua, X., Hökfelt, T., and Fischer, J. A. (1985). Coexistence of substance P and calcitonin gene-related peptide-like immunoreactivities in sensory nerves in relation to cardiovascular and bronchoconstrictor effects of capsaicin. *Eur. J. Pharmacol. 108,* 315–319.
Lundberg, J. M., Rudehill, A., Sollevi, A., Theodorsson-Norheim, E., and Hamberger, B. (1986). Frequency- and reserpine-dependent chemical coding of sympathetic transmission: Differential release of noradrenaline and neuropeptide Y from pig spleen. *Neurosci. Lett. 63,* 96–100.
Lüscher, T. F. (1988). "Endothelial Vasoactive Substances and Cardiovascular Disease." Karger, Basel.
Machalay, M., Dalziel, H. H., and Sneddon, P. (1988). Evidence for ATP as a cotransmitter in dog mesenteric artery. *Eur. J. Pharmacol. 147,* 83–91.
Mackenzie, I., Manzini, S., and Burnstock, G. (1988). Regulation of voltage-dependent excitatory responses to α,β-methylene ATP, ATP and nonadrenergic nerve stimulation by dihydropyridines in the guinea-pig vas deferens. *Neuroscience 27,* 317–332.
Maggi, C. A., and Meli, A. (1988). The sensory-efferent function of capsaicin-sensitive sensory nerves. *Gen. Pharmacol. 19,* 1–43.
Mangiarua, E. I., and Bevan, R. D. (1986). Altered endothelium-mediated relaxation after denervation of growing rabbit ear artery. *Eur. J. Pharmacol. 122,* 149–152.
Miller, M. J., Pinto, A., and Mullane, K. M. (1987). Impaired endothelium-dependent relaxations in rabbits subjected to aortic coarctation hypertension. *Hypertension 10,* 164–170.
Milner, P., Ralevic, V., Hopwood, A. M., Fehér, E., Lincoln, J., Kirkpatrick, K. A., and Burnstock, G. (1989). Ultrastructural localisation of substance P and choline acetyltransferase in endothelial cells of

rat coronary artery and release of substance P and acetylcholine during hypoxia. *Experientia 45,* 121–125.

Milner, P., Kirkpatrick, K. A., Ralevic, V., Toothill, V., Pearson, J. D., and Burnstock, G. (1990a). Endothelial cells cultured from human umbilical vein release ATP, substance P and acetylcholine in response increased flow. *Proc. Roy. Soc. B 241,* 245–248.

Milner, P., Bodin, P., Loesch, A., and Burnstock, G. (1990b). Rapid release of endothelin and ATP from isolated aortic endothelial cells exposed to increased flow. *Biochem. Biophys. Res. Commun. 170,* 649–656.

Mione, M. C., Dhital, K. K., Amenta, F., and Burnstock, G. (1988a). An increase in the expression of neuropeptidergic vasodilator, but not vasoconstrictor, cerebrovascular nerves in aging rats. *Brain Res. 460,* 103–113.

Mione, M. C., Cavallotti, C., Burnstock, G., and Amenta, F. (1988b). The peptidergic innervation of the guinea pig uterine artery in pregnancy. *Basic Appl. Histochem. 32,* 153–159.

Mione, M. C., Ralevic, V., and Burnstock, G. (1990). Peptides and vasomotor mechanisms. *Pharmacol. Ther. 46,* 429–468.

Muir, T. C., and Wardle, K. A. (1988). The electrical and mechanical basis of co-transmission in some vascular and non-vascular smooth muscles. *J. Auton. Pharmacol. 8,* 203–218.

Mulvany, M. J., and Aalkjaer, C. (1990). Structure and function of small arteries. *Physiol. Rev. 70,* 921–961.

Muramatsu, I. (1986). Evidence for sympathetic, purinergic transmission in the mesenteric artery of the dog. *Br. J. Pharmacol. 87,* 478–480.

Muramatsu, I., and Kigoshi, S. (1987). Purinergic and non-purinergic innervation in the cerebral arteries of the dog. *Br. J. Pharmacol. 92,* 901–908.

Osborne, N. N. (1983). "Dale's Principle and Communication between Neurones." pp. 204. Pergamon Press, Oxford.

Owman, C. (1990). Peptidergic vasodilator nerves in the peripheral circulation and in the vascular beds of the heart and brain. *Blood Vessels 27,* 73–93.

Oyama, Y., Kawasaki, H., Hattori, Y., and Kanno, M. (1986). Attenuation of endothelium-dependent relaxation in aorta from diabetic rats. *Eur. J. Pharmacol. 132,* 75–78.

Paddle, B. M., and Burnstock, G. (1974). Release of ATP from perfused heart during coronary vasodilatation. *Blood Vessels 11,* 110–119.

Palmer, R. M. J., Ferrige, A. G., and Moncada, S. (1987). Nitric oxide release accounts for the biological activity of endothelium-derived relaxing factor. *Nature 327,* 524–526.

Panek, R. L., Dixon, W. R., and Rutledge, C. O. (1985). Modification of sympathetic neuronal function in the rat tail artery by dietary lipid treatment. *J. Pharmacol. Exp. Ther. 233,* 578–583.

Parnavelas, J. G., Kelly, W., and Burnstock, G. (1985). Ultrastructural localization of choline acetyltransferase in vascular endothelial cells in rat brain. *Nature 316,* 724–725.

Pernow, J., Saria, A., and Lundberg, J. M. (1986). Mechanisms underlying pre- and postjunctional effects of neuropeptide Y in sympathetic vascular control. *Acta Physiol. Scand. 126,* 239–249.

Power, R. F., Wharton, J., Salas, S. P., Kanse, S., Ghatei, M., Bloom, S. R., and Polak, J. M. (1989). Autoradiographic localisation of endothelin binding sites in human and porcine coronary arteries. *Eur. J. Pharmacol. 160,* 199–200.

Ralevic, V., Kristek, F., Hudlická, O., and Burnstock, G. (1989). A new protocol for removal of the endothelium from the perfused rat hindlimb preparation. *Circ. Res. 64,* 1190–1196.

Ramme, D., Regenold, J. T., Starke, K., Busse, R., and Illes, P. (1987). Identification of the neuroeffector transmitter in jejunal branches of the rabbit mesenteric artery. *Naunyn Schmiedebergs Arch. Pharmacol. 336,* 267–273.

Rubanyi, G. M., and Vanhoutte, P. M. (1985). Hypoxia releases a vasoconstrictor substance from the canine vascular endothelium. *J. Physiol. (Lond.) 364,* 45–56.

Rubenstein, A. E., Rudansky, M. C., and Yahr, M. D. (1983). Autonomic failure due to amyloid. *In* "Autonomic Failure. A Textbook of Clinical Disorders of the Autonomic Nervous System." (R. Bannister, ed.), pp. 574–595. Oxford University Press, Oxford.

Salt, T. E., and Hill, R. G. (1983). Excitation of single sensory neurones in the rat caudal trigeminal nucleus by iontophoretically applied adenosine 5′-triphosphate. *Neurosci. Lett. 35,* 53–57.

Saville, V. L., and Burnstock, G. (1988). Use of reserpine and 6-hydroxydopamine supports evidence for purinergic cotransmission in the rabbit ear artery. *Eur. J. Pharmacol. 155,* 271–277.

Saville, V. L., Maynard, K. I., and Burnstock, G. (1990). Neuropeptide Y potentiates purinergic as well as adrenergic responses of the rabbit ear artery. *Eur. J. Pharmacol. 176,* 117–126.

Schlicker, E., Urbanek, E., and Göthert, M. (1989). ATP, α,β-methylene ATP and suramin as tools for characterization of vascular P_{2X} receptors in the pithed rat. *J. Auton. Pharmacol. 9,* 371–380.

Schon, F., Ghatei, M., Allen, J. M., Mulderry, P. K., Kelly, J. S., and Bloom, S. R. (1985). The effect of sympathectomy on calcitonin gene-related peptide levels in the rat trigemino-vascular system. *Brain Res. 348,* 197–200.

Schwartz, D. D., and Malik, K. U. (1989). Renal periarterial nerve stimulation-induced vasoconstriction at low frequencies is primarily due to release of a purinergic transmitter in the rat. *J. Pharmacol. Exp. Ther. 250,* 764–771.

Shimada, S. G., and Stitt, J. T. (1984). An analysis of the purinergic component of active muscle vasodilatation obtained by electrical stimulation of the hypothalamus in rabbits. *Br. J. Pharmacol. 83,* 577–589.

Sneddon, P., and Burnstock, G. (1984). ATP as a co-transmitter in rat tail artery. *Eur. J. Pharmacol. 106,* 149–152.

Stjärne, L. (1989). Basic mechanisms and local modulation of nerve-impulse-induced secretion of neurotransmitters from individual sympathetic nerve varicosities. *Rev. Physiol. Biochem. Pharmacol. 112,* 1–137.

Stjärne, L., Lundberg, J. M., and Åstrand, P. (1986). Neuropeptide Y—a cotransmitter with noradrenaline and adenosine 5′-triphosphate in the sympathetic nerves of the mouse vas deferens? A biochemical, physiological and electropharmacological study. *Neuroscience 18,* 151–166.

Stone, T. W. (1981). Differential blockade of ATP, noradrenaline and electrically evoked contractions of the rat vas deferens by nifedipine. *Eur. J. Pharmacol. 74,* 373–376.

Suzuki, H. (1985). Electrical responses of smooth muscle cells of the rabbit ear artery to adenosine triphosphate. *J. Physiol. (Lond.) 359,* 401–415.

Taguchi, T., Ishii, Y., Matsubara, F., and Tenaka, K. (1986). Intimal thickening and the distribution of vasomotor nerves in the mechanically injured dog coronary artery. *Exp. Mol. Pathol. 444,* 138–146.

Taylor, E. M., and Parsons, M. E. (1989). Adrenergic and purinergic neurotransmission in arterial resistance vessels of the cat intestinal circulation. *Eur. J. Pharmacol. 164,* 23–33.

Terenghi, G., Polak, J. M., Rodrigo, J., Mulderry, P. K., and Bloom, S. R. (1986). Calcitonin gene-related peptide-immunoreactive nerves in the tongue, epiglottis and pharynx of the rat: Occurrence, distribution and origin. *Brain Res. 365,* 1–14.

Uddman, R., and Edvinsson, L. (1989). Neuropeptides in the cerebral circulation. *Cerebrovasc. Brain Metab. Rev. 1,* 230–252.

Uddman, R., Edvinsson, L., Ekman, R., Kingman, T., and McCulloch, J. (1985). Innervation of the feline cerebral vasculature by nerve fibres containing calcitonin gene-related peptide: Trigeminal origin and co-existence with substance P. *Neurosci. Lett. 62,* 131–136.

Vanhoutte, P. M. , and Rimele, T. J. (1983). Role of endothelium in the control of vascular smooth muscle function. *J. Physiol. (Paris) 78,* 681–686.

Vidal, M., Hicks, P. E., and Langer, S. Z. (1986). Differential effects of α,β-methylene ATP on responses to nerve stimulation in SHR and WKY tail arteries. *Naunyn Schmiedebergs Arch. Pharmacol. 332,* 384–390.

Von Kügelgen, I., and Starke, K. (1985). Noradrenaline and adenosine triphosphate as co-transmitters of neurogenic vasoconstriction in rabbit mesenteric artery. *J. Physiol. (Lond.) 367,* 435–455.

Wahlestedt, C., Yanaihara, N., and Håkanson, R. (1986). Evidence for different pre- and postjunctional receptors for neuropeptide Y and related peptides. *Regul. Pept. 13,* 307–318.

Warland, J. J. I., and Burnstock, G. (1987). Effects of reserpine and 6-hydroxydopamine on the adrenergic and purinergic components of sympathetic nerve responses of the rabbit saphenous artery. *Br. J. Pharmacol. 92,* 871–880.

Warner, T., de Nucci, G., and Vane, J. R. (1988). Release of EDRF by endothelin in the rat isolated perfused mesentery. *Br. J. Pharmacol. Proc. 95* (Suppl.), 723P.

Yamamoto, H., Bossaller, C., Cartwright, J., Jr., and Henry, P. D. (1988). Videomicroscopic demonstration of defective cholinergic arteriolar vasodilation in atherosclerotic rabbit. *J. Clin. Invest. 81,* 1752–1758.

Yanagisawa, M., Kurihara, H., Kimura, S., Tomobe, Y., Kobayashi, M., Mitsui, Y., Yazaki, Y., Goto, K., and Masaki, T. (1988). A novel potent vasoconstrictor peptide produced by vascular endothelial cells. *Nature 332,* 411–415.

Part I

Basic Studies

CHAPTER 2

Regulation of the Autonomic Innervation of Blood Vessels during Development and Aging

T. COWEN

I. Introduction

Recent information has emphasized the regionally specific nature of the autonomic innervation of different blood vessels. Patterns of chemical and morphological coding have been described suggesting specialized functions that are more highly localized than was previously thought. Local differences in innervation may allow the modulation of general homeostatic activity to suit the requirements of the particular region. We need to understand how these locally specific patterns of innervation are established during development. Furthermore, with the realization that neuromuscular relationships may remain in a state of dynamic equilibrium throughout life, it becomes important to understand how specific local equilibria are maintained through life. Finally, because diseases of the cardiovascular system are clinically important in an ever-increasing population of elderly people, we need to investigate whether changes in old age in the relationship between nerves and muscles contribute to cardiovascular pathology.

II. The Establishment of Locally Specific Patterns of Innervation during Early Development

A. *Noradrenergic Nerves*

In small laboratory animals, the noradrenergic sympathetic nerves begin to grow out over the vessel surface about halfway through gestation and by the perinatal period nerve plexuses are present in the majority of those vessels that will be innervated in maturity (Cowen *et al.*, 1982; De Champlain *et al.*, 1970; Read and Burnstock, 1970). Nerve varicosities, the sites of transmitter release, are sparse at this stage and, in general, the nerve trunks are smooth, nonvaricose, and oriented parallel to the vessel axis. Rapid growth of plexus density and number of varicosities occurs during the perinatal period with increases of 50% or more being recorded in some vessels (Cowen *et al.*, 1982). Maximal nerve density is achieved in most of the nerve plexuses studied between 3 and 8 weeks postnatum (Cowen *et al.*, 1982; Lundberg *et al.*, 1976; Su *et al.*, 1977; Todd, 1980; Todd and Tokito, 1981) and during the initial period after hatching in chicks (Rickenbacher and Ruflin, 1974). (For a review, see Cowen and Burnstock, 1986.)

Similar patterns of postnatal growth of adrenergic fibers have been observed in the coronary vessels and myocardium of the dog (Dolezel *et al.*, 1974), rabbit (Friedman *et al.*, 1968), and rat (Lipp and Rudolph, 1972). Interestingly, the catecholaminergic innervation of the heart of guinea pigs (Lipp and Rudolph, 1972) and sheep (Lebowitz *et al.*, 1972), which are more able to fend for themselves at birth than many other mammals, develops earlier than in other species and by birth appears almost mature.

Quantitative studies of catecholaminergic nerve density around blood vessels have shown that following a relatively homogeneous period of postnatal nerve growth, plexuses show heterogeneous changes with marked reductions and increases of density being recorded in different regions and species (Todd, 1980; Cowen *et al.*, 1982). Localized reductions of nerve density in the renal (Gallen *et al.*, 1982) and femoral (Dolezel *et al.*, 1974) arteries during early postnatal development may be the result of growth in body size affecting the vessel, causing it to change from a small artery with vasomotor functions to a larger conduit vessel. As further evidence for this suggestion, more peripheral regions of the femoral artery supplying muscles of the lower limb show increases in nerve density over the same period (Dolezel *et al.*, 1974). Thus, during postnatal development a process of diversification occurs that results in the development of the mature, locally specific pattern of innervation.

B. Nonadrenergic Nerves

While the development of perivascular noradrenergic nerves has been investigated in some depth, there are few studies of nonadrenergic nerves. Peptide-containing perivascular nerves appear to develop earlier than the catecholaminergic nerve plexuses to the same blood vessels (Dhall *et al.*, 1986). Substance P-, vasoactive intestinal polypeptide- (VIP), and calcitonin gene-related peptide- (CGRP) immunoreactive nerves appear around blood vessels of fetal guinea pigs in the second half of gestation and are relatively mature by birth. The greatest density of peptidergic nerve plexuses is generally found in vessels that also exhibit dense noradrenergic nerves. 5-Hydroxytryptamine (5-HT) immunoreactivity has been demonstrated around submucosal blood vessels of human fetal intestine at about half-term (Griffith and Burnstock, 1983). These nerves may be axons of either enteric or extrinsic, probably sympathetic, neurons.

C. Morphological and Physiological Maturation

Morphological and physiological maturation of nerve–muscle relationships may not follow the same time course, although there are few studies in this area. Combined histochemical and electrophysiological investigations of the developing sympathetic innervation in rat mesenteric arteries (Hill *et al.*, 1983) have shown that excitatory junction potentials resembling those seen in maturity are not elicited until near the end of the postnatal period of nerve outgrowth. The same study also provided evidence that an increase in neurotransmitter content of terminal nerve fibers occurred after the completion of axon outgrowth. Physiological maturation of the neurogenic responses of the hepatic portal vein of the rat has been shown to depend on the development of myogenic propagation mechanisms between innervated and noninnervated smooth muscle cells (Ljung and Stage, 1975). While responses to nerve stimulation can be elicited before this occurs, they are smaller and slower than mature neurogenic responses.

III. Regulation of Autonomic Neuronal Phenotype during Early Development

A. *Extraneuronal Trophic Influences*

There is little information about how blood vessels achieve their phenotypically diverse patterns of innervation during development. However, because sympathetic neurons have been widely studied as models for developmental regulation of neuronal phenotype, some inferences can be drawn regarding the innervation of blood vessels.

Considerable plasticity exists in sympathetic neurons during early development, in which the local environment plays a vital role. Extraneuronal influences appear to be important in determining survival and connectivity of sympathetic neurons during development. Nerve growth factor (NGF) (Levi-Montalcini and Hamburger, 1953), produced by effectors including vascular smooth muscle (Shelton and Reichard, 1984), is the principal trophic messenger molecule in the sympathetic nervous system on which sympathetic neurons depend for their survival (Thoenen and Barde, 1980; Levi-Montalcini, 1987; Yankner and Shooter, 1982; Barde, 1989). Competition between adjacent axons for target-produced NGF is responsible for the modeling of axonal branching patterns during development (Purves, 1980). Furthermore, this trophic relationship between target tissues and their innervation appears to continue into maturity because a correlation has been shown between levels of NGF mRNA in different target tissues and the density of their adrenergic innervation in adult animals (Korsching and Thoenen, 1983; Heumann *et al.*, 1984; Shelton and Reichard, 1984). NGF affects the maturation of some aspects of sympathetic neurotransmitter phenotype (Thoenen *et al.*, 1971) and the expression of catecholaminergic enzymes in maturity. However, the actions of NGF appear to be principally permissive rather than instructive as regards transmitter phenotype (Patterson, 1978).

While NGF regulates aspects of connectivity in sympathetic neurons during development, other extraneuronal factors are important in establishing neurotransmitter phenotype. Cultured sympathetic neurons taken from early postnatal rats can be "switched" to a cholinergic phenotype by exposure to a culture medium conditioned by heart muscle cells (Patterson, 1978). These extraneuronal influences appear to be instructive in nature in that neurons from older animals appear to have lost this plasticity (Patterson, 1978). The sympathetic nerves that supply sweat glands undergo a developmental switch from a catecholaminergic to a cholinergic phenotype (Potter *et al.*, 1983), which has been shown to result from contact with the appropriate glandular target tissue (Stevens and Landis, 1990). Although the sympathetic neurons involved in these situations do not supply blood vessels, the examples indicate the importance of the local environment and the contact between axons and their peripheral target tissues in the determination of locally specific transmitter phenotype during development.

B. Neuronal Trophic Influences

Not all trophic effects act retrogradely from the target to the nerve. Denervation of the rabbit ear artery during development results in structural and functional impairment of vascular smooth muscle (Bevan and Tsuru, 1981). The medial component of the vessel wall was significantly reduced in thickness when denervation was carried out on 4-week-old animals but not when denervation was carried out later in life. Similar results were obtained when the superior cervical ganglion (SCG), which supplies sympathetic nerves to the ear artery, was decentralized rather than removed. Thus, nerves may exert trophic effects on muscle cells that are specific to particular stages of development.

Little is known about the regulation of nonadrenergic characteristics in autonomic neurons supplying blood vessels although there is evidence from *in vitro* studies that the electrical activity of sympathetic neurons influences their pattern of peptide expression (Black *et al.*, 1984).

IV. Changes in the Nerve Supply to Blood Vessels during Maturity

Blood vessels exhibit local specificity in their patterns of innervation (Burnstock, 1990; Bevan and Bevan, 1981; Mione *et al.*, 1990). Developmental studies (see Section II.A,B) suggest that these patterns of innervation are also temporally specific, showing substantial quantitative variations in the density of different nerve populations. During the period between puberty and the achievement of maximum body size, some noradrenergic perivascular nerve plexuses continue to increase in density or remain stable. However, the nerves of the femoral artery show sharp decreases over the same period (Cowen *et al.*, 1982). Staining for neuropeptide-containing nerves also changes during mature life (Dhall *et al.*, 1986).

A. 5-Hydroxytryptamine (5-HT) in Perivascular Nerves

The presence of 5-HT in perivascular nerves has been demonstrated in intestinal (Dahlstrom and Ahlman, 1983; Griffith and Burnstock, 1983), cerebral (Chang *et al.*, 1988; Edvinsson *et al.*, 1984; Cowen *et al.*, 1986), mesenteric (Gale and Cowen, 1988), and femoral (Gale *et al.*, 1989) arteries. It has been shown to be present in noradrenergic nerve fibers by studies that demonstrated its colocalization with catecholamines or their synthesizing enzymes (Jackwoski *et al.*, 1989; Chang *et al.*, 1988), or its disappearance following surgical sympathectomy (Cowen *et al.*, 1986).

5-HT has been considered to be present in sympathetic nerves largely as a result of uptake$_1$ (Iversen, 1971; Chang *et al.*, 1988), the uptake system for noradrenaline. Immunohistochemical studies of cerebrovascular nerves showed that exsanguination resulted in the abolition or substantial reduction of 5-HT staining (Levitt and Duckles, 1986; Jackowski *et al.*, 1989), confirming that the majority of neuronal 5-HT was present as a result of uptake, probably of 5-HT released from platelets. However, *in vitro* studies have suggested that 5-HT may be synthesized in sympathetic neurons (Sah and Matsumoto, 1987). Stronger evidence that sympathetic, perivascular neurons can synthesize 5-HT has come from a recent immunohistochemical study demonstrating staining for tryptophan hydroxylase, the rate-limiting enzyme in the synthesis of 5-HT, in nerve fibers around brain vessels of the rat and cat (Chedotal and Hamel, 1990).

Developmental studies of 5-HT-immunoreactive perivascular nerves have shown some differences compared with noradrenergic nerves (Gale *et al.*, 1989). For example, the density of 5-HT-immunoreactive nerves in the femoral artery of the rabbit shows no decrease between 6 weeks and 6 months of age, whereas the noradrenergic nerves show a sharp decline (Cowen *et al.*, 1982). These observations support the notion that coexisting neuroactive substances can be differentially regulated during maturity as well as during earlier development (Dhall *et al.*, 1986).

B. Extraneuronal and Neuronal Influences on Plasticity in Maturity

Recent work on sympathetic neurons has indicated that plasticity persists into maturity (Voyvodic, 1989; Adler and Black, 1984). Experimental manipulations of target size have shown that sympathetic neurons adapt their axonal (Todd, 1986) and dendritic (Voyvodic, 1989) branching patterns to the new target dimensions and that these adaptations are largely independent of the afferent (preganglionic) nervous input (i.e., the trophic relationship controlling connectivity between neuron and target is dominated by retrograde influences from the target tissue) (Voyvodic, 1987; Voyvodic, 1989; Yawo, 1987). Perhaps trophic support of neurons by their target tissues persists into maturity, providing mechanisms whereby injury or changes in dynamic features, such as body size, can evoke appropriate growth responses from the nervous system (Voyvodic, 1989; Yawo, 1987; Purves, 1988; Purves and Lichtman, 1985; Purves *et al.*, 1986).

Nerves may continue to exert trophic influences on their vascular muscle effectors during maturity despite evidence from earlier studies (Bevan and Tsuru, 1981) that these "orthograde" effects were limited to the immediate postnatal period. While denervation during early development causes structural impairment of the blood vessel wall, sympathectomy of blood vessels during maturity appears to have an opposite effect, resulting in hypertrophy and signs of increased metabolic activity in smooth muscle cells and fibroblasts (Fronek *et al.*, 1978; Dimitriadou *et al.*, 1988; Branco *et al.*, 1984). These hypertrophic effects can be prevented by infusion over several days of purine agonists but not by infusion of noradrenaline (Albino-Teixeira *et al.*, 1990). Thus, purines, which are cotransmitters with noradrenaline in a number of vascular beds (Burnstock, 1990), may also have a trophic role in regulating smooth muscle activity. As a result, nerve injury may cause disinhibition of smooth muscle metabolic activity and the stimulation of nerve regeneration, perhaps through increased production of growth factors.

C. Effects of Chronic Sympathectomy

Evidence for plasticity of neurotransmitter expression comes from studies of the effects of surgical and long-term chemical sympathectomy of cerebral perivascular nerves (Mione *et al.*, 1990; Gibbins and Morris, 1988). Chronic sympathectomy causes substantial changes in the transmitter expression of remaining, nonsympathetic nerves, one feature of which is an increase in neuropeptide Y (NPY) expression in VIP-containing parasympathetic nerve fibers. Chronic sympathectomy of the iris results in increased substance P immunoreactivity (Kessler *et al.*, 1983; Cole *et al.*, 1983) in presumptive afferent fibers. While some of these studies have been carried out on tissues other than blood vessels, the demonstration that mature autonomic and sensory nerves retain plasticity of their neurotransmitter phenotype may be helpful in understanding the changes that occur in the innervation of blood vessels during old age (see Section VII).

V. Changes during Aging in Perivascular Nerves

A. Changes in Sympathetic Ganglia

A discussion about aging in the innervation of blood vessels should begin by stating that aging appears to involve the loss of only a minority of the neurons supplying autonomic perivascular fibers. In young and old rats, counts of neuronal cell bodies in autonomic sympathetic ganglia, which supply postganglionic axons to many peripheral blood vessels, show decreased packing density with age (Baker and Santer, 1988; Partanen *et al.*, 1982) but no evidence of cell loss (Santer, 1991). However, extensive cell death of autonomic enteric neurons supplying, amongst other tissues, intestinal blood vessels has been indicated from studies of the small and large intestine of aging rats (Santer and Baker, 1988) and the small intestine of guinea pigs (Gabella, 1989).

Metabolic indices of neuronal function during aging have shown apparently contradictory results in sympathetic ganglia. Glucose utilization increases in the superior cervical ganglion in old age (Partanen *et al.*, 1982) while succinate dehydrogenase activity decreases in SCG and celiac-mesenteric ganglia (CSMG) (Baker and Santer, 1990). Studies of tyrosine hydroxylase, the noradrenaline-synthesizing enzyme, have shown increases in its mRNA (Kedzierski and Porter, 1990) and in enzyme activity (Yurkewicz *et al.*, 1981; Reis *et al.*, 1977) in the SCG in old age.

However, previous studies using fluorescence histochemistry have shown reduced catecholamine synthesis in the CSMG (Santer, 1979) and reduced catecholamine levels in the SCG of aging rats (Partanen *et al.*, 1982) and humans (Hervonen *et al.*, 1978). At present, it is not clear whether these divergent results are due to the different sympathetic ganglia studied or the different methods used.

B. Changes in Perivascular Nerve Fibers

While in sympathetic ganglia the general tendency is for an increase in activity and expression of noradrenergic enzymes with age (see Section V.A), the perivascular terminal nerve fibers of these neurons rarely exhibit obvious increases in noradrenaline levels in old age, and decreases in noradrenergic indices are more common (Cowen and Burnstock, 1986). Neurotransmitter expression in autonomic nerves to blood vessels varies during aging in a regionally specific way. The noradrenergic nerves supplying coronary, internal carotid and cerebral arteries, different levels of the renal arterial circulation, and the abdominal aorta decline markedly in old rats and rabbits, whereas the innervation of coronary veins, mesenteric arteries, and pancreatic arterioles is not impaired (Amenta and Mione, 1988; Mione *et al.*, 1988; Vega *et al.*, 1990; Santer, 1982; Cowen *et al.*, 1982; Saba *et al.*, 1984; Cowen and Thrasivoulou, 1990; Dhall *et al.*, 1986). A single report has documented increases in noradrenergic nerve density around vasa nervorum in the sciatic nerve of aged rats (Mione *et al.*, 1987), while in this laboratory, increases of noradrenergic nerve density on the tail vein of aged rats have been observed (Andrews and Cowen, unpublished data).

Nonadrenergic nerves may not follow the same patterns of age change as noradrenergic fibers. Thus, CGRP-immunoreactive nerves to cerebral arteries increase in old age (Mione *et al.*, 1988) but decline in mesenteric vessels (Dhall *et al.*, 1986). VIP nerves around pulmonary blood vessels decline in old age (Geppetti *et al.*, 1988). Choline esterase staining around cerebral arteries declines in old age (Saba *et al.*, 1984) as do choline and acetyl choline levels (Marchi and Giacobini, 1980) and choline acetyl transferase activity (Marchi *et al.*, 1980) in the iris of old chickens.

Age changes are temporally specific, thus noradrenergic nerves decline markedly in nerves supplying the renal artery in old (>3.5 yr) rabbits, whereas noradrenergic nerves in the femoral artery also decline rapidly but at an earlier period of life (see Section IV). Age changes may also be species specific: noradrenergic nerves to the basilar and carotid arteries decline in old age in the rat (Santer, 1982; Mione *et al.*, 1988) but not in the rabbit (Cowen *et al.*, 1982).

C. Changes of Amine Uptake in Perivascular Nerves

Studies of changes in noradrenaline uptake during aging have mainly been carried out using *in vitro* procedures in combination with pharmacological blocking agents, such as cocaine (Duckles, 1987; Shibata *et al.*, 1971). No obvious age-related changes have been demonstrated (Duckles *et al.*, 1985) although there may be some reductions in noradrenaline uptake with age in some blood vessels (see Docherty,

1986) and in the iris (Hoffman *et al.*, 1980). 5-HT-immunoreactivity in perivascular nerves from old animals is reduced in all the vessels studied (Gale *et al.*, 1989) whereas noradrenergic nerves show reduced, stable, or, occasionally, increased nerve plexuses in different vascular beds (see Section V.B). Noradrenaline content of nerves during aging is likely to be affected by changes in synthesis, uptake, and, possibly, storage whereas 5-HT content is probably controlled mainly by uptake (see Section IV.A).

In this laboratory, we have attempted to study changes in amine uptake during aging using microscopical techniques. Age changes in immunohistochemical staining are often seen as alterations in brightness of fluorescence labeling as well as in the frequency of stained nerve fibers or cell bodies. A densitometric application of image analysis has been used in an attempt to quantify these changes in immunohistochemical stain intensity (Cowen and Thrasivoulou, 1991). A model system (Schipper and Tilders, 1983) has been used to establish the relationship between stain intensity and antigen concentration. Preliminary results suggest reductions in net accumulation of 5-HT (Cowen and Thrasivoulou, 1990) (Fig. 1) and in the sensitivity of 5-HT-uptake (Violet and Cowen, 1990) in old age. The microscopical

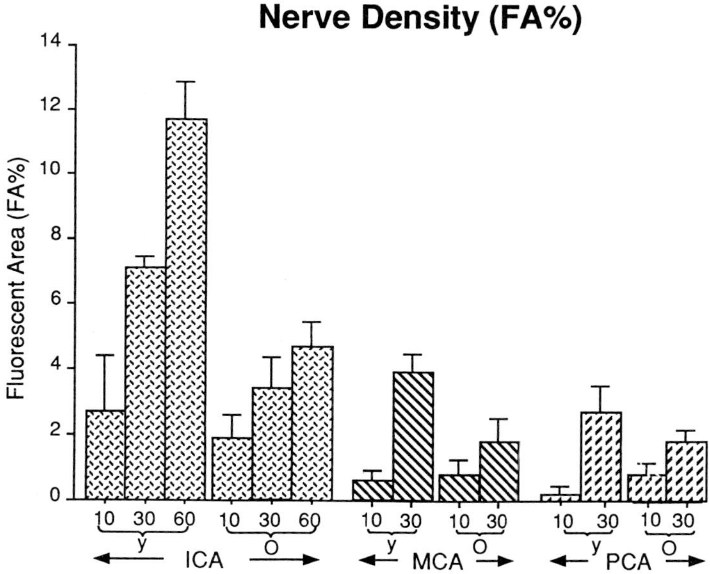

Figure 1 Histogram showing mean fluorescent area (% scanned surface area) of 5-HT-immunoreactive nerves in internal carotid (ICA), middle (MCA), and posterior (PCA) cerebral arteries from young (y) and old (o) rats, following incubation with 10^{-7} M 5-HT for 10, 30, or 60 min. Measurements made using image analysis. Error bars indicate SEM. Values for 30 and 60 min incubations of all vessels from old animals were significantly different from equivalent values from young animals (ICA, 30 min: $P < 0.01$; ICA, 60 min: $P < 0.001$; MCA, 30 min: $P < 0.05$; PCA, 30 min: $P < 0.05$). (Modified with permission from Cowen and Thrasivoulou, 1990.)

assay system used for these experiments (Cowen and Thrasivoulou, 1991) may have advantages over conventional assay procedures involving tissue homogenization. In heterogeneous tissues, such as blood vessels, amine-handling capacity exists in several cell types, including nerves, endothelium (Burnstock, 1990), smooth muscle (Iversen, 1971), and fibroblasts (Jacobowitz, 1972). In homogenized tissue samples, amine uptake by nonneuronal tissues could prevent sensitive discrimination of amine uptake in nerve fibers because of their relatively small volume compared to that of other tissue components of the vessel wall.

Studies of age changes in the peripheral processes of autonomic neurons have generally utilized histochemical or immunohistochemical techniques at the light microscopical level. A general problem with these studies is the difficulty of distinguishing structural from biochemical changes in the particular nerve population under investigation. For this reason, it should not be assumed that the disappearance of a particular neurotransmitter during aging provides evidence that the particular group of axons has degenerated.

The general neuronal marker PGP 9.5 (Thompson *et al.*, 1983) has been used in an attempt to demonstrate changes in the total population of perivascular nerves, irrespective of transmitter phenotype (Cowen and Thrasivoulou, 1990). In cerebral arteries, there was a significant decrease in density of PGP-immunoreactive perivascular fibers, indicating a loss of some axons to cerebral blood vessels in old age (Fig 2). However, because PGP stains the majority of autonomic terminal fibers, this study was not able to identify which particular population of axons was affected. The surviving nerve fibers showed evidence of reduced amine uptake during aging, indicating that the regulation of neurotransmitter phenotype may be separate from the regulation of growth or degeneration in perivascular nerves. Future studies may be able to show whether neurochemical changes precede structural degeneration during aging in sympathetic perivascular nerve plexuses.

VI. Regulatory Factors in Aging of Perivascular Nerves

An investigation of age changes in a range of neurotransmitter substances in nerves supplying cerebral blood vessels has shown that not all the changes that occur in old age are degenerative (Mione *et al.*, 1988). While deficits in staining for noradrenaline and 5-HT were demonstrated in vessels from old animals, immunoreactivity for VIP and CGRP increased in old age. NPY immunoreactivity remained constant. Following chronic sympathectomy, substance P expression in presumptive sensory nerves (Cole *et al.*, 1983; Kessler *et al.*, 1983) and NPY-expression in parasympathetic nerves (Gibbins and Morris, 1988; Mione *et al.*, 1990) to autonomically innervated tissues have both been shown to increase (see Section IV.C). Experiments using NGF and anti-NGF antibodies (Kessler *et al.*, 1983) have suggested that competition between sympathetic and sensory fibers for target-derived growth factors could explain these apparently compensatory interactions, although the participation of parasympathetic nerves is such a competitive relationship has not been demonstrated.

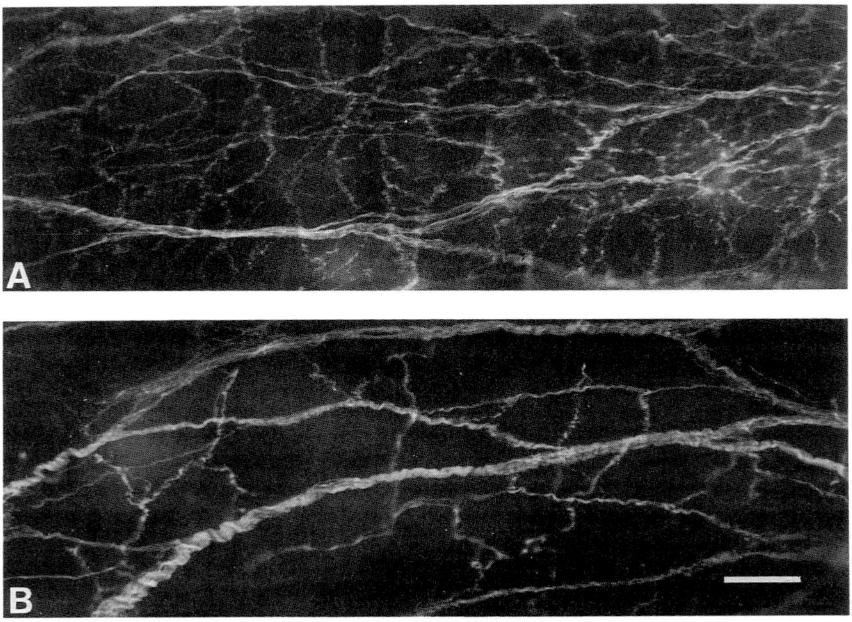

Figure 2 Photomicrographs of PGP-immunoreactive nerves in posterior cerebral artery from young (A) and old (B) rats. Note the loss of fine fibers in old age. Scale bar, 50 μm. Measurements of nerve intercept density using image analysis showed significant ($P < 0.01$) reductions in nerve density in the vessels from old rats. (Reproduced with permission from Cowen and Thrasivoulou, 1990.)

Since changes in old age show some similarities with those following chronic sympathectomy, it is tempting to consider whether alterations in one group of nerves in tissues with multiple innervation trigger reciprocal changes in other populations of nerves, perhaps through the mechanism of competition for common, target-produced growth factors. The nature of these changes is such that they could be nonadaptive and even destabilizing of cardiovascular homeostasis. The discovery of new growth factors with localized patterns of distribution in the nervous system (Thoenen, 1991) may help to explain the regulation and aging of localized patterns of neurochemical coding in the nerve supply to blood vessels and other tissues.

I have described the establishment of an interrelationship between neurons and their target tissues during early development and maturity. This may provide a means whereby neurons can respond to long-term changes such as growth or perhaps, in the case of cardiovascular homeostatic mechanisms, seasonal and other alterations in the environment. In addition, sympathetic nerves appear to be able to influence their targets through electrical activity or other undiscovered mechanisms. It is only a short step to understanding changes that occur in old age as the result of disturbances of these regulatory equilibria established earlier in development. The lack of any obvious separation between the processes of development and aging in the autonomic (Cowen *et al.*, 1982) and other parts of the nervous system (see (Haugh, 1984) provides support for this "developmental approach" (Russell, 1987) to aging.

We have now carried out experiments that support the hypothesis that age changes in nonneuronal tissues of the blood vessel wall regulate aging in their innervating nerve fibers (Gavazzi *et al.*, 1992). Cerebral arteries show a reduction of about 50% in their overall density of innervation as revealed by PGP 9.5 in old age (see Section V.C). Transplantation of these vessels taken from young and old donors *in oculo* into young hosts results in a pattern of reinnervation appropriate to the donor's age rather than that of the host. Thus, old vessels become reinnervated with about 50% fewer nerves than young vessels, showing that some feature of the target tissue determines the extent of reinnervation. Donor Schwann cells appear not to be involved in regulation of nerve regrowth in this model. Further experiments in our and other laboratories have shown that old sympathetic neurons from the superior cervical ganglion, which supply fibers to cerebral arteries, have unimpaired capacity for regrowth and plasticity in old age (Stieg *et al.*, 1991; Gavazzi and Cowen, 1991), supporting the concept that age changes in target tissues determine neuronal phenotype in old age.

VII. Conclusions

It appears that target-derived trophic influences may be important in the development and aging of cardiovascular innervation. During development, an interdependency is established between autonomic nerves and their blood vessel and other effector target tissues. In this relationship, while growing smooth muscle cells may be dependent on nerves for their maturation, nerves are dependent on target-produced substances and other products of their cellular environment for their survival and expression of an appropriate transmitter phenotype. During later periods of postnatal development, expression in nerves of the phenotype appropriate to particular targets continues to depend on contact with the target tissue, perhaps through the constant access of axons to target-produced substances. These substances are retrogradely transported back to the nerve cell body where they influence a wide range of cellular activities. Axons continue to grow and retract, probably throughout life, and it appears possible that the same target-produced factors determine the point of balance between these two processes. The changes that occur in perivascular nerves during old age may result from a shift in the balance between growth and retraction, initiated by changes in the target tissues. The changes induced in one group of nerves in this way may, through unknown mechanisms, have reciprocal effects on other groups of nerves innervating the same tissue. In this way, the delicate equilibria established during development for the maintenance of cardiovascular homeostasis may be disrupted in old age. It is interesting to consider whether this disruption may extend to higher centers of nervous control of the cardiovascular system through similar retrograde, transsynaptic trophic influences.

Acknowledgments

I thank my colleagues, Isabella Gavazzi, Chris Thrasivoulou, and Tim Andrews, for their discussion and helpful comments and for their part in the studies on which this chapter is based. This work was supported by the Wellcome Trust and the Peter Samuels Royal Free Fund. My thanks to my wife, Liz, for valuable editorial assistance.

References

Adler, J. E., and Black, I. B. (1984). Plasticity of substance P in mature and aged sympathetic neurons in culture. *Science 225,* 1499–1500.

Albino-Teixeira, A., Azevedo, I., Branco, D., and Osswald, W. (1990). Purine agonists prevent trophic changes caused by sympathetic denervation. *Eur. J. Pharmacol. 179,* 141.

Amenta, F., and Mione, M. C. (1988). Age-related changes in the noradrenergic innervation of the coronary arteries in old rats: A fluorescent histochemical study. *J. Auton. Nerv. Syst. 22,* 247–251.

Baker, D. M., and Santer, R. M. (1988). Morphometric studies on pre- and paravertebral sympathetic neurons in the rat: Changes with age. *Mech. Ageing Dev. 42,* 139–145.

Baker, D. M., and Santer, R. M. (1990). Development of a quantitative histochemical method for determination of succinate dehydrogenase activity in autonomic neurons and its application to the study of aging in the autonomic nervous system. *J. Histochem. Cytochem. 38,* 525–531.

Barde, Y.-A. (1989). Trophic factors and neuronal survival. *Neuron 2,* 1525–1534.

Bevan, J. A., and Bevan, R. D. (1981). Developmental influences on vascular structure and function. *In* "CIBA Foundation Symposium 83: Development of the Autonomic Nervous System." (K. Elliot and G. Lawrenson, eds.), pp. 101–107. Pitman Books Ltd, London.

Bevan, R. D., and Tsuru, H. (1981). Functional and structural changes in the rabbit ear artery after sympathetic denervation. *Circ. Res. 49,* 478–485.

Black, I. B., Adler, J. E., Dreyfus, C. F., Jonakait, G. M., Katz, D. M., LaGamma, E. F., and Markey, K. M. (1984). Neurotransmitter plasticity at the molecular level. *Science 225,* 1266–1270.

Branco, D., Albino-Teixeira, A., Azevedo, I., and Osswald, W. (1984). Structural and functional alterations caused at the extraneuronal level by sympathetic denervation of blood vessels. *Naunyn-Schmiedeberg's Arch. Pharmacol. 326,* 302–312.

Burnstock, G. (1990). Local mechanisms of blood flow control by perivascular nerves and endothelium. *J. Hypertension 8,* S95–S106.

Chang, J.-Y., Owman, C., and Steinbusch, H. W. M. (1988). Evidence for coexistence of serotonin and noradrenaline in sympathetic nerves supplying brain vessels of guinea-pig. *Brain Res. 438,* 237–246.

Chedotal, A., and Hamel, E. (1990). Serotonin-synthesizing nerve fibers in rat and cat cerebral arteries and arterioles: Immunohistochemistry of tryptophan-5-hydroxylase. *Neurosci. Lett. 116,* 269.

Cole, D. F., Bloom, S. R., Burnstock, G., Butler, J. M., McGregor, G. P., Saffrey, M. J., Unger, W. G., and Zhang, S. Q. (1983). Increase in SP-like immunoreactivity in nerve fibers of rabbit iris and ciliary body one to four months following sympathetic denervation. *Exp. Eye Res. 37,* 191–197.

Cowen, T., and Burnstock, G. (1986). Development, aging, and plasticity of perivascular autonomic nerves. *In* "Developmental Neurobiology of the Autonomic Nervous System." (P. M. Gootman, ed.), pp. 211–231. Humana Press, New York.

Cowen, T., and Thrasivoulou, C. (1990). Cerebrovascular nerves in old rats show reduced accumulation of 5-hydroxytryptamine and loss of nerve fibres. *Brain Res. 513,* 237–243.

Cowen, T., and Thrasivoulou, C. (1991). Methods of image analysis for the measurement of densitometric data on immunofluorescence-stained preparations of nerves. *In* "Quantitative Methods of Neuroanatomy." (M. J. Stewart, ed.), pp. 85–96. Open University Press, Milton Keynes.

Cowen, T., Haven, A. J., Wen Qin, C., Gallen, D. D., Franc, F., and Burnstock, G. (1982). Development and ageing of perivascular adrenergic nerves in the rabbit. A quantitative fluorescence histochemical study using image analysis. *J. Auton. Nerv. Syst. 5,* 317–336.

Cowen, T., Alafaci, C., Crockard, H. A., and Burnstock, G. (1986). 5-HT-containing nerves to major cerebral arteries of the gerbil originate in the superior cervical ganglion. *Brain Res. 384,* 51–59.

Dahlstrom, A., and Ahlman, H. (1983). Immunocytochemical evidence for the presence of tryptaminergic nerves of blood vessels, smooth muscle and myenteric plexus in the rat small intestine. *Acta Physiol. Scand. 117,* 589–591.

De Champlain, J., Malmfors, T., Olson, L., and Sachs, C. (1970). Ontogenesis of peripheral adrenergic neurons in the rat: Pre- and postnatal observations. *Acta Physiol. Scand. 80,* 276–288.

Dhall, U., Cowen, T., Haven, A. J., and Burnstock, G. (1986). Perivascular noradrenergic and peptide-containing nerves show different patterns of change during development and ageing in the guinea-pig. *J. Auton. Nerv. Syst. 16,* 109–126.

Dimitriadou, V., Aubineau, P., Taxi, J., and Seylaz, J. (1988). Ultrastructural changes in the cerebral artery wall induced by long-term sympathetic denervation. *Blood Vessels 25,* 122–143.

Docherty, J. R. (1986). Aging and the cardiovascular system. *J. Auton. Pharmac. 6,* 77–84.

Dolezel, S., Gerova, M., and Gero, J. (1974). Postnatal development of the sympathetic innervation in skeletal muscles of the dog. *Physiol. bohemoslov. 23,* 138–139.

Duckles, S. P. (1987). Effect of norepinephrine uptake blockade on contractile responses to adrenergic nerve stimulation of isolated rat blood vessels: Influence of age. *J. Pharm. Exp. Ther. 243,* 521–526.

Duckles, S. P., Carter, B. J., and Williams, C. L. (1985). Vascular adrenergic neuroeffector function does not decline in aged rats. *Circ. Res. 56,* 109–116.

Edvinsson, L., Birath, E., Uddman, R., Lee, T. J.-F., Duverger, D., MacKenzie, E. T., and Scatton, B. (1984). Indoleaminergic mechanisms in brain vessels: Localization, concentration, uptake and *in vitro* responses of 5-hydroxytryptamine. *Acta Physiol. Scand. 121,* 291–299.

Friedman, W. F., Pool, P. E., Jacobowitz, D., Seagren, S. C., and Braunwald, E. (1968). Sympathetic innervation of the developing rabbit heart. *Circ. Res. 23,* 25–31.

Fronek, K., Bloor, C. M., Amiel, D., and Chvapil, M. (1978). Effect of long-term sympathectomy on the arterial wall in rabbits and rats. *Exp. Mol. Path. 28,* 279–289.

Gabella, G. (1989). Fall in the number of myenteric neurons in aging guinea pigs. *Gastroenterology 96,* 1487–1493.

Gale, J. D., and Cowen, T. (1988). The origin and distribution of 5-hydroxytryptamine-like immunoreactive nerve fibres to major mesenteric blood vessels of the rat. *Neurosci. 24,* 1051–1059.

Gale, J. D., Alberts, J. C. J., and Cowen, T. (1989). A quantitative study of changes in old age of 5-hydroxytryptamine-like immunoreactivity in perivascular nerves of the rabbit. *J. Auton. Nerv. Syst. 28,* 51–60.

Gallen, D. D., Cowen, T., Griffith, S. G., Haven, A. J., and Burnstock, G. (1982). Functional and non-functional nerve-smooth muscle transmission in the renal arteries of new born and adult rabbit and guinea pig. *Blood Vessels 19,* 237–246.

Gavazzi, I., and Cowen, T. (1991). Use of transplants to investigate target-neuron interaction in old age. *Eur. J. Neurosci.* Suppl. 4, 187.

Gavazzi, I., Andrews, T., Thrasivoulou, C., and Cowen, T. (1992). Influence of target tissues on their innervation in old age: A transplantation study. *NeuroReport 3,* 117–120.

Geppetti, P., De Rossi, M., Mione, M. C., Renzi, D., and Amenta, F. (1988). Age-related changes in vasoactive intestinal polypeptide levels and distribution in the rat lung. *J. Neural. Transm. 74,* 1–10.

Gibbins, I. L., and Morris, J. L. (1988). Coexistence of immunoreactivities to neuropeptide Y and vasoactive intestinal peptid in non-noradrenergic axons innervating guinea pig cerebral arteries after sympathectomy. *Brain Res. 444,* 402–406.

Griffith, S. G., and Burnstock, G. (1983). Serotoninergic neurons in human fetal intestine: An immunohistochemical study. *Gastroenterology 85,* 929–937.

Haugh, H. (1984). Macroscopic and microscopic morphometry of the human brain and cortex. A survey in the light of new results. *Brain Path. 1,* 123–146.

Hervonen, A., Vaalasti, A., Partanen, M., Kanerva, L., and Hervonen, H. (1978). Effects of ageing on the histochemically demonstrable catecholamines and acetylcholinesterase of human sympathetic ganglia. *J. Neurocytol. 7,* 11–23.

Heumann, R., Korsching, S., Scott, J., and Thoenen, H. (1984). Relationship between levels of nerve growth factor (NGF) and its messenger RNA in sympathetic ganglia and peripheral target tissue. *EMBO J. 3,* 3183–3189.

Hill, C. E., Hirst, G. D. S., and van Helden, D. F. (1983). Development of sympathetic innervation to proximal and distal arteries of the rat mesentery. *J. Physiol. 338,* 129–147.

Hoffman, D., Salzman, S. K., Marchi, M., and Giacobini, E. (1980). Norepinephrine levels in peripheral nerve terminals during development and aging. *J. Neurochem. 34,* 1785–1787.

Iversen, L. L. (1971). Role of transmitter uptake mechansism in synaptic neurotransmission. *Br. J. Pharmacol. 41,* 571–591.

Jackowski, A., Crockard, A., Burnstock, G., and Lincoln, J. (1989). Alterations in serotonin and neuropeptide Y content of cerebrovascular sympathetic nerves following experimental subarachnoid hemorrhage. *J. Cerebr. Blood. Flow. Metabol. 9,* 271.

Jacobowitz, D. M. (1972). Localization of catechol-O-methyl-transferase and monoamine oxidase in fibroblasts in tissue culture. *Life Sci. 11,* 965–974.

Kedzierski, W., and Porter, J. C. (1990). Quantitative study of tyrosine hydroxylase mRNA in catecholaminergic neurons and adrenals during development and aging. *Mol. Brain Res. 7,* 45.

Kessler, J. A., Bell, W. O., and Black, I. B. (1983). Interactions between the sympathetic and sensory innervation of the iris. *J. Neurosci. 3,* 1301–1307.

Korsching, S., and Thoenen, H. (1983). Nerve growth factor in sympathetic ganglia and corresponding target organs of the rat: Correlation with density of sympathetic innervation. *Proc. Natl. Acad. Sci. U.S.A. 80,* 3513–3516.

Lebowitz, E. A., Novick, J. S., and Rudolph, A. M. (1972). Development of myocardial sympathetic innervation in the fetal lamb. *Pediatr. Res. 6,* 887–893.

Levi-Montalcini, R. (1987). The nerve growth factor: 35 years later. *Science 237,* 1154–1162.

Levi-Montalcini, R., and Hamburger, V. (1953). A diffusible agent of mouse sarcoma producing hyperplasia of sympathetic ganglia and hyperneurotization of viscera in the chick embryo. *J. Exp. Zool. 123,* 233–287.

Levitt, B., and Duckles, S. P. (1986). Evidence against serotonin as a vasoconstrictor neurotransmitter in the rabbit basilar artery. *J. Pharm. Exp. Ther. 238,* 880–885.

Lipp, J. A. M., and Rudolph, A. M. (1972). Sympathetic nerve development in the rat and guinea-pig heart. *Biol. Neonate 21,* 76–82.

Ljung, B., and Stage, D. (1975). Postnatal ontogenetic development of neurogenic and myogenic control in the rat portal vein. *Acta Physiol. Scand. 94,* 112–127.

Lundberg, J., Ljung, B., Stage, D., and Dahlstrom, A. (1976). Postnatal ontogenic development of the adrenergic innervation pattern in the rat portal vein. *Cell Tiss. Res. 172,* 15–27.

Marchi, M., and Giacobini, E. (1980). Development and aging of cholinergic synapses II. continuous growth of acetylcholine and choline levels in autonomic ganglia and iris of the chick. *Dev. Neurosci. 3,* 39–48.

Marchi, M., Hoffman, D. W., and Giacobini, E. (1980). Development and aging of cholinergic synapses IV. Acetylcholinesterase and choline acetyltransferase activities in autonomic ganglia and iris of the chick. *Dev. Neurosci. 3,* 235–247.

Mione, M. C., Cavallotti, C., Collier, W. L., and Amenta, F. (1987). The noradrenergic innervation of the vasa nervorum in old rats: A fluorescence histochemical study. *J. Auton. Nerv. Syst. 18,* 177–180.

Mione, M. C., Dhital, K. K., Amenta, F., and Burnstock, G. (1988). An increase in the expression of neuropeptidergic vasodilator, but not vasoconstrictor, cerebrovascular nerves in aging rats. *Brain Res. 460,* 103–113.

Mione, M., Ralevic, V., and Burnstock, G. (1990). Peptides and vasomotor mechanisms. *Pharmacol. Ther. 46,* 429–468.

Mione, M. C., Cavanagh, J. F. R., Lincoln, J., Milner, P., and Burnstock, G. (1990). Long-term chemical sympathectomy leads to an increase of neuropeptide Y immunoreactivity in cerebrovascular nerves and iris of the developing rat. *Neurosci 34,* 369–378.

Partanen, M., Hervonen, A., and Rapoport, S. I. (1982). Microspectrofluorimetric quantitation of histochemically demonstrable catecholamines in peripheral and brain catecholamine-containing neurons in male Fischer-344 rats at different ages. *In* "Mechanisms of Aging of the Nervous System." (E. Giacobini, ed.), pp. 161–171. Raven Press, New York.

Patterson, P. H. (1978). Environmental determination of autonomic neurotransmitter functions. *Ann. Rev. Neurosci. 1,* 1–18.
Potter, D. D., Furshpan, E. J., and Landis, S. C. (1983). Transmitter status in cultured rat sympathetic neurons: Plasticity and multiple function. *Fed. Proc. 42,* 1626–1632.
Purves, D. (1980). Neuronal competition. *Nature 287,* 585–586.
Purves, D. (1988). "Body and Brain, A Trophic Theory of Neural Connections." Harvard University Press, Cambridge, Massachusetts.
Purves, D., and Lichtman, J. (1985). Geometrical differences among homologous neurons in mammals. *Science 228,* 298–302.
Purves, D., Hadley, R. D., and Voyvodic, J. T. (1986). Dynamic changes in the dendritic geometry of individual neurons visualized over periods of up to three months in the superior cervical ganglion of living ice. *J. Neurosci. 6,* 1051–1060.
Read, J. B., and Burnstock, G. (1970). Development of the adrenergic innervation and chromaffin cells in the human fetal gut. *Develop. Biol. 22,* 513–524.
Reis, D. J., Ross, R. A., and Joh, T. H. (1977). Changes in the activity and amounts of enzymes synthesisizing catecholamines and acetylcholine in brain, adrenal medulla, and sympathetic ganglia of aged rat and mouse. *Brain Res. 136,* 465–474.
Rickenbacher, J., and Ruflin, G. (1974). Zur entwicklung der inervation der extremitatengefasse beim huhnchen. *VASA 3,* 5–9.
Russell, R. L. (1987). Modern biological theories of aging. *In* "Aging." (H. R. Warner, R. N. Butler, R. L. Sprott, and E. L. Schneider, eds.), Vol. 33, pp. 35–61. Raven Press, New York.
Saba, H., Cowen, T., Haven, A. J., and Burnstock, G. (1984). Uneven reduction of noradrenergic perivascular nerve densities in the left and right cerebral arteries of old rabbits. *J. Cereb. Blood Flow Metab. 4,* 283–289.
Sah, D. W. Y., and Matsumoto, S. G. (1987). Evidence for serotonin synthesis, uptake and release in dissociated rat sympathetic neurones in culture. *J. Neurosci. 7,* 391–399.
Santer, R. M. (1979). Fluorescence histochemical evidence for decreased noradrenaline synthesis in sympathetic neurones of aged rats. *Neurosci. Letts. 15,* 177–180.
Santer, R. M. (1982). Fluorescence histochemical observations on the adrenergic innervation of the cardiovascular system in the aged rat. *Brain Res. Bull. 9,* 667–672.
Santer, R. M. (1991). Morphological evidence for the maintenance of the cervical sympathetic system in aged rats. *Neurosci. Letts. 130,* 248–250.
Santer, R. M., and Baker, D. M. (1988). Enteric neuron numbers and sizes in Auerbach's plexus in the small and large intestine of young adult and aged rats. *J. Auton. Nerv. Syst. 25,* 59–67.
Schipper, J., and Tilders, F. J. H. (1983). A new technique for studying specificity of immunocytochemical procedures: Specificity of serotonin immunostaining. *J. Histochem. Cytochem. 31,* 12–18.
Shelton, D. L., and Reichard, L. F. (1984). Expression of the b-nerve growth factor gene correlates with the density of sympathetic innervation in effector organs. *Proc. Natl. Acad. Sci. U.S.A. 81,* 7951–7955.
Shibata, S., Hattori, K., Sakurai, I., Mori, J., and Fujiwara, M. (1971). Adrenergic innervation and cocaine-induced potentiation of adrenergic responses of aortic strips from young and old rabbits. *J. Pharm. Exp. Ther. 177,* 621–632.
Stevens, L. M., and Landis, S. C. (1990). Target influences on transmitter choice by sympathetic neurons developing in the anterior chamber of the eye. *Dev. Biol. 137,* 109–124.
Stieg, P., Strömberg, I., and Olson, L. (1991). Effects of donor age on superior cervical ganglion transplants: Evaluation by Falck–Hillarp histochemistry and immunocytochemistry. *Exp. Brain Res. 85,* 55–65.
Su, C., Bevan, J. A., Assali, N. S., and Brinkman, C. R. (1977). Development of neuroeffector mechanisms in the carotid artery of the fetal lamb. *Blood Vessels 14,* 12–24.
Thoenen, H. (1991). The changing scene of neurotrophic factors. *TINS 14,* 165–170.
Thoenen, H., and Barde, Y.-A. (1980). Physiology of nerve growth factor. *Physiol Rev. 60,* 1284–1335.

Thoenen, H., Angeletti, P. U., Levi-Montalcini, R., and Kettler, R. (1971). Selective induction by nerve growth factor of tyrosine hydroxylase and dopamine-B-hydroxylase in the rat superior cervical ganglia. *Proc. Natl. Acad. Sci. U.S.A. 68,* 1598–1602.

Thompson, R. J., Doran, J. F., Jackson, P., Dhillon, A. P., and Rode, J. (1983). PGP 9.5—a new marker for vertebrate neurons and neuroendocrine cells. *Brain Res. 278,* 224–228.

Todd, M. E. (1980). Development of adrenergic innervation in rat peripheral vessels: A fluorescence microscopic study. *J. Anat. 131,* 121–133.

Todd, M. E. (1986). Trophic interactions between rat nerves and blood vessels in denervated peripheral arteries and in anterior eye chamber transplants. *Circ. Res. 58,* 641–652.

Todd, M. E., and Tokito, M. K. (1981). An ultrastructural investigation of developing vasomotor innervation in rat peripheral vessels. *Amer. J. Anat. 160,* 195–212.

Vega, J. A., Ricci, A., and Amenta, F. (1990). Age-dependent changes on the sympathetic innervation of the rat kidney. *Mech. Age. Dev. 54,* 185.

Violet, J., and Cowen, T. (1990). Changes in uptake of serotonin into sympathetic perivascular nerves in old age. *J. Anat. 173,* 220P.

Voyvodic, J. T. (1987). Development and regulation of dendrites in the rat superior cervical ganglion. *J. Neurosci. 7,* 904–912.

Voyvodic, J. T. (1989). Peripheral target regulation of dendritic geometry in the rat superior cervical ganglion. *J. Neurosci. 9,* 1997–2010.

Yankner, B. A., and Shooter, E. M. (1982). The biology and mechanism of action of nerve growth factor. *Ann. Rev. Biochem. 51,* 845–868.

Yawo, H. (1987). Changes in the dendritic geometry of mouse superior cervical ganglion cells following postganglionic axotomy. *J. Neurosci. 7,* 3703–3711.

Yurkewicz, L., Marchi, M., Lauder, J. M., and Giacobini, E. (1981). Development and aging of noradrenergic cell bodies and axon terminals in the chicken. *J. Neurosci. Res. 6,* 621–641.

CHAPTER 3

Endothelial-Derived Relaxing Factors: Focus on Nitric Oxide

S. M. GARDINER T. BENNETT

I. Introduction

The seminal observations of Furchgott and Zawadzski (1980) on the role of endothelial cells in the relaxant responses of vascular preparations gave birth to what is now a burgeoning area of research activity. Since the original demonstration of the existence of "endothelium-derived relaxing factor" (EDRF), it has become apparent that endothelial cells may synthesize and release several factors that could be involved in vasodilator responses; these include prostacyclin, endothelium-derived hyperpolarizing factor (EDHF), and nitric oxide (NO). Since recent reviews on prostacyclin exist (e.g., Dusting and MacDonald, 1990), and little is known of EDHF (e.g., Nagao and Vanhoutte, 1991), this review will deal with NO, which, in many instances, appears to be the major EDRF.

II. General Considerations

The concept of endothelium-dependent vasodilatation is based on *in vitro* experiments in which removal of the endothelium from various vascular preparations renders them unresponsive to agents such as acetylcholine. However, the demonstration of the endothelial dependence of vasodilator phenomena *in vivo* is much less straightforward due to several reasons. A major problem, until recently, was the lack of availability of drugs to inhibit endothelium-dependent vasodilatation *in vivo*. With the demonstration that the major EDRF is NO, and the important finding that the synthesis of NO can be inhibited by various analogs of L-arginine (see Moncada *et al.*, 1991a), this problem has been partly, but not completely, solved.

In the majority of studies, the demonstration of vasodilatation *in vivo* has hung on measurement of systemic arterial blood pressure alone. However, it does not follow that a fall in systemic arterial blood pressure is necessarily due to vasodilatation, or that similar hypotensive responses under different conditions are due to similar underlying hemodynamic changes. Although measurement of cardiac and regional hemodynamic responses to administration of vasodilator agents *in vivo* may be helpful in this regard, interpretation of results in the presence of L-arginine analogs is not straightforward because the latter cause substantial changes in resting cardiovascular variables (Gardiner *et al.*, 1989f,g, 1990d,e,f,g,h,i, 1991b,c,f,g,h,j). In various studies, responses to glyceryl trinitrate or sodium nitroprusside have been included as "endothelium-independent," internal standards. However, it is now clear that the responses to such agents are "endothelium-modulated," inasmuch as pretreatment with L-arginine analogs causes specific sensitization to their effects (Moncada *et al.*, 1991b; Gardiner *et al.*, 1991g). Thus far, the only vasodilator agent we have found whose actions are largely unaffected by N^G-monomethyl-L-arginine (L-NMMA) or N^G-nitro-L-arginine methyl ester (L-NAME) is the K^+-channel opener, BRL 38227 (Gardiner *et al.*, 1991h). Thus, the inclusion of this agent in any protocol concerned with assessing the effects of L-arginine analogs on hemodynamic responses to vasodilator agents could be helpful. However, in our experience, the detailed regional hemodynamic profile of different vasodilator agents *in vivo* differs markedly. For example, BRL 38227 causes an increase in mesenteric blood flow only (Gardiner *et al.*, 1991h), whereas acetylcholine increases renal and internal and common carotid flows, but not mesenteric or hindquarters flows (Gardiner *et al.*, 1991g), and salbutamol causes a marked increase in hindquarters flow, a slight increase in renal flow, but no increase in mesenteric or carotid flows (Gardiner *et al.*, 1991h). Hence, unless one could find a series of vasodilator agents whose hemodynamic effects were uninfluenced by L-arginine analogs, and whose profiles of action in different vascular beds were matched with those of the vasodilator agents under investigation, it could be claimed the experiments were not fully controlled.

This line of argument raises another fundamental problem. If any vasoactive agent is administered *in vivo,* it could, by dint of changing flow characteristics, influence release of NO and other endothelium-derived factors (Holtz *et al.*, 1984; Hutcheson and Griffith, 1991; Kelm *et al.*, 1991). Hence, one might suggest that any agent given *in vivo* would, unavoidably, exert effects that were endothelium-modulated. Therefore, the finding that the hemodynamic actions of agents such as BRL 38227 are relatively unaffected by NO synthesis inhibitors, such as L-NAME, may require more explanation than simply to say they act in an endothelium-independent manner.

In the past, particularly in experiments in which blood pressure was the only hemodynamic variable monitored, attempts were made to control for any putative effects of the pressor action of L-arginine analogs by comparing responses to vasodilator agents in the presence of other pressor agents at doses selected to give the same rise in mean arterial blood pressure as seen with the L-arginine analogs. However, just because pressor effects were similar, it does not follow that the cardiac and regional hemodynamic changes evoked by the NO synthesis inhibitor and the other pressor agent would be, and this calls into question the value of such experiments. Although we have found that the pressor and regional hemodynamic effects of appropriate doses of L-NMMA and vasopressin are reasonably well matched (Gardiner *et al.*, 1989g, 1990d), we have not compared the cardiac hemodynamic effects of the two agents formally.

Another complication of *in vivo* studies is that agents, such as acetylcholine, that exert clear, endothelium-dependent effects *in vitro* can elicit hypotensive vasodilator responses *in vivo* that may not be inhibited completely by L-NAME (Gardiner *et al.*, 1990i, 1991g). Finally, there is evidence that agents, such as β_2-adrenoceptor agonists, which are said to cause endothelium-independent vasodilatation *in vitro*, exert vasodilator effects *in vivo* that are, to an extent, sensitive to L-NAME (Gardiner *et al.*, 1991g,h).

As a generalization, it appears that the responses to any vasodilator agent *in vivo* may vary in magnitude from one vascular bed to another, may involve different factors in different vascular beds, and in different segments of the same vascular bed, and these factors may vary with the dose of agonist used. Moreover, even in the case of classical endothelium-dependent vasodilators, large components of the response seen may be due to indirect mechanisms, such as adrenal medullary adrenaline release, as in the hindquarters vasodilator response to bradykinin (Gardiner *et al.*, 1991i, 1992).

Although these complications could lead one to assert that all studies of endothelium-dependent vasodilator phenomena should be carried out in *in vitro* systems, it is undeniable that the question one is trying to answer, ultimately, is what happens *in vivo*? Indeed, since one is attempting to understand the workings of the system in order to manipulate it to therapeutic benefit, one could argue that what happens *in vitro* is irrelevant. For example, *in vitro,* both atrial natriuretic peptide (ANP) and calcitonin gene-related peptide (CGRP) are potent mesenteric vasodilators, yet *in vivo* they can cause vasoconstriction (Gardiner *et al.*, 1988a,b, 1989a,b, 1991e). Thus, in conditions such as splanchnic ischemia these agents, or analogs thereof, may not be helpful; but this conclusion is the opposite to that which one would have reached on the basis of *in vitro* studies. However, it should always be remembered that the effect of any *in vivo* intervention under normal conditions may not be the same as that under abnormal conditions.

Another factor that is likely to cause confusion is the interspecies variation in the role of NO as the major EDRF. Already it is clear that rats, cats, mice, rabbits, dogs, and man may show differences with, for example, rats, rabbits, dogs, and man having NO as a major EDRF, but cats and mice being less dependent on this system. Furthermore, these differences may vary from vascular bed to vascular bed.

In vivo, the synthesis and release of NO (and other factors) by the endothelium are likely affected by patterns of pulsatile flow through changes in shear stress (Hutcheson and Griffith, 1991). For this reason, it seems unlikely that *in vitro* studies of isolated segments, often of conduit vessels, will produce results that mimic the *in vivo* condition. Moreover, studying isolated vascular beds perfused at constant flow could give results quite different from the *in vivo* situation in which flow is variable and pulsatile. We anticipate that *in vitro* studies in which the dynamics of flow and pressure simulate those *in vivo* will produce more meaningful results.

With these various points in mind, this review deals with some aspects of the biosynthesis of NO and the factors that influence it, and then goes on to cover the systemic effects of interfering with NO production, and, finally, considers the putative influence of NO in different regions of the cardiovascular system.

III. Biosynthesis of Nitric Oxide

Although suggested independently by Furchgott (1988) and Ignarro *et al.* (1988) in 1986, the first direct demonstration that EDRF was NO came from the work of Palmer *et al.* (1987). This group went on to demonstrate that NO was synthesized from the terminal guanidino nitrogen of L-arginine (Palmer *et al.*, 1988a,b). Subsequently it was shown that analogs of L-arginine, such as L-NMMA, N^G-nitro-L-arginine (L-NNA), and L-NAME, inhibited production of NO from L-arginine (see Moncada *et al.*, 1991a). At an early stage it became clear that the production of NO from L-arginine was not confined to endothelial cells, and that the L-arginine-NO pathway probably constituted a mechanism of fundamental importance in mediating cellular signaling (see Moncada *et al.*, 1991a). It is known now that the production of NO from L-arginine can involve NO synthases with different cofactor and ionic requirements in different tissues. Furthermore, it is apparent that NO synthases exist in constitutive and inducible forms that also differ in these respects (see Moncada *et al.*, 1991a).

Recent work from Snyder's laboratory (Bredt *et al.*, 1991) has shown that the brain-endothelial NO synthase has recognition sites for NADPH, FAD, flavin mononucleotide, and calmodulin, in addition to phosphorylation sites. It is intriguing that this enzyme shows close homology with cytochrome P-450 reductase (Bredt *et al.*, 1991). It seems there is a family of NO synthases, and it is feasible that drugs will be developed that can selectively inhibit the different types of NO synthase. This would be of particular value in the context of disease states in which the inducible form of NO synthase is produced. A good example of this is in endotoxemia (see Moncada *et al.*, 1991a), where it appears that abnormal NO synthesis may be responsible for the cardiovascular sequelae. Indeed, it has been suggested that treatment with NO synthesis inhibitors may be of therapeutic benefit in endotoxemia (Kilbourn *et al.*, 1990). However, it does not follow that the apparent improvement in cardiovascular status, as judged by recovery of systemic arterial blood pressure following administration of inhibitors of NO synthesis in endotoxic shock, would be clinically beneficial. For instance, although blood pressure may be "improved," this may be at the expense of diminished regional blood flows and cardiac output (Gardiner *et al.*, 1990h; Klabunde and Ritger, 1991), which could further exacerbate the problem. Nonetheless, a compound that selectively inhibited NO synthesis by the inducible enzyme, while leaving the constitutive NO synthase functional, could prove useful. It appears that the efficacious actions of steroids, such as dexamethasone, in endotoxemia may be due to the inhibition of synthesis of the inducible form of NO synthase (see Moncada *et al.*, 1991a). In other models of shock there is evidence that administration of exogenous NO may be beneficial (e.g., Aoki *et al.*, 1990).

IV. Whole Animal Studies with L-Arginine Analogs

It is now well established that administration of L-NMMA, L-NNA, or L-NAME *in vivo* causes hypertension in rats, guinea pigs, rabbits, sheep, and dogs (e.g., Rees *et al.*, 1989b; Aisaka *et al.*, 1989; Whittle *et al.*, 1989; Gardiner *et al.*, 1989f,g, 1990d,e,f,g,h,i, 1991b,c,f,g,h,i; Hecker *et al.*, 1990; Fozard and Part, 1991; Chu *et al.*, 1991; Humphries *et al.*, 1991; Vargas *et al.*, 1990; Klabunde *et al.*, 1991; Zambetis *et al.*, 1991; Tresham *et al.*, 1991; Yamazaki *et al.*, 1991). However, it has been reported that cats do not show an increase in mean arterial blood pressure (MAP) following L-NAME (van Gelderen *et al.*, 1991), and that indomethacin inhibits the pressor effects of L-NMMA in anesthetized dogs (Klabunde *et al.*, 1991).

In conscious, unrestrained rats, intravenous administration of L-NMMA or L-NAME caused constriction in renal, mesenteric, hindquarters, and internal and external carotid vascular beds, and these effects were variably susceptible to reversal by L-arginine (Gardiner *et al.*, 1990d,e,h). The hypertensive effect was offset by substantial reductions in cardiac function (Gardiner *et al.*, 1990h); although the bradycardia was largely reflex (i.e., almost completely abolished in the presence of atropine and atenolol), the reductions in other cardiac functions following L-NAME still occurred under those conditions (Widdop *et al.*, 1992).

Fozard and Part (1991) studied anesthetized, ganglion-blocked, normotensive and hypertensive rats and found that cumulative bolus injections of L-NMMA caused dose-related, and similar, pressor effects in both groups. These effects were accompanied by dose-related reductions in conductance in renal, mesenteric, hindquarters, and common carotid vascular beds. However, in contrast to our findings in conscious rats (Gardiner *et al.*, 1990d), the reductions in conductance in renal and carotid vascular beds were greater than those in mesenteric and hindquarters vascular beds, and all effects were completely reversed by L-arginine. As Fozard and Part (1991) pointed out, it is feasible that the regional differences in maximal vasoconstrictor effect of L-NMMA seen in their experiments reflected differential contributions of NO to vascular tone. However, we do not think it is likely that our failure to observe such differences was due to baroreflex buffering, as suggested by Fozard and Part (1991), since the mesenteric vascular bed is an important target for such effects, while flow through the internal carotid bed is not markedly influenced by baroreflexes, yet, in our studies, L-NMMA induced changes in conductance that were similar in these vascular beds. Furthermore, L-arginine caused almost complete reversal of the mesenteric vasoconstrictor effect of L-NMMA, whereas its influence on the internal carotid bed was relatively resistant to reversal by L-arginine (Gardiner *et al.*, 1990d). There is a theoretical possibility that the hemodynamic consequences of inhibiting NO synthesis are due to factors other than, or in addition to, loss of endothelial-derived NO (see later). However, we have not been able to show any change in baroreflex sensitivity in the presence of L-NAME (Gardiner *et al.*, 1991a,c).

Our observations of substantial pressor and vasoconstrictor responses to L-NAME in conscious rats rendered areflexic and treated with inhibitors of renin-angiotensin- and vasopressin-mediated processes (Gardiner *et al.*, 1990i, 1991h) are at variance with those of others in anesthetized rats (Vargas *et al.*, 1990; Lacolley *et al.*, 1991). However, there is evidence that anesthesia has effects on responses to NO synthesis inhibitors although the reports are contradictory (Wang *et al.*, 1991; Aisaka *et al.*, 1991). Thus, we would argue that the apparent dependence of the pressor effects of L-arginine analogs on sympathoadrenal function is a reflection of the experimental model studied and is not representative of the situation *in vivo* in conscious rats.

It has been reported that NO inhibits endothelin release (Boulanger and Lüscher, 1990), so it is feasible that the substantial hemodynamic effects of NO synthesis inhibition are not due simply to the removal of vasodilator tone but are contributed to by disinhibition of endothelin release. Data obtained from anesthetized dogs support this proposition in that L-NMMA caused a modest increase in plasma endothelin (Perrella *et al.*, 1991). However, administration of L-NNA to conscious sheep caused no change in plasma endothelin (Tresham *et al.*, 1991) in spite of the fact that in the latter study there was a sustained increase in mean arterial blood pressure, whereas in the dogs studied by Perrella *et al.* (1991), L-NMMA was without pressor effect. But it could be argued that plasma levels of endothelin are a poor index of the abluminal release of endogenous endothelin. In this connection it is notable that the initial response to both exogenous endothelin (Gardiner *et al.*, 1989e,g, 1990a,b,c) and L-NMMA or L-NAME (Gardiner *et al.*, 1990d,e,h, 1991b) in conscious rats includes hindquarters vasodilatation. However, it is not known if these effects are due to the same mechanisms and, certainly in the case of endothelin-1, it does not appear that NO release is responsible for the initial hindquarters vasodilatation (Gardiner *et al.*, 1989g; Ohlstein *et al.*, 1990).

There is evidence that the production of endothelin-1 from proendothelin-1 is mediated by a phosphoramidon-sensitive enzyme (Matsumura *et al.*, 1990; Fukuroda *et al.*, 1990), and we found pretreatment with phosphoramidon markedly reduced the regional hemodynamic effects of proendothelin-1 in conscious rats (Gardiner *et al.*, 1991d). However, pretreatment with phosphoramidon does not inhibit the hemodynamic effects of L-NAME (Gardiner, S. M., and Bennett, T., unpublished observations). While this observation might indicate that endothelin-1 is not involved in the hemodynamic effects of L-NAME, other interpretations are possible. Thus, treatment with L-NAME might disinhibit endothelin release from a preformed pool, and/or phosphoramidon treatment *in vivo* might inhibit extracellular conversion of exogenous proendothelin-1 to endothelin while leaving intracellular conversion of endogenous proendothelin-1 to endothelin-1 intact. A definitive assessment of the involvement of endothelin-1 in the responses to L-NAME or other NO synthesis inhibitors should come from experiments using nonpeptide antagonists of endothelin-1, but these are not yet widely available.

Considering the evidence that NO exerts potent vasodilator effects, inhibits endothelin release, and has antiproliferative and antimitogenic actions (see Moncada *et al.*, 1991a), we examined the possibility that chronic administration of NO synthesis inhibitors might cause sustained and irreversible hypertension. In our initial studies with L-NMMA, it appeared that intravenous administration for 6 h had hypertensive and vasoconstrictor effects that persisted for some time after the infusion of L-NMMA was stopped (Gardiner *et al.*, 1990f). Subsequently, we found that L-NMMA and L-NAME were orally active and that adding them to the drinking water of polydipsic, Brattleboro rats caused sustained hypertension, but this was associated with variable changes in renal, mesenteric, and hindquarters hemodynamics over a period of 9 h (Gardiner *et al.*, 1990g). In fact, at the end of this period, the hindquarters vascular bed was the only one in which flow was reduced below control levels, raising the possibility that the renal and mesenteric vascular beds were better able to counteract the loss of NO-mediated vasodilator tone.

In a later study, Brattleboro rats were given L-NMMA to drink for a period of 7 days and hemodynamics were monitored throughout this time (Gardiner *et al.*, 1991j). The sustained hypertension was associated with maintained reductions in conductance only in the hindquarters vascular bed. Moreover, within 9 h of withdrawal of L-NMMA, the animals were no longer hypertensive (Gardiner *et al.*, 1991j). In order to determine if longer periods of NO synthesis inhibition could produce irreversible effects, Brattleboro rats were given L-NAME in their drinking water for 5–6 months. Although at the end of this time the surviving animals were grossly hypertensive, all became normotensive within 48 h of withdrawing L-NAME (Gardiner *et al.*, 1991j). While we have yet to determine the structural correlates of these changes, they appear to question the involvement of vascular remodeling in the maintenance of hypertension.

V. Influence of Nitric Oxide in Different Regions of the Cardiovascular System

A. Aorta

The aorta has been used widely for the *in vitro* study of endothelial cell function and its relation to vascular smooth muscle activity. Although it is clear that aortic endothelial cells can produce NO, which mediates aortic relaxation, there are no data illustrating the significance of this phenomenon *in vivo*. As we have indicated elsewhere (Kiff *et al.*, 1991), it is feasible that endothelial cells have an important effect on aortic compliance and, thereby, influence both cardiac function and regional hemodynamics downstream. In this context it is notable that Hutcheson and Griffith (1991) recently demonstrated that release of NO from perfused segments of the descending aorta of rats or rabbits was increased by pulse frequency and decreased by pulse amplitude. It is likely that this phenomenon contributed to the control of hydraulic resistance of arteries by blood flow as described by Melkumyants *et al.* (1987). Hutcheson and Griffith (1991) also obtained evidence for factors, besides inhibition of NO release, contributing to the effects of increased pulse pressure; one candidate is endothelin (Boulanger and Lüscher, 1990; Milner *et al.*, 1990).

It would be of interest to know whether or not aortic endothelial NO production shows regional variations, but there appear to be no studies available in which a systematic, *in vitro* comparison has been made of endothelial-mediated vasodilator phenomena in segments of aorta from different levels. Furthermore, it is not known to what extent increases in aortic compliance and/or volume contribute to the hypotensive effects of vasodilator agents *in vivo*.

Palmer *et al.* (1988b) were the first to report the *in vitro* effects of L-NMMA on preparations of the thoracic aorta from rabbits. L-NMMA caused small, dose-dependent contractions whereas L-arginine caused small, dose-dependent relaxations when preparations were precontracted with phenylephrine. Acetylcholine, substance P, and A23187 caused endothelium-dependent relaxations; the maximal inhibitory effects of L-NMMA on these responses were 64, 55, and 76%, respectively. These effects were reversed by washing or, dose dependently, by L-arginine (Palmer *et al*,. 1988b, Rees, *et al.*, 1989a). Rees *et al.* (1989a) suggested that the inability of L-NMMA completely to inhibit endothelium-dependent responses could have been due to an inability of L-NMMA to gain access to all sites of synthesis of NO, and/or to the involvement of factors, other than NO, in endothelium-dependent vasodilatation. However, L-NNA was shown subsequently to be a more potent inhibitor than L-NMMA of acetylcholine-induced, endothelium-dependent relaxation of rabbit and rat thoracic aortic rings (Moore *et al.*, 1990; Rees *et al.*, 1990), possibly because L-NMMA, but not L-NNA, is metabolized to L-arginine (Hecker *et al.*, 1990).

It is notable that inhibition of NO synthesis had no effect on the sensitivity of rabbit aortic rings to glyceryl trinitrate or sodium nitroprusside (Rees *et al.*, 1989a,b; Moore *et al.*, 1990; Kobayashi and Hattori, 1990), whereas the same intervention caused supersensitivity to these agents in rat aortic rings (Moncada *et al.*, 1991b). While this difference could be due to species differences in the basal rate of formation of NO, Schmidt *et al.* (1990) found that the sensitivity to glyceryl trinitrate of aortic rings from both rabbits and rats was unaffected by L-NMMA, and Dubbin *et al.* (1990) reported that responses of rat aortic rings to sodium nitroprusside were unaffected by L-NNA. However, both groups studied aortic rings precontracted with phenylephrine, and it is feasible that the contractions were supramaximal in the presence of the NO synthesis inhibitors, thus obscuring any enhanced sensitivity to nitrovasodilators (D. D. Rees, personal communication).

Thomas *et al.* (1989) claimed that L-NMMA was not a specific antagonist of endothelium-mediated relaxation because they found this compound "antagonized" the relaxation evoked by various compounds applied to rat aortic rings with the endothelium intact. However, Thomas *et al.* (1989) failed to report the effects of L-

NMMA applied before the other agents, or in the absence of the endothelium. Hence, it is likely that the antagonism they saw was due to inhibition of ongoing NO effects by L-NMMA, as suggested by Fasehun *et al.* (1990).

Apart from classical, endothelium-dependent vasodilators, peptides, such as CGRP, have been shown to cause relaxation of rat aortic rings by a mechanism that was sensitive to L-NMMA or L-NNA (Gray and Marshall, 1990, 1992; Wang *et al.*, 1991; Fiscus *et al.*, 1991). In addition, contractile substances, such as endothelin and α-adrenoceptor agonists, exerted effects on the aorta that were modulated by endothelial activity (Topouzis *et al.*, 1991a,b).

Given the potent aortic effects that NO exerts, it is logical to consider the consequences that might result when NO synthesis becomes disordered. Since atherosclerosis impairs endothelium-mediated vasodilatation, Minor *et al.* (1990) investigated the possibility that rabbits fed a high-cholesterol diet would show impaired production of nitrogen oxides in the aorta. In fact, they found a marked increase in production of nitrogen oxides by the atherosclerotic aorta and concluded that impaired endothelium-dependent vasodilatation was due neither to reduced NO synthase activity nor to arginine deficiency. In contrast, Cooke *et al.* (1991) found that the impaired acetylcholine-mediated, endothelium-dependent relaxations of aortic rings from hypercholesterolemic rabbits were normalized by *in vivo* exposure to L-arginine. The explanation of this disparity is elusive at present, but there is some evidence for diminished NO-mediated effects in other abnormal conditions, such as congestive heart failure (Ontkean *et al.*, 1991; Elsner *et al.*, 1991).

Clearly, there is much work to be done to characterize properly the biochemical defects and *in vivo* sequelae associated with abnormalities of the L-arginine-NO pathway as it impinges on cardiovascular function.

B. Cerebral Circulation

There is growing evidence that NO might be involved in various ways in the control of the cerebral circulation in several species, and that these processes might become disordered in pathological conditions (e.g., Katusic *et al.*, 1989; Gardiner *et al.*, 1989f, 1990d,e, 1991g; Faraci, 1990, 1991; Mayhan, 1990; Rosenblum *et al.*, 1990; Busija *et al.*, 1990; Fischer-Nakielski and Schrör, 1990; Rossitch *et al.*, 1991; Fozard and Part, 1991).

Marshall *et al.* (1988) reported that acetylcholine-induced vasodilatation of the cerebral microcirculation in the cat was inhibited by methylene blue, and that this effect was antagonized by superoxide dismutase, catalase, or deferoxamine. Marshall *et al.* (1988) suggested, therefore, that the inhibitory effect of methylene blue on acetylcholine-induced vasodilatation was due to hydroxyl radical-mediated oxidation of EDRF. However, it appeared the latter was not NO since NO- and sodium nitroprusside-induced vasodilatations were not inhibited by methylene blue. This finding is odd since methylene blue is generally considered to be an inhibitor of guanylyl cyclase, and hence would be expected to inhibit the effects of NO and compounds acting through the release of NO.

In the canine basilar artery, Katusic *et al.* (1989) found that responses to bradykinin and exogenous NO were indistinguishable on the basis of similar responses to various pharmacological interventions. They suggested, therefore, that the major EDRF in this system was NO. A similar conclusion was reached by Ohta *et al.* (1990) who found that relaxations of the canine basilar artery by substance P were reduced by L-NNA and this effect was partly reversed by L- but not by D-arginine. Ohta *et al.* (1990) observed that the vasorelaxant responses to glyceryl trinitrate (GTN) were not inhibited by L-NNA but did not say whether or not they were augmented (see above).

In conscious, unrestrained rats, Gardiner *et al.* (1989f, 1990d,e) reported that L-NMMA caused dose-dependent pressor effects and reductions in internal carotid blood flow and vascular conductance; D-NMMA had no pressor or vasoconstrictor effects. Although L-arginine caused dose-dependent reversal of the pressor effects of L-NMMA, the internal carotid vasoconstrictor effects of L-NMMA were relatively resistant to reversal by L-arginine; the reasons for this have not been established. In a subsequent study, Gardiner *et al.* (1991g) showed that L-NAME caused reductions in internal and common carotid blood flows and vascular conductances. Under these conditions the hypotensive and internal and common carotid vasodilator effects of sodium nitroprusside were enhanced. Interestingly, the internal and common carotid vasodilator effects of a low dose of acetylcholine were more clearly attenuated than were those of a higher dose (Gardiner *et al.*, 1991g), so raising the possibility that the ease with which NO can be shown to be involved in vasodilator responses by inhibiting its synthesis might depend on the magnitude of the agonist stimulus as well as the vascular bed considered and the species studied. In the latter regard, it is notable that, contrary to findings in other species, vasopressin did not exert carotid vasodilator effects in conscious rats (Gardiner *et al*, 1989c).

The results of Faraci (1990) on anesthetized rats corroborate those of Gardiner *et al.* (1989f, 1990d,e), inasmuch as Faraci (1990) produced evidence that topical application of L-NMMA on to the basilar artery caused a dose-dependent reduction in vessel diameter that was opposed by L-arginine. Although Faraci (1990) found that L-NMMA abolished the vasodilator effects of acetylcholine, he only considered a single dose of the latter, so it is not clear if the dose-dependent effects described by Gardiner *et al.* (1991g), for the interaction between acetylcholine and L-NAME in the internal carotid arterial bed, apply also to the basilar artery. Moreover, it is feasible that systemic administration of NO synthesis inhibitors would have cerebral microcirculatory effects that were contributed to by actions upstream of the microvessels. Furthermore, it does not follow that the marked internal carotid vasoconstrictor effects of L-NMMA or L-NAME would necessarily be reflected in changes in cerebral blood flow. However, this latter possibility has been excluded by two recent studies. One, by Tanaka *et al.* (1991), examined the effects of L-NMMA on regional brain blood flow in conscious rats; they found uniform reductions in blood flow throughout all regions of the brain in spite of the fact that L-NMMA caused a rise in systemic arterial blood pressure. A subsequent study by Macrae *et al.* (1991) obtained similar results with L-NAME.

Recently, Faraci (1991) extended his earlier observations by comparing the responses of large (174–326-μm-diameter) and small (27–115-μm-diameter) cerebral vessels to various agents, including L-NMMA. He found that L-NMMA applied to cerebral vessels evoked a constriction that was linearly, and positively, related to baseline diameter. Separation of vessels into large (275 ± 10-μm-diameter) and small (62 ± 6-μm-diameter) categories showed a significant difference between their constrictor responses. In contrast to the differential effects of L-NMMA, Faraci (1991) found that large and small vessels showed similar constrictor responses to U-46619 and similar dilator responses to sodium nitroprusside. Hence, the difference between large and small vessels in their constrictor responses to L-NMMA was not due to differences in vasoconstrictor capacity or sensitivity to NO. Acetylcholine caused dilatation of large and small cerebral vessels, and these effects were inhibited to similar extents by L-NMMA, whereas the vasodilator effects of sodium nitroprusside were not reduced. Indeed, the effects of the latter drug on large vessels tended to be augmented in the presence of L-NMMA (Faraci 1991). Collectively, these results indicate a lesser resting production of NO affecting small, compared to large, vessels in the cerebral circulation. Whether or not this difference was due to different synthetic abilities of endothelial cells and/or segmental differences in hemodynamic variables that influence NO synthesis remains to be determined. However, these findings and our earlier observations are consistent with an important role of NO in the maintenance of cerebral vascular conductance. But this does not constitute direct evidence that the NO concerned is endothelial in origin since NO synthase occurs in the brain tissue, and some cerebral vasodilator nerves act via NO-mediated mechanisms that are not endothelium dependent.

An example of the latter phenomenon was provided by Toda and Okamura (1990a,b,c) who found that transmural stimulation of cerebral (basilar and middle cerebral) arteries from dogs and monkeys elicited vasorelaxant responses that were inhibited by L- but not by D-NMMA. The inhibition was reversed by L- but not by D-arginine; furthermore, the effect of L-NMMA was still seen in the absence of the endothelium. Toda and Okamura (1990a,b,c) suggested, therefore, that the results indicated an involvement of NO in neurotransmission. Recently, Toda and Okamura (1991) considered the possibility that the expression of inducible NO synthase in canine cerebral arterial strips might explain the involvement of NO in the responses to transmural stimulation or nicotine. However, they concluded this was unlikely although there was no direct evidence that NO synthase was absent from the vascular smooth muscle in their preparations. Despite questions remaining about the origin of NO, the findings with L-NMMA indicate a role for NO in the mediation of neural vasodilatation in the cerebral vasculature. Toda *et al.* (1990) confirmed these findings using L-NNA and also suggested that, in canine temporal arteries, NO release in response to transmural stimulation opposed vasoconstriction elicited by stimulation of noradrenergic nerves (Toda *et al.*, 1991b), by a mechanism other than inhibition of noradrenaline release. As in dogs, Gaw *et al.* (1991) reported that neurogenic vasodilatation of the ovine middle cerebral artery was not dependent on the endothelium but was inhibited by L-NMMA. In this case the vasodilatation appeared to be due to vasoactive intestinal polypeptide causing stimulation of NO release.

The results of Toda and Okamura (1990a,b,c) and Gaw *et al.* (1991) differ from those of González and Estrada (1991) who studied bovine pial arterioles. Although González and Estrada (1991) reported that relaxant responses to transmural stimulation were inhibited by L-NMMA or L-NAME, these responses were abolished also by the removal of the endothelium. Thus, it appeared that the NO responsible for vasodilatation of bovine pial arterioles was endothelial in origin and released by an agent arising from perivascular nerves. What this agent is and how it reaches endothelial cells remain to be established, as does the role of NO in neurogenic vasodilatation of the cerebral vasculature in species such as rats and man.

One aspect of the control of the cerebral circulation that is of particular interest is the etiology of vasospasm following events such as subarachnoid hemorrhage. Although it is likely that the genesis of this problem is multifactorial, the ability of hemoglobin to bind NO and thereby annul its vasodilator influence could be a major contributor to the problem. However, the susceptibility of oxyhemoglobin-induced basilar artery spasm to inhibition by nicardipine (Toda *et al.*, 1991a) indicates that Ca^{2+} flux may be an important step in the development of the vasoconstriction. In this regard the ability of NO to inhibit the release of endothelin (Boulanger and Lüscher, 1990) may be relevant since binding of NO by hemoglobin could disinhibit endothelin release in addition to removing the vasodilator influence of NO.

The possibility that endothelin might be involved in cerebral vascular dysfunction following subarachnoid hemorrhage has been strengthened recently by the finding that the vasospasm in this condition is inhibited by phosphoramidon (Matsumura *et al.*, 1991; Shinyama *et al.*, 1991), a substance that suppresses the hemodynamic effects of proendothelin-1 *in vivo* (Gardiner *et al.*, 1991d; Shinyama *et al.*, 1991) by preventing its conversion to endothelin-1. However, it is feasible that endothelin release following subarachnoid hemorrhage is not due to the removal of a braking action of NO but to an effect of oxyhemoglobin independent of an interaction with NO to stimulate endothelin release (Cocks *et al*,. 1991).

The temporal sequence of vascular and neurological disorders following subarachnoid hemorrhage is still not delineated; vasospasm may underlie the neuropathy, or neuronal dysfunction may trigger the vasospasm. Either way, the neurological sequelae of subarachnoid hemorrhage could involve NO-mediated processes because there is growing evidence that NO is involved in the mechanisms whereby excitotoxic amino acids evoke neuronal death (Dawson *et al.*, 1991a,b).

C. Coronary Circulation

The coronary circulation in the perfused rabbit heart was the first intact vascular bed in which NO was shown to mediate acetylcholine-induced vasodilatation (Amezcua *et al.*, 1988). However, since it is now known that endocardial cells may synthesize and release NO (Schulz *et al.*, 1991), one cannot preclude a contribution from this source to NO-mediated effects on the coronary vasculature. Nonetheless, it is clear that, in the perfused rabbit heart, L-NMMA, but not D-NMMA, caused dose-dependent increases in coronary perfusion pressure, and L-arginine reversed these effects (Amezcua *et al.*, 1989). Furthermore, L-NMMA inhibited the coronary vasodilator effects of acetylcholine (Amezcua *et al.*, 1989). Similar effects were obtained with L-NNA by Lamontagne *et al.* (1991) who also found the coronary vasodilator responses to 5-hydroxytryptamine (5-HT) were inhibited by L-NNA under conditions in which the responses to sodium nitroprusside were unaffected. Lamontagne *et al.* (1991) did not comment on the absence of supersensitivity to sodium nitroprusside in the presence of L-NNA although this observation is surprising in the light of findings in other systems (Moncada *et al.*, 1991b; Gardiner *et al.*, 1991g).

An involvement of NO in the maintenance of a coronary vasodilator tone and the contribution of NO to the vasodilator responses to various agents in several *in vitro* systems have now been demonstrated by the use of various arginine analogs (e.g., Chester *et al.*, 1990; Kelm and Schrader, 1990; Richard *et al.*, 1990; Cowan and Cohen, 1991; Tschudi *et al.*, 1991; Mankad *et al.*, 1991). However, studies *in vivo* have demonstrated variable coronary vasoconstrictor effects of inhibiting NO biosynthesis.

Humphries *et al.* (1991) found that intravenous infusion of L-NNA caused a pressor effect accompanied by bradycardia and an increase in coronary vascular resistance in conscious rabbits, thereby corroborating the data obtained in the isolated rabbit heart. Although Humphries *et al.* (1991) reported that L-NNA caused significant increases in coronary vascular resistance in subendocardial and mid-myocardial layers, they did not comment on changes in myocardial blood flow. However, from the data in their Table 1 (Humphries *et al.*, 1991), it appears cardiac blood flow was about 2.3 ml min^{-1} g^{-1} at rest and about 1.98 ml min^{-1} g^{-1} during infusion of L-NNA, but we do not know if this was a significant change. This point is an important one since Richard *et al.* (1991) have suggested that a rise in coronary vascular resistance may not result in reduced myocardial perfusion (see below), and van Gelderen (1991) failed to find a reduction in myocardial blood flow following administration of L-NAME to anesthetized cats, in contrast to the observations of Benyó *et al.* (1991) who, however, used L-NNA.

In an earlier study, Chu *et al.* (1990) had observed that intraatrial infusion of L-NMMA caused decreases in the diameter of epicardial coronary arteries in conscious dogs although under those conditions there was no change in phasic coronary flow. Chu *et al.* (1990) suggested that either vasoconstriction of the distal autoregulatory vessels following inhibition of NO biosynthesis was overridden by cardiac effects influencing coronary conductance, or NO was exerting a preferential effect in the proximal, compared to the distal, coronary vasculature. However, in a subsequent study, Chu *et al.* (1991) did find a significant, albeit modest (about 19%), reduction in resting phasic coronary blood flow following L-NMMA. Under the same conditions, they observed that L-NMMA was without influence on left ventricular end diastolic pressure or dP/dt, thus indicating that L-NMMA was not affecting myocardial contractility (see below). Doses of L-NMMA less than 120 mg/kg had inconsistent effects on epicardial arterial responses to acetylcholine although in a later study, Woodman and Dusting (1991) found that a lower dose of L-NNA (5 mg/kg) caused significant attenuation of the coronary vascular effects of acetylcholine. However, it is notable that this dose of L-NNA had no significant effects on coronary blood flow or vascular resistance in the anesthetized greyhounds studied by Woodman and Dusting (1991).

Richard *et al.* (1991) assessed coronary vascular tone and myocardial perfusion following administration of L-NMMA or L-NAME in anesthetized, open-chest, mongrel dogs. Although they found clear evidence of coronary vasoconstriction and inhibition of coronary vasodilator effects of acetylcholine following L-NMMA or L-NAME, they saw no evidence of changes in myocardial tissue perfusion (either subendocardial, subepicardial, or transmural). Richard *et al.* (1991) suggested that the failure of coronary vasoconstriction to cause compromised myocardial perfusion might have been due to "compensatory mechanisms occurring at the level of small arterioles and/or capillaries." Klabunde *et al.* (1991) monitored circumflex artery flow continuously in anesthetized beagles and reported a significant reduction in cardiac output, unaccompanied by any reduction in coronary flow, following administration of L-NMMA (30 mg/kg). Likewise, Cappelli-Bigazzi *et al.* (1991) found that intracoronary injections of L-NMMA and L-NNA were without effects on coronary flow or diameter in anesthetized, open-chest dogs, and, recently, Ishizaka *et al.* (1991) reported that a dose of L-NMMA, which caused significant inhibition of the coronary vasodilator effects of acetylcholine, was without influence on resting coronary hemodynamics. In contrast, Bassenge and Huckstorf (1992) found that L-NAME injected into the coronary arteries of conscious dogs reduced vessel diameter by 4% and myocardial perfusion by 37%.

These disparities highlight, once again, the possibility that anesthesia interferes with the normal, physiological mechanisms mediated by NO. However, the difficulty of demonstrating a cardinal role for NO in the control of myocardial perfusion *in vivo* is, presumably, also a reflection of the multifactorial nature of the control systems and the major input myocardial activity has on local flow. In this regard, the observations by Chu *et al.* (1991) and Klabunde *et al.* (1991) showing that L-NMMA had no effect on indices of myocardial contractility are noteworthy since they indicate that the large reduction in cardiac function that occurs following NO biosynthesis inhibitors (Gardiner *et al.*, 1990h), and which is independent of reflex mechanisms (Widdop *et al.*, 1992), could be due entirely to the increase in afterload (Gardiner *et al.*, 1990h). Klabunde *et al.* (1991) tried to exclude this possibility by showing that the reduction in cardiac output following administration of L-NMMA or L-NNA was much more than that after a dose of phenylephrine, which had a much greater pressor effect. However, the regional hemodynamic effects of L-NMMA and L-NNA could have been quite different from those of phenylephrine; in addition, phenylephrine could have exerted direct α-adrenoceptor-mediated positive inotropic effects. An interesting observation that adds further complexity to the manner whereby NO synthesis inhibition could affect the heart is that, in the conscious rat, L-NAME inhibits the tachycardic effects of circulating adrenaline, and this could influence its inotropic action by increasing diastolic filling time (Gardiner *et al.,* 1991h). In addition, there is some evidence that NO mediates vagally induced, coronary vasodilatation (Broten *et al.,* 1992).

D. Pulmonary Circulation

The pulmonary vascular bed is notable for its large surface area and the important contribution that the associated endothelial cells make to the metabolic activation and/or degradation of a variety of vasoactive compounds. As might be expected, then, there is evidence indicating an involvement of NO in control of the pulmonary circulation under a variety of conditions. Nevertheless, as in other systems, one cannot dismiss the possibility that the NO concerned could derive from sources other than, or in addition to, endothelial cells.

Crawley *et al.,* (1990) carried out *in vitro* studies on pulmonary artery rings from rats and humans and found that L-NMMA caused dose-dependent inhibition of the acetylcholine-induced relaxations of precontracted preparations. This effect was not seen with D-NMMA and was inhibited by L- but not by D-arginine. Furthermore, L-histidine was without influence on the effects of L-NMMA. These results indicated NO biosynthesis was involved in pulmonary arterial relaxant responses to acetylcholine. Interestingly, L-NMMA alone was without effect on tone in the preparations, indicating absence of basal release of NO, in contrast to the findings with aortic ring preparations (Rees *et al.,* 1989a,b). Crawley *et al.* (1990) found that L-NMMA and removal of the endothelium increased the sensitivity of pulmonary arterial preparations to phenylephrine to the same extent, consistent with phenylephrine-induced NO release acting to oppose its contractile effects. Gold *et al.* (1990), studying bovine pulmonary artery and vein preparations, made similar observations.

In addition to any direct effects of NO on pulmonary vascular tone, Cederqvist *et al.* (1991) showed that L-NMMA enhanced contractile responses of rings of guinea pig pulmonary artery to transmural nerve stimulation. This effect was inhibited by L-arginine but not by the removal of the endothelium. L-NMMA enhanced contractions to exogenous noradrenaline without influencing stimulation-induced release of (^{3}H)-noradrenaline. Cederqvist *et al.* (1991) also found that L-NMMA augmented nonadrenergic, noncholinergic responses to transmural stimulation and concluded that a nonendothelial source of NO caused postjunctional enhancement of the actions of noradrenaline and pre- and/or postjunctional facilitation of nonadrenergic, noncholinergic effects in the pulmonary artery of the guinea pig. Very similar results were reported recently by Liu *et al.* (1991) although these workers considered the NO was of endothelial origin.

In artificially ventilated, open-chest rabbits under normoxic conditions, Persson *et al.* (1990b) found that intravenous infusion of L-NAME caused systemic hypertension but no significant bradycardia. Cardiac output and stroke volume were significantly reduced, while mean pulmonary arterial pressure increased; there was no change in mean left atrial pressure. Although there was a marked increase in pulmonary vascular resistance, blood gases did not change; all the effects of L-

NAME were inhibited by L-arginine. In contrast to these findings, Persson *et al.* (1990b) observed that L-NAME caused a reduction in P_{O_2} and a rise in central venous pressure in spontaneously breathing rabbits. In spite of a marked increase in respiratory frequency, there was no change in P_{CO_2}. This, together with the hypoxemia, indicates that L-NAME caused ventilation-perfusion mismatching. Pulmonary vasoconstrictor responses to infusion of L-NMMA have been reported in anesthetized dogs (Perrella *et al.,* 1991), but no information was given about blood gases.

In connection with the pulmonary effects of L-NAME it is notable that oral administration of this compound to conscious Brattleboro rats can cause sudden death (Gardiner *et al.,* 1990g) and the premonitory signs of this event include severe respiratory distress (Gardiner, S. M., and Bennett, T., unpublished observations).

Important findings regarding the possible contribution of NO to the changes in the pulmonary circulation at birth have been reported by Abman *et al.* (1990). These workers studied the effects of L-NNA on pulmonary vascular tone and vasodilator responses in the late-gestation ovine fetus and also investigated the influence of L-NNA on changes in the pulmonary circulation following delivery by caesarian section. *In utero,* L-NNA inhibited the pulmonary vasodilator effects of acetylcholine but not those of atrial natriuretic peptide. L-NNA also caused rises in pulmonary arterial and systemic arterial blood pressures and a reduction in pulmonary arterial blood flow. Furthermore, in the presence of L-NNA, the increase in pulmonary arterial blood flow usually seen on delivery and ventilation was markedly reduced. Although Abman *et al.* (1990) found that arterial blood gas tensions in the two groups were not different 1 h after delivery, the F_{iO_2} necessary to achieve this was significantly higher in the L-NNA-treated group. If this was due to ventilation-perfusion mismatching, as suggested by the experiments of Persson *et al.* (1990b), it is not clear why the L-NNA-treated lambs in the study of Abman *et al.* (1990) were not hypercapnic since respiratory rate was the same in control and treated animals. However, the balance of evidence indicates a substantial role for NO in the pulmonary vascular adaptations at birth and, as pointed out by Abman *et al.* (1990), the marked changes in the stimuli impinging on the pulmonary endothelial cells in this circumstance are likely to have an important influence on NO biosynthesis and release.

There is evidence for an involvement of NO-dependent mechanisms in the control of pulmonary vascular resistance and the mediation of acetylcholine-induced pulmonary vasodilator responses in conscious postnatal lambs (Fineman *et al.,* 1991). But, in the experiments of Fineman *et al.* (1991), L-NNA was reported to have no significant effects on cardiac output, right or left atrial pressures, or blood gases. However, as far as cardiac output is concerned, the apparent lack of effect of L-NNA seems to have been due to marked variations in the baseline values.

The pulmonary circulation is distinguished by responding to hypoxia with a vasoconstriction, rather than a vasodilatation, as seen in other vascular beds, and several studies have considered the possibility that NO might influence this hypoxic pulmonary vasoconstriction. The first experiments to address this question directly were reported by Archer *et al.* (1989) who investigated the effects of L-NMMA on isolated, perfused lungs and on pulmonary arterial rings from the rat. In the isolated lung, L-NMMA had no effect on perfusion pressure under normoxic conditions, but it increased perfusion pressure under hypoxic conditions. At the peak of the vasoconstrictor response induced by hypoxia alone, L-arginine reduced perfusion pressure; this vasodilator effect of L-arginine was enhanced in the presence of L-NMMA. Interestingly, although L-NMMA augmented the vasoconstrictor response to angiotensin II and inhibited the vasodilator response to bradykinin, it had no significant effect on the vasodilator response to acetylcholine. In contrast to the lack of influence of L-NMMA on perfusion pressure in the normoxic lung, Archer *et al.* (1989) found that L-NMMA caused sustained contraction of pulmonary arterial rings but only in the presence of the endothelium. These results differ from the findings of Crawley *et al.* (1990) (see above) who found no effect of L-NMMA on the basal tone of pulmonary arterial rings from rats; the reason for this disparity is not obvious.

Archer *et al.* (1989) suggested that the difference between the effects of L-NMMA on pulmonary arterial rings and on the normoxic, perfused lung could have been due to the latter having a "large vascular reserve," such that significant pulmonary vasoconstriction did not cause a rise in pulmonary arterial pressure. If this was the case, it appears to pertain specifically to the *in vitro* setting since *in vivo* experiments indicate that inhibition of NO biosynthesis does cause pulmonary vasoconstriction under normoxic conditions (see above).

Similar findings to those of Archer *et al.* (1989) were reported by Robertson *et al.* (1990) who studied lungs of rats, perfused with blood *in situ*. Under hyperoxic conditions, L-NMMA had no effect, but it markedly enhanced pulmonary vasoconstriction in response to hypoxia. The study of Persson *et al.* (1990b) also showed that L-NAME enhanced hypoxic pulmonary vasoconstriction in open-chest rabbits and provided evidence that this was not an unspecific effect since hypoxic pulmonary vasoconstriction was not enhanced during endothelin infusion at a dose that increased pulmonary arterial pressure to an extent similar to that seen with L-NAME.

While these various studies are consistent, and indicate that NO may act to offset hypoxic pulmonary vasoconstriction, they do not reveal the nature of the mediator of this event. It is feasible that substances such as endothelin are responsible for the vasoconstriction since there is evidence for release of endothelin in response to hypoxia (Horio *et al.*, 1991). However, Thomas *et al.* (1991) found that hypoxic pulmonary vasoconstriction was inhibited, selectively, by diphenyleneiodinium (DPI). DPI is an inhibitor of NADPH oxidase, but its mode of action in preventing pulmonary vasoconstriction in response to hypoxia was not defined by Thomas *et al.* (1991). They did, however, exclude the possibility that DPI was interfering with NO production by carrying out their experiments in the presence of L-NNA. This was an important control since DPI inhibits NO synthase (Stuehr *et al.*, 1991) although, as previously indicated, this would not be expected to suppress hypoxic pulmonary vasoconstriction.

In addition to a role in opposing pulmonary vasoconstriction in response to acute hypoxia, there is some evidence that NO might offset the development of pulmonary hypertension in rats subjects to chronic hypoxia (Barer *et al.*, 1990). This is an interesting experimental model because the structure of the pulmonary vessels changes markedly, with a substantial increase in smooth muscle mass. It is not known if this new muscle expresses inducible NO synthase, but there is evidence that endothelium-dependent relaxation is impaired in the pulmonary vasculature of rats with chronic hypoxemia (Adnot *et al.*, 1990). Furthermore, a recent investigation by Dinh-Xuan *et al.* (1991) found impaired endothelium-dependent pulmonary vasorelaxation in vessels from patients with chronic obstructive lung disease. Pulmonary arterial rings from patients developed a greater tension than those from controls in response to phenylephrine, and this difference was abolished by L-NMMA. In control preparations, L-NMMA itself increased tension, but it had no effect on the preparations from patients. Pulmonary arterial rings from controls and patients, when precontracted with phenylephrine, showed dose-dependent relaxations to acetylcholine or adenosine diphosphate but only when the endothelium was intact. However, the maximum responses of preparations from patients with chronic obstructive lung disease were substantially less than those of the controls (41 and 49% for acetylcholine and adenosine diphosphate, respectively). In contrast, preparations from both groups showed similar relaxations to sodium nitroprusside. As mentioned elsewhere, one would expect supersensitivity to sodium nitroprusside if synthesis of endogenous NO were impaired (Moncada *et al.*, 1991b; Gardiner *et al.*, 1991g). Yet, when Dinh-Xuan *et al.* (1991) compared responses of both sets of preparations to sodium nitroprusside in the absence and presence of L-NMMA, there was no evidence of supersensitivity to sodium nitroprusside in the presence of L-NMMA in either group. Nevertheless, the results of Dinh-Xuan *et al.* (1991) indicate selective diminution of endothelium-dependent responses in pulmonary arterial preparations from patients with chronic obstructive lung disease.

Since Dinh-Xuan *et al.* (1991) found there was an inverse correlation between the degree of intimal thickening and the degree of relaxation to acetylcholine or adenosine diphosphate, it is feasible that intimal thickening impaired access of NO to the underlying smooth muscle. However, Dinh-Xuan *et al.* (1991) suggested that while such a "barrier" effect might have delayed onset of relaxation, it should not have reduced its magnitude; thus, they felt diminished endothelium-dependent responses in preparations from patients with chronic obstructive lung disease were more likely due to diminished NO production. This argument is only sustainable if the thickened intima did not sequester NO by means such as production of superoxide radicals, for example. Although in concluding, Dinh-Xuan *et al.* (1991) made the noteworthy point that the pulmonary vascular remodeling that occurs in chronic obstructive lung disease could be due to loss of the antiproliferative and antimitogenic effects of NO, such abnormalities also could be accounted for by diminished access of NO to the tissues, rather than to reduced NO synthesis.

In regard to access of NO to pulmonary vascular tissues, the lung circulation is unique because inhaled, gaseous NO could influence pulmonary vascular smooth muscle before being bound by hemoglobin. This raises the possibility that NO might be delivered by inhalation to influence, selectively, pulmonary vascular tone. Frostell *et al.* (1991) tested this hypothesis in conscious, chronically instrumented lambs. Under normal conditions, inhalation of NO (80 ppm) had no effects on pulmonary or systemic vascular resistance, cardiac output, or left atrial or central venous pressures. However, inhalation of NO (40 ppm) reversed pulmonary hypertension due to hypoxia or to U46619 infusion; under these conditions, NO did not cause systemic vasodilatation. In addition, there were no signs of short-term tolerance to the pulmonary vasodilator effects of NO inhalation during 1-h infusions of U46619 and no changes in methemoglobin levels, or lung water, or histology (Frostell *et al.*, 1991).

These important results immediately raised the possibility that inhaled NO might be of therapeutic benefit in respiratory disorders. Indeed, Falke *et al.* (1990) have reported that, in a patient with severe adult respiratory distress syndrome (ARDS), inhaled NO selectively reduced pulmonary hypertension and improved gas exchange and right heart ejection fraction. These workers subsequently (Falke *et al.*, 1992) reported on their first three patients with severe ARDS, treated with inhaled NO. Falke *et al* (1992) measured pulmonary and systemic hemodynamics and blood gases before, during, and after interruption of NO inhalation (20 ppm) for periods up to 6 days after initiating NO therapy. Over the first 3 days, NO inhalation caused reproducible reductions in pulmonary arterial pressure and increases in Pa_{O_2}. Only two of the patients were examined over the next 3 days, and in both the effect of NO inhalation on pulmonary arterial pressure was sustained. However, in one of these patients pulmonary arterial pressure remained reduced during withdrawal of NO from the inspired gas. Falke *et al.* (1992) did not report on the final clinical outcome or on the possibility that NO therapy might have allowed damaged pulmonary endothelial cells to recover their normal function and thereby contribute to the amelioration of pulmonary hypertension in this patient.

Pepke-Zaba *et al.* (1991) confirmed the ability of inhaled NO to cause selective pulmonary vasodilatation, both in patients with pulmonary hypertension, and also in cardiac patients with normal pulmonary function. Inhalation of NO had no systemic hemodynamic effects in either group of patients, or in normal volunteers. In contrast, infusion of prostacyclin caused pulmonary and systemic vasodilatation in the patients with pulmonary hypertension.

The exciting prospect that inhaled NO might be of substantial significance in the therapy of respiratory disorders is particularly ironic since NO has long been known as an environmental pollutant!

E. Renal Circulation

Actions of NO on the renal vasculature could have profound effects on fluid and electrolyte balance, as could influences of NO on other renal mechanisms. For example, bradykinin caused cyclic GMP levels in cultured mesangial cells to rise, but only when they were coincubated with endothelial cells (Shultz *et al.*, 1990); this effect was inhibited by L-NMMA and enhanced by superoxide dismutase.

Incubation of mesangial cells directly with NO increased their cyclic GMP levels, and hemoglobin inhibited this effect. Together, these results indicate that mesangial cells were influenced by NO from the endothelial cells. However, in an extension of their earlier work, Shultz *et al.* (1991) showed that mesangial cells in culture expressed an inducible form of NO synthase when challenged with endotoxin (see also Pfeilshifter, 1991). Under these conditions, mesangial cell cyclic GMP levels were increased and this effect was markedly inhibited by L-NMMA. Thus, the results of Shultz *et al.* (1990, 1991) could be taken to indicate that cultured endothelial cells can express inducible NO synthase but do not possess the constitutive form of the enzyme. In contrast, Springall *et al.* (1992) have provided immunohistochemical evidence for the existence of the constitutive form of NO synthase in cultured mesangial cells from rats, in support of their earlier functional findings. Whatever the resolution of this contradiction, Shultz *et al.* (1990) also found that NO inhibited the reduction in mesangial cell cross sectional area evoked by angiotensin II, thus raising the possibility that NO might be involved in determining the glomerular ultrafiltration coefficient *in vivo,* particularly if NO affects mesangial cell mitogenesis (Garg and Hassid, 1989).

Wilcox *et al.* (1992) produced evidence for the existence of a single nephron, L-arginine–NO system, which regulated glomerular capillary hydrostatic pressure and set the level at which the tubuloglomerular feedback mechanism operated. In addition, there is indirect evidence that NO from endocardial or coronary endothelial cells might exert a tonic inhibitory influence on release of ANP (Sanchez-Ferrer *et al.*, 1990), and that EDRF might inhibit renin release (Vidal *et al.*, 1988). However, it is not immediately obvious how NO acting to inhibit both ANP and renin release would be integrated to achieve fluid homeostasis since the outcome of these effects would be conflicting (i.e., antinatriuresis and natriuresis, respectively). Furthermore, there is circumstantial evidence that pretreatment with L-NAME caused a relative suppression of the activation of the renin–angiotensin system usually seen following administration of CGRP (Gardiner *et al.*, 1991f); this effect is the opposite of that which would be expected if NO was inhibiting renin release *in vivo.* Our misgivings on this point are supported by recent experiments of Kurtz *et al.* (1991) involving cocultures of mouse juxtaglomerular cells with bovine microvascular endothelial cells. In this system, indomethacin caused significant inhibition of spontaneous renin release but only in the presence of endothelial cells; these findings are consistent with the known stimulatory effect of prostacyclin on renin release. More interesting, in the context of the present discussion, is the observation that L-NMMA attenuated spontaneous renin secretion from cocultures and did not prevent the inhibitory effect of endothelial cells on renin secretion. Hence, these results of Kurtz *et al.* (1991) are in line with our *in vivo* findings (Gardiner *et al.*, 1991f), indicating that NO might stimulate renin release.

Bhardwaj and Moore (1989) were the first to assess the putative role of NO in the control of the renal circulation, which they did by investigating the actions of arginine and NO on the perfused rat kidney. They found that acetylcholine, sodium nitroprusside, and NO caused dose-related vasodilatations when the vasculature had been preconstricted with noradrenaline. Removal of endothelial cells abolished the effect of acetylcholine but did not affect responses to sodium nitroprusside or NO. In preconstricted kidneys, L-arginine base caused vasoconstriction followed by vasodilatation, whereas L-arginine salts caused only vasodilatation, but D-arginine was without effect. The vasodilator actions of L-arginine base or salts were endothelium dependent, but preloading kidneys with L-arginine did not enhance vasodilator responses to acetylcholine. Bhardwaj and Moore (1989) concluded that L-arginine was not converted into NO in the kidney, but that L-arginine and its salts might act to release EDRF from rat renal endothelial cells.

Radermacher *et al.* (1990), also using isolated perfused kidneys from rats, found that L-NMMA caused renal vasoconstriction that was attenuated by L-arginine, and these effects were accompanied by directionally similar changes in glomerular filtration rate (GFR) (see also El Sayed *et al.*, 1990a). L-NNA attenuated the vasodilator effects of ATP, but not those of sodium nitroprusside; in equimolar doses, L-NNA caused a 2-fold greater renal vasoconstriction than did L-NMMA. Although these results are consistent with generation of NO from L-arginine by a pathway susceptible to L-NMMA or, particularly, L-NNA, Farhat *et al.* (1990) argued that L-arginine is a poor substrate for EDRF generation in the isolated kidney of the rat and suggested, as had Bhardwaj and Moore (1989), that it might act by releasing EDRF. However, given the weight of evidence showing the major EDRF is NO, it is not at all clear, if L-arginine only acts by releasing NO, why it should reverse the renal vasoconstrictor effects of L-NMMA, L-NNA, or L-NAME.

In conscious, unrestrained rats, Gardiner *et al.* (1989f, 1990d,e,f,g) found that L-NMMA caused dose-related reductions in renal blood flow and vascular conductance. In contrast to the findings of Bhardwaj and Moore (1989), L-arginine salts alone had little effect on renal flow or vascular conductance. However, L-arginine acetate and L-arginine hydrochloride caused significant attenuation of the renal vasoconstrictor effects of L-NMMA, without influencing those seen in response to vasopressin. D-NMMA had no pressor or renal vasoconstrictor effects, and D-arginine did not attenuate the effects of L-NMMA (Gardiner *et al.*, 1990d,e). In subsequent studies, L-NAME was found to cause dose-related renal vasoconstriction that was attenuated by L-arginine although L-NAME was about 10-fold more potent than L-NMMA in causing renal vasoconstriction (Gardiner *et al.*, 1990h). Fozard and Part (1991) also observed that L-NMMA caused renal vasoconstriction in anesthetized, ganglion-blocked, normotensive and hypertensive rats. However, this effect was greater than that in the mesenteric vascular bed, whereas the opposite was the case in the studies of Gardiner *et al.* (1989f, 1990d,e,f) on conscious rats.

Baylis *et al.* (1990) studied the effects of L-NMMA and L-NNA on renal hemodynamics in conscious rats and they confirmed a marked renal vasoconstrictor response to administration of both compounds; they also observed an increase in filtration fraction because GFR was reduced less than renal plasma flow. Baylis *et al.* (1990), in addition, noted diuresis and natriuresis, the latter being due to diminished tubular reabsorption; this might have been a "pressure natriuresis" due to the marked hypertensive effects of L-NMMA and L-NNA.

In contrast to the findings of Gardiner *et al.* (1989f, 1990d,e,f,g) and Baylis *et al.* (1990), Tolins *et al.* (1990) reported that L-NMMA had no effect on renal blood flow in anesthetized rats. However, as mentioned earlier, there is evidence that anesthesia interferes with NO-mediated mechanisms. But Fozard and Part (1991) also studied anesthetized rats and found marked renal vasoconstrictor responses to L-NMMA. Unfortunately, they did not report changes in renal blood flow, so the possibility that the renal vasoconstriction they observed was not associated with a reduction in renal flow cannot be excluded. That such an event occurred is feasible, considering the profound pressor effect of L-NMMA in their ganglion-blocked rats (Fozard and Part, 1991). If there was a renal vasoconstriction, but no change in flow, then the results of Fozard and Part (1991) and those of Tolins *et al.* (1990) may not differ, inasmuch as Tolins *et al.* (1990) also found L-NMMA caused renal vasoconstriction, with no change in renal blood flow.

Tolins *et al.* (1990) observed that L-NMMA reduced urinary cyclic GMP excretion, and this and the other actions of L-NMMA were reversed by L-arginine. Moreover, all the effects of acetylcholine infusion (hypotension, renal vasodilatation, and increased urinary cyclic GMP excretion) were reversed by L-NMMA even though renal blood flow was not significantly increased during infusion of acetylcholine. However, in conscious rats, infusion of acetylcholine caused substantial increases in renal blood flow and vascular conductance that were significantly inhibited in the presence of L-NAME although increases in both variables still occurred under those conditions (Gardiner *et al.*, 1990i, 1991g).

A lack of effect of acetylcholine on renal blood flow in anesthetized rats also contrasts with the picture seen by Lahera *et al.* (1990) in anesthetized dogs. But, in the latter case, acetylcholine was administered intrarenally, rather than intravenously, so the different effects in the two studies could have been due simply to differences in effective concentrations of the agonist, without the need to invoke species differences.

In the studies of Lahera *et al.* (1990), intrarenal administration of acetylcholine in the presence of meclofenamate caused renal hyperemia, diuresis, and increased sodium excretion, with no change in GFR. After administration of L-NMMA, acetylcholine did not alter renal blood flow, urine flow, or GFR, whereas sodium excretion increased almost as much as it did with acetylcholine in the absence of L-NMMA. While it is feasible that this dissociation was due to differences in susceptibility of vascular and tubular actions of NO to inhibition by L-NAME (see earlier), other possibilities, including changes in intrarenal hemodynamics (see below), cannot be excluded. Although the results of Lahera *et al.* (1990) are entirely consistent with the renal hemodynamic effects of acetylcholine being due to release of NO through mechanisms inhibited by L-NMMA, it is notable that they, like Tolins *et al.* (1990), found L-NMMA alone was without effects on renal blood flow. However, in line with our observations in conscious rats (Gardiner *et al.*, 1990d,e,f,g), Walder *et al.* (1991) reported that L-NMMA and L-NNA caused marked reductions in renal cortical blood flow in anesthetized rats. In addition, these compounds inhibited the renal cortical hemodynamic effects of acetylcholine without affecting those of dopamine. Perella *et al.* (1991) also found that L-NMMA caused significant reductions in renal blood flow and vascular conductance, with no changes in GFR or sodium excretion in anesthetized dogs. These workers, like Walder *et al.* (1991), noted that renal hemodynamic effects of L-NMMA could occur in the absence of changes in systemic arterial blood pressure. Walder *et al.* (1991) also found that infusions of L- but not D-arginine caused increases in renal cortical blood flow. This observation differs from that of Gardiner *et al.* (1990d,h) who measured total renal blood flow, but, as Walder *et al.* (1991) commented, it is feasible that this effect of L-arginine was not mediated through NO (see later).

Welch *et al.* (1991) reported that, in anesthetized rats, L-NMMA decreased renal plasma flow and GFR, but only if renal perfusion pressure was kept constant by an aortic snare. When the systemic pressor effects of L-NMMA were allowed to cause elevation of renal perfusion pressure, GFR and renal plasma flow remained constant; calculated renal vascular resistance increased by 44%. Welch *et al.* (1991) concluded that L-NMMA most likely had a predominant preglomerular action. In anesthetized rabbits, acetylcholine caused dose-dependent hypotension and increases in glomerular filtration rate, and L-NMMA reversed these effects (Cairns *et al.*, 1991); hence, these results are consistent with those of Welch *et al.* (1991).

The effects of L-NMMA on the glomerular microcirculation in anesthetized Munich Wistar rats were investigated by Zatz and De Nucci (1991). The systemic pressor effects of L-NMMA were accompanied by a 34% decrease in renal plasma flow, indicating the occurrence of marked vasoconstriction. Since GFR did not change, there was a significant increase in filtration fraction, yet there was a substantial polyuria. Zatz and De Nucci (1991) found that single nephron GFR did not change significantly in the presence of L-NMMA in spite of an increase in glomerular capillary hydrostatic pressure because there was a significant decrease in calculated glomerular ultrafiltration coefficient. The rise in glomerular capillary hydrostatic pressure was due to the fact that efferent arteriolar resistance increased more than afferent arteriolar resistance. Since there is little evidence that efferent arterioles contribute to renal autoregulation, Zatz and De Nucci (1991) suggested that the effects of L-NMMA were probably due to a direct action on the microvasculature of the kidney. The explanation of the disparity between the results of Zatz and De Nucci (1991), showing a preferential postglomerular vasoconstrictor action of L-NMMA, and those of Welch *et al.* (1991), showing a preglomerular action of the same compound, is not obvious. Neither is the reason why L-NMMA produced a polyuria in the rats studied by Zatz and De Nucci (1991) although it is feasible this was due to the systemic hypertension, as suggested by Baylis *et al.* (1990), or to an intrarenal action of L-NMMA. However, in preliminary experiments we have been unable to show any change in urine flow rate in conscious Brattleboro rats ingesting L-NMMA chronically in their drinking water.

Recently, Brezis *et al.* (1991) investigated the possibility that NO was involved in controlling renal medullary oxygenation. In anesthetized rats, L-NMMA caused a pressor effect accompanied by a reduction in renal blood flow, as observed by Gardiner *et al.* (1990d,e,f,g) in conscious rats. Both effects of L-NMMA were reversed by L-arginine, as was the reduction in medullary P_{O_2} caused by L-NMMA; the latter had no effect on cortical P_{O_2}. Noradrenaline, at a dose that caused a reduction in renal blood flow similar to that seen with L-NMMA, produced significant reductions in cortical as well as medullary P_{O_2}; endothelin-3 caused a reduction in cortical P_{O_2} and a small increase in medullary P_{O_2} (Brezis *et al.*, 1991). In isolated, perfused kidneys, Brezis *et al.* (1991) found that L-NMMA exacerbated the damage to medullary thick ascending limbs caused by hypoxia. Moreover, the effect of hypoxia was ameliorated by sodium nitroprusside at a dose that had no effect on GFR, total renal perfusate flow, or fractional excretion of sodium or potassium. In an *in vivo* model of acute renal failure, induced by administration of radiocontrast medium, Brezis *et al.* (1991) found that L-NMMA augmented both renal failure and tubular damage.

Clearly, these results are of great importance and interest from several points of view. As Brezis *et al.* (1991) pointed out, their findings indicate a role for NO in the selective control of renal medullary perfusion. There is evidence that the kidney is a major site of arginine synthesis (Dhanakoti *et al.*, 1990; Levillain *et al.*, 1990), although medullary tubules do not appear to make a significant contribution to this process (Levillain *et al.*, 1990). However, it is feasible that medullary recycling of arginine occurs as it does for other amino acids (Dantzler and Silbernagl, 1988), and that this leads to its concentration within the medulla, which could explain the high concentration of NO in this region (Biondi *et al.*, 1990). Brezis *et al.* (1991) raised the possibility that their results could bear upon the etiology of renal failure associated with hemolysis or rhabdomyolysis since the elevated circulating heme levels in these conditions could inactivate NO. In support of this proposal, they cited unpublished data showing that infusion of myoglobin caused a selective decrease in medullary P_{O_2}.

One particular way in which L-arginine might impinge on renal function is in the mediation of the hyperperfusion and hyperfiltration responses to a protein meal or challenge with intravenous amino acids. In isolated, perfused rat kidneys, El Sayed *et al.* (1990b) noted that individual amino acids differed in their ability to reverse the decline in GFR, with an order of potency L-arginine > L-glutamate > glycine > L-leucine. However, this order differed from that of their effects on perfusate flow (L-glutamate > L-arginine > glycine > L-leucine). In a subsequent study, El Sayed *et al.* (1991) found that the relative hyperperfusion and hyperfiltration, caused by increasing the mixed amino acid concentration perfusing the isolated kidney, could be affected selectively by different interventions. Thus, indomethacin caused 50% reduction of the effect on GFR with no influence on the change in renal perfusate flow, sulpiride inhibited the effects on GFR by 60% and on perfusate flow by 50%, whereas L-NAME completely blocked the change in GFR without preventing the rise in renal perfusate flow. This latter observation is notable because L-NAME caused a substantial reduction in baseline renal perfusate flow.

Tolins and Raij (1991) studied anesthetized rats and investigated the effects of L-NMMA on the renal responses to intravenous infusion of a mixture of amino acids (with L-arginine representing 9.8% of the total amino acids). L-NMMA given intrarenally caused reductions in GFR and renal plasma flow (with a rise in filtration fraction). Although these effects appeared to be blunted by an angiotensin II antagonist, examination of the data (see Table 1 in Tolins and Raij, 1991) shows that this was due to marked differences in the baseline values in the different conditions. While infusion of a mixture of amino acids caused significant hyperperfusion and hyperfiltration, there was no significant renal vasodilatation (Tolins and Raij, 1991). However, in the presence of L-NMMA, the same challenge elicited no increase in GFR, and there was a significant reduction of renal plasma flow accompanied by renal vasoconstriction. The ability of L-NMMA to prevent the renal effects of mixed amino acids in the experiments of Tolins and Raij (1991), when L-NAME had no effect on the hyperperfusion response in the experiments of El Sayed *et al.* (1991), is puzzling, particularly in view of the more potent effects of L-NAME on renal hemodynamics (Gardiner *et al.,* 1990h). However, it is not clear to what extent effects of L-arginine, independent of NO (e.g., Battle and Chan, 1989) or due to changes in pH (e.g., Mitchell *et al.,* 1991), might contribute to its renal effects.

F. Splanchnic Circulation

In the isolated, perfused, superior mesenteric vascular bed of the rat, exogenous NO caused vasodilatation (Warner *et al.,* 1989a), and L-NMMA and L-NNA caused vasoconstriction of the noradrenaline-preconstricted preparation (Moore *et al.,* 1990). Interestingly, although Moore *et al.* (1990) found the vasoconstrictor effects of L-NMMA and L-NNA faded within 5–10 min, at that time the vasodilator effects of acetylcholine were dose-dependently inhibited. These inhibitory effects of L-NMMA or L-NAME were not reversed by washing, but they were by the addition of L-, but not D-arginine, to the perfusion medium (Moore *et al.,* 1990).

Moore *et al.* (1990) did not comment on the effects of L-NMMA or L-NNA on baseline perfusion pressure in the isolated mesenteric vascular bed, but Ebeigbe *et al.* (1990) found L-NMMA was without effect under those conditions; moreover, L-NMMA did not influence the vasoconstrictor response to noradrenaline. In addition, L-NMMA pretreatment had no influence on the vasodilator response to acetylcholine. Ebeigbe *et al.* (1990) interpreted these results to mean that L-NMMA was only effective in conditions of an "activated EDRF release state." If by this they meant basal NO synthesis was not inhibited by L-NMMA, but acetylcholine-stimulated NO synthesis was, it is not clear why L-NMMA pretreatment did not have any effect on acetylcholine-mediated vasodilatation since its presence throughout the experiment should have interfered with acetylcholine-induced NO synthesis (Ebeigbe *et al.,* 1990).

The results of Ebeigbe *et al.* (1990) are difficult to reconcile also with the findings of Dohi *et al.* (1990) who reported that L-NMMA augmented noradrenaline- or phenylephrine-induced contractions of rat, isolated, perfused, pressurised mesenteric arterioles. Dohi *et al.* (1990) further found that L-NMMA partly inhibited vasodilator responses to acetylcholine in this system. Thus, although NO may not have mediated the entire vasodilator response to acetylcholine, L-NMMA pretreatment had inhibitory effects on acetylcholine-mediated responses of a sort not seen by Ebeigbe *et al.* (1990).

Earlier studies (Gardiner *et al.*, 1989f, 1990d,e,f,g,h) and subsequent experiments (Gardiner *et al.*, 1991b) in conscious rats showed marked superior mesenteric vasoconstrictor responses to L-NMMA or L-NAME. D-NMMA and D-NAME were without vasoconstrictor effect, and the action of L-NMMA was reversed by L-, but not by D-arginine (Gardiner *et al.*, 1990d,e, 1991b). In this experimental model, the initial vasoconstrictor response to L-NMMA or L-NAME was more marked in the mesenteric than in the other vascular beds studied. Indeed, it was possible to elicit reductions of mesenteric blood flow with doses of L-NMMA that were insufficient to cause a rise in mean systemic arterial blood pressure (Gardiner *et al.*, 1990e). However, in the steady state, the vasoconstrictor responses to L-NMMA were similar in renal, mesenteric hindquarters and internal carotid vascular beds (Gardiner *et al.*, 1990d), whereas following administration of L-NAME, the mesenteric vasoconstriction was earlier and more rapid in onset and reached a lower nadir than those in the renal or hindquarters vascular beds (Gardiner *et al.*, 1990h). In all respects, except this, the findings of Fozard and Part (1991), who studied anesthetized, ganglion-blocked rats, were similar.

As pointed out by Gardiner *et al.* (1990h), it is feasible that the differences they observed in the regional hemodynamic responses to inhibition of NO synthesis were due to different degrees of involvement of neural mechanisms in the control of conductances in the different vascular beds. However, in a separate study, in which conscious rats were given L-NAME in the presence of pentolinium, captopril, and a vasopressin V_1-receptor antagonist (Gardiner *et al.*, 1990i), the vasoconstrictor effects of L-NAME were also greater in the mesenteric, than in the renal, vascular bed, both in absolute and in percentage terms. Notably, although the absolute changes in renal, mesenteric, and hindquarters vascular conductances evoked by L-NAME were all less in the blocked, than in the unblocked, state, the percentage changes in the two conditions were the same (Gardiner *et al.*, 1990i). Hence, these results do not provide evidence for a simple contribution from baroreflex buffering to the regional hemodynamic changes evoked by L-NAME. However, this is not surprising since NO-mediated processes act to oppose noradrenergic vasomotor tone (Toda *et al.*, 1991b; Cederqvist *et al.*, 1991; Vo *et al.*, 1991), and there are differences in resting vasomotor tone in different vascular beds (Gardiner *et al.*, 1989d), and different vascular beds differ in the extent to which they are affected by baroreflexes (Sagawa, 1983). Thus, all of these factors could influence the effect that ganglion blockade has on the regional hemodynamic responses to inhibition of NO synthesis. In addition, ganglion blockade normally causes activation of the renin–angiotensin system and stimulates vasopressin release; these affect cardiovascular status (Gardiner and Bennett, 1985) and, thereby, could influence responses to compounds such as L-NMMA or L-NAME in experiments in which angiotensin II- and vasopressin-mediated effects were not blocked (e.g., Fozard and Part, 1991). Moreover, regional hemodynamic responses to the L-arginine analogs would inevitably involve different degrees of autoregulatory vasoconstriction in different vascular beds (Meininger and Trzeciakowski, 1988; Hinojosa-Laborde *et al.*, 1991). Hence, various combinations of these factors could contribute to the regional hemodynamic changes following administration of NO synthesis inhibitors under different experimental conditions. As mentioned earlier, an additional complication in the experiments of Fozard and Part (1991) was the presence of an anesthetic.

A further confounding element in the assessment of the role of NO in the control of the superior mesenteric vascular bed has been highlighted by the observations of Ahlner *et al.* (1991), indicating the occurrence of an endothelium-independent, neurogenic vasodilator response in bovine mesenteric arteries. Although this phenomenon had many biochemical and pharmacological attributes consistent with NO being the mediator of the vasodilatation, L-NMMA and L-NNA were without effect. Ahlner *et al.* (1991) had no ready explanation for this anomaly, although they suggested the existence of different isoenzymes of NO synthase, and/or diffusional barriers to L-NMMA or L-NNA could explain their results.

There is *in vitro* evidence for the existence of a vasodilator innervation of the rat superior mesenteric vascular bed that utilizes CGRP as a neurotransmitter (Kawasaki *et al.*, 1988, 1990; Fujimori *et al.*, 1989, 1990; Han *et al.*, 1990). However, there is no convincing data from *in vivo* studies for the existence of CGRP-mediated vasodilator tone in the superior mesenteric vascular bed of the rat (Gardiner *et al.*, 1990j). Furthermore, in conscious rats the mesenteric vasodilator effects of human α-CGRP were similar whether that vascular bed had been preconstricted with endothelin-1 or with L-NAME (Gardiner *et al.*, 1991f). Thus, even if CGRP acts as a neurotransmitter in the mesenteric vascular bed of the rat, it appears unlikely that NO is involved in the vasodilator responses.

In contrast, there is substantial support for NO being involved in the mesenteric vasodilator responses to endothelins in the isolated, preconstricted, vascular bed (e.g., De Nucci *et al.*, 1988; Warner *et al.*, 1989b; Randall *et al.*, 1989; Douglas and Hiley, 1990, 1991; Fukuda *et al.*, 1990; Douglas *et al.*, 1991). However, in conscious rats, endothelins do not produce clear mesenteric vasodilatation (Gardiner *et al.*, 1989e, 1990a,b,c), possibly because the initial hypotensive effect due to hindquarters vasodilatation triggers baroreflex-mediated mesenteric vasoconstriction.

Although *in vitro* (see earlier) and some *in vivo* (e.g., Fortes *et al.*, 1990) studies indicate an involvement of NO in mesenteric vasodilator responses to agents such as acetylcholine, results of experiments in conscious rats are less clear-cut (Gardiner *et al.*, 1990i, 1991g); this could be due to a variety of reasons, including an involvement of NO-independent processes (Gardiner *et al.*, 1990i, 1991g).

Apart from its effects on the superior mesenteric vascular bed as a whole, there is evidence for an involvement of NO in the control of the microcirculation of the gut. Pique *et al.* (1989) observed that treatment with L-NMMA caused a rise in mean systemic arterial blood pressure accompanied by an increase in gastric mucosal blood flow in anaesthetized rats. These effects were prevented by concurrent administration of L-arginine and were not seen with D-NMMA. Pique *et al.* (1989) pointed out that a low dose of L-NMMA, which had a significant pressor effect, did not change gastric mucosal blood flow. Considered together with the reductions in superior mesenteric blood flow seen after nonpressor doses of L-NMMA in conscious rats (Gardiner *et al.*, 1990e), these results indicate that, within the tissues supplied by the celiac and superior mesenteric arteries, there are likely to be marked variations in the involvement of NO in the control of local blood flow.

Although Pique *et al.* (1989) observed no gastric mucosal damage following administration of L-NMMA, they suggested that NO might be involved in preserving gastric mucosal integrity. Subsequently, Whittle *et al.* (1990) produced evidence that NO, alone, was not essential for protecting the gastric mucosa, but under conditions in which sensory neuropeptides had been depleted by pretreatment with capsaicin, or prostacyclin synthesis was prevented by indomethacin, then administration of L-NMMA caused dose-dependent, gastric mucosal damage. This effect was inhibited by L-arginine and was not found with D-NMMA. Whittle *et al.* (1990) also observed the most extensive hemorrhagic necrosis after L-NMMA was administered to capsaicin-treated rats that had received indomethacin. Thus, these findings indicate that NO, together with prostacyclin and sensory neuropeptides, are responsible for maintaining gastric mucosal integrity (Whittle *et al.*, 1990; Whittle, 1990). These results pertain to the resting state; in addition, there are data indicating that exogenous NO may protect against ethanol-induced gastric mucosal damage (MacNaughton *et al.*, 1989), and that endogenous NO may be involved in the acute, capsaicin-induced protection of the gastric mucosa in the same model (Peskar *et al.*, 1991). The origin of NO in the study by Peskar *et al.* (1991) was not determined.

Walder *et al.* (1990) have shown that the gastric mucosal hyperemia induced by pentagastrin, in indomethacin-treated rats, is inhibited by L-NMMA or L-NAME. In these experiments, gastric acid secretion was not measured, although it is likely to have been increased, at least in the absence of L-NMMA or L-NAME. It would, therefore, be interesting to know if the latter compounds inhibit the gastric acid secretory response to pentagastrin since this could influence the functional hyperemia. However, it is feasible also that any diminution in hyperemia, due to L-NMMA or L-NAME, might act to limit the secretory response. Thus, determining whether or not NO was involved directly in the latter phenomenon would be difficult.

Apart from the data previously cited, there is little information regarding the contribution of NO to the control of other regions of the splanchnic circulation. However, Mathie *et al.* (1991) have observed that L-NMMA or L-NAME inhibited vasodilator responses to acetylcholine and ATP in the perfused liver of the rabbit. Notably, neither compound evoked significant vasoconstriction in this preparation, possibly indicating low basal release of NO in the hepatic vascular bed.

G. Skeletal Muscle and Cutaneous Circulation

Skeletal muscle and cutaneous circulations have been grouped here because many studies involve experiments in which the relative contributions of these vascular beds to the results obtained were not delineated.

Systemic administration of L-NMMA or L-NAME in conscious rats caused an initial hyperemic vasodilatation in the hindquarters, followed by reductions in flow and vascular conductance (Gardiner *et al.*, 1990d,e,h). Although the latter effects were less marked than in other vascular beds, this may have been due, partly, to the initial vasodilatation acting to oppose vasoconstriction. Some support for this proposition comes from the finding that infusion or oral administration of L-NMMA or L-NAME caused only a slowly developing hindquarters vasoconstriction, but eventually this exceeded that in the renal and mesenteric vascular beds (Gardiner *et al.*, 1990f,g, 1991j).

As mentioned earlier, the initial increase in hindquarters flow seen following administration of L-NMMA or L-NAME resembled that observed after injection of exogenous endothelins (Wright and Fozard, 1988; Gardiner *et al.*, 1989e, 1990a,b,c), and one explanation for this could be that inhibition of NO synthesis caused disinhibition of endothelin release (Boulanger and Lüscher, 1990). However, it has been claimed that the initial hypotensive vasodilator response to exogenous endothelin-1 was due to release of NO (Whittle *et al.*, 1989); the mutual exclusivity of these arguments is obvious. Moreover, we (Gardiner *et al.*, 1989g), and others (Ohlstein *et al.*, 1989g), have found no evidence for involvement of NO in the early hindquarters vasodilator effects of exogenous endothelin-1. Thus, it could be that the initial hindquarters effects of L-NMMA and L-NAME were due to disinhibition of endothelin release, triggering activation of an, as yet, unidentified vasodilator mechanism. It is feasible that such a mechanism is susceptible to anesthetic agents and/or ganglion blockade since it was not observed in the experiments of Fozard and Part (1991), but if this is the case, it is unlikely that endothelin-1 release underlies the event since hindquarters vasodilator responses to exogenous endothelin-1 occur in anesthetized, ganglion-blocked rats (Wright and Fozard, 1988).

With chronic administration of L-NMMA or L-NAME, the sustained systemic hypertension was accompanied by maintained constriction only in the hindquarters vascular bed (Gardiner *et al.*, 1990g, 1991j), possibly because in this region counter-regulatory mechanisms protecting flow are the least well developed. Nevertheless, even here, vascular conductance returned to normal within a relatively short time of withdrawal of the NO synthesis inhibitor (Gardiner *et al.*, 1991j). These results contrast somewhat with those of Du *et al.* (1991) in conscious rabbits since they observed that L-NNA caused hypertension and hindquarters vasoconstriction, but the latter persisted for at least 48 h after administration of L-NNA.

Ekelund and Mellander (1990) carried out a detailed study of the effects of L-NMMA on the vasculature of skeletal muscle in the acutely sympathectomized hindlimb of the cat. L-NMMA increased total vascular resistance by about 99%, large-bore arterial resistance by about 138%, small arteriolar vessel resistance by about 18%, and venous resistance by about 23%. The relatively small effect of L-NMMA on arterioles in the experiments of Ekelund and Mellander (1990) contrasts with the 36% reduction in diameter of transverse arterioles seen in rabbit tenuissimus muscle following topical application of L-NMMA or L-NAME (Persson *et al.*, 1990a, 1991) although terminal arteriolar diameter decreased by only 20% in the same experiment. Furthermore, Nakamura and Prewitt (1991a,b) also saw maximal reductions of the order of 20% in the diameter of arcade arterioles in the spinotrapezius muscles of anesthetized rats following superfusion of L-NMMA. Hence, on balance, these results and others (Bellan *et al.*, 1991; Ross *et al.*, 1991; Mügge *et al.*, 1991) are consistent with an important influence of NO on arterial resistance in the hindlimb.

The lack of an initial vasodilator response to L-NMMA in the experiments of Ekelund and Mellander (1990) might indicate that the effect observed in the hindquarters of conscious rats (Gardiner *et al.*, 1990d,e) was confined to the cutaneous vascular bed. However, although Ekelund and Mellander (1990) indicated that L-NMMA had no effect on myogenic vessel reactivity, there is some evidence that NO might be involved in this process, at least under some conditions (Johnsson *et al.*, 1991), so it is feasible that L-NMMA or L-NAME could also influence hindquarters flow by affecting myogenic tone. Once again, anesthesia might be a confounding variable.

In conscious rats, the hindquarters vasodilator effects of infused acetylcholine were unaffected by L-NAME whereas those to 5′-N-ethylcarboxamidoadenosine, salbutamol, adrenaline, and bradykinin were inhibited (Gardiner *et al.*, 1991g,h,i). In the case of bradykinin, there appeared to be a contribution from β_2-adrenoceptors, probably due to adrenaline released from the adrenal medulla (Gardiner *et al.*, 1991i, 1992).

The finding that the acetylcholine-mediated hindquarters vasodilatation was resistant to L-NAME should be considered against the background that the dose of acetylcholine administered did not increase hindquarters flow; thus, the change in vascular conductance could have been autoregulatory (Gardiner *et al.*, 1991g). However, it does not follow that doses of acetylcholine sufficient to cause hindquarters hyperemia would not stimulate NO release. Indeed, Ekelund and Mellander (1990) found that the hindlimb vasodilator effects of acetylcholine in the cat were inhibited by L-NMMA, but the effect was not impressive. Nakamura and Prewitt (1991a,b) reported that the inhibitory effects of L-NMMA on arteriolar responses to acetylcholine in the rat spinotrapezius muscle were more marked at lower than at higher doses of the latter although Persson *et al.* (1990a, 1991) found the opposite with both L-NMMA and L-NAME in arterioles of the rabbit tenuissimus muscle.

While it could be argued that some of these differences were due to the extensive surgical interventions necessary to make the *in vivo* microscopical observations, similar variations exist between studies in which surgical preparation was relatively restricted. Thus, Bellan *et al.* (1991) found that local infusion of L-NNA, into the hindlimb of the anesthetized cat, caused vasoconstriction and inhibition of acetylcholine-mediated vasodilatation. However, the inhibitory effect of L-NNA diminished with increasing dose of acetylcholine, as it did with bradykinin. Although Bellan *et al.* (1991) noted a slight enhancement of the vasodilator effects of sodium nitroprusside in the presence of L-NNA, neither this nor the reduction of the vasodilator effects of isoprenaline under the same conditions reached significance.

In order to avoid reflex effects influencing responses to L-NMMA, Ross *et al.* (1991) studied its actions in the denervated hindlimb of the anesthetized cat. L-NMMA given intravenously caused hypertension, bradycardia, and hindlimb vasoconstriction; the latter observation is consistent with our assertion that the vasoconstrictor effects of NO synthesis inhibitors are not dependent on vasomotor tone (see earlier) but is contrary to other findings (Lacolley *et al.*, 1991; Vargas *et al.*, 1990). However, Ross *et al.* (1991) saw no effect of L-NMMA, injected intravenously, on vasodilator responses to acetylcholine given into the femoral artery. When L-NMMA was administered intraarterially it increased hindlimb vascular resistance, but still with no effect on the maximum reduction in vascular resistance elicited by acetylcholine, although the duration of the vasodilatation was decreased.

The disparity between the results of Bellan *et al.* (1991) and Ross *et al.* (1991) is puzzling although the former studied the hindlimb perfused at constant flow and the latter's experiments were under free-flow conditions. Since Ross *et al.* (1991) did not give raw data for flow and vascular resistance, it is not possible to judge if their percentage transforms obscured an effect of L-NMMA on responses to acetylcholine. In addition, the use of resistance, rather than conductance, as an index of vasomotion can give rise to difficulties of interpretation *in vivo* (Stark, 1967; Rodbard, 1971; Lautt, 1989; O'Leary, 1991).

The possibility that differences in the interpretation of results have impinged on reported data is heightened by the publication from Mügge *et al.* (1991). These workers found that L-NMMA and L-NNA inhibited acetylcholine-mediated relaxation of rings of rabbit femoral arteries *in vitro,* whereas the same substances caused vasoconstriction of skeletal muscle in the hindlimb and enhanced the vasodilator effects of acetylcholine *in vivo*. This enhancement was the same as that seen with sodium nitroprusside, which is surprising since supersensitivity to the effects of the latter might have been expected under those conditions. Whatever the final resolution of these anomalies, it seems likely that, *in vivo,* responses to classical endothelium-dependent vasodilators involve factors besides NO (Gardiner et al., 1990i, 1991g). However, it is notable that the impaired hindlimb vasodilator responses to acetylcholine, seen in cholesterol-fed rabbits, were normalized by L-arginine (Girerd *et al.*, 1990).

It appears NO plays a major role in the control of resting hindquarters vascular conductance and there is some evidence that neurogenic mechanisms may be involved (Persson *et al.*, 1991). It is likely also that NO-mediated events contribute to the control of the cutaneous circulation (Hughes *et al.*, 1990).

The limb circulation was the first vascular terriotory in which the effects of L-NMMA were investigated in normal man (Vallance *et al.*, 1989a). Infusion of L-NMMA into the brachial artery caused dose-dependent reductions in forearm blood flow, with the highest dose causing a 60% fall; however, blood flow had returned to normal by 60 min after the end of the L-NMMA infusion. L-NMMA caused a significant reduction of the hyperemic effects of acetylcholine but not those of glyceryl trinitrate (Vallance *et al.*, 1989a). L-NMMA also inhibited the venodilator and augmented the venoconstrictor effects of acetylcholine in normal subjects, without influencing basal venous tone (Vallance *et al.*, 1989b). It is of interest that the forearm vasodilator effects of acetylcholine have been shown to be attenuated in patients with heart failure (Kubo *et al.*, 1991) although the effects of NO synthesis inhibitors were not investigated in that study.

Nakaki *et al.* (1990) observed that intravenous infusion of L-arginine monochloride caused hypotension and tachycardia in normotensive and hypertensive subjects; they suggested the most likely explanation of their results was increased NO release. However, Paton (1990) pointed out that L-arginine might have been releasing histamine, thereby eliciting hypotension. Both Hishikawa *et al.* (1991) and Petros *et al.* (1991) attempted to counter this argument by demonstrating that L-arginine still exerted its hypotensive effects in the presence of a dose of chlorpheniramine sufficient to inhibit the depressor action of H_1-receptor stimulation. But Paton (1991) came back with the salutary reminder that it is much more difficult to block the effects of endogenous than of exogenous histamine, and that the hypotensive effects of histamine may also involve H_2-receptors. Another problem is that the hypotensive effects of L-arginine were not demonstrated to be selective for the L-enantiomer. Calver *et al.* (1990, 1991) investigated this question, by comparing forearm arterial and venous responses to L- and D-arginine free base and hydrochloride. At an infusion rate of 160 μmol/min, all compounds caused increases in forearm blood flow, while infusion of all compounds at 5 μmol/min caused increases in the diameters of superficial hand veins. These results do not support the claim that the hypotensive effects of L-arginine *in vivo* are due to increased NO production.

Investigations of NO-mediated mechanisms in blood vessels isolated from human subjects are sparse. Those that have been done indicate L-NMMA inhibits acetylcholine-mediated vasorelaxation and can cause endothelium-dependent contraction of a variable magnitude in different vessels (Woolfson and Poston, 1990; Yang *et al.*, 1990; Lin *et al.*, 1991).

VI. Conclusions

The staggering rate at which our knowledge about the involvement of the L-arginine–NO pathway in biological processes is developing is due to several factors:

1. its ubiquity
2. the magnitude of its effects, even in systems subject to multifactorial control
3. the ease with which the system can be manipulated, and at little financial cost

It is now clear that NO is a major EDRF. However, it is not the only one, and from the foregoing review, it should be obvious that findings *in vitro* cannot be extrapolated to the *in vivo* setting with impunity.

In these days of increasing activity on the molecular biological and animal liberation fronts, those who work *in vivo* on integrated systems are threatened in different ways. But the day will come when it will be appreciated that such work is indispensable if we are to understand the role of mechanisms such as the L-arginine–NO pathway in normal and pathological conditions, to the extent that we can manipulate it to clinical benefit.

Acknowledgments

Work from the author's laboratory was supported by grants from the MRC, BHF, Wellcome Trust, and Wellcome Research Laboratories. We are indebted to Michelle Neville for her unstinting help in the preparation and typing of this review.

References

Abman, S. H., Chatfield, B. A., Hall, S. L., and McMurtry, I. F. (1990). Role of endothelium-derived relaxing factor during transition of pulmonary circulation at birth. *Am. J. Physiol. 258,* H1921–1927.

Adnot, S., Raffestin, B., Eddahibi, S., Braquet, P., and Chabrier, P-E. (1990). Loss of endothelium-dependent relaxant activity in the pulmonary circultion of rats exposed to chronic hypoxia. *J. Clin. Invest. 87,* 155–162.

Ahlner, J., Ljusegren, M. E., Grundström, N., and Axelsson, K. L. (1991). Role of nitric oxide and cyclic GMP as mediators of endothelium-independent neurogenic relaxation in bovine mesenteric artery. *Circ. Res. 68,* 756–762.

Aisaka, K., Gross, S. S., Griffith, O. W., and Levi, R. (1989). N^G-methylarginine, an inhibitor of endothelium-derived nitric oxide synthesis, is a potent pressor agent in the guinea pig: Does nitric oxide regulate blood pressure *in vivo*? *Biochem. Biophys. Res. Comm. 160,* 881–886.

Aisaka, K., Mitani, A., Kitajima, Y., Ohno, T., and Ishihara, T. (1991). Difference in pressor responses to N^G-monomethyl-L-arginine between conscious and anesthetized rats. *Jpn. J. Pharmacol. 56,* 245–248.

Amezcua, J. L., Dusting, G. J., Palmer, R. M. J., and Moncada, S. (1988). Acetylcholine induces vasodilatation in the rabbit isolated heart through the release of nitric oxide, the endogenous nitrovasodilator. *Br. J. Pharmacol. 95,* 830–834.

Amezcua, J. L., Palmer, R. M. J., de Souza, B. M., and Moncada, S. (1989). Nitric oxide synthesized from L-arginine regulates vascular tone in the coronary circulation of the rabbit. *Br. J. Pharmacol. 97,* 1119–1124.

Aoki, N., Johnson, G., and Lefer, A. M. (1990). Beneficial effects of two forms of NO administration in feline splanchnic artery occlusion shock. *Am. J. Physiol. 258,* G275–G281.

Archer, S. L., Tolins, J. P., Raij, L., and Weir, E. K. (1989). Hypoxic pulmonary vasoconstriction is enhanced by inhibition of the synthesis of an endothelium derived relaxing factor. *Biochem. Biophys. Res. Comm. 164,* 1198–1205.

Barer, G., Bee, D., and Emery, D. (1990). An arginine analogue exacerbates hypertension in normoxic lungs isolated from chronically hypoxic rats. *J. Physiol. 430,* 43P.

Bassenge, E., and Huckstorf, C. (1992). Endothelium-derived nitric oxide adjusts coronary tone and regulates myocardial perfusion. *In* "Biology of Nitric Oxide." (S. Moncada, M. A. Marletta, and J. B. Hibbs, eds.) (in press).

Battle, D. C., and Chan, Y. L. (1989). Effect of L-arginine on renal tubular bicarbonate reabsorption by the rat kidney. *Min. Electro. Metab. 15,* 187–194.

Baylis, C., Harton, P., and Engels, K. (1990). Endothelial derived relaxing factor controls renal hemodynamics in the normal rat kidney. *J. Am. Soc. Nephrol. 1,* 875–881.

Bellan, J. A., Minkes, R. K., McNamara, D. B., and Kadowitz, P. J. (1991). N^ω-nitro-L-arginine selectively inhibits vasodilator responses to acetylcholine and bradykinin in cats. *Am. J. Physiol. 260,* H1025–H1029.

Benyó, Z., Kiss, G., Szabó, C., Csáki, C., and Kovách, A. G. B. (1991). Importance of basal nitric oxide synthesis in regulation of myocardial blood flow. *Cardiovasc. Res. 25,* 700–703.

Bhardwaj, R., and Moore, P. K. (1989). The effect of arginine and nitric oxide on resistance blood vessels of the perfused rat kidney. *Br. J. Pharmacol. 97,* 739–744.

Biondi, M. L., Vanhoutte, P. M., and Romero, J. C. (1990). Endothelium-derived relaxing factor (EDRF) in renal medulla. *Kidney Intl. 37,* 364a.

Boulanger, C., and Lüscher, T. F. (1990). Release of endothelin from the porcine aorta: Inhibition by endothelium-derived nitric oxide. *J. Clin. Invest. 85,* 587–590.

Bredt, D. S., Hwang, P. M., Glatt, C. E., Lowenstein, C., Reed, R. R., and Snyder, S. H. (1991). Cloned and expressed nitric oxide synthase structurally resembles cytochrome P-450 reductase. *Nature (London) 351,* 714–717.

Brezis, M., Heyman, S. N., Dinour, D., Epstein, F. H., and Rosen, S. (1991). Role of nitric oxide in renal medullary oxygenation: Studies in isolated and intact rat kidneys. *J. Clin. Invest. 88,* 390–395.

Broten, T. P., Miyashiro, J. K., Moncada, S., and Feigl, E. O. (1992). Role of endothelium-derived relaxing factor in parasympathetic cholinergic coronary vasodilation. *Am. J. Physiol. 262,* H1579–H1584.

Busija, D. W., Leffler, C. W., and Wagerle, L. C. (1990). Mono-L-arginine-containing compounds dilate piglet pial arterioles via an endothelium-derived relaxing factor-like substance. *Circ. Res. 67,* 1374–1380.

Cairns, H. S., Rogerson, M. E., Westwick, J., and Neild, G. H. (1991). Regional heterogeneity of endothelium-dependent vasodilatation in the rabbit kidney. *J. Physiol. 436,* 431–429.

Calver, A., Collier, J., and Vallance, P. (1990). L-arginine-induced hypotension. *Lancet 336,* 1016–1017.

Calver, A., Collier, J., and Vallance, P. (1991). Dilator actions of arginine in human peripheral vasculature. *Clin. Sci. 81,* 695–700.

Cappelli-Bigazzi, M., Nuno, D. W., and Lamping, K. G. (1991). Evidence of a role for compounds derived from arginine in coronary response to serotonin *in vivo. Am. J. Physiol. 261,* H404–H409.

Cederqvist, B., Wiklund, N. P., Persson, M. G., and Gustafsson, L. E. (1991). Modulation of neuroeffector transmission in the guinea pig pulmonary artery by endogenous nitric oxide. *Neurosci. Letts. 127,* 67–69.

Chester, A. H., O'Neil, G. S., Tadjkarimi, S., Palmer, R. M. J., Moncada, S., and Yacoub, M. H. (1990). The role of nitric oxide in mediating endothelium dependent relaxations in the human epicardial coronary artery. *Int. J. Cardiol. 29,* 305–309.

Chu, A., Chambers, D. E., Lin, C.-C., Kuehi, W. D., and Cobb, F. R. (1990). Nitric oxide modulates epicardial coronary basal vasomotor tone in awake dogs. *Am. J. Physiol. 258,* H1250–H1254.

Chu, A., Chambers, D. E., Lin, C-C., Kuehl, W. D., Palmer, R. M. J., Moncada, A., and Cobb, F. R. (1991). Effects of inhibition of nitric oxide formation on basal vasomotion and endothelium-dependent responses of the coronary arteries in awake dogs. *J. Clin. Invest. 87,* 1964–1968.

Cocks, T. M., Malta, E., King, S. J., Woods, R. L., and Angus, J. A. (1991). Oxyhaemoglobin increases the production of endothelin-1 by endothelial cells in culture. *Eur. J. Pharmacol. 196,* 177–182.

Cooke, J. P., Andon, N. A., Girerd, X. J., Hirsch, A. T., and Creager, M. A. (1991). Arginine restores cholinergic relaxation of hypercholesterolemic rabbit thoracic aorta. *Circulation 83,* 1057–1062.

Cowan, C. L., and Cohen, R. A. (1991). Two mechanisms mediate relaxation by bradykinin of pig coronary artery: NO-dependent and -independent responses. *Am. J. Physiol. 261,* H830–H835.

Crawley, D. E., Liu, S. F., Evans, T. W., and Barnes, P. J. (1990). Inhibitory role of endothelium-derived relaxing factor in rat and human pulmonary arteries. *Br. J. Pharmacol. 101,* 166–170.

Dantzler, W. H., and Silbernagl, S. (1988). Amino acid transport by juxtamedullary nephrons: Distal reabsorption and recycling. *Am. J. Physiol. 255,* F397–F407.

Dawson, T. M., Dawson, V. L., Bredt, D. S., Fotuhi, M., Hwang, P. M., Uhl, G. R., and Snyder, S. H. (1991a). Nitric oxide synthase/NADPH diaphorase neurons: Role in neurotoxicity. *Soc. Neurosci. Abst. 17,* 784.

Dawson, V. L., Dawson, T. M., London, E. D., Bredt, D. S., and Snyder, S. H. (1991b). Glutamate neurotoxicity is mediated by nitric oxide. *Soc. Neurosci. Abst. 17,* 783.

De Nucci, G., Thomas, R., D'Orléans-Juste, P., Antunes, E., Walder, C., Warner, T. D., and Vane, J. R. (1988). Pressor effects of circulating endothelin are limited by its removal in the pulmonary circulation and by the release of prostacyclin and endothelium-derived relaxing factor. *Proc. Natl. Acad. Sci. U.S.A. 85,* 9797–9800.

Dhanakoti, S. N., Brosnan, J. T., Herzberg, G. R., and Brosnan, M. E. (1990). Renal arginine synthesis: Studies *in vitro* and *in vivo*. *Am. J. Physiol. 259,* E437–E442.

Dinh-Xuan, A. T., Higenbottam, T. W., Clelland, C. A., Pepke-Zaba, J., Cremona, G., Butt, A. Y., Large, S. R., Wells, F. C., and Wallwork, J. (1991). Impairment of endothelium-dependent pulmonary-artery relaxation in chronic obstructive lung disease. *N. Engl. J. Med. 324,* 1539–1547.

Dohi, Y., Thiel, M. A. Bühler, F. R., and Lüscher, T. F. (1990). Activation of endothelial L-arginine pathway in resistance arteries: Effect of age and hypertension. *Hypertension 15,* 170–179.

Douglas, S. A., and Hiley, C. R. (1990). Endothelium-dependent vascular activities of endothelin-like peptides in the isolated superior mesenteric arterial bed of the rat. *Br. J. Pharmacol. 101,* 81–88.

Douglas, S. A., and Hiley, C. R. (1991). Endotheium-dependent mesenteric vasorelaxant effects and systemic actions of endothelin (16–21) and other endothelin-related peptides in the rat. *Br. J. Pharmacol. 104,* 311–320.

Douglas, S. A., James, S., and Hiley, C. R. (1991). Endothelial modulation and changes in endothelin pressor activity during hypoxia in the rat isolated perfused superior mesenteric arterial bed. *Br. J. Pharmacol. 103,* 1441–1448.

Du, Z. Y., Dusting, G. J., and Woodman, O. L. (1991). Haemodynamic responses to N-nitro-L-arginine in conscious rabbits. *Clin. Exp. Pharmacol. Physiol. 18,* 371–374.

Dubbin, P. N., Zambetis, M., and Dusting, G. J. (1990). Inhibition of endothelial nitric oxide biosynthesis by *N*-nitro-L-arginine. *Clin. Exp. Pharmacol. Physiol. 17,* 281–286.

Dusting, G. J., and MacDonald, P. S. (1990). Prostacyclin and vascular function: Implications for hypertension and atherosclerosis. *Pharmacol. Therap. 48,* 323–344.

Ebeigbe, A. B., Cressier, F., Konneh, M. K., Luu, T. D., and Criscione, L. (1990). Influence of N^G-monomethyl-L-arginine on endothelium-dependent relaxations in the perfused mesenteric vascular bed of the rat. *Biochem. Biophys. Res. Comm. 169,* 873–879.

Ekelund, U., and Mellander, S. (1990). Role of endothelium-derived nitric oxide in the regulation of tonus in large-bore arterial resistance vessels, arterioles and veins in cat skeletal muscle. *Acta Physiol. Scand. 140,* 301–309.

El Sayed, A. A., El Nahas, A. M., Towers, J. D., and Haylor, J. (1990a). Effect of nitro-arginine methyl ester on renal function in isolated perfused rat kidney. *Br. J. Pharmacol. 100,* 327P.

El Sayed, A. A., Haylor, J., and El Nahas, A. M. (1990b). Differential effects of amino acids on the isolated perfused rat kidney. *Clin. Sci. 79,* 381–386.

El Sayed, A. A., Haylor, J., and El Nahas, A. M. (1991). Mediators of the direct effects of amino acids on the rat kidney. *Clin. Sci. 81,* 427–432.

Eisner, D., Müntze, A., Kromer, E. P., and Riegger, G. A. J. (1991). Systemic vasoconstriction induced by inhibition of nitric oxide synthesis is attenuated in conscious dogs with heart failure. *Cardiovasc. Res. 25,* 438–440.

Falke, K., Rossaint, R., Pison, U., Slama, K., Lopez, F., Santak, B., and Zapol, W. M. (1990). Inhaled nitric oxide selectively reduces pulmonary hypertension in severe ARDS and improves gas exchange as well as right heart ejection fraction—a case report. *Am. Rev. Respir. Dis. 143,* (suppl.) A248.

Falke, K. J., Rossaint, R., Keitel, M., Pison, U. K., Slama, K., Lopez, F., Grüning, T., and Zapol, W. M. (1992). Successful treatment of severe adult respiratory distress syndrome with nitric oxide—the first three patients. *In* "Biology of Nitric Oxide." (S. Moncada, M. A. Marletta and J. B. Hibbs, eds.) (in press).

Faraci, F. M. (1990). Role of nitric oxide in regulation of basilar artery tone *in vivo*. *Am. J. Physiol. 259,* H1216–H1221.

Faraci, F. M. (1991). Role of endothelium-derived relaxing factor in cerebral circulation: Large arteries vs microcirculation. *Am. J. Physiol. 261,* H1038–H1042.

Farhat, M. Y., Ramwell, P. W., and Thomas, G. (1990). Endothelium-mediated effects of N-substituted arginines on the isolated perfused rat kidney. *J. Pharmacol. Exp. Ther. 255,* 473–477.

Fasehun, O. A., Gross, S. S., Rubin, S. E., Jaffe, E. A., Griffith, O. W., and Levi, R. (1990). L-

Arginine, but not Nα-benzoyl-L-arginine ethyl ester, is a precursor of endothelium-derived relaxing factor. *J. Pharmacol. Exp. Ther. 255,* 1348–1353.

Fineman, J. R., Heymann, M. A., and Soifer, S. J. (1991). N^W-nitro-L-arginine attenuates endothelium-dependent pulmonary vasodilation in lambs. *Am. J. Physiol. 260,* H1299–H1306.

Fischer-Nakielski, H., and Schrör, K. (1990). Nitric oxide is the endothelium-derived relaxing factor in bovine pial arteroles. *Stroke 21* (suppl. iv), IV-46–IV-48.

Fiscus, R. R., Zhou, H-L., Wang, X., Han, C., Ali, S., Joyce, C. D., and Murad, F. (1991). Calcitonin gene-related peptide (CGRP)-induced cyclic AMP, cyclic GMP and vasorelaxant responses in rat thoracic aorta are antagonized by blockers of endothelium-derived relaxant factor (EDRF). *Neuropeptides 20,* 133–143.

Fortes, Z. B., Oliveira, M. A., Scivoletto, R., Carvalho, M. H. C., De Nucci, G., and Nigro, D. (1990). Nitric oxide release may be involved in the microcirculatory response to acetylcholine. *Eur. J. Pharmacol. 182,* 143–147.

Fozard, J. R., and Part, M-L. (1991). Haemodynamic responses to N^G-monomethyl-L-arginine in spontaneously hypertensive and normotensive Wistar-Kyoto rats. *Br. J. Pharmacol. 102,* 823–826.

Frostell, C., Fratacci, M.-D., Wain, J. C., Jones, R., and Zapol, W. M. (1991). Inhaled nitric oxide: A selective pulmonary vasodilator reversing hypoxic pulmonary vasoconstriction. *Circulation 83,* 2038–2047.

Fujimori, A., Saito, A., Kimura, S., Watanabe, T., Uchiyama, Y., Kawasaki, H., and Goto, K. (1989). Neurogenic vasodilation and release of calcitonin gene-related peptide (CGRP) from perivascular nerves in the rat mesenteric artery. *Biochem. Biophys. Res. Comm. 165,* 1391–1398.

Fujimori, A., Saito, A., Kimura, S., and Goto, K. (1990). Release of calcitonin gene-related peptide (CGRP) from capsaicin-sensitive vasodilator nerves in the rat mesenteric artery. *Neurosci. Letts. 112,* 183–178.

Fukuda, N., Izumi, Y., Soma, M., Watanabe, Y., Watanabe, M., Hatano, M., Sakuma, I., and Yasuda, H. (1990). L-N^G-monomethyl arginine inhibits the vasodilating effects of low dose of endothelin-3 on rat mesenteric arteries. *Biochem. Biophys. Res. Comm. 167,* 739–745.

Fukuroda, T., Noguchi, K., Tsuchida, S., Nishikibe, M., Ikemoto, F., Okada, K., and Yano, M. (1990). Inhibition of biological actions of big endothelin-1 by phosphoramidon. *Biochem. Biophys. Res. Comm. 172,* 390–395.

Furchgott, R. F. (1988). Studies on relaxation of rabbit aorta by sodium nitrite: The basis for the proposal that the acid-activatable inhibitory factor from retractor penis is inorganic nitrite and the endothelium-derived relaxing factor is nitric oxide. *In* "Vasodilatation: Vascular Smooth Muscle, Peptides, Autonomic Nerves and Endothelium." (P. M. Vanhoutte, ed.), pp. 401–414. Raven Press, New York.

Furchgott, R. F., and Zawadzki, J. V. (1980). The obligatory role of endothelial cells in the relaxation of arterial smooth muscle by acetylcholine. *Nature (London) 288,* 373–376.

Gardiner, S. M., and Bennett, T. (1985). Interactions between neural mechanisms, the renin–angiotensin system and vasopressin in the maintenance of blood pressure during water deprivation: Studies in Long Evans and Brattleboro rats. *Clin. Sci. 68,* 647–657.

Gardiner, S. M., Compton, A. M., and Bennett, T. (1988a). Regional haemodynamic effects of depressor neuropeptides in conscious, unrestrained, Long Evans and Brattleboro rats. *Br. J. Pharmacol. 95,* 197–208.

Gardiner, S. M., Compton, A. M., and Bennett, T. (1988b). Regional hemodynamic effects of atrial natriuretic peptide or captopril in Brattleboro rats. *Am. J. Physiol. 255,* R737–R743.

Gardiner, S. M., Compton, A. M., and Bennett, T. (1989a). Regional haemodynamic effects of calcitonin gene-related peptide. *Am. J. Physiol. 256,* R332–R338.

Gardiner, S. M., Compton, A. M., and Bennett, T. (1989b). Regional haemodynamic effects of human α- and β-calcitonin gene-related peptide in conscious Wistar rats. *Br. J. Pharmacol. 98,* 1225–1232.

Gardiner, S. M., Compton, A. M., and Bennett, T. (1989c). The effects of infusions of arginine vasopressin or 1-deamino-8-D-arginine vasopressin on common carotid vascular resistance in conscious, Long Evans rats. *Br. J. Pharmacol. 97,* 153–157.

Gardiner, S. M., Compton, A. M., and Bennett, T. (1989d). Regional hemodynamic changes following hypovolemia in conscious rats. *Am. J. Physiol. 256,* R1076–R1083.

Gardiner, S. M., Compton, A. M., and Bennett, T. (1989e). Regional hemodynamic effects of endothelin-1 in conscious, unrestrained, Wistar rats. *J. Cardiovasc. Pharmacol. 13* (suppl. 5), S202–S204.

Gardiner, S. M., Compton, A. M., Bennett, T., Palmer, R. M. J., and Moncada, S. (1989f). Haemodynamic effects of N^G-monomethyl-L-arginine (L-NMMA) in conscious Long Evans rats. *Br. J. Pharmacol. 98* (suppl.), 623P.

Gardiner, S. M., Compton, A. M., Bennett, T., Palmer, R. M. J., and Moncada, S. (1989g). N^G-monomethyl-L-arginine does not inhibit the hindquarters vasodilator action of endothelin-1 in conscious rats. *Eur. J. Pharmacol. 171,* 237–240.

Gardiner, S. M., Compton, A. M., and Bennett, T. (1990a). Regional haemodynamic effects of endothelin-1 and endothelin-3 in conscious Long Evans and Brattleboro rats. *Br. J. Pharmacol. 99,* 107–112.

Gardiner, S. M., Compton, A. M., and Bennett, T. (1990b). Regional hemodynamic effects of endothelin-2 and sarafotoxin-S6b in conscious rats. *Am. J. Physiol. 258,* R912–R917.

Gardiner, S. M., Compton, A. M., and Bennett, T. (1990c). Effects of indomethacin on the regional haemodynamic responses to low doses of endothelins and sarafotoxin. *Br. J. Pharmacol. 100,* 158–162.

Gardiner, S. M., Compton, A. M., Bennett, T., Palmer, R. M. J., and Moncada, S. (1990d). Control of regional blood flow by endothelium-derived nitric oxide. *Hypertension 15,* 486–492.

Gardiner, S. M., Compton, A. M., Bennett, T., Palmer, R. M. J., and Moncada, S. (1990e). Regional haemodynamic effects of inhibiting endothelial cell production of nitric oxide with N^G-monomethyl-L-arginine in conscious rats. *In* "Nitric Oxide from L-arginine: A Bioregulatory System." (S. Moncada and E. A. Higgs, eds.), pp. 81–88. Excerpta Medica, Amsterdam.

Gardiner, S. M., Compton, A. M., Bennett, T., Palmer, R. M. J., and Moncada, S. (1990f). Persistent haemodynamic changes following prolonged infusions of N^G-monomethyl-L-arginine (L-NMMA) in conscious rats. *In* "Nitric Oxide from L-arginine: A Bioregulatory System." (S. Moncada and E. A. Higgs, eds.), pp. 489–491. Excerpta Medica, Amsterdam.

Gardiner, S. M., Compton, A. M., Bennett, T., Palmer, R. M. J., and Moncada, S. (1990g). Regional haemodynamic changes during oral ingestion of N^G-monomethyl-L-arginine or N^G-nitro-L-arginine methyl ester in conscious Brattleboro rats. *Br. J. Pharmacol. 101,* 10–12.

Gardiner, S. M., Compton, A. M., Kemp, P. A., and Bennett, T. (1990h). Regional and cardiac haemodynamic effects of N^G-nitro-L-arginine methyl ester in conscious, Long Evans rats. *Br. J. Pharmacol. 101,* 625–631.

Gardiner, S. M., Compton, A. M., Kemp, P. A., and Bennett, T. (1990i). Regional and cardiac haemodynamic responses to glyceryl trinitrate, acetylcholine, bradykinin and endothelin-1 in conscious rats: Effects of N^G-nitro-L-arginine methyl ester. *Br. J. Pharmacol. 101,* 632–639.

Gardiner, S. M., Compton, A. M., Kemp, P. A., Bennett, T., Bose, C., Foulkes, R., and Hughes, B. (1990j). Antagonistic effect of human α-CGRP [8-37] on the *in vivo* regional haemodynamic actions of human α-CGRP. *Biochem. Biophys. Res. Comm. 171,* 938–943.

Gardiner, S. M., Compton, A. M., and Bennett, T. (1991a). N^G-nitro-L-arginine methyl ester (L-NAME) does not impair baroreflex control of pulse interval acutely in conscious rats. *Clin. Auto. Res. 1,* 93.

Gardiner, S. M., Compton, A. M., Kemp, P. A., and Bennett, T. (1991b). Hemodynamic effects of N^G-monomethyl-L-arginine and N^G-nitro-L-arginine methyl ester in conscious, unrestrained rats. *J. Cardiovasc. Pharmacol. 17* (suppl. 3), S173–S181.

Gardiner, S. M., Compton, A. M., Kemp, P. A., and Bennett, T. (1991c). Effects of N^G-nitro-L-arginine methyl ester or indomethacin on differential regional and cardiac haemodynamic actions of arginine vasopressin and lysine vasopressin in conscious rats. *Br. J. Pharmacol. 102,* 65–72.

Gardiner, S. M., Compton, A. M., Kemp, P. A., and Bennett, T. (1991d). The effects of phosphoramidon on the regional haemodynamic responses to human proendothelin [1-38] in conscious rats. *Br. J. Pharmacol. 103,* 1009–2015.

Gardiner, S. M., Compton, A. M., Kemp, P. A., Bennett, T., Hughes, B., and Foulkes, R. (1991e). Differential effects of (±)-dobutamine and human α-CGRP on cardiac and on regional haemodynamics in conscious Long Evans rats. *Br. J. Pharmacol. 103,* 1500–1508.

Gardiner, S. M., Compton, A. M., Kemp, P. A., Bennett, T., Foulkes, R., and Hughes, B. (1991f). Haemodynamic effects of human α-calcitonin gene-related peptide following administration of endothelin-1 or N^G-nitro-L-arginine methyl ester in conscious rats. *Br. J. Pharmacol. 103,* 1256–1262.

Gardiner, S. M., Kemp, P. A., and Bennett, T. (1991g). Effects of N^G-nitro-L-arginine methyl ester on vasodilator responses to acetylcholine, 5′-N-ethylcarboxamidoadenosine or salbutamol in conscious rats. *Br. J. Pharmacol. 103,* 1725–1732.

Gardiner, S. M., Kemp, P. A., and Bennett, T. (1991h). Effects of N^G-nitro-L-arginine methyl ester on vasodilator responses to adrenaline or BRL 38227 in conscious rats. *Br. J. Pharmacol. 104,* 731–737.

Gardiner, S. M., Kemp, P. A., Bennett, T., Bose, C., Foulkes, R., and Hughes, B. (1991i). Involvement of β_2-adrenoceptors in the regional haemodynamic responses to bradykinin in conscious rats. *Br. J. Pharmacol. 104* (suppl.), 123P.

Gardiner, S. M., Kemp, P. A., Bennett, T., Palmer, R. M. J., and Moncada, S. (1991j). Sustained hypertension in Brattleboro rats during chronic oral ingestion of N^G-monomethyl-L-arginine. *Br. J. Pharmacol. 104* (suppl.), 10P.

Gardiner, S. M., Kemp, P. A., Bennett, T., Bose, C., Foulkes, R., and Hughes, B. (1992). Involvement of β_2-adrenoceptors in the regional haemodynamic responses to bradykinin in conscious rats. *Br. J. Pharmacol. 105,* 839–898.

Garg, U. C., and Hassid, A. (1989). Inhibition of rat mesangial cell mitogenesis by nitric oxide-generating vasodilators. *Am. J. Physiol. 257,* F60–F66.

Gaw, A. J., Aberdeen, J., Humphrey, P. P. A., Wadsworth, R. M., and Burnstock, G. (1991). Relaxation of sheep cerebral arteries by vasoactive intestinal polypeptide and neurogenic stimulation: Inhibition by L-N^G-monomethyl arginine in endothelium-denuded vessels. *Br. J. Pharmacol. 102,* 567–572.

Girerd, X. J., Hirsch, A. T., Cooke, J. P., Dzau, V. J., and Creager, M. A. (1990). L-Arginine augments endothelium-dependent vasodilation in cholesterol-fed rabbits. *Circ. Res. 67,* 1301–1308.

Gold, M. E., Wood, K. S., Byrns, R. E., Fukuto, J., and Ignarro, L. J. (1990). N^G-methyl-L-arginine causes endothelium-dependent contraction and inhibition of cyclic GMP formation in artery and vein. *Proc. Natl. Acad. Sci. U.S.A. 87,* 4430–4434.

González, C., and Estrada, C. (1991). Nitric oxide mediates the neurogenic vasodilation of bovine cerebral arteries. *J. Cereb. Blood Flow Metab. 11,* 366–370.

Gray, D. W., and Marshall, I. (1990). Calcitonin gene-related peptide (CGRP) endothelium-dependent relaxation in rat aorta is inhibited by L-NMMA. *Br. J. Pharmacol. 99* (suppl), 104P.

Gray, D. W., and Marshall, I. (1992). Nitric oxide synthesis inhibitors attenuate calcitonin gene-related peptide endothelium-dependent vasorelaxation in rat aorta. *Eur. J. Pharmacol.* (in press).

Han, S-P., Naes, L., and Westfall, T. C. (1990). Calcitonin gene-related peptide is the endogenous mediator of nonadrenergic-noncholinergic vasodilation in rat mesentery. *J. Pharmacol. and Exp. Ther. 255,* 423–427.

Hecker, M., Mitchell, J. A., Harris, H. J., Katsura, M., Thiemermann, C., and Vane, J. R. (1990). Endothelial cells metabolize N^G-monomethyl-L-arginine to L-citrulline and subsequently to L-arginine. *Biochem. Biophys. Res. Comm. 167,* 1037–1043.

Hinojosa-Laborde, C., Frohlich, B. H., and Cowley, A. W. (1991). Contribution of regional vascular responses to whole body autoregulation in conscious areflexic rats. *Hypertension 17,* 1078–1084.

Hishikawa, K., Nakaki, T., Suzuki, H., Saruta, T., and Kato, R. (1991). L-arginine induced hypotension. *Lancet 337,* 683–684.

Holtz, J., Förstermann, U., Pohl, U., Giesler, M., and Bassenge, E. (1984). Flow-dependent, endothelium-mediated dilation of epicardial coronary arteries in conscious dogs: Effects of cyclooxygenase inhibition. *J. Cardiovasc. Pharmacol. 6,* 1161–1169.

Horio, T., Kohno, M., Yokokawa, K., Murakawa, K., Yasunari, K., Fujiwara, H., Kurihara, N., and

Takeda, T. (1991). Effect of hypoxia on plasma immunoreactive endothelin-1 concentration in anesthetized rats. *Metabolism 40,* 999–1001.

Hughes, S. R., Williams, T. J., and Brain, S. D. (1990). Evidence that endogenous nitric oxide modulates oedema formation induced by substance P. *Eur. J. Pharmacol. 191,* 481–484.

Humphries, R. G., Carr, R. D., Nicol, A. K., Tomlinson, W., and O'Connor, S. E. (1991). Coronary vasoconstriction in the conscious rabbit following intravenous infusion of L-N^G-nitro-arginine. *Br. J. Pharmacol. 102,* 565–566.

Hutcheson, I. R., and Griffith, T. M. (1991). Release of endothelium-derived relaxing factor is modulated both by frequency and amplitude of pulsatile flow. *Am. J. Physiol. 261,* H257–H262.

Ignarro, L. J., Byrns, R. E., and Wood, K. S. (1988). Biochemical and pharmacological properties of endothelium-derived relaxing factor and its similarity to nitric oxide radical. *In* "Vasodilatation: Vascular Smooth Muscle, Peptides, Autonomic Nerves and Endothelium." (P. M. Vanhoutte, ed.), pp. 427–436. Raven Press, New York.

Ishizaka, H., Okumua, K., Yamabe, H., Tsuchiya, T., and Yasue, H. (1991). Endothelium-derived nitric oxide as a mediator of acetylcholine-induced coronary vasodilation in dogs. *J. Cardiovasc. Pharmacol. 18,* 665–669.

Johnsson, E., Folkow, B., and Karlström, G. (1991). Myogenic responsiveness in rat hindquarter vessels during constant-flow and constant-pressure perfusion *in vitro;* effects of various potassium concentrations and of endothelial nitrous oxide blockade. *Acta Physiol. Scand. 142,* 319–328.

Katusic, Z. S., Marshall, J. J., Kontos, H. A., and Vanhoutte, P. M. (1989). Similar responsiveness of smooth muscle of the canine basilar artery to EDRF and nitric oxide. *Am. J. Physiol. 257,* H1235–H1239.

Kawasaki, H., Takasaki, K., Saito, A., and Goto, K. (1988). Calcitonin gene-related peptide acts as a novel vasodilator neurotransmitter in mesenteric resistance vessels of the rat. *Nature (London) 335,* 164–167.

Kawasaki, H., Saito, A., and Takasaki, K. (1990). age-related decrease of calcitonin gene-related peptide-containing vasodilator innervation in the mesenteric resistance vessel of the spontaneously hypertensive rat. *Circ. Res. 67,* 733–743.

Kelm, M., and Schrader, J. (1990). Control of coronary vascular tone by nitric oxide. *Circ. Res. 66,* 1561–1575.

Kelm, M., Feelisch, M., Deussen, A., Strauer, B. E., and Schrader, J. (1991). Release of endothelium derived nitric oxide in relation to pressure and flow. *Cardiovasc. Res. 25,* 831–836.

Kiff, R. J., Gardiner, S. M., Compton, A. M., and Bennett, T. (1991). Selective impairment of hindquarters vasodilator responses to bradykinin in conscious Wistar rats with streptozotocin-induced diabetes mellitus. *Br. J. Pharmacol. 103,* 1357–1362.

Kilbourn, R. G., Jubran, A., Gross, S. S., Griffith, O. W., Levi, R., Adams, J., and Lodao, R. F. (1990). Reversal of endotoxin-mediated shock by N^G-methyl-L-arginine, an inhibitor of nitric oxide synthesis. *Biochem. Biophys. Res. Comm. 172,* 1132–1138.

Klabunde, R. E., and Ritger, R. C. (1991). N^G-Monomethyl-L-arginine (NMA) restores arterial blood pressure but reduces cardiac output in a canine model of endotoxic shock. *Biochem. Biophys. Res. Comm. 178,* 1135–1140.

Klabunde, R. E., Ritger, R. C., and Helgren, M. C. (1991). Cardiovascular actions of inhibitors of endothelium-derived relaxing factor (nitric oxide) formation/release in anesthetized dogs. *Eur. J. Pharmacol. 199,* 51–59.

Kobayashi, Y., and Hattori, K. (1990). Nitroarginine inhibits endothelium-derived relaxation. *Jpn. J. Pharmacol. 52,* 167–169.

Kubo, S. H., Rector, T. S., Bank, A. J., Williams, R. E., and Heifetz, S. M. (1991). Endothelium-dependent vasodilation is attenuated in patients with heart failure. *Circulation 84,* 1589–1596.

Kurtz, A., Kaissling, B., Busse, R., and Baier, W. (1991). Endothelial cells modulate renin secretion from isolated mouse juxtaglomerular cells. *J. Clin. Invest. 88,* 1147–1154.

Lacolley, P. J., Lewis, S. J., and Brody, M. J. (1991). Role of sympathetic nerve activity in the generation of vascular nitric oxide in urethane-anesthetized rats. *Hypertension 17,* 881–887.

Lahera, V., Salom, M. G., Fiksen-Olsen, M. J., Raij, L., and Romero, J. C. (1990). Effects of N^G-monomethyl-L-arginine and L-arginine on acetylcholine renal response. *Hypertension 15,* 659–663.

Lamontagne, D., Pohl, U., and Busse, R. (1991). N^G-Nitro-L-arginine antagonizes endothelium-dependent dilator responses by inhibiting endothelium-derived relaxing factor release in the isolated rabbit heart. *Eur. J. Physiol. 418,* 266–270.

Lautt, W. W. (1989). Resistance or conductance for expression of arterial vascular tone. *Microvasc. Res. 37,* 230–236.

Levillain, O., Hus-Citharel, A., Morel, R., and Bankir, L. (1990). Localization of arginine synthesis along rat nephron. *Am. J. Physiol. 259,* F916–F923.

Lin, P. J., Pearson, P. J., and Schaff, H. V. (1991). Endothelium-dependent contraction and relaxation of the human and canine internal mammary artery: Studies on bypass graft vasospasm. *Surgery 110,* 127–135.

Liu, S. F., Crawley, D. E., Evans, T. W., and Barnes, P. J. (1991). Endogenous nitric oxide modulates adrenergic neural vasoconstriction in guinea-pig pulmonary artery. *Br. J. Pharmacol. 104,* 565–569.

MacNaughton, W. K., Cirino, G., and Wallace, J. L. (1989). Endothelium-derived relaxing factor (nitric oxide) has protective actions in the stomach. *Life Sci. 45,* 1869–1876.

Macrae, I. M., Dawson, D. A., Reid, J. L., and McCulloch, J. (1991). Effect of nitric oxide synthetase inhibition on cerebral blood flow (CBF) in the conscious rat. *Soc. Neurosci. Abst. 17,* 475.

Mankad, P. S., Chester, A. H., and Yacoub, M. H. (1991). 5-Hydroxytryptamine mediates endothelium dependent coronary vasodilatation in the isolated rat heart by the release of nitric oxide. *Cardiovasc. Res. 25,* 244–248.

Marshall, J. J., Wei, E. P., and Kontos, H. A. (1988). Independent blockade of cerebral vasodilation from acetylcholine and nitric oxide. *Am. J. Physiol. 255,* H847–H854.

Mathie, R. T., Ralevic, V., Alexander, B., and Burnstock, G. (1991). Nitric oxide is the mediator of ATP-induced dilatation of the rabbit hepatic arterial vascular bed. *Br. J. Pharmacol. 103,* 1602–1606.

Matsumura, Y., Hisaki, K., Takaoka, M., and Morimoto, S. (1990). Phosphoramidon, a metalloproteinase inhibitor, suppresses the hypertensive effect of big endothelin-1. *Eur. J. Pharmacol. 185,* 103–106.

Matsumura, Y., Ikegawa, R., Suzuki, Y., Takaoka, M., Uchida, T., Kido, H., Shinyama, H., Hayashi, K., Watanabe, M., and Morimoto, S. (1991). Phosphoramidon prevents cerebral vasospasm following subarachnoid hemorrhage in dogs: The relationship to endothelin-1 levels in the cerebrospinal fluid. *Life Sci. 49,* 841–848.

Mayhan, W. G. (1990). Impairment of endothelium-dependent dilatation of basilar artery during chronic hypertension. *Am. J. Physiol. 259,* H1455–H1462.

Meininger, G. A., and Trzeciakowski, J. P. (1988). Vasoconstriction is amplified by autoregulation during vasoconstrictor-induced hypertension. *Am. J. Physiol. 254,* H709–H718.

Melkumyants, A. M., Balashov, S. A., Veselova, E. S., and Khayutin, V. M. (1987). Continuous control of the lumen of feline conduit arteries by blood flow rate. *Cardiovasc. Res. 21,* 863–870.

Milner, P., Bodin, P., Loesch, A., and Burnstock, G. (1990). Rapid release of endothelin and ATP from isolated aortic endothelial cells exposed to increased flow. *Biochem. Biophys. Res. Comm. 170,* 649–656.

Minor, R. L., Myers, P. R., Guerra, R., Bates, J. N., and Harrison, D. G. (1990). Diet-induced atherosclerosis increases the release of nitrogen oxides from rabbit aorta. *J. Clin. Invest. 86,* 2109–2116.

Mitchell, J. A., De Nucci, G., Warner, T. D., and Vane, J. R. (1991). Alkaline buffers release EDRF from bovine cultured aortic endothelial cells. *Br. J. Pharmacol. 103,* 1295–1302.

Moncada, S., Palmer, R. M. J., and Higgs, E. A. (1991a). Nitric oxide: Physiology, pathophysiology, and pharmacology. *Pharmacol. Rev. 43,* 109–142.

Moncada, S., Rees, D. D., Schulz, R., and Palmer, R. M. J. (1991b). Development and mechanism of a specific supersensitivity to nitrovasodilators after inhibition of vascular nitric oxide synthesis *in vivo*. *Proc. Natl. Acad. Sci. U.S.A. 88,* 2166–2170.

Moore, P. K., al-Swayeh, O. A., Chong, N. W. S., Evans, R. A., and Gibson, A. (1990). L-N^G-nitro-arginine (L-NOARG), a novel, L-arginine-reversible inhibitor of endothelium-dependent vasodilatation *in vitro*. *Br. J. Pharmacol. 99,* 408–412.

Mügge, A., Lopez, J. A. G., Piegors, D. J., Breese, K. R., and Heistad, D. D. (1991). Acetylcholine-induced vasodilatation in rabbit hindlimb *in vivo* is not inhibited by analogues of L-arginine. *Am. J. Physiol. 260,* H242–H247.

Nagao, T., and Vanhoutte, P. M. (1991). Hyperpolarization contributes to endothelium-dependent relaxations to acetylcholine in femoral veins of rats. *Am. J. Physiol. 261,* H1034–H1037.

Nakaki, T., Hishikawa, K., Suzuki, H., Saruta, T., and Kato, R. (1990). L-arginine-induced hypotension. *Lancet 336,* 696.

Nakamura, T., and Prewitt, R. L. (1991a). Effect of N^G-monomethyl-L-arginine on arcade arterioles of rat spinotrapezius muscles. *Am. J. Physiol. 261,* H46–H52.

Nakamura, T., and Prewitt, R. L. (1991b). Effect of N^G-monomethyl L-arginine on endothelium-dependent relaxation in arterioles of one-kidney, one clip hypertensive rats. *Hypertension 17,* 875–880.

Ohlstein, E. H., Vickery, L., Sauermelch, C., and Willette, R. N. (1990). Vasodilation induced by endothelin: Role of EDRF and prostanoids in rat hindquarters. *Am. J. Physiol. 259,* H1835–H1841.

Ohta, F., Kobayashi, Y., Shimoura, K., Hattori, K., and Moritake, K. (1990). Nitro-L-arginine inhibits endothelium-derived relaxation in canine basilar artery. *Eur. J. Pharmacol. 183,* 1794.

O'Leary, D. S. (1991). Regional vascular resistance vs conductance: Which index for baroreflex responses? *Am. J. Physiol. 260,* H632–H637.

Ontkean, M., Gay, R., and Greenberg, B. (1991). Diminished endothelium-derived relaxing factor activity in an experimental model of chronic heart failure. *Circ. Res. 69,* 1088–1096.

Palmer, R. M. J., Ferrige, A. G., and Moncada, S. (1987). Nitric oxide release accounts for the biological activity of endothelium-derived relaxing factor. *Nature (London) 327,* 524–526.

Palmer, R. M. J., Ashton, D. S., and Moncada, S. (1988a). Vascular endothelial cells synthesize nitric oxide from L-arginine. *Nature (London) 333,* 664.

Palmer, R. M. J., Rees, D. D., Ashton, D. S., and Moncada, S. (1988b). L-Arginine is the physiological precursor for the formation of nitric oxide in endothelium-dependent relaxation. *Biochem. Biophys. Res. Comm. 153,* 1251–1256.

Paton, W. D. M. (1990). L-arginine-induced hypotension. *Lancet 336,* 1016.

Paton, W. D. M. (1991). L-arginine-induced hypotension. *Lancet 337,* 1045.

Pepke-Zaba, J., Higenbottam, T., Dinh-Xuan, A. T., Stone, D., and Wallwork, J. (1991). Inhaled nitric oxide as a cause of selective pulmonary vasodilatation in pulmonary hypertension. *Lancet 338,* 1173–1174.

Perrella, M. A., Hildebrand, F. L., Margulies, K. B., and Burnett, J. C. (1991). Endothelium-derived relaxing factor in regulation of basal cardiopulmonary and renal function. *Am. J. Physiol. 261,* R323–R328.

Persson, M. G., Gustafsson, L. E., Wiklund, N. P., Hedqvist, P., and Moncada, S. (1990a). Endogenous nitric oxide as a modulator of rabbit skeletal muscle microcirculation *in vivo*. *Br. J. Pharmacol. 100,* 463–466.

Persson, M. G., Gustafsson, L. E., Wiklund, N. P., Moncada, S., and Hedqvist, P. (1990b). Endogenous nitric oxide as a probable modulator of pulmonary circulation and hypoxic pressor response *in vivo*. *Acta Physiol. Scand. 140,* 449–457.

Persson, M. G., Wiklund, N. P., and Gustafsson, L. E. (1991). Nitric oxide requirement for vasomotor nerve-induced vasodilatation and modulation of resting blood flow in muscle microcirculation. *Acta Physiol. Scand. 141,* 49–56.

Peskar, B., Respondek, M., Müller, K. M., and Peskar, B. A. (1991). A role for nitric oxide in capsaicin-induced gastroprotection. *Eur. J. Pharmacol. 198,* 113–114.

Petros, A. J., Hewlett, A. M., Bogle, R. G., and Pearson, J. D. (1991). L-arginine-induced hypotension. *Lancet 337,* 1044–1045.

Pfeilschifter, J. (1991). Anti-inflamatory steroids inhibit cytokine induction of nitric oxide synthase in rat renal mesangial cells. *Eur. J. Pharmacol. 195,* 179–180.

Pique, J. M., Whittle, B. J. R., and Esplugues, J. V. (1989). The vasodilator role of endogenous nitric oxide in the rat gastric microcirculation. *Eur. J. Pharmacol. 174,* 293–296.

Radermacher, J., Förstermann, U., and Frölich, J. C. (1990). Endothelium-derived relaxing factor influences renal vascular resistance. *Am. J. Physiol. 259,* F9–F17.

Randall, M. D., Douglas, S. A., and Hiley, C. R. (1989). Vascular activities of endothelin-1 and some alanyl substituted analogues in resistance beds of the rat. *Br. J. Pharmacol. 98,* 685–699.

Rees, D. D., Palmer, R. M. J., Hodson, H. F., and Moncada, S. (1989a). A specific inhibitor of nitric oxide formation from L-arginine attenuates endothelium-dependent relaxation. *Br. J. Pharmacol. 96,* 418–424.

Rees, D. D., Palmer, R. M. J., and Moncada, S. (1989b). Role of endothelium-derived nitric oxide in the regulation of blood pressure. *Proc. Natl. Acad. Sci. U.S.A. 86,* 3375–3378.

Rees, D. D., Palmer, R. M. J., Schulz, R., Hodson, H. F., and Moncada, S. (1990). Characterization of three inhibitors of endothelial nitric oxide synthase *in vitro* and *in vivo. Br. J. Pharmacol. 101,* 746–752.

Richard, V., Tanner, F. C., Tshudi, M., and Lüscher, T. F. (1990). Different activation of L-arginine pathway by bradykinin, serotonin, and clonidine in coronary arteries. *Am. J. Physiol. 259,* H1433–H1439.

Richard, V., Berdeaux, A., Drieu la Rochelle, C., and Giudicelli, J.-F. (1991). Regional coronary haemodynamic effects of two inhibitors of nitric oxide synthesis in anaesthetized, open-chest dog. *Br. J. Pharmacol. 104,* 59–64.

Robertson, B. E., Warren, J. B., and Nye, P. C. G. (1990). Inhibition of nitric oxide synthesis potentiates hypoxic vasoconstriction in isolated rat lungs. *Exp. Physiol. 75,* 255–257.

Rodbard, S. (1971). The burden of the resistance vessels. *Circ. Res. 28,* (suppl. I), I-2–I-8.

Rosenblum, W. I., Nishimura, H., and Nelson, G. H. (1990). Endothelium-dependent L-Arg- and L-NMMA-sensitive mechanisms regulate tone of brain microvessels. *Am. J. Physiol. 259,* H1396–H1401.

Ross, G., Chaudhuri, G., Ignarro, L. J., and Chyu, K.-Y. (1991). Acetylcholine vasodilation of resistance vessels *in vivo* may not entirely depend on newly synthesized nitric oxide. *Eur. J. Pharmacol. 195,* 291–293.

Rossitch, E., Alexander, E. Black, P., and Cooke, J. P. (1991). L-Arginine normalizes endothelial function in cerebral vessels from hypercholesterolemic rabbits. *J. Clin. Invest. 87,* 1295–1299.

Sagawa, K. (1983). Baroreflex control of systemic arterial pressure and vascular bed. *In* "The Cardiovascular System: Peripheral Circulation and Organ Blood Flow." (J. T. Shepherd and F. M. Abboud, eds.), Handbook of Physiology Sect. 2, Vol. III, Pt. 2, pp. 453–496. American Physiological Society, Bethesda.

Sanchez-Ferrer, C. F., Burnett, J. C., Lorenz, R. R., and Vanhoutte, P. M. (1990). Possible modulation of release of atrial natriuretic factor by endothelium-derived relaxing factor. *Am. J. Physiol. 259,* H982–H986.

Schmidt, H. H. W., Baeblich, S. E., Zernikow, B. C., Klein, M. M., and Böhme, E. (1990). L-Arginine and arginine analogues: Effects on isolated blood vessels and cultured endothelial cells. *Br. J. Pharmacol. 101,* 145–151.

Schulz, R., Smith, J. A., Lewis, M. J., and Moncada, S. (1991). Nitric oxide synthase in cultured endocardial cells of the pig. *Br. J. Pharmacol. 104,* 21–24.

Shinyama, H., Uchida, T., Kido, H., Hayashi, K., Watanabe, M., Matsumura, Y., Ikegawa, R., Takaoka, M., and Morimoto, S. (1991). Phosphoramidon inhibits the conversion of intracisternally administered big endothelin-1 to endothelin-1. *Biochem. Biophys. Res. Comm. 178,* 24–30.

Shultz, P. J., Schorer, A. E., and Raij, L. (1990). Effects of endothelium-derived relaxing factor and nitric oxide on rat mesangial cells. *Am. J. Physiol. 258,* F162–F167.

Shultz, P. J., Tayeh, M. A., Marletta, M. A., and Raij, L. (1991). Synthesis and action of nitric oxide in rat glomerular mesangial cells. *Am. J. Physiol. 261,* F600–F606.

Springall, D. R., Unwin, R., Ganz, M. B., Buttery, L. D. K., Ciani, F., Moncada, S., and Polak, J. M. (1992). Rat mesangial cells in culture contain nitric oxide synthase immunoreactivity. *In* "Biology of Nitric Oxide." (S. Moncada, M. A. Marletta, and J. Hibbs, eds.) (in press).

Stark, R. D. (1967). Conductance or resistance? *Nature (London) 217,* 779.

Stuehr, D. J., Fasehun, O. A., Kwon, N. S., Gross, S. R., Gonzalez, J. A., Levi, R., and Nathan, C. F. (1991). Inhibition of macrophage and endothelial cell nitric oxide synthase by diphenyleneiodonium and its analogs. *FASEB J. 5,* 98–103.

Tanaka, K., Gotoh, F., Gomi, S., Takashima, S., Mihar, B., Shirai, T., Nogawa, S., and Nagata, E. (1991). Inhibition of nitric oxide synthesis induces a significant reduction in local cerebral blood flow in the rat. *Neurosci. Letts. 127,* 129–132.

Thomas, G., Cole, E. A., and Ramwell, P. W. (1989). N^G-monomethyl L-arginine is a non-specific inhibitor of vascular relaxation. *Eur. J. Pharmacol. 170,* 123–124.

Thomas, H. M., Carson, R. C., Fried, E. D., and Novitch, R. S. (1991). Inhibition of hypoxic pulmonary vasocontriction by diphenyleneiodonium. *Biochem. Pharmacol. 42,* R9–R12.

Toda, N., and Okamura, T. (1990a). Mechanism underlying the response to vasodilator nerve stimulation in isolated dog and monkey cerebral arteries. *Am. J. Physiol. 259,* H1511–H1517.

Toda, N., and Okamura, T. (1990b). Modification by L-N^G-monomethyl arginine (L-NMMA) of the response to nerve stimulation in isolated dog mesenteric and cerebral arteries. *Jpn. J. Pharmacol. 52,* 170–173.

Toda, N., and Okamura, T. (1990c). Possible role of nitric oxide in transmitting information from vasodilator nerve to cerebroarterial muscle. *Biochem. Biophys. Res. Comm. 170,* 308–313.

Toda, N., and Okamura, T. (1991). Role of nitric oxide in neurally induced cerebroarterial relaxation. *J. Pharmacol. Exp. Ther. 258,* 1027–1031.

Toda, N., Minami, Y., and Okamura, T. (1990). Inhibitory effects of L-N^G-nitro-arginine on the synthesis of EDRF and the cerebroarterial response to vasodilator nerve stimulation. *Life Sci. 47,* 345–351.

Toda, N., Kawakami, M., and Kodama, N. (1991a). Prevention by nicardipine of dog basilar artery spasm caused by oxyhemoglobin *in vivo* and *in vitro*. *Eur. J. Pharmacol. 195,* 175–177.

Toda, N., Yoshida, K., and Okamura, T. (1991b). Analysis of the potentiating action of N^G-nitro-L-arginine on the contraction of the dog temporal artery elicited by transmural stimulation of noradrenergic nerves. *Naunyn-Schmiedeberg's Arch. Pharmacol. 343,* 221–224.

Tolins, J. P., and Raij, L. (1991). Effects of amino acid infusion on renal hemodynamics: Role of endothelium-derived relaxing factor. *Hypertension 17,* 1045–1051.

Tolins, J. P., Palmer, R. M. J., Moncada, S., and Raij, L. (1990). Role of endothelium-derived relaxing factor in regulation of renal hemodynamic responses. *Am. J. Physiol. 258,* H655–H662.

Topouzis, S., Huggins, J. P., Pelton, J. T., and Miller, R. C. (1991a). Modulation by endothelium of the responses induced by endothelin-1 and by some of its analogues in rat isolated aorta. *Br. J. Pharmacol. 102,* 545–549.

Topouzis, S., Schott, C., and Stoclet, J.-C. (1991b). Participation of endothelium-derived relaxing factor and role of cyclic GMP in inhibitory effects of endothelium on contractile responses elicited by α-adrenoceptors agonists in rat aorta. *J. Cardiovasc. Pharmacol. 18,* 670–678.

Tresham, J. J., Dusting, G. J., Coghlan, J. P., and Whitworth, J. A. (1991). Haemodynamic and hormonal effects of N-nitro-L-arginine, an inhibitor of nitric oxide biosynthesis, in sheep. *Clin. Exp. Pharmacol. Physiol. 18,* 327–330.

Tschudi, M., Richard, V., Bühler, F. R., and Lüscher, T. R. (1991). Importance of endothelium-derived nitric oxide in porcine coronary resistance arteries. *Am. J. Physiol. 260,* H13–H20.

Vallance, P., Collier, J., and Moncada, S. (1989a). Effects of endothelum-derived nitric oxide on periopheral anteriolar tone in man. *Lancet 2,* 997–999.

Vallance, P., Collier, J., and Moncada, S. (1989b). Nitric oxide synthesised from L-arginine mediates endothelium dependent dilatation in human veins *in vivo*. *Cardiovasc. Res. 23,* 1053–1057.

van Gelderen, E. M., Heiligers, J. P. C., and Saxena, P. R. (1991). Haemodynamic changes and acetylcholine-induced hypotensive responses after N^G-nitro-L-arginine methyl ester in rats and cats. *Br. J. Pharmacol. 103,* 1899–1904.

Vargas, H. M., Ignarro, L. J., and Chaudhuri, G. (1990). Physiological release of nitric oxide is dependent on the level of vascular tone. *Eur. J. Pharmacol. 190,* 393–397.

Vidal, M. J., Romero, J. C., and Vanhoutte, P. M. (1988). Endothelium-derived relaxing factor inhibits renin release. *Eur. J. Pharmacol. 149,* 401–402.

Vo, P. A., Reid, J. J., and Rand, M. J. (1991). Endothelial nitric oxide attenuates vasoconstrictor responses to nerve stimulation and noradrenaline in the rat tail artery. *Eur. J. Pharmacol. 199,* 123–125.

Walder, C. E., Thiemermann, C., and Vane, J. R. (1990). Endothelium-derived relaxing factor participates in the increased blood flow in response to pentagastrin in the rat stomach mucosa. *Proc. Roy. Soc. London 241,* 195–200.

Walder, C. E., Thiemermann, C., and Vane, J. R. (1991). The involvement of endothelium-derived relaxing factor in the regulation of renal cortical blood flow in the rat. *Br. J. Pharmacol. 102,* 967–973.

Wang, X., Han, C., and Fiscus, R. R. (1991). Calcitonin gene-related peptide (CGRP) causes endothelium-dependent cyclic AMP, cyclic GMP and vasorelaxant responses in rat abdominal aorta. *Neuropeptides 20,* 115–124.

Wang, Y.-X., Zhou, T., Chua, T. C., and Pang, C. C. Y. (1991). Effects of inhalation and intravenous anaesthetic agents on pressor response to N^G-nitro-L-arginine. *Eur. J. Pharmacol. 198,* 183–188.

Warner, T. D., De Nucci, G., and Vane, J. R. (1989a). Comparison of the survival of endothelium-derived relaxing factor and nitric oxide within the isolated perfused mesenteric arterial bed of the rat. *Br. J. Pharmacol. 97,* 777–782.

Warner, T. D., De Nucci, G., and Vane, J. R. (1989b). Rat endothelin is a vasodilator in the isolated perfused mesentery of the rat. *Eur. J. Pharmacol. 159,* 325–326.

Welch, W. J., Wilcox, C. S., Aisaka, K., Gross, S. S., Griffith, O. W., Fontoura, B. M. A., Maack, T., and Levi, R. (1991). Nitric oxide synthesis from L-arginine modulates renal vascular resistance in isolated perfused and intact rat kidneys. *J. Cardiovasc. Pharmacol. 17* (suppl. 3), S165–S168.

Whittle, B. J. R. (1990). Role of endogenous nitric oxide in the mechanisms underlying gastric mucosal integrity. *In* "Nitric oxide from L-arginine: A Bioregulatory System." (S. Moncada and E. A. Higgs, eds.), pp. 365–371. Excerpta Medica, Amsterdam.

Whittle, B. J. R., Lopez-Belmonte, J., and Rees, D. D. (1989). Modulation of the vasodepressor actions of acetylcholine, bradykinin, substance P and endothelin in the rat by a specific inhibitor of nitric oxide formation. *Br. J. Pharmacol. 98,* 646–652.

Whittle, B. J. R., Lopez-Belmonte, J., and Moncada, S. (1990). Regulation of gastric mucosal integrity by endogenous nitric oxide: Interactions with prostanoids and sensory neuropeptides in the rat. *Br. J. Pharmacol. 99,* 607–611.

Widdop, R. E., Gardiner, S. M., Kemp, P. A., and Bennett, T. (1992). The influence of atropine and atenolol on the cardiac haemodynamic effects of N^G-nitro-L-arginine methyl ester in conscious, Long Evans rats. *Br. J. Pharmacol. 105,* 653–656.

Wilcox, C. S., Welch, W. J., Gross, S. S., Levi, R., and Taylor, G. (1992). A single nephron arginine-NO system regulating glomerular capillary pressure in the rat. *In* "Biology of Nitric Oxide." (S. Moncada, M. A., Marletta, and J. Hibbs, eds.) (in press).

Woodman, O. L., and Dusting, G. J. (1991). N-nitro L-arginine causes coronary vasoconstriction and inhibits endothelium-dependent vasodilatation in anaesthetized greyhounds. *Br. J. Pharmacol. 103,* 1407–1410.

Woolfson, R. G., and Poston, L. (1990). Effect of N^G-monomethyl-L-arginine on endothelium-dependent relaxation of human subcutaneous resistance arteries. *Clin. Sci. 79,* 273–278.

Wright, C. E., and Fozard, J. R. (1988). Regional vasodilation is a prominent feature of the haemodynamic response to endothelin in anesthetized spontaneously hypertensive rats. *Eur. J. Pharmacol. 155,* 201–203.

Yamazaki, J., Fujita, N., and Nagao, T. (1991). N^G-monomethyl-L-arginine-induced pressor response at developmental and established stages in spontaneously hypertensive rats. *J. Pharmacol Exp. Ther.* *259*, 52–57.

Yang, Z., von Segesser, L., Bauer, E., Stulz, P., Turina, M., and Lüscher, T. F. (1990). Different activation of the endothelial L-arginine and cyclooxygenase pathway in the human internal mammary artery and saphenous vein. *Circ. Res.* *68*, 52–60.

Zambetis, M., Dusting, G. J., Rajanayagam, S., and Woodman, O. L. (1991). Mechanism of the hypertension produced by inhibition of nitric oxide biosynthesis in rats. *J. Cardiovasc. Pharmacol.* *17* (suppl. 3), S191–S197.

Zatz, R., and De Nucci, G. (1991). Effects of acute nitric oxide inhibition on rat glomerular microcirculation. *Am. J. Physiol.* *261*, F360–F363.

CHAPTER 4

The Effect of Endothelin on the Circulation System

TOMOH MASAKI

I. Introduction

Endothelin (ET) is a potent vasoconstrictive peptide with 21 amino acid residues isolated initially from the conditioned medium of the cultured vascular endothelial cells (Yanagisawa *et al.*, 1988). It includes four cysteine residues forming two disulfide bonds between 1–15 and 3–11 positions (Fig. 1). There are successive hydrophobic amino acid residues at the carboxyl-terminus, which were very important in the activation of the ET-receptor.

There were three ET isoforms predicted initially by the three different ET-related genes (Inoue *et al.*, 1989). The first of those peptides, isolated from cultured medium of porcine aortic endothelial cells, was termed ET-1. The other two were named ET-2 and ET-3 (Fig. 1). Interestingly, interspecies differences between those three ETs were barely observable. All mammalian species are likely to have those three ETs. We could not detect the similar sequence to this peptide when it was discovered. However, a sequence of a rare snake venom, sarafotoxin, was shown later to be very similar, suggesting an occurrence of an evolutional divergence of ET family genes a very long time ago.

Likewise at least two types of ET-receptor have been reported, designated ET_A and ET_B (Arai *et al.*, 1990; Sakurai *et al.*, 1990). Both are about 46-kDa proteins and belong to seven-turned G-protein-coupled receptor family (i.e., the rhodopsin-type superfamily).

ET is known to be produced in the neuron, glia cell, intestinal cell, amnion cell, kidney cell, neutrophil, parathyroid cell (Fujii *et al.*, 1991), and epithelial cell (Ohta *et al.*, 1990) as well as in endothelial cell (Masaki *et al.*, 1992). The ET-receptor was also shown to distribute widely in various tissues, not only in circulating organ, but also in other organs, including the brain, thyroid gland, adrenal gland, kidney, lung, intestine, liver, spleen, uterus, and placenta (Hosoda *et al.*, 1991; Sakamoto *et al.*, 1991). These results were compatible with the fact that ET has many functions in various tissues.

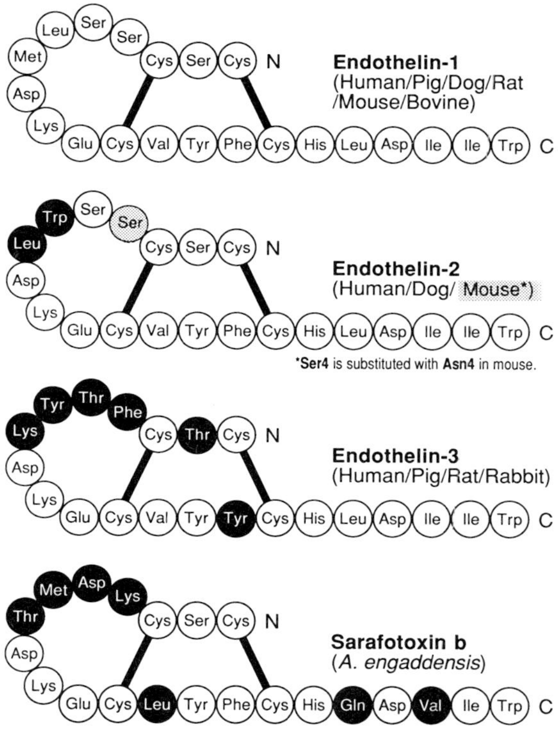

Figure 1 Structure of endothelins and sarafotoxin b.

Since ET induces slow-rising and long-lasting vasoconstriction in isolated blood vessels, and pressor response *in situ,* it has been thought to play an important role in maintenance of blood pressure, hypertension, or vasospasm. The purpose of this article is to review the current knowledge of the pharmacological action of ET in the circulation system and to discuss the physiological and pathophysiological significance of ET in this system.

II. Cellular Mechanism of Endothelin-Induced Vasoconstriction

A subtype of ET-receptor, ET_A, had a high affinity to ET-1 and ET-2 but a low affinity to ET-3 (Arai *et al.*, 1990). ET_B had an equally potent affinity to all of the three isotypes of ET (Sakurai *et al.*, 1990) (Table 1). ET_A was distributed in a smooth muscle and a wide variety of other human tissues, including the lung, heart, brain, kidney, and intestine. On the other hand, ET_B was found in endothelial cell but not in vascular smooth muscle. ET_B was distributed also densely in the brain, lung, and kidney. ET_A probably played an important role in vasoconstriction, while ET_B was important in releasing endothelium-derived relaxing factor (EDRF) or prostacyclin from endothelial cells.

Table 1 Comparison of Cloned Endothelin Receptors

Receptor subtype[a]	Affinity rank order for endothelins	Origin	Amino acid residues (incl. signal sequences)	Tissue distribution
ET_A	ET−1=ET−2>>ET−3	Bovine lung	427	Heart, lung, intestine, brain
		rat A10 cell	426	Heart, lung, intestine, brain
		human	427	Aorta, heart, lung, intestine, brain
ET_B	ET−1=ET−2=ET−3	Bovine lung,	415	Lung
		rat lung,	441	Brain, lung, kidney
		human jejunum/ liver	442	Brain, lung, kidney, heart, endothelial cells

[a]Both subtypes have a hepatahelical tertiary structure and couple to phospholipase C via G protein.

Activation of both ET-receptor subtypes was coupled to an activation of phospholipase C via G-protein in all of the tissues that had an ET-receptor, and produced phosphoinositol and diacylglycerol (N. Takuwa *et al.*, 1990; Y. Takuwa *et al.*, 1990). In isolated cells, for instance A10 or 3T3 cells, ET-1 elicited a rapid and transient increase in cytosolic free calcium level followed by a prompt decrease to a lower level, which was significantly higher than the original basal level (Goto *et al.*, 1989). The later higher level was sustained for a long time. The transient increase in cytosolic free calcium ion was due to the release of calcium ion from intracellular free calcium ion pool stimulated by inositoltrisphosphate. The later sustained increase in cytosolic free calcium ion was ascribed to the exterior calcium ion. The mechanism of the opening of calcium channel at the later stage is still unclear. Concomitantly produced diacylglycerol activated protein kinase C. Activated protein kinase C in turn activated Na^+-K^+ATPase and Na^+-H^+ exchanger (Gupta *et al.*, 1991). The resulting intracellular alkalinization was thought to play a role in excitation–contraction coupling (Danthuluri and Deth, 1989).

ET-induced vasoconstriction was long lasting. The sustained constriction was likely due to the exterior calcium ion. The activation of phospholipase C may not be important in physiological constriction since circulating ET in the blood stream was low and the amount of ET released from endothelial cell was small. Indeed, low concentration of ET stimulated calcium influx without elevation of phosphoinstide breakdown (Muldoon *et al.*, 1991). This mechanism was probably more important in physiological ET-induced vasoconstriction.

It was also demonstrated that ET-induced vasoconstriction was observed in the absence of exterior calcium ion even after complete depletion of intracellular calcium store (Kodama *et al.*, 1989; Kasuya *et al.*, 1989b). This constriction was completely abolished by protein kinase C inhibitor. Protein kinase C was activated by diacylglycerol. The activation was also long lasting. The ET-induced sustained vasoconstruction was associated with this activation of protein kinase C (Danthuluri and Deth, 1989). However, the further detailed mechanism was unclear. One possibility was that contractile apparatus was sensitized to calcium ion via activation of protein kinase C (Nishimura *et al.*, 1990).

ET-1 induced a phasic depolarization, which depended on exterior calcium ion and was easily inactivated. The depolarization was due to activation of calcium-dependent nonselective ion channel (Chen and Wagoner, 1991).

III. Hemodynamic Effect of Endothelin

When a single intravenous bolus of ET-1 was administered, for instance, into a rat, a significant hypotensive response occurred in the first few minutes, which was followed by a sustained long-lasting pressor response (DeNucci *et al.*, 1988; Miyauchi *et al.*, 1989; LeMonnier de Gouville *et al.*, 1990a) (Fig. 2). The initial transient depressor response and the later sustained pressor response was mediated by different mechanisms.

The prolonged systemic hypertension after the initial transient hypotension can be entirely explained by the direct vasoconstrictor actions of ET on peripheral vessels. Concurrent with the sustained pressor response, cardiac output and heart rate decreased at the high dose of ET-1 (Goetz *et al.*, 1988; Mortensen and Fink, 1990; Rohmeiss *et al.*, 1990; Hinojosa-Laborde *et al.*, 1989). A fall in right atrial pressure and significant hemoconcentration were also associated with this pronounced vasoconstrictor effect (King *et al.*, 1990). A decrease in venous return probably caused a reduction in right atrial pressure and in cardiac output.

The ET-induced pressor response was observed both in anesthetized, chemically denervated animals and in conscious animals (Goetz *et al.*, 1988; Rohmeiss *et al.*, 1990; Miyauchi *et al.*, 1989). However, the pressor effect of ET appeared to be buffered by neurohumoral system. Indeed, in rats without reflex, infusion with 1 ng/kg-min of ET caused significant increases in blood pressure, while rats with intact reflexes did not respond significantly until a dose of 100 ng/kg-min (Hinojosa-Laborde *et al.*, 1989).

The ET-induced bradycardia is likely to be a result of the sustained increases in

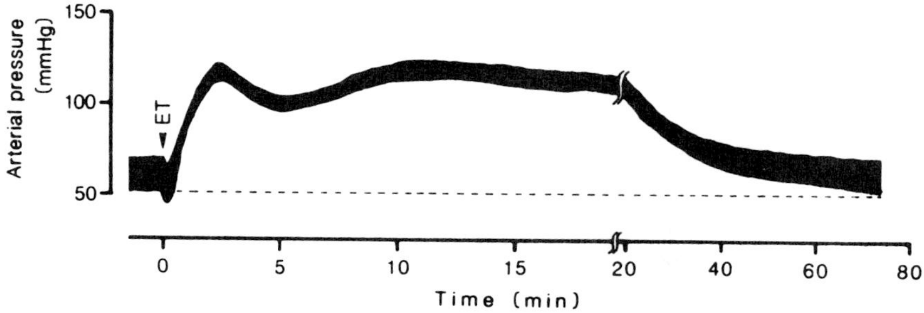

Figure 2 Effect of ET-1 on blood pressure. One ng/kg ET-1 was injected in bolus intravenously into a rat. The rat was anesthetized by urethane and pretreated by atropine, propranolol, and bunazosin (Yanagisawa *et al.*, 1988).

arterial pressure. ET did not affect baroreflex function directly (Knuepfer *et al.*, 1989). However, an intracisternal administration of the higher dose of ET-1 induced a significant increase in vagal component of baroreflex sensitivity (Itoh and van den Busse, 1991).

The central nervous system was sensitive to ET and mediated cardiovascular response to ET. The results were compatible with the results that ET/ETmRNA and ET-receptor widely distributed in the central nervous system. However, experimental results of neural action of ET on circulation in central nervous system were disparate depending on the region. Intrathecal administration of ET elicited a decrease in blood pressure associated with hindquarters vasodilation in rats (Han *et al.*, 1991a). On the other hand, intracisternal administration of ET-1 induced a marked increase in blood pressure (Ouchi *et al.*, 1989). Indeed, several specific areas of the ventral surface of medulla responded to ET-1, which induced transient but significant increases in mean arterial pressure, heart rate, and renal sympathetic nerve activity (Kuwaki *et al.*, 1991a).

As previously described, an intravenous administration of ET elicited elevation of plasma levels of atrial natriuretic peptide, aldosteron, renin, and vasopressin. These factors may function as regulatory factors of the cardiovascular system. However, this problem still remains to be investigated.

As previously described, the hypotensive effect may be mediated by ET_B receptor on the endothelial cells, whereas pressor response was mediated by ET_A receptor on the smooth muscle (Sakurai *et al.*, 1992). The fact that the hypotensive effects of ETs underwent a tachyphylaxis phenomenon supported the above concept (LeMonnier de Gouville *et al.*, 1990a).

ET-1 produced markedly different hemodynamic responses that depended on region (Wright and Fozard, 1988; Gardiner *et al.*, 1989a; Minkes and Kadowitz, 1989). The initial hypotension in conscious rats was associated with marked vasodilations in hindquarters and carotid vascular beds, whereas the subsequent hypertension was associated with vasoconstriction in renal, mesenteric, hindquarters, and carotid vascular beds (Gardiner *et al.*, 1989a). In squirrel monkeys, the vasoconstrictor effect of ET-1 was most remarkable in splanchnic vascular beds. The coronary blood flow also decreased in response to ET-1. In contrast, the blood flow to the brain and stomach did not decrease in the presence of ET-1 (Clozel and Clozel, 1989). In general, skeletal muscle vascular bed showed vasodilation in response to ET at relatively low doses. The regional heterogeneity of the responsiveness to ET-1 was demonstrated not only between various arteries but also within the venous system. Substantial differences of hemodynamic effects induced by various isotypes of ETs also existed both with *in vitro* and *in vivo* experiments (Leadly *et al.*, 1991; Gardiner *et al.*, 1990a,b; Han *et al.*, 1991b; Minkes *et al.*, 1990). The diversity of the responses to ETs as seen in regional blood flow may be ascribed to varied proportions of the number of the two receptors on the respective blood vessels.

Since intravenously injected bolus ET-1 disappeared promptly in the circulating blood stream, prolonged vasoconstrictor effect of ET-1 may be ascribed to its high affinity for binding to vascular smooth muscle (Shiba *et al.*, 1989). Probably the ET-1 that was bound to the receptor kept the calcium channel open for a long time. The sustained systemic vasoconstriction was clearly dependent on extracellular calcium ion (Cao and Banks, 1990; Madeddu *et al.*, 1990). Dihydropyridines partially prevented systemic vasoconstriction. Potassium channel opener markedly antagonized the pressor response.

The initial transient depressor response was observed only when the vascular tonus was elevated, for instance, when the blood pressure in pithed rat was supported by vasopressin (Le Monnier de Gouville *et al.*, 1990). Bilateral vagotomy or glibenclamide, a potassium channel blocker, failed to abolish the ET-induced depressor response, which indicated that systemic vasodilation was not mediated by acetylcholine or by the opening of potassium channels (LeMonnier de Gouville *et al.*, 1990b). The initial transient hypotension was primarily due to a change in systemic vascular resistance since cardiac index remained unchanged in the early phase after the administration (King *et al.*, 1990). The transient hypotension was remarkable at the low dose of ET-1, while it was counteracted by the pressor response at the high dosage. The ET-1-induced hypotension was probably elicited by EDRF or prostacyclin released from endothelial cell by ET itself, while the vasoconstriction occurred only if sufficient quantities of the peptide penetrated the endothelial monolayer to bind to receptors on vascular smooth muscle cells. In conscious rats, the vasodilation was blunted by methylene blue and was also attenuated in the presence of N^G-nitro-L-arginine methyl ester (L-NAME) (DeNucci *et al.*, 1988; Gardiner *et al.*, 1990d). L-NAME was an inhibitor of nitric oxide synthesis. ET-3 was more a potent vasodilator than ET-1 in this respect (Inoue *et al.*, 1989; Warner *et al.*, 1989). However, recently it was demonstrated that in the rat hindquarter artery, EDRF nitric oxide as well as archidonic acid metabolites were not involved in the transient increase in the vascular conductance (Gardiner *et al.*, 1989b; Ohlstein *et al.*, 1990). Additionally, it was also demonstrated, in this case, that vasodilator response in hindquarter vessels was not dependent on neurogenic mechanisms (Ohlstein *et al.*, 1990; Minkes *et al.*, 1989).

In this context, ET-induced responses of the trout vascular bed were interesting. Trout did not appear to have an endothelium-derived nitric oxide system (Olson and Villa, 1991). However, when a bolus of ET-1 was injected into conscious trout, transient depressor response was induced as seen in mammals. Interestingly, continuous intravenous infusion of low dose of ET-1 in trout produced a dose-dependent hypotensive effect. This depressor response to ET-1 was not mediated by local mechanisms and probably required an intact cardiovascular reflex system (Olson *et al.*, 1991).

The pulmonary artery was sensitive to ET-1. ET-1 elicited both dilator and constrictor effects in pulmonary artery depending on the condition of vascular beds and species. The increased resistance of pulmonary circulation was primarily arterial in rabbit and rat and venous in the canine lung (Barnard *et al.*, 1991). In cat, fetal rat, and juvenile lamb, low dose of ET-1 elicited vasodilation of pulmonary artery (Lippton *et al.*, 1989; Hasunuma *et al.*, 1990; Chatfield *et al.*, 1991). Mechanism of the ET-induced vasodilation in pulmonary artery was unclear. However, it was suggestive that the ET-induced vasodilation was abolished by the potassium channel blockers tetraethylammonium chloride and glybenclamide, and by ouabain, a sodium potassium pump inhibitor, which suggested the involvement of an ATP-sensitive potassium channel (Hasunuma *et al.*, 1990).

Cyclooxygenase products seemed to be, more or less, involved in the regulatory mechanism of ET-induced vascular responses. However, ET-elicited release of cyclooxygenase products that modified the vascular response to ET depended on the type of blood vessel and species. In canine and rat lungs, ET-induced increased resistance of pulmonary circulation was reversed by cyclooxygenase inhibitor (Barnard *et al.*, 1991). It may not have played a role in ET-mediated vasodilation of pulmonary artery (Hasunuma *et al.*, 1990; O'Donnell, 1990). In renal artery, ET induced vasoconstrictor prostanoid, while in mesenteric artery, ET induced vasodilator prostanoid. Those prostanoids counteracted the ET-induced initial vasodilation and the vasoconstriction, respectively (Gardiner *et al.*, 1990f; Miura *et al.*, 1991).

IV. Regional System of Regulation of Vasoconstriction by Endothelin

Vascular endothelium produced ET-1 exclusively (Masaki *et al.*, 1991, 1992). In conscious dog coronary artery, a significant decrease of blood flow in large coronary artery was obtained by infusion rate of endothelin at 10 ng/kg·min. At this rate, plasma concentration of ET was about 10–20 times higher than that of the normal plasma concentration of ET (Wang *et al.*, 1991b). The normal concentration of plasma ET-1 was around 1 pg/ml (Suzuki *et al.*, 1989). Accordingly, the circulating ET-1 was unlikely to be involved in the regulatory mechanism of vascular tonus. Assuming that ET was involved in the maintenance of vascular tonus, it may act on the underlying smooth muscle cells in a paracrine manner after being released from the endothelial cell. In fact, Yoshimoto *et al.* (1991) demonstrated that the cultured endothelial cell monolayer released twice as much of the amount of ET-1 to basal side than to apical side. To this effect, the shorter the distance between the endothelium and smooth muscle layer was, the more effective the ET-induced constriction became. In other words, microvessels and small blood vessels appeared to be more important than the large blood vessels in the regulation of vasoconstriction. ET in picomolar doses was probably released from endothelial cells and bound to the underlying smooth muscle that elicited arteriolar constriction and to endothelial cells that released EDRF in the intima. Indeed, it has been reported that rat coronary microvessels arterioles of less than 20 μm in diameter were very sensitive to ET-1 (Homma *et al.*, 1991). The arterioles were more sensitive to venules in the rat mesenteric and in the hamster cheek pouch arterioles (Homma *et al.*, 1992; Brain *et al.*, 1989). On the contrary, small veins, such as the saphenus vein, were more sensitive than small mesenteric arteries (Miller *et al.*, 1989). ET-induced pressor response may be ascribed to the ET-induced constriction of those arterioles since the total cross sectional area of the arterioles was about 60 times larger than that of the large arteries.

Subsequently, the question arose as to what kind of stimuli regulated the production and release of ET in endothelial cells. Recent studies from many laboratories have demonstrated that numerous factors, including platelet comprising factors (transforming growth factor-β and thrombin) (Ohlstein *et al.*, 1991), cytokines [tumor-necrotizing factor (TNF)], and interleukin-1 (IL-1) (Yoshizumi *et al.*, 1990)), and vasoactive substances, such as angiotensin II, arginine-vasopressin, and histamine, as well as shear stress, induced the expression of ET mRNA and endothelin biosynthesis (Emori *et al.*, 1991; Yoshizumi *et al.*, 1989). Platelet-activating factor or cyclooxygenase products may not have been a stimulated factor of ET-biosynthesis. However, endotoxin stimulated the biosynthesis of ET but only in the presence of serum, which suggested a role for serum factors in priming endothelial cells for ET release by endotoxin (Ohlstein *et al.*, 1991). In addition to those known factors, an unknown factor(s) in calf serum was suggested to have an ability to produce and release of ET-1 (Brown *et al.*, 1990). However, as yet, no sufficient evidence has been demonstrated that indicates these factors are involved in physiological regulators of production and release of ET-1.

V. Conversion from Big Endothelin to Endothelin

In porcine endothelium, a precursor of ET-1, proET-1, was cleaved in a vesicle by a regular endopeptidase specific for a pair of dibasic amino acids at the initial step in order to produce an intermediate 39-amino acid peptide, designated as big ET-1, which was in turn cleaved into mature ET-1 by a putative endothelin-converting enzyme, one of the metalloproteinase (Okada *et al.*, 1990). Human big ET-1 was determined to be 38 amino acid residues long. Both big ET-1 and ET-1 were excreted into the extracellular space.

The conversion of big ET-1 into ET-1 is likely to be essential for the pressor activity since big ET-1 exerted only weak vasoconstrictor effects (D'Orlean-Juste *et al.*, 1990; Pollack and Opgenorth, 1991). In fact, the hypertensive effect of big ET-1 was markedly attenuated in animals administered with phosphoramidon, an inhibitor of the conversion (McMahon *et al.*, 1991). In addition, phosphoramidon infused intravenously lowered the mean arterial pressure in conscious spontaneously hypertensive rats. However, Modin *et al.* (1991) demonstrated recently that when big ET-1 was injected into pig or rat, the plasma levels of ET-1 increased similarly in the absence and presence of phosphoramidon, whereas the vasoconstrictor effect of big ET-1 in the splenic blood vessels was significantly reduced after phosphoramidon. The results suggested two possibilities. First, ET-1 that was converted from big ET-1 on endothelium effectively induced vasoconstriction of the underlying smooth muscle alone. Second, the degradation of other vasoactive peptides was reduced by phosphoramidon, probably via inactivation of phosphoramidon-sensitive endopeptidase.

VI. Interaction of Endothelin to Other Factors

ET-induced vascular response was influenced by the presence of endothelium. Removal of endothelium augmented the ET-1 induced vasoconstriction. EDRF blunted the ET-induced vasoconstriction. Conversely, ET-1 inhibited endothelium-dependent relaxation of vein and arteries. As mentioned earlier, ET-1 stimulated the release of EDRF (DeNucci *et al.*, 1988). Conversely, EDRF inhibited the production of ET-1 (Boulanger and Lüscher, 1990). Decrease in production of ET-1 appeared to be coupled to increase in production of EDRF, probably via activation of guanylate cyclase. The balance of the effects of EDRF and ET-1 may be of great importance in ET-1-induced vasoconstriction (Lüsher *et al.*, 1990).

Interaction of ET-1 and α-calcitonin gene-related peptide (α-CGRP) was also observed. The ET-induced vasoconstriction was attenuated by α-CGRP in hindquarter artery but not remarkably in renal artery (Gardiner *et al.*, 1991).

Atrial natriuretic peptide (ANP) counteracted ET-induced vasoconstriction although production of ET-1 in parathyroid cells was stimulated by ANP (De Fes *et al.*, 1991). ET enhanced the responses to noradrenaline (NA) and 5-hydroxytryptamine (5-HT) in isolated arteries of rat and human (Tabuchi *et al.*, 1989; Nakayama *et al.*, 1991). It has been demonstrated that endothelium of dog coronary artery released a noneicosanoid vasoconstrictive substance under hypoxia condition. In fact, hypoxia increased maximum response to ET-1 probably due to the increase in number of ET-receptor. However, Douglas *et al.* (1991) suggested that the responses cannot be explained by endothelin.

VII. Effect of Endothelin on Cardiac Muscle

ET has been shown to have positive inotropic effects on isolated spontaneously beating right atria and in electrically driven left atrium of guinea pig, rabbit, and rat but not of dog (Ishikawa *et al.*, 1988a,b; Moravec *et al.*, 1989; Vigne *et al.*, 1989). The effect varied from species to species. The ET-induced positive inotropic response was slow developing and long lasting and attenuated partially by a dihydrophytidine calcium antagonist, nicardipine, in guinea pig atria. However, in rabbit and rat papillary muscle, activation of voltage-dependent calcium channel appeared to be not important (Li *et al.*, 1991). In those cases, ET-induced contraction of cardiac muscle was mediated by other factors.

ET potentiated the calcium sensitivity of the contractile apparatus similarly to smooth muscle (Wang *et al.*, 1991; Kelly *et al.*, 1990). The same level of tension was developed by a lesser amount of intracellular free calcium ion in the presence of ET. In rat ventricular myocyte amiloride, an inhibitor of Na^+-H^+ exchanger prevented the increase in intracellular pH in response to ET and reduced the inotropic response in rat ventricular myocytes, and inhibitors of protein kinase C diminished the alkalinization, which suggested that activation of sarcolemmal Na^+-H^+ exchanger by protein kinase C-mediate pathway was a cause of the inotropic action of ET (Kramer *et al.*, 1991).

Since ET-induced increase in contraction was also attenuated either by propranolol or by depletion of catecholomine with reserpine, ET also appeared to act by causing endogenous release of catecholamine (Li *et al.*, 1991).

It was demonstrated that ET increased the sinoatrial nodal rate in a dose-dependent manner on the isolated right atria of guinea pigs. However, Kim (1991) demonstrated recently that in isolated right and left atrial tissue from 1 to 2-day-old newborn rats, each isotype of ET showed potent negative chronotropic action. This phenomenon could be explained by the opening of the muscarinic-gated potassium channels. The failure of ET to activate the potassium current in pertussis toxin-treated cells indicated that ET receptors were coupled to the inhibitory G-protein, similar to that of muscarinic receptors.

VIII. Physiological and Pathophysiological Significance of Endothelin

The unique pharmacological action of ET and production and release of ET in response to physiological stimuli lead to the concept that ET had a significant role in endothelial modulation of vascular tone. However, as described earlier, the plasma ET level in normal subject was determined to be around 1–2 pg/ml (Suzuki *et al.*, 1989). Systemic administration of such a low dose of ET-1 showed no effect on systemic blood pressure. In addition, the picomolar doses of ET caused vasodilation rather than vasoconstriction. Therefore, the circulating ET is unlikely to maintain the vascular tonus.

However, as described earlier, it might be possible that ET regulated vascular tonus in a paracrine manner (Fig. 3). The picomolar doses of ET released from endothelial cells bound to the underlying smooth muscle that elicited arteriolar constriction. Probably, some vasoactive substances in plasma may have regulated the production and release of ET in endothelial cells. In order to induce vasoconstriction, the concentration of ET-1 in intima must be more than 1 ng/ml. The experimental result that ET-1 was preferentially released to basal side rather than to apical side (Yoshimoto *et al.*, 1991) supported this idea. This tendency was enhanced when the distance between endothelium and smooth muscle layer was closer, in other words, in the smaller vessels. In accordance with this fact, the mesenteric microvessels of rat, with arterioles of 15–40 μm in diameter, were very sensitive to ET-1 (Homma *et al.*, 1991). The arterioles were more sensitive than the venules. Since microvessels have proved to be more important in the maintenance of blood pressure than the large arteries, these facts favorably lead to the idea of paracrine regulation of ET in vascular tonus.

However, very recently the result of an experiment using a specific antagonist to ET_A receptor did not consistently support this idea. Ihara *et al.* (1992) demonstrated that pretreatment with the specific antagonist BQ-123 did not reduce the basal level of blood pressure of spontaneously hypertensive rat, while the pretreatment markedly antagonized ET-1-induced sustained pressor response.

Although there were several reports on an increased plasma ET-1 level in hyper-

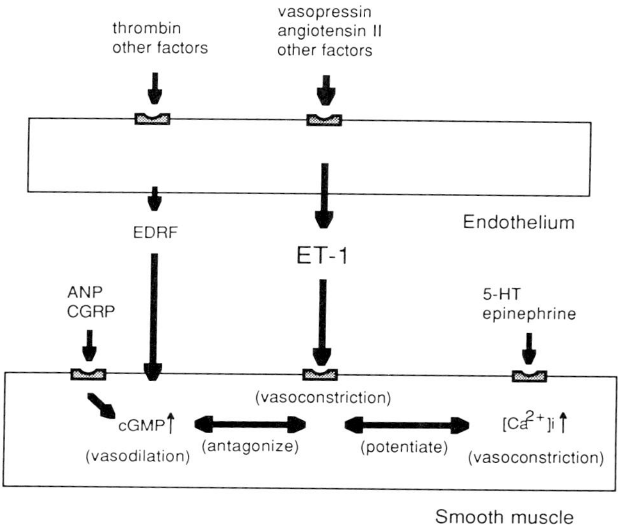

Figure 3 Mechanisms of vasoconstriction and vasodilation triggered by endogenous endothelin (ET). ET interacts with the other local vasoactive factors. (ANP, atrial natiuretic peptide; EDRF, endothelium-derived relaxing factor.)

tensive patients (Schichiri *et al.*, 1990; Kohno *et al.*, 1990; Saito *et al.*, 1990), we reported that the plasma level of ET-1 was not significantly different between normal and hypertensive subjects but increases with age in both states of male and female groups (Miyauchi *et al.*, 1992b). In addition, plasma levels of ET-1 in females were significantly lower than those in males in both normal and hypertensive subjects. These results suggested that ET-1 was not the primary cause of hypertension. Nevertheless, there remains a possibility that it played some secondary role in generation of the hypertension. An increase in plasma level of ET-1, in patients with chronic hemodialysis, suggested a significant correlation between blood pressure and production of ET-1 (Miyauchi *et al.*, 1991). On the other hand, in patients with pulmonary hypertension with congenital heart defect or with congestive heart failure, plasma concentration of ET-1 increased. ET was likely to be produced in pulmonary circulation (Yoshibayashi *et al.*, 1991; Cody *et al.*, 1992). Similarly, an increased level of plasma ET-1 was also observed in experimental congestive heart failure in anesthetized dog (Cavero *et al.*, 1990).

Although plasma level of ET in spontaneously hypertensive rat (SHR) was not high as compared with that in Wister Kyoto rat (WKY), vasocontractile sensitivity to ET-1 was enhanced in the isolated arteries of SHR. This may have been partly due to the depolarization of the resting membrane potential and increase in the number of ET-receptors in vascular smooth muscle of SHR (Tomobe *et al.*, 1991). These results also suggested that ET was not the major factor of maintenance of blood pressure.

Coronary artery was sensitive to ET-1. An ischemic episode potentiated further the sensitivity because of increase in the density of ET-1 binding site in the heart muscle (Chen *et al.*, 1990), which suggested that ET-1 may have acted as an aggravating factor in the ischemic damage. In fact, in experimental cardiac infarction, the plasma and cardiac tissue concentration of endogenous ET were increased. Antibody against ET reduced the infarction size (Watanabe *et al.*, 1990).

On the other hand, in variant angina, plasma ET-1 levels in peripheral venous and coronary sinus blood were higher in patients with vasospastic angina (Toyo-oka *et al.*, 1991). Although the plasma level of ET-1 was not sufficient to induce coronary spasm, the increase in ET-1 level may have enhanced the response to other vasoconstrictor substances.

Mechanism of the delayed cerebral vasospasm after subarachnoid hemorrhage was not clarified. Many factors have been implicated in this phenomenon, such as the increasing amount of ET-1 induced by dysfunction of endothelium after the hemorrhage elicits constriction of basilar artery from adventitial side but not from luminal side (Mima *et al.*, 1989). Indeed, after experimental subarachnoid hemorrhage in dog, ET was produced in endothelium of vasospastic basilar artery. In addition, treatment of the experimental animal with actinomycin D completely inhibited the development of vasospasm (Shigeno *et al.*, 1991). These results strongly suggested that ET played an important role in the cerebral vasospasm.

It has been known that plasma concentration of ET increases in patients with acute myocardial infarction, cardiogenic shock, or in patients who have undergone surgical operation (Miyauchi *et al.*, 1989; Onizuka *et al.*, 1992; Cernacek and Stewart, 1989). The increase in plasma ET level can possibly be ascribed to thrombin release. Thrombin is known to act on periferal tissues, for instance forearm, to generate ET-1. The production of ET-1 might differ between various organs under surgical stress, implying a compensatory response of ET-1 to acute stress (Onizuka *et al.*, 1992) (e.g., the reduction of local blood flow in nonvital organs in contrast to the increase in the blood flow in vital organs). Output of ET from forearm in response to cold stress was also reported (Fyhrguist *et al.*, 1990).

In renal tissues, ET displayed a profound effect in two ways. ET affected renal blood flow and glomerular filtration rate at the relatively high dose. On the other hand, a low dose of ET reduced renal vascular resistance and affected tubulus, which induced diuresis. Experimental bilateral renal artery occlusion induced production of ET in the renal parenchyma in accordance with the renal lesion. Treatment with antibody against ET or ET-antagonist BQ-123 remarkably ameliorated the development of the renal lesion (Shibouta *et al.*, 1990; Mino *et al.*, personal communication). On the other hand, increase in plasma level of ET-1 was demonstrated in ischemic diseases. These results strongly suggested that ET was an aggravating factor of the renal failure.

Finally, recent results also demonstrated that ET-1 may have played some causal role in generation of atherosclerosis. One of the characteristic features of atherosclerosis is the proliferation of arterial smooth muscle cells. ET-1 has also been found to be a cell growth factor (Takuwa *et al.*, 1989). In patients with symptomatic atherosclerosis, plasma concentrations of ET-1 significantly increased. In experimentally cholesterol-fed rats, the phasma concentration of ET-1 increased as compared with those of normal rats, which suggested a contribution of ET-1 in atherogenesis (Miyauchi *et al.*, 1991).

References

Arai, H., Hori, S., Aramori, I., Ohkubo, H., and Nakanishi, S. (1990). Cloning and expression of a cDNA encoding an endothelin receptor. *Nature* (London) *348,* 730–732.

Barnard, J. W., Barman, S. A., Adkins, W. K., Longenecker, G. L., and Taylor, A. E. (1991). Sustained effects of endothelin-1 on rabbit, dog and rat pulmonary circulation. *Am. J. Physiol. 261,* H479–H486.

Boulanger, C., and Lüscher, T. F. (1990). Release of endothelium from porcine aorta inhibition by endothelium-derived nitric oxide. *J. Clin. Invest. 85,* 587–590.

Brain, S. D. (1989). The direct observation of arteriolar constriction induced by endothelin *in vivo*. *Eur. J. Pharmacol. 160,* 401–403.

Brown, M. R., Vanghan, J., Walsh, J., Jimenez, L., Hexum, T. D., Baird, A., and Vale, W. (1990). Endothelin releasing activity in calf serim and porcine follicular fluid. *Biochem. Biophys. Res. Comm. 173,* 807–815.

Cao, L., and Banks, R. O. (1990). Cardiovascular and renal actions of endothelin: effects of calcium-channel blockers. *Am. J. Physiol. 258,* F254–F258.

Cavero, P. G., Miller, W. L., Heublein, D. M., Margulies, K. B., and Burnett, J. C. J. (1990). Endothelin in experimental congestive heart failure in the anesthetized dog. *Am. J. Physiol. 259,* F312–F317.

Cernacek, P., and Stewart, D. J. (1989). Immunoreactive endothelin in human plasma: Marked elevations in patients in cardiogenic shock. *Biochem. Biophys. Res. Comm. 161,* 562–567.

Chatfield, B. A., Mcmurtry, I. F., Hall, S. L., and Abman, S. H. (1991). Hemodynamic effects of endothelin-1 on ovine fetal pulmonary circulation. *Am. J. Physiol. 261,* R182–R187.

Chen, C., and Wagoner, P. K. (1991). Endothelin induces a nonselective cation current in vascular smooth muscle cells. *Circul. Res. 69,* 447–454.

Chen, J. L. R., Casley, D. J., and Nayler, W. G. (1990). Ischemia and resperfusion increase 125I-labeled endothelin-1 binding in rat cardiac membranes. *Am. J. Physiol. 258,* H829–H835.

Clozel, M., and Clozel, J. (1989). Effects of endothelin on regional flood flows in squirrel monkeys. *J. Pharmacol. Exp. Ther. 250,* 1125–1131.

Cody, R. J., Haas, G. J., Binkley, P. F., Capers, Q., and Kelley, R. (1992). Plasma endothelin correlates with the extent of pulmonary hypertension in patients with chronic congestive heart failure. *Circulation 85,* 504–509.

D'Orléans-Juste, Lidbury, P. S., Warner, T. D., and Vane, J. R. (1990). Intravascular big endothelin increases circulating levels of endothelin-1 and prostanoids in the rabbit. *Biochem. Pharmacol. 29,* 821–822.

Danthuluri, N. R., and Deth, R. C. (1989). Effect of intracellular alkalinization on resting and agonist-induced vascular tone. *Am. J. Physiol. 256,* H867–H875.

De Fes, M. A., Bartoliri, O., Orlands, C., Maggi, M., Seris, M., Pines, M., Hurwitz, S., Fujii, Y., Sakaguchi, K., Aurbach, G. D., and Brandi, M. L. (1991). Natriuretic peptide receptors regulate endothelin synthesis and release from parathyroid cells. *Proc. Natl. Acad. Sci. U.S.A. 88,* 6496–6500.

DeNucci, G., Thomas, R., D'Orleans-Juste, P., Antunes, E., Walder, C., Warner, T. D., and Vane, J. R. (1988). Pressor effects of circulating endothelin are limited by its removal in the pulmonary circulation and by the release of prostacyclin and endothelium-derived relaxing factor. *Proc. Natl. Acad. Sci. U.S.A. 85,* 9797–9800.

Douglas, S. A., James, S., and Hiley, C. R. (1991). Endothelial modulation and changes in endothelin pressor activity during hypoxia in the rat isolated perfused superior mesenteric arterial bed. *Br. J. Pharmacol. 103,* 1441–1448.

Emori, T., Hirata, Y., Ohta, K., Kanno, K., Eguchi, S., Imai, T., Schichiri, M., and Marumo, F. (1991). Cellular mechanism of endothelin-1 release by angiotensin and vasopressin. *Hypertension 18,* 165–170.

Fujii, Y., Moreira, J. E., Orlands, C., Maggi, M., Aurbach, G. D., Brandi, M. L., and Sakaguchi, K. (1991). Endothelin as an autocrine factor in the regulation of parathyroid cells. *Proc. Natl. Acad. Sci. U.S.A. 88,* 4235–4239.

Fyhrquist, F., Saijonmaa, O., Metsärinne, K., Tikkanen, I., Rosenlöf, K., and Tikkanen, T. (1990). Raised plasma endothelin-1 concentration following cold pressor test. *Biochem. Biophys. Res. Comm. 169,* 217–221.

Gardiner, S. M., Compton, A. M., and Bennett, T. (1989a). Regional-hemodynamic effects of endothelin-1 in conscious, unrestrained Wistar rats. *J. Cardiovascul. Pharmacol. 13* (Suppl. 5), S202–S204.

Gardiner, S. M., Compton, A. M., Bennett, T., Palmer, R. M. J., and Moncada, S. (1989b). NG-monomethyl-L-arginine does not inhibit the hindquarters vasodilator action of endothelin-1 in conscious rats. *Eur. J. Pharmacol. 171,* 237–240.

Gardiner, S. M., Compton, A. M., and Bennett, T. (1990a). Regional haemodynamic effects of endothelin-1 and endothelin-3 in conscious Long Evans and Brattleboro rats. *Br. J. Pharmacol. 99,* 107–112.

Gardiner, S. M., Compton, A. M., and Bennett, T. (1990b). Regional hemodynamic effects of endothelin-2 and sarafotoxin-S6b in conscious rats. *Am. J. Physiol. 258,* R912–R917.

Gardiner, S. M., Compton, A. M., Kemp, P. A., and Bennett, T. (1990c). Regional and cardiac haemodynamic responses to glyceryl trinitrate, acetylcholine, bradykinin and endothelin-1 in conscious rats: Effects of NG-nitro-L-arginine methyl ester. *Br. J. Pharmacol. 101,* 632–639.

Gardiner, S. M., Compton, A. M., and Bennett, T. (1990d). Effects of indomethacin on the regional haemodynamic responses to low doses of endothelins and sarafotoxin. *Br. J. Pharmacol. 100,* 158–162.

Gardiner, S. M., Compton, A. M., Kemp, P. A., Bennett, T., Foulkes, R., and Hughes, B. (1991). Haemodynamic effects of human α-calcitorin gene-related peptide following administration of endothelin-1 or N^G-nitro-L-arginine methyl ester in conscious rats. *Br. J. Pharmacol. 103,* 1256–1262.

Giaid, A., Gibson, S. J., Ibrahim, N., Legon, S., Bloom, R., Yanagisawa, M., Masaki, T., Varudell, I. M., and Polak, I. M. (1989). Endothelin-1, an endothelium-derived peptide is expressed in neurons of the human spinal cord and dorsal root ganglia. *Proc. Natl. Acad. Sci. U.S.A. 86,* 7634–7638.

Goetz, K. L., Wang, B. C., Madwed, J. B., Zhu, J. L., and Leadley, R. J. J. (1988). Cardiovascular, renal, and endocrine responses to intravenous endothelin in conscious dogs. *Am. J. Physiol. 255,* R1064–R1068.

Goto, K., Kasuya, Y., Matsuki, N., Takuwa, Y., Kurihara, H., Ishikawa, T., Kimura, S., Yanagisawa, M., and Masaki, T. (1989). Endothelin activates the dihydrophyridine-sensitive, voltage-dependent Ca^{2+} channel in vascular smooth muscle. *Proc. Natl. Acad. Sci. U.S.A. 86,* 3915–3918.

Gupta, S., Ruderman, N. B., Cragoe, E. J. J., and Sussman, I. (1991). Endothelin stimulates Na^+-K^+-ATPase activity by a protein kinase C-dependent pathway in rabbit aorta. *Am. J. Physiol. 261,* H38–H45.

Han, S. P., Trapani, A. J., Fok, K. F., Westfall, T., and Knuepfer, M. (1989). Effects of endothelin or regional hemodynamics in conscious rats. *Eur. J. Pharmacol. 159,* 303–305.

Han, S., Chen, X., Westfall, T. C., and Knuepfer, M. M. (1991a). Characterization of the depressor effect of intrathecal endothelin in anesthetized rats. *Am. J. Physiol. 260,* H1685–H1691.

Han, S., Trapani, A. J., Fok, K. F., Westfall, T. C., and Knuepfer, M. M. (1991b). A comparison of the hemodynamic effects of endothelin-1 and sarafotoxin S66 in conscious rats. *J. Pharmacol. Exp. Ther. 256,* 183–188.

Hasunuma, K., Rodman, D. M., O'Brien, R. F., and Mcmurtay, I. F. (1990). Endothelin-1 causes nulmonary vasodilation in rats. *Am. J. Physiol. 259,* H48–H54.

Hinojosa-Laborde, C., Osborn, J. W. J., and Cowley, A. W. J. (1989). Hemodynamic effects of endothelin in conscious rats. *Am. J. Physiol. 256,* H1742–H1746.

Homma, S., Miyauchi, T., Goto, K., Sugishita, Y., Sato, M., and Ohshima, N. (1991). Effects of

endothelin-1 on coronary microcirculation in isolated beating hearts of rats. *J. Cardiovascul. Pharmacol. 17* (Suppl. 7), S276–S278.
Hosoda, K., Nakao, K., Arai, H., Suga, S., Ogawa, Y., Mukoyama, M., Shirakami, G., Saito, Y., Nakanishi, S., and Imura, H. (1991). Cloning and expression of human endothelin-1 receptor cDNA. *FEBS Lett. 287,* 23–26.
Ihara, M., Noguchi, K., Saeki, T., Fukuroda, T., Tsuchida, S., Kimura, S., Fukami, T., Ishikawa, K., Nishikibe, M., and Yano, M. (1992). Biological profiles of highly potent novel endothelin antagonists selective for the ETA receptor. *Life Science 50,* 247–255.
Inoue, A., Yanagisawa, M., Kimura, S., Kasuya, Y., Miyauchi, T., Goto, K., and Masaki, T. (1989). The human endothelin family: Three structurally and pharmacologically distinct isopeptides predicted by three separate genes. *Proc. Natl. Acad. Sci. U.S.A. 86,* 2863–2867.
Ishikawa, T., Yanagisawa, M., Kimura, S., Goto, K., and Masaki, T. (1988a). Positive chronotropic effects of endothelin, a novel endothelium-derived vasoconstrictor peptide. *Pflügers Arch 413,* 108–110.
Ishikawa, T., Yanagisawa, M., Kimura, S., Goto, K., and Masaki, T. (1988b). Positive inotropic action of novel vasoconstrictor peptide endothelin on guinea pig atria. *Am. J. Physiol. 255,* H970–H973.
Itoh, S., and van den Busse, M. (1991). Sensitization of baroreceptor reflex by central endothelin in conscious rats. *Am. J. Physiol. 260,* H1106–H1112.
Kasuya, Y., Ishikawa, T., Yanagisawa, M., Kimura, S., Goto, K., and Masaki, T. (1989a). Mechanism of contraction to endothelin in isolated porcine coronary artery. *Am. J. Physiol. 257,* H1828–H1835.
Kasuya, Y., Takuwa, Y., Yanagisawa, M., Kimura, S., Goto, K., and Masaki, T. (1989b). Endothelin-1 induces vasoconstriction through two functionally distinct pathways in porcine coronary artery: Contribution of phosphoinositide turnover. *Biochem. Biophys. Res. Comm. 161,* 1049–1055.
Kelly, R. A., Eid, H., Kramer, B. K., O'Neill, M., Liang, T., Rears, M., and Smith, T. W. (1990). Endothelin enhances the contractile responsiveness of adult rat ventricular myocytes to calcium by a pertussis toxin-sensitive pathway. *J. Clin. Invest. 86,* 1164–1171.
Kim, D. (1991). Endothelin activation of an inwardly rectifying K^+ current in atrial cells. *Circulation Research 69,* 250–255.
King, A. J., Pfeffer, J. M., Peffer, M. A., and Brenner, B. M. (1990). Systemic hemodynamic effects of endothelin in rats. *Am. J. Physiol. 258,* H787–H792.
Knuepfer, M. M., Han, S. P., Trapani, A. J., Fok, K. F., and Westfall, T. C. (1989). Regional hemodynamic and baroreflex effects of endothelin in rats. *Am. J. Physiol. 257,* H918–H926.
Kodama, M., Kanaide, H., Abe, S., Hirano, K., Kai, H., and Nakamura, M. (1989). Endothelin-induced Ca-independent contraction of the porcine coronary artery. *Biochem. Biophys. Res. Commun. 160,* 1302–1308.
Kohno, M., Yasunari, K., Murakawa, K., Yokokawa, K., Horie, T., Fukui, T., and Takeda, T. (1990). Plasma immunoreactive endothelin in essential hypertension. *Am. J. Med. 88,* 614–618.
Kohno, M., Murakawa, K., Horio, T., Yokokawa, K., Yasunari, K., Fukui, T., and Takeda, T. (1991). Plasma immunoreactive endothelin-1 in experimental malignant hypertension. *Hypertension 18,* 93–100.
Koseki, C., Imai, M., Hirata, Y., Yanagisawa, M., and Masaki, T. (1989). Autoradiographic distribution in rat tissues of binding sites for endothelin: A neuropeptide? *Am. J. Physiol. 256,* R858–R866.
Krämer, B. K., Smith, T. W., and Kelly, R. A. (1991). Endothelin and increased contractibility in adult rat ventricular myocytes role of intracellular alkalosis induced by activation of the protein kinase C-dependent Na^+-H^+ exchanger. *Circulation Research 68,* 269–279.
Kuwaki, T., Cao, W., Yamamoto, M., Terui, N., and Kumada, K. (1991a). Cardiorespiratory effects of topical application of endothelin-1 to the ventral surface of the rat medulla. *J. Cardiovascul. Pharmacol. 17* (Suppl. 7), S343–S345.
Kuwaki, T., Koshiya, N., Terui, N., and Kumada. (1991b). Endothelin-1 modulates cardiorespiratory control by the central nervous system. *Neurochem. Int. 18,* 19–24.
Leadley, R. J. J., Zhu, J. L., and Goetz, K. L. (1991). Effects of endothelin-1 and sarafotoxin S66 on regional hemodynamics in the conscious dog. *Am. J. Physiol. 260,* R1210–R1217.
Lee, T., Chao, T., Hu, K., and King, G. L. (1989). Endothelin stimulates a sustained 1, 2-diacylglycerol

increase and protein kinase C activation in bovine aortic smooth muscle cells. *Biochem. Biophys. Res. Comm. 162,* 381–386.

LeMonnier de Gouville, A., Lippton, H., Cohen, G., Cavero, I., and Hyman, A. (1990a). Vasodilator activity of endothelin-1 and endothelin-3: Rapid development of cross-tachyphylaxis and dependence on the rate of endothelin administration. *J. Pharmacol. Exp. Ther. 254,* 1024–1028.

LeMonnier de Gouville, A. C., Mondot, S., Lippton, H., Hyman, A., and Cavero, I. (1990b). Hemodynamic and pharmacological evaluation of the vasodilator and vasoconstrictor effects of endothelin-1 in rats. *J. Pharmacol. Exptl. Ther. 252,* 300–311.

Lerman, A., Edwards, B. S., Hallett, J. W., Heublein, D. M., Sandberg, S. M., and Burnett, J. C. J. (1991). Circulating and tissue endothelin immunoreactivity in advanced atherosclerosis. *N. Engl. J. Med. 325,* 997–1001.

Li, K., Stewart, J., and Rouleau, J. (1991). Myocardial contractile actions of endothelin-1 in rat and rabbit papillary muscle: Role of endocardial endothelium. *Circulation Research 69,* 301–312.

Lippton, H. L., Hauth, T. A., Summer, W. R., and Hyman, A. L. (1989). Endothelin produces pulmonary vasoconstriction and systemic vasodilation. *J. Appl. Physiol. 66,* 1008–1012.

Lüscher, T. F., Yang, Z., Tschudi, M., von Segesser, L., Stulz, P., Boulanger, C., Siebenmann, R., Turina, M., and Bühler, F. R. (1990). Interaction between endothelin-1 and endothelium-derived relaxing factor in human arteries and veins. *Circulation Res. 66,* 1088–1094.

Madeddu, P., Yang, X., Anania, V., Troffa, C., Pazzola, A., Soro, A., Manunta, P., Tonolo, G., Demontis, M. P., Varoni, M. V., Melis, M. G., and Glorioso, N. (1990). Efficacy of nifedipine to prevent systemic and renal vasoconstrictor effects of endothelin. *Am. J. Physiol. 259,* F304–311.

Masaki, T., Kimura, S., Yanagisawa, M., and Goto, K. (1991). Molecular and cellular mechanism of endothelin regulation, Implications for vascular function. *Circulation 84,* 1457–1468.

Masaki, T., Yanagisawa, M., and Goto, K. (1992). Physiology and pharmacology of endothelins. *Medicinal Res. Rev. 12,* 391–421.

McMahon, E. G., Palomo, M. A., Moore, W. M., McDonald, J. F., and Stern, M. K. (1991). Phosphoramidon blocks the pressor activity of porcine big endothelin-1-(1–39) in vivo and conversions of big endothelin-1-(1–39) to endothelin-1 (1–21) *in vivo. Proc. Natl. Acad. Sci. U.S.A. 88,* 703–707.

Miller, V. M., Komori, K., Burnett, J. C., and Vanhoutte, P. M. (1989). Differential sensitivity to endothelin in canine arteries and veins. *Am. J. Physiol. 257,* H1127–H1131.

Mima, T., Yanagisawa, M., Shigeno, T., Saito, A., Goto, K., Takakura, K., and Masaki, T. (1989). Endothelin acts in feline and canine cerebral arteries from the adventitial side. *Stroke 20,* 1553–1556.

Minkes, R. K., and Kadowitz, P. J. (1989). Influence of endothelin on systemic arterial pressure and regional blood flow in the cat. *Eur. J. Pharmacol. 163,* 163–166.

Minkes, R. K., MacMillan, L. A., Bellan, J. A., Kerstein, M. D., McNamara, D. B., and Kadowitz, P. J. (1989). Analysis of regional responses to endothelin in hindquarters vascular bed of cats. *Am. J. Physiol. 256,* H598–H602.

Minkes, R. K., Bellan, J. A., Saroyan, R. M., Kerstein, M. D., Coy, D. H., Murphy, W. A., Nossaman, B. D., Mcnamara, D. B., and Kadowritz, P. J. (1990). Analysis of cardiovascular and pulmonary responses to endothelin-1 and endothelin-3 in the anesthetized cat. *J. Pharmacol. Exper. Ther. 253,* 1118–1125.

Miura, K., Yukimura, T., Yamashita, Y., Shimmen, T., Okumura, M., Yamanaka, S., Imanishi, M., and Yamamoto, K. (1991). Renal and femoral vascular responses to endothelin-1 in dogs: Role of prostaglandins. *J. Pharmacol. Exp. Thr. 256,* 11–17.

Miyauchi, T., Ishikawa, T., Tomobe, Y., Yanagisawa, M., Kimura, S., Sugishita, Y., Ito, I., Goto, K., and Masaki, T. (1989). Characteristics of pressor response to endothelin in spontaneously hypertensive and Wistar-Kyoto rats. *Hypertension 14,* 427–434.

Miyauchi, T., Suzuki, N., Kurihara, T., Yamaguchi, I., Sugishita, Y., Matsumoto, H., Goto, K., and Masaki, T. (1991). Endothelin-1 and endothelin-3 play different roles in acute and chronic alterations of blood pressure in patients with chronic hemodialysis. *Biochem. Biophys. Res. Comm. 178,* 276–281.

Miyauchi, T., Sugishita, Y., Matsuda, M., Sakai, H., Suzuki, N., Masaki, T., and Goto, K. (1992a). Increased plasma concentration of endothelin-1 in cholesterol-fed rats. *Atherosclerosis 93,* 257–259.

Miyauchi, T., Yanagisawa, M., Iida, K., Ajisaka, R., Suzuki, N., Fujino, M., Goto, K., Masaki, T., and Sugishita, Y. (1992b). Age- and sex-related variation of plasma endothelin-1 concentration in normal and hypertensive subjects. *Am. Heart J. 123,* 1092–1093.

Modin, A., Pernow, J., and Lundberg, J. M. (1991). Phosphoramidon inhibits the vasoconstrictor effects evoked by big endothelin-1 but not the elevation of plasma endothelin-1 *in vivo. Life Science 49,* 1619–1625.

Moravec, C. S., Reynolds, E. E., Stewart, R. W., and Bond, M. (1989). Endothelin is a positive inotropic agent in human and rat heart *in vitro. Biochem. Biophys. Res. Comm. 159,* 14–18.

Mortensen, L. H., and Fink, G. D. (1990). Hemodynamic effects of human and rat endothelin administration into conscious rats. *Am. J. Physiol. 258,* H362–H368.

Muldoon, L. L., Enslen, H., Rodland, K. D., and Magun, B. E. (1991). Stimulation of Ca^{2+} influx by endothelin-1 is subject to negative feedback by elevated intracellular Ca^{2+}. *Am. J. Physiol. 260,* C1273–C1281.

Nakayama, K., Ishigai, Y., Uchida, H., and Tanaka, Y. (1991). Potentiation by endothelin-1 of 5-hydroxytryptamine-induced contraction in coronary artery of the pig. *Br. J. Pharmacol. 104,* 978–986.

Nishimura, J., Khalil, R. A., Drenth, J. P., and Van Breeman, C. (1990). Evidence for increased myofilament Ca^{2+} sensitivity in norepinephrine-activated vascular smooth muscle. *Am. J. Physiol. 259,* H2–H8.

O'Donnell, S. R., Wanstall, J. C., and Zang, X. (1990). Pinacidil antagonism of endothelin-induced contractions of smooth muscle in the lungs: Differences between tracheal and pulmonary artery preparations. *J. Pharmacol. Exp. Ther. 252,* 1318–1323.

Ohlstein, E. H., Vickery, L., Sauermelch, C., and Willette, R. N. (1990). Vasodilation induced by endothelin: Role of EDRF and prostanoids in rat hindquarters. *Am. J. Physiol. 259,* H1835–H1841.

Ohlstein, E. H., Storer, B. L., Butcher, J. A., Debouck, C., and Feuerstein, G. (1991). Platelets stimulate expression of endothelin mRNA and endothelin biosynthesis in cultured endothelial cells. *Circulation Research 69,* 832–841.

Ohta, K., Hirata, Y., Imai, T., Kanno, K., Emori, T., Schichiri, M., and Marumo, F. (1990). *Biochem. Biophys. Res. Comm. 169,* 578–584.

Okada, K., Miyazaki, Y., Takada, J., Matsuyama, K., Yamaki, T., and Yano, M. (1990). Conversion of big endothelin-1 by membrane-bound methalloendopeptidase in cultured bovine endothelial cells. *Biochem. Biophys. Res. Commun. 171,* 1192–1198.

Olson, K. R., and Villa, J. (1991). Evidence against nonprostanoid endothelium-derived relaxing factor(s) in trout vessels. *Am. J. Physiol. 260,* R925–R933.

Olson, K. R., Duff, D. W., Farrell, A. P., Keen, J., Kellogg, M. D., Kullman, D., and Villa, J. (1991). Cardiovascular effects of endothelin in trout. *Am. J. Physiol. 260,* H1214–H1223.

Onizuka, M., Miyauchi, T., Mitsui, K., Suzuki, N., Ueno, H., Goto, K., Masaki, T., and Hori, M. (1992). Plasma levels of endothelin-1 and thrombin-antithrombin III complex in patients undergoing open-chest surgery. *J. Thorac Cardiovasc. Surg.* (in press).

Onizuka, M., Miyauchi, T., Mitsui, K., Suzuki, N., Masaki, T., Goto, K., and Hori, M. (1992). Endothelin-1 mediates regional blood flow during and after pulmonary surgery. *J. Thoracic Cardiovasc. Surgery* (in press).

Ouchi, Y., Kim, S., Souza, A. C., Iijima, S., Hattori, A., Orimi, H., Yoshizumi, M., Kurihara, H., and Yazaki, Y. (1989). Central effect of endothelin on blood pressure in conscious rats. *Am. J. Physiol. 256,* H1747–H1751.

Pollack, D. M., and Opgenorth, T. J. (1991). Evidence for metalloprotenase involvement in the *in vivo* effects of big endothelin-1. *Am. J. Physiol. 261,* R257–R263.

Rohmeiss, P., Phitiadis, J., Rohmeiss, S., and Unger, T. (1990). Hemodynamic actions of intravenous endothelin in rats: Comparison with sodium nitroprusside and methoxamine. *Am. J. Physiol. 258,* H337–H346.

Saito, Y., Nakao, K., Mukoyama, M., and Imura, H. (1990). Increased plasma endothelin level in patients with essential hypertension. *N. Engl. J. Med. 322,* 205.

Sakamoto, A., Yanagisawa, M., Sakurai, T., Takuwa, Y., Yanagisawa, H., and Masaki, T. (1991). Cloning and functional expression of human cDNA for the ETB endothelin receptor. *Biochem. Bipohys. Res. Comm. 178,* 656–663.

Sakurai, T., Yanagisawa, M., Takuwa, Y., Miyazaki, H., Kimura, S., Goto, K., and Masaki, T. (1990). Cloning of a cDNA encoding a non-isopeptide-selective subtype of the endothelin receptor. *Nature* (London) *348,* 732–735.

Schichiri, M., Hirata, Y., Ando, K., Emori, T., Ohta, K., Kimoto, S., Ogura, M., Inoue, A., and Marumo, F. (1990). Plasma endothelin levels in hypertension and chronic renal failure. *Hypertension 15,* 493–496.

Shiba, R., Yanagisawa, M., Miyauchi, T., Ishii, Y., Kimura, S., Uchiyama, Y., Masaki, T., and Goto, K. (1989). Elimination of intravenously injected endothelin-1 from the circulation of the rat. *J. Cardiovascul. Pharmacol. 13* (Suppl. 5), S98–S101.

Shibouta, Y., Suzuki, N., Shino, A., Matsumoto, H., Terashita, Z., Kondo, K., and Nishikawa, K. (1990). Pathophysiological role of endothelin in acute renal failure. *Life Sciences 46,* 1611–1618.

Shigeno, T. Mima, T., Yanagisawa, M., Saito, A., Goto, K., Yamashita, K., Takenouchi, T., Matsurura, N., Yamasaki, Y., Yamada, K., Masaki, T., and Takakure, K. (1991). Prevention of cerebral vasospasm by actinomycin D. *J. Neurosurg. 74,* 940–943.

Suzuki, N., Matsumoto, H., Kitada, C., Masaki, T., and Fujino, M. (1989). A sensitive sandwich-enzyme immunoassay for human endothelin. *J. Immunol. Meth. 118,* 245–250.

Tabuchi, Y., Nakamura, M., Rakugi, H., Nagano, M., and Ogihara, T. (1989). Endothelin-1 enhances adnergic vasoconstriction in perfused rat mesenteric arteries. *Biochem. Biophys. Res. Comm. 159,* 1304–1308.

Takuwa, N., Takuwa, Y., Yanagisawa, M. Yamashita, K., and Masaki, T. (1989). A novel vasoactive peptide endothelin stimulates mitogenesis through inositol lipid turnover in Swiss 3T3 fibroblasts. *J. Biol. Chem. 264,* 7856–7861.

Takuwa, Y., Kasuya, Y., Takuwa, N., and Kudo, M. (1990). Endothelin receptor is coupled to phospholipase C via a pertussis toxin-insensitive guanine vascular smooth muscle cells. *J. Clin. Invest. 85,* 653–658.

Tomobe, Y., Ishikawa, T., Yanagisawa, M., Kimura, S., Masaki, T., and Goto, K. (1991). Mechanism of altered sensitivity to endothelin-1 between aortic smooth muscles of spontaneously hypetensive and wistar-Kyoto rats. *J. Pharmacol. Exp. Ther. 257,* 555–561.

Totsune, K., Mouri, T., Takahashi, K., Ohneda, M., Sone, M., Saito, T., and Yoshinaga, K. (1989). Detection of immunoreactive endothelin in plasma of hemodialysis patients. *FEBS Lett. 249,* 239–239.

Toyo-oka, T., Aizawa, T., Suzuki, N., Hirata, Y., Miyauchi, T., and Shin, W. S. (1991). Increased plasma level of endothelin-1 and coronary spasm induction in patients with vasospastic angina pectoris. *Circulation 83,* 476–483.

Vigne, P., Lazdunski, M., and Frelin, C. (1989). The inotropic effect of endothelin-1 on rat atria involves hydrolysis of phosphatidylinositol. *FEBS Lett. 249,* 143–146.

Wagner-Mann, C., and Sturek, M. (1991). Endothelin mediates Ca influx and release in porcine coronary smooth muscle cells. *Am. J. Physiol. 260,* C771–C777.

Wang, J., Paik, G., and Morgan, J. P. (1991a). Endothelin-1 enhances myofilament Ca^{2+} responsiveness in aequorin-loaded ferret myocardium. *Circulation 69,* 582–589.

Wang, J., Zeballos, G. A., Kaley, G., and Hintze, T. H. (1991b). Dilation and costriction of large coronary arteries in conscious dogs by endothelin. *Am. J. Physiol. 261,* H1379–H1386.

Warner, T. D. G., DeNucci, and Vane, J. R. (1989). Rat endothelin is a vasodilator in the isolated perfused mesentery of the rat. *Eur. J. Pharmacol. 159,* 325–326.

Watanabe, T., Suzuki, N., Shimamoto, N., Fujino, M., and Imada, A. (1990). Endothelin in myocardial infarction. *Nature 344,* 114.

Wright, C. E., and Fozard, J. R. (1988). Regional vasodilation is a prominent feature of the haemodynamic response to endothelin in anaesthetized, spontaneously hypertensive rats. *Eur. J. Pharmacol. 155,* 210–203.

Xuan, Y., Watkins, W. D., and Whorton, A. R. (1991). Regulation of endothelin-mediated calcium mobilization in vascular smooth muscle cells by isoproterenol. *Am. J. Physiol. 260,* C492–C502.

Yanagisawa, M., Kurihara, H., Kimura, S., Tomobe, Y., Kobayashi, M., Mitsui, Y., Yazaki, Y., Goto, K., and Masaki, T. (1988). A novel potent vasoconstrictor peptide produced by vascular endothelial cells. *Nature* (London) *332,* 411–415.

Yoshibayashi, M., Nishioka, K., Nakao, K., Saito, Y., Matsumura, M., Ueda, T., Temma, S., Shirakami, G., Imura, H., and Mikawa, H. (1991). Plasma endothelin concentrations in patients with pulmonary hypertension associated with congenital heart defects. *Circulation 84,* 2280–2285.

Yoshimoto, S., Ishizaki, Y., Sasaki, T., and Murota, S. (1991). Effect of carbon dioxide and oxygen on endothelin production by cultured porcine cerebral endothelial cells. *Stroke 22,* 378–383.

Yoshizumi, M., Kurihara, H., and Sugiyama, T. (1989). Hemodynamic shear stress stimulates endothelin production by cultured cells. *Biochem. Biophys. Res. Comm. 161,* 859–864.

Yoshizumi, M., Kurihara, H., and Morita, T. (1990). Interleukin-1 increases the production of endothelin-1 by cultured endothelial cells. *Biochem. Biophys. Res. Comm. 166,* 324–329.

CHAPTER 5

Interaction between Endothelium-Derived Vasoactive Factors and Neurohumoral Substances in the Regulation of Vascular Tone

GABOR M. RUBANYI

I. Introduction

The caliber of blood vessels, an important determinant of tissue perfusion, is dependent on vascular smooth muscle tone. Vascular smooth muscle tone is influenced by adrenergic innervation (i.e., by norepinephrine released from adrenergic nerve endings in the blood vessel wall), which provides the input from the central nervous system. Circulating (e.g., catecholamines, vasopressin, angiotensin) or locally produced humoral substances (e.g., serotonin, adenine nucleotides, thrombin) can also modify vascular tone in a significant way. Discoveries in the last decade revealed that vascular smooth muscle tone is modulated also by the endothelium, which responds to changes in the local hemodynamic and humoral environment. Although the majority of studies on endothelium dealt with the chemical nature and mechanisms of synthesis/release and biological action of endothelium-derived vasoactive substances, evidence has accumulated for potential interactions between these three regulatory systems (i.e., neural, humoral, and endothelial) in the control of vascular tone. Such interactions may occur in several ways (Fig. 1): (1) the endothelium exhibits a physical and biochemical barrier (i.e., uptake and metabolic degradation of various neurohumoral vasoactive substances); (2) neurohumoral substances can act on the endothelium via specific receptors and stimulate the synthesis/release of endothelium-derived vasoactive factors; (3) endothelium-derived vasoactive factors can alter the responsiveness of vascular smooth muscle to neurohumoral substances; (4) although anatomically separated, a bidirectional interaction was proposed to exist between substances released from the endothelium and adrenergic neurotransmission in the vascular wall; and (5) the endothelium may store, synthesize, and release neurohumoral substances.

This chapter briefly reviews the present knowledge on endothelium-derived vas-

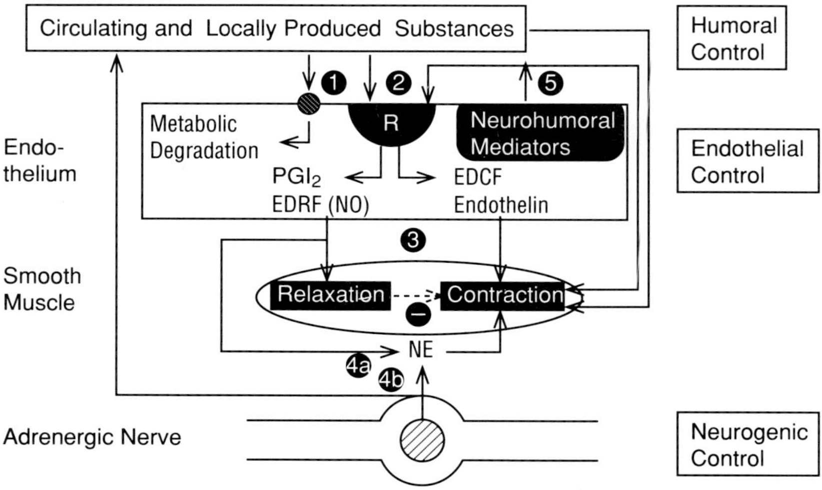

Figure 1 Possible interactions (1–5) between neural, humoral, and endothelial control mechanisms in the regulation of vascular tone. PGI_2, prostacyclin; EDRF(NO), endothelium-derived relaxing factor (nitric oxide); EDCF, endothelium-derived contracting factor; NE, norepinephrine; R, receptors on endothelial cells.

oactive factors and subsequently gives a summary of the evidence supporting the various forms of interaction between these factors and neurohumoral vasoactive substances. Finally, an attempt is made to speculate on the potential importance of these interactions in the control of vascular tone in physiological and pathological conditions.

II. Endothelium-Derived Vasoactive Factors

A. *Relaxing Factors*

The first endothelial vasoactive substance was discovered in the late 1970s, when it was demonstrated that human vascular endothelial cells synthesize prostacyclin, a potent vasorelaxant and platelet inhibitory metabolite of arachidonic acid [1] (Fig. 2).

After the observation of Furchgott and Zawadski in 1980 [2] that the presence of endothelium is obligatory for acetylcholine-induced vasorelaxation, bioassay studies demonstrated that a very labile, diffusible, nonprostanoid substance (called endothelium-derived relaxing factor, or EDRF) mediates endothelium-dependent vasorelaxation [3] (Fig. 2). EDRF is released under basal conditions and upon stimulation [4]. Among the many physiologic stimuli that can elicit the release of EDRF are platelet products, thrombin, hormones, neurotransmitters, local autacoids, changes in oxygen tension, and increase in flow (shear stress) [4].

After the discovery that, similar to nitric oxide (NO), the effect of EDRF is mediated by an increase in smooth muscle cyclic GMP content [5], and hemoglobin [6] and superoxide anions [7] inactivate both NO and EDRF, it was proposed that EDRF and NO are the same [8,9]. The demonstration of NO production by endo-

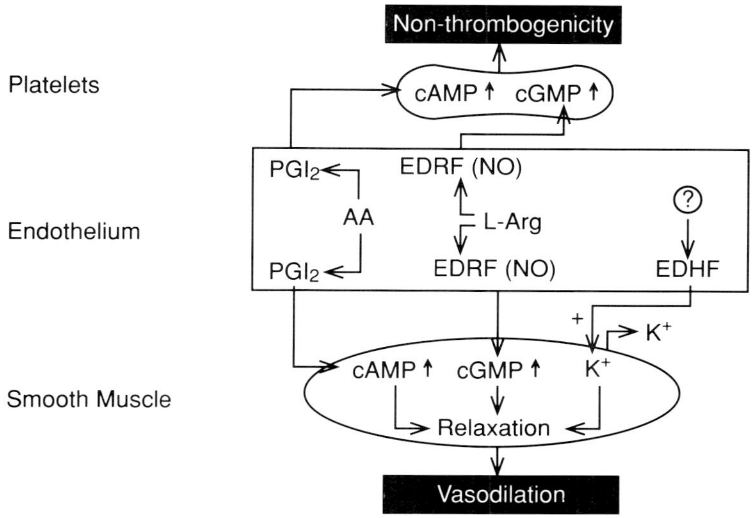

Figure 2 Endothelium-derived relaxing factors (EDRFs). Prostacyclin (PGI_2) and EDRF [identical with nitric oxide (NO) or a labile nitroso compound] relax vascular smooth muscle and inhibit platelet aggregation by elevating tissue levels of cyclic AMP (cAMP) and cyclic GMP (cGMP), respectively. Endothelium-derived hyperpolarizing factor (EDHF) relaxes vascular smooth muscle by membrane hyperpolarization presumably via activating K^+ channels. These factors contribute to the "thromboresistant" and "vasodilator function" of the endothelium.

thelial cells provided more direct evidence for this hypothesis [10]. Although it is still uncertain whether EDRF is free NO or a labile nitroso compound from which NO is liberated (or both)[11–13], it is widely accepted that L-arginine is the substrate required for formation of NO and EDRF (Fig. 2).

Vascular endothelial cells in culture synthesize NO from the terminal guanido nitrogen atom(s) of L-arginine [14]. This reaction is specific since other analogs of L-arginine, including its D-enantiomer, are not substrates. Furthermore, N^G-monomethyl-L-arginine (L-NMMA) and other arginine analogs inhibit the reaction in a dose-dependent and enantiomerically specific manner. This inhibition can be reversed by L-arginine in a way that suggests a competition between the inhibitor and the substrate. The enzyme (NO-synthase) responsible for this reaction was isolated and characterized [15]. Several isoforms of NO-synthase were also identified: a constitutive enzyme present in endothelial cells and the brain, which requires calcium and calmodulin, and an inducable form, which is Ca^{2+}-independent, and can be observed in numerous cell types after treatment with endotoxin, tumor-necrotizing factor (TNF), and other cytokines [16]. It is postulated that the arginine-NO signaling pathway is ubiquitous, and may contribute to a variety of biological functions.

In addition to the "classical" EDRF(NO), the existence of other, chemically different, nonprostanoid relaxing mediators [17,18], including an endothelium-derived hyperpolarizing factor (EDHF) [19], was proposed (Fig. 2).

B. Contracting Factor

The observation that in canine veins arachidonic acid causes endothelium-dependent contraction and that removal of the endothelium from canine femoral arteries significantly reduces the contractions evoked by several agonists and hypoxia was the first indication that the endothelium not only mediates relaxation of the underlying vascular smooth muscle but can also facilitate smooth muscle contraction [20]. Ample experimental evidence, accumulated over the past years, has clearly established the existence of endothelium-dependent vasoconstriction [21], which can be stimulated by naturally occurring substances (e.g., acetylcholine, arachidonic acid, norepinephrine, prostaglandin H_2, thrombin), pharmacological agents (e.g., calcium ionophores, nicotine, high K^+), physical forces (stretch, pressure), and hypoxia [21].

The endothelium-derived contracting factor(s) (EDCF) mediating endothelium-dependent vasoconstriction belongs to two categories (Fig. 3): (1) vasoconstrictor metabolites of arachidonic acid (TXA_2, endoperoxides) and (2) a potent peptidergic vasoconstrictor substance produced by endothelial cells in culture that in contrast to the prompt and reversible contraction triggered by the arachidonic acid metabolites causes slowly developing, and long-lasting vasoconstriction [22]. This peptidergic EDCF has been isolated, purified, sequenced, and termed endothelin [23]. The molecular biology, biosynthesis, diverse biological actions, and potential pathological significance of this 21 amino acid potent vasoconstrictor peptide has been recently reviewed [24].

Classically, the synthesis of angiotensin II (AII) has been described as a series of biochemical reactions occurring within the circulation. This renin-angiotensin system (RAS) functions as a systemic endocrine unit. Renin, secreted by the kidney, converts angiotensinogen, secreted by the liver, to angiotensin I (AI). Angiotensin-converting enzyme (ACE) breaks AI down to its vasoactive product, AII [25,26]. This last reaction occurs primarily at the luminal surface of endothelial cells in the lung [27], but ACE has been localized in endothelial cells of capillaries, large arteries, and veins throughout the vasculature [28]. An additional pathway for the synthesis of AII has also been described that takes place within the vascular wall [25]. Bovine aortic endothelial cells in culture express renin and are able to synthesize AII intracellularly [26,29]. Thus, AII can be regarded as an endothelium-derived contracting factor (Fig. 3).

C. Interaction between Endothelium-Derived Relaxing and Contracting Factors

The two major endothelium-derived relaxing factors, prostacyclin (PGI_2) and EDRF(NO), cause vasodilation and inhibit platelet aggregation via different cellular mechanisms (elevation of cAMP and cGMP, respectively) (Fig. 2). They act synergistically in different biological systems, which may represent a physiologically important interaction between these two mediators [30].

Several interactions between the potent vasoconstrictor endothelin and endothelium-derived relaxing factors have been proposed (Fig. 4). Endothelin can stimulate the synthesis/release of PGI_2 and EDRF (NO) [31] by acting on ET_B endothelin receptors present in endothelial cells [32]. These vasorelaxant substances can in turn

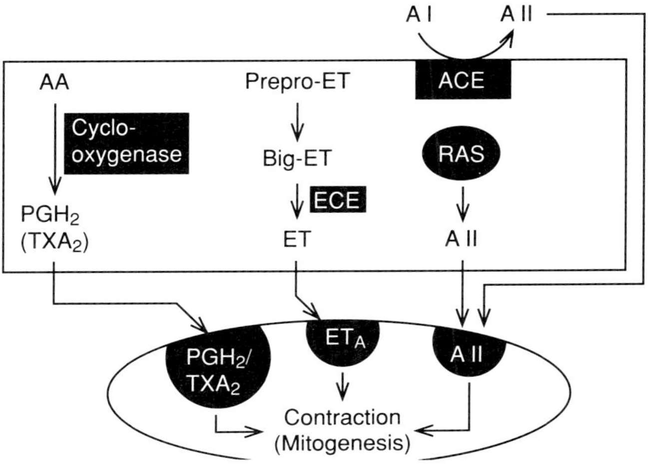

Figure 3 Endothelium-derived contracting factors (EDCFs). Endothelium-dependent contractions evoked by stretch and agonists can be prevented by inhibition of cyclooxygenase, suggesting that the mediator is probably a vasoconstrictor metabolite of arachidonic acid (TXA_2, endoperoxide). The recently discovered potent vasoconstrictor polypeptide endothelin can also be regarded as an endothelium-derived contracting factor; however, the stimuli that trigger its synthesis/release under physiological or pathological conditions are still unknown. The renin–angiotensin system (RAS) has been localized in endothelial cells, thus angiotensin II (AII) can be regarded as an EDCF. AA, arachidonic acid; TXA_2, thromboxane A_2; ET, endothelin; ET_A, ET receptor subtype on smooth muscle; ECE, endothelin-converting enzyme; ACE, angiotensin-converting enzyme.

oppose (at the level of smooth muscle) the direct vasoconstrictor action of endothelin. EDRF (NO) was shown to suppress endothelin biosynthesis in endothelial cells [33] representing an other potentially important interaction between these factors. Recent studies demonstrated that smooth muscle cells exhibit protease activity, which effectively breaks down (and inactivates) ET-1 [34]. In addition, smooth muscle cells release a still unidentified diffusible factor(s) (smooth muscle–derived inhibitory factor; SMIF) that inhibits endothelin biosynthesis in endothelial cells [34,35]. Thus, the blood vessel wall appears to have "developed" several protective mechanisms against the potent vasoconstrictor peptide endothelin.

III. Neurohumoral Substances and the Endothelium

A. *Metabolic Degradation of Vasoactive Substances by the Endothelium*

It has been known for some time that endothelial cells, particularly in the pulmonary vascular bed, have a marked capacity to degrade and/or take up vasoactive substances, such as bradykinin, norepinephrine, serotonin, and adenine nucleotides [27,36–39]. Norepinephrine and serotonin are metabolically degraded predomi-

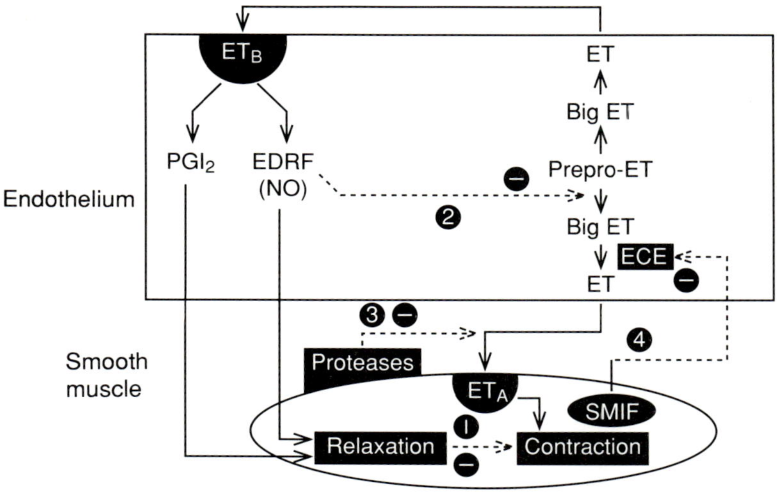

Figure 4 Modulation of endothelin (ET) biosynthesis and biological activity in the blood vessel wall. The vasoconstrictor action of ET (via acting on ET_A receptors on vascular smooth muscle) is reduced by the release of EDRF(NO) and PGI_2 by ET itself acting on ET_B receptor subtype present in endothelial cells (1). EDRF(NO) can also suppress ET synthesis in endothelial cells (2). Vascular smooth muscle cells can reduce the bioactivity (via proteases degrading ET to inactive products) (3) and biosynthesis of ET via a still unidentified factor (smooth muscle-derived inhibitory factor, SMIF) (4).

nantly by the enzymes monoamine oxidase (MAO) and to a lesser extent by catechol-ortho-methyltransferase (COMT). Endothelial cells can also convert angiotensin I to angiotensin II by the angiotensin-converting enzyme (ACE) localized on the luminal surface of these cells [27]. The same enzyme degrades bradykinin to an inactive peptide [27].

Contractions of rabbit carotid arteries with endothelium caused by adrenergic nerve stimulation remain less than those without endothelium even after blocking the action of EDRF [40,41]. Inhibition of norepinephrine disposition with a combination of cocaine, hydrocortisone, and pargyline prevents this remaining endothelial cell-mediated inhibition. Thus, the reduction in norepinephrine concentration at its postjunctional receptors by endothelial cell metabolism may be sufficient to reduce neurogenic contractions.

B. Neurohumoral Substances Stimulate the Release of Endothelium-Derived Vasoactive Factors

A growing list of substances for which a neurotransmitter role has been described was shown to have endothelium-dependent actions. These include acetylcholine (ACh), adenosine 5′-triphosphate (ATP), adenosine 5′-diphosphate (ADP), bradykinin (BK), epinephrine (E), histamine (H), norepinephrine (NE), substance P

(SP), serotonin (5-HT), and vasopressin (VP) [42–44]. Evidence of a pharmacological action has been supported by autoradiographic demonstration of specific receptors on the endothelium [45].

1. Adrenergic Neurotransmitter

In coronary arteries, exogenously added norepinephrine causes relaxation of the smooth muscle primarily by way of β-adrenoceptors [46, 47] (Fig. 5). In epicardial coronary arteries of the pig, relaxations caused by norepinephrine are slightly greater in arteries with, compared to those without, endothelium [47]. The increased relaxations to norepinephrine in arteries with endothelium are prevented by an α_2-adrenoceptor antagonist, indicating that stimulation of endothelial cell α_2-adrenoceptors increases EDRF release, leading to the greater relaxation [47]. If β-adrenoceptors are blocked, α-adrenergic coronary artery contractions (mediated by α_1-adrenoceptors in response to norepinephrine) are observed that are less in arteries with than in those without endothelium. Even in the presence of an α_2-adrenoceptor antagonist, norepinephrine-induced contractions remain markedly attenuated by the endothelium, presumably by the spontaneous release of EDRF. Thus, in coronary arteries, spontaneous as well as α_2-adrenoceptor-stimulated EDRF inhibits adrenergic contractions.

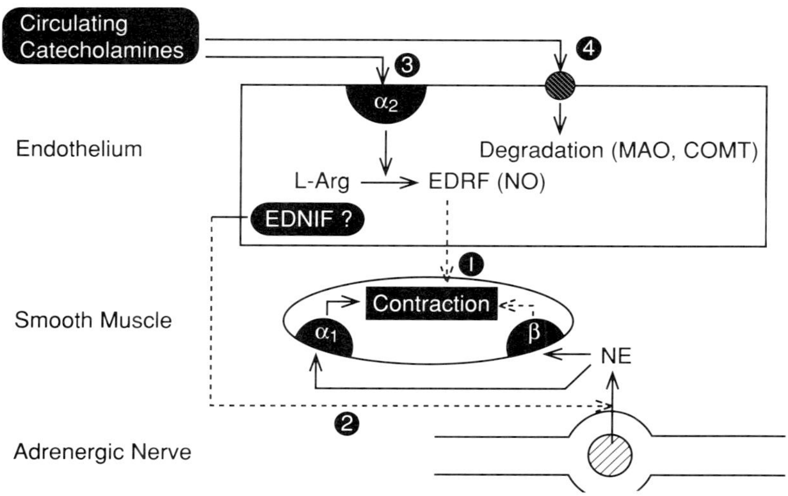

Figure 5 Mechanisms involved in suppression of adrenergic vasoconstriction by the endothelium. Basal release of EDRF(NO) decreases the reactivity of vascular smooth muscle cells to the constrictor action (via α_1-adrenergic receptor) of the adrenergic neurotransmitter norepinephrine (NE) (1). A still unidentified endothelial substance(s) (endothelium-derived nerve inhibitory factor, EDNIF) suppresses NE release from adrenergic nerve terminals (2). Circulating or neurogenic NE can act on endothelial α_2-receptors and stimulates the release of EDRF(NO) (3). Circulating catecholamines are taken up and metabolically degraded by the endothelium (4).

2. Platelet Products

Similar to several vasoactive substances, aggregating platelets cause vasoconstriction in canine, porcine, and human coronary arteries [48–51], which is inhibited by the presence of endothelium. In precontracted coronary arterial preparations, platelets evoke endothelium-dependent relaxation [48–51], which is mediated by EDRF (i.e., methylene blue and hemoglobin prevent the relaxations) and not by prostacyclin (i.e., inhibition of cyclooxygenase has no effect on the response) [51] (Fig. 6). Thus, aggregating platelets release substances that cause endothelium-independent vasoconstriction and in parallel, trigger endothelium-dependent inhibition of smooth muscle contraction.

Among the substances contained in platelets and released upon aggregation, the role of adenosine di- and triphosphate (ADP, ATP), 5-hydroxytryptamine (5-HT, serotonin), platelet-activating factor (PAF), vasopressin, and thromboxane A_2 has been analyzed (Fig. 6).

Adenosine diphosphate and ATP cause endothelium-dependent relaxation that can be inhibited by apyrase, an enzyme that hydrolyzes ADP and ATP to adenosine monophosphate [52]. The receptor on endothelial cells that mediates the response to ATP is the P_{2y} purinergic receptor subtype [52]. Platelet-induced release of EDRF is mediated mainly by this purinergic receptor in canine [49] and human coronary arteries [51].

In precontracted coronary arteries, 5-HT causes endothelium-dependent relaxation that is abolished by the 5-HT_1-serotonergic antagonists methiothepin and metergoline but not by the 5-HT_2-antagonist ketanserin [49,53]. Platelet-induced endothelium-dependent relaxation is mediated mainly by 5-HT_1-receptors in isolated porcine coronary arteries [54]. Although both PAF and vasopressin cause endothelium-dependent relaxation in isolated coronary arteries, these substances probably do not contribute to platelet-induced release of EDRF because selective antagonists of these agents (CV-3988 for PAF and $d[CH_2]_5$ Tyr[Me]AVP for vasopressin) do not affect endothelium-dependent relaxation related to aggregating platelets [55,56].

C. The Endothelium Inhibits Vascular Contraction Evoked by Adrenergic Nerve Stimulation, Aggregating Platelets, and Circulating Vasoactive Substances

Contraction of isolated blood vessels caused by adrenergic nerve stimulation can be augmented after the removal of the endothelium [40,57,58]. Vasoconstriction of perfused segments of the rabbit carotid artery caused by adrenergic nerve stimulation is also greater if the endothelium is removed or after treatment with methylene blue [59]. In many isolated blood vessels, contractions caused by aggregating platelets and platelet products [48–54] and other circulating vasocontrictor agonists, including norepinephrine [60–63], are greater after the removal of the endothelium. Thus, the formation of cGMP in smooth muscle caused by spontaneous or shear

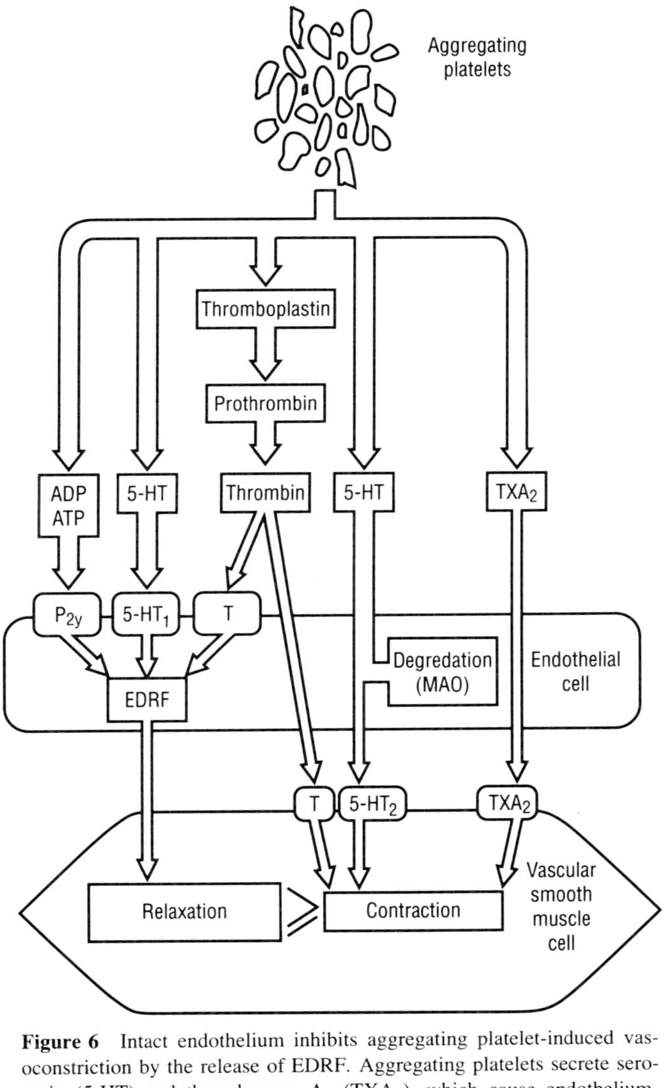

Figure 6 Intact endothelium inhibits aggregating platelet-induced vasoconstriction by the release of EDRF. Aggregating platelets secrete serotonin (5-HT) and thromboxane A_2 (TXA_2), which cause endothelium-independent contraction (*right*). These contractions are well balanced by endothelium-dependent relaxation, achieved primarily by adenine nucleotides (ADP and ATP)- and serotonin (5-HT)-induced release of EDRF by acting on endothelial receptors (P_{2y} and 5-HT_1, respectively) (*left*). Aggregating platelets can initiate the coagulation cascade with the production of thrombin, which also stimulates the release of EDRF from intact endothelium (*middle*). Serotonin is taken up by the endothelial cells and degraded by monoamine oxidase (MAO). T—thrombin receptor. (Reproduced with permission from *Current Science*.)

stress-induced release of EDRF inhibits contractions evoked by these vasoactive substances.

D. Endothelium-Derived Vasoactive Factors Modulate Adrenergic Neurotransmission in Blood Vessels

1. Relaxing Factors

Evidence for the modulation of neurotransmission by factors released from the endothelium comes from the observation that contractions and release of tritriated norepinephrine evoked by electrical stimulation in mammalian arteries are inhibited by the endothelial cells [41]. For example, the release of radiolabeled norepinephrine from dog pulmonary artery and vein is less in segments with endothelium compared to those without endothelium [58].

The identity of substances synthesized and released by the endothelium having prejunctional actions is still unknown, but the involvement of endothelium-derived relaxing and contracting factors was postulated (Fig. 5). Prostaglandins are released from the endothelium, and endogenous prostaglandins have also been shown to inhibit [64] or facilitate (prostacyclin) (G. M. Rubanyi, unpublished observations) norepinephrine release from adrenergic nerve terminals. However, it is unlikely that prostaglandings formed by the endothelium of rabbit carotid artery inhibit norepinephrine release because blockade of prostaglandin synthesis with indomethacin fails to influence neurogenic contractions or the release of norepinephrine [40,65].

The inhibitory factor(s) is also unlikely to be identical with EDRF(NO). For example, methylene blue does not prevent the inhibition of norepinephrine release by the endothelium, suggesting that its mechanism is not via a rise in cGMP in the nerve endings [65]. Maximally relaxing concentrations of nitric oxide have no effect on norepinephrine release from the rabbit cortid artery [41], indicating that endothelial cell-derived nitric oxide is unlikely to affect nerve endings. These findings suggest that the existence of a still unidentified "endothelium-derived nerve inhibitory factor" (EDNIF) with a mechanism of action and chemical nature different from EDRF (Fig. 5).

The putative endothelium-derived hyperpolarizing factor (EDHF) may also inhibit norepinephrine release from nerve endings. However, the observation that the endothelium inhibits norepinephrine release during depolarization with 80 mM potassium chloride argues against this possibility [60].

Several other substances, including acetylcholine, serotonin, and adenine nucleotides, can inhibit norepinephrine release that has been shown to be synthesized by some endothelial cells [66–68] (see Section III.E). However, evidence is still lacking that they are produced by endothelial cells in sufficient concentrations to have a physiological prejunctional action.

2. Contracting Factors

Vasoactive peptides synthesized by the endothelium could influence norepinephrine release. Angiotension II facilitates the release of norepinephrine from adrenergic nerves [69], making it unlikely that the peptide plays a role as an endothelium-derived nerve inhibitory factor. The potent vasoconstrictor peptide endothelin has been shown to inhibit norepinephrine release from guinea pig femoral artery [70]. However, in the rabbit carotid artery, a maximal contractile concentration of endothelin failed to influence norepinephrine release [41], indicating that in this artery it is unlikely that endothelium-derived endothelin has a prejunctional action.

E. Neurohumoral Substances in the Endothelium

Although endothelium-dependent responses to many neurotransmitter and neurosecretory substances have been extensively documented, the origin of many of these neurohumoral substances is still unknown. The structure of most of the blood vessels makes it unlikely that the source of substances acting on the endothelium is the perivascular nerves (although evidences presented in Sections III.A and III.B argue for such a possibility in some blood vessels). Activated and aggregating platelets in the vicinity of endothelial cells can release 5-HT and ATP (see Sections III.A,B), but they cannot be the source of other substances (e.g., ACh, BK, and SP) since their rapid breakdown keeps the plasma concentration of them very low. Several lines of evidence suggest (primarily provided by the laboratory of G. Burnstock) that for some of these substances the source is the endothelial cell itself [71].

1. Localization

Choline acetyltransferase (ChAT), the enzyme responsible for the synthesis of ACh, was shown to be present in the endothelial cells of small vessels and capillaries in the rat cortex [66]. Endothelial cells in pial vessels have been able to take up choline [72], and endothelial cells in microvessels of the rat cortex can synthesize ACh [73,74].

Substance P (SP) has been measured in freshly isolated bovine aortic endothelial cells [75]. In bovine cerebral microvessel endothelial cells, SP was still detectable in cultured cells [75].

Adenosine, produced by the breakdown of adenine nucleotides, is rapidly taken up by endothelial cells [35,76]. It has been shown that adenosine taken up by porcine aortic endothelial cells in culture is rapidly synthesized into adenine nucleotides, mainly ATP [76].

Vasopressin immunoreactivity can be localized in the cytoplasm of endothelial cells in the rat mesenteric and renal arteries [77]. In both of these vessels, about 10% of the endothelial cells were immunoreactive for vasopressin.

Synthesis of histamine has been demonstrated within endothelial cells. Histidine decarboxylase, the enzyme responsible for the synthesis of histamine, is present in most tissues [78]. It was proposed that histamine synthesized in the microcirculation was a primary mechanism in intrinsic microcirculatory regulation [79].

The kallikrein–bradykinin synthetic pathway has been shown recently to be present in endothelial cells [80]. This finding suggests that bradykinin, together with the other substances listed above, which are potent agonists of prostacyclin and

EDRF(NO) synthesis/release from endothelial cells [81], may be regarded as autocrine/paracrine local mediators.

2. Release by Increased Flow and Hypoxia

The neurohumoral substances ACh, ATP, BK, 5-HT, and SP were shown to be released from endothelial cells in response to stimulation with increased flow or hypoxia.

In the perfused rat hindlimb preparation, increased flow evokes the release of SP [82], which can be prevented by the removal of endothelial cells following air treatment. Similarly, in the isolated perfused rat mesenteric arterial bed, the release of SP during high flow was significant although variable between preparations. In contrast, ATP was released during increased flow. After the removal of the endothelium by sodium deoxycholate, increased flow caused no significant release of ATP, suggesting that it was released from endothelial cells.

Flow-induced endothelium-dependent vasodilation can be prevented by a selective antagonist of bradykinin receptor (R. Busse, personal communication) and by inhibitors of EDRF(NO) [83], suggesting that increased flow stimulates the release of BK from endothelial cells, which in turn triggers the release of EDRF in an autocrine/paracrine fashion.

Hypoxia triggers the release of ATP from isolated guinea pig heart preparations [84]. In the rat heart, 5-HT has also been shown to be released during the initial rapid phase of hypoxic vasodilatation [67]. Similarly, hypoxia causes the release of SP and ACh from the rat heart [85].

Thus, there is increasing evidence that endothelial cells contain and under certain conditions (e.g., increased flow or hypoxia) release a variety of vasoactive substances that can act via endothelial receptors to modify vascular tone. It has also been shown that the endothelium is capable of synthesizing its own stores of these substances. With regard to the endothelial content and release of these vasoactive agents, there is considerable heterogeneity both within and between different vascular beds. It is also uncertain, whether under physiological conditions, that this mechanism plays a significant role in the modulation of vascular tone.

IV. Physiological Significance

A. Vascular Resistance and Blood Pressure

Treatment of animals with analogs of L-arginine that inhibit NO-synthase causes significant elevation of systemic arterial blood pressure [86]. In contrast, cyclooxygenase inhibitors (e.g., aspirin, indomethacin, etc.) have no significant effect on blood pressure, ruling out prostacyclin as an important modulator of vascular smooth muscle tone *in vivo*. These findings indicate that endothelial cells continuously produce EDRF (NO), which exerts a tonic vasodilator influence on resistance vessels *in vivo*. As discussed earlier, the decrease of vascular tone by EDRF (NO) can be the result of direct action on smooth muscle combined with inhibition of adrenergic vasoconstriction and suppression of vasoactivity of circulating substances (Fig. 1). Inhibition by the endothelium of adrenergic neurogenic contractions provides a mechanism for the local environment of the blood vessel to modulate adrenergic input. In addition, continuous synthesis of EDRF(NO) can effectively antagonize vasoconstriction triggered by circulating (e.g., vasopressin) and locally produced (e.g., platelet-derived substances, AII, endothelin) agents. It also plays a role in suppressing myogenic (pressure-induced) vascular tone. Furthermore, EDRF(NO) together with PGI_2 contribute in a significant way to the physiological "nonthrombogenic" property of the vessel wall. This balance between vasodilator and vasoconstrictor influences can be disturbed by acute (evoked by rapid changes in flow, pressure, and pO_2) and chronic changes in endothelial synthesis/release of EDRF(NO). Acute or chronic decrease in EDRF(NO) synthesis (endothelial dysfunction) can shift this physiological balance and can lead to local (vasospasm) or generalized (hypertension) increase in vascular tone and thrombosis (see also Section V).

B. Flow-Induced Vasodilation and the Endothelium

Increments in blood flow velocity or blood viscosity through arteries *in situ* or *in vitro* induce vasodilatation that can be abolished by the removal of the endothelium [83,87–90]. These observations suggest that endothelial cells release vasodilator mediator(s) in response to increased shear stress. An increase in steady flow rate or the introduction of pulsatile flow enhances the release of prostacyclin from endothelial cells in culture [91,92] and from the endothelium of perfused arteries [88,91]. However, inhibitors of cyclooxygenase did not affect flow-induced vasodilation [83,93]. Therefore, it was postulated that increased shear stress stimulates the release of a relaxing mediator from endothelial cells, which is not prostacyclin. Experiments using the bioassay technique developed for the detection of endothelium-derived relaxing factor [94] showed that a sudden increase in steady flow rate or the introduction of pulsatile flow (without change in mean flow rate) increased the release of PGI_2 and induced relaxations of the bioassay ring. Indomethacin significantly depressed the release of PGI_2 during increases in flow but had no significant effect on the relaxing activity of the effluent. Superoxide dismutase restored the relaxing activity of the effluent during increases in flow. These data demonstrate that in addition to PGI_2, flow triggers the release of another relaxing substance (or substances) from vascular endothelial cells that has characteristics similar to the endothelium-derived relaxing factor released by acetylcholine (Fig. 7). The mechanism of shear stress-sensor function may involve activation of a K^+-channel in endothelial cell membrane [101]. Increased K^+-outflow hyperpolarizes the cell membrane that enhances Ca^{2+}-influx and triggers synthesis/release of EDRF(NO) [101] (Fig. 7).

The demonstration that shear stress may influence the release of neurohumoral substances endogenous to the endothelium, for example, BK, SP, and ATP (see Section III.E), may also play a role in vascular responses to changes in flow. Since it has been shown that EDRF participates in this response (see above), it is possible

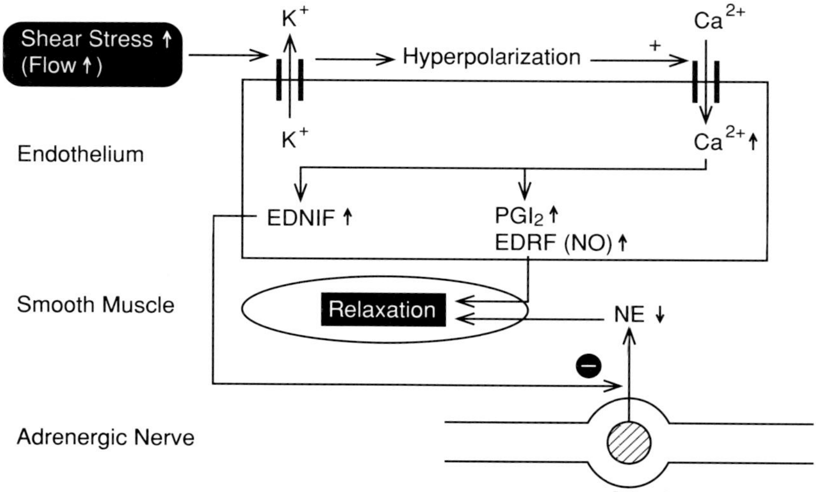

Figure 7 Flow-induced endothelium-dependent vasodilation. By virtue of releasing vasoactive substances, endothelial cells can transduce changes in these physical forces into changes in vascular tone. Shear force (increases in blood flow) stimulates the release of PGI_2 and EDRF(NO) and causes vasodilation. Increase in shear stress hyperpolarizes the membrane of endothelial cells by activation of the inward rectifying K^+-channel. Membrane hyperpolarization increases the driving force of Ca^{2+}-entry, resulting in elevation of free cytoplasmic calcium ($[Ca^{2+}]_i$) and stimulated synthesis/release of EDRF.

that a neurohumoral substance, such as BK, may be involved in a primary step of the mechanism. Increased flow may thus trigger an initial release of BK, which would subsequently act on endothelial cell receptors locally or downstream to cause vasodilatation via EDRF. Indeed, not only inhibitors of EDRF (NO) but also BK receptor antagonists were shown to selectively prevent flow-induced vasodilation (R. Busse, personal communication). While this mechanism is described in terms of BK, it is likely to be a general one applicable to any of the other neurohumoral substances localized within endothelial cells and capable of inducing EDRF production. For example, flow-induced endothelium-dependent relaxation in some, but not all, isolated canine femoral artery segments tested can be inhibited by pirenzipine, a muscarinic receptor antagonist, suggesting the involvement of locally produced muscarinic receptor agonists (e.g., ACh) in the response (Rubanyi, unpublished observation).

Although these studies have specifically addressed the role of endothelial cells in the regulation of blood vessel tone in response to changes in blood flow, it is important to remember that the situation *in vivo* is far more complex. Because the release of EDRF is responsive to shear stress, one might expect that when vascular resistance falls, as in response to metabolic demand, increased blood flow dilates the conductance vessels not only by a direct action on the smooth muscle but also by inhibiting vasoconstrictor nerves (see Fig. 5). An increase in perfusate viscosity resulting in increased shear stress inhibited neurogenic vasocontriction only if the endothelium was present [95]. This was attributed to a shear stress-induced release of a still unidentified factor (termed endothelium-derived nerve inhibitory factor, EDNIF), as demonstrated in bioassay experiments. Thus, changes in shear stress may affect neurogenic vasoconstriction by modulating EDRF release. This demonstration of a modulatory effect of shear stress on adrenergic vasoconstriction emphasizes the fact that vessel tone is the result of effects exerted at both the adventitia and the intima [95].

C. Role of Endothelium in Pressure-Induced Vasoconstriction: Myogenic Autoregulation of Blood Flow

Rapid stretch of canine basilar artery rings [96] and sudden increase of transmural pressure in canine carotid arteries [97] increase wall tension, which could be prevented by mechanical removal of the endothelium [96,97].

Elevation of pressure in isolated segments of cerebral arteries from cats [98] depolarizes the vascular smooth muscle cells, and the internal diameter of the vessel is either maintained or reduced. When the endothelial lining of cerebral vessels is disrupted (either by collagenase perfusion or introduction of an air bolus), the vessels dilate as transmural pressure is elevated and membrane potential does not change [98].

Bioassay studies were used to determine whether increases in pressure trigger the release of diffusable factor(s) from the endothelium of canine carotid arteries. These studies revealed that, in parallel with the pressure-induced contraction of the donor carotid artery segment, the bioassay canine coronary artery ring (not exposed to pressure and superfused by the perfusate from the donor segment) contracted as well [97]. These results indicated that increases in transmural pressure modulate the release of diffusible and bioassayable vasoactive factor(s) from the endothelium of carotid arteries.

Similar bioassay studies showed that pressure-induced endothelium-dependent contraction and membrane depolarization in feline cerebral arteries are also mediated by diffusible factor(s) [99]. There appears to be a difference between these two preparations as far as the nature of the diffusible mediator(s) is concerned. In contrast to canine carotid arteries, in feline cerebral arteries inhibitors of EDRF (e.g., hemoglobin) have no effect on the phenomenon [99]. It is postulated that a vasoconstrictor cytochrome P450 metabolite of arachidonic acid mediates pressure-induced vasoconstriction in this preparation [100]. Stretch-induced contraction in canine basilar arteries appears to be mediated by cyclooxygenase product (PGH_2/TXA_2) of arachidonic acid [96]. Either suppression of production of EDRF (NO) or augmented release of EDCF enhances vascular smooth muscle tone and reactivity to neural and humoral vasoconstrictor stimuli.

Thus, a sudden increase in transmural pressure can lead to vasoconstriction via several mechanisms (Fig. 8): (1) direct activation of smooth muscle cells (so-called myogenic response); (2) suppression of EDRF(NO) synthesis/release via (i) reducing shear stress-induced release of EDRF [97], (ii) generation of free radicals that

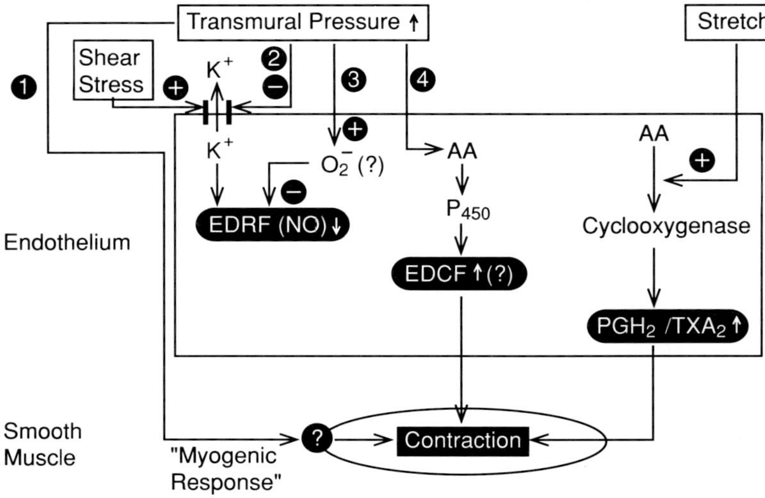

Figure 8 Mechanoreception by the endothelium. The endothelial cell (EC) serves as a unique sensor of transmural pressure and rapid stretch. Increase in transmural pressure evokes vasoconstruction via direct activation of smooth muscle, suppression of the synthesis, or destruction of EDRF (e.g., canine carotid artery), and facilitation of the release of a still unidentified EDCF (e.g., feline cerebral artery). Rapid stretch of canine basilar arteries causes endothelium-dependent contraction by the release of a cyclooxygenase product of arachidonic acid (AA) (PGH_2/TXA_2). Elevated transmural pressure decreases Ca^{2+}-entry and suppresses synthesis/release of EDRF, presumably by inactivation of inward rectifying K^+-channel.

inactive EDRF(NO) [3], and (iii) inhibition of a K^+-current, which leads to a decrease of intracellular Ca^{2+} [101]; and (3) stimulation of synthesis/release of EDCFs (PGH_2 and cytochrome P_{450} metabolite of arachidonic acid) [96,100].

D. Metabolic Vasodilation: Potential Role of Endothelium in pO_2-, pH-, and K^+-Induced Changes in Vascular Tone

Increased tissue metabolic demands (e.g., exercise) or decreased supply of nutrients or oxygen (e.g., hypoxia, ischemia) triggers potent vasodilator mechanisms. Several factors were proposed to mediate this "metabolic vasodilation," including adenosine [102], and changes in partial pressure of oxygen (pO_2), hydrogen ion concentration (pH), and potassium ion concentration (K^+) [103]. The potential involvement of the endothelium in this important physiological autoregulatory mechanism will be briefly discussed.

1. Oxygen

a. HYPOXIA. Changes in the partial pressure of oxygen (pO_2) of arterial blood are potent stimuli for changes in vessel caliber (vascular smooth muscle tone) in a number of organs [104] (Fig. 9). While hypoxia generally causes relaxation of most vascular beds, the response of the pulmonary vascular bed [105,106] and of several isolated rat, canine, porcine, and monkey arteries [107–111] is vasoconstriction. Hypoxic vasoconstriction in the pulmonary circulation serves the important physiologic function of excluding poorly ventilated parts of the lung from the circulation.

Endothelial cells are ideally situated to function as sensors of changes in oxygen content of the blood. Busse *et al.* showed that intraluminal hypoxia of rat tail and femoral arteries elicited a dilatation that was abolished after the removal of the endothelium [112].

Oxygen appears to be crucial to the production of EDRF since it has been demonstrated that, under anoxic conditions, EDRF is no longer produced [111].

In isolated precontracted rings of canine femoral and coronary arteries [110,111] and rat aorta [113], severe hypoxia (oxygen pressure below 5 mmHg) induced transient relaxations followed by contractions only in preparations with endothelium. Bioassay studies indicated that both the hypoxic relaxation and contraction are mediated by diffusable factor(s) released from the endothelium [110,112]. Indomethacin (a cyclooxygenase inhibitor) had no effect on these responses, ruling out the involvement of vasodilator prostanoids [110]. Methylene blue and LY 83583 (inhibitors of soluble guanylate cyclase) and nitro-L-arginine (an inhibitor of nitric oxide synthase) inhibited both hypoxia-induced relaxations and contractions [113]. In rings without endothelium, exogenous donors of nitric oxide (SIN-1 and sodium nitroprusside) and dibutyril cGMP unmasked hypoxic contractions but not hypoxic relaxations [114]. Hypoxia significantly decreased cGMP levels only in vascular preparations with endothelium [114]. These results suggest that hypoxic relaxation is due to the release of EDRF (NO) from endothelium whereas hypoxic contraction

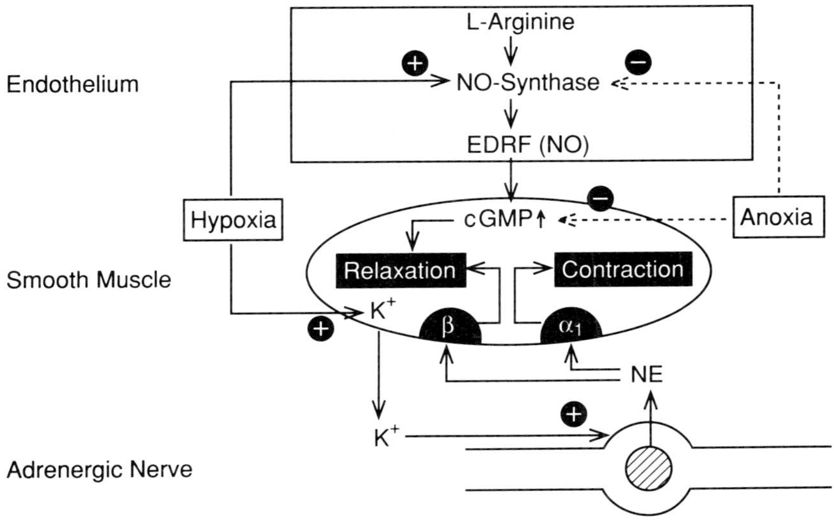

Figure 9 Contribution of endothelial (*top*) and neurogenic (*bottom*) control mechanisms to hypoxia (*left*) and anoxia (*right*) induced changes in vascular tone. (See text for details.)

is the consequence of inhibition of the effect of endothelium-derived nitric oxide in vascular smooth muscle. In contrast, in porcine and bovine coronary arteries, hypoxia-induced contractions can be inhibited by cyclooxygenase blockers, suggesting that vasoconstrictor archidonic acid metabolites mediate the response in these preparations [109].

It has been suggested that ATP and 5-HT may be released in response to hypoxia from endothelial cells [67,115]. These substances may also contribute to the stimulated release of EDRF (NO) during hypoxia.

Hypoxia can also change smooth muscle tone by modulating adrenergic neurotransmission via the release of K^+ in certain vascular preparation [107].

b. HYPEROXIA AND REOXYGENATION. Elevated pO_2, either during hyperoxic conditions usually found in most organ chamber experiments (95 to 100% O_2) or during reoxygenation of tissues after a period of hypoxia/ischemia, can also have significant effects on vascular tone. Both conditions are associated with excessive generation of oxygen-derived free radicals [116]. Superoxide anion production elevates vascular tone by virtue of inactivating EDRF (NO) [3]. Hydrogen peroxide causes vasodilation either by stimulating EDRF(NO) production [7] or by stimulating soluble guanylate cyclase in vascular smooth muscle [117]. Reoxygenation after a short period of hypoxia triggers vasoconstriction in porcine and canine coronary veins, which is mediated by cyclooxygenase product(s) of arachidonic acid [118].

2. Hydrogen Ions

In spite of its potential importance, very little is known on the effect of changes in pH on endothelial cells. This may be particularly relevant for the skeletal muscle since exercise is often accompanied by a local accumulation of lactic acid. The effect of hypercapnia (increased CO_2), which causes vasodilation primarily of cerebral blood vessels, is likely to be dependent on changes in the hydrogen ion concentration. The increase in cerebral blood flow due to hypercapnia has been shown to be attenuated by indomethacin [119]. Since endothelial cells are known sources of prostaglandins [30], they may be involved in the response. Hypercapnic acidosis has also been shown to increase coronary blood flow, and the involvement of mediators other than adenosine was suggested by the fact that this was accompanied by only a transient increase in cardiac adenosine production [120].

3. Potassium Ions

Recent bioassay studies on canine femoral arteries demonstrated, that in addition to direct action on vascular smooth muscle [121] and changes of the release of adrenergic neurotransmitter [107], elevation in extracellular potassium on concentration ($[K^+]$) can modulate blood vessel tone via the endothelium [122] (Fig. 10). These studies showed that the bioassay rings progressively relaxed when increasing concentrations of KCl (10–40 mM) were added extraluminally to the organ bath where the perfused donor segment was mounted. That the relaxing mediator is released from endothelial cells and not from other constituents of the vessel wall

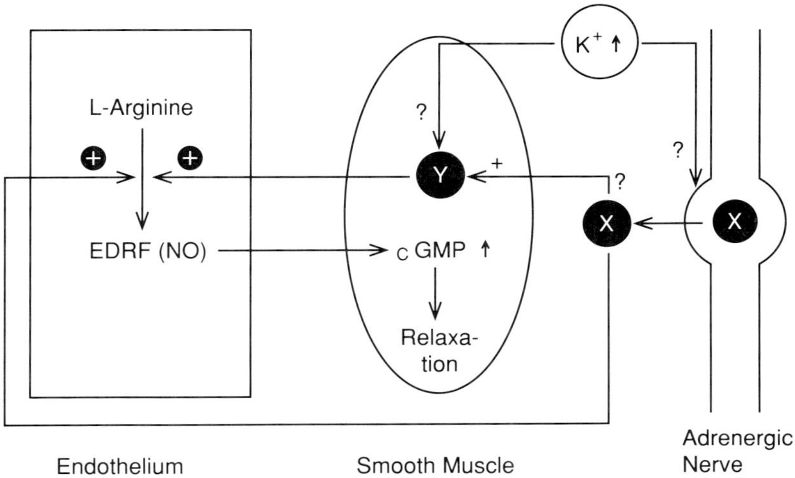

Figure 10 Contribution of endothelial (*left*) and neurogenic (*right*) control mechanisms to potassium ion (K^+)-induced endothelium-dependent vasorelaxation. X and Y: putative endogenous mediators released from adrenergic nerves and smooth muscle cells by K^+, respectively, which stimulate the synthesis/release of EDRF(NO) from endothelial cells. (See text for further details.)

was demonstrated by the prevention of relaxations by the removal of the endothelium from the femoral arteries. Because the relaxations were obtained in the presence of indomethacin, they cannot be mediated by vasodilator prostaglandins. Similar to EDRF released by acetylcholine [13], prolongation of the transit time from 1 to 6 s significantly depressed the biological activity of the endothelial relaxing factor released by K^+. Superoxide dismutase restored the biological activity of the mediator, probably by preventing its destruction during the prolonged transit, which is similar to that observed with EDRF [3]. Therefore, it can be postulated that K^+ and acetylcholine stimulate the release of a similar labile nonprostaglandin relaxing factor from the endothelium of canine femoral arteries. It is unlikely that elevation of extracellular K^+ concentration stimulates the release of the relaxing mediator by a direct action on endothelial cells because the intraluminal infusion of the ion did not evoke relaxations [122]. It is likely that K^+ stimulates the release of a substance (or substances) in the vessel wall that diffuses to the endothelial cells and triggers the release of EDRF. The relatively long latency period between the application of KCl and the onset of relaxation may underly this assumption. Although the identity of this endogenous substance(s) is still unknown, the potential involvement of acetylcholine, norepinephrine, adenosine diphosphate and triphosphate, and products of cyclooxygenase were ruled out by the use of pharmacological antagonists [122]. Since K^+ triggered the release of a relaxing factor from the endothelium only when added extraluminally, the source of the mediator substance is presumably a structure(s) at or near the adventitial site of the vessel wall (e.g., perivascular nerve endings, smooth muscle) [122] (Fig. 10).

V. Pathological Consequences of Endothelial Dysfunction

Endothelial dysfunction (characterized by an imbalance between the production of prostacyclin and thromboxane A_2, EDRFs, EDCFs, and pro- and anticoagulation factors) could lead to hyperreactivity of underlying vascular smooth muscle and pathologically elevated vascular tone (vasospasm), and to activation, adhesion, and aggregation of platelets (thrombosis). Some of these potential consequences have been demonstrated experimentally in a variety of animal models of endothelial injury and dysfunction and also in human coronary artery disease. In contrast to the sudden changes in endothelial function during increased flow and pressure or changes in pO_2, pH, or $[K^+]_o$ (see above), various risk factors (e.g., hypertension, diabetes, elevated cholesterol) cause chronic alteration in endothelial function with several cardiovascular consequences (vasospasm, thrombosis, atherosclerosis).

It is also possible that vascular diseases can interfere with normal endothelial cell–neuroeffector interactions. Following a 12-week high-cholesterol diet, the adrenergic neurogenic contractions of the rabbit carotid artery were increased in rings with endothelium but not without endothelium [123]. Endothelial cells remained intact, and no atherosclerotic intimal thickening occurred. These studies suggest that hypercholesterolemia impairs endothelial cell–neuroeffector interactions selectively. It is also possible that by adversely affecting endothelial cell function, such diseases as diabetes mellitus or hypertension could result in exaggerated adrenergic contractions.

Pathological changes in endothelial cells have been implicated in a wide variety of cardiovascular disorders. These include atherosclerosis, diabetic vasculopathy, tumor angiogenesis, thrombosis, hypertension, coronary artery spasm, cerebral vasospasm, and migraine [81,124]. As a general rule, endothelium-dependent responses to vasodilator substances are diminished and responses to vasoconstrictor substances are increased in clinical and experimental models of hypertension [125,126], hypercholestereamia and atherosclerosis [127–131], and diabetes [132–136].

In a study of porcine coronary arteries with endothelium that had regenerated 4 weeks after its removal, impaired responses to 5-HT and aggregating platelets were reported. Responses to BK, adenosine diphosphate, and thrombin were unaltered [50]. After longer time intervals of 8 and 16 weeks, responses of previously denuded arteries with regenerated endothelium to all of these agents were depressed [137]. Thus, endothelium-dependent relaxations progressively worsen after regeneration of the endothelium. Evidence has been presented that dysfunction of a pertussis-toxin-sensitive G protein may account in part for dysfunction of chronically regenerated endothelium [137]. As has been discussed previously, the expression of neurohumoral substances within endothelial cells may be altered during culture. Similarly, endothelial cells that regenerate after injury or removal may not recover all their original vasoactive components. This has implications on the long-term control of vascular tone following percutaneous angioplasty.

Under normal conditions, in the adult, the endothelial cells of both capillaries and larger blood vessels are a stable population with very low mitotic activity [138]. However, in pathological states, such as hyperlipidemia, hypertension, and mechanical stress, the turnover of endothelial cells is increased. Further, capillary growth, of which endothelial cell proliferation is the initiating process, can occur as a result of exercise, hypoxia, and low pH, all of which stimulate vasodilatation and increased blood flow [138]. The factors involved in capillary growth have been divided into two groups; (1) mechanical changes in flow and/or blood pressure and (2) local chemical factors. It has been suggested that chemical factors may act by the changes they cause in blood flow or permeability rather than by trophic actions on the endothelial cells per se [138]. Damage of endothelial cells by increased shear stress has been implicated in the induction of endothelial cell proliferation in the aorta [139] and in culture [140]. It is interesting to note that there is evidence that increased flow, hypoxia, and low pH may all stimulate the release of a variety of substances from the endothelium. Thus, the two groups of factors considered above may, in fact, be interdependent. Increased blood flow, often mediated by an endothelium-dependent process, may in turn cause the further release of endothelial agents, which, under chronic conditions, could promote endothelial proliferation.

VI. Summary and Conclusions

Modulation of intrinsic vascular tone by neural and humoral mechanisms has been well studied and documented. The newly discovered role of endothelium in the control of vascular tone added a third, equally important regulatory pathway. This chapter reviewed experimental observations in the last years arguing for a potential interaction between these three regulatory mechanisms. It is by now clearly established that vascular tone is the result of effects exerted both at the adventitia (neural control) and at the intima (humoral agents and endothelium-derived vasoactive factors). Although several forms of interaction (see Fig. 1) were documented in certain vascular preparations, their role in the physiological regulation of tissue perfusion or their contribution to pathophysiological conditions remains to be determined. The endothelium appears to exert tonic suppression of vascular reactivity under physiological conditions both by directly affecting vascular smooth muscle tone and by suppressing adrenergic neurotransmission and inhibiting the release of vasoactive products from platelets. The endothelium also contributes in a significant way to adjustment of vascular tone to the changing physiological environment. It plays an important role in sensing and transducing changes in hemodynamic forces (shear stress and pressure) and also contributes to metabolic autoregulation via responding with altered synthesis release of vasoactive factors to change in pO_2 and K^+. Disturbance of this delicate balance in various cardiovascular diseases leads to localized (spasm) or generalized (hypertension) increase in vascular reactivity and tone. Endothelial dysfunction also plays a pivotal role in the pathogenesis of atherosclerosis. A better understanding of the complexities of local regulation of vascular function by these three mechanisms in health and disease is a challenge for future investigations. The rewards of such efforts will certainly be the unraveling of the many mysteries of cardiovascular homeostasis both under normal conditions and in pathological processes.

References

1. Moncada, S., and Vane, J. R. (1979). Pharmacology and endogenous roles of prostaglandin endoperoxides, thromboxane A_2, and prostacyclin. *Pharmacol. Rev. 30,* 293–331.

2. Furchgott, R. F., and Zawadski, J. V. (1980). The obligatory role of endothelial cells in the relaxation of arterial smooth muscle by acetylcholine. *Nature 288,* 373–376.

3. Rubanyi, G. M., and Vanhoutte, P. M. (1986). Superoxide anions and hyperoxia inactivate endothelium-derived relaxing factor. *Am. J. Physiol. 250,* H822–H827.

4. Vanhoutte, P. M., Rubanyi, G. M., Miller, V. M., and Houston, D. S. (1986). Modulation of vascular smooth muscle contraction by the endothelium. *Ann. Rev. Physiol. 48,* 307–320.

5. Rapoport, R. M., Draznin, M. B., and Murad, F. (1983). Endothelium-dependent relaxation in rat aorta may be mediated through cyclic GMP-dependent protein phosphorylation. *Nature 306,* 174–176.

6. Martin, W., Villani, G. M., Jothianandan, D., and Furchgott, R. F. (1985). Selective blockade of endothelium-dependent and glyceryl trinitrate-induced relaxation by hemoglobin and by methylene blue in the rabbit aorta. *J. Pharmacol. Exp. Ther. 232,* 708–716.

7. Rubanyi, G. M., and Vanhoutte, P. M. (1986). Oxygen-derived free radicals endothelium and responsiveness of vascular smooth muscle. *Am. J. Physiol. 250,* H815–H821.

8. Furchgott, R. F. (1988). Studies on relaxation of rabbit aorta by sodium nitrite. The basis for the proposal that the acid-activatable inhibitory factor from bovine retractor penis is inorganic nitrite and the endothelium-derived relaxing factor is nitric oxide. *In* "Mechanisms of Vasodilatation." (P. M. Vanhoutte, ed.), pp. 401–414. Raven Press, New York.

9. Ignarro, L. J., Byrns, R. E., and Wood, K. S. (1988). Biochemical and pharmacological properties of endothelium-derived relaxing factor and its similarity to nitric oxide radical. *In* "Mechanisms of Vasodilatation." (P. M. Vanhoutte, ed.), pp. 427–435. Raven Press, New York.

10. Palmer, R. M., Ferrige, A. G., and Moncada, S. (1987). Nitric oxide release accounts for the biological activity of endothelium-derived relaxing factor. *Nature 327,* 524–526.

11. Rubanyi, G. M., Greenberg, S. S., and Wilcox, D. E. (1990). Endothelium-derived relaxing factor cannot be identified as free nitric oxide by electron paramagnetic resonance spectorscopy. *In* "Endothelium-Derived Relaxing Factors." (G. M. Rubanyi, P. M. Vanhoutte, eds.), pp. 32–38. Karger, Basel.

12. Myers, P. R., Minor, R. L. Jr., Guerra, R. Jr., Bates, J. N., and Harrison, D. G. (1990). Vasorelaxant properties of the endothelium-derived relaxing factor more closely resemble S-introscysteine than nitric oxide. *Nature 345,* 161–163.

13. Rubanyi, G. M., Johns, A., Wilcox, D., Bates, F. N., and Harrison, D. (1991). Evidence that a S-nitrosothiol, but not nitric oxide, may be identical with endothelium-derived relaxing factor. *J. Cardiovascular Pharmacol. 17,* S41–S45.

14. Palmer, R. M., Ashton, D. S., and Moncada, S. (1988). Vascular endothelial cells synthesize nitric oxide from L-arginine. *Nature 333,* 664–666.

15. Schmidt, H. H. W., Pollock, J. S., Nakane, M., Gorsky, L. D. Förstermann, U., and Murad, F. (1991). Purification of a soluble isoform of guanylyl cyclase-activating-factor synthase. *Proc. Natl. Acad. Sci. U.S.A. 88,* 365–369.

16. Förstermann, U., Gorsky, L. D., Pollock, J. S., Schmidt, H. H. W., Heller, M., and Murad, F. (1990). Regional distribution of EDRF/NO-synthesizing enzyme(s) in rat brain. *Biochem. and Biophysical Res. Comm. 168,* 727–732.

17. Rubanyi, G. M., and Vanhoutte, P. M. (1987). Nature of endothelium-derived relaxing factor: Are there two relaxing mediators? *Circulation Res. 61,* 1161–1167.

18. Rubanyi, G. M., McKinney, M., and Vanhoutte, P. M. (1987). Biphasic release of endothelium-derived relaxing factor(s) by acetylcholine from perfused canine femoral arteries. Characterization of muscarinic receptors. *J. Pharmacol. Exp. Ther. 240,* 802–808.

19. Suzuki, H., and Chen, G. (1990). Endothelium-derived hyperpolarizing factor EDHF an endogenous potassium-channel activator. *NIPS 5,* 212–215.

20. DeMey, J. G., and Vanhoutte, P. M. (1982). Heterogenous behavior of the canine arterial and venous wall. Importance of the endothelium. *Circulation Res. 51,* 439–447.

21. Rubanyi, G. M. (1988). Endothelium-derived vasoconstrictor factors. *In* "Endothelial cells." (U. Ryan, ed.), Vol. II, pp. 61–74. CRC Press, Boca Raton, FL.

22. Hickey, K. A., Rubanyi, G. M., Paul, R. J., and Highsmith, R. F. (1985). Characterization of a coronary vasoconstrictor produced by cultured endothelial cells. *Am. J. Physiol. 248,* C550–C556.

23. Yanagisawa, M., Kurhara, H., Kimura, S., Tomobe, Y., Kobayashi, M., Mitsui, Y., Yazaki, Y., Goto, K., and Masaki, T. (1988). A novel potent vasoconstrictor peptide produced by vascular endothelial cells. *Nature 332,* 411–415.

24. Rubanyi, G. M. (1992). "Endothelin." Oxford University Press, New York.

25. Dzau, V. J. (1984). Vascular wall renin angiotensin pathway in control of the circulation. *Am. J. Med. 77,* 31–36.

26. Kifor, I., and Dzau, V. J. (1987). Endothelial renin-angiotension pathway: Evidence for intracellular synthesis and secretion of angiotensin. *Circulation Res. 60,* 422–428.

27. Ryan, J. W. (1984). The metabolism of angiotensin I and bradykinin by endothelial cells. *In* "Biology of Endothelial Cells." (E. A. Jaffe ed.), pp. 317–329. Martinus Nijhoff, Boston.

28. Caldwell, P. R. B., Seegal, B. C., Hsu, K. C., Das, M., and Soffer, R. L. (1976). Angiotensin-converting enzyme: Vascular endothelial localization. *Science 191,* 1050–1051.

29. Lilly, L. S., Pratt, R. E., Alexander, R. W., Larson, D. M., Ellison, K. E., Gimbrone, M. A. Jr., and Dzau, V. J. (1985). Renin expression by vascular endothelial cells in culture. *Circulation Res. 47,* 312–318.

30. Moncada, S., Palmer, R. M. J., and Higgs, E. A. (1987). Prostacyclin and endothelium-derived relaxing factor: Biological interactions and significance. *In* "Thrombosis and Haemostasis. International Society of Thrombosis and Haemostasis." (V. Verstraete, J. Vermylen, and H. R. Lijnen, *et al.* eds.), pp. 597–618. Leuven University Press, Leuven.

31. DeNucci, G. F., Thomas, F., D'Orleans-Juste, P., Antunes, E., Walder, C., Warner, T. D., Vane, J. R. (1989). Pressor effects of circulating endothelin are limited by its removal in the pulmonary circulation and by the release of prostacyclin and endothelium-derived relaxing factor. *Proc. Natl. Acad. Sci. U.S.A. 85,* 9797–9800.

32. Sakurai, T., Yanagisawa, M., Takuwa, Y., Miyazaki, H., Kimura, S., Goto, K., and Masaki, T. (1990). Cloning of a cDNA encoding a non-isopeptide-selective subtype of the endothelin receptor. *Nature 348,* 732–735.

33. Boulanger, C., and Luscher, T. F. (1990). Endothelin is released from the porcine aorta: Inhibition by endothelium-derived nitric oxide. *J. Clin. Invest. 85,* 587–590.

34. Rubanyi, G. M., Parker-Botelho, L. H. (1991). Smooth muscle cells release a diffusible factor(s) that inhibits ET-1 production by cultured endothelial cells. *J. Cardiovasc. Pharmacol. 17* (Suppl. 7), 71–75.

35. Stewart, D. J., Langleben, D., Cernacek, P., and Cianflone, K. (1990). Endothelin release is inhibited by coculture of endothelial cells with cells of vascular media. *Am. J. Physiol. 259* (Heart Circ. Physiol. 28), H1928–1932.

36. Catravas, J. D., Bassingthwaighte, J. B., and Sparks, H. V. Jr. (1988). Adenosine transport and uptake by cardiac and pulmonary endothelial cells. *In* "Endothelial Cells." (U.S. Ryan ed.), Vol. I, pp. 65–82. CRC Press, Boca Raton, FL.

37. Junod, A. F., and Ody, C. (1977). Amine uptake and metabolism by endothelium of pig pulmonary artery and aorta. *Am. J. Physiol. 232,* C88–C94.

38. Pearson, J. D., and Gordon, J. L. (1984). Metabolism of serotonin and adenosine. *In* "Biology of Endothelial Cells." (E. A. Jaffe ed.), pp. 330–342. Martinus Nijhoff, Boston.

39. Robinson-White, A., Peterson, S., Hechtman, H. B., and Shepro, D. (1981). Serotonin uptake by isolated adipose capillary endothelium. *J. Pharmacol. Exp. Ther. 216,* 125–128.

40. Tesfamariam, B., Weisbrod, R. M., and Cohen, R. A. (1987). Endothelium inhibits responses of rabbit carotid artery to adrenergic nerve stimulation. *Am. J. Physiol. 22,* H792–H798.

41. Cohen, R. A., Tesfamariam, B., and Weisbrod, R. M. (1990). The endothelium inhibits adrenergic neurotransmission. *In* "Endothelium-Derived Relaxing Factors." (G. M. Rubanyi and P. M. Vanhoutte eds.), pp. 206–212. Karger, Basel.

42. Angus, J. A., and Cocks, T. M. (1989). Endothelium-derived relaxing factor. *Pharmacol. Ther. 41,* 303–351.

43. Katusic, Z. S., Shepherd, J. T., and Vanhoutte, P. M. (1984). Vasopressin causes endothelium-dependent relaxations of the canine basilar artery. *Circulation Res. 55,* 575–579.

44. Toda, M. (1984). Endothelium-dependent relaxation induced by angiotensin II and histamine in isolated arteries of dog. *Br. J. Pharmacol. 81,* 301–307.

45. Stephenson, J. A., and Summers, J. R. (1987). Autoradiographic analysis of receptors on vascular endothelium. *Eur. J. Pharmacol. 134,* 35–43.

46. Cohen, R. A., Shepherd, J. T., and Vanhoutte, P. M. (1983). Prejunctional and postjunctional actions of norepinephrine at the sympathetic neuroeffector junction in canine coronary arteries. *Cirs. Res. 52,* 16–25.

47. Cohen, R. A., Zitnay, K. M., Weisbrod, R. M., and Tesfamariam, B. (1988). Influence of the endothelium on tone and the response of isolated pig coronary artery to norepinephrine. *J. Pharamcol. Exp. Ther. 244,* 550–555.

48. Cohen, R. A., Shepherd, J. T., and Vanhoutte, P. M. (1983). Inhibitory role of the endothelium in the response of isolated coronary arteries to platelets. *Science 221,* 273–274.

49. Houston, D. S., Shepherd, J. T., and Vanhoutte, P. M. (1986). Aggregating human platelets cause direct contraction and endothelium-dependent relaxation of isolated canine coronary arteries: Role of serotonin, thromboxane A_2 and adenine nucleotides. *J. Clin. Invest. 78,* 539–544.

50. Shimokawa, H., Aarhus, L. L., and Vanhoutte, P. M. (1987). Porcine coronary arteries with regenerated endothelium have a reduced endothelium-dependent responsiveness to aggregating platelets and serotonin. *Circ. Res. 61,* 256–270.

51. Förstermann, U., Mügge, A., Alheid, U., Bode, S. M., and Fröhlich, J. C. (1989). Endothelium-derived relaxing factor (EDRF): A defense mechanism against platelet aggregation and vasospasm in human coronary arteries. *Eur. Heart J.* (Suppl. F, 36–43).

52. Houston, D. S., Burnstock, G., and Vanhoutte, P. M. (1987). Different P_2-purinergic receptor subtypes on endothelium and smooth muscle in canine blood vessels. *J. Pharmacol. Exp. Ther. 241,* 501–506.

53. Houston, D. S., and Vanhoutte, P. M. (1987). Comparison of serotonergic receptor subtypes on the smooth muscle and endothelium of the canine coronary artery. *J. Pharmacol. Exp. Ther. 244,* 1–10.

54. Cohen, R. A., Shepherd, J. T., and Vanhoutte, P. M. (1983). 5-Hydroxytryptamine can mediate endothelium-dependent relaxation of coronary arteries. *Am. J. Physiol. 245,* H1077–H1080.

55. Vanhoutte, P. M., and Houston, D. S. (1985). Platelets, endothelium and vasospasm. *Circulation 72,* 728–734.

56. Vanhoutte, P. M., and Shimokawa, H. (1989). Endothelium-derived relaxing factor and coronary vasospasm. *Circulation 80,* 1–9.

57. Hynes, M. R., Dang, H., and Duckles, S. P. (1988). Contractile responses to adrenergic nerve stimulation are enhanced with removal of endothelium in rat caudal artery. *Life Sci. 42,* 338–365.

58. Greenberg, S., Diecke, F. P. J., Peevy, K., and Tanaka, T. P. (1989). The endothelium modulates adrenergic neurotransmission to canine pulmonary arteries and veins. *Eur. J. Pharmacol. 162,* 67–80.

59. Tesfamariam, B., and Cohen, R. A. (1988). Inhibition of adrenergic vasoconstriction by endothelial cell shear stress. *Circulation Res. 63,* 720–725.

60. Tesfamariam, B., Weisbrod, R. M., and Cohen, R. A. (1989). The endothelium inhibits activation by calcium of vascular neurotransmission. *Am. J. Physiol. 257,* H1871–H1877.

61. Murakami, K., Hideaki, K., and Urakawa, N. (1985). Role of endothelium in the contractions induced by norepinephrine and clonidine in rat aorta. *Jpn. J. Pharmacol. 39,* 357–364.

62. Martin, W., Furchgott, R. F., Villani, G. M., and Jothianandan, D. (1986). Depression of contractile responses in rat aorta by spontaneously released endothelium-derived relaxing factor. *J. Pharamcol. Exp. Ther. 237,* 529–538.

63. Chelly, J., Tsao, C., Nath, R., Doursout, M. F., and Buckley, J. (1986). Effects of norepinephrine on the dog common carotid, coronary and renal arteries: *In vitro* studies. *Eur. J. Pharmacol. 120,* 137–139.

64. Hedqvist, P. (1977). Basic mechanisms of prostaglandin action on autonomic neurotransmission. *Annu. Rev. Pharmacol. Toxicol. 17,* 259–279.

65. Cohen, R. A., and Weisbrod, R. M. (1988). The endothelium inhibits norepinephrine release from adrenergic nerves of the rabbit carotid artery. *Am. J. Physiol. 254,* H871–878.

66. Parnavelas, J. G., Kelly, W., and Burnstock, G. (1985). Ultrastructural localization of choline acetyltransferase in vascular endothelial cells in rat brain. *Nature 316,* 724–725.

67. Burnstock, G., Lincoln, J., Fehér, E., Hopwood, A. M., Kirkpatrick, K., Milner, P., and Ralevic, V. (1988). Serotonin is localized in endothelial cells of coronary arteries and released during hypoxia: A possible new mechanism for hypoxia-induced vasodilatation of the rat heart. *Experientia 44,* 705–707.

68. Pearson, J. D., and Gordon, J. L. (1979). Vascular endothelial and smooth muscle cells in culture selectively release adenine nucleotides. *Nature 281,* 384–386.

69. Story, D. F., and Ziogas, J. (1986). Role of the endothelium on the facilitatory effects of angiotensin I and angiotensin II on noradrenergic transmission in the caudal artery of the rat. *Br. J. Pharmacol. 87,* 249–255.

70. Wiklund, N. P., Ohlen, A., and Cederqvist, B. (1988). Inhibition of adrenergic neuroeffector transmission by endothelin in the guinea-pig femoral artery. *Acta Physiol. Scand. 134,* 311–312.

71. Lincoln, J., Ralevic, V., and Burnstock, G. (1991). Neurohumoral substances and the endothelium. *In* "Cardiovascular Significance of Endothelium-Derived Vasoactive Factors." (G. M. Rubanyi, ed.), pp. 83–110. Futura Publ. Co., Inc. Mount Kinsco, N.Y.

72. Hamel, E., Assumel-Lurdin, C., Edvinsson, L., Fage, D., and MacKenzie, E. T. (1987). Neuronal versus endothelial origin of vasoactive acetylcholine in pial vessels. *Brain Res. 420,* 391–396.

73. Arneric, S. P., Honig, M. A., Milner, T. A., Greco, S., Iadecola, C., and Reis, D. J. (1988). Neuronal and endothelial sites of acetylcholine synthesis and release associated with micro-vessels in rat cerebral cortex: Ultrastructural and neurochemical studies. *Brain Res. 454,* 11–30.

74. Gonzalez, J. L., and Santos-Benito, F. F. (1987). Synthesis of acetylcholine by endothelial cells isolated from rat brain cortex capillaries. *Brain Res. 412,* 148–150.

75. Linnik, M. D., and Moskowitz, M. A. (1989). Identification of immunoreactive substance P in human and other mammalian endothelial cells. *Peptides 10*(5), 957–962.

76. Pearson, J. D., Carleton, J. S., Hutchings, A., and Gordon, J. L. (1978). Uptake and metabolism of adenosine by pig aortic endothelial and smooth-muscle cells in culture. *Biochem. J. 170,* 265–271.

77. Lincoln, J., Loesch, A., and Burnstock, G. (1990). Localization of vasopressin, sertotonin and angiotensin II in endothelial cells of the renal and mesenteric arteries of the rat. *Cell Tissue Res. 259,* 341–344.

78. Kahlson, G., and Rosengren, E. (1968). New approaches to the physiology of histamine. *Physiol. Rev. 48,* 155–196.

79. Schoefl, G. I. (1963). Studies on inflammation III. Growing capillaries: Their structure and permeability. *Virchows Arch. Pathol. Anat. Physiol. 337,* 97–114.

80. Busse, R.–Personal communication (1991).

81. Rubanyi, G. M. (1991). "Cardiovascular Significance of Endothelium-Derived Vasoactive Factors." pp. 1–357. Futura Publ. Co. Inc., Mount Kisco, N.Y.

82. Ralevic, V., Milner, P., Hudlicka, O., Kristek, F., and Burnstock, G. (1990). Substance P is released from the endothelium of normal and capsaicin-treated rat hindlimb vasculature, *in vivo,* by increased flow. *Circ. Res. 66*(5), 1178–1183.

83. Rubanyi, G. M., Romero, J. C., and Vanhoutte, P. M. (1986). Flow-induced release of endothelium-derived relaxing factor. *Am. J. Physiol. 250,* H1145–H1149.

84. Paddle, B. M., and Burnstock, G. (1974). Release of ATP from perfused heart during coronary vasodilation. *Blood Vessels 11,* 110–119.

85. Milner, P., Ralevic, V., Hopwood, A. M., Fehér, E., Lincoln, J., Kirkpatrick, K. A., and Burnstock, G. (1989). Ultrastructural localization of substance P and choline acetyltransferase in endothelial cells of rat coronary artery and release of substance P and acetylcholine during hypoxia. *Experientia 45,* 121–125.

86. Rees, D. D., Palmer, R. M. J., and Moncada, S. (1989). Role of endothelium-derived nitric oxide in the regulation of blood pressure. *Proc. Natl. Acad. Sci. U.S.A. 86,* 3375–3378.

87. Busse, R., Pohl, U., Forstermann, U., and Bassenge, E. (1985). Endothelium-dependent modulation of arterial smooth muscle tone and PGI_2-release: Pulsatile versus steady flow. *In* "Prostaglandins and Other Eicosanoids in the Cardiovascular System." (U. Pohl, U. Forstermann, R. Busse, E. Bassenge, and E. Schror, ed.), 2nd International Symposium Nurnberg-Furth, 1984, pp. 553–558 (Karger, Basel).

88. Holtz, J., Forstermann, U., Pohl, U., Giesler, M., and Bassenge, E. (1984). Flow-dependent, endothelium-mediated dilation of epicardial coronary arteries in conscious dogs: Effects of cyclooxygenase inhibition. *J. Cardiovasc. Pharmacol. 6,* 1161–1169.

89. Hutcheson, I. R., and Griffith, T. M. (1991). Modulation of EDRF release by the frequency and amplitude of pulsatile flow. *Blood Vessels 28,* 296–297.

90. Smiesko, V., Kozik, J., and Dolezel, S. (1986). Role of endothelium in the control of arterial diameter by blood flow. *Blood Vessels 22,* 247–251.

91. Frangos, J. A., Eskin, S. G., McIntire, L. V., and Ives, C. L. (1985). Flow effects on prostacyclin production by cultured human endothelial cells. *Science 227,* 1477–1479.

92. van Grondelle, A., Worthen, G. S., Ellis, D., Mathias, M. M., Murphy, R. C., Striefl, R. J., Reeves, J. T., and Voelkes, N. F. (1984). Altering hydrodynamic variables influences PGI_2 production by isolated lungs and endothelial cells. *J. Appl. Physiol. 57,* 388–395.

93. Hintze, T. H., and Vatner, S. F. (1984). Reactive dilation of larger coronary arteries in conscious dogs. *Circulation Res. 54,* 50–57.

94. Rubanyi, G. M., Lorenz, R. R., and Vanhoutte, P. M. (1985). Bioassay of endothelium-derived relaxing factor(s). Inactivation by catecholamines. *Am. J. Physiol. 249,* H95–H101.

95. Tesfamariam, B., and Cohen, R. A. (1988). Inhibition of adrenergic vasoconstriction by endothelial cell shear stress. *Circulation Res. 63,* 720–725.

96. Katusic, Z. S., Shepherd, J. T., and Vanhoutte, P. M. (1987). Endothelium-dependent contraction to stretch in canine basilar arteries. *Am. J. Physiol. 252,* H671–H673.

97. Rubanyi, G. M. (1988). Endothelium-dependent pressure-induced contraction of isolated canine carotid arteries. *Am. J. Physiol. 255,* H783–H788.

98. Harder, D. R. (1987). Pressure-induced myogenic activation of cat cerebral arteries is dependent on intact endothelium. *Circulation Res. 60,* 102–107.

99. Harder, D. R., Sanchez-Ferrer, C., Kauser, K., Stekiel, W. J., and Rubanyi, G. M. (1989). Pressure releases a transferable endothelial contractile factor in cat cerebral arteries. *Circulation Res. 65,* 193–198.

100. Kauser, K., Clark, J. E., Masters, B. S., Ortiz de Montellano, P. R., Ma, Y. H., Harder, D. R., and Roman, R. J. (1991). Inhibitors of cytochrome p-450 attenuate the myogenic response of dog renal arcuate arteries. *Circulation Res. 68,* 1154–1163.

101. Rubanyi, G. M., Freay, A. D., Kauser, K., Johns, A., and Harder, D. R. (1990). Mechanoreception by the Endothelium: Mediators and Mechanisms of Pressure- and Flow-Induced Vascular responses. *Blood Vessels 27,* 246–257.

102. Berne, R. M., Gidday, J. M., Hill, H. E., Curnish, R. R., and Rubio, R. (1987). Adenosine in the local regulation of blood flow: Some controversies. *In* "Topics and Perspectives in Adenosine Research." (E. Gerlach and B. F. Becker, eds.), pp. 395–405. Springer Verlag, Berlin.

103. Sparks, H. V. (1980). Effect of local metabolic factors on vascular smooth muscle. *In* "Handbook of Physiology." (D. F. Bohr, A. P. Somlyo, H. V. Sparks, Jr., eds.), Section 2: The Cardiovascular System, Vol. II: Vascular Smooth Muscle, pp. 475–513. American Physiological Society, Bethesda, MD.

104. King, C. E., and Cain, S. M. (1985). Peripheral vascular response to hypoxic hypoxia after aortic denervation. *Can. J. Physiol. Pharmacol. 63,* 1197–1201.

105. Rounds, S. S., McMurtry, I. F., and Reeves, J. T. (1981). Glucose metabolism accelerates the decline of hypoxic vasoconstriction in rat lungs. *Resp. Physiol. 44,* 239–249.

106. Kivity, S., and Souhrada, J. F. (1981). Plasma and platelets potentiate a hypoxic vascular response of the isolated lung. *J. Appl. Physiol. 51,* 875–880.

107. Borda, L. J., Shuchleib, R., and Henry, P. D. (1980). Hypoxic contraction of isolated canine coronary artery. Mediation by potassium-dependent exocytosis of norepinephrine. *Circulation Res. 46,* 870–879.

108. Kwan, Y. W., Wadsworth, R. M., and Kane, K. A. (1989). Hypoxia- and endothelium-mediated changes in the pharmacological responsiveness of circumflex coronary artery rings from the sheep. *Br. J. Pharmacol. 96,* 857–863.

109. Rubanyi, G. M., and Paul, R. J. (1985). Two distinct effects of oxygen on vascular tone in isolated porcine coronary arteries. *Circulation Res. 56,* 1–10.

110. Rubanyi, G. M., and Vanhoutte, P. M. (1985). Hypoxia releases a vasoconstrictor substance from the canine vascular endothelium. *J. Physiol. 364,* 45–46.

111. DeMey, J. G., and Vanhoutte, P. M. (1983). Anoxia and endothelium-dependent reactivity of the canine femoral artery. *J. Physiol. Lond. 335,* 65–74.

112. Busse, R., Pohl, U., Kellner, C., and Klemm, U. (1983). Endothelial cells are involved in the vasodilatory response to hypoxia. *Pflügers Arch. 397,* 78–80.

113. Rubanyi, G. M., and Gräser, T. (1991). Hypoxic contraction in rat aorta is due to inhibition of EDRF action on vascular smooth muscle. *Circulation*—in press.

114. Gräser, T., and Vanhoutte, P. M. (in press). Contraction of canine coronary arteries to severe hypoxia: Role of endothelium, contraction level, and cyclic GMP. *Am. J. Physiol.*

115. Hopwood, A. M., Lincoln, J., Kirkpatrick, K. A., Burnstock, G. (1986). Adenosine 5′-triphosphate, adenosine and endothelium-derived relaxing factor in hypoxic vasodilatation of the heart. *Eur. J. Pharmacol. 165,* 323–326.

116. Rubanyi, G. M. (1988). Vascular effects of oxygen-derived free radicals. *Free Radical Biology and Medicine 4,* 107–120.

117. Wolin, M. S., and Burke, T. M. (1987). Hydrogen peroxide elicits activation of bovine pulmonary arterial soluble guanylate cyclase by a mechanism associated with its metabolism by catalase. *Biochem. Biophys. Res. Comm. 143,* 20–25.

118. Vedernikov, Y. P., Gräser, T., and Li, D. S. (1990). Hypoxic and posthypoxic responses in isolated coronary arteries and veins: Role of endothelium. *Biomed. Biochim. Acta 49,* 1177–1184.

119. Phillis, J. W., and Delong, R. E. (1987). An involvement of adenosine in cerebral blood flow regulation during hypercapnia. *Gen. Pharmacol. 18,* 133–139.

120. Deussen, A., Möser, G., and Schrader, J. (1986). Contribution of coronary endothelial cells to cardiac adenosine production. *Pflügers Arch. 406,* 608–614.

121. Bohr, D. F. (1978). Vascular Smooth Muscle. *In* "Peripheral Circulation." (P. C. Johnson, ed.), pp. 13–44. Wiley, N.Y.

122. Rubanyi, G. M., and Vanhoutte, P. M. (1988). Potassium-induced release of EDRF from canine femoral arteries. *Circulation Res. 62,* 1098–1103.

123. Tesfamariam, B., Weisbrod, R. M., and Cohen, R. A. (1989). Augmented adrenergic contractions of carotid arteries from cholesterol-fed rabbits due to endothelial cell dysfunction. *J. Cardiovasc. Pharamcol. 13,* 820–825.

124. Petty, R. G., and Pearson, J. D. (1989). Endothelium: the axis of vascular health and disease. *J.R. Coll. Physicians,* London *23,* 92–102.

125. Vanhoutte, P. M., and Lüscher, T. F. (1987). Peripheral mechanisms in cardiovascular regulation: Transmitters, receptors and the endothelium. *In* "Handbook of Hypertension." Vol. 8, Physiology and Pathophysiology of Hypertension Regulatory Mechanism, (R. C. Tarazi and A. Zanchetti, eds.), pp. 96–123. Elsevier, Amsterdam.

126. Lüscher, T. F. (1988). Endothelial vasoactive substances and cardiovascular disease." Karger, London.

127. Aksulu, H. E., Cellek, S., and Türker, R. K. (1986). Cholesterol feeding attenuates endo-

thelium-dependent relaxation response to acetylcholine in the main pulmonary artery of chickens. *Eur. J. Pharmacol. 129,* 397–400.

128. Ganz, P., Ludmer, P. L., Leopold, J. A., Hollenberg, N. K., Shook, T. L., Wayne, R. R., Mudge, G. H., Alexander, R. W., and Selwyn, A. P. (1988). "Endothelial dysfunction *in vivo:* Studies in animals and in patients with coronary atherosclerosis. *In* "Vasodilatation: Vascular Smooth Muscle, Peptides, Autonomic Nerves, and Endothelium." (P. M. Vanhoutte ed)., pp. 543–549. Raven Press, New York.

129. Heistad, D. D., Armstrong, M. S., Marcus, M. L., Piegors, D. J., and Mark, A. L. (1984). Augmented responses to vasoconstrictor stimuli in hypercholesterolemic and atherosclerotic monkeys. *Circulation Res. 54,* 711–718.

130. Verbeuren, T. J., Joraens, F. H., Zonnekeyn, L. L., Van Hove, C. E., Coene, M. C., and Herman, A. G. (1986). Effect of hypercholesterolemia on vascular reactivity in the rabbit I. Endothelium-dependent and endothelium-independent contractions and relaxations in isolated arteries of control and hypercholesterolemic rabbits. *Circulation Res. 58,* 552–564.

131. Yamamoto, H., Bossaller, C., Cartwright, J. Jr., and Henry, P. D. (1988). Videomicroscopic demonstration of defective cholinergic arteriolar vasodilation in atherosclerotic rabbit. *J. Clin. Invest. 81,* 1752–1758.

132. Deussen, A., Möser, G., and Schrader, J. (1986). Contribution of coronary endothelial cells to cardiac adenosine production. *Pflügers Arch. 406,* 608–614.

133. Durante, W., Sen, A. K., and Sunahara, F. A. (1988). Impairment of endothelium-dependent relaxation in aortae from spontaneously diabetic rats. *Br. J. Pharmacol. 94,* 463–468.

134. Fortes, Z. B., Leme, J. G., and Scivoletto, R. (1983). Vascular reactivity in diabetes mellitus: role of the endothelial cell. *Br. J. Pharmacol. 79,* 771–781.

135. Petty, R. G., and Pearson, J. D. (1989). Endothelium: The axis of vascular health and disease. *J. R. Coll. Physicians London 23,* 92–102.

136. Pieper, G. M., and Gross, G. J. (1991). Endothelial dysfunction in diabetes. *In* "Cardiovascular Significance of Endothelium-Derived Vasoactive Factors." (G. M. Rubanyi, ed.), pp. 223–249. Futura Publ. Co., Inc., Mount Kisco, N.Y.

137. Shimokawa, H., Flavahan, N. A., Vanhoutte, P. M. (1989). Natural course of the impairment of endothelium-dependent relaxations after balloon endothelium-dependent removal in porcine coronary arteries: Possible dysfunction of a pertussis toxin-sensitive G protein. *Circulation Res. 65,* 740–753.

138. Hudlicka, O. (1984). Development of microcirculation: Capillary growth and adaptation. *In* "Handbook of Physiology: The Cardiovascular System IV. American Physiological Society. pp. 165–216. Bethesda, MD.

139. Fry, D. L. (1968). Acute vascular endothelial changes associated with increased blood velocity gradients. *Circulation Res. 22,* 165–197.

140. Gimbrone, M. A. Jr., Cotran, R. S., and Folkman, J. (1974). Human vascular endothelial cells in culture: Growth and DNA synthesis. *J. Cell. Biol. 60,* 673–684.

CHAPTER 6

Modulation of Endothelial–Inflammatory Cell Interactions by Primary Afferent Nerves

ZEINAB KHALIL ANTHONY EGLEZOS ROBERT D. HELME

I. Introduction

A. General Background

Endothelium and the vascular basement membrane form important barriers between the circulation and extravascular compartment. Endothelium forms the inner lining of all blood vessels and as a consequence, is in direct contact with the blood. Because of this and its ability to synthesize and secrete regulatory substances, the endothelium is able to modulate blood cell functions and play an active role in hemostasis, inflammatory, and immune reactions.

In the 1970s the role of endothelium in inflammatory processes was primarily considered passive; the role of endothelial cells (ECs) was to form a physical barrier, and injury merely exposed the thrombogenic subendothelium to the blood and thereby initiated the vascular response to injury (Willms-Kretschmer *et al.*, 1967). In the 1980s, however, this view was altered after the discovery that injured ECs can express new molecules and biologic functions. During the last decade there has been rapid growth in the recognition and understanding of factors released by the endothelium as well as understanding of the physiology and pharmacology of mediators released from circulating blood cells (Pober and Grimbrone, 1982). Evidence for a role of neural elements in these processes, particularly of primary afferent nerves, has also accumulated during this time. As a consequence, a clearer understanding of the integrated functional interaction between endothelium and blood cells has become more apparent during both normal and pathological conditions.

This chapter will briefly review mechanisms involved in the interaction between endothelium and blood cells, particularly leukocytes and platelets, and their role in the control of vascular reactivity, inflammation, and immunity. The role of primary afferent nerves in modulating this interaction will then be discussed. The role of the sympathetic nervous system in this interaction is not addressed because of the limited data available for review at this time. The earliest component of the inflammatory response consists of vasodilatation (VD) and plasma extravasation (PE). There is an extensive literature on the modulatory effects of the nervous system on this process (Holzer, 1988). VD and PE are clearly important to the subsequent series of events that leads to the involvement of cellular defense mechanisms. We do not intend to review this interaction here (see, however, Chapter 15).

B. Role of EDRF

Under normal physiological conditions, the vascular endothelium generates and releases endothelium-derived relaxing factor (EDRF) that is closely related to nitric oxide (NO) (Palmer *et al.*, 1987). EDRF accounts for the vasodilation induced by a number of local mediators and consequently plays an important role in the control of vascular reactivity, particularly at the level of distributing arteries (Dusting *et al.*, 1989); the microvasculature has rarely been examined.

NO can be synthesized by a variety of cells and tissues other than endothelium, including macrophages (Marletta *et al.*, 1988; Stuehr *et al.*, 1989), polymorphonuclear leukocytes (McCall *et al.*, 1989), mast cells (Salvemini *et al.*, 1990), and neuronal tissue (Garthwaite *et al.*, 1988; Moncada *et al.*, 1989). Vascular ECs (Palmer *et al.*, 1988) and macrophages (Marletta *et al.*, 1988) synthesize NO from the terminal guanidino nitrogen atom of the amino acid L-arginine. This synthesis is inhibited by N^G-monomethyl-L-arginine (L-NMMA), which also inhibits both endothelium-dependent vasodilation *in vitro* and *in vivo*. Studies by Rimele *et al.* (1988) showed that neutrophils release a factor with the pharmacological profile of NO, which relaxes vascular smooth muscle and inhibits platelet aggregation (Cynk *et al.*, 1988). Further evidence by McCall *et al.* (1989) has shown that the release of NO from neutrophils is inhibited by L-NMMA, indicating that NO is formed from L-arginine in these cells via an enzymic mechanism similar to that in vascular endothelium (Palmer and Moncada, 1989) and macrophages (Marletta *et al.*, 1988). The substrate and the inhibitor profile for NO synthesis in neutrophils (McCall *et al.*, 1989) are similar to that in macrophages (Hibbs *et al.*, 1987) but differs from that in ECs (Palmer *et al.*, 1988). The ability of these blood cells to release NO, which in turn is capable of relaxing smooth muscles, suggests a physiological role for NO of cellular origin in the control of vascular reactivity. In addition, Moncada *et al.* (1989) suggested a role of NO in macrophage-mediated cytotoxicity. It is therefore likely that NO of different cell origins could have various physiological roles that will be clarified with the use of specific and selective inhibitors.

C. Endothelium–Platelet Interaction

ECs are known to synthesize platelet-activating factor (PAF) when stimulated with agonists that bind to cell surface receptors. The process of PAF accumulation in ECs is regulated by Ca^{2+} and also by protein kinase C (Whatley *et al.*, 1989). PAF released from endothelial cells can influence both endothelial and platelet functions. Renkonen *et al.* (1990) have provided evidence for the ability of PAF to induce lymphocyte binding to ECs via a mechanism that also involves activation of protein kinase C. On the other hand, PAF induces platelet aggregation via activating PAF receptors located on platelets (Ashby *et al.*, 1990). Activation of these receptors results in an increase in platelet intracellular Ca^{2+} with a subsequent release of a number of mediators that can influence endothelial function. Serotonin (5-HT), a biogenic amine that is released from platelets, can activate receptors located on ECs in small blood vessels, resulting in smooth muscle relaxation via EDRF release from ECs (Moncada *et al.*, 1989). The released NO, or EDRF by implication, can subsequently inhibit platelet aggregation by an acceleration of Ca^{2+} removal from the cytosol leading to a decrease in intracellular Ca^{2+} and by inhibiting PAF-induced increases in platelets intracellular Ca^{+} (Whatley *et al.*, 1989).

Another mechanism by which ECs can inhibit platelet aggregation involves the synthesis of prostacyclin. ECs can be activated by other mediators of inflammation, such as bradykinin, to produce prostacyclin (Bartha *et al.*, 1989). This involves a mechanism that is related to receptor-linked activation of inositide hydrolysis.

On the other hand, platelets can indirectly play a role in EC injury. On the basis of *in vitro* observations, a synergy exists between platelets and neutrophils, resulting in enhanced oxygen formation by neutrophils and this could result in endothelial injury. Cytokines released from macrophages have both direct stimulatory effects on oxygen radical formation in neurophils and can prime macrophages for enhanced oxygen radical responses to other agonists. Cytokines can also alter ECs rendering them more susceptible to oxygen radical mediated injury by neutrophils (Ward *et al.*, 1988).

D. Role of Vascular Endothelium in Immunity

The orderly movement of lymphocytes between blood, lymph nodes, bone marrow, spleen, and thymus ensures maintenance of immunological competence (Trnka and Cahill, 1980). Lymphocytes recirculate from the blood to the lymphatics soon after they leave the primary lymphoid organs (Gowans, 1959).

Leukocyte–EC interaction regulates the entry of immune effectors into the tissues. This interaction occurs in lymphoid tissues and inflammatory sites, via postcapillary venules (PCVs) lined with high ECs (Butcher *et al.*, 1986), as opposed to large capacitance vessels lined with flat ECs (O'Connell and Edidin, 1990). Adherence of leukocytes to the endothelium is a requirement for infiltration and accumulation of cells at inflammatory sites. Several sites of contact exist between macrophages and EC membranes (Azzali *et al.*, 1990) that stress a possible functional role played by cellular interactions between the two.

ECs lining the walls of the PCVs are the preferred sites of leukocyte entry into the lymphoid organs (Duijvestijn and Herman, 1989). These high endothelial venules (HEV) characterized by plump or high endothelial morphology (Butcher *et al.*, 1986) allow sustained cell traffic without a large loss of fluid. The nonrandom distribution of lymphocytes into different lymphoid organs is mediated by the selective adherance of lymphocytes to these ECs (Butcher *et al.*, 1986), an action mediated by the presence of specific glycoprotein adhesion molecules called homing receptors. Functionally distinct and organ specific homing receptors control lymphocyte migration to different lymphoid tissues. HEV are normally present in the paracortex of lymph nodes, and mucosal-associated lymphoid tissues and vessels with HEV-like morphology can also arise at sites of inflammation.

The interleukins (ILs) and colony-stimulating factors (CSFs) are a group of regulatory molecules that function to prime and regulate the immune and hemopoietic systems (for reviews see: Dinarello, 1988; Balkwill and Burke, 1989; Metcalf, 1989; Kelso, 1990; Kelso and Metcalf, 1990; Habenicht, 1990; Smith, 1989, 1990). Immunologic factors, such as IL-1 (Cavender *et al.*, 1986) and γ-interferon (γ-IFN) (Yu *et al.*, 1985), have also been shown to be involved in the development and maintenance of HEV (Hendricks and Estermans, 1983). These factors, which are produced at the site of an inflammatory response (Bendtzen, 1988), may act to attract and direct lymphocytes to the sites of immunologic activity. IL-1 is a potent inflammation promoting cytokine that induces multiple biological effects on a variety of cells (Dinarello, 1988). Several studies have demonstrated that a number of factors, including IL-1, enhance lymphocyte binding to ECs. Issekutz and Stoltz (1989) have also shown that IL-1 combined with γ-IFN produce a synergistic increase in lymphocyte migration.

The EC has recently been shown to be an important target for IL-1. The IL-1-stimulated EC plays an essential role in inflammatory reactions, and IL-1 and IL-2 are important biological response modifiers of inflammatory and immunological reactions. IL-1 released by macrophages can activate ECs to increase total cell glycosphinogolipids content without altering their composition on cell surface binding. Immunoprecipitation and chemical cross-linking analysis have demonstrated that the 78-kDa human EC IL-1 receptor is similar to the human T lymphocyte and fibrobast IL-1 receptor (Thieme and Wagner, 1989). Although ILs are produced primarily by activated lymphocytes and macrophages, the capacity to elaborate IL-1 and IL-6 in particular is not confined to these cells. Studies have shown that proteoglycans can induce the production of IL-1 α/β by ECs and that activation *in vitro* of ECs results in the production of IL-1 and IL-6 (for review see Bendtzen, 1988). In addition, EC-derived IL-1 was found to have a chemotactic activity for human tumor cells (Orr *et al.*, 1988.)

II. Neural Receptor Mechanisms

A. Neurogenic Modulation of EDRF Release

In recent years it has become apparent that many agents, including sensory-derived neuropeptides, relax blood vessels by endothelium-dependent mechanisms. Of these neuropeptides, substance P (SP) is the most studied in this regard. As a putative sensory neurotransmitter, SP is believed to be responsible for mediating nociception and peripheral neurogenic responses, including vasodilatation and plasma extravasation. These effects on the vasculature are thought to be partially mediated by the release of histamine from mast cells (Johnson and Erdos, 1973). SP-containing perivascular nerves, however, are present in the adventitial-medial border of almost all the arteries and veins of guinea pigs (Furness *et al.*, 1982) and rats (Barja *et al.*, 1983). In addition, autoradiographic analysis has demonstrated cell surface receptors for SP (Stephenson and Summers, 1987) on endothelial cells. Activation of these receptors causes vascular relaxation that is coupled to the formation of cyclic GMP (Bolton and Clapp, 1986) and is mediated via the release of EDRF (Christie *et al.*, 1989; Smith and Lang, 1990). It is possible, therefore, that the effects of SP on the vasculature could be partially mediated via an action on the endothelium. Although the relaxing effect of SP on veins appears to be endothelium independent (Regoli *et al.*, 1984a,b), its effect in arteries usually requires the presence of an intact endothelium (Furchgott, 1983; D'Orleans-Juste *et al.*, 1985; Edvinsson *et al.*, 1985). A recent study at the level of the rat skin microvasculature using the NO synthetase inhibitor (NOLA) has confirmed that the endothelially derived mediator that is in part responsible for SP-induced relaxation is EDRF or NO derived from L-arginine (manuscript in preparation). In this later study, using a blister model of inflammation in the rat hind footpad, we investigated possible contribution of NO derived from the microvascular endothelium to the plasma extravasation and vasodilatation responses induced by perfusion of SP over the blister base. In the presence of NOLA, both responses were significantly reduced, suggesting a role of NO in the inflammatory response to SP and conversely the ability of a sensory neuropeptide, SP, to release EDRF from the endothelium. The specificity of action of NOLA was established by its inability to inhibit the vasodilator response to sodium nitroprusside, an endothelium independent vasodilator that was also perfused over the blister base.

B. Neurogenic Modulation of the Endothelium–Platelet Interaction

There is insufficient data on the role of neural elements in the interaction between ECs and platelets to allow for review. However, neural modulation can be postulated based on pharmacological studies that have examined neural responses to mediators released from platelets. Serotonin released from platelets can activate receptors that are located on primary afferents (Richardson *et al.*, 1985) that in turn leads to the release of sensory neuropeptides. These peptides can subsequently modulate endothelium function since studies by Maggi *et al.* (1990) have shown the presence of receptors for sensory peptides on EC.

C. Neurogenic Modulation of the Immune Response

Activation and regulation of the immune system involve complex cellular interactions associated with and/or including the production of soluble regulatory factors by leukocytes and other cells. Recently a network of external regulatory messages arising from the nervous system has been shown to modulate immune responses. Shared ligand-receptor and related peptide receptors also exist as a possible route of communication between the nervous and immune systems. Certain antigens, such as Thy 1, class I, and II MHC antigens (Wong *et al,* 1984), and receptors for hormones (Blalock *et al.*, 1985) and sensory neuropeptides (Payan *et al.*, 1984a) are shared by lymphocytes and cells of the nervous system. It is therefore clear that the potential exists for neuropeptides to act at a number of different levels of the immune response. The widespread innervation of lymphoid organs and the close association of nerves with lymphocytes and structural components of lymphoid organs can be a mechanism for the transfer of messages between the nervous and immune systems (for review see Payan *et al.*, 1986; Stanisz *et al.*, 1987b).

Lymphokines are known to affect the functions of many cells involved in immune responses, including lymphocytes, fibroblasts, and ECs. Other factors affecting these cells, but not of immune origin, include the tachykinins substance P (SP), neurokinin A (NKA) and neurokinin B (NKB) (Payan *et al.*, 1983; Nilsson *et al.*, 1985; Stanisz *et al.*, 1986; Helme *et al.*, 1987a; Kimball *et al.*, 1988; Scicchitano *et al.*, 1988a; Tanaka *et al.*, 1988; Casini *et al.*, 1989; Eglezos *et al.*, 1991a; Eglezos, 1990), calcitonin gene-related peptide (CGRP) (Casini *et al.*, 1989), somatostatin (SOM) (Payan *et al.*, 1984b; Stanisz *et al.*, 1986; Eglezos *et al.*, 1991b), and vasoactive intestinal peptide (VIP) (Ottaway *et al.*, 1984; O'Dorisio *et al.*, 1985). The interactions between these neuropeptides and immunologically active cells provide one of the basic building blocks of "neuroimmunomodulation."

Much of the evidence for involvement of primary afferent nerves in inflammatory and immunologic responses has depended on experiments in rats pretreated with the sensory neurotoxin capsaicin (see Buck and Burks, 1986; Bevan and Szolcsanyi, 1990; Eglezos, 1990). Capsaicin, 8-methyl-*N*-vanillyl-6-nonenamide, the pungent comonent of certain hot red peppers, causes a dose-dependent destruction of primary afferent tachykinin-containing nerves when injected into neonatal rats (Nagy *et al.*, 1983) with associated decreases in *in vitro* (Nilsson and Ahlstedt, 1988) and *in vivo* (Eglezos, 1990) immune reactivity.

1. Neuropeptide Receptor Expression on Cells of Immune System

Neuropeptides such as SP, SOM, and VIP have been shown to bind to specific membrane receptors on the surface of cells important in immune responses (see Table 1). These lymphocyte and mononuclear leukocyte receptors have been characterized using both radioiodinated and fluoresceinated neuropeptides (for review see Payan *et al.*, 1986; McGillis *et al.*, 1987; Stanisz *et al.*, 1987a; O'Dorisio, 1988). Receptors for other neuropeptides, such as NKA, NKB, and CGRP, are yet to be described.

a. SUBSTANCE P. SP binds to a specific cell surface receptor on human circulating lymphocytes (Payan *et al.*, 1984a). SP-R are expressed on both Th/i and Tc/s subclasses, with the Th/i binding being predominant (Payan *et al.*, 1984a). The human T cell SP-R is of a single affinity with a binding constant (K_D) of 18 μM, and approximately 7000 receptor sites/T cell. Murine spleen and peyer's patch (PP) T and B cells have also been shown to possess a high-affinity SP-R (Stanisz *et al.*, 1987b). This receptor has a similar K_D on both cell types from both tissues with a

Table 1. Role of Tachykinins in the Immune System

	SP[a]	NKA	NKB
A. Receptor expression			
T cell			
human	+		
murine	+		
B cell			
human	–		
murine	+		
Macrophage			
guinea pig	+		
Monocyte			
human	+		
PMN			
rabbit	+		
Cell lines			
IM-9	+		
Molt-4b	+		
Hut 78	–		
Jurkat	–		
B. Cellular functions			
Proliferation			
T cell	+	+	
B cell	+		
fibroblast	+	+	
sm. muscle	+	+	
keratinocyte	+		
NK cell activity	+		
Ig production	+	+	
Phagocytosis	+		
IL-1 production	+	+	+
IL-6 production	+	+	
TNF-α production	+		
IFN-γ production	+		
Chemotaxis			
PMN	+		
monocyte	+		
Lymphocyte recirculation	+		
Mast cell degranulation	+		
Thymic involution	–		

[a] +, receptor present; –, no receptor (A).

mean of 0.68 nM. Other cells that possess receptor-binding sites for SP include rabbit polymorphonuclear leukocytes (Maracso *et al.*, 1981), human monocytes (Ruff *et al.*, 1985), and guinea pig macrophages (Hartung *et al.*, 1986) with similar K_D values to that of the lymphocyte SP-R.

b. Somatostatin. Murine lymphocytes from spleen and PP have specific binding sites for SOM (SOM-R), which are found predominantly on Tc/s cells

(Scicchitano *et al.*, 1987a). In murine B cells, SOM binding was predominantly found in sIgA-positive B cells (Scicchitano *et al.*, 1987a). The murine IgA-secreting plasmacytoma cell-line MOPC-315 expresses high-affinity SOM-R (Scicchitano *et al.*, 1988b). Human blood mononuclear cells also have high-affinity SOM-R, which have a K_D of 0.5 μM and express 300–500 receptor sites/cell (Bhathena *et al.*, 1981).

c. VASOACTIVE INTESTINAL PEPTIDE. Specific binding sites for VIP have been detected on a wide variety of cells, including human mononuclear cells and lymphocytes, Molt-4b lymphoblasts, murine adherent and nonadherent mononuclear cells (for review see O'Dorisio *et al.*, 1985), and more recently human SUP-T1 lymphoblasts (Christophe *et al.*, 1990). Human lymphocytes bear high- and low-affinity receptors for VIP (O'Dorisio *et al.*, 1985; Guerrero *et al.*, 1981). VIP-receptor expression in T cells is predominantly on $L3T4^+/Lyt2^-$ ($CD4^+$) Th cells (Ottaway, 1987). Murine mesenteric lymph node cells express a single class of a high-affinity VIP-R. Receptors for VIP have also been shown on spleen T cells and T cells from other lymph node groups but not on thymocytes (Ottaway and Greenberg, 1984).

2. Neuropeptide Regulation of Cellular Responses

Studies both *in vivo* and *in vitro* have shown that neuropeptides influence the functions of acute inflammatory cells, lymphocytes, and cells involved in repair mechanisms.

a. THE TACHYKININS AND CGRP. Cellular chemotaxis of rabbit neutrophils (Marasco *et al.*, 1981), human monocytes (Ruff *et al.*, 1985), and rat neutrophils (Roch-Arvellier *et al.*, 1986; Helme *et al.*, 1987b) is modulated by SP. In all of these studies the C-terminal end of the undecapeptide SP was the active component.

Pharmacological (Mazurek *et al.*, 1981; Piotrowski and Foreman, 1985) and histochemical (Skofitsch *et al.*, 1985) studies provide a role for SP in hypersensitivity. SP was found to be capable of degranulating mast cells leading to histamine release in a dose-dependent manner with an EC_{50} of 10 μM (Mazurek *et al.*, 1981; Piotrowski and Foreman, 1985); this effect was mediated by the N-terminal end of the peptide and is maximal at approximately 1 μM (Mazurek *et al.*, 1981).

SP as well as NKA and CGRP (and to a lesser extent NKB) have been shown to modulate macrophage function. Guinea pig macrophages primed with Corynebacterium parvum have increased chemiluminescence responses and arachadonic acid metabolism in the presence of SP (Hartung *et al.*, 1986). Cellular chemotaxis was maximal at 0.6 μM and mediated by the C-terminal end of the peptide (Hartung and Toyka, 1983). However, the stimulatory effects of SP on phagocytosis of yeast were maximal at 40 nM and mediated by N-terminal residues. It appears that the effects of SP are mediated by the whole peptide with some selectivity of function for C- and N-terminal residues. Human alveolar macrophages can be induced by SP to synthesize and release both IL-1 and tumor necrosis factor-α (TNF-α) (Cozens and Rowe, 1987). In fact SP, SP(4–11) and NKA are capable of inducing human macrophages to generate IL-1, IL-6, and TNF-α (Lotz *et al.*, 1988), whereas NKB is only able to stimulate macrophages to generate IL-1 (Kimball *et al.*, 1988). SP(4–11) is the most potent in these effects with a maximal response at 7.4 pM; SP at levels below 22 pM is inactive. NKA and NKB were more potent than SP but less so than NKA(4–10) (Kimball *et al.*, 1988). Human bronchoalveolar macrophges, on the other hand, are not affected by SP (Pujol *et al.*, 1989), showing a possible tissue or species specificity for the actions of the tachykinins.

The stimulatory effects of SP on lymphocyte proliferation first described by Payan *et al.* (1983) showed a capacity for SP to stimulate proliferation of human T lymphocytes (by 60–70% at 10 nM) in the presence or absence of mitogenic stimulation. SP is also stimulatory for murine PP, mesenteric lymph node (MLN), and spleen lymphocytes at between 1 nM and 0.1 μM, increasing the response by over 60% with maximal stimulation occurring in the murine system at 10 nM (Stanisz *et al.*, 1986). Similar results have also been obtained *in vivo*. Infusion of SP (10 μM) has been shown to increase the *in vivo* PFC response (production of antigen-specific immunoglobulin) of neonatally capsaicin-treated rats from 25% of normal to over 70% of normal (Helme *et al.*, 1987a). At a similar concentration, Scicchitano *et al.* (1988a) also showed that SP stimulated lymphocyte proliferation after *in vivo* infusion. SP enhances the pokeweed mitogen (PWM) induced release of IFN-γ from human mononuclear cells (Wagner *et al.*, 1987) and the release of both Ig and IFN-γ from human duodenal mucosal cells (Hart *et al.*, 1990). SP-containing capsaicin-sensitive nerves have also been shown to be involved in lymphocyte production of IL-2 to Con A stimulation. Following neonatal capsaicin treatment, rats produced 60% less IL-2 following activation with optimal Con A levels (Eglezos, 1990).

NKA, NKA(4–10), and CGRP have a dose-related stimulatory effect on PHA-induced proliferation of murine mononuclear leukocytes (Casini *et al.*, 1989), with NKA(4–10) being 16 times more potent than NKA. The effects of the two peptides were evident at concentrations as low as 1 fM; EC_{50} values were 0.6 pM and 0.04 pM for NKA and NKA(4–10), respectively. Due to the high potency of NKA(4–10), it was concluded that the receptor involved may be an NK-2 type tachykinin receptor (Casini *et al.*, 1989). Recently, an *in vivo* immunomodulatory role for NKA in modulation of the primary antibody response to SRBC was reported. NKA was 10 times more potent than SP, with NKA active at concentrations of 10 nM (Eglezos *et al.*, 1991a). This *in vivo* study reinforces the *in vitro* results of Kimball *et al.* (1988) and Casini *et al.*, (1989) of NK-2 receptors involved in the immunomodulatory effects of the tachykinins (Eglezos *et al.*, 1991b). NKA has also been shown to modulate human B-cell IgG production, once again being slightly more effective than SP (Laurenzi *et al.*, 1989). CGRP stimulates Con A-induced lymphocyte proliferation, maximal at 1 μM with an ED_{50} of 4.6 nM (Casini *et al.*, 1989).

Other investigators have suggested a role for SP in control of lymphocyte recirculation (Moore, 1984; Moore *et al.*, 1989), reversal of stress-induced thymic involution (Oehme *et al.*, 1987) and repair mechanisms (Nilsson *et al.*, 1985; Tanaka *et al.*, 1988). Fifty μg/ml SP increased lymphatic drainage and the output of small cells and blast cells from sheep popliteal lymph nodes. These increases (20–80%) occur within the first 2–4 h following peptide infusion into the node (Moore, 1984; Moore *et al.*, 1989) and is within the same time frame of the effects of the tachykinins on the *in vivo* PFC response (Eglezos *et al.*, 1990). SP injected i.p. at 125 μg/kg 1 h prior to immobilization stress decreased the stress-induced involution of the thymus from 40 to 12%. When SP(4–11) was used (52.8 mg/kg, i.p.), thymic weight was not decreased but increased by 50% following stress (Oehme *et al.*, 1987). The reasons for this protective effect of SP are unknown, but possibilities raised include SP-mediated inhibition of adrenal function, and SP protection of the thymus and/or lymphocytes from steroid hormone. The role of SP and SP-containing nerves on adrenal function has been described in detail by Khalil (1987).

Another protective role for SP has been observed in wound healing. SP stimulation of proliferative responses of smooth muscle cells, skin fibroblasts (Nilsson *et al.*, 1985), and keratinocytes (Tanaka *et al.*, 1988) implicates SP in the wound healing response during tissue damage and inflammation. In addition, Nilsson *et al.* (1985) provided further evidence to show that NKA was more potent than SP, suggesting another possible role for NK-2 receptor–mediated activity. It therefore appears that not only immunomodulatory but other actions of the tachykinins could be mediated by NK-2 receptors. These results need to be validated by the use of the recently available second generation NK-2 receptor agonists (Rovero *et al.*, 1990; Maggi *et al.*, 1990).

b. SOMATOSTATIN. Contrary to the stimulatory effects of SP, SOM generally has an inhibitory effect. SOM inhibits T- and B-cell proliferation, (Payan *et al.*, 1984b; Stanisz *et al.*, 1986) and Ig secretion (Stanisz *et al.*, 1986; Eglezos, 1990). *In vivo* infusion of SOM immediately following antigen injection decreased the capacity of rats to respond to antigenic stimulation by over 65% (Eglezos, 1990). This effect was related to the concentration of SOM infused, and reversed by coinfusion of NKA but not SP. This result provides evidence that the mode of action of SOM may be via an interaction with a receptor activated by NKA (possibly an NK-2 receptor) (Eglezos, 1990).

c. VASOACTIVE INTESTINAL PEPTIDE. As for SOM, VIP has a net inhibitory effect on many components of the immune response. VIP exerts its inhibitory effect on lymphocyte function (O'Dorisio *et al.*, 1981; Ottaway *et al.*, 1983), being confined to T cells (Ottaway and Greenberg, 1984; Krco *et al.*, 1986; Ottaway, 1987). VIP also inhibits murine Ig secretion (Stanisz *et al.*, 1986) and lymphocyte recirculation (Moore, 1984) but not rat Ig production (Eglezos, 1990). VIP-receptors have been postulated to facilitate lymphocyte homing and entry into specific tissues. VIP-receptor positive cells accumulate preferentially in PP and MLN (Ottaway, 1984). It appears that receptors for VIP affect the migration of T cells *in vivo*, or that VIP may act as the ligand that enhances T-cell binding to the vascular endothelium of MLN and PP (Ottaway, 1984).

III. Interaction between Endothelial Cells and Leukocytes in Disease States

A. *Inflammatory Disease*

The adherence of circulating neutrophils to the microvascular endothelium is the initial step in diapedesis, the process by which leukocytes migrate through blood vessels to accumulate at sites of cutaneous disease or injury. Neutrophil adherence to EC can be enhanced by a variety of mediators of inflammation, and both the neutrophil and the EC can actively contribute to the adhesive interaction. Leukocyte–EC adhesion is mediated through specific adhesion molecules. The leukocyte adhesion molecule-1 (LAM-1) is principally involved in the initial interactions between leukocytes and ECs (Spertini *et al.*, 1991). Other adhesion molecules, including CD-11/CD-18 complexes (on the leukocyte surface), the endothelial leukocyte adhesion molecule-1 (ELAM-1), and intercellular adhesion molecule-1 (ICAM-1) (on the EC surface) (Luscinskas *et al.*, 1989), have a role in subsequent adhesion and diapedsis into tissues.

Recent evidence by Leeuwenberg *et al.* (1990) has suggested that the binding of leukocytes to ECs can be activated by a number of different mechanisms, including inflammatory mediators that might act in concert to strengthen this binding. These mechanisms include activation by TNF-α, IL-1, and lipopolysaccharide (LPS). These factors strengthen cellular binding and the expression of the adhesion molecules ELAM-1 and CD-18 antigen, while inhibiting leukocyte adhesion inhibitor (LAI) production by EC. Further evidence by Doukas and Pober (1990) showed that treatment of EC with either TNF or IL-1 caused a transient induction of ELAM-1 and a sustained increase in ICAM-1 expression.

Granule membrane protein-140 (GMP-140), known as PADGEM protein, is a membrane glyceroprotein of platelet and EC secretory granules and is structurally related to ELAM-1 (McEver, 1990). When expressed by activated endothelium, PADGEM promotes rapid neutrophil targeting to sites of acute inflammation (Geng *et al.*, 1990), and recruit of platelets to the sites of tissue injury. It also mediates adhesion of stimulated platelets to neutrophils (Hamburger and McEver, 1990) and monocytes (Larsen *et al.*, 1989). It was therefore suggested by these same authors that PADGEM, ELAM-1, and MEL-14 have not only structural but also functional homology.

Evidence from both *in vitro* and *in vivo* experiments has suggested that SP is able to function as an immunomodulator in disease states associated with inflammation. This possibility was supported by the findings of specific high-affinity receptors for SP on lymphocytes (Payan *et al.*, 1986) and macrophages (Hartung *et al.*, 1986). In addition, SP has been detected *in vivo* at locations and in concentrations that would allow the interaction between immunocytes and SP (Stead *et al.*, 1987). Thus, the essential elements for a contribution of neuronal systems to inflammatory and immune responses associated with disease states exist. There are increased levels of SP in inflammatory exudates in response to physical injury in rats (Tissot *et al.*, 1988) and in humans with inflammatory bowel diseases (Koch *et al.*, 1987). Not only are the levels of SP increased in patients with inflammatory bowel diseases, but receptor binding sites for SP are also increased (Mantyh *et al.*, 1988).

Recent evidence by Matis *et al.* (1990) has suggested that SP induces the expression of an ELAM-1 by microvascular endothelium at the level of postcapillary dermal venules as a direct consequence of degranulation of mast cells. It was therefore suggested that SP release by dermal nerve fibres upon physiologic or electrical stimulation may be important in the regulation of endothelial–leukocyte interactions *in vivo*.

The origin of SP in inflammatory exudates has been attributed to the secretion of the peptide from peripheral neurons (Nakanishi, 1987) and also from eosinophils (Weinstock *et al.*, 1988) and macrophages (Pascual and Bost, 1990) that have been shown to synthesize IR-SP during acute inflammation. SP released by leukocytes or neuronally derived SP can subsequently, via activation of NK1 receptors on ECs (Maggi *et al.*, 1990) and NK2 receptors in the immune system (Eglezos *et al.*, 1991a), modulate both endothelial and macrophage function by increasing secretion of factors, such as EDRF and IL-1. In addition, Kimball *et al.* (1988) have demonstrated that one of the functions of SP as well as NKA is to maintain chronic inflammation by promoting IL-1 production by macrophages. As mentioned earlier, the EC has recently been shown to be an important target for IL-1 and consequently, IL-1-stimulated ECs play an essential role in the inflammatory reaction of disease states. Activation of macrophages by SP could also result in the release of other inflammatory factors that could subsequently modulate EC function, including the arachidonic acid derived mediators, prostaglandins, and leukotrienes. One of the mechanisms underlying the actions of prostaglandin (PG) is the stabilization of endothelial lining and decrease in circulating ECs (Sinzinger *et al.*, 1988). Leukotrienes, on the other hand, can increase vascular permeability via an action on ECs at the level of PCV (Peck *et al.*, 1981). Our current knowledge of the interactions of the peripheral nervous system in disease states is clearly sketchy and, even in those diseases studied, incomplete.

B. Atherosclerosis

Endothelial cells and macrophages play a key role in the development of atherosclerosis. The formation of the early proliferative lesion of atherosclerosis is the result of chronic interaction of ECs, monocytes/macrophages, smooth muscle cells, and potentially other cells, such as T lymphocytes and platelets. The precise manner in which these cells interact during the early stage of the disease is unclear. Evidence was recently provided by Rajovashisth *et al.* (1990) to show exposure of ECs, which have been identified in atherosclerotic plaques, to oxidized lipoproteins results in a rapid induction of the expression of granulocyte macrophage-CSF (GM-CSF), macrophage-CSF (M-CSF), and granylocyte-CSF (G-CSF). These growth factors effect proliferation and migration of cells of the macrophage/granulocyte lineages (Metcalf, 1989) and in addition G-CSF also effects the migration and proliferation of ECs.

Further evidence has shown that exposure of ECs to these lipoproteins stimulates monocyte binding (Berliner *et al.*, 1990) as well as the production of chemotactic factors for monocytes (Berliner *et al.*, 1986). It is therefore evident that exposure of ECs to these lipoproteins is a potent inducer of three CSFs. These factors contribute to the initiation and progression of atherosclerotic lesions by affecting macrophage function. Macrophages do not only take up cholesterol to give rise to the foam cells (Rajavashisth *et al.*, 1990), but they also produce growth factors that can affect smooth muscle cell proliferation (Ross, 1986). Therefore, the link between ECs and macrophages is important for the initiation and progression of atherosclerosis. The initial mechanisms involved in the pathophysiological interaction of these cells with neuronal system identified in arterial walls represent a major challenge for future research in this disease. An involvement of noradrenaline (NA)-containing perivascular nerves in hypertension has been suggested based on changes in the pattern of perivascular innervation, including an increase in NA-containing nerves in several blood vessels (Abboud, 1982). In addition, it was demonstrated that the trophic role exerted by sympathetic nerves can reduce lipid accumulation in rabbit aorta (Fronek and Alexander, 1986). The relationship between the sympathetic nervous system and the development of hypertension and atherosclerosis has been further examined (Albert and Campbell, 1990) and reviewed (Campbell and Campbell, 1989). The role of primary afferent, however, has not been examined despite the demonstration of the innervation of the arteries where the lesions occur (Franco-Cereceda, 1988). As shown by immunocytochemistry, the peripheral nerve endings of primary sensory neurons are extensively associated with the cardiovascular system (Dalsgaard *et al.*, 1986, Gibbins *et al.*, 1988). These have been identified in the heart, around large arteries and veins, and in smaller vessels supplying vascular beds (Weihe, 1987). In particular, SP-LI nerve fibers were identified in perivascular nerves of the heart (Reinecke *et al.*, 1980), and coexistence of SP and CGRP was also found to occur extensively in the cardiovascular system (Lundberg *et al.*, 1985; Gibbins *et al.*, 1985). Evidence was also provided by Beny *et al.* (1986) that the relaxing effects of SP on the coronary artery are mediated by an endothelium-dependent mechanism involving the release of EDRF. In addition and in contrast to the vasodilator action of CGRP in many vessels, its relaxing effects in rat aortic rings are entirely endothelium dependent (Grace *et al.*, 1987). This effect is mediated via an increase in cyclic AMP and prostacyclin (Crossman *et al.*, 1987).

Although Foote *et al.* (1986) have shown no changes in SP-LI nerves in hypertensive rats, changes in the pattern of sensory perivascular innervation during atherosclerosis are largely unknown. The fact remains, however, that endothelium-mediated processes are highly attenuated in this disease (Harrison, 1988) and that sensory neuropeptides, including SP (Beny *et al.*, 1986), VIP (Ignarro *et al.*, 1987), and CGRP (Crossman *et al.*, 1987), exert their effects on vascular tone of cardiovascular blood vessels largely through the endothelium. These together with the extensive innervation of the cardiovascular system with primary afferent sensory nerves (Gibbins *et al.*, 1988) suggest that the role of these nerves and their sensory-derived peptides in the development of atherosclerosis will be important.

IV. Summary

In summary, it is clear that over recent years there have been many new developments in our knowledge of the tissue responses to injury that culminates in the development of inflammation and an immune response. However, much work still remains to be done before we understand the complex interaction between circulating inflammatory and immune cells, the microvascular endothelium and neural systems during the inflammatory reaction *in vivo*. Evidence has been accumulating for neural modulation for those components of the immune response that were described 10 years ago. It remains to be seen whether neural modulation also plays a role in the more recently described phenomena. It would be our view that such modulation will be demonstrated during the next few years. As we learn more about these interactions, it is suggested that new strategies of therapy to prevent or ameliorate inflammatory disease will also be developed. To delineate the role of the nervous system in the pathogenesis of astherosclerosis is an immediate and exciting challenge.

References

Abboud, F. M. (1982). The sympathetic system in hypertension. State of the art review. *Hypertension. 4* (Suppl. II), 11208–11225.

Albert, V., and Campbell, G. R. (1990). Relationship between the sympathetic nervous system and vascular smooth muscle: A morphometric study of adult and juvenile spontaneously hypertensive rat/wistar-kyoto rat caudal artery. *Heart vessels 5,* 129–139.

Ashby, B., Daniel, J. L., and Smith, J. B. (1990). Mechanisms of platelet activation and inhibition. *Hematol. Oncol. Clin. North Am. 4,* 1–26.

Azzali, G., Gatti, R., and Orlandini, G. (1990). Macrophage migration through the endothelium in the absorbing peripheral lymphatic vessel of the small intestine. *J. Submicrosc. Cytol. Pathol. 22,* 273–280.

Balkwill, F. R., and Burke, F. (1989). The cytokine network. *Immunol. Today 10,* 299–304.

Barja, F., Mathison, R., and Huggel, H. (1983). Substance P containing nerve fibres in large peripheral blood vessels of the rat. *Cell tissue Res. 229,* 411–422.

Bartha, K., Muller-Peddinghaus, R., and Van Rooifen, L. A. (1989). Bradykinin and thrombin effects on polyphosphoinositide hydrolysis and prostacyclin production in endothelial cells. *Biochem. J. 263,* 149–155.

Bendtzen, K. (1988). Interleukin 1, interleukin 6 and tumor necrosis factor in infection, inflammation and immunity. *Immunol. Lett. 19,* 183–191.

Bendtzen, K., Sevenson, M., Jonsson, V., and Hippe, E. (1990). Autoantibodies to cytokines—friends or foes? *Immunol. Today 11,* 167–169.

Beny, J. L., Brunet, P. C., and Huggel, H. (1986). Effect of mechanical stimulation, substance P and vasoactive intestinal polypeptide on the electrical and mechanical activities of circular smooth muscles from pig coronary arteries contracted with acetylcholine: Role of endothelium. *Pharmacol. 33,* 61–68.

Berliner, J. A., Territo, M., Almada, L., Carter, A., Shafonsky, E., and Fogelman, A. M. (1986). Monocyte chemotactic factor produced by large vessel endothelial cells *in vitro*. *Arteriosclerosis 6,* 254–258.

Berliner, J. A., Territo, M. C., Sevanian, A., Ramin, S., Kim, J. A., Bamshad, B., Esterson, M., and Fagelman, A. N. (1990). *J. Clin. Invest. 85,* 1260–1266.

Bevan, S., and Szolcsanyi, J. (1990). Sensory neuron-specific actions of capsaicin: Mechanisms and applications. *TIPS 11,* 330–333.

Bhathena, S. J., Louie, J., Schechter, G. P., Redman, R. S., Wahl, L., and Recant, L. (1981). Identification of human mononuclear leukocytes bearing receptors for somatostatin and glucagon. *Diabetes 30,* 127–131.

Blalock, J. E., Harbour-McMenamin, D., and Smith, E. M. (1985). Peptide hormones: Shared by the neuroendocrine and immune systems. *J. Immunol. 129,* 856s–861s.

Bolton, T. B., and Clapp, L. H. (1986). Endothelial-dependent relaxant actions of carbochol and substance P in the arterial smooth muscle. *Br. J. Pharmacol. 87,* 713–723.

Buck, S. H., and Burks, T. F. (1986). The neuropharmacology of capsaicin: Review of some recent observations. *Pharmacol. Rev. 38,* 179–226.

Butcher, E. C., Lewinsohn, D., Duijvestijn, A., Bargatze, R., Wu, N., and Jalkaren, S. (1986). Interactions between endothelial cells and leukocytes. *J. Cell. Biochem. 30,* 121–131.

Campbell, J. H., and Campbell, G. R. (1989). Biology of the vessel wall and atherosclerosis. *Clin. Exp. Hypertons. (A)11,* 901–913.

Casini, A., Geppetti, P., Maggi, C. A., and Surrenti, C. (1989). Effects of calcitonin gene-related peptide (CGRP), neurokinin A and neurokinin A (4–10) on the mitogenic response of human peripheral blood mononuclear cells. *Naunyn-Schmiedeberg's Arch. Pharmacol. 339,* 354–358.

Cavender, D. E., Haskard, D. O., Joseph, B., and Ziff, M. (1986). Interleukin-1 increases the binding of human B and T-lymphocytes to endothelial cell monolayers. *J. Immunol. 136,* 203–207.

Christie, M. L., Griffith, T. M., and Lewis, M. J. (1989). A comparison of basal and agonist-stimulated release of endothelium-derived relaxing factor from different arteries. *Br. J. Pharmacol. 98,* 397–406.

Christophe, J., Cauvin, A., Vervisch, E., Buscail, L., Damien, C., Abello, J., Gourlet, P., and Robberecht, P. (1990). Vasoactive intestinal peptide receptors on human SUP-T1 lymphoblasts. *Digestion 46,* 148–156.

Cozens, P. J., and Rowe, F. M. (1987). Substance P is a potent inducer of tumor necrosis factor and interleukin-1 secretion by macrophages. A potential role for tumor necrosis factor in the pathogenesis of asthma. *Immunobiol. 175,* 7.

Crossman, D., Dollery, C. T., MacDermot, J., MacIntyre, I., and MacEwan, J. (1987). Effects of human calcitonin gene-related peptide on human endothelial cells. *Br. J. Pharmacol. 90,* 38P.

Cynk, E., Kondo, K., Salvemini, D., Sneddon, J. M., and Vane, J. R. (1988). *J. Physiol.* (London) *407,* 28P.

Dalsgaard, C. J., Franco-Cereceda, A., Saria, A., Lundberg, J. M., Theodrsson-Norheim, E., and Hokfelt, J. M. (1986). Distribution and origin of substance P and neuropeptide y—immunoreactive nerves in the guinea pig heart. *Cell Tissue Res. 243,* 477–485.

Dinarello, C. A. (1988). Interleukin-I and its biologically related cytokines. *Adv. Immunol. 44,* 153–205.

D'Orleans-Juste, P., Dion, S., Mizrahi, J., and Regoli, D. (1985). Effects of peptides and non-peptides on isolated arterial smooth muscle: Role of endothelium. *Eur. J. Pharmacol. 114,* 9–21.

Doukas, J., and Pober, J. S. (1990). IFN-gamma enhances endothelial activation induced by tumor necrosis factor but not IL-1. *J. Immunol. 145,* 1727–1733.

Duijvestijn, A., and Haman, A. (1989). Mechanisms and regulation of lymphocyte migration. *Immunol. Today 10,* 23–28.

Dusting, G. J., Higgs, E. A., Macdonald, P. S., and Moncada, S. (1989). The endogenous nitro vasodilator produced by the vascular endothelium. *Aust. NZ. J. Med. 19,* 493–498.

Edvinsson, L., Fredholm, B. B., Hamel, E., Jansen, I., and Verrecchia, C. (1985). Perivascular peptides relax cerebral arteries concomitant with stimulation of cyclic adenosine monophosphate accumulation or release of an endothelium-derived relaxing factor in the cat. *Neurosci. Lett. 56,* 213–217.

Eglezos, A. (1990). Neurogenic modulation of the immune response. Ph.D. Thesis. University of Melbourne, Melbourne, Australia.

Eglezos, A., Andrews, P. V., Boyd, R. L., and Helme, R. D. (1990). Effects of capsaicin treatment on immunoglobulin secretion in the rat: Evidence for involvement of tachykinin containing afferent nerves. *J. Neuroimmunol. 26,* 131–138.

Eglezos, A., Andrews, P. V., Boyd, R. L., and Helme, R. D. (1991a). Tachykinin mediated modulation of the primary antibody response in rats: Evidence for mediation by an NK-2 receptor. *J. Neuroimmunol. 32,* 11–18.

Eglezos, A., Andrews, P. V., Boyd, R. L., Helme, R. D. (1991b). Modulation of the immune response by tachykinins: A review. *Immunol. Cell Biol. 69,* 285–294.

Foote, J., Robinson, J., and Scott, T. M. (1986). A comparison of substance P innervation of the mesenteric vascular bed in hypertensive and normotensive rats. *J. Physiol. Lond. 377,* 55P.

Franco-Cereceda, A. (1988). Calcitonin gene-related peptide and tachykinins in relation to local sensory control of cardiac contractility and coronary vascular tone. *Acta physiol. Scand. 569,* 1–63.

Fromek, K., and Alexander, M. (1986). Sympathetic activity, lipids accumulation and arterial wall morphology in rabbits at high altitude. *Am. J. Physiol. 250,* R485–R492.

Furchgott, R. F. (1983). Role of endothelium responses of vascular smooth muscle. *Circ. Res. 35,* 557–573.

Furness, J. B., Papka, R. E., Della, N. G., Costa, M., and Eskay, R. L. (1982). Substance P like immunoreactivity in nerves associated with the vascular system of guinea pigs. *Neurosci. 7,* 447–459.

Garthwaite, J., Charles, S. L., and Chess-Williams, R. (1988). Endothelium derived relaxing factor release on activation of NMDA receptors suggests role as intracellular messenger in the brain. *Nature 326,* 387–395.

Geng, J. G., Bevilacqua, M. P., Moore, K. L., McIntyre, T. M., Prescott, S. M., Kim, J. M., Bliss, G. A., Zimmerman, G. A., and McEver, R. P. (1990). Rapid neutrophile adhesion to activated endothelium mediated by GMP. 140. *Nature 343,* 757–760.

Gibbins, I. L., Furness, J. B., Costa, M., MacIntyre, I., Hillyard, C. J., and Girgis, S. (1985). Colocalization of calcitonin gene-related peptide-like immunoreactivity with substance P in cutaneous, vascular and visceral sensory neurons of guinea pigs. *Neurosci. Lett. 57,* 125–130.

Gibbins, I. L., Morris, J. L., Furness, J. B., and Costa, M. (1988). Innervation of systemic blood vessels. *In* Non-Adrenergic Innervation of Blood Vessels." G. Burnstock and S. G. Griffith, eds., Vol. II, pp. 1–36. CRC Press, Boca Raton, Florida.

Gowans, J. L. (1959). The recirculation of lymphocytes from blood to lymph in the rat. *J. Physiol.* (Lond.) *146,* 54–69.

Grace, G. C., Dusting, G. J., Kemp, B. E., and Martin, T. J. (1987). Endothelium and the vasodilator action of rat calcitonin gene-related peptide (CGRP). *Br. J. Pharmacol. 91,* 729–733.

Guerrero, J. M., Prieto, J. C., Elorza, F. L., Ramirez, R., and Goberna, R. (1981). Inaction of vasoactive intestinal peptide with human blood mononuclear cells. *Mol. Cell. Endocrinol. 21,* 151–160.

Habenicht, A. (1990). "Growth Factors, Differentiation Factors and Cytokines." Springer Verlag, Berlin.

Hamburger, S. A., and McEver, R. O. (1990). GMP-140 mediates adhesion of stimulated platelets to neutrophils. *Blood. 75,* 550–554.

Harrison, D. G. (1988). Artherosclerosis alters endothelium dependent responses to a variety of neurohumoral agents: Potential mechanisms and pathophysiological consequences. *In* "Vasodilatation, Vascular Smooth Muscle, Peptides, Autonomic Nerves and Endothelium." (P. M. Vanhoutte, ed.), pp. 517–521. Raven Press, New York.

Hart, R., Dancygier, H., Wagner, F., Lersch, C., and Classen, M. (1990). Effect of substance P on immunoglobulin and interferon-gamma secretion by human cultured duodenal mucosa. *Immunol. Lett. 23,* 199–204.

Hartung, H.-P., and Toyka, K. V. (1983). Activation of macrophages by substance P: Induction of oxidative burst and thromboxane release. *Eur. J. Immunol. 89,* 301–305.

Hartung, H.-P., Wolters, K., and Toyka, K. V. (1986). Substance P: Binding properties and studies on cellular responses in guinea-pig macrophages. *J. Immunol. 136,* 3856–3863.

Helme, R. D., Eglezos, A., and Andrews, P. V. (1987a). The effects of capsaicin denervation on leukocytes and complement components of the inflammatory response. *Clin. Exp. Neurol. 24,* 207–211.

Helme, R. D., Eglezos, A., and Hosking, C. S. (1987b). Substance P induces chemotaxis of neutrophils in normal and capsaicin treated rats. *Immunol. Cell Biol. 65,* 267–269.

Hendricks, H. R., and Estermans, I. L. (1983). Disappearance and reappearance of high endothelial venules and immigrating lymphocytes in lymph nodes deprived of afferent lymphatic vessels: A possible regulatory role of macrophages in lymphocyte migration. *Eur. J. Immunol. 13,* 663–669.

Hibbs, J. B., Tainter, R. R., and Varrin, Z. (1987). Macrophage cyto-toxicity: Role for L-arginine deiminase and imino nitrogen oxidation to nitrite. *Science. 235,* 473–476.

Holzer, P. (1988). Local effector functions of capsacin sensitive sensory nerve endings: Involvement of tachykinins, calcitonin gene-related peptide and other neuropeptides. *Neuroscience. 24,* 739–768.

Ignarro, L. J., Burns, R. E., Buga, G. M., and Wood, K. S. (1987). Mechanisms of endothelium dependent vascular smooth muscle relaxation elicited by bradykinin and VIP. *Am. J. Physiol. 253,* H1074–H1082.

Issekutz, T. B., and Stoltz, J. M. (1989). Stimulation of lymphocyte migration by endotoxin, tumor necrosis factor, and interferon. *Cell Immunol. 120,* 165–173.

Johnson, A. R., and Erdos, E. G. (1973). Release of histamine from mast cells by vasoactive peptides. *Proc. Soc. Exp. Biol. Med. 142,* 1252–1256.

Kelso, A. (1990). Cytokines in infectious disease. *Aust. Microbiol. 11,* 372–376.

Kelso, A., and Metcalf, D. (1990). T lymphocyte derived colony stimulating factors. *Adv. Immunol. 48,* 69–105.

Khalil, Z. (1987). Role of tachykinins in catecholamine secretion. Ph.D. thesis. University of Melbourne.

Kimball, E. S., Persico, F. J., and Vaught, J. L. (1988). Substance P, neurokinin A and neurokinin B induce generation of IL-1 like activity in P388DI cells. Possible relevance to arthritic disease. *J. Immunol. 141,* 3564–3569.

Koch, T. R., Carney, J. A., and Go, V. L. (1987). Distribution and quantitation of gut neuropeptides in normal intestine and inflammatory bowel disease. *Dig. Dis. Sci. 32,* 369–376.

Krco, C. J., Gores, A., and Go, V. L. W. (1986). Gastrointestinal regulatory peptides modulate *in vitro* immune reactions of mouse lymphoid cells. *Clin. Immunol. Immunopathol. 39,* 308–318.

Larsen, E., Celi, A., Gilbert, G. E., Furie, B. C., Erban, J. K., Bonfanti, R., Wagner, D. D., and Furie, B. (1989). PADGEM protein: A receptor that mediates the interaction of activated platelets with neutrophils and monocytes. *Cell. 59,* 305–312.

Laurenzi, M. A., Persson, M. A. A., Dalsgaard, C. J., and Ringden, O. (1989). Stimulation of human B lymphocyte differentiation by the neuropeptide substance P and neurokinin A. *Scand. J. Immunol. 30,* 695–701.

Leeuwenberg, J. F., Von Asnuth, E. J., Jeunhomme, T. M., and Buurman, W. A. (1990). IFN-gamma regulates the expression of the adhesion molecule ELAM-1 and IL-6 production by human endothelial cells *in vitro. J. Immunol. 145,* 2110–2114.

Lotz, M., Vaughan, J. H., and Carson, D. A. (1988). Effects of neuropeptides on production of inflammatory cytokines by human monocytes. *Science 241,* 1218–1221.

Lundberg, J. M., Franco-Cereceda, A., Hua, X., Hokfelt, T., and Fischer, J. A. (1985). Co-existence of substance P and calcitonin gene related peptide-like immunoreactivities in sensory nerves in relation to cardiovascular and bronchoconstrictor effects of capsaicin. *Eur. J. Pharmacol. 108,* 315–319.

Luscinskas, F. W., Brock, A. F., Arnout, M. A., and Gimbrone, M. A. Jr. (1989). Endothelial leukocyte adhesion molecule-1-dependent and leukocyte (CD11/CD18)-dependent mechanisms contribute to polymorphonuclear leukocyte adhesion to cytokine-activated human vascular endothelium. *J. Immunol. 142,* 2257–2263.

McCall, T. B., Boughton-Smith, N. K., Palmer, R. M. J., Whittle, B. J. R., and Moncada, S. (1989). Synthesis of nitric oxide from L-arginine by neutrophils. *Biochem. J. 261,* 293–296.

McEver, R. P. (1990). Properties of GMP-140, an inducible granule membrane protein of platelets and endothelium. *Blood Cells. 16,* 73–80.

McGillis, J. P., Organist, M. L., Scriven, K. H., and Payan, D. G. (1987). Purification of the 33,000-dalton ligand binding-protein constituent of the lymphoblast substance P receptor. *J. Neurosci. Res. 18,* 190–194.

Maggi, C. A., Patacchini, R., Perretti, F., Tramontana, M., Manzini, S., Geppetti, P., and Santicioli, P. (1990). Sensory nerves, vascular endothelium and neurogenic relaxation of the guinea-pig isolated pulmonary artery. *Naunyn Schmiedebergs Arch. Pharmacol. 342,* 78–84.

Maggi, C. A., Patacchini, R., Giuliani, S., Rovero, P., Dion, S., Regoli, D., Giachetti, A., and Meli, A. (1990). Competitive antagonists discriminate between NK2 tachykinin receptor subtypes. *Br. J. Pharmacol. 100,* 589–592.

Mantyh, C. R., Gates, T. S., Zimmerman, R. P., Welton, M. L., Passearo, Jr. E. D., Vigna, S. R., Maggio, J. E., Kruger, L., and Mantyh, P. W. (1988). Receptor binding sites for substance P but not substance K or neuromedin K, are expressed in high concentrations by arterioles, venules and lymph nodules in surgical specimens obtained from patients with ulcerative colitis and Crohn disease. *Proc. Natl. Acad. Sci. U.S.A. 85,* 3235–3239.

Marasco, W. A., Showell, H. J., and Becker, E. L. (1981). Substance P binds the formyl-peptide chemotaxis receptor on the rabbit neutrophil. *Biochem. Biophys. Res. Commun. 99,* 1065–1072.

Marletta, M. A., Yoon, P. S., Iyengar, R., Leaf, C. D., and Wishnok, J. S. (1988). Macrophage oxidation of L-arginine to nitrite and nitrate nitric oxide is an intermediate. *Biochem. 27,* 8706–8711.

Matis, W. L., Lavker, R. M., and Murphy, G. F. (1990). Substance P induces the expression of an endothelial-leukocyte adhesion molecule by microvascular endothelium. *J. Invest. Dermatol. 94,* 492–495.

Mazurek, N., Pecht, I., Teichburg, V. I., and Blumberg, S. (1981). The role of the N-terminal tetrapeptide in the histamine releasing action of substance P. *Neuropharmacol. 20,* 1025–1027.

Metcalf, D. (1989). The molecular control of cell division, differentiation commitment and maturation in haemopoietic cells. *Nature 339,* 27–30.

Moncada, S., Palmer, R. M. J., and Higgs, E. A. (1989). Biosynthesis of nitric oxide from L-arginine. *Biochem. Pharmacol. 38,* 1709–1715.

Moore, T. C. (1984). Modification of lymphocyte traffic by vasoactive neurotransmitter substances. *Immunol. 52,* 511–518.

Moore, T. C., Lami, J. L., and Spruck, C. H. (1989). Substance P increases lymphocyte traffic and lymph flow through peripheral lymph nodes of sheep. *Immunol. 67,* 109–114.

Nagy, J. I., Iversen, L. L., Goedert, M., Chapman, D., and Hunt, S. P. (1983). Dose-dependent effects of capsaicin on primary sensory neurones in the neonatal rat. *J. Neurosci. 3,* 339–406.

Nakanishi, S. (1987). Substance P precursor and kininogen: Their structures, gene organizations, and regulation. *Physiol. Rev. 67,* 1117–1142.

Nilsson, G., and Ahlstedt, S. (1988). Altered lymphocyte proliferation of immunized rats after neurological manipulation with capsaicin. *Int. J. Immunopharmacol. 10,* 747–751.

Nilsson, J., von Euler, A. M., and Dalsgaard, C.-J. (1985). Stimulation of connective tissue cell growth by substance P and substance K. *Nature* (Lond.) *315,* 61–63.

O'Connell, K. A., and Edidin, M. (1990). A mouse lymphoid endothelial cell line immortalized by simian virus 40 binds lymphocytes and retains functional characteristics of normal endothelial cells. *J. Immunol. 144,* 521–525.

O'Dorisio, M. S. (1988). Neuropeptide modulation of the immune response in gut associated lymphoid tissue. *Intern. J. Neurosci. 38,* 189–198.

O'Dorisio, M. S., Hermina, N. S., O'Dorisio, T. M., and Balcerzak, S. P. (1981). Vasoactive intestinal peptide modulation of lymphocyte adenylate cyclase. *J. Immunol. 127,* 2551–2554.

O'Dorisio, M. S., Wood, C. L., and O'Dorisio, T. M. (1985). Vasoactive intestinal peptide and neuropeptide modulation of the immune response. *J. Immunol. 135,* 792s–796s.

Oehme, P., Hecht, K., Jumatov, J., Repke, H., Hilse, H., and Cordova, A. (1987). Prevention of stress-induced involution of the thymus in rats by substance P (SP_{1-11}) and its N-terminal fragment SP_{1-4}. *Pharmazie 42,* 34–36.

Orr, F. W., Buchanan, M. R., Tron, V. A., Guy, D., Lauri, D., and Sauder, D. N. (1988). Chemotactic activity of endothelial cell-derived interleukin I for human tumor cells. *Cancer Res. 48,* 6758–6763.

Ottaway, C. A. (1984) *In vitro* alteration of receptors for vasoactive intestinal peptide changes the *in vivo* localization of mouse T cells. *J. Exp. Med. 160,* 1054–1069.

Ottaway, C. A. (1987). Selective effects of vasoactive intestinal peptide on the mitogenic response of murine T cells. *Immunobiol. 62,* 291–297.

Ottaway, C. A., and Greenberg, G. R. (1984). Interaction of vasoactive intestinal peptide with mouse lymphocytes: Specific binding and the modulation of mitogen responses. *J. Immunol. 132,* 417–423.

Ottaway, C. A., Bernaerts, C., Chan, B., and Greenberg, G. R. (1983). Specific binding of vasoactive intestinal peptide to human circulating mononuclear cells. *Can. J. Phys. Pharm. 61,* 664–671.

Ottaway, C. A., Lewis, D. L., and Asa, S. L. (1987). VIP-containing nerves in peyers patches. *Brain Beh. Immun. 1,* 148–158.

Palmer, R. M., and Moncada, S. (1989). A novel citrulline-forming enzyme implicated in the formation of nitic oxide by vascular endothelial cells. *Biochem. Biophys. Res. Commun. 156,* 345–352.

Palmer, R. M. J., Ferrige, A. G., and Moncada, S. (1987). Nitric oxide release accounts for the biological activity of endothelium-derived relaxing factor. *Nature 327,* 524–526.

Palmer, R. M. J., Ashton, D. S., and Moncada, S. (1988). Vascular endothelial cells synthesize nitric oxide from L-arginine. *Nature 33,* 664–666.

Pascual, D. W., and Bost, K. L. (1990). Substance P production by P388D1 macrophages: A possible autocrine function for this neuropeptide. *Immunol. 71,* 52–56.

Payan, D. G., Brewster, D. R., and Goetzl, E. J. (1983). Specific stimulation of human T lymphocytes by substance P. *J. Immunol. 131,* 1613–1615.

Payan, D. G., Brewster, D. R., Missirian-Bastien, A., and Goetzl, E. J. (1984a). Substance P recognition of a subset of human T-lymphocytes. *J. Clin. Invest. 74,* 1532–1539.

Payan, D. G., Brewster, D. R., and Goetzl, E. J. (1984b). Stereospecific receptors for substance P on cultured human IM-9 lymphoblasts. *J. Immunol. 133,* 3260–3265.

Payan, D. G., McGillis, J. P., and Goetzl, E. J. (1986). Neuroimmunology. *Adv. Immunol. 30,* 299–323.

Peck, M. J., Piper, P. J., and Williams, T. J. (1981). The effect of leukotrinenes C_4 and D_4 on the microvasculature of guinea pig skin. *Prostaglandins. 21,* 315–317.

Piotrowski, W., and Foreman, J. C. (1985). On the actions of substance P, somatostatin and vasoactive intestinal polypeptide on rat peritoneal mast cells in human skin. *Naunyn-Schmiedeberg's Arch. Pharmacol. 331,* 364–368.

Pober, J. S., and Grimbrone, M. A. Jr. (1982). Expression of Ia-like antigens by human vascular endothelial cells is inducible *in vitro:* Demonstration by monoclomal antibody binding and immune precipitation. *Proc. Nat. Acad. Sci. (U.S.A.) 79,* 6641–6645.

Pujol, J.-L., Bousquet, J., Grenier, J., Michel, F., Godard, P., Chanez, P., De Vos, C., De Paulet, C., and Michel, F.-B. (1989). Substance P activation of bronchoalveolar macrophages from asthmatic patients and normal subjects. *Clin. Exp. Allergy 19,* 625–628.

Rajavashisth, T. B., Eng, R., Shadduck, R. K., Waheed, A., Ben-Avram, C. M., Shively, J. E., Lusis, A. J. (1987). Cloning and tissue-specific expression of mouse macrophage colony-stimulating factor in RNA. *Proc. Natl. Acad. Sci. (U.S.A.). 84,* 1157–1161.

Rajavashisth, T. B., Andalibi, A., Territo, M. C., Berliner, J. A., Navab, M., Fogelman, A. M., and Lusis, A. J. (1990). Induction of endothelial cell expression of granulocyte and macrophage colony-stimulating factors by modified low-density lipoproteins. *Nature 344,* 254–257.

Regoli, D., Escher, E., Drapeau, G., D'Orleans-Juste, P., and Mizrahi, J. (1984a). Receptors for substance P. III. Classification by competitive antagonists. *Eur. J. Pharmacol. 97,* 179–189.

Regoli, D., Escher, E., and Mizrahi, J. (1984b). Substance P structure-activity studies and the developments of antagonists. *Pharmacol. 28,* 301–32.

Reinecke, M., Weihe, E., and Forsmann, W. C. (1980). Substance P-immunoreactive nerve fibres in the heart. *Neurosci. Lett. 20,* 265–269.

Renkonen, R., Mattila, P., and Ustinov, J. (1990). Signal transduction during platelet-activation factor-induced lymphocyte binding to endothelial cells. *Scand. J. Immunol. 31,* 523–527.

Richardson, B. P., Engel, G., Donatsch, P., and Stadler, P. A. (1985). Identification of serotonin M-receptor sub types and their specific blockade by a new class of drugs. *Nature* (Lond.) *316,* 126–131.

Rimele, T. J., Sturm, R. J., Adams, L. M., Henry, D. E., Heaslip, R. J., Weichman, B. M., and Grimes, D. (1988). Interaction of neutrophils with vascular smooth muscle: Identification of a neutrophil-derived relaxing factor. *J. Pharmacol. Exp. Ther. 245,* 102–111.

Roch-Arvellier, M., Regoli, D., Chanaud, B., Lenoir, M., Muntaner, O., Stalzko, S., and Giroud, J.-P. (1986). Tachykinins: Effects on motility and metabolism of rat polymorphonuclear leukocytes. *Pharmacol. 33,* 266–273.

Ross, R. (1986). The pathogenesis of arthersclerosis an update. *N. Engl. J. Med. 314,* 488–500.

Rovero, P., Pestellini, V., Patacchini, R., Giuliani, S., Maggi, C. A., Meli, A., and Giachetti, A. (1990). Synthesis and biological activity of NK-2 selective tachykinin antagonists containing D-tryptophan. *Peptides 11,* 619–620.

Ruff, M. R., Wahl, S. M., and Pert, C. B. (1985). Substance P mediated chemotaxis of human monocytes. *Peptides 6,* (suppl. 2), 107–111.

Salvemini, S., Masini, E., Anggard, E., Mannaioni, P. F., and Vane, J. (1990). Synthesis of a nitric oxide-like factor from L-arginine by rat serosal mast cells: Stimulation of guanylate cyclase and inhibition of platelet aggregation. *Biochem. Biophys. Res. Comm. 169,* 596–601.

Scicchitano, R., Dazin, P., Bienenstock, J., Payan, D. G., and Stanisz, A. M. (1987a). Distribution of somatostatin receptors on murine spleen and peyer's patch T and B lymphocytes. *Brain Beh. Immun. 1,* 173–184.

Scicchitano, R., Bienenstock, J., and Stanisz, A. M. (1987b). The differential effect with time of neuropeptides on the proliferative responses of murine peyer's patch and splenic lymphocytes. *Brain. Beh. Immun. 1,* 231–237.

Scicchitano, R., Bienenstock, J., and Stanisz, A. M. (1988a). *In vivo* immunomodulation by the neuropeptide substance P. *Immunol. 63,* 733–735.

Sinzinger, H., Virgolini, I., Fitscha, P., Rauscha, F., and Kaliman, J. (1988). Stabilization of endothelial lining and decrease in circulating endothelial cells—one mechanism underlying the clinical action of PGE1?. *Br. J. Clin. Pharmacol. 25,* 775–776.

Skofitsch, G., Savvit, J. M., and Jacobwitz, D. M. (1985). Suggestive evidence for a functional unit between macrophages and substance P fibres in the rat diaphragm and mesentery. *Histochem. 82,* 5–8.

Smith, K. A. (1989). The interleukin 2 receptor. *Ann. Rev. Cell Biol. 5,* 397–425.

Smith, K. A. (1990). Interleukin-2. *Sci. American 262,* 26–33.

Smith, J. A., and Lang, D. (1990). Release of endothelium-derived relaxing factor from pig cultured aortic endothelial cells, as assessed by changes in endothelial cell cyclic GMP content, is inhibited by a pharbol ester. *Br. J. Pharmacol. 99,* 565–571.

Spertini, O., Kansas, G. S., Munro, J. M., Griffin, J. D., and Tedder, T. F. (1991). Regulation of leukocyte migration by activation of the leukocyte adhesion molecule-1 (LAM-1) selection. *Nature 349,* 691–694.

Stanisz, A. M., Befus, D., and Bienenstock, J. (1986). Differential effects of vasoactive intestinal peptide, substance P and somatostatin on immunoglobulin synthesis and proliferation by lymphocytes from peyer's patches, mesenteric lymph nodes and spleen. *J. Immunol. 136,* 152–156.

Stanisz, A. M., Scicchitano, R., Dazin, P., Bienenstock, J., and Payan, D. G. (1987a). Distribution of substance P receptors on murine spleen and peyer's patch T and B cells. *J. Immunol. 139,* 749–754.

Stanisz, A., Scicchitano, R., Stead, R., Matsuda, H., Tomioka, M., Denburg, J., and Bienenstock, J. (1987b). Neuropeptides and immunity. *Am. Rev. Resp. Dis. 136,* S48–S51.

Stead, R. H., Bienenstock, J., and Stanisz, A. M. (1987). Neuropeptide regulation of mucosal immunity. *Immunol. Rev. 100,* 333–359.

Stephenson, J. A., and Summers, R. J. (1987). Autoradiographic analysis of receptors on vascular endothelium. *Eur. J. Pharmacol. 134,* 35–43.

Stuehr, D. J., Gross, S. S., Sakuma, I., Levi, R., and Nathan, C. F. (1989). Activated murine macrophages secrete a metabolite of arginine with the bioactivity of endothelium derived relaxing and the chemical reactivity of nitric oxide. *J. Exp. Med. 169,* 1011–1020.

Tanaka, T., Danno, K., Ikai, K., and Imamura, S. (1988). Effects of substance P and substance K on the growth of cultured keratinocytes. *J. Invest. Dermatol. 90,* 399–401.

Thieme, T. R., and Wagner, C. R. (1989). The molecular weight of the endothelial cell IL-1 receptor is 78,000. *Mol. Immunol. 26,* 249–253.

Tissot, M., Pradelles, P., and Giroud, J. P. (1988). Substance P-like levels in inflammatory exudates. *Inflammation 12,* 25–35.

Trnka, Z., and Cahill, R. N. P. (1980). Aspects of the immune response in single lymph nodes. Monogr. *Allergy 16,* 245–259.

Vanhoutte, P. M. (1987). Serotonin and the vascular wall. *Int. J. Cardiol. 14,* 189–203.

Wagner, F., Fink, R., Hart, R., and Dancygier, H. (1987). Substance P enhances interferon-γ production by human peripheral blood mononuclear cells. *Reg. Pept. 19,* 355–364.

Ward, P. A., Warren, J. S., and Johnson, K. J. (1988). Oxygen radicals, inflammation, and tissue injury. *Free Radic. Biol. Med. 5,* 403–408.

Weihe, E. (1987). Peripheral innervation of the heart. *In* "Silent Ischemia." (T. H. Von Armin and A. Maseri, eds.), pp. 7–18. Steinkoff, Darmstadt.

Weinstock, J. V., Blum, A., Walder, J., and Walder, R. (1988). Eosinophils from granulomas in murine schistosomiasis mansoni produce substance P. *J. Immunol. 141,* 961–966.

Whatley, R. E., Nelson, P., Zimmerman, G. A., Stevens, D. L., Parker, C. J., McIntyre, T. M., and Prescott, S. M. (1989). The regulation of platelet activating factor production in endothelial cells. The role of calcium and protein kinase C. *J. Biol. Chem. 264,* 6325–6333.

Willms-Kretschmer, K., Flax, M. H., and Cotran, R. S. (1967). The fine structure of the vascular response in hapten specific delayed hypersensitivity and contact dermatitis. *Lab. Invest. 17,* 334–349.

Wong, G. H., Bartlett, P. F., Clark-Lewis, I., Battye, F., and Schrader, J. W. (1984). Inducible expression of H2 and Ia antigens on brain cells. *Nature* (Lond.) *310,* 688–691.

Yu, C. L., Haskard, D. O., Cavender, D., Johnson, A. R., and Ziff, M. (1985). Human gamma interferon increases the binding of T lymphocytes to endothelial cells. *Clin. Exp. Immunol. 62,* 554–560.

PART II

The Sympathetic System

CHAPTER 7

Noradrenaline and Neuropeptide Y in the Sympathetic System: Prejunctional Interactions and Regulation of Release

PIA LINTON-DAHLÖF   CARL DAHLÖF

I. Introduction

Neuropeptide Y (NPY) is a 36 amino acid peptide present in the brain, the adrenal medulla, and peripheral sympathetic nerves. NPY-like immunoreactivity (LI) has been found in abundance in the perivascular nerve fibers of many organs, particularly around the cerebral, renal, and femoral arteries. NPY-LI-containing fibers are more abundant around arteries than veins. Fibers around the small arteries show more immunoreactivity than nerve fibers around large vessels. The sympathetic origin of these nerves has been established by the presence of the enzymes involved in the synthesis of catecholamines, tyrosine-hydroxylase, and dopamine-β-hydroxylase (Lundberg *et al.*, 1982, 1983). In addition, the NPY-LI-containing nerves are sensitive to surgical sympathectomy as well as chemical destruction of the sympathetic nerves with the neurotoxin 6-hydroxy-dopamine (Lundberg *et al.*, 1982, 1983; Edvinsson *et al.*, 1983; Ekblad *et al.*, 1984).

Cotransmission by additional neurotransmitters has been suggested to be the most likely explanation for cardiovascular effector responses to sympathetic nerve stimulation that are resistant to classical adrenoceptor antagonists. Peptides, such as NPY (Lundberg *et al.*, 1982, 1983), purine nucleotides, such as adenosine triphosphate (ATP) (Burnstock, 1976), and also amines, such as adrenaline (Guimaraes, 1978; Rand *et al.*, 1979; Dahlöf, 1981) and dopamine may accompany noradrenaline (NA) as neuronal transmitter substances in postganglionic sympathetic neurons. ATP is co-stored with NA in small and large synaptic vesicles, whereas NPY is co-stored with NA in large dense-cored vesicles within the sympathetic nerves (Fried, 1980; Fried *et al.*, 1985). Results of experimental research provide strong support for the proposal that NPY is coreleased with NA from postganglionic sympathetic nerve terminals and involved in the regional effector control of sympathetic neurotransmission (Pernow, 1988; Linton-Dahlöf, 1989). High-frequency electrical stimulation of sympathetic nerve fibers causes secretion of both NPY and NA, whereas continuous low-frequency stimulation results in the preferential secretion of NA from small vesicles (Lundberg *et al.*, 1986a). The location and mode of release of NPY led to the proposal that this peptide plays an important role in modulating the contribution of the sympathetic nervous system to the control of vascular tone and/or cardiac function (Linton-Dahlöf, 1989).

The nerve terminals of the peripheral sympathetic nerves appear to be endowed with a vast number of prejunctional receptor types, each capable of mediating a modulatory effect on the amount of the transmitter released per nerve impulse. The existence of prejunctional receptors raises the question of their activation by an endogenous agonist. The prejunctional receptors can either be categorized as prejunctional *autoreceptors* (mainly affected by their neuronal transmitters) or prejunctional *heteroreceptors* ("autacoidreceptors," which respond to neurotransmitters released by adjacent neurons and to endogenous "naturally occurring" substances other than neuronal transmitters, e.g., hormones, circulating in the bloodstream). As yet, the differential physiologic role and/or impact of the modulatory mechanisms that are triggered by activation of these prejunctional receptors is not fully understood.

This review will mainly focus on modulatory interactions at the prejunctional level of action potential-evoked release of NA and/or NPY from sympathetic nerve terminals. Most of the referred results have been obtained in animal experimental studies carried out *in vitro* or *in vivo*.

II. Modulatory Mechanisms of Noradrenaline Release

The action potential-mediated release of NA from sympathetic neurons is modulated by a variety of prejunctional receptors including adrenergic receptors, such as inhibitory α_2-adrenoceptors and facilitatory β-adrenoceptors (Dahlöf, 1981; Langer, 1981; Starke, 1977, 1987). Furthermore, these prejunctional receptors include inhibitory muscarinic cholinergic, dopaminergic, prostaglandin, serotonin, GABA, and histamine receptors. In addition, there are prejunctional facilitatory angiotensin and nicotinic cholinergic receptors. The nonadrenergic prejunctional receptors known at present are inhibitory NPY-receptors, opioid receptors, and P_1-purinoceptors (adenosine) (Fig. 1).

A. NA Release—Prejunctional Autoreceptors

1. Prejunctional α-Adrenoceptors

The autoinhibition caused by released NA via prejunctional α-adrenoceptors and the interaction of drugs with this regulatory mechanism has been extensively studied. The negative feedback control exerted by prejunctional α-adrenoceptors has been thoroughly documented (Gillespie, 1980; Langer, 1977; Starke, 1987). Stimulation of these prejunctional receptors thus diminishes the NA release per impulse, whereas α-adrenoceptor blockade greatly enhances transmitter release. For the sake

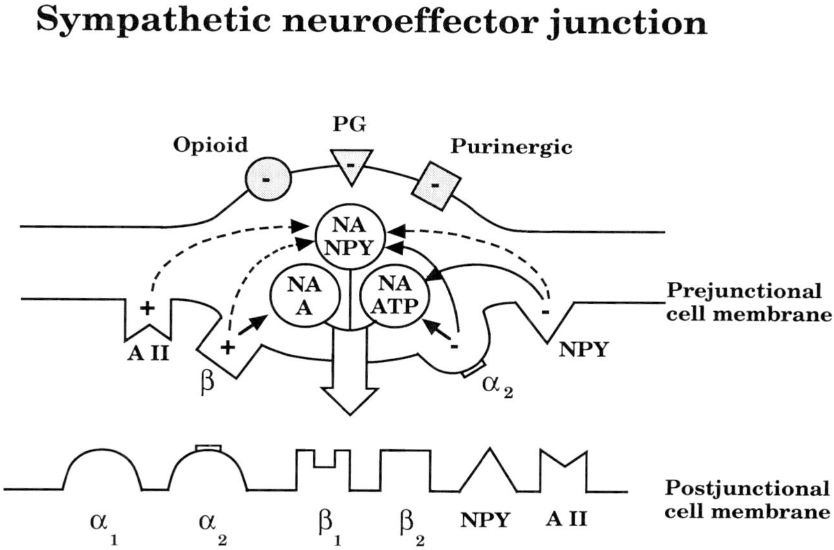

Figure 1 Schematic illustration of some pre- and postjunctional receptors involved in sympathetic neurotransmission. (+ and − indicate prejunctional receptor-mediated facilitation and inhibition of transmitter release, respectively.)

of simplicity, the α-adrenoceptors will only be subdivided into α_1 and α_2 subtypes here. With regard to the α_1 subtype, there is no firm evidence that α_1-adrenoceptors are active prejunctionally and capable of modulating the release of the transmitter from sympathetic nerve terminals. For example, the selective α_1-adrenoceptor antagonist prasozin appears not normally to affect the action potential-evoked release of NA (Starke, 1987).

The overflow radioactivity evoked by a single pulse is not facilitated in the presence of the α_2-adrenoceptor antagonist rawolscine or idazoxan (Valenta *et al.*, 1989). However, both antagonists display marked facilitatory effects *in vitro* when brain slices are stimulated with a short train of pulses delivered at 1 Hz. The results are in agreement with the view that stimulation with a single pulse—in contrast to two or more pulses—does not elicit the autoinhibitory mechanism due to the fact that the amount of NA in the biophase around the prejunctional α_2-adrenoceptors is too low at the time of stimulation (Langer, 1981). The α_2-adrenocepter agonist clonidine appears to decrease stimulation-evoked release of NA independently of the number of pulses employed. It appears that the full development of the autoinhibitory feedback mechanism has a certain time requirement since on two-pulse stimulation at 1 Hz, the amount of NA released by the second pulse is only a fraction (approximately 15%) of that released by the first (Singer, 1988). However, when the two stimuli are repeated within <100 ms (>10 Hz), the overflow evoked by this second pulse sharply increases and reaches values approximately equal to the amount released by a single stimulus.

2. Prejunctional β-Adrenoceptors

Another control mechanism for NA release seems to be mediated by prejunctional facilitatory β-adrenoceptors (Dahlöf, 1981; Langer, 1981; Misu and Kubo, 1986). Isoprenaline and other β-adrenoceptor agonists acting on prejunctional mainly β_2-adrenoceptors increase the action potential-evoked release of NA in most of the tissues that have been studied, and pretreatment with β-adrenoceptor antagonists abolishes their facilitatory effect. Adrenaline (A) in concentrations occurring in the circulation in stress also exerts this effect and appears to serve as a humoral modulator when it is secreted from the adrenal medulla. In addition, A is taken up into sympathetic noradrenergic nerve terminals, incorporated into transmitter stores together with NA, and subsequently released as a cotransmitter. When a sufficient concentration of A has accumulated, the cotransmitter A mediates a positive feedback loop, facilitating transmitter release by acting on the prejunctional β-adrenoceptors (Majewski and Rand, 1986).

B. NA Release—Prejunctional Heteroreceptors

1. Prejunctional NPY-Receptors

The proposal of a prejunctional inhibitory effect of NPY on the nerve stimulation-induced release of NA arises from *in vitro* studies on vas deferens preparations where NPY was found to inhibit the contraction elicited by electrical field stimulation without attenuating the contractile response to NA (Allen *et al.*, 1982; Lundberg *et al.*, 1982; Lundberg and Stjärne, 1984). This inhibitory effect of NPY has now been studied in many *in vitro* vascular preparations. Thus, NPY at a concentration of 500 nM has been shown to reduce the transmural nerve stimulation-evoked efflux of ^{3}H-NA from rat portal vein (Dahlöf *et al.*, 1985), rat femoral artery (Lundberg *et al.*, 1985c), portal and femoral veins, superior mesenteric and basilar arteries (Pernow *et al.*, 1986b), and human submandibular arteries (Lundberg *et al.*, 1985d) as well as the nerve stimulation-evoked release of endogenous NA from the mesenteric arterial bed (Westfall *et al.*, 1987). The threshold concentration for the inhibitory effect of NPY on the rat femoral vein and mesenteric arterial bed has been found to be 10^{-9} M and 10^{-10} M, respectively (Pernow *et al.*, 1986b; Westfall *et al.*, 1987). The inhibitory effect of NPY appears to be resistant to treatment with calcium antagonists (nifedipine), α_1-(prazosin), and α_2-(yohimbine) adrenoceptor blockade (Lundberg *et al.*, 1985c; Pernow *et al.*, 1986b; Westfall *et al.*, 1987). The degree of inhibition varies with the vascular preparations between 40% (portal vein) and 80% (basilar and mesenteric arteries) (Dahlöf *et al.*, 1985; Pernow *et al.*, 1986b). In addition to species differences, this discrepancy might be due to differences in the methods used. A similar inhibitory effect of NPY on the nerve stimulation-induced release of NA has also been demonstrated in isolated guinea pig heart preparations (Franco-Cereceda *et al.*, 1985; Haas *et al.*, 1989) and in other tissues, such as rabbit oviduct (Chernaeva and Charakchieva, 1988) and mouse vas deferens (Stjärne *et al.*, 1986; Serfozo *et al.*, 1986). On the other hand, no inhibitory effect of NPY on NA release has been observed in isolated vessels from cats and rabbits (Edvinsson and Skärby, 1984; Ekblad *et al.*, 1984).

The prejunctional inhibitory effect of exogenous NPY on the nerve stimulation-induced release of NA has also been demonstrated *in situ* as well as *in vivo*. Thus, intravenous administration of NPY, in a dose (470 pmol $\times$ kg^{-1} $\times$ min^{-1}) that gives a plasma concentration of approximately 20 nM, to the pithed rat has been shown to reduce preganglionic nerve stimulation-induced release of NA by 20% (Dahlöf *et al.*, 1986). This inhibitory effect of exogenous NPY on the nerve stimulation-induced release of NA in the pithed rat model is in agreement with findings obtained in studies on the canine gracilis muscle (*in situ*) preparation (Pernow *et al.*, 1988a; Kahan *et al.*, 1988). Thus, it was found that local intraarterial infusion of NPY inhibited the nerve stimulation-induced outflow of endogenous NA by about 30% at plasma concentrations of NPY of about 1 μM (Pernow *et al.*, 1988a). In addition, infusion of NPY into the blood-perfused pig spleen *in vivo* reduced the nerve stimulation-evoked overflow of ^{3}H-NA by 25% (Lundberg *et al.*, 1989a). The inhibitory effect of NPY on the nerve stimulation-evoked increase of NA could be due to a vasoconstriction elicited by exogenous NPY (Lundberg and Tatemoto, 1982; Edvinsson *et al.*, 1983; Allen *et al.*, 1983b) that impairs the diffusion of NA, but this is less likely since it has been shown that moderate variations of vascular tone in the gracilis muscle preparation do not influence nerve stimulation-evoked NA overflow (Kahan *et al.*, 1985). The magnitude of the inhibition *in vivo* appears to be smaller than that observed *in vitro*. This difference could be due to the fact that changes in ^{3}H-NA release might be more difficult to demonstrate due to a more restricted diffusion and a more marked degradation of NPY *in vivo*.

NPY has also been shown to reduce the preganglionic nerve stimulation (PNS)-induced secretion of A from the adrenal medulla by about 40% (Dahlöf *et al.*, 1986). In accordance with this finding, an inhibitory effect of NPY on nicotine-mediated release of catecholamines from cultured bovine adrenal chromaffin cells has been reported (Higuchi *et al.*, 1987). In the same preparation, peptides structurally related to NPY were shown to have a similar inhibitory effect on catecholamine release. In addition, binding to membranes of the bovine adrenal medulla indicated a correlation between their ability to inhibit catecholamine release and to displace the radioligand *N*-propionyl-[^{3}H]-NPY (Higuchi *et al.*, 1987).

The results obtained in studies with NPY-agonists suggest that at least three subtypes of NPY receptors exist, Y_1-like, Y_2-like, and Y_3-like (Mitchel, 1991). These NPY receptors belong to the family of G-protein-coupled receptors, and accordingly, most effects of NPY are sensitive to pertussis toxin (Balasubramaniam and Sheriff, 1990; Ewald *et al.*, 1988; Mitchel *et al.*, 1990; Wahlestedt *et al.*, 1990). The results of these studies support the suggestion that inhibition of adenylate cyclase is a common signaling mechanism of NPY receptors. Thus, the signal construction mechanism most frequently associated with NPY is the inhibition of adenylate cyclase activity, via an inhibitory G-protein. Activation of NPY-receptors can also influence free cytosolic calcium (Ca^{2+}) concentrations by regulating both the influx of extracellular Ca^{2+} and mobilization of intracellular Ca^{2+} (Mitchel, 1991). Inhibition of *N*-type Ca^{2+}-channels in neuronal preparations is of particular interest because it has been suggested that this may be the mechanism of prejunctional inhibition of transmitter release. The prejunctional inhibitory NPY-receptors appear to be of the Y_2-like subtype. Accordingly, NPY has convincingly been shown to be capable of inhibiting NA release from the sympathetic nerve terminals via activation of prejunctional NPY receptors. However, the scarcity of NPY-receptor antagonists has hampered definitive demonstration of the physiologic role of NPY in the sympathetic neurotransmission.

2. Prejunctional Purinergic Receptors (Adenosine)

The ATP (adenosine triphosphate) co-released with NA or its breakdown product adenosine is known to act on prejunctional P_1-purinoceptors to inhibit NA release (Hedqvist and Fredholm, 1976; Kügelgen and Starke, 1991). These receptors are blocked by methylxanthines like theophylline but not by the P_2-purinoceptor blocker α,β-methylene-ATP. In this sense, adenosine receptors on sympathetic nerve terminals can be regarded as prejunctional inhibitory autoreceptors, but there are also other sources of ATP and adenosine (Fredholm, 1988).

3. Prejunctional Opioid Receptors

The opioid peptides are another group of compounds that have been demonstrated to be co-stored and co-released with NA. Met- and leuenkephalin are two of these opioids that interact with at least three opioid receptor subtypes (μ, κ, and δ). There is, at present, no strong evidence that the opioid receptors on sympathetic noradrenergic terminals act as prejunctional inhibitory autoreceptors under physiologic conditions. However, electrically evoked NA release is depressed by opioid receptor agonists. Morphine inhibits sympathetic neurotransmission by acting on opioid receptors to inhibit transmitter release in only a few sites: the cat nictitating membrane (Dubocovich and Langer, 1980), the mouse vas deferens (Henderson *et al.*, 1972), and the cat spleen (Gaddis and Dixon, 1982).

4. Prejunctional Prostaglandin Receptors

Prostaglandins (PG), particularly PGE_2, are released from many effector tissues when they are activated by sympathetic nerve stimulation or other stimuli. PGE_2 mediates prejunctional inhibition of NA release, thus providing a brake on excessive excitation. On the other hand, inhibition of PG synthesis results in an increase in NA release (Westfall, 1977; Langer, 1981). The PGE_2-mediated pre- and/or transjunctional feedback loop operates independently of the NA-mediated autoinhibition (Rump and Majewski, 1987). Prostaglandins of the F series, particularly PGF_2, have a facilitatory action on NA release in some tissues (Vanhoutte *et al.*, 1980).

5. Prejunctional Angiotensin II Receptors

Angiotensin II (AII) has been claimed to have two sites of action on noradrenergic neurons: one at the cell bodies, where it elicits action potentials (Trendelenburg, 1967), and one at the terminal axons, where it facilitates the action potential-evoked NA release (Zimmerman, 1978; Starke, 1977; Westfall, 1977). In addition, the AII-induced potentiation of sympathetic neurotransmission can be blocked by competitive AII antagonists like saralasin, which demonstrates that the prejunctional facilitation is mediated by AII-receptors.

III. Modulatory Mechanisms of NPY Release

The release of neuronal NPY can be modulated by different sympatholytic drugs. Electrical stimulation of postganglionic nerves to the cat spleen has been shown to evoke the concomitant release of NPY-LI and NA by a mechanism sensitive to the neuronal blocking agent guanethidine (Lundberg *et al.*, 1984). Other sympatholytic drugs seem to affect the release of NPY-LI. Thus, reserpine, which interferes with vesicular monoamine storage (Carlsson, 1965), appears to deplete NPY from certain sympathetically innervated tissues (Lundberg *et al.*, 1985a, 1986c). This effect of reserpine can be prevented by the ganglionic blocking agent chlorisondamine, which indicates that the NPY depletion by reserpine is due to enhanced neuronal activity and consequent release of the neuropeptide (Lundberg *et al.*, 1986c). The NPY-LI secretion from the adrenal medulla upon preganglionic nerve stimulation seems to be parallel to A overflow, resistant to guanethidine but sensitive to chlorisondamine (Lundberg *et al.*, 1986c). The reserpine-induced depletion of medullary NPY-LI, as in nerve terminals, can also be prevented by the ganglionic blocking agent (Lundberg *et al.*, 1986c).

For various reasons, NPY release is normally evaluated by comparing levels of plasma NPY-like immunoreactivity (LI) before and after a specific intervention. Circulating levels of plasma NPY-LI vary widely between species, being low in humans, about 20 pM (Lundberg *et al.*, 1986a) and high in rats, from about 200 pM (Pernow *et al.*, 1988c) to 1–4 nM, (Morris *et al.*, 1987). The reason for the very high plasma levels of NPY-LI in the rat is most likely related to the presence of NPY in thrombocytes in this species (Ericsson *et al.*, 1987). Physiological activation of the sympathetic nervous system has been shown to enhance plasma NPY-LI levels in parallel with plasma catecholamines both in rats, after exposure to stress (Zukowska-Grojec *et al.*, 1988; Zukowska-Grojec and Vaz, 1988), and in man, after physical exercise (Lundberg *et al.*, 1985b). In both species, a neural rather than adrenal medullary origin of the NPY-LI levels has been suggested since, in these situations, plasma NPY-LI seems to correlate better with the plasma levels of NA than with A. Further evidence suggesting the neuronal origin of plasma NPY-LI has been obtained in experiments where the time course of plasma A, NA, and NPY-LI release after preganglionic nerve stimulation was studied in the pithed guinea pig (Dahlöf *et al.*, 1991). The peak plasma A level occurred immediately after the initiation of nerve stimulation, which is compatible with A being secreted directly into the circulating blood. In contrast to this, NA and NPY-LI reached their peak plasma levels 20–40 s later. This pattern suggests that NA and NPY are mainly released from the sympathetic nerve terminals, from where the two transmitters diffuse into the systemic circulation.

Furthermore, it has been suggested that NPY-LI is released during strong but probably not mild sympathetic activation under physiological conditions in man (Pernow *et al.,* 1986a). Accordingly, NPY-LI has been shown to be preferentially released at high frequencies whereas NA can already be detected at low frequencies of stimulation (Lundberg *et al.,* 1986a).

In patients with pheochromocytoma, a tumor of adrenal medullary origin with high (up to nM) plasma levels of NPY-LI has been detected (Adrian *et al.,* 1983; Corder *et al.,* 1985; Lundberg *et al.,* 1986b). Other conditions in humans that have been demonstrated to be associated with increased NPY-LI plasma levels are thoracotomy and surgery involving cardiopulmonary bypass (Lundberg *et al.,* 1985d) and vaginal delivery, where, in fact, the highest plasma levels of NPY-LI observed under physiologic conditions in humans have been found in the newborn infant (Lundberg *et al.,* 1986d). Thus, stressful situations appear to be associated with increased plasma NPY-LI levels in different species, including man. In addition, the release of NPY into the circulation apparently varies with the intensity of the stress stimulus. A progressive increase in plasma NPY levels was observed in rats exposed to a graded series of stress (e.g., cold, insulin, handling, and electric shock) (Castagné *et al.,* 1987). Plasma NPY-LI levels have also been shown to be increased in conscious rats after hemorrhage (Morris *et al.,* 1987), exposure to cold water or transfer to a new environment (Zukowska-Grojec and Vaz, 1988), in the anesthetized pig during hemorrhagic hypovolemia (Rudehill *et al.,* 1987), and in man during physical exercise (Lundberg *et al.,* 1985b; Pernow *et al.,* 1986a; Morris *et al.,* 1986a), an acute asthma attack (Dahlöf *et al.,* 1988), insulin-induced hypoglycemia (Takahashi *et al.,* 1988), and congestive heart failure (Maisel *et al.,* 1989). Although the increase in plasma NPY-LI levels correlates with that of NA during an acute asthma attack (Dahlöf *et al.,* 1988) and during exercise in humans, the neuropeptide appears to be preferentially released at high work loads (i.e., at high levels of sympathetic nerve activity) (Pernow *et al.,* 1986a). In that study, neither handgrip nor tilting, which were considered to be low degrees of sympathetic activation, altered plasma NPY-LI resting levels. With regard to the half-life of NPY, the NPY-LI disappearance curve after intravenous infusion of NPY has been found, at least in human plasma, to be biphasic, with half-lives of 4 and 20 min (see Pernow, 1988).

A. NPY Release—Prejunctional Autoreceptors

1. NPY Receptors

For several reasons, but mainly due to the lack of specific NPY antagonists, it has not yet been possible to elucidate whether neuronally released NPY is capable of modifying its own release. However, there is one piece of information available suggesting that neuronal NPY can act prejunctionally on NPY autoreceptors (Y_2-like?) to inhibit its own release. It was namely shown that exposure to the NPY analog PYY reduced action-potential evoked release of NPY in an *in situ* preparation of pig kidney (Pernow *et al.*, 1988c).

B. NPY Release—Prejunctional Heteroreceptors

1. Prejunctional α-Adrenoceptors

In vitro studies on the isolated blood-perfused cat and pig spleen have shown that the nerve stimulation-mediated overflow of NPY-LI can be enhanced in the presence of the α-adrenoceptor antagonist phenoxybenzamine. This finding has been thought to be due to a prejunctional regulation of NPY release (Lundberg *et al.*, 1984, 1986a) apparently similar to the α-adrenoceptor-mediated modulation of NA release (Langer 1981; Starke, 1977). In accordance with these data, *in vivo* studies on the pithed guinea pig support the existence of an α-adrenoceptor-mediated inhibitory mechanism of nerve stimulation-induced increase of plasma NPY-LI in this whole animal preparation. In addition, this mechanism has been suggested to be mediated by the α_2-adrenoceptor subtype since prazosin was not able to elicit any changes of plasma NPY-LI levels after electrical nerve stimulation (Dahlöf *et al.*, 1986). Furthermore, it was shown in this study that guanethidine pretreatment reduced the yohimbine-induced enhancement of stimulation-evoked increase of plasma NPY-LI, which suggests that the contribution of adrenal medullary NPY-LI to this increase is low. Studies in the isolated guinea pig heart (Archelos *et al.*, 1987), dog gracilis muscle (Pernow *et al.*, 1988a; Kahan *et al.*, 1988), and perfused dog spleen (Schoups *et al.*, 1988) confirm the existence of this α-adrenoceptor-mediated inhibition. This mechanism has been further elucidated in the blood-perfused pig spleen (Lundberg *et al.*, 1986a, 1989b). The nerve stimulation-induced release of NA and NPY-LI was enhanced after phentolamine as well as after treatment with reserpine, when the degree of prejunctional α_2-adrenoceptor activation is supposed to be reduced (Lundberg *et al.*, 1986a). The enhancement was greater at low stimulation frequencies (2 Hz). Pernow *et al.* (1988b) recently suggested that the release of NPY and NA in man seems to be influenced by both peripheral and central α-adrenoceptors since the increase in plasma levels during physical exercise was found to be enhanced and suppressed by phentolamine and clonidine, respectively. The prejunctional nature of this inhibition cannot be distinguished from a compensatory increase in the sympathetic nerve activity owing to the decrease in peripheral resistance induced by these drugs. However, the findings are in line with the results obtained in the pithed animal model (Dahlöf *et al.*, 1986) and lend further support to the existence of a prejunctional α-adrenoceptor-mediated inhibitory control mechanism of neuronal NPY release, which is apparently operating under physiological *in vivo* conditions.

2. Prejunctional β-Adrenoceptors

In addition to prejunctional α-adrenoceptors, the NPY-release also appears to be modulated by a prejunctional β-adrenoceptor-mediated facilitatory mechanism

(Dahlöf *et al.*, 1991) similar to that controlling the release of NA (Dahlöf, 1981; Misu and Kubo, 1986). Experiments in the pithed guinea pig have shown that treatment with the β-adrenoceptor agonist isoprenaline enhanced plasma NPY-LI in response to preganglionic nerve stimulation and this effect could be blocked by the β-blocker propranolol, which *per se* did not alter plasma NPY-LI (Dahlöf *et al.*, 1991). The magnitude of this facilitation (37%), however, is small compared to the inhibitory α_2-adrenoceptor-mediated control mechanism (about three- to fourfold). In yohimbine-pretreated animals, however, the enhanced plasma NPY-LI in response to preganglionic nerve stimulation was not further increased but, on the contrary, was partially inhibited by isoprenaline. However, propranolol was demonstrated to reduce the yohimbine-induced enhancement of plasma NPY-LI during sympathetic nerve stimulation in the blood-perfused pig spleen (Lundberg *et al.*, 1989a).

3. Angiotensin II Receptors

Beside the prejunctional α- and β-adrenoceptor-mediated modulatory mechanisms, the NPY release also appears to be affected by the angiotensin-converting enzyme inhibitor captopril. In the pithed guinea pig, it was recently demonstrated that captopril treatment *per se* reduced both basal plasma NPY-LI values and the increase of plasma NPY-LI in response to preganglionic nerve stimulation (by 38% and 46%, respectively) (Dahlöf *et al.*, 1991). This finding would indicate that captropril affects a prejunctional facilitatory mechanism. The observed effects of captopril on the plasma NPY-LI levels, however, appeared not to be due to inhibited formation of circulating angiotensin II (AII) since infusion of exogenous AII did not reverse this effect. Nor did treatment with the angiotensin-receptor antagonist saralasin have any significant effect on plasma NPY-LI levels. On the other hand, Pernow and coworkers (1988c) demonstrated a facilitatory effect of AII on the nerve stimulation-evoked release of NPY-LI in the pig kidney that could be blocked by saralasin. In these experiments, however, captopril did not significantly affect this parameter. This discrepancy leaves the field open for speculation. Nevertheless, an angiotensin-dependent mechanism appears to be involved in the modulatory control of NPY release.

IV. Conclusions

Strong experimental support has recently been provided for the concept that NPY is co-stored and co-released with NA from postganglionic sympathetic nerve terminals and involved in the regional effector control of sympathetic neurotransmission. At low sympathetic nerve activity the release of NA dominates, whereas NPY is preferentially released during conditions of moderate or high sympathetic nerve activity. Opioid peptides, like met- and leuenkephalin, purine nucleotides, such as ATP, as well as amines, like adrenaline and dopamine, may also accompany NA as neuronal transmitter substances in postganglionic sympathetic neurons.

NA and NPY constitute a well-studied example of the modulatory interactions between an adrenergic and a nonadrenergic transmitter component of the sympathetic nervous system, respectively. Activation of prejunctional α_2-adrenoceptors and/or Y_2-like NPY receptors leads to a reduction of the release of NA and NPY, whereas stimulation of prejunctional β-adrenoceptors appears to enhance the release of these transmitters. In addition to that, the release of NA and in some cases NPY can also be modulated via prejunctional P_1-purinoceptors (adenosine) and receptors for acetylcholine, angiotensin, opioids, prostaglandin, and serotonin. The cotransmission of NA and NPY apparently represents a functional mechanism employed by the sympathetic nervous system to achieve the control of vascular tone, cardiac function, and/or other functional demands. In this delicate interplay, NA mainly accounts for the postjunctional immediate/intermediate responses, whereas NPY is more likely to be responsible for the intermediate/slow responses of the sympathetic neurotransmission.

References

Adrian, T. E., Allen, J. M., Terenghi, G., Bacarese-Hamilton, A. J., Brown, M. J., Polak, J. M., and Bloom, S. R. (1983). Neuropeptide Y in phaeochromocytomas and ganglioneuroblastomas. *Lancet ii,* 540–542.

Allen, J. M., Adrian, T. E., Tatemoto, K., Polak, J. M., Hughes, J., and Bloom, S. R. (1982). Two novel related peptides, neuropeptide Y and peptide YY (PYY) inhibit the contraction of the electrically stimulated mouse vas deferens. *Neuropeptides 3,* 71–77.

Allen, J. M., Adrian, T. E., Polak, J. M., and Bloom, S. R. (1983a). Neuropeptide Y (NPY) in the adrenal gland. *J. Auton. Nerv. Syst. 9,* 559–563.

Allen, J. M., Bircham, P. M. M., Edwards, A. V., Tatemoto, K., and Bloom, S. R. (1983b). Neuropeptide Y (NPY) Reduces Myocardial Perfusion and inhibits the force of contraction of the isolated perfused rabbit Heart. *Regul. Pept. 6,* 247–253.

Archelos, J., Xiang, J. Z., Reinecke, M., and Lang, R. E. (1987). Regulation of release and function of neuropeptides in the heart. *J. Cardiovasc. Pharmacol. 10* (Supp 12), 45–50.

Balasubramaniam, A., and Sheriff, S. (1990). Neuropeptide Y (18–36) is a competitive antagonist of neuropeptide Y in rat cardiac membranes. *J. Biol. Chem. 265,* 14724–14727.

Burnstock, G. (1976). Do some nerve cells release more than one transmitter? *Neurosci. 1,* 239–248.

Carlsson, A. (1965). Drugs Which Block the Storage of 5-hydroxytryptamine and Related Amines. *In* "Handbook of Experimental Pharmacology." (O. Eichler and A. Farah, eds.), Vol. 19, pp. 529–592. Springer-Verlag, Berlin.

Castagné, V., Corder, R., Gaillard, R., and Morméde, P. (1987). Stress-induced changes of circulating neuropeptide Y in the rat: Comparison with catecholamines. *Regul. Pept. 19,* 55–63.

Chernaeva, L., and Charakchieva, S. (1988). Leucine-enkephalin- and neuropeptide Y-modulation of (^{3}H)-noradrenaline release in the oviduct of mature and juvenile rabbits. *Gen. Pharmacol. 19,* 137–142.

Corder, R., Lowry, P. J., Emson, P. C., and Gaillard, R. C. (1985). Chromatographic characterisation of the circulating neuropeptide Y immunoreactivity from patients with phaeochromocytoma. *Regul. Pept. 10,* 91–93.

Dahlöf, C. (1981). Studies on β-adrenoceptor mediated facilitation of sympathetic neurotransmission. *Acta Physiol. Scand.* (Suppl 500), 1–147.

Dahlöf, C., Dahlöf, P. Tatemoto, K., and Lundberg, J. M. (1985). Neuropeptide Y (NPY) reduces field stimulation-evoked release of noradrenaline and enhances force of contraction in the rat portal vein. *Naunyn. Schmiedebergs. Arch. Pharmacol. 328,* 327–330.

Dahlöf, C., Dahlöf, P., and Lundberg, J. M. (1986). Alpha-2-adrenoceptor-mediated inhibition of nerve stimulation-evoked release of neuropeptide Y (NPY)-like immunoreactivity in the pithed guinea-pig. *Eur. J. Pharmacol. 131,* 279–283.

Dahlöf, C., Dahlöf, P., Lundberg, J. M., and Strömbom, U. (1988). Elevated plasma concentration of neuropeptide Y and low level of circulating adrenaline in elderly asthmatics during rest and acute severe asthma. *Pulmonary pharmacol. 1,* 3–6.

Dahlöf, P., Lundberg, J. M., and Dahlöf, C. (1991). Effect of captopril on plasma levels of neuropeptide Y(NPY)-like immunoreactivity in the pithed guinea-pig. *Neuropeptides 18,* 171–180.

Dahlöf, P., Tarizzo, V. I., Lundberg, J. M., and Dahlöf, C. (1991). Alpha- and beta-adrenoceptor-mediated effects on nerve stimulation-evoked release of neuropeptide Y(NPY)-like immunoreactivity in the pithed guinea-pig. *J. Autonom. Nervous System 35,* 199–210.

Dubocovich, M. L., and Langer, S. Z. (1980). Pharmacological differentiation of presynaptic inhibitory α-adrenoceptors and opiate receptors in the cat nictitating membrane. *Br. J. Pharmacol. 70,* 383–393.

Edvinsson, L., and Skärby, T. (1984). Interaction of peptides present in perivascular nerves on the smooth muscle of cat cerebral arteries and on the stimulation-evoked efflux of 3H-noradrenaline. *J. Auton. Pharmacol. 4,* 193–198.

Edvinsson, L., Emson, P., Mcculloch, J., Tatemoto, K., and Uddman, R. (1983). Neuropeptide Y: Cerebrovascular innervation and vasomotor effects in the cat. *Neurosci. Lett. 43,* 79–84.

Edvinsson, L., Håkanson, R., Steen, S., Sundler, F., Uddman, R., and Wahlestedt, C. (1985). Innervation of human omental arteries and veins and vasomotor response to noradrenaline, neuropeptide Y, substance P and vasoactive intestinal peptide. *Regul. Pept. 12,* 67–79.

Ekblad, E., Edvinsson, L., Wahlestedt, C., Uddman, R., Håkanson, R., and Sundler, F. (1984). Neuropeptide Y co-exists and co-operates with noradrenaline in perivascular nerve fibers. *Regul. Pept. 8,* 225–235.

Ericsson, A., Schalling, M., Mcintyre, K. R., Lundberg, J. M., Larhammar, D., Seroogy, K., Hökfelt, T., and Persson, H. (1987). Expression of Neuropeptide Y mRNA and peptide in megakaryocytes: Enhanced levels in certain autoimmune mice. *Proc. Natl. Acad. Sci. 84,* 5585–5590.

Ewald, D. A., Sternweis, P. C., and Miller, R. J. (1988). Guanine nucleotide-binding protein G_0-induced coupling of neuropeptide Y receptors to Ca^{2+} channels in sensory neurones. *Proc. Natl. Acad. Sci. U.S.A. 85,* 3633–3637.

Franco-Cereceda, A., Lundberg, J. M., and Dahlöf, C. (1985). Neuropeptide Y and sympathetic control of heart contractility and coronary vascular tone. *Acta Physiol. Scand. 124,* 361–369.

Fredholm, B. B. (1988). Presynaptic adenosine receptors. *ISI Atlas of Sciences: Pharmacol. 2,* 257–260.

Fried, G. (1980). Small noradrenergic storages vesicles isolated from rat vas deferens—biomedical and morphological characterization. *Acta Phys. Scand. 111* (Suppl. 493), 1–28.

Fried, G., Terenius, L., Hökfelt, T., and Goldstein, M. (1985). Evidence for differential localization of noradrenaline and neuropeptide Y (NPY) in neuronal storage vesicles isolated from rat vas deferens. *J. Neurosci. 5,* 450–458.

Fried, G., Terenius, L., Hökfelt, T., and Goldstein, M. (1985a). Evidence for differential localization of noradrenaline and neuropeptide Y in neuronal storage vesicles isolated from rat vas deferens. *J. Neurosci. 5,* 450–458.

Fried, G., Lundberg, J. M., and Theodorsson-Norheim, E. (1985b). Subcellular storage and axonal transport of neuropeptide Y (NPY) in relation to catecholamines in the cat. *Acta Physiol. Scand. 125,* 145–154.

Gaddis, R. R., and Dixon, W. R. (1982). Modulation of peripheral adrenergic neurotransmission by methionine-enkephalin. *J. Pharmacol. Exp. Ther. 221,* 282–288.

Gillespie, J. S. (1980). Presynaptic receptors in the autonomic nervous system. *In* "Handbook of Experimental Pharmacology" Adrenergic activators and inactivators. (L. Szekeres, ed.), Vol 54, Part I, pp. 352–425. Springer-Verlag, Berlin.

Guimaraes, S., Brandao, F., and Paiva, M. Q. (1978). A study of the adrenoceptor-mediated feedback mechanisms by using adrenaline as a false transmitter. *Naunyn. Schmiedeberg's Archives of Pharmacol. 305,* 185–188.

Haas, M., Cheng, B., Richardt, G., Lang, R., and Schömig, A. (1989). Characterization and presynaptic modulation of stimulation-evoked exocytotic co-release of noradrenaline and neuropeptide Y in guinea-pig heart. *Naunyn Schmiederberg'S Arch. Pharmacol. 339,* 71–78.

Hedquist, P. (1977). Basics mechanisms of prostaglandins on autonomic neurotransmission. *Ann. Rev. Pharmacol. Toxicol. 17,* 259–279.

Hedqvist, P., and Fredholm, B. B. (1976). Effects of adenosine on adrenergic neurotransmission: Prejunctional inhibition and postjunctional enhancement. *Naunyn Schmiedeberg's Archives of Pharmacology 293,* 217–223.

Henderson, G., Hughes, J., and Kosterlitz, H. W. (1972). A new example of a morphine-sensitive neuroeffector junction: Adrenergic transmission in the mouse vas deferens. *Br. J. Pharmacol. 46,* 764–766.

Higuchi, H., Costa, E., and Yang, H. Y. (1987). Neuropeptide Y inhibits the nicotine-mediated release of catecholamines from bovine adrenal chromaffin cells. *J. Pharmacol. Exp. Ther. 244,* 468–474.

Kahan, T., Dahlöf, C., and Hjemdahl, P. (1985). Influence of acetylcholine, peptides and other vasodilators on endogenous noradreneline overflow in canine blood perfused gracilis muscle. *Acta Physiol. Scand. 124,* 457–465.

Kahan, T., Pernow, J., Schwieler, J., Lundberg, J. M., Hjemdahl, P., and Wallin, B. G. (1988). Involvement of neuropeptide Y in sympathetic vascular control of skeletal muscle in vivo. *J. Hypertens. 6*(Suppl. 4), 532–534.

Kügelgen, I., and Starke, K. (1991). Noradrenaline-ATP co-transmission in the sympathetic nervous system. *TIPS 12,* 319–324.

Langer, S. Z. (1977). Presynaptic receptors and their roles in the regulation of transmitter release. *Br. J. Pharmacol. 60,* 481–497.

Langer, S. Z. (1981). Presynaptic regulation of the release of Catecholamines. *Pharmacol. Rev. 32,* 337–362.

Linton-Dahlöf, P. (1989). Modulatory interactions of neuropeptide Y (NPY) on sympathetic neurotransmission. *Acta Phys. Scand. 137*(suppl. 586), 1–85.

Lundberg, J. M., and Tatemoto, K. (1982). Pancreatic polypeptide family (APP, BPP, NPY and PYY) in relation to sympathetic vasoconstriction resistent to alpha-adrenoceptor blockade. *Acta Physiol. Scand. 116,* 393–402.

Lundberg, J. M., and Stjärne, L. (1984). Neuropeptide Y (NPY) Depresses the secretion of ^{3}H-noradrenaline and the contractile response evoked by field stimulation, in rat vas deferens. *Acta Physiol. Scand. 120,* 477–479.

Lundberg, J. M., Hamberger, B., Schultzberg, M., Hökfelt, T., Granberg, P.-O., Efendic, S., Terenius, L., Godstein, M., and Luft, R. (1979). Enkephalin- and somatostatin-like immunoreactivities in human adrenal medulla and pheochromocytoma. *Proc. Natl. Acad. Sci. 76,* 4079–4083.

Lundberg, J. M., Terenius, L., Hökfelt, T., Martling, C., Tatemoto, K., Mutt, V., Polak, J., Bloom, S., and Goldstein, M. (1982). Neuropeptide Y (NPY)-like immunoreactivity in peripheral noradrenergic neurons and effects of NPY on sympathetic function. *Acta Physiol. Scand. 116,* 477–480.

Lundberg, J. M., Terenius, L., Hökfelt, T., and Goldstein, M. (1983). High levels of neuropeptide Y in peripheral noradrenergic neurons in various mammals including man. *Neurosci. Lett. 42,* 167–172.

Lundberg, J. M., Änggård, A., Theodorsson-Norheim, E., and Pernow, J. (1984). Guanethidine-sensitive release of Neuropeptide Y-like immunoreactivity in the cat spleen by sympathetic nerve stimulation. *Neurosci. Lett. 52,* 175–180.

Lundberg, J. M., Saria, A., Franco-cereceda, A., Hökfelt, T., Terenius, L., and Goldstein, M. (1985a). Differential effects of reserpine and 6-hydroxydopamine on neuropeptide Y (NPY) and noradrenaline in peripheral neurons. *Naunyn-Schmiederberg's Arch. Pharmacol. 328,* 331–340.

Lundberg, J. M., Martinsson, A., Hemsén, A., Theodorsson-Norheim, E., Svedenhag, J., Ekblom, B., and Hjemdahl, P. (1985b). Co-release of neuropeptide Y and catecholamines during physical exercise in man. *Biochem. and Biophys. Res. Comm. 133,* 30–36.

Lundberg, J. M., Pernow, J., Tatemoto, K., and Dahlöf, C. (1985c). Pre- and postjunctional effects of NPY on sympathetic control of rat femoral artery. *Acta Physiol. Scand. 123,* 511–513.

Lundberg, J. M., Torssell, L., Sollevi, A., Pernow, J., Theodorsson-Norheim, E., Änggård, A., and Hamberger, B. (1985d). Neuropeptide Y and sympathetic vascular control in man. *Regul. Pept. 13,* 41–52.

Lundberg, J. M., Rudehill, A., Sollevi, A., Theodorsson-Norheim, E., and Hamberger, B. (1986a). Frequency- and reserpine-dependent chemical coding of sympathetic transmission: Differential release of noradrenaline and neuropeptide Y from pig spleen. *Neurosci. Lett. 63,* 96–100.

Lundberg, J. M., Hökfelt, T., Hemsén, A., Theodorsson-Norheim, E., Pernow, J., Hamberger, B., and Goldstein, M. (1986b). Neuropeptide Y-like immunoreactivity in adrenaline cells of adrenal medulla and in tumors and plasma of pheochromocytoma patients. *Regul. Pept. 13,* 169–182.

Lundberg, J. M., Al-Saffar, A., Saria, A., and Theodorsson-Norheim, E. (1986c). Reserpine-induced depletion of neuropeptide Y from cardiovascular nerves and adrenal gland due to enhanced release. *Naunyn-Schmiederberg'S Pharmacol. 332,* 163–169.

Lundberg, J. M., Hemsén, A., Fried, G., Theodorsson-Norheim, E., and Lagercrantz, H. (1986d). Co-release of neuropeptide Y (NPY)-like immunoreactivity and catecholamines in newborn infants. *Acta Physiol. Scand. 126,* 471–473.

Lundberg, J. M., Rudehill, A., and Sollevi, A. (1989a). Pharmacological characterization of neuropeptide Y and noradrenaline mechanisms in sympathetic control of pig spleen. *Eur. J. Pharmacol. 163,* 103–113.

Lundberg, J. M., Rudehill, A., Sollevi, A., Fried, G., and Wallin, G. (1989b). Co-release of neuropeptide Y and noradrenaline from pig spleen *in vivo:* Importance of subcellular storage, nerve impulse frequency and pattern, feedback regulation and resupply by axonal transport. *Neurosci. 28,* 475–486.

Maisel, A. S., Scott, N. A., Motulsky, H. J., Michel, M. C., Boublik, J. H., Rivier, J. E., Ziegler, M., Allen, R. S., and Brown, M. R. (1989). Elevation of plasma neuropeptide Y levels in congestive heart failure. *Am. J. Med. 86,* 43–48.

Majewski, H., and Rand, M. J. (1986). A possible role of epinephrine in the development of hypertension. *Med. Res. Rev. 6,* 467–486.

Michel, M. C. (1991). Receptors for neuropeptide Y: Multiple subtypes and multiple and multiple second messengers. *TIPS 12,* 389–394.

Michel, M. C., Schlicker, E., Fink, K., Boublik, J. H., Göthert, M., Willette, R. N., Daly, R. N., Hieble, J. P., Rivier, J. E., Motulsky, H. J. (1990). Distinction of NPY receptor subtypes *in vitro* and *in vivo. Am. J. Physiol. 259,* E131–139.

Misu, Y., and Kubo, T. (1986). Presynaptic β-adrenoceptors. *Med. Res. Rev. 6,* 197–225.

Morris, M. J., Elliott, J. M., Cain, M. D., Kapoor, V., West, M. J., and Chalmers, J. P. (1986a). Plasma neuropeptide Y levels rise in patients undergoing exercise tests for the investigation of chest pain. *Clin. Exp. Pharmacol. Physiol. 13,* 437–440.

Morris, M., Kapoor, V., and Chalmers, J. (1987). Plasma neuropeptide Y concentration is increased after hemorrhage in conscious rats: Relative contributions of sympathetic nerves and the adrenal medulla. *J. Cardiovasc. Pharmacol. 9,* 541–545.

Pernow, J. (1988). Co-release and functional interactions of neuropeptide Y and noradrenaline in peripheral sympathetic Vascular Control. *Acta Physiol. Scand.* (Suppl 568), 1–56.

Pernow, J., and Lundberg, J. M. (1989). Modulation of noradrenaline and neuropeptide Y (NPY) Release in the pig kidney *in vivo:* involvement of alpha$_2$, NPY and Angiotensin II Receptors. *Naunyn-Schmiederberg's Arch. Pharmacol. 340,* 379–385.

Pernow, J., Lundberg, J. M., Kaijser, L., Hjemdahl, P., Theodorsson-Norheim, E., Martinsson, A., and Pernow, B. (1986a). Plasma neuropeptide Y-like immunoreactivity and catecholamines during various degrees of sympathetic activation in man. *Clin. Physiol. 6,* 561–578.

Pernow, J., Saria, A., and Lundberg, J. M. (1986b). Mechanisms underlying pre- and postjunctional effects of neuropeptide Y in sympathetic vascular control. *Acta Physiol. Scand. 126,* 239–249.

Pernow, J., Kahan, T., Hjemdahl, P., and Lundberg, J. M. (1988a). Possible involvement of neuropeptide Y in sympathetic vascular control of canine skeletal muscle. *Acta Physiol. Scand. 132,* 43–50.

Pernow, J., Lundberg, J. M., and Kaijser, L. (1988b). Alpha-adrenoceptor influence on plasma levels of neuropeptide Y-like immunoreactivity and catecholamines during rest and sympathoadrenal activation in humans. *J. Cardiovasc. Pharmacol. 12,* 593–599.

Pernow, J., Thorén, P., Millberg, B. I., and Lundberg, J. M. (1988c). Renal sympathetic nerve activation in relation to reserpine-induced depletion of neuropeptide Y in the kidney of the Rat. *Acta Physiol. Scand. 134,* 53–59.

Rand, M. J., Majewski, H., McCulloch, M. W., and Story, D. F. (1979). An adrenaline-mediated positive feedback loop in sympathetic transmission and its possible role in hypertension. *In* "Advances in the biosciences. Presynaptic receptors." (S. Z. Langer, K. Starke, and M. L. Dubocovich, eds.), Vol 18, pp. 263–269. Pergamon Press, New York.

Rudehill, A., Sollevi, A., Franco-Cereceda, A., and Lundberg, J. M. (1987). Neuropeptide Y (NPY) and the pig heart: Release and coronary vasoconstrictor effects. *Peptides 7,* 821–826.

Rump, L. C., and Majewski, H. (1987). Modulation of noradrenaline release through α_1- and α_2-adrenoceptors in rat isolated kidney. *J. Cardiovasc. Pharmacol. 9,* 500–507.

Schoups, A. A., Saxena, V. K., Tombeur, K., and De Potter, W. P. (1988). Facilitation of the release of noradrenaline and neuropeptide Y by the α_2-adrenoceptor blocking agents idazoxan and hydergine in the dog spleen. *Life Sci. 42,* 517–523.

Serfozo, P., Bartfai, T., and Vizi, E. S. (1986). Presynaptic effects of neuropeptide Y on (^{3}H)-noradrenaline and (^{3}H)-acetylcholine release. *Regul. Pept. 16,* 117–123.

Singer, E. A. (1988). Transmitter release from brain slices elicited by single pulses: A powerful method to study presynaptic mechanisms. *TIPS 9,* 274–276.

Sneddon, P., and Burnstock, G. (1985). ATP as a co-transmitter in rat tail artery. *Eur. J. Pharmacol. 106,* 149–152.

Starke, K. (1977). Regulation of noradrenaline release by presynaptic receptor systems. *Rev. Physiol. Biochem. Pharmacol. 77,* 1–124.

Starke, K. (1987). Presynaptic α-adrenoceptors. *Rev. Physiol. Biochem. Pharmacol. 107,* 73–146.

Stjärne, L., Lundberg, J. M., and Åstrand, P. (1986). Neuropeptide Y- a cotransmitter with noradrenaline and adenosine 5′-triphosphate in the sympathetic nerves of the mouse vas deferens? A biochemical, physiological and electropharmacological study. *Neurosci. 18,* 151–166.

Takahashi, K., Mouri, T., Murakami, O., Itoi, K., Sone, M., Ohneda, M., Nozuki, M., and Yoshinaga, K. (1988). Increases of neuropeptide Y-like immunoreactivity in plasma during insulin-induced hypoglycemia in man. *Peptides 9,* 433–435.

Trendelenburg, U. (1967). Some aspects of the pharmacology of autonomic ganglion cells. *Erg. Physiol. 59,* 1–85.

Valenta, B., Drobny, H., and Singer, E. A. (1989). Presynaptic autoinhibition of central noradrenaline release *in vitro:* Operational characteristics and effects of drugs acting at alpha-2 adrenoceptors in the presence of uptake inhibition. *J. Pharmacol. Exp. Ther. 245,* 944–949.

Vanhoutte, P. M., Webb, R. C., and Collins, M. G. (1980). Pre- and postjunctional adrenergic mechanisms and hypertension. *Clin. Sci. 59,* 211–223.

Wahlestedt, C., Grandemar, L., Håkanson, R., Heilig, M., Shen, G. H., Zukowska-Grojec, Z., and Reis, D. J. (1990). Neuropeptide Y receptor subtypes, Y_1 and Y_2. *Ann. New York Acad. Sci. 611,* 7–26.

Westfall, T. C. (1977). Local regulation of adrenergic neuro-transmission. *Physiol. Rev. 57,* 659–728.

Westfall, T. C., Carpentier, S., Chen, X., Beinfeld, M. C., Naes, L., and Meldrum, M. (1987). Prejunctional and postjunctional effects of neuropeptide Y at the noradrenergic neuroeffector junction of the perfused mesenteric arterial bed of the rat. *J. Cardiovasc. Pharmacol. 10,* 716–722.

Zimmermann, B. G. (1978). Actions of angiotensin on Adrenergic nerve endings. *Fed. Proc. 37,* 199–202.

Zukowska-Grojec, Z., Konarska, M., and McCarty, R. (1988). Differential plasma catecholamine and neuropeptide Y responses to acute stress in rats. *Life Sci 42,* 1615–1624.

Zukowska-Grojec, Z., and Vaz, A. C. (1988). Role of neuropeptide Y (NPY) in cardiovascular responses to stress. *Synapse 2,* 293–298.

CHAPTER 8

Neuropeptide Y Receptor Subtypes and Putative Receptor Ligands

THOMAS HEDNER XIANG-YING SUN LARS EDVINSSON

I. Introduction

Neuropeptide Y (NPY), which is a tyrosine-rich 36 amino acid residue peptide, was initially isolated by Tatemoto in 1982 from porcine brain (Tatemoto *et al.*, 1982). NPY is a member of the larger "PP" family of structurally related peptides that includes peptide YY (PYY) and the pancreatic polypeptide (PP) (Kimmel *et al.*, 1975; Tatemoto, 1982a). These bioactive peptides share a common secondary structure that is termed the PP fold (Michel, 1991). NPY is widely distributed in the central as well as the peripheral nervous systems and is one of the most abundant neuropeptides found in the mammalian brain (Tatemoto *et al.*, 1982; De Quid and Emson, 1986; Sundler *et al.*, 1986).

II. NPY Localization, Release, and Mechanisms in the Cardiovascular System

By the use of immunohistochemical techniques, the extensive central as well as the peripheral distribution of NPY has been revealed.

In the blood vessels of many mammals, including man, NPY is costored with noradrenaline (NA) in the sympathetic nerve fibers (Lundberg *et al.*, 1982b, 1983, 1985a; Edvinsson *et al.*, 1983; Ekblad *et al.*, 1984a). Most abundantly, NPY is present in the perivascular nerve fibers of many organs, such as the cerebral, renal, coronary, mesenteric, and femoral arteries (Edvinsson *et al.*, 1987a; Walker *et al.*, 1991). NPY-containing fibers are generally also more densely located around arteries than veins (Ekblad *et al.*, 1984; Sundler *et al.*, 1986), and small arteries are also likely to have more NPY immunoreactivity compared to the nerve fibers supplying large vessels (Walker *et al.*, 1991).

There are numerous NPY-immunoreactive nerve fibers around the cerebral arteries (Edvinsson *et al.*, 1983) and in particular, the rostral part of the circle of

Willis appears to be more densely supplied with NPY-containing fibers when compared to the caudal part and the basilar and vertebral arteries (Edvinsson *et al.*, 1987a). In the heart, there are also numerous NPY-containing fibers around coronary arteries (Ekblad *et al.*, 1984a; Allen *et al.*, 1986). The NPY-containing sympathetic fibers in the heart are preferentially found in the atria and the auricles whereas they appear in less numbers in the ventricles and septum (Lundberg *et al.*, 1982b, 1983; Gu *et al.*, 1984; Allen *et al.*, 1985; Uddman *et al.*, 1985). There is also a nonsympathetic system of NPY-containing fibers in the conduction system throughout the heart (Hassall and Burnstock, 1984; Dalsgaard *et al.*, 1986; Forsgren, 1989).

NPY-immunoreactive nerve fibers that appear to be identical to those containing adrenaline are also found in cells of the adrenal medulla of the rat (Lundberg *et al.*, 1986b; Sundler *et al.*, 1986). In the capsule, NPY immunoreactivity is found in a nerve plexus that penetrates the cortex into the medulla. Only a portion of these fibers seems to be of perivascular origin (Varndell *et al.*, 1984; Sundler *et al.*, 1986).

It is clear that NPY acts through several receptor subtypes to elicit its function through multiple ways in the central and peripheral nervous system. The peripheral mechanisms of action of NPY at the sympathetic neuroeffector junction include (Edvinsson *et al.*, 1987b; Walker *et al.*, 1991; Michel, 1991):

I. Direct vascular effects
II. Indirect vascular effects by interaction with vasoactive agents
  a. Vasoconstrictor potentiation
  b. Interaction with cyclooxygenase products
  c. Interaction with vasodilators
III. Prejunctional suppression of transmitter release
IV. Induction of histamine release

These aspects, however, will not be dealt with in the present review, which aims to focus on NPY receptor subtypes and putative receptor ligands.

III. Structures of NPY, Peptide YY, and Pancreatic Polypeptide

The amino acid sequence of NPY is highly conserved among species. Within the PP family, Peptide YY (PYY) is most closely related to NPY and displays 68% homology (Table 1). PYY occurs mainly in endocrine cells in the lower gastrointestinal tract (Tatemoto, 1982a; Lundberg *et al.*, 1982a; Böttcher *et al.*, 1984). Pancreatic polypeptide (PP) is present in a subpopulation of pancreatic islet cells (Kimmel *et al.*, 1975; Larsson *et al.*, 1975, 1976; Alumets *et al.*, 1978) and shares approximately 50% sequence identity with NPY (Table 1).

The PP family (i.e., NPY, PYY, and PP) all display the so-called PP fold that consists of a polyproline helix comprising residues 1–8, followed by an amphiphilic

Table 1 Amino Acid Sequences for the PP Family of Peptides: Neuropeptide Y (NPY), Peptide YY (PYY), and Pancreatic Polypeptide (PP)[a]

1   5   10   15   20   25   30   35
NPY *human*
Y-P-S-K-P-D-N-P-G-E-D-A-P-A-E-D-M-A-R-Y-Y-S-A-L-R-H-Y-I-N-L-I-T-R-Q-R-Y-*
NPY *porcine*
Y-P-S-K-P-D-N-P-G-E-D-A-P-A-E-D-**L**-A-R-Y-Y-S-A-L-R-H-Y-I-N-L-I-T-R-Q-R-Y-*
PYY *porcine*
Y-P-**A**-K-P-**E-A**-P-G-E-D-A-**S-P**-E-**E-L-S**-R-Y-Y-**A-S**-L-R-H-Y-I-N-L-**V**-T-R-Q-R-Y-*
PP *rat*
A-P-**L-E**-P-**M-Y**-P-G-**D-Y**-A-**T-H**-E-**Q-R**-A-**Q**-Y-**E-T-Q**-L-R-**R**-Y-I-N-**T-L**-T-R-**P**-R-Y-*

[a]The underlined amino acid residues differ from the human NPY sequence.

alpha-helix in residues 15–30. These helices are joined by a beta turn in residues 9–12. Furthermore, the configuration is stabilized by hydrophobic interactions between the alpha helix and the N-terminal part of the peptide. The C-terminal end of the peptides (residues 30–36) projects away from the PP fold at an angle of approximately 90° (Glover *et al.*, 1985; Taylor and Kaiser, 1986; Allen *et al.*, 1986; Schwartz *et al.*, 1990).

In most biological assays evaluating peripheral effects of the peptides, PYY appears to be approximately equipotent compared to NPY, while PP is virtually inactive (see Wahlestedt *et al.*, 1990a). Like many other regulatory peptides, NPY and PYY are amidated (Eipper *et al.*, 1985) and the C-terminal amido group seems to be vital in order to induce binding (MacKerell *et al.*, 1989) and to confer bioactivity. This can be demonstrated by the lack of biological effects of for example, desamido-NPY in various assays (Wahlestedt *et al.*, 1986; MacKerell *et al.*, 1989; Grundemar and Håkanson, 1990). One possible explanation for this phenomenon could be that when NPY-related peptides are deamidated, they cannot form a stable helix (Tonan *et al.*, 1990). The functional importance of the carboxy amide group may also be illustrated by the fact that the C-terminally extended 39 amino acid residue peptide, NPY-Gly-Lys-Arg, does not exhibit any biological effects (Grundemar and Håkanson, 1990). Thus, an intact C-terminus seems to be essential for the bioactivity of NPY/PYY. There are several C-terminal fragments of NPY and PYY that display both binding and bioactivity. However, several N-terminal NPY fragments, such as NPY_{1-12} and NPY_{1-24} (Danho *et al.*, 1988), NPY_{1-15} (Danger *et al.*, 1987), NPY_{1-21} (MacKerell *et al.*, 1989), or the midfragments NPY_{6-14}, NPY_{7-20}, NPY_{13-24}, and NPY_{20-30}, were without biological effects in the assays (Danho *et al.*, 1989).

In the NPY holopeptide, the C-terminal (residues 25–36) end has a rigid configuration within the intact parent molecule but seems to become increasingly more flexible with progressive N-terminal truncation. With the N-terminal truncation, there is a corresponding progressive loss of bioactivity, which may suggest that the C-terminal rigid configuration is of importance for the receptor recognition (Mac-

Kerell *et al.*, 1989; Grundemar and Håkanson, 1990). A centrally truncated NPY analog has recently been synthesized in which the 8–17 residues were deleted. In this structure, the N- and C-termini were joined together by an 8-aminooctanic acid residue and stabilized with a disulphid bridge ([D-Cys^7-Aoc^{8-17}-Cys^{20}]NPY). The [D-Cys^7-Aoc^{8-17}-Cys^{20}]NPY analog retained full NPY-binding affinity and suppressed the stimulated formation of cAMP (Krstenansky *et al.*, 1989; Gordon *et al.*, 1990). Another analog, [Aca^{5-24}]NPY, was formed by the linking of NPY_{1-4} to NPY_{25-36} via an epsilon aminocaproic acid sequence. [Aca^{5-24}]NPY was approximately 10 times less potent compared to NPY at the Y_1 and Y_2 receptor sites (Beck *et al.*, 1989). From these experiments it seems likely that both the C- and N-termini of NPY are essential for the full recognition and binding to the Y_1/Y_2 receptors. The PP fold, however, may be of importance in order to bring the C- and N-terminal ends in close and optimal steric relation (Krstenansky *et al.*, 1989; Fuhlendorff *et al.*, 1990b; Schwartz *et al.*, 1990).

IV. NPY Receptor Subtypes

NPY can elicit a number of different physiological responses by acting on specific pre- and postsynaptic receptors that have been detected and characterized in several different tissue and cell types (Edvinsson *et al.*, 1987b; Walker *et al.*, 1991). Such specific binding sites for NPY have been demonstrated in, for example, the porcine spleen (Lundberg *et al.*, 1988), rabbit (Chang *et al.*, 1988) and porcine aortic membranes (Shigeri *et al.*, 1991), rat cardiac ventricular membranes (Balasubramanian *et al.*, 1990), bovine adrenal medulla (Cherdchu *et al.*, 1989), rabbit pulmonary arteries (Wahlestedt *et al.*, 1990b), and rat vascular smooth muscle cells in the pancreas (Sheikh *et al.*, 1991).

A subclassification of NPY receptors was first suggested by Wahlestedt *et al.* (1986). Based on binding experiments using a series of C-terminal NPY fragments, Wahlestedt *et al.* proposed the existence of two different receptor subtypes of which one was postsynaptic (Y_1) and the other presynaptic (Y_2) (Wahlestedt *et al.*, 1990a; Sheikh *et al.*, 1990). By action on these binding sites within the synaptic cleft, NPY caused a presynaptic inhibition of norepinephrine release and a postsynaptic contraction of the vasculature (see below). Further studies by Wahlestedt *et al.* in the vas deferens preparation revealed that the C-terminal fragments NPY_{13-36} and PYY_{13-36} were about equipotent compared to NPY and PYY in suppressing the electrically (sympathetically) evoked twitches by interaction with prejunctional sympathetic activity. In contrast, on the postjunctional site, where NPY and PYY induced a concentration-dependent contraction of the guinea pig iliac vein, NPY_{13-36} or PYY_{13-36} seemed to be inactive. Furthermore, PYY_{13-36}, and desamido-NPY, which lacks the C-terminal amido group, did not demonstrate any antagonistic properties. These findings suggested that NPY/PYY acted on at least two different receptor subtypes, referred to as Y_1 and Y_2, respectively, since NPY_{13-36} and PYY_{13-36} only showed activity at the prejunctional receptor site (Wahlestedt *et al.*, 1986, 1987).

Further support for the existence of at least two NPY/PYY receptor subtypes stems from studies of the human neuroblastoma cell lines MC-IXC and SK-N-MC, as well as the rat pheochromocytoma cell line PC-12. These cell lines exhibit receptor binding to the $NPY_{1\text{-}36}$ and PYY_{1-36} structures but not the NPY_{13-36} or PYY_{13-36} sequences, indicating the presence of a Y_1-type receptor (Sheikh *et al.*, 1989; Schwartz *et al.*, 1990; Wahlestedt *et al.*, 1990a). In contrast, the human neuroblastoma cell lines SMS-MSN or SK-N-BE2 and porcine hippoocampal membranes bind NPY_{13-36} and PYY_{13-36} almost as effectively as NPY_{1-36} and PYY_{1-36}, consistent with the existence of a Y_2-type receptor (Sheikh *et al.*, 1989; Schwartz *et al.*, 1990; Wahlestedt *et al.*, 1990a).

There is also other evidence supporting the existence of NPY receptor heterogeneity. Fuhlendorff *et al.*, (1990a) demonstrated that a Y_1-specific agonist can be synthesized by exchanging an isoleucine to leucine at position 31 and a glutamine to proline at position 34 of the NPY sequence. Moreover, the substitution of Tyr^{20} resulted in a break of the α-helix and in a loss of affinity to the Y_1 receptor, while $[Pro^{20}]$NPY still recognized the Y_2 receptor (Fuhlendorff *et al.*, 1990b). Moreover, the linkage of PP_{1-30} to NPY_{31-36} lead to a hybrid sequence that was a selective Y_2 receptor ligand (Fuhlendorff *et al.*, 1990b).

One characteristic for the Y_1 receptor subtype is that $[Pro^{34}]$NPY is equipotent with NPY in all vascular segments, and the deletion of the first C-terminal amino acid, resulting in NPY_{2-36}, is characterized by a marked loss of potency. Further N-terminal truncation of the NPY sequence results in fragments that are even less potent compared to the intact NPY molecule. The importance of an intact N-terminus is also illustrated by the observation that a single substitution of the Y_1 receptor agonist $[Pro^{34}]$NPY from Lys^4 to Glu^4 ($[Glu^4,Pro^{34}]$NPY results in a marked loss of affinity to the Y_1 receptor (Schwartz *et al.*, 1990). Moreover, if there is systematic D-substitutions in either of the five first amino acid residues in the N-terminal there is a marked loss of potency at the Y_1 receptor (Boublik *et al.*, 1990).

Taken together, these results suggest that an intact C-terminal of the NPY sequence is the most important part of the molecule for recognition of the Y_2 receptor, since several N-terminally truncated NPY/PYY fragments down to NPY_{22-36} are rather potent, and that the 12 C-terminal amino acid residues of NPY/PYY are the minimum length required to activate the Y_2 receptor (Grundemar and Håkanson, 1990).

The NPY-evoked pressor response in coronary resistance arteries and contraction of isolated vessels as well as the specific binding characteristics of NPY to vascular smooth muscle cells are consistent with the presence of Y_1 receptors in the vascular system. It is also likely that Y_2 receptors in some way may contribute to the vasoconstriction in some vascular beds (Grundemar *et al.*, 1992) since recent experiments indicate that the C-terminal NPY fragments appear to be relatively more potent in increasing the arterial pressure or the coronary resistance compared to their ability to displace radioligand binding in the aortic smooth muscle cells, which is suggested to display Y_1 receptors.

From a structural point of view, cross-linking experiments indicate that the Y_1 receptor is a glycoprotein with a 70-kDa binding subunit, whereas the Y_2 receptor seems to be a glycoprotein with a 50-kDa binding subunit (Sheikh and Williams, 1990). In addition to the prejunctional sympathetic site, the Y_2 receptor seems to be prejunctionally located also at the sensory C fiber and the parasympathetic neurons. Thus, NPY_{13-36} induced a prejunctional suppression of sensory C fiber-mediated contractions in the guinea pig bronchi (Grundemar *et al.*, 1990), as well as parasympathetic activity in the rat uterine cervix (Stjernquist *et al.*, 1990). Indeed, these results suggest the presence of Y_2 receptors on C fibers and parasympathetic fibers, respectively. In addition to the neuronal location, binding studies and the use of autoradiography techniques have suggested the presence of Y_2 receptors on tubular cells in the rabbit kidney (Sheikh *et al.*, 1989; Nielsen *et al.*, 1991).

In contrast to NPY/PYY, PP binds neither to the Y_1 nor the Y_2 receptor subtype (Schwartz *et al.*, 1989, 1990).

A new third type of postsynaptic NPY receptor has been postulated based on binding experiments in rat cardiac ventricular membranes. Unlike the Y_1 and Y_2 receptors that bind NPY and PYY with similar affinities, this third type of NPY receptor discriminates between these two peptides (Balasubramanian and Sheriff, 1990).

A 65-kDa surface receptor was recently demonstrated by electrophoresis experiments in the pancreatic islets and acini, on hepatocytes and on the epithelium in the gastrointestinal tract. By affinity cross-linking it was further demonstrated that NPY, PYY, and PP were equally potent in displacing radiolabeled PP, which would suggest that this receptor structure represented a common receptor for NPY-related peptides (Nata *et al.*, 1990). Moreover, recent experiments have shown that PP but not NPY binds with high affinity to PC12 cells, suggesting the presence of specific PP receptors not recognizing NPY (Schwartz *et al.*, 1987, 1989, 1990; Jörgensen *et al.*, 1990). This is further supported by experiments showing that PP suppressed the electrically evoked twitches in the rat vas deferens via receptors distinct from the Y_1 and Y_2 sites (Jörgensen *et al.*, 1990).

In addition to these actions by NPY, recent experiments on rat peritoneal mast cells may suggest that the NPY-induced histamine release is dependent on a mechanism that involves neither Y_1 nor Y_2 receptors. There are in fact reasons to believe that the positively charged amino acids at the C-termini of NPY/PYY activate G proteins in the mast cell membrane by a nonreceptor mechanism, as has been suggested for tachykinins and other basic peptides (Mously *et al.*, 1990).

To what extent the NPY-induced histamine release from mast cell represents a physiologically relevant process or not is currently unclear. Since mast cells are particularly numerous around blood vessels and nerves (Stead *et al.*, 1989; Nilsson *et al.*, 1990), it may be hypothesized that the NPY-evoked histamine release from the mast cells may in some way contribute to the sympathetic control of microcirculation in certain vascular beds.

V. NPY Signal Transduction

The molecular mechanisms leading to the cellular actions of NPY are complex and differ according to the experimental system studied. One signal transduction mechanism most frequently associated with NPY receptor activation is the inhibition of adenylyl cyclase activity via an inhibitory G protein (Edvinsson *et al.*, 1987b; Walker *et al.*, 1991).

NPY may also influence the intracellular Ca^{2+} concentration by regulating both the influx of extracellular Ca^{2+} and the mobilization of free cytosolic Ca^{2+}. Using intracellular single-electrode and whole-cell patch-clamp techniques, Wiley *et al.* (1990) showed that NPY produced concentration-dependent reduction in Ca^{2+} currents in rat vagal afferent neurons, which was blocked by pertussis toxin. Perney and Miller (1989) demonstrated that NPY also inhibited voltage-dependent Ca^{2+} channels in primary cultures of neurons prepared from rat dorsal root ganglia. This action was again abolished by pertussis toxin and restored when pertussis toxin-treated cells were perfused with a combination of GTP and the α-subunit of the inhibitory G protein G_0. Several other investigators have also demonstrated that NPY induces a decrease in Ca^{2+} currents in both peripheral and central neurons. Mobilization of intracellular Ca^{2+} stores by NPY is also linked to the phosphoinositol pathway (Perney and Miller, 1989; Sun *et al.*, 1991). However, NPY appears to mobilize Ca^{2+} in cultured smooth muscle cells from porcine aorta without a detectable change in IP_3 concentration (Mihara *et al.*, 1989). Recent experiments have demonstrated that D-myo-1,2,6-inositol-trisphosphate (PP56), which seems to interfere with the Ca^{2+} mobilization, selectively blocks the NPY-induced vascular contraction and pressor responses (Edvinsson *et al.*, 1990; Sun *et al.*, 1991) as well as the potentiating effect of NPY on other vasoconstrictors (Sun *et al.*, 1992). Moreover, thapsigargin, a drug that selectively depletes the Ca^{2+} store of the endoplasmic reticulum, abolishes the rise in $[Ca^{2+}]_i$ produced by NPY (Aakerlund *et al.*, 1990) in neuroblastoma cells. This may suggest that some of the Y_1-related actions of NPY may depend on $[Ca^{2+}]$ release from the endoplasmic reticulum.

VI. NPY Receptor Ligands

The elucidation of the potential physiological role of NPY has been hampered by the lack of potent and selective receptor ligands at the Y_1 and Y_2 receptor sites. Hitherto, most of the work relating to NPY receptor subtype selective agents has been focused around studies relating to truncated NPY or PYY fragments. Several centrally truncated NPY analogs with intact C- and N-termini have been shown to act at both the Y_1 and Y_2 receptors (Beck *et al.*, 1989; Krstenansky *et al.*, 1989; Baeza and Undén, 1990; Schwartz *et al.*, 1990).

The substituted analog ([Leu^{31},Pro^{34}]NPY is a selective Y_1 receptor agonist that is at least equipotent with NPY_{1-36} (Fuhlendorff *et al.*, 1990). In fact, the single substitution to proline in position 34 confers Y_1 receptor selectivity (Krstenansky *et al.*, 1989; Dumont *et al.*, 1991).

On the other hand, the linking of PP_{1-30} with N_{31-36} results in a hybrid structure that is about equipotent with NPY in terms of Y_2 receptor selectivity in hippocampal membranes, while this sequence displayed a low affinity to a neuroblastoma cell line that selectively expressed Y_1 receptors (Fuhlendorff *et al.*, 1990b).

There are as yet only some nonpeptidergic compounds that have been claimed to act as antagonists at the NPY receptor. The irreversible α-adrenoceptor antagonist benextramine exhibits a weak antagonism of NPY binding in the rat brain and in higher doses functionally attenuates NPY-induced pressor effects (Doughty *et al.*, 1990). Finally, another compound (He 90481, Heuman Pharma, Nürnberg) has been reported to act as a low-potency NPY receptor antagonist. In addition, it was found to be a potent H_2 receptor agonist (Michel and Motulsky, 1990). Recently an inositol derivative, D-myo-inositol-1,2,6-trisphosphate (PP56), has been shown to attenuate NPY-evoked vasoconstriction in a noncompetitive manner (Edvinsson *et al.*, 1990; Sun *et al.*, 1991). At higher concentrations of PP56, the contractions evoked by histamine, serotonin and prostaglandin $F_{2\alpha}$, were also suppressed (Edvinsson *et al.*, 1990). Furthermore, in the pithed rat, PP56 has recently been demonstrated to act as a selective antagonist of exogenously infused NPY within a defined dose range (Sun *et al.*, 1991). Furthermore, it has been shown that PP56 antagonizes not only the direct vasoconstrictor actions of NPY *in vivo* in the conscious rat but also the potentiating effect of NPY on a variety of vasoconstrictor agents (Sun *et al.*, 1992).

Acknowledgments

This research was supported by grants from the Swedish Medical Research Council (Project no. 8642).

References

Aakerlund, L., Gether, U., Fuhlendorff, J., Schwartz, T. W., and Thastrup, O. (1990). Y_1 receptors for neuropeptide Y are coupled to mobilization of intracellular calcium and inhibition of adenylate cyclase. *FEBS Lett. 260,* 73–78.

Allen, J. M., Polak, J. M., Rodrigo, J., Darcy, K., and Bloom, S. R. (1985). Localization of neuropeptide Y in nerves in rat cardiovascular system and the effect of 6-hydroxydopamine. *Cardiovasc. Res. 19,* 570–577.

Allen, J. M., Gjörstrup, P., Björkman, J.-A., Ek, L., Abrahamsson, T., and Bloom, S. R. (1986). Studies of cardiac distribution and function of neuropeptide Y. *Acta Physiol. Scand. 126,* 405–411.

Alumets, J., Håkanson, R., and Sundler, F. (1978). Distribution, ontogeny and ultrastructure of pancreatic polypeptide (PP) cells in pancreas and gut of the chicken. *Cell. Tissue Res. 194,* 377–386.

Baeza, C. R., and Undén, A. (1990). Studies of the structural requirement for ligand binding to the neuropeptide Y (NPY) receptor from rat cerebral cortex. *FEBS Lett. 277,* 23–25.

Balasubramaniam, A., Sheriff, S., Rigel, D. F., and Fischer, J. E. (1990). Characterization of neuropeptide Y binding sites in rat cardiac ventricular membranes. *Peptides 11,* 545–550.

Balasubramaniam, A., and Sheriff, S. (1990). Neuropeptide Y (18–36) is a competitive antagonist of neuropeptide Y in rat cardiac membranes. *J. Biol. Chem. 265,* 14724–14727.

Beck, A., Jung, G., Gaida, W., Köppen, H., Lang, R., and Schnorrenberg, G. (1989). Highly potent and small neuropeptide Y agonist obtained by linking NPY 1–4 via spacer to α-helical NPY 25–36. *FEBS Lett. 244,* 119–122.

Böttcher, G., Sjölund, K., Ekblad, E., Håkanson, R., Schwartz, T. W., and Sundler, F. (1984). Coexistence of peptide YY in glicentin immunoreactivity in endocrine cells of the gut. *Regul. Pept. 8,* 261–273.

Boublik, J. H., Spicer, M. A., Scott, N. A., Brown, M. R., and Rivier, J. E. (1990). Biologically active neuropeptide Y analogs. *Ann. N.Y. Acad. Sci. 611,* 27–34.

Chang, R., and Lotti, V. J. (1988). Specific [^{3}H] proprionyl-neuropeptide Y (NPY) binding in rabbit aortic membranes: comparison with binding in rat brain and biological responses in rat vas deferens. *Biochem. Biophys. Res. Commun. 151,* 1213–1219.

Cherdchu, C., Deupree, J. D., and Hexum, T. D. (1989). Biding sites for ^{125}I-neuropeptide Y (NPY) on membranes from bovine adrenal medulla. *Eur. J. Pharmacol. 173,* 115–119.

Chronwall, B. M., DiMagio, D. A., Massari, V. J., Picker, V. M., Ruggiero, D. A., O'Donohue, T. L. (1985). The anatomy of neuropeptide Y-containing neurons in rat brain. *Neuroscience 15,* 1159–1181.

Dalsgaard, C.-J., Franco-Cereceda, A., Saria, A., Lundberg, J. M., Theodorsson-Norheim, E., and Hökfelt, T. (1986). Distribution and origin of substance P and neuropeptide Y-immunoreactive nerves in the guinea-pig heart. *Cell Tissue Res. 243,* 477–485.

Danger, J. M., Tonon, M. C., Lamacz, M., Martel, J. C. Saint-Pierre, S., Pellitier, G., and Vaudry, H. (1987). Melanotropin release inhibiting activity of neuropeptide Y: Structure-activity relationships. *Life Sci. 40,* 1875–1880.

Danho, W., Triscari, J., Vincent, G., Nakajima, T., Taylor, J., and Kaiser, E. T. (1988). Synthesis and biological evaluation of pNPY fragments. *Int. J. Peptide Protein Res. 32,* 496–505.

de Quidt, M.E., and Emson, P. C. (1986). Distribution of neuropeptide Y-like immunoreactivity in the rat central nervous system. II. Immunohistochemical analysis. *Neuroscience 18,* 545–618.

Doughty, M. B., Chu, S. S., Miller, D. W., Li, K., and Tessel, R. E. (1990). Benextramine: a long-lasting neuropeptide Y receptor antagonist. *Eur. J. Pharmacol. 185,* 113–114.

Dumont, Y., Fournier, A., St-Pierre, S., Schwartz, T. W., and Quirion, R. (1991). Differential distribution of neuropeptide Y_1 and Y_2 receptors in the rat brain. *Eur. J. Pharmacol. 191,* 501–503.

Edvinsson, L., Emson, P. C., McCulloch, K., Tatemoto, K., and Uddman, R. (1983). Neuropeptide Y: Cerebrovascular innervation and vasomotor effect in the cat. *Neurosci. Lett. 43,* 79–84.

Edvinsson, L., Copeland, J. R., Emson, P. C., McCulloch, J., and Uddman, R. (1987a). Nerve fibers containing neuropeptide Y in the cerebrovascular bed: Immunocytochistry, radioimmunoassay and vasomotor effects. *J. Cer. Blood Flow Metab. 7,* 45–57.

Edvinsson, L., Håkanson, R., Wahlestedt, C., and Uddman, R. (1987b). Effects of neuropeptide Y on the cardiovascular system. *Trends Pharmacol. Sci. 8,* 231–235.

Edvinsson, L., Adamsson, M., and Jansen, I. (1990). Neuropeptide Y antagonistic properties of D-myo-inositol-1,2,6-trisphosphate in guinea pig basilar arteries. *Neuropeptides 17,* 99–105.

Ekblad, E., Edvinsson, L., Wahlestedt, C., Uddman, R., and Håkanson, R. (1984a). Neuropeptide Y co-exists and co-operates with noradrenaline in perivascular fibres. *Regul. Pept. 8,* 225–235.

Eipper, B. A., Mains, R. E., and Herbert, E. (1985). Peptides in the nervous system. *Trends in Neurosci. 9,* 463–468.

Forsgren, S. (1989). Neuropeptide Y-like immunoreactivity in relation to the distribution of sympathetic nerve fibers in the heart conduction system. *J. Mol. Cell. Cardiol. 21,* 279–290.

Fuhlendorff, J., Gether, U., Aakerlund, L., Langeland-Johansen, N., Thogersen, H., Melberg, S. G., Bang-Olsen, U., Thastrup, O., and Schwartz, T. W. (1990). [Leu^{31}, Pro^{34}] Neuropeptide Y—a specific Y_1 receptor agonist. *Proc. Natl. Acad. Sci. U.S.A. 187,* 182–186.

Glover, I. D., Barlow, D. J., Pitts, J. E., Wood, S. P., Tickle, I. J., Blundell, T. L., Tatemoto, K., Kimmel, J. R., Wollmer, A., Strassburger, W., and Zhang, Y.-S. (1985). Conformal studies of the pancreatic polypeptide hormone family. *Eur. J. Biochem. 142,* 379–385.

Gordon, E. A., Krstenansky, J. L., and Fishman, P. H. (1990). Centrally truncated neuropeptide Y analog acts as an agonist for Y_1 receptors on SK-N-MC cells. *Neurosci. Lett. 119,* 187–190.

Grundemar, L., and Håkanson, R. (1990). Effects of various neuropeptide Y/peptide YY fragments on electrically-evoked contractions of the rat vas deferens. *Br. J. Pharmacol. 100,* 190–192.

Grundemar, L., Grundström, N., Johansson, I. G. M., Andersson, R. G. G., and Håkanson, R. (1990). Suppression by neuropeptide Y of capsaicin-sensitive sensory nerve-mediated contraction in guinea-pig airways. *Br. J. Pharmacol. 99,* 473–476.

Grundemar, L., Mörner, J., Högeslätt, E., Wahlestedt, C., and Håkanson, R. (1992). Characterization of vascular receptors for neuropeptide Y. *Br. J. Pharmacol. 105,* 45–50.

Gu, J., Polak, J. M., Allen, J. M., Huang, W. M., Sheppard, M. N., Tatemoto, K., and Bloom, S. R.

(1984). High concentrations of a novel peptide, neuropeptide Y, in the innervation of mouse and rat heart. *J. Histochem. Cytochem. 52,* 467–472.

Hassall, C. J. S., and Burnstock, G. (1984). Neuropeptide Y-like immunoreactivity in cultured intrinsic neurones of the heart. *Neurosci. Lett. 52,* 111–115.

Jörgensen, J. Ch., Fuhlendorff, J., and Schwartz, T. W. (1990). Structure–function studies on neuropeptide Y and pancreatic polypeptide—evidence for two PP-fold receptors in vas deferens. *Eur. J. Pharmacol. 186,* 105–114.

Kimmel, J. R., Hayden, L. J., and Pollock, H. G. (1975). Isolation and characterization of a new pancreatic polypeptide hormone. *J. Biol. Chem. 250,* 9369–9376.

Krstenansky, J. L., Owen, T. J., Buck, S. H., Hagaman, K. A., and Lean, L. R. (1989). Centrally truncated and stabilized porcine neuropeptide Y analogs. Design, synthesis, and mouse brain receptor binding. *Proc. Natl. Acad. Sci. U.S.A. 86,* 4377–4381.

Larsson, L. I., Sundler, F., and Håkanson, R. (1975). Immunohistochemical localization of human pancreatic polypeptide (HPP) to a population of islet cells. *Cell Tissue Res. 156,* 167–171.

Larsson, L. I., Sundler, F., and Håkanson, R. (1976). Pancreatic polypeptide—a postulated new hormone. Identification of its cellular storage site by light and electron microscopic immunocytochemistry. *Diabetologia 12,* 211–226.

Lundberg, J. M., Tatemoto, K., Terenius, L., Hellström, P. M., Mutt, V., Hökfelt, T., and Hamberger, B. (1982a). Localization of peptide YY (PYY) in gastrointestinal endocrine cells and effects on intestinal blood flow and motility. *Proc. Natl. Acad. Sci. U.S.A. 79,* 4471–4475.

Lundberg, J. M., Terenius, L., Hökfelt, T., Martling, C. R., Tatemoto, K., Mutt, V., Polak, J., Bloom, S. R., and Goldstein, M. (1982b). Neuropeptide Y (NPY)-like immunoreactivity in peripheral noradrenergic neurons and effects of NPY on sympathetic function. *Acta Physiol. Scand. 116,* 477–480.

Lundberg, J. M., Terenius, L., Hökfelt, L., Hökfelt, T., and Goldstein, M. (1983). High levels of neuropeptide Y (NPY) in peripheral noradrenergic neurons in various mammals including man. *Neurosci. Lett. 42,* 167–172.

Lundberg, J. M., Ängård, A., Pernow, J., and Hökfelt, T. (1985a). Neuropeptide Y-, substance P and VIP-immunoreactive nerves in cat spleen in relation to autonomic vascular and volume control. *Cell. Tiss. Res. 239,* 9–18.

Lundberg, J. M., Hökfelt, T., Hemsén, A., Theodorsson-Norheim, E., Pernow, J., Hamberger, B., and Goldstein, M. (1986b). Neuropeptide Y-like immunoreactivity in adrenaline cells of adrenal medulla and in tumors and plasma of pheochromocytoma patients. *Regul. Pept. 13,* 169–182.

Lundberg, J. M., Hemsen, A., Larsson, O., Rudehill, A., Saria, A., and Fredholm, B. (1988). Neuropeptide Y receptor in pig spleen: Binding characteristics, reduction of cyclic AMP formation and calcium antagonist inhibition of vasoconstriction. *Eur. J. Pharmacol. 145,* 21–29.

MacKerell, A. D., Hemsén, A., Lacroix, J. S., and Lundberg, J. M. (1989). Analysis of structure–function relationships of neuropeptide Y using molecular dynamics simulations and pharmacological activity and binding measurements. *Regul. Pept. 25,* 295–313.

Michel, M. C. (1991). Receptors for neuropeptide Y: Multiple subtypes and multiple second messengers. *Trends Pharmacol. Sci. 12,* 389–394.

Michel, M. C., and Motulsky, H. J. (1990). He 90481: A competitive non-peptidergic antagonist at NPY receptors. *Ann. N.Y. Acad. Sci. 611,* 392–394.

Mihara, S., Shigeri, Y., and Fujimoto, M. (1989). Neuropeptide Y-induced intracellular Ca^{2+} increase in vascular smooth muscle cells. *FEBS Lett. 259,* 79–82.

Mously, M., Bueb, J.-L., Bronner, C., Rouot, B., and Landry, (1990). Y. G protein activation: A receptor-independent mode of action for cationic amphiphilic neuropeptides and venom peptides. *Trends Pharmacol. Sci. 11,* 358–362.

Nata, K., Yonekura, H., Yamamoto, H., and Okamoto, H. (1990). Identification of a novel 65-kDa cell surface receptor common to pancreatic polypeptide, neuropeptide Y and peptide YY. *Biochem. Biophys. Res. Commun. 171,* 330–335.

Nielsen, S., Sheikh, S. P., Sheikh, M. I., and Christensen, E. I. (1991). Axial heterogeneity of peptide YY processing in renal proximal tubules. *Am. J. Physiol. 260,* F359–367.

Nilsson, G., Alving, K., Ahlstedt, S., Hökfelt, T., and Lundberg, J. M. (1990). Peptidergic innervation of rat lymphoid tissue and lung: Relation to mast cells and sensitivity to capsaicin and immunization. *Cell Tissue Res. 262,* 125–133.

Perney, T. M., and Miller, R. J. (1989). Two different G-protein mediate neuropeptide Y and bradykinin-stimulated phospholipid breakdown in cultured sensory neurons. *J. Biol. Chem. 264,* 7317–7327.

Schwartz, T. W., Sheikh, S. P., and O'Hara, M. M. T. (1987). Receptors on pheocromocytoma cells for two members of the PP-fold family—NPY and PP. *FEBS Lett. 225,* 209–214.

Schwartz, T. W., Fuhlendorff, J., Langeland, N., Thogersen, H., Jörgensen, J. Ch., and Sheikh, S. P. (1989). Y_1 and Y_2 receptors for NPY—The evolution of PP-fold peptides and their receptors. *In* "Neuropeptide Y–Nobel Conference Series." (V. Mutt, T. Hökfelt, K. Fuxe, and J. M. Lundberg, eds.), pp. 143–151. Raven Press, New York, N.Y.

Schwartz, T. W., Fuhlendorff, J., Kjems, L. L., Kristensen, M. S., Vervelde, M., O'Hara, M., Kretemansky, J. L., and Björnholm, B., (1990). Signal epitopes in the three-dimensional structure of NPY interaction with Y_1, Y_2 and PP receptors. *Ann. N.Y. Acad. Sci. 611,* 35–47.

Sheikh, S. P., and Williams, J. A. (1990). Structural characterization of Y_1 and Y_2 receptors for neuropeptide Y and peptide YY by affinity cross-linking. *J. Biol. Chem. 265,* 8304–8310.

Sheikh, S. P., Håkanson, R., and Schwartz, T. W. (1989). Y_1 and Y_2 receptors for neuropeptide Y. *FEBS Lett. 245,* 209–214.

Sheikh, S. P., Roach, E., Fuhlendorff, J., and Williams, J. A. (1991). Localization of Y_1 receptors for NPY and PYY on vascular smooth muscle cells in rat pancreas. *Am. J. Physiol. 260,* G250–257.

Sheik, S. P., Sheikh, M. I., and Schwartz, T. W. (1989). Y_2-type receptors for peptide YY on renal proximal tubular cells in the rabbit. *Am. J. Physiol. 257,* (Renal Fluid Electrolyte Physiol. 26), F978–F984.

Shigeri, Y., Mihara, S.-I., and Fujimoto, M. (1991). Neuropeptide Y receptor in vascular smooth muscle. *J. Neurochem. 56,* 852–859.

Stead, R. H., Dixon, F., Bramwell, N. H., Riddel, R. H., and Bienenstock, J. (1989). Mast cells are closely apposed to nerves in the human gastrointestinal mucosa. *Gastronenterol. 97,* 575–585.

Stjernquist, M., and Owman, Ch. (1990). Further evidence for a prejunctional action of neuropeptide Y on cholinergic motor neurons in the rat uterine cervix. *Acta Physiol. Scand. 138,* 95–96.

Sun, X. Y., Dahlöf, C., Edvinsson, L., and Hedner, T. (1991). D-myo-inositol-1.2.6-trisphosphate (PP 56) is a selective antagonist of neuropeptide Y (NPY) induced pressor response in the pithed rat. *Eur. J. Pharmacol. 204,* 281–286.

Sun, X. Y., Edvinsson, L., and Hedner, T. (1992). Effect of D-myo-inositol-1.2.6-trisphosphate (PP 56) on neuropeptide Y (NPY) induced potentiation of various vasoconstrictor agents in the rat. *J. Pharmacol. Exp. Ther. 261,* 1147–1152.

Sundler, F., Håkanson, R., Ekblad, E., Uddman, R., and Wahlestedt, C. (1986). Neuropeptide Y in peripheral and enteric nervous systems. *Ann. Rev. Cytol. 102,* 234–269.

Tatemoto, K., Carlquist, M., and Mutt, V. (1982). Neuropeptide Y—a novel brain peptide with structural similarities to peptide YY and pancreatic polypeptide. *Nature* (Lond.) *296,* 659–660.

Tatemoto, K. (1982a). Isolation and characterization of peptide YY (PYY), a candidate gut hormone that inhibits pancreatic exocrine function. *Proc. Natl. Acad. Sci. U.S.A. 79,* 2514–2518.

Taylor, J. N., and Kaiser, E. T. (1986). The structural characterization of β-endorphine and related peptide hormones and neurotransmitters. *Pharmacol. Rev. 38,* 291–320.

Tonan, K., Kawata, Y., and Hamaguchi, K. (1990). Conformations of isolated fragments of pancreatic polypeptide. *Biochem. 29,* 4424–4429.

Uddman, R., Ekblad, E., Edvinsson, L., Håkansson, R., and Sundler, F. (1985). Neuropeptide Y immunoreactivity in peripheral vascular nerve fibers of the guinea pig. *Regul. Pept. 10,* 243–257.

Varndell, I. M., Polak, J. M., Allen, J. M., Terenghi, G., and Bloom, S. R. (1984). Neuropeptide tyrosine (NPY)-immunoreactivity in noradrenaline-containing cells and nerves of the mammalian adrenal gland. *Endocrinol. 114,* 1460–1462.

Wahlstedt, C., Yanaihara, N., and Håkanson, R. (1986). Evidence for different pre- and postjunctional receptors for neuropeptide Y and related peptides. *Regul. Pept. 13,* 307–318.

Wahlestedt, C., Edvinsson, L., Ekblad, E., and Håkanson, R. (1987). Effects of neuropeptide Y at sympathetic neuroeffector junctions: Existence of Y_1- and Y_2-receptors. *In* "Neuronal Messengers in Vascular Function." (A. Nobin, C. Owman, and B. Arneklo-Nobin, eds.), pp. 231–241.

Wahlestedt, C., Grundemar, L., Håkanson, R., Heilig, M., Shen, G. H., Zukowska-Grojec, Z., and Reis, D. J. (1990a). Neuropeptide Y receptor subtypes, Y_1 and Y_2. *Ann. N.Y. Acad. Sci. 611,* 7–26.

Wahlestedt, C., Håkanson, R., Vaz, C., and Zukowska-Grojec, Z. (1990b). Norepinephrine and neuropeptide Y: Vasoconstrictor cooperation *in vivo* and *in vitro. Am. J. Physiol. 258,* R736–R742.

Walker, M. W., and Miller, M. J. (1988). ^{125}-Neuropeptide Y and ^{125}I-peptide YY bind to multiple receptor sites in rat brain. *Mol. Pharmacol. 34,* 779–792.

Walker, P., Grouzmann, E., Burnier, M., and Waeber, B. (1991). The role of neuropeptide Y in cardiovascular regulation. *Trends Pharmacol. Sci. 12,* 111–115.

Wiley, J. W., Gross, R. A., Lu, Y., and Macdonald, R. L. (1990). Neuropeptide Y reduces calcium current and inhibits acetylcholine release in noclose neuros via a pertussis toxin-sensitive mechanism. *J. Neurophysiol. 63,* 1499–1507.

CHAPTER 9

ATP as a Cotransmitter with Noradrenaline in Sympathetic Perivascular Nerves

CHARLES KENNEDY

I. Introduction

The possibility that adenosine triphosphate (ATP) is a neurotransmitter has long been controversial. However, rather than acting on its own as was originally proposed, ATP may be stored and released with other neurotransmitters. Thus, in recent years many studies have supported a role for ATP as a cotransmitter with noradrenaline (NA) in sympathetic nerves and with acetylcholine in parasympathetic nerves innervating visceral and vascular tissues. In this chapter, I will consider the evidence that ATP is an excitatory cotransmitter with NA in the sympathetic nerves innervating blood vessels and discuss the functional implications for the regulation of blood pressure.

II. Vascular Actions of Purines

The potent and widespread vascular actions of purine nucleosides and nucleotides were first demonstrated over 60 years ago (Drury and Szent-Györgi, 1929) and have since been extensively studied. In general, ATP elicits either vasodilation alone or a transient vasoconstriction followed by vasodilation. These actions are seen both *in vivo* following systemic administration (Gillespie, 1934; Green and Stoner, 1950; Haddy and Scott, 1968) and *in vitro* on isolated preparations (see Burnstock, 1980; Burnstock and Kennedy, 1986; Kennedy, 1988). For many years it was assumed that these effects resulted from a direct action of ATP at P_2-purinoceptors (Burnstock, 1978) located in vascular smooth muscle and since the main action of ATP was vasodilation, there was little reason to consider ATP as an excitatory transmitter.

This changed following the discovery of endothelium-derived relaxing factor (EDRF) (Furchgott and Zawadzki, 1980) when it became clear that, in most cases, ATP induces vasodilation via P_2-purinoceptors located on the endothelial cells lining the vascular lumen. Removal of the endothelial cells abolishes these relaxations, revealing contractions due to an effect of ATP directly on the vascular smooth muscle. Thus, in the isolated rat femoral artery Kennedy *et al.* (1985) showed that in the presence of an intact endothelium low concentrations of ATP produced relaxation while higher concentrations induced transient contractions followed by relaxation. Removal of the endothelium abolished the relaxations, and contractions were seen instead, often at concentrations that produced no change in vessel tone in the presence of endothelium.

Clearly, the responses produced by ATP here are the sum of an inhibitory action via P_2-purinoceptors located on the endothelium and an excitatory action at P_2-purinoceptors located on the smooth muscle. P_2-purinoceptors have been divided into several subtypes on the basis of the rank order of potency of a number of P_2-agonists (Burnstock and Kennedy, 1985; Kennedy, 1990). The receptor mediating the excitatory actions of ATP on many blood vessels has been characterized as being of the P_{2x}-subtype and the receptor present on the endothelium is of the P_{2y}-subtype.

Thus, in the absence of the inhibitory influence of the endothelium the contractile activity of ATP becomes apparent and the possibility that ATP is an excitatory transmitter can be considered.

III. ATP as an Excitatory Cotransmitter in Sympathetic Nerves

A. *Co-storage and Release of ATP and NA*

An essential condition required to support the cotransmission hypothesis is that ATP be shown to be co-stored with NA in sympathetic nerves and that their release can be measured following nerve stimulation. Certainly, the existence of nerves that can synthesize, store, and release more than one pharmacologically active substance is now widely accepted (Cuello, 1982; Osborne, 1983) and NA and ATP have been shown to be stored together in vesicles in sympathetic neurons (Stjärne and Lishajko, 1966; Geffen and Livett, 1971; Smith, 1972). Recently, brief exposure to X-rays was found to selectively impair the ATP component of neurotransmission in the rabbit central ear artery (Stewart-Lee *et al.*, 1991). This could be due to a change in the synthesis and/or the vesicular uptake and storage of ATP. Alternatively, it may indicate that as well as being co-stored in one population of vesicles with NA, ATP can also be stored separately from NA in another vesicular pool.

Tetrodotoxin- and guanethidine-sensitive release of ATP following sympathetic nerve stimulation have been demonstrated in several blood vessels (Su, 1975; Levitt and Westfall, 1982; Katsuragi and Su, 1980). ATP appeared to be released from a neuronal rather than from a postjunctional source. In the rabbit pulmonary artery KCl-induced [^{3}H] purine efflux was unaffected by reserpine or phentolamine but was greatly decreased by the removal of external calcium, 6-hydroxydopamine pretreatment or cold storage, whereas [^{3}H] purine efflux induced by exogenous adrenaline was inhibited by phentolamine but unaffected by 6-hydroxydopamine or cold storage (Katsuragi and Su, 1980).

B. Electrophysiological Actions of Endogenously Released ATP

ATP was first shown to be a cotransmitter with NA in the early 1980s in the guinea pig vas deferens. In this preparation, nerve stimulation evokes an initial transient, phasic contraction followed by a maintained, tonic contraction. Both phases are inhibited by guanethidine, but only the maintained portion is sensitive to α-adrenoceptor antagonists. However, the phasic contraction can be selectively abolished by $ANAPP_3$, a photoaffinity analog of ATP (Fedan *et al.*, 1981), or by α,β-methyleneATP (α,β-meATP) (Meldrum and Burnstock, 1983), a selective and potent P_2-agonist that induces desensitization of the P_{2x}-purinoceptor (Kasakov and Burnstock, 1983). The excitatory junction potentials (EJPs) that underlie the phasic contraction are also sensitive to these compounds (Sneddon *et al.*, 1982; Sneddon and Burnstock, 1984a). Thus, the EJPs and phasic contraction appear to be mediated by neuronally released ATP.

EJPs that are resistant to α-adrenoceptor blockade but abolished by guanethidine and tetrodotoxin are seen in many blood vessels (see Burnstock and Kennedy, 1986; Kennedy, 1988). They were first shown to be mediated by ATP by Sneddon and Burnstock (1984b) in the rat tail artery where a train of nerve impulses induced rapid, transient EJPs superimposed on a slowly developing, maintained depolarization. The α-adrenoceptor antagonist phentolamine inhibited the maintained depolarization but had no effect on the EJPs. In contrast, desensitization of the P_{2x}-purinoceptor by α,β-meATP inhibited the EJPs but not the maintained depolarization, indicating that α,β-meATP was not acting at a prejunctional site to depress the release of NA. Addition of both phentolamine and α,β-meATP inhibited the entire neurogenic response. Thus, the electrical responses in smooth muscle evoked by nerve stimulation appear to be mediated by both ATP and NA.

C. Mechanical Actions of Endogenously Released ATP

The contractile response to nerve stimulation was first shown to be mediated by both ATP and NA in the mesenteric artery of the rabbit (Kugelgen and Starke, 1985) and dog (Muramatsu, 1986) and in the rabbit central ear artery (Kennedy *et al.*, 1986). These studies showed that a portion of the neurogenic contraction was resistant to α-adrenoceptor blockade but could be selectively inhibited by α,β-meATP. Again α,β-meATP had no effect on the release of NA, excluding a prejunctional site of action. Subsequently, a great number of isolated blood vessels and vascular beds have been shown to exhibit EJPs and/or neurogenic contractions that are resistant to α-adrenoceptor antagonists but are sensitive to desensitization of the P_{2x}-purinoceptor by α,β-meATP (see Kennedy, 1988).

Unlike the vas deferens, the contractile response of most blood vessels to nerve stimulation does not consist of two clearly defined components but rather tends to be monophasic. One study in which a clearly biphasic neurogenic contraction was seen was that of Machaly *et al.*, (1988) in the dog mesenteric artery where α-adrenoceptor blockade had little effect on a rapid, transient phase but inhibited a slower, maintained phase. α,β-meATP greatly depressed the nonadrenergic contraction.

The relative contribution of NA and ATP to neurogenic contractions is both vessel dependent and species dependent. For example, at low frequencies of nerve stimulation ATP appears to be the exclusive excitatory transmitter in the isolated kidney of the rat, but NA is the exclusive transmitter in the rabbit kidney under the same conditions (Schwartz and Malik, 1989).

α,β-meATP also selectively inhibits neurogenic vasoconstriction in the whole animal (Bulloch and McGrath, 1988a), implicating ATP in the control of blood pressure *in vivo*. An early study could not find an effect of α,β-meATP in the absence of α-adrenoceptor blockade (Flavahan *et al.*, 1985), but this has since been shown to be due to the α,β-meATP-induced desensitization being more transient *in vivo* (Bulloch and McGrath, 1988a). Also, within a single animal the various vascular beds will utilize ATP as a cotransmitter to different extents and so inhibition of the action of neuronally released ATP may have a larger effect on some beds than on others.

As yet, cotransmission has not been studied in human blood vessels. However, exogenously administered ATP has been shown to elicit vasoconstriction via P_{2x}-purinoceptors in human cerebral artery (Hardebo *et al.*, 1987), pulmonary artery (Liu *et al.*, 1989), and subcutaneous and omental arteries (Martin *et al.*, 1991). Further studies are required to determine whether these receptors are involved in a cotransmitter role for ATP.

D. *Frequency Dependence of Cotransmission*

The relative contributions of ATP and NA to neurogenic contractions are also dependent on the frequency of stimulation and on the number of pulses applied. Kennedy *et al.* (1986) used two patterns of nerve stimulation to induce contractions of the rabbit central ear artery. Stimulation was applied for (a) 1 s or (b) until a maximum response was reached (Fig. 1). Contractions evoked by continuous stimulation were significantly greater than those by a 1-s train. The α-adrenoceptor antagonist prazosin (10^{-6} M) significantly reduced, but did not abolish, all neurogenic contractions, such that contractions to both types of stimulation were of a similar size (Figs. 1, 2). A higher concentration of prazosin (10^{-5} M) had no additional inhibitory effect. Desensitization of the postjunctional P_{2x}-purinceptor present in this tissue (Kennedy and Burnstock, 1985) by α,β-meATP inhibited the remaining nonadrenergic neurogenic contractions and contractions to exogenous ATP but had no effect on contractions to exogenous NA.

When expressed as a percentage of the maximum contraction to NA, the prazosin-resistant contractions appear to be quite small (Fig. 2). However, if they are instead expressed as a percentage of the control response in the absence of prazosin, then the importance of the purinergic response becomes clearer (Fig. 3). The greatest resistance to α-adrenceptor blockade was seen with a 1-s train at low frequencies when the nonadrenergic component of contraction was 60% of the total contraction. The relative contribution of ATP decreased as the number of pulses delivered was increased, either by increasing the frequency of the 1-s bursts or by stimulating until a maximum was reached. Thus, the relative contribution of NA

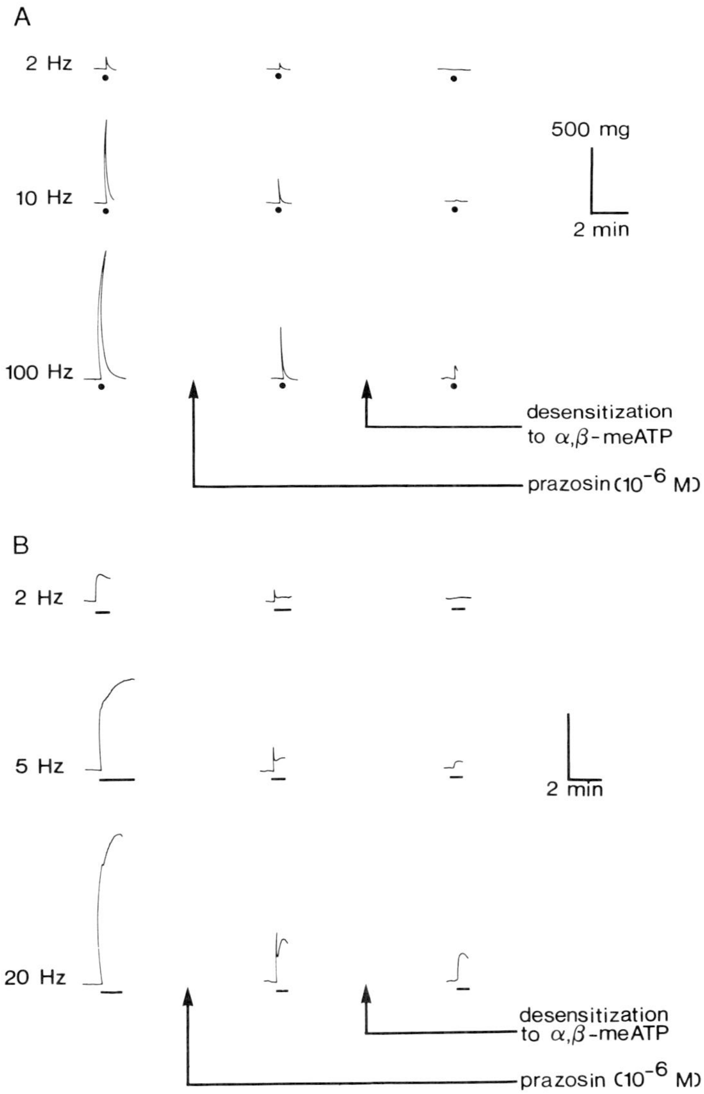

Figure 1 Isolated rabbit central ear artery: contractions to perivascular nerve stimulation (0.1-ms pulse width, supramaximal voltage) with either (A) a 1-s train (●) (2, 10, 100 Hz) or (B) with continuous stimulation until a maximum response was reached (−) (2, 5, 20 Hz). The inhibition of control responses by prazosin (10^{-6} M) and subsequently by desensitization of the P_{2x}-purinoceptor by α,β-meATP is shown. Note that in (B) the vertical scale represents 1000 mg in the left-hand column but 500 mg in the center and right-hand columns. (Reprinted with permission from Kennedy *et al.*, 1986.)

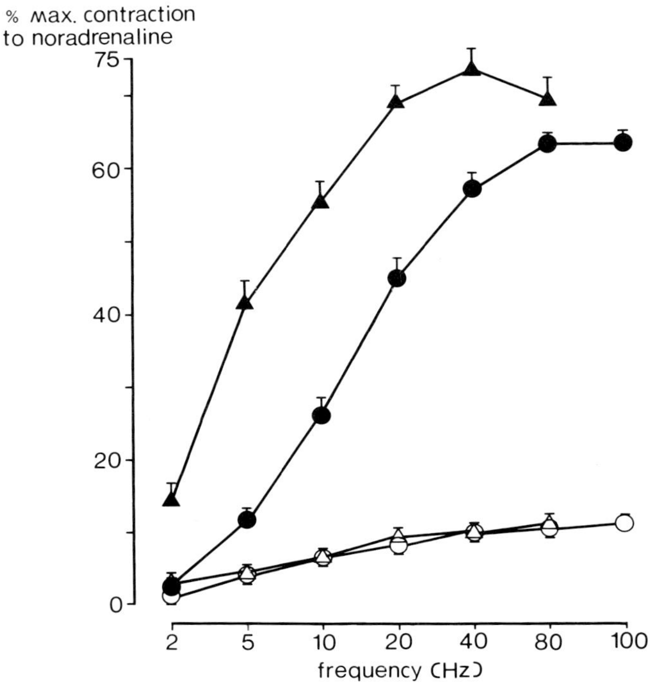

Figure 2 Isolated rabbit central ear artery: perivascular nerve stimulation (2–100 Hz, 0.1-ms pulse width, supramaximal voltage) with a 1-s train (○) or with continuous stimulation until a maximum response was reached (△) prior to (filled symbols) or following 30 min preincubation with prazosin (10^{-6} M) (open symbols) (n = 6). Vertical bars show s.e.m. (Reprinted with permission from Kennedy *et al.*, 1986.)

and ATP to neurogenic contractions is highly dependent on the parameters of stimulation used. This may explain why some studies using long trains of stimulation find that the contractions are virtually abolished by α-adrenoceptor antagonists. Purinergic contractions will be favored by short bursts of neuronal activity and noradrenergic by longer trains. Recordings of nerve activity *in vivo* suggest that short bursts may be more physiologically relevant than long bursts of activity.

E. Sympathetic Nerve Activity in Vivo

It has long been held that sympathetic nerves rarely fire at greater than 6–8 Hz under physiological conditions. This is based on experiments by Folkow (1952) in which it was shown that long trains of stimulation of the sympathetic chain produced vasoconstriction in the splanchnic bed that was maximal at about 6–8 Hz. However, it has since been shown that, rather than being regular, sympathetic nerve activity *in vivo* is in fact highly irregular (see Nilsson *et al.*, 1985). Although the average firing frequency may be low, neuronal action potentials tend to fire in short bursts and

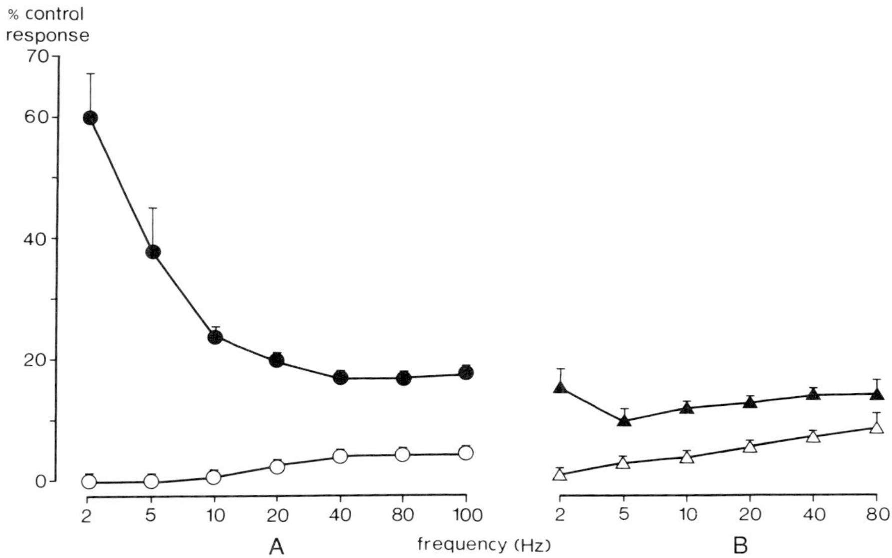

Figure 3 Isolated rabbit central ear artery: contractions to perivascular nerve stimulation (2–100 Hz, 0.1-ms pulse width, supramaximal voltage) with either (A) a 1-s train (○) or (B) with continuous stimulation until a maximum response was reached (△). Contractions in the presence of prazosin (10^{-6} M) (filled symbols) and subsequently following desensitization of the P_{2x}-purinoceptor by α,β-meATP (open symbols) are expressed by a percentage of control contractions at each frequency prior to the addition of prazosin (n = 6). Vertical bars show s.e.m. (Reprinted with permission from Kennedy *et al.*, 1986.)

often at frequencies much higher than 6–8 Hz. Also, vasoconstriction is greater when a given number of pulses are delivered in bursts rather than continuously (Andersson, 1983; Nilsson *et al.*, 1985). Finally, irregular stimulation favors the transmitter evoking transient responses, rather than that producing maintained contractions (i.e., ATP is favored over NA) (Sjöblom-Widfeldt *et al.*, 1990). Thus, the short trains of stimulation used by Kennedy *et al.* (1986) may have a greater physiological relevance than the long trains of stimulation used in previous studies.

F. Cellular Mechanism of the Action of ATP

The dependency of neurogenic vasoconstriction on the frequency and number of pulses delivered may be explained if we consider the cellular mechanisms underlying the action of ATP. Application of discrete amounts of ATP evokes a rapid, transient depolarization in vascular smooth muscle (see Burnstock and Kennedy, 1986; Kennedy, 1988). In the rabbit ear artery ATP opens ion channels that are nonselective for cations (Benham and Tsien, 1987; Benham *et al.*, 1987). The resulting current is inward at negative membrane potentials and so depolarizes the muscle cell. This triggers the opening of voltage-dependent calcium channels leading to influx of calcium and muscle contraction. The ATP-induced current displays rapid desensitization in the continued presence of ATP such that the membrane potential repolarizes to below threshold for activation of calcium channels. Thus, a postjunctional mechanism appears to limit the effects of ATP. Noradrenaline, on the other hand, tends to produce a slower, more maintained response that is not subject to desensitization over the same time scale (see Sneddon and Burnstock, 1984b).

If postjunctional responses to endogenously released ATP and NA are similar to those of exogenously administered drug, then increasing the number of stimulation pulses applied to a nerve is likely to lead to a progressively smaller action of ATP but an increasing action of NA. This difference in the response to ATP and NA may therefore explain the frequency dependence of neurogenic vasoconstriction described earlier.

The channels activated by ATP in the rabbit central ear artery display a 3 to 1 selectivity for calcium over sodium ions, which means that at physiological ion concentrations a significant influx of calcium will occur during channel opening (Benham, 1989). However, the main route of calcium entry is likely to be through voltage-dependent calcium channels, the opening of which will be triggered by the ATP-induced depolarization. Antagonists at L-type calcium channels inhibit the α,β-meATP-sensitive component of neurogenic vasoconstriction but have little or no effect on the prazosin-sensitive component (Bulloch and McGrath, 1988b; Omote *et al.*, 1989; Bulloch *et al.*, 1991).

ATP (acting via the P_{2x}-purinoceptor) appears to open the nonselective, cationic channels independent of readily diffusable secondary messengers (see Bean and Friel, 1990). Thus, the P_{2x}-purinoceptor may be part of a receptor–ion channel complex, similar to the nicotinic acetylcholine receptor. Specific binding sites thought to be the P_{2x}-purinoceptor have been found in the rabbit ear artery and other smooth muscle preparations using [^{3}H] α,β-meATP (Bo and Burnstock, 1989).

IV. Role of ATP in Hypertension

The spontaneously hypertensive rat (SHR) derived from the normotensive Wystar-Kyoto (WKY) strain is widely used as a model of hypertension. In the isolated tail artery of these two strains Vidal *et al.* (1986) found that the contribution of ATP to neurogenic vasoconstriction was significantly increased in SHR vessels. This suggested that an increased purinergic response could be involved in hypertension. However, in the same preparation Dalziel *et al.* (1989) found no such increase in the contribution of ATP. Furthermore, they found no enhanced action of ATP in the pithed SHR animal compared with the WKY.

The rat tail artery may not be the best tissue in which to study changes in cotransmission since the role of ATP in neurogenic contractions is not as clear here as in other vessels. Although, as discussed earlier, the postjunctional electrical response to nerve stimulation clearly consists of noradrenergic and purinergic components, the same is not true for the mechanical response. Prazosin abolishes contractions in some preparations, but in others a prazosin-resistant component remains (Cheung, 1984). While Vidal *et al.* (1986) reported that neurogenic contractions are unaffected by α,β-meATP, Bao *et al.* (1990) reported complex effects.

α,β-meATP abolished the rapid, prazosin-resistant phasic contraction but also caused the appearance of a larger, more slowly developing contraction. These effects occur with no change in the release of endogenous NA or on the action of exogenous NA. The slow decay of the depolarization induced by α,β-meATP can lead to an increase in neurogenic noradrenergic contractions (Neild and Kotecha, 1986). Finally, the effects of α,β-meATP and prazosin are dependent on both the age of the animal (Bao *et al.*, 1989a) and the region of the tail artery from which the preparation was taken (Bao *et al.*, 1989b). Clearly the actions of α,β-meATP are not simple in this issue. Thus, the role of ATP in hypertension remains unclear.

V. Functional Significance of Cotransmission

The functional significance of cotransmission remains to be determined, but several possible reasons for this arrangement can be considered. First, the presence in sympathetic nerves of two neurotransmitters displaying different temporal profiles of action may allow the whole animal a flexible response to different conditions. ATP acts rapidly, producing a response that is not maintained. This could be involved in short-term (pulse to pulse or second to second) regulation of blood pressure. In contrast NA is relatively slow acting and elicits a maintained vasoconstriction, which may be more useful when a maintained rise in blood pressure is advantageous. Second, corelease of two transmitters may be linked to a postjunctional synergistic interaction between these two substances. ATP and NA show synergistic potentiation in several blood vessels (Krishnamurty and Kadowitz, 1983; Burnstock *et al.*, 1984; Kennedy *et al.*, 1985; Kennedy and Burnstock, 1986; Ralevic and Burnstock, 1990). This may reflect an intimate association between α-adrenoceptors and P_2-purinoceptors or may simply be due to a general sensitization of the vascular smooth muscle cell membrane by these compounds. Finally, the functional significance of co-storage of transmitters may depend on the functional activity of the nerve terminal (i.e., the relative rates of release of cotransmitters may change according to the number, frequency, and pattern of impulses reaching the nerve terminal). Parasympathetic nerves innervating blood vessels associated with cat submaxillary gland appear to release both acetylcholine and vasoactive intestinal polypeptide (VIP) as inhibitory transmitters. However, acetylcholine is optimally released at frequencies well below those required to give effective release of VIP. Optimal release and response to endogenous VIP are seen when short bursts of high frequency of stimulation are applied (Edwards and Bloom, 1982). If such a differential release of ATP and NA were to occur in sympathetic nerves, then this would imply that ATP and NA can be stored in separate vesicles.

VI. Conclusions

That ATP acts as a cotransmitter with NA in perivascular sympathetic nerves appears to be widely accepted. ATP has been known for a number of years to be stored in and released from these nerves. However, ATP was only conclusively shown to be a cotransmitter when it was demonstrated that selective desensitization of the vascular P_{2X}-purinoceptor by α,β-meATP inhibits a component of both the electrical and the mechanical response to nerve stimulation, α,β-meATP has no effect on the evoked release of NA. The relative contribution of ATP and NA to the overall response varies greatly not only from vessel to vessel within a species but also from species to species for a given vessel. The relative contribution also depends on the parameters of stimulation used to evoke transmitter release. ATP is favored by short bursts of release whereas NA is favored by longer maintained periods of stimulation. This difference in the temporal profile of action of ATP and NA may allow greater flexibility in the control of blood pressure under different conditions.

References

Andersson, P. O. (1983). Comparative vascular effects of stimulation continuously and in bursts of the sympathetic nerves to cat skeletal muscle. *Acta Physiol. Scand. 118*, 343–348.

Bao, J. X., Eriksson, I. E., and Stjarne, L. (1989a). Age-related variations in the relative importance of noradrenaline and ATP as mediators of the contractile response of rat tail artery to sympathetic nerve stimulation. *Acta Physiol. Scand. 136*, 287–288.

Bao, J. X., Eriksson, I. E., and Stjarne, L. (1989b). Neurogenic contractions in caudal artery from young rats: Receptor–receptor interaction and regional differences in effects of noradrenaline and ATP via α_1-, α_2- and P_{2x}-receptors. *Acta Physiol. Scand. 136*, 619–620.

Bao, J. X., Eriksson, I. E., and Stjarne, L. (1990). Neurotransmitters and pre- and post-junctional receptors involved in the vasoconstrictor response to sympathetic nerve stimulation in rat tail artery. *Acta Physiol. Scand. 140*, 467–479.

Bean, B. P., and Friel, D. D. (1990). ATP-activated channels in excitable cells. *In* "Ion Channels." (T. Narahashi, ed.), Vol. 2, pp. 169–203. Plenum Press, New York.

Benham, C. D. (1989). ATP-activated channels gate calcium entry in single smooth muscle cells dissociated from rabbit ear artery. *J. Physiol. 419*, 689–701.

Benham, C. D., and Tsien, R. W. (1987). A novel receptor-operated Ca^{2+}-permeable channel activated by ATP in smooth muscle. *Nature 328*, 275–278.

Benham, C. D., Bolton, T. B., Byrne, N. G., and Large, W. A. (1987). Action of externally applied adenosine triphosphate on single smooth muscle cells dispersed from rabbit central ear artery. *J. Physiol. 387*, 473–488.

Bo, X., and Burnstock, G. (1989). [^{3}H]-α,β-methyleneATP, a radioligand labelling P_2-purinoceptors. *J. Autonom. Nerv. Syst. 28*, 85–88.

Bulloch, J. M., and McGrath, J. C. (1988a). Blockade of vasopressor and vas deferens responses by α,β-methyleneATP in the pithed rat. *Br. J. Pharmacol. 94*, 103–108.

Bulloch, J. M., and McGrath, J. C. (1988b). Selective blockade of nifedipine of purinergic rather than adrenergic nerve-mediated vasopressor responses in the pithed rat. *Br. J. Pharmacol. 95*, 695–700.

Bulloch, J. M., McDonald, A., and McGrath, J. C. (1991). Different sensitivities of rabbit isolated blood vessels exhibiting co-transmission to the slow calcium channel blocker, nifedipine. *Br. J. Pharmacol. 103*, 1685–1690.

Burnstock, G. (1978). A basis for distinguishing two types of purinergic receptor. *In* "Cell Membrane Receptors for Drugs and Hormones: A Multidisciplinary Approach." (R. W. Straub and L. Bolis, eds), pp. 107–118. Raven Press, New York.

Burnstock, G. (1980). Cholinergic and purinergic regulation of blood vessels. *In* "Handbook of Physiology, Sec 2, The Cardiovascular System, Vol II, Vascular Smooth Muscle." (D. F. Bohr, A. P. Somlyo, H. V. Sparks, and S. R. Geiger, eds.), pp. 567–612. American Physiological Society, Washington, D.C.

Burnstock, G., and Kennedy, C. (1985). Is there a basis for distinguishing two types of P_2-purinoceptor? *Gen. Pharmacol. 16*, 433–440.

Burnstock, G., and Kennedy, C. (1986). A dual function for adenosine 5′-triphosphate in the regulation of vascular tone: Excitatory cotransmitter with noradrenaline from perivascular nerves and locally released inhibitory intravascular agent. *Circ. Res. 58,* 319–330.

Burnstock, G., Crowe, R., Kennedy, C., and Török, J. (1984). Indirect evidence that purinergic modulation of perivascular adrenergic neurotransmission in the portal vein is a physiological process. *Br. J. Pharmacol. 82,* 359–368.

Cheung, D. W. (1984). Neural regulation of electrical and mechanical activities in the rat tail artery. *Pflugers Arch. 400,* 335–337.

Cuello, A. C. (1982). "Cotransmission." Macmillan, London.

Dalziel, H. H., Machaly, M., and Sneddon, P. (1989). Comparison of the purinergic contribution to sympathetic vascular responses in SHR and WKY rats *in vitro* and *in vivo*. *Eur. J. Pharmacol. 173,* 19–26.

Drury, A. N., and Szent-Gyorgi, A. (1929). The physiological activity of adenine compounds with special reference to their action upon the mammalian heart. *J. Physiol. 68,* 213–237.

Edwards, A. V., and Bloom, S. R. (1982). Recent physiological studies of the alimentary autonomic innervation. *Scand. J. Gastroenterol. 17,* [Suppl], 77–89.

Fedan, J. S., Hogaboom, G. K., O'Donnell, J. P., Colby, J., and Westfall, D. P. (1981). Contribution by purines to the neurogenic response to the vas deferens of the guinea pig. *Eur. J. Pharmacol. 69,* 41–53.

Flavahan, N. A., Grant, T. L., Greig, J., and McGrath, J. C. (1985). Analysis of the α-adrenoceptor-mediated and other components in the sympathetic vasopressor responses of the pithed rat. *Br. J. Pharmacol. 86,* 265–274.

Folkow, B. (1952). Impulse frequency in sympathetic vasomotor fibres correlated to the release and elimination of the transmitter. *Acta Physiol. Scand. 25,* 49–76.

Furchgott, R. F., and Zawadzki, J. V. (1980). The obligatory role of endothelial cells in the relaxation of arterial smooth muscle by acetylcholine. *Nature 288,* 373–376.

Geffen, L. B., and Livett, B. G. (1971). Synaptic vesicles in sympathetic neurons. *Physiol. Rev. 51,* 98–157.

Gillespie, J. H. (1934). The biological significance of the linkages in adenosine triphosphoric acid. *J. Physiol. 80,* 345–359.

Green, H. N., and Stoner, H. B. (1950). "Biological actions of the adenine nucleotides". Lewis, London.

Haddy, F. J., and Scott, J. B. (1968). Metabolically linked vasoactive chemicals in local regulation of blood flow. *Physiol. Rev. 48,* 688–707.

Hardebo, J. E., Kåhström, J., and Owman, C. (1987). P_1- and P_2-purine receptors in brain circulation. *Eur. J. Pharmacol. 144,* 343–352.

Kasakov, L., and Burnstock, G. (1983). The use of the slowly degradable analog, α,β-methylene ATP, to produce desensitisation of the P_2-purinoceptor: Effect on non-adrenergic, non-cholinergic responses of the guinea-pig urinary bladder. *Eur. J. Pharmacol. 86,* 291–294.

Katsuragi, T., and Su, C. (1980). Purine release from vascular adrenergic nerves by high potassium and a calcium ionophore A-23187. *J. Pharmacol. Exp. Ther. 215,* 685–690.

Kennedy, C. (1988). Possible roles for purine nucleotides in perivascular neurotransmission. *In* "Non-adrenergic Innervation of Blood Vessels, Vol. 1, Putative Neurotransmitters." (G. Burnstock and S. G. Griffith), pp. 65–76. CRC Press, Boca Raton, FL.

Kennedy, C. (1990). P_1- and P_2-purinoceptor subtypes—an update. *Arch. Int. Pharmacodyn. 303,* 30–50.

Kennedy, C., and Burnstock, G. (1985). ATP produces vasodilation via P_1-purinoceptors and vasoconstriction via P_2-purinoceptors in the isolated rabbit central ear artery. *Blood Vessels 22,* 145–155.

Kennedy, C., and Burnstock, G. (1986). ATP potentiates noradrenergic contractions in the portal vein of guinea-pig and rat. *J. Pharm. Pharmacol. 38,* 307–309.

Kennedy, C., Delbro, D., and Burnstock, G. (1985). P_2-purinoceptors mediate both vasodilation (via the endothelium) and vasoconstriction of the isolated rat femoral artery. *Eur. J. Pharmacol. 107,* 161–168.

Kennedy, C., Saville, V. L., and Burnstock, G. (1986). The contributions of noradrenaline and ATP to the responses of the rabbit central ear artery to sympathetic nerve stimulation depends on the parameters of stimulation. *Eur. J. Pharmacol. 122,* 291–300.

Krishnamurty, V. S. R., and Kadowitz, P. J. (1983). Influence of adenosine triphosphate on the isolated perfused mesenteric artery of the rabbit. *Can. J. Physiol. Pharmacol. 61,* 1409–1417.

Kugelgen, I. V., and Starkê, K. (1985). Noradrenaline and adenosine triphosphate as cotransmitters of neurogenic vasoconstriction in rabbit mesenteric artery. *J. Physiol. 367,* 435–455.

Levitt, B., and Westfall, D. P. (1982). Factors influencing the release of purines and norepinephrine in the rabbit portal vein. *Blood Vessels 19,* 30–40.

Liu, S. F., McCormack, D. G., Evans, T. W., and Barnes, P. J. (1989). Characterisation of P_2-purinoceptors in human small pulmonary arteries. *Br. J. Pharmacol. 97,* 498P.

Machaly, M., Dalziel, H. H., and Sneddon, P. (1988). Evidence for ATP as a cotransmitter in dog mesenteric artery. *Eur. J. Pharmacol. 147,* 83–91.

Martin, G. N., Thom, S. A. McG., and Sever, P. S. (1991). The effects of adenosine triphosphate (ATP) and related purines on human isolated subcutaneous and omental resistance arteries. *Br. J. Pharmacol. 102,* 645–650.

Meldrum, L. A., and Burnstock, G. (1983). Evidence that ATP acts as a co-transmitter with noradrenaline in sympathetic nerves supplying the guinea-pig vas deferens. *Eur. J. Pharmacol. 29,* 161–163.

Muramatsu, I. (1986). Evidence for sympathetic, purinergic transmission in the mesenteric artery of the dog. *Br. J. Pharmacol. 87,* 478–480.

Neild, T. O., and Kotecha, N. (1986). Effects of α,β-methyleneATP on membrane potential, neuromuscular transmission and smooth muscle contraction in the rat tail artery. *Gen. Pharmacol. 17,* 461–464.

Nilsson, H., Ljung, B., Sjoblom, N., and Wallin, B. G. (1985). The influence of the sympathetic impulse pattern on contractile responses of rat mesenteric arteries and veins. *Acta Physiol. Scand. 123,* 303–309.

Omote, S., Kigoshi, S., and Muramatsu, I. (1989). Selective inhibition by nifedipine of the purinergic component of neurogenic vasoconstriction in the dog mesenteric artery. *Eur. J. Pharmacol. 160,* 239–245.

Osborne, N. N. (1983). "Dale's Principle and Communication between Neurones." Pergamon Press, Oxford.

Ralevic, V., and Burnstock, G. (1990). Postjunctional synergism of noradrenaline and adenosine 5′-triphosphate in the mesenteric arterial bed of the rat. *Eur. J. Pharmacol. 175,* 291–299.

Schwartz, D. D., and Malik, K. U. (1989). Renal periarterial nerve stimulation-induced vasoconstriction at low frequencies is primarily due to release of a purinergic transmitter in the rat. *J. Pharmacol. Exp. Ther. 250,* 764–771.

Sjöblom-Widfeldt, N., and Nilsson, H. (1990). Sympathetic transmission in small mesenteric arteries from the rat: Influence of impulse pattern. *Acta Physiol Scand. 138,* 523–528.

Smith, A. D. (1972). Subcellular localisation of noradrenaline in sympathetic neurons. *Pharmacol. Rev. 24,* 435–457.

Sneddon, P., and Burnstock, G. (1984a). Inhibition of excitatory junction potentials in guinea-pig vas deferens by α, β-methylene-ATP: Further evidence for ATP and noradrenaline as cotransmitters. *Eur. J. Pharmacol. 100,* 85–90.

Sneddon, P., and Burnstock, G. (1984b). ATP as a cotransmitter in rat tail artery. *Eur. J. Pharmacol. 106,* 149–152.

Sneddon, P., Westfall, D. P., and Fedan, J. S. (1982). Cotransmitters in the motor nerves of the guinea-pig vas deferens: Electrophysiological evidence. *Science 218,* 693–695.

Stewart-Lee, A. L., Maynard, K. I., Lincoln, J., and Burnstock, G. (1991). Sympathetic neurotransmission in the rabbit isolated central ear artery is affected as early as one week following a single dose of irradiation. *Br. J. Pharmacol. 102,* 23–26.

Stjärne, L., and Lishajko, F. (1966). Comparison of spontaneous loss of catecholamines and ATP *in vitro* from isolated bovine adrenomedullary, vesicular gland, vas deferens and splenic nerve granules. *J. Neurochem. 13,* 1213–1216.

Su, C. (1975). Neurogenic release of purine compounds in blood vessels. *J. Pharmacol. Exp. Ther. 195,* 159–166.

Vidal, M., Hicks, P. E., and Langer, S. Z. (1986). Differential effects of α,β-methylene ATP on responses to nerve stimulation in SHR and WKY tail arteries. *Naunyn Schmeidebergs Arch. Pharmacol. 332,* 384–390.

PART III

Parasympathetic Innervation of Circulation

CHAPTER 10

Regulatory Peptides of the Parasympathetic Vascular Innervation

ROLF UDDMAN ROLF HÅKANSON
ANDERS LUTS FRANK SUNDLER

I. Introduction

The automatic nervous system plays an important role in the regulation of regional blood flow as suggested by the rich supply of autonomic nerve fibers around most blood vessels. The adrenergic vascular innervation has been mapped out in several species using catecholamine histofluorescence as well as using immunocytochemistry for enzymes involved in noradrenaline (NA) synthesis. The demonstration of an anticipated cholinergic innervation of blood vessels has been more difficult due to the shortcomings of existing methods. The "classical" postganglionic neurotransmitter in the parasympathetic nervous system is acetylcholine (ACh). Acetylcholinesterase (AChE) staining has been used to delineate cholinergic fibers. AChE-positive nerve fibers are numerous around both cerebral and peripheral blood vessels (Edvinsson *et al.*, 1972; Burnstock, 1980; Hara *et al.*, 1985). It is generally assumed that such fibers represent cholinergic, postganglionic parasympathetic fibers. However, AChE may occur also in sympathetic and sensory fibers (Cauna and Naik, 1963; Kyösola and Rechardt, 1974; Burnstock, 1978; Lehman and Fibiger, 1979). Recently, more specific methods for the demonstration of cholinergic fibers have been introduced [e.g. immunocytochemistry for choline acetyltransferase (ChAT) responsible for the synthesis of ACh] (Dauphin *et al.*, 1991). Direct evidence for the existence of perivascular cholinergic fibers has come from the demonstration of ChAT in nerve fibers of cerebral arteries (Hardebo *et al.*, 1977; Florence and Bevan, 1979; Duckles, 1981; Estrada *et al.*, 1983; Suzuki *et al.*, 1990). It has further been demonstrated that cerebral blood vessels possess an uptake mechanism for ^{3}H-choline (Florence and Bevan, 1979). Although the distribution of AChE-positive fibers around blood vessels suggests a role in the regulation of local blood flow, the function of cholinergic fibers has not been extensively studied. Cholinomimetic agents, such as ACh and carbachol, mediate relaxation in isolated cerebral arteries. This relaxation is antagonized by the muscarinic receptor antagonist atropine (Edvinsson *et al.,* 1977). Transmural nerve stimulation (TNS) evokes relaxation. Whether this relaxation depends on the release of ACh from functioning cholinergic nerve fibers is debated (Lee, 1980; Duckles 1981; Bevan *et al.,* 1982).

It has been known for a long time that neuronally mediated responses in the periphery may be resistant to both adrenergic and cholinergic blocking agents (Langley, 1898; Burnstock, 1980; Lundberg, 1981). During recent years, various neuropeptides have been promoted to the rank of candidate neurotransmitters. Some of the peptides have well-defined physiological effects, either excitatory or inhibitory, whereas other peptides do not appear to display such effects and have instead been proposed to act as neuromodulators, a term that is intended to mean that they modulate the effects of other transmitters rather than exerting effects of their own (which of course is a transmitter property). Neuropeptides often coexist with classical neurotransmitters, such as NA and ACh. Moreover, more than one peptide may coexist in one and the same neuron (Hökfelt *et al.,* 1980; Sundler *et al.,* 1985, 1989; Hökfelt *et al.,* 1986; Håkanson *et al.,* 1990). Such coexistence can be expected in situations where the coexisting peptides arise from the same precursor. However, there are several instances on record where peptides generated from different precursors are stored together in the same neuron. The discovery that neurons can produce, store, and release more than one putative transmitter has attracted considerable attention over the past decade (Burnstock 1986; Hökfelt *et al.,* 1986). If several messenger compounds can be released from the same nerve ending, it is conceivable, perhaps even likely, that they cooperate. Moreover, if several messenger compounds are involved in the transmission of nerve impulses, it may be assumed not only that they cooperate but also that the proportion of the different messengers released may vary during different conditions. In the present review we will describe and discuss many of the neuropeptides that seem to be involved in parasympathetic vascular innervation.

II. Vasoactive Intestinal Peptide/Peptide Histidine Isoleucine

Vasoactive intestinal polypeptide (VIP), originally isolated from the porcine gut (Said and Mutt, 1970), belongs to a family of structurally related peptides, such as glucagon and secretin. Immunocytochemical studies have revealed a varying supply of VIP-containing nerve fibers in the cardiovascular system. Generally, arteries have a rich supply whereas veins have relatively few (Uddman *et al.,* 1981; Della *et al.,* 1983). Cerebral blood vessels, arteries as well as veins, receive a rich supply of VIP-containing fibers (Larsson *et al.,* 1976; Edvinsson *et al.,* 1980; Kobayashi *et al.,* 1983; Matsuyama *et al.,* 1983) (Fig. 1). A moderate number of VIP-containing fibers can be seen in the choroid plexus of some mammals (Lindvall *et al.,* 1978; Nilsson *et al.,* 1990). Intestinal blood vessels (Furness and Costa, 1982; Håkanson *et al.,* 1982; Hökfelt *et al.,* 1982) and blood vessels of the genitourinary tract (Alm *et al.,* 1977; Larsson *et al.,* 1977a,b; Fahrenkrug *et al.,* 1989) are supplied with

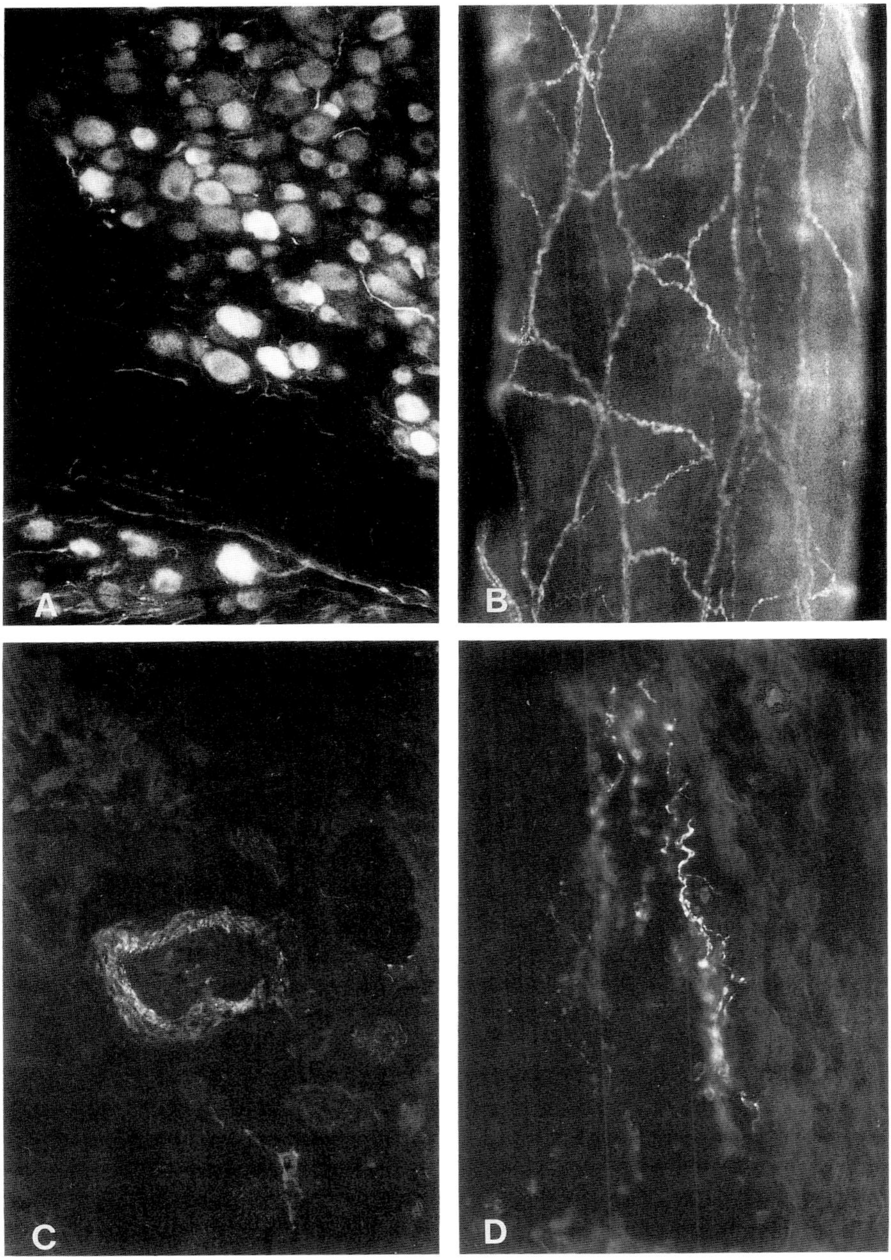

Figure 1 (A) Cat sphenopalatine ganglion. Virtually all nerve cell bodies are immunoreactive for VIP. (B) Whole mount spread of a blood vessel from the circle of Willis of cat. A rich supply of VIP-containing nerve fibers encompasses the vessel. (C) Nasal mucosa of dog. A dense supply of PHI-containing nerve fibers surrounds a small artery. (D) PACAP-containing nerve fibers in the adventitia of a pulmonary artery of guinea pig (×150).

numerous VIP-containing fibers (Fig. 2) whereas blood vessels in striated muscle, liver, and kidney have comparatively few such fibers. Ultrastructural studies have revealed the presence of VIP in large dense-cored vesicles (Johansson and Lundberg, 1981; Leander *et al.*, 1981).

In the cardiovascular system many of the VIP-containing neurons seem to be identical with cholinergic neurons and there is evidence for the coexistence of VIP and ACh in various parasympathetic ganglia (Lundberg *et al.*, 1979). In several tissues, low-frequency nerve stimulation gives rise to responses that mainly are atropine sensitive, whereas high-frequency stimulation, particularly when applied in bursts, induces predominantly atropine-resistant responses (Bloom and Edwards, 1980; Lundberg *et al.*, 1984). Under these latter conditions the release of VIP is greatly enhanced (Bloom and Edwards; 1980; Lundberg *et al.*, 1984). VIP is a powerful coronary vasodilator (Forssmann *et al.*, 1988; Smitherman and Dehmer, 1988) and dilates splanchnic and other vascular beds. As a consequence, exogenous VIP induces systemic hypotension.

High levels of VIP have been measured by radioimmunoassay in cerebral arteries of several species (Edvinsson *et al.*, 1980; Duckles and Said, 1982). The origin of VIP- and AChE-positive fibers to cerebral blood vessels has been an enigma for a long time. It has been suggested that some of the fibers, terminating on cortical surface arterioles, may derive from cortical VIP-containing neurons (Loren *et al.*, 1979). In addition, it was observed that VIP-containing and ChAT-positive nerve fibers in the cortex were closely associated with intraparenchymal blood vessels (Eckenstein and Baughman, 1984). Some of the VIP-containing fibers surrounding cerebral arteries in the cat seem to derive from local microganglia in, for example, the cavernous plexus and external rete (Gibbins *et al.*, 1984). In some species, VIP-containing fibers belonging to the rostral part of the circle of Willis emanate from the sphenopalatine ganglia (Hara *et al.*, 1985; Walters *et al.*, 1986; Suzuki *et al.*, 1988; Hara *et al.*, 1989). Lesioning of the sphenopalatine ganglia in the cat and rat causes only partial loss of VIP-immunoreactive fibers in the circle of Willis (Edvinsson *et al.*, 1980; Hara *et al.*, 1985, 1989). The findings indicate that approximately one-third of the cholinergic/VIP innervation of the circle of Willis (the rostral part) originates in the sphenopalatine ganglion. Other sources include local microganglia (Gibbins *et al.*, 1984; Hara and Kobayashi, 1987; Suzuki *et al.*, 1988) and the otic ganglia (Walters *et al.*, 1986; Edvinsson *et al.*, 1989). The VIP innervation of cerebral blood vessels seems to return 5 weeks after sphenopalatine ganglionectomy (Suzuki *et al.*, 1988; Hara *et al.*, 1989), probably due to the expression of VIP in other neurons. Parasympathetic nerve fibers supplying blood vessels in the head and neck region derive mainly from the sphenopalatine (Fig. 1) and otic ganglia.

Binding sites for VIP have been characterized in bovine cerebral arteries by the use of radiolabeled VIP (Suzuki *et al.*, 1985). The binding of ^{125}I-VIP to cerebral arteries was saturable, specific, and reversible and Scatchard analysis suggested the presence of both high- and low-affinity binding sites for ^{125}I-VIP. TNS releases VIP from the cat cerebral artery segments (Toda, 1982, Bevan *et al.*, 1986). VIP causes relaxation of isolated cerebral arteries (Larsson *et al.*, 1976; Toda, 1982; Bevan *et*

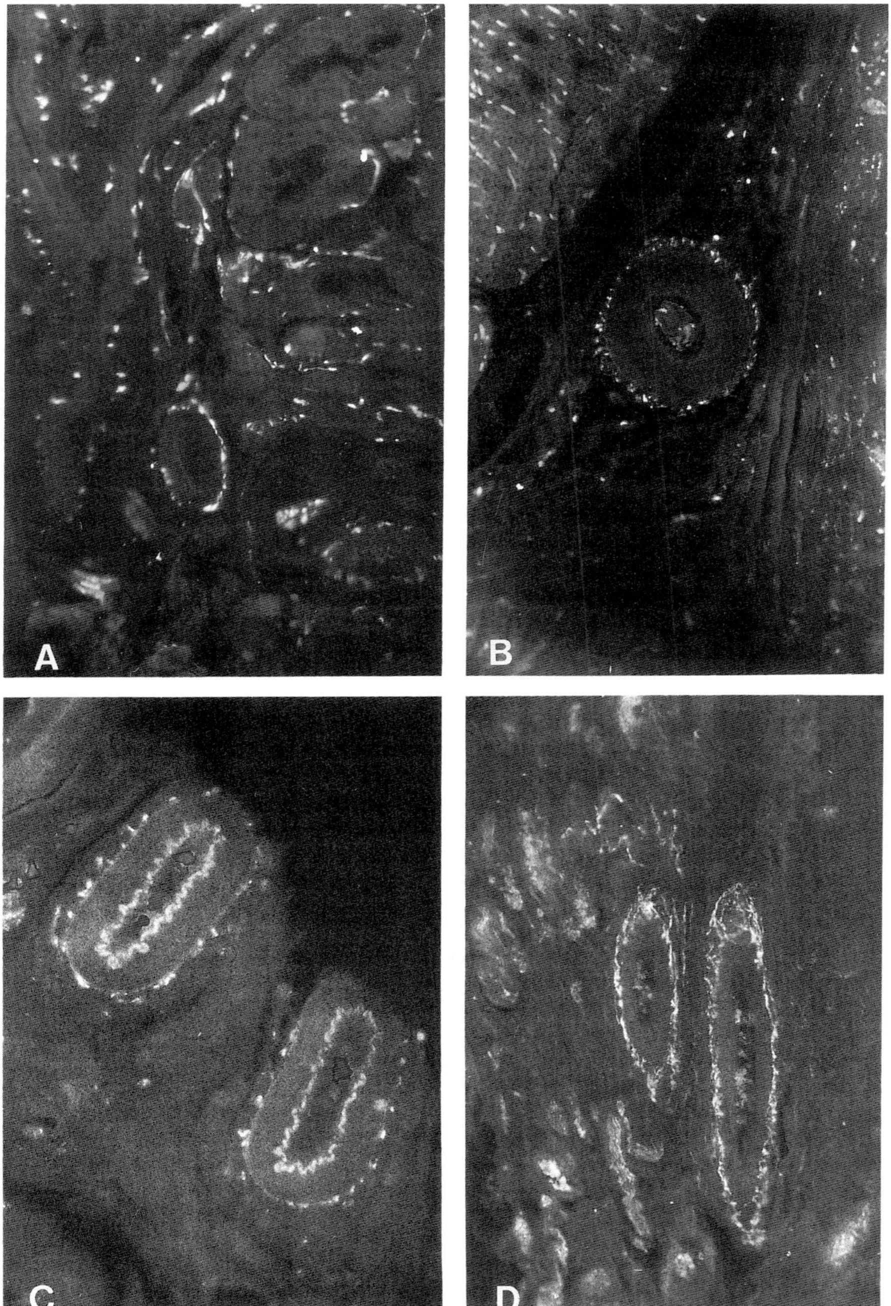

Figure 2 (A) Cat submaxillary gland. Numerous VIP-containing nerve fibers around glandular acini and small blood vessels. (B) Chicken proventrikel. VIP-containing nerve fibers surround a small blood vessel. (C) Numerous VIP-immunoreactive nerve fibers surrounding small blood vessels in the vas deferens of cat. (D) Prostate gland with VIP fibers around small blood vessels (×160).

al., 1984). These results have later been supplemented by *in vivo* experiments that indicate that VIP dilates cerebral blood vessels and increases cerebral blood flow (Heistad *et al.*, 1980; McCulloch and Edvinsson, 1980; Wei *et al.*, 1980). The effects of VIP are unaffected by cholinoceptor blockade and by β-adrenoceptor blockade, indicating that VIP exerts its action directly on the vascular smooth muscle. The relaxation occurs in parallel with activation of vascular adenylate cyclase (Edvinsson *et al.*, 1985). An intact endothelial layer is known to be essential for the vasodilatory response to ACh (Furchgott, 1981). When the endothelium is removed, cerebral and pulmonary arteries still relax in response to TNS and VIP (Duckles and Said, 1982; Lee *et al.*, 1984; Edvinsson *et al.*, 1985; Sata *et al.*, 1986).

Microapplication or superfusion of VIP elicits a concentration-dependent dilation of pial arterioles and veins *in situ*. Studies on the relationship between the caliber of the arterioles and the response to VIP have revealed that the response of smaller arterioles is greater than that of larger vessels (McCulloch and Edvinsson, 1980). It has also been shown that the intracarotid infusion of VIP results in an increase of cerebral blood flow and metabolism in primates and other species (McCulloch and Edvinsson, 1980), and it has been suggested that VIP may be involved in the interrelation between neuronal activity, energy generation, and local blood flow (McCulloch and Kelly, 1983; McCulloch *et al.*, 1983).

VIP-containing fibers to the upper respiratory tract derive from the sphenopalatine ganglia. The lower respiratory tract is supplied with fibers from the jugular-nodose ganglia and local ganglia along the vagus nerve and in the tracheobronchial wall (Martling *et al.*, 1990; Dey *et al.*, 1991). VIP dilates both resistance and capacitance vessels of the nasal mucosa in cats (Malm *et al.*, 1980). Autoradiography using ^{125}I-VIP as ligand has shown a moderate to dense labeling over the smooth muscle of the major airways and pulmonary vessels (Barnes *et al.*, 1986; Power *et al.*, 1988). In isolated pulmonary blood vessels VIP induces relaxation (Altiere and Diamond, 1983; Hand *et al.*, 1984; Obara *et al.*, 1989).

In the female genital tract VIP induces a concentration-dependent increase in vaginal, endometrial, myometrial, and total uterine blood flow (Ottesen and Fahrenkrug, 1981; Ottesen *et al.*, 1983; Ottesen and Einer-Jensen, 1984). In the male genital tract VIP causes a concentration-dependent relaxation of smooth muscle *in vitro* (Willis *et al.*, 1983; Steers *et al.*, 1984). By close intraarterial infusion in dogs, VIP elicits a vasodilation in the penis causing erection (Andersson *et al.*, 1984). In man, intracavernous injection of VIP provokes penile erection (Ottesen *et al.*, 1984).

The VIP precursor contains also peptide histidine isoleucine/methionine (PHI/PHM), a VIP-like peptide, sharing several biological activities with VIP (Tatemoto and Mutt, 1981; Itoh *et al.*, 1983). PHI, discovered by virtue of its C-terminal α-amide, was first isolated from the porcine duodenum (Tatemoto and Mutt, 1981). PHI resembles VIP by having 13 amino acid residues in identical positions. VIP and PHI have an identical distribution in central and peripheral neurons (Fig. 1). In fact, double immunostaining has revealed their coexistence in parasympathetic ganglia and in perivascular nerve fibers (Christofides *et al.*, 1982; Ekblad *et al.*, 1984a; Edvinsson and McCulloch, 1985). However, there appears to be differences in the relative amounts of VIP and PHI present in various regions, probably reflecting differential processing of the precursor (Fahrenkrug *et al.*, 1985). In particular, the presence of a C-terminally extended form of PHI has been demonstrated in the upper gut and respiratory tract (Yiangou *et al.*, 1987).

Like VIP, PHI induces relaxation of bovine and human pulmonary arteries. In both species VIP alone causes an almost complete relaxation of precontracted arteries. PHI appears to be less potent than VIP in relaxing blood vessels (Greenberg *et al.*, 1987). Both VIP and PHI produce vasodilatory responses on the myometrial blood flow in rabbits (Bardrum *et al.*, 1986). Although PHI generally seems to have a lower vasodilatory potency than VIP and although the posttranslational processing of the precursor sometimes favors the expression of VIP over that of PHI (Fahrenkrug *et al.*, 1985), both peptides might be regarded as candidates for causing the atropine-resistant dilatation of blood vessels in the respiratory tract (Lundberg *et al.*, 1988). Helodermin and helospectins are VIP-like peptides isolated from the venom of the lizard Gila monster (Hoshino *et al.*, 1984; Parker *et al.*, 1984; Cauvin *et al.*, 1986). There is recent evidence that helodermin/helospectin-like peptides are present also in mammals, where they seem to occur in certain endocrine cells (thyroid C cells, adrenomedullary cells, and endocrine-like cells in the airways) (Robberecht *et al.*, 1985; Bjartell *et al.*, 1989; Grunditz *et al.*, 1989; Luts *et al.*, 1991) and in neurons in the peripheral and central nervous system (Absood *et al.*, 1992; Kivipelto *et al.*, 1992). Like VIP, helospectin reduces blood pressure and relaxes precontracted femoral arteries *in vitro* (Grundemar and Högestätt, 1990). There is a moderate number of helodermin- and helospectin-immunoreactive nerve fibers around cerebral and peripheral blood vessels. Recently, still another member of the VIP/secretin family was discovered, namely pituitary adenylate cyclase activating peptide (PACAP) (Miyata *et al.*, 1989). PACAP occurs in two forms: PACAP 27 and PACAP 38 (Miyata *et al.*, 1989, 1990). Both peptides are amidated and PACAP 27 constitutes the N-terminal (VIP-like) portion of PACAP 38. A moderate supply of PACAP-containing nerve fibers can be seen around peripheral blood vessels in several laboratory animals (Sundler *et al.*, 1992; Uddman *et al.*, 1991) (Fig. 2). In guinea pigs, PACAP causes concentration-dependent relaxation of isolated precontracted basilar and pulmonary arteries. These dilatory effects are unaffected by pretreatment with atropine or propranolol. Removal of the endothelium abolishes the PACAP-induced relaxation (Cardell *et al.*, 1991). Double immunostaining has shown that helospectin-like peptides as well as PACAP coexist with VIP in a subpopulation of the VIP-containing neurons in the central and peripheral nervous system (Absood *et al.*, 1992; Sundler *et al.*, 1991; Uddman *et al.*, 1991). The coexistence of VIP, PACAP, and helospectin-like peptides is intriguing and may imply that the same neurons produce several homologous peptides. The structural homologies of these peptides and recent *in vitro* data suggest that they have similar biological activities.

Recently, C-terminally extended precursors of VIP (VIP-Gly-Lys and VIP-Gly-

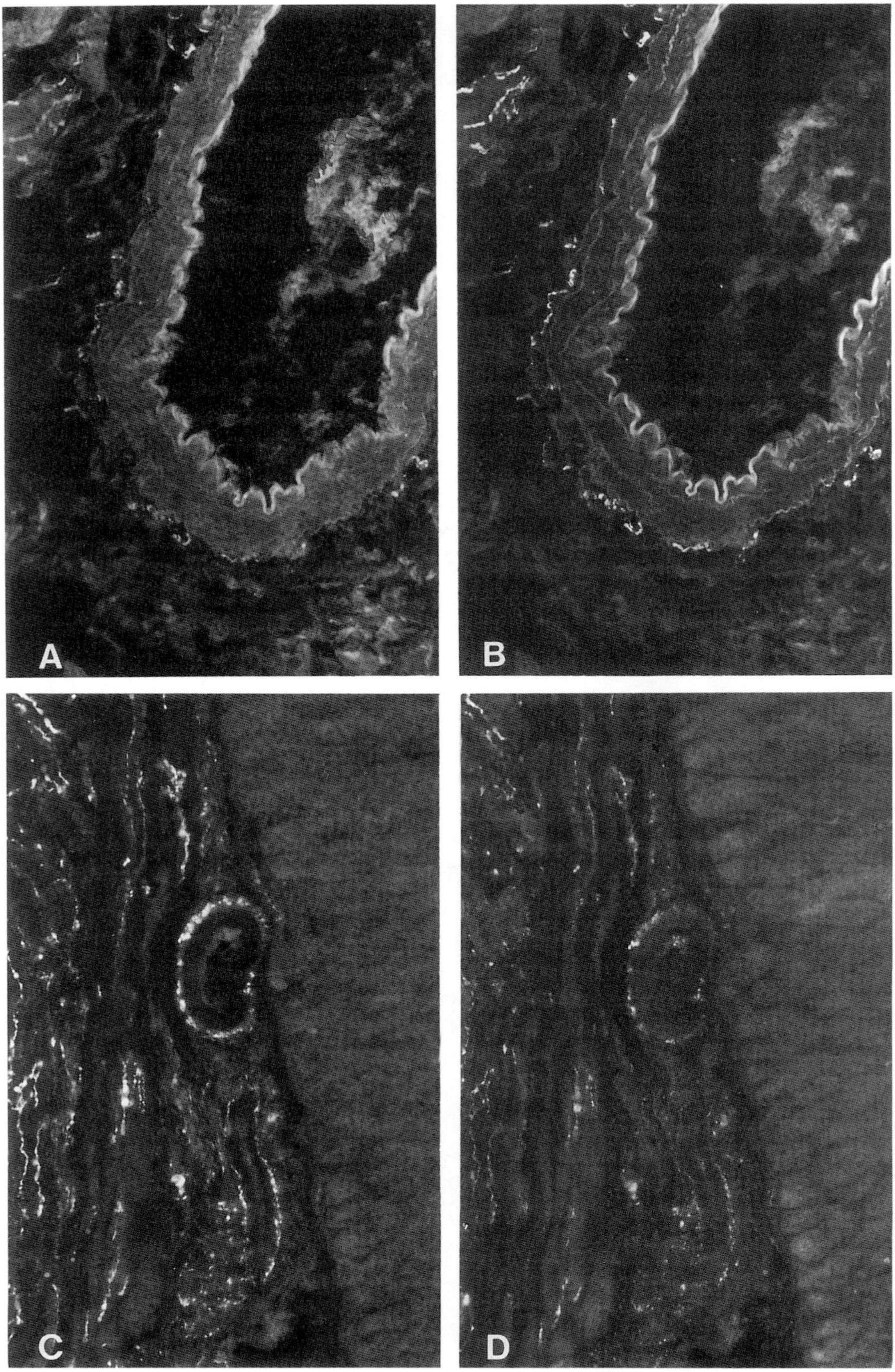
A
B
C
D

Lys-Arg) were synthesized. They showed high bronchodilatory activity and an increased depressor activity when compared with VIP itself (Ito and Tachibana, 1991). In addition, it has been shown that there are relatively large amounts of glycine-extended VIP in certain mammals (Fahrenkrug, 1991).

III. Neuropeptide Y

Neuropeptide Y (NPY) is a 36 amino acid peptide with N- and C-terminal tyrosine residues (Tatemoto *et al.*, 1982). NPY was first isolated from porcine brain but has since then been found to be widely distributed also outside the central nervous system. Immunocytochemical studies have revealed a particularly rich supply of NPY-containing nerve fibers in association with blood vessels (Lundberg *et al.*, 1982; Edvinsson *et al.*, 1983, 1984; Ekblad *et al.*, 1984b; Uddman *et al.*, 1985; Edvinsson *et al.*, 1987; Sundler *et al.*, 1989). NPY-containing nerve cell bodies are numerous in the sympathetic ganglia. Extirpation of the ganglia leads to the near complete disappearance of perivascular NA-containing fibers (Edvinsson *et al.*, 1972). The NPY-containing fibers are markedly reduced in number and there is a reduction in the tissue NPY content in vessels from the circle of Willis (Edvinsson *et al.*, 1987). After chemical sympathectomy (treatment with 6-hydroxydopamine), there is a reduction in the number of NPY-immunoreactive nerve fibers around cerebral blood vessels and a reduction of the NPY concentration by 70% in cerebral and cranial blood vessels (Edvinsson *et al.*, 1987). Simultaneous immunostaining with antibodies to NPY and to tyrosine hydroxylase or dopamine-β-hydroxylase (DBH) (enzymes found more or less exclusively in adrenergic neurons) has provided evidence that NPY and NA coexist in a major subpopulation of nerve cell bodies in sympathetic ganglia and in numerous cardiovascular nerve fibers (Lundberg *et al.*, 1983; Uddman *et al.*, 1985; Sundler, *et al.*, 1989). In the guinea pig, rat, and pig a population of nerve cell bodies in the sphenopalatine and otic ganglia contains NPY emphasizing the presence of NPY also in nonadrenergic neurons (Lundberg *et al.*, 1988). In the guinea pig uterine artery NPY has been found to coexist with NA in one population of nerve fibers and with VIP in another (Morris *et al.*, 1985; Fallgren *et al.*, 1989). Since very few VIP-immunoreactive axons contained DBH, it was concluded that the NPY/VIP axons were distinct from the noradrenergic ones. Also in the gut, respiratory tract, and thyroid, subpopulations of perivascular NPY-containing fibers are known to contain VIP and to be nonadrenergic (Ekblad *et al.*, 1984a; Grunditz *et al.*, 1987; Luts and Sundler, 1989) (Fig. 3). In addition, a population of the nonadrenergic NPY-containing fibers that surround pial arteries contains VIP (Gibbins and Morris, 1988). The number of neurons that contain both NPY and VIP is increased upon surgical sympathectomy.

Figure 3 Double immunostaining for VIP (A and C) and NPY (C and D). Note apparent identical immunostaining for the two peptides. Guinea pig pulmonary artery (A and B). Cryostat section first immunostained for VIP (A) and then for NPY (B). Proventrikel of chicken (C and D). A rich supply of VIP-containing fibers is seen in the smooth muscle and around a small blood vessel (×150).

The functional significance of the NPY and VIP colocalization remains to be clarified. It has been shown for the uterine artery that a low dose of NPY may antagonize the vasodilator effect of ACh and substance P (Fallgren *et al.*, 1989). This also seems to be the case for the brain circulation (Gibbins and Morris, 1988).

IV. Conclusions

Over the past few years, evidence has accumulated that indicates that blood vessels are innervated by different populations of nerve fibers that contain catecholamines, ACh, and neuropeptides, respectively. The origin and innervation pattern of the perivascular neuronal systems remain poorly defined in several tissues.

Most neuropeptides have strong and distinct effects on blood vessels. VIP/PHI are potent vasodilators whereas NPY acts as a vasoconstrictor. The physiological significance of these multiple neuropeptides will be fully revealed only after the development of specific receptor blocking agents.

The fact that some peptides derive from the same precursor molecule explains why they coexist in the same neuron (e.g., VIP/PHI, substance P/neurokinin A). In addition, peptides generated from distinct precursors have also been found to coexist in neurons. There are reasons to assume that they also coexist within the same vesicles in the cytoplasm and that they are coreleased upon nerve stimulation. The functional significance of the coexistence and possible corelease of a peptide and a classical transmitter (NPY/NA) or of several peptides (VIP/PHI, NPY/VIP/PHI, dynorphin/VIP/NPY) from perivascular nerve fibers is unclear. Conceivably, several bioactive substances may be released simultaneously from the same nerve endings, and each neuron may therefore represent a multimessenger system.

Much work will be needed in the years to come to define the chemical coding of the neurons that control the cardiovascular system and its functional significance.

Acknowledgment

Supported by grants from the Swedish Medical Research Council (17X-6859 and 014X-5958).

References

Absood, A., Ekblad, E., Ekelund, M., Håkanson, R., and Sundler, F. (1992). Helospectin-like peptides in the gastrointestinal tract: Immunocytochemical localization and immunochemical characterization. *Neuroscience 46,* 431–438.

Alm, P., Alumets, J., Håkanson, R., and Sundler, F. (1977). Peptidergic (VIP) nerves in the genitourinary tract. *Neuroscience 2,* 751–754.

Andersson, P.-O., Bloom, S. R., and Mellander, S. (1984). Haemodynamics of pelvic nerve induced penile erection in the dog: Possible mediation by vasoactive intestinal polypeptide. *J. Physiol. 350,* 209–224.

Altiere, R. J., and Diamond, L. (1983). Relaxant effects of vasoactive intestinal peptide and bethanechol in cat intrapulmonary artery. *European J. Pharmacol. 93,* 121–124.

Bardrum, B., Ottesen, B., and Fahrenkrug, J. (1986). Peptides PHI and VIP: Comparison between vascular and non-vascular smooth muscle effect in rabbit uterus. *Am. J. Physiol. 251,* E48–E51.

Barnes, P. J., Cadieux, A., Carstairs, J. R., Greenberg, B., Polak, J. M., and Rhoden, K. (1986). VIP in bovine artery. Localization, function and receptor autoradiography. *Br. J. Pharmacol. 89,* 157–162.

Bevan, J. A., Buga, G. M., Florence, V. M., Gonsalves, A., and Snowden, A. (1982). Distribution of choline acetyltransferase in cerebral and extracerebral cranial arteries of the cat. Its relationship to neurogenic atropine-sensitive dilation. *Circ. Res. 50,* 470–476.

Bevan, J. A., Buga, G. M., Moskowitz, M. A., and Said, S. I. (1986). *In vitro* evidence that vasoactive intestinal peptide is a transmitter of neuro-vasodilation in the head of the cat. *Neuroscience 19,* 597–604.

Bevan, J. A., Moskowitz, M. A., Said, S. I., and Buga, G. M. (1984). Evidence that vasoactive intestinal polypeptide is a dilator transmitter to some cerebral and extracerebral arteries. *Peptides 5,* 385–388.

Bjartell, A., Persson, P., Absood, A., Sundler, F., and Håkanson, R. (1989). Helodermin-like peptides in noradrenaline cells of adrenal medulla. *Regul. Pept. 26,* 27–34.

Bloom, S. R., and Edwards, A. V. (1980). Vasoactive intestinal polypeptide in relation to atropine resistant vasodilation in the submaxillary gland of the cat. *J. Physiol. 300,* 41–53.

Burnstock, G. (1978). Do some sympathetic neurones synthesize and release both noradrenaline and acetylcholine? *Prog. Neurobiol. 11,* 205–222.

Burnstock, G. (1980). Cholinergic and purinergic regulation of blood vessels. *In* "Handbook of Physiology." Sect. 2 "The Cardiovascular System" Vol. II, Vascular smooth muscle (D. F. Bohr, A. D. Somlyo, H. W. Sparks, and S. R. Geiger, eds.), pp. 567–612. Am. Physiol. Soc., Waverly Press, Baltimore.

Burnstock, G. (1986). Purines as cotransmitters in adrenergic and cholinergic neurones. *In* "Coexistence of neuronal messengers: A new principle in chemical transmission." (T. Hökfelt, K. Fuxe, and B. Pernow, eds.), Progr. Brain Res. Vol. 68, pp. 193–203. Elsevier, Amsterdam.

Cardell, L. O., Uddman, R., Luts, A., and Sundler, F. (1991). Pituitary adenylate cyclase activating peptide (PACAP) in guinea pig lung: Distribution and dilatory effects. *Regul. Pept. 36,* 379–390.

Cauna, N., and Naik, N. T. (1963). The distribution of cholinesterase in the sensory ganglia of men and some mammals. *J. Histochem. Cytochem. 11,* 129–138.

Cauvin, A., Vandermeers, A., Vandermeers-Piret, M. C., de Neef, P., Yanaihara, C., Yanaihara, N., Robberecht, P., and Christophe, J. (1986). Multiple variants of the PHI/VIP/helodermin family in rat brain. *Arch. Int. Physiol. Biochem. 94,* B69.

Christofides, N., Yiangou, Y., McGregor, G., Aarons, E., Woodhams, P., Tatemoto, K., and Bloom, S. R. (1982). Distribution of PHI in the rat brain. *Biomed. Res. 3,* 573–574.

Dauphin, F., Richard, J. W., Seylaz, J., Quirion, R., and Hamel, E. (1991). Acetylcholine levels and choline acetyltransferase activity in rat cerebrovascular bed after uni- or bilateral sphenopalatine ganglionectomy. *J. Cereb. Blood Flow Metab. 11,* 253–260.

Della, N. G., Papka, R. E., Furness, J. B., and Costa, M. (1983). Vasoactive intestinal peptide-like immunoreactivity in nerves associated with the cardiovascular system of guinea-pigs. *Neuroscience 9,* 605–619.

Dey, R. D., Altemus, J. B., and Michalkiewicz, M. (1991). Distribution of vasoactive intestinal peptide- and substance P-containing nerves originating from neurons of airway ganglia. *J. Comp. Neurol. 304,* 330–340.

Duckles, S. P. (1981). Evidence for a functional cholinergic innervation of cerebral arteries. *J. Pharmacol. Exptl. Ther. 217,* 544–548.

Duckles, S. P., and Said, S. I. (1982). Vasoactive intestinal polypeptide as a neurotransmitter in the cerebral circulation. *Eur. J. Pharmacol. 78,* 371–374.

Eckenstein, F., and Baughman, S. H. (1984). Two types of cholinergic innervation in cortex, one co-localized with vasoactive intestinal polypeptide. *Nature* (London) *309,* 153–155.

Edvinsson, L., and McCulloch, J. (1985). Distribution and vasomotor effects of peptide HI (PHI) in feline cerebral blood vessels *in vitro* and *in situ*. *Regul. Pept. 10,* 345–356.

Edvinsson, L., Nielsen, K. C., Owman, C., and Sporrong, B. (1972). Cholinergic mechanisms in pial vessels. Histochemistry, electron microscopy and pharmacology. *Z. Zellforsch. 134,* 311–325.

Edvinsson, L., Owman, C., Rosengren, E., and West, K. A. (1972). Concentration of noradrenaline in pial vessels, choroid plexus, and iris during two weeks after sympathetic ganglionectomy or decentralization. *Acta Physiol. Scand. 85,* 201–206.

Edvinsson, L., Falck, B., and Owman, C. (1977). Possibilities for a cholinergic action on smooth musculature and on sympathetic axons in brain vessels mediated by muscarinic and nicotinic receptors. *J. Pharmacol. Exptl. Ther. 200,* 117–126.

Edvinsson, L., Fahrenkrug, J., Hanko, J., Owman, C., Sundler, F., and Uddman, R. (1980). VIP (vasoactive intestinal polypeptide)-containing nerves of intracranial arteries in mammals. *Cell Tissue Res. 208,* 135–142.

Edvinsson, L., Emson, P., McCulloch, J., Tatemoto, K., and Uddman, R. (1983). Neuropeptide Y: Cerebrovascular innervation and vasomotor effect in the cat. *Neurosci. Lett. 43,* 79–84.

Edvinsson, L., Emson, P. C., McCulloch, J., Tatemoto, K., and Uddman, R. (1984). Neuropeptide Y: Immunocytochemical localization to and effect upon feline pial arteries and veins *in vitro* and *in situ*. *Acta Physiol. Scand. 122,* 155–163.

Edvinsson, L., Fredholm, B. B., Hamel, E., Jansen, I., and Verrecchia, C. (1985). Perivascular peptides relax cerebral arteries concomitant with stimulation of cyclic adenosine monophosphate accumulation or release of an endothelium-derived relaxing factor in the cat. *Neurosci. Lett. 58,* 213–217.

Edvinsson, L., Copeland, J. R., Emson, P. C., McCulloch, J., and Uddman, R. (1987). Nerve fibers containing neuropeptide Y in the cerebrovascular bed: Immunocytochemistry, radioimmunoassay, and vasomotor effects. *J. Cereb. Blood Flow Metab. 7,* 45–57.

Edvinsson, L., Håkanson, R., Wahlestedt, C., and Uddman, R. (1987c). Effects of neuropeptide Y on the cardiovascular system. *Trends Pharmacol. Sci. 8,* 231–235.

Edvinsson, L., Hara, H., and Uddman, R. (1989). Retrograde tracing of nerve fibers to the rat middle cerebral artery with True Blue: Colocalization with different peptides. *J. Cereb. Blood Flow Metab. 9,* 212–228.

Ekblad, E., Håkanson, R., and Sundler, F. (1984a). VIP and PHI coexist with a NPY-like peptide in intramural neurones of the small intestine. *Regul. Pept. 10,* 47–55.

Ekblad, E., Edvinsson, L., Wahlestedt, C., Uddman, R., Håkanson, R., and Sundler, F. (1984b). Neuropeptide Y co-exists and co-operates with noradrenaline in perivascular nerve fibers. *Regul. Pept. 8,* 225–235.

Estrada, C., Hamel, E., and Krause, D. N. (1983). Biochemical evidence for cholinergic innervation of intracerebral blood vessels. *Brain Res. 266,* 261–270.

Fahrenkrug, J. (1991). Glycine-extended processing intermediate of proVIP: A new form of VIP in the rat. *Biochem. Biophys. Res. Commun. 178,* 173–177.

Fahrenkrug, J., Bek, T., Lundberg, J. M., and Hökfelt, T. (1985). VIP and PHI in cat neurons: Colocalization but variable tissue content possibly due to differential processing. *Regul. Pept. 12,* 21–34.

Fahrenkrug, J., Palle, C., Jörgensen, J., and Ottesen, B. (1989). Regulatory peptides in the mammalian urogenital system. *In* "Regulatory Peptides." (J. Polak, ed.), pp. 362–381. Birkhauser Verlag, Basel.

Fallgren, B., Ekblad, E., and Edvinsson, L. (1989). Co-existence of neuropeptides and differential inhibition of vasodilator responses by neuropeptide Y in guinea-pig uterine arteries. *Neurosci. Lett. 100,* 71–76.

Florence, V. M., and Bevan, J. A. (1979). Biochemical determination of cholinergic innervation in cerebral arteries. *Circ. Res. 45,* 212–218.

Forssmann, W. G., Triepel, J., Daffner, C., and Heym, C. H. (1988). Vasoactive intestinal peptide in the heart. *In* "Vasoactive intestinal peptide and related peptides." (S. I. Said and V. Mutt, eds.), *Ann. N. Y. Acad. Sci. 527,* 405–420.

Furchgott, R. F. (1981). The requirement for endothelial cells in the relaxation of arteries by acetylcholine and some other vasodilators. *Trends Pharmacol. Sci. 2,* 173–176.

Furness, J. B., and Costa, M. (1982). Enteric inhibitory nerves and VIP. *In* "Vasoactive intestinal peptide. Advances in peptide hormone research series." (S. I. Said, ed.), pp. 391–406. Raven Press, New York.

Gibbins, I. L., and Morris, J. L. (1988). Co-existence of immunoreactivity to neuropeptide Y and vasoactive intestinal peptide in non-noradrenergic axons innervating guinea pig cerebral arteries after sympathectomy. *Brain Res. 444,* 402–406.

Gibbins, I. L., Brayden, J. E., and Bevan, J. A. (1984). Perivascular nerves with immunoreactivity to vasoactive intestinal polypeptide in cephalic arteries of the cat: Distribution, possible origin and functional implications. *Neuroscience 13,* 1327–1346.

Greenberg, B., Rhoden, K., and Barnes, P. J. (1987). Relaxant effects of vasoactive intestinal peptide and peptide histidine isoleucine in human and bovine pulmonary arteries. *Blood Vessels 24,* 45–50.

Grundemar, L., and Högestätt, E. (1990). Vascular effects of helodermin, helospectin I and helospectin II: A comparison with vasoactive intestinal peptide (VIP). *Br. J. Pharmacol. 99,* 526–528.

Grunditz, T., Ekman, R., Håkanson, R., Sundler, F., and Uddman, R. (1987). Neuropeptide Y and vasoactive intestinal peptide coexist in rat thyroid nerve fibers emanating from the thyroid ganglion. *Regul. Pept. 23,* 193–208.

Grunditz, R., Persson, P., Håkanson, R., Absood, A., Böttcher, G., Rerup, C., and Sundler, F. (1989). Helodermin-like peptides in thyroid C cells. Stimulation of thyroid hormone secretion and suppression of calcium incorporation in bone. *Proc. Natl. Acad. Sci. U.S.A. 86,* 1357–1361.

Håkanson, R., Sundler, F., and Uddman, R. (1982). Distribution and topography of peripheral VIP nerve fibers: Functional implications. *In* "Vasoactive intestinal peptide. Advances in peptide hormone research series." (S. I. Said, ed.), pp. 121–145. Raven Press, New York.

Håkanson, R., Böttcher, G., Ekblad, E., Grunditz, T., and Sundler, F. (1990). Functional implications of messenger coexpression in neurons and endocrine cells. *In* "Neuropeptides and their receptors." (T. W. Schwartz, L. M. Hilsted, and F. G. Rehfeld, eds.), pp. 211–232. *Proc. A. Benzon Symp. 29.*

Hand, J. M., Laravuso, R. B., and Will, J. A. (1984). Relaxation of isolated guinea pig trachea, bronchi and pulmonary arteries by vasoactive intestinal peptide. *European J. Pharmacol. 98,* 279–284.

Hara, H., and Kobayashi, S. (1987). Vasoactive intestinal polypeptide (VIP)-like immunoreactive cells in the skull base of rats. A combined study with acetylcholinesterase histochemistry. *Histochem. 87,* 217–221.

Hara, H., Hamill, G. S., and Jacobowitz, D. M. (1985). Origin of cholinergic nerves to the rat major cerebral arteries: Coexistence with vasoactive intestinal polypeptide. *Brain Res. Bull. 14,* 179–188.

Hara, H., Jansen, I., Ekman, R., Hamel, E., MacKenzie, E. T., Uddman, R. and Edvinsson, L. (1989). ACh and vasoactive intestinal peptide in cerebral blood vessels: Effect of extirpation of the sphenopalatine ganglion. *J. Cereb. Blood Flow Metab. 9,* 204–211.

Hardebo, J. E., Edvinsson, L., Emson, P. C., and Owman, C. (1977). Isolated brain microvessels: Enzymes related to adrenergic and cholinergic functions. *In* "Neurogenic Control of the Brain Circulation." (C. Owman, and L. Edvinsson, eds.), pp. 105–113. Pergamon, London.

Heistad, D. D., Marcus, M. L., Said, S. I., and Gross, P. M. (1980). Effect of ACh and vasoactive intestinal peptide on cerebral blood flow. *Am. J. Physiol. 239,* H73–H89.

Hökfelt, T., Johansson, O., Ljungdahl, Å. Lundberg, J. M., and Schultzberg, M. (1980). Peptidergic neurons. *Nature* (London) *284,* 515–521.

Hökfelt, T., Schultzberg, M., Lundberg, J. M., Fuxe, K., Mutt, V., Fahrenkrug, J., and Said, S. I. (1982). Distribution of vasoactive intestinal polypeptide in the central and peripheral nervous system as revealed by immunocytochemistry. *In* "Vasoactive intestinal peptide. Advances in peptide hormone research series." pp. 65–91. Raven Press, New York.

Hökfelt, T., Holets, V. R., Staines, W., Meister, B., Melander, T., Schalling, M., Schultzberg, M., Freedman, J., Björklund, H., Olson, L., Lindh, B., Elfvin, L.-G., Lundberg, J. M., Lindgren, J. Å., Samuelsson, B., Pernow, B., Terenius, L., Post, C., Everitt, B., Goldstein, M. (1986). Coexistence of neuronal messengers—an overview. *In* "Coexistence of neuronal messengers: A new principle in chemical transmission." (Hökfelt, T., Fuxe, K., and Pernow, B. eds.), "Progress in brain research." Vol. 68, pp. 33–70. Elsevier, Amsterdam.

Hoshino, M., Yanaihara, C., Hong, Y.-M., Kishida, S., Katsumaru, Y., Vandermeers, A., Vandermeers-Piret, M.-C. Robberecht, P., Christophe, J., and Yanaihara, N. (1984). Primary structure of helodermin, a VIP-secretin-like peptide isolated from Gila monster venom. *FEBS Lett. 178,* 233–239.

Ito, O., and Tachibana, S. (1991). Vasoactive intestinal polypeptide precursors have highly potent bronchodilatory activity. *Peptides 12,* 131–137.

Itoh, N., Obata, K., Yanaihara, N., and Okamoto, H. (1983). Human preprovasoactive intestinal polypeptide contains a novel PHI-27-like peptide, PHM-27. *Nature* (London) *304,* 547–549.

Johansson, O., and Lundberg, J. M. (1981). Ultrastructural localization of VIP-like immunoreactivity in large dense core vesicles of "cholinergic-type" nerve terminals in cat exocrine glands. *Neuroscience 5,* 847–862.

Kivipelto, L., Absood, A., Håkanson, R., Sundler, F., and Panula, P. (1992). Helodermin- and helospectin-like immunoreactivities in the rat brain: An immunochemical and immunohistochemical study. *Neuroscience 47,* 135–153.

Kobayashi, S., Kyoshima, K., Olschowka, J. A., and Jacobowitz, D. M. (1983). Vasoactive intestinal polypeptide immunoreactive and cholinergic nerves in the whole mount preparation of the major cerebral arteries of the rat. *Histochem. 79,* 377–381.

Kyösola, K., and Rechardt, L. (1974). The anatomy and innervation of the sphincter Oddi in the dog and cat. *Am. J. Anat. 140,* 497–522.

Langley, J. N. (1898). On inhibitory fibres in the vagus for the end of the oesophagus and stomach. *J. Physiol. 23,* 407–414.

Larsson, L.-I., Edvinsson, L., Fahrenkrug, J., Håkanson, R., Owman, C., Schaffalitzky de Muckadell, O. B., and Sundler, F. (1976). Immunohistochemical localization of a vasodilatory polypeptide (VIP) in cerebrovascular nerves. *Brain Res. 113,* 400–404.

Larsson, L.-I., Fahrenkrug, J., and Schaffalitzky de Muckadell, O. B. (1977a). Occurrence of nerves containing vasoactive intestinal polypeptide immunoreactivity in the male genital tract. *Life Sci. 21,* 503–508.

Larsson, L.-I., Fahrenkrug, J., and Schaffalitzky de Muckadell, O. B. (1977b). Vasoactive intestinal polypeptide occurs in nerves of the female genitourinary tract. *Science, 197,* 1374–1375.

Leander, S., Håkanson, R., and Sundler, F. (1981). Nerves containing substance P, vasoactive intestinal polypeptide, enkephalin or somatostatin in the guinea-pig taenia coli. Distribution, ultrastructure and possible functions. *Cell Tissue Res. 215,* 21–39.

Lee, T. (1980). Direct evidence against acetycholine as the dilator transmitter in the cat cerebral artery. *Eur. J. Pharmacol. 68,* 393–394.

Lee, T., Saito, A., and Beresin, I. (1984). Vasoactive intestinal polypeptide: The potential transmitter for cerebral vasodilation. *Science, 224,* 898–901.

Lehmann, J., and Fibiger, H. C. (1979). Acetylcholinesterase and the cholinergic neuron. *Life Sci. 25,* 1939–1947.

Lindvall, M., Alumets, J., Edvinsson, L., Fahrenkrug, J., Håkanson, R., Hanko, J., Owman, C., Schaffalitzky de Muckadell, O. B., and Sundler, F. (1978). Peptidergic (VIP) nerves in the mammalian choroid plexus. *Neurosci. Lett. 9,* 77–82.

Lorén, I., Emson, P., Fahrenkrug, J., Björklund, A., Alumets, J., Håkanson, R., and Sundler, F. (1979). Distribution of vasoactive intestinal polypeptide in the rat and mouse brain. *Neuroscience 4,* 1953–1976.

Lundberg, J. M. (1981). Evidence for coexistence of vasoactive intestinal polypeptide (VIP) and acetylcholine in neurons of cat exocrine glands. Morphological, biochemical and functional studies. *Acta Physiol. Scand. suppl. 496,* 1–57.

Lundberg, J. M., Hökfelt, T., Schultzberg, M., Uvnäs-Wallensten, K., Köhler, C., and Said, S. I. (1979). Occurrence of vasoactive intestinal polypeptide (VIP)-like immunoreactivity in certain cholinergic neurones of the cat: Evidence from combined immunohistochemistry and acetylcholinesterase staining. *Neuroscience 4,* 1539–1559.

Lundberg, J. M., Terenius, L., Hökfelt, T., Martling, C. R., Tatemoto, K., Mutt, V., Polak, J., Bloom, S. R., and Goldstein, M. (1982). Neuropeptide Y (NPY)-like immunoreactivity in peripheral noradrenergic neurons and effects of NPY on sympathetic function. *Acta Physiol. Scand. 116,* 477–480.

Lundberg, J. M., Terenius, L., Hökfelt, T., and Goldstein, M. (1983). High levels of neuropeptide Y in peripheral noradrenergic neurons in various mammals including man. *Neurosci. Lett. 42,* 167–172.

Lundberg, J. M., Fahrenkrug, J., Larsson, O., and Änggård, A. (1984). Corelease of vasoactive intestinal polypeptide and peptide histidine isoleucine in relation to atropine resistant vasodilation in cat submandibular salivary gland. *Neurosci. Lett.* *52,* 37–42.

Lundberg, J. M., Martling, C.-R., and Hökfelt, T. (1988). Airways, oral cavity and salivary glands: Classical transmitters and peptides in sensory and autonomic neurons. *In* "Handbook of Chemical Neuroanatomy." Vol. 6, "The peripheral nervous system" (A. Björklund, T. Hökfelt, and C. Owman, eds.), pp. 391–444. Elsevier, Amsterdam.

Luts, A., and Sundler, F. (1989). Peptide-containing nerve fibers in the respiratory tract of the ferret. *Cell Tissue Res.* *258,* 259–267.

Luts, A., Uddman, R., Absood, A., Håkanson, R., and Sundler, F. (1991). Chemical coding of endocrine cells of the airways: Presence of helodermin-like peptides. *Cell Tissue Res.* *265,* 425–433.

Malm, L., Sundler, F., and Uddman, R. (1980). Effects of vasoactive intestinal polypeptide VIP on resistance and capacitance vessels in the nasal mucosa. *Acta Otolaryngol.* *90,* 304–308.

Martling, C.-R., Matran, R., Alving, K., Hökfelt, T., and Lundberg, J. M. (1990). Innervation of lower airways and neuropeptide effects on bronchial and vascular tone in the pig. *Cell Tissue Res.* *260,* 223–233.

Matsuyama, T., Shiosaka, S., Matsumoto, M., Yoneda, S., Kimura, K., Abe, H., Hayakawa, T., Inoue, H., and Tohyama, M. (1983). Overall distribution of vasoactive intestinal polypeptide-containing nerves in the wall of the cerebral arteries: An immunohistochemical study using whole mounts. *Neuroscience* *10,* 89–96.

McCulloch, J., and Edvinsson, L. (1980). Cerebral circulatory and metabolic effects of vasoactive intestinal polypeptide. *Am. J. Physiol.* *238,* H449–H456.

McCulloch, J., Kelly, P. A. T. (1983). A functional role for vasoactive intestinal polypeptide in anterior cingulate cortex. *Nature* *304,* 438–440.

McCulloch, J., Kelly, P. A. T., Uddman, R., and Edvinsson, L. (1983). Functional role for vasoactive intestinal polypeptide in the caudate nucleus: A 2-deoxy(^{14}C) glucose investigation. *Proc. Natl. Acad. Sci.* *80,* 1472–1476.

Miyata, A., Arimura, A., Dahl, R. R., Minamino, N., Uehara, A., Jiang, L., Culler, M. D., and Coy, D. H. (1989). Isolation of a novel 38-residue hypothalamic polypeptide which stimulates adenylate cyclase in pituitary cells. *Biochem. Biophys. Res. Commun.* *164,* 567–574.

Miyata, A., Jiang, L., Dahl, R. R. Kitada, C., Kubo, K., Fujino, M., Minamino, N., and Arimura, A. (1990). Isolation of a neuropeptide corresponding to the N-terminal 27 residues of the pituitary adenylate cyclase activating polypeptide with 38 residues (PACAP 38). *Biochem. Biophys. Res. Commun.* *170,* 643–648.

Morris. J. L., Gibbins, I. L., Furness, J. B., Costa, M., and Murphy, R. (1985). Co-localization of neuropeptide Y, vasoactive intestinal polypeptide and dynorphin in non-noradrenergic axons of the guinea-pig artery. *Neurosci. Lett.* *62,* 31–37.

Nilsson, C., Ekman, R., Lindvall-Axelsson, M., and Owman, C. (1990). Distribution of peptidergic nerves in the choroid plexus, focusing on coexistence of neuropeptide Y, vasoactive intestinal polypeptide and peptide histidine isoleucine. *Regul. Pept.* *27,* 11–26.

Obara, H., Kusunoki, M., Mori, M., Mikawa, K., and Iwai, S. (1989). The effect of various peptides on the isolated pulmonary artery. *Peptides* *10,* 242–243.

Ottesen, B., and Fahrenkrug, J. (1981). Effect of vasoactive intestinal polypeptide (VIP) upon myometrial blood flow in non-pregnant rabbit. *Acta Physiol. Scand.* *112,* 195–201.

Ottesen, B., and Einer-Jensen, N. (1984). Increased endometrial clearance of krypton-85 evoked by VIP in rabbits. *Acta Physiol. Scand.* *121,* 185–187.

Ottesen, B., Gram, B. R., and Fahrenkrug, J. (1983). Neuropeptides in the female genital tract: Effect on vascular and non-vascular smooth muscle. *Peptides,* *4,* 387–392.

Ottesen, B., Wagner, G., Viraq, R., and Fahrenkrug, J. (1984). Penile erection: Possible role for vasoactive intestinal polypeptide as a neurotransmitter. *Br. Med. J.* *28,* 9–11.

Parker, D. S., Raufman, J.-P., O'Donohue, T. L., Bledsoe, M., Yoshida, H., and Pisano, J. J. (1984). Amino acid sequences of helospectins, new members of the glucagon superfamily, found in Gila monster venom. *J. Biol. Chem.* *259,* 11751–11755.

Power, R. F., Bishop, A. E., Wharton, J., Inyama, C. O., Jackson, R. H., Bloom, S. R., and Polak, J.

M. (1988). Anatomical distribution of vasoactive intestinal peptide binding sites in peripheral tissues investigated by *in vitro* autoradiography. *In* "Vasoactive intestinal peptide and related peptides." (S. I. Said and V. Mutt, eds.), *Ann. N. Y. Acad. Sci. 527,* 314–325.

Robberecht, P., de Graef, J., Woussen, M.-C., Vandermeers-Piret, M.-C., Yanaihara, C., Yanaihara, N., and Christophe, J. (1985). Immunoreactive helodermin-like peptides in rat: A new class of mammalian neuropeptides related to secretin and VIP. *Biochem. Biophys. Res. Commun. 130,* 333–342.

Said, S. I., and Mutt, V. (1970). Polypeptide with broad biological activity: Isolation from small intestine. *Science 169,* 1217–1218.

Sata, T., Misra, H. P., Kubota, E., and Said, S. I. (1986). Vasoactive intestinal polypeptide relaxes pulmonary artery by an endothelium-independent mechanism. *Peptides* (suppl. 1), *7,* 225–227.

Smitherman, T. C., and Dehmer, G. J. (1988). Vasoactive intestinal peptide as a coronary vasodilator *In* "Vasoactive intestinal peptide and related peptides." (S. I. Said and V. Mutt, eds.), *Ann. N. Y. Acad. Sci. 527,* 421–430.

Steers, W. D., McConnell, J., and Benson, G. S. (1984). Anatomical localization and some pharmacological effects of vasoactive intestinal polypeptide in human and monkey corpus cavernosum. *J. Urol. 132,* 1048–1053.

Sundler, F., Alumets, J., Ekblad, E., Böttcher, G., and Håkanson, R. (1985). Coexistence of peptides in the neuroendocrine system. *In* "Biogenetics of Neurohormonal Peptides." (R. Håkanson and J. Thorell, eds.), pp. 213–243. London, Academic Press.

Sundler, F., Ekblad, E., Grunditz, T., Håkanson, R., Luts, A., and Uddman, R. (1989). NPY in peripheral non-adrenergic neurons. *In* "Neuropeptide Y." (V. Mutt, T. Hökfelt, and K. Fuxe, eds.), pp. 93–102. Raven Press, New York.

Sundler, F., Ekblad, E., Absood, A., Håkanson, R., Köves, K., and Arimura, A. (1992). Pituitary adenylate cyclase activating peptide (PACAP): A novel VIP-like neuropeptide in the gut. *Neuroscience 46,* 439–454.

Suzuki, Y., McMaster, D., Huang, M., Lederis, K., and Rorstad, O. P. (1985). Characterization of functional receptors for vasoactive intestinal peptide in bovine cerebral arteries. *J. Neurochem. 45,* 890–899.

Suzuki, N., Hardebo, J.-E., and Owman, C. (1988). Origins and pathways of cerebrovascular vasoactive intestinal polypeptide-positive nerves in rat. *J. Cerebr. Blood Flow Metab. 8,* 697–712.

Suzuki, N., Hardebo, J.-E., Kåhrström, J., and Owman, C. (1990). Origins and pathways of choline-acetyltransferase-positive parasympathetic nerve fibers to cerebral blood vessels in rat. *J. Cereb. Blood Flow Metab. 10,* 399–408.

Suzuki, N., Hardebo, J.-E., Kåhrström, J., and Owman, C. (1990). Neuropeptide Y coexists with vasoactive intestinal polypeptide and acetylcholine in parasympathetic cerebrovascular nerves originating in the sphenopalatine, otic, and internal carotid ganglia of the rat. *Neuroscience 36,* 507–519.

Tatemoto, K., and Mutt, V. (1981). Isolation and characterization of the intestinal peptide porcine PHI (PHI-27), a new member of the glucagon-secretin family. *Proc. Natl. Acad. Sci. U.S.A. 78,* 6603–6607.

Tatemoto, K., Carlquist, M., and Mutt, V. (1982). Neuropeptide Y—a novel brain peptide with structural similarities to peptide YY and pancreatic polypeptide. *Nature 296*, 659–660.

Toda, N. (1982). Relaxant responses to transmural stimulation and nicotine of dog and monkey cerebral arteries. *Am. J. Physiol. 243,* H145–H153.

Uddman, R., Alumets, J., Edvinsson, L., Håkanson, R., and Sundler, F. (1981). VIP nerve fibres around peripheral blood vessels. *Acta Physiol. Scand. 112,* 65–70.

Uddman, R., Ekblad, E., Edvinsson, L., Håkanson, R., and Sundler, F. (1985). Neuropeptide Y-like immunoreactivity in perivascular nerve fibres of the guinea-pig. *Regul. Pept. 10,* 243–257.

Uddman, R., Luts, A., Arimura, A., and Sundler, F. (1991). Pituitary adenylate cyclase activating peptide (PACAP), a new vasoactive intestinal peptide (VIP)-like peptide in the respiratory tract. *Cell Tissue Res. 265,* 197–201.

Walters, B. B., Gillespie, S. A., and Moskowitz, M. A. (1986). Cerebrovascular projections from the sphenopalatine and otic ganglia to the middle cerebral artery of the cat. *Stroke 17,* 488–494.

Wei, E. P., Kontos, H. A., and Said, S. I. (1980). Mechanism of action of vasoactive intestinal polypeptide on cerebral arteries. *Am. J. Physiol. 239,* H765–H768.

Willis, E. A., Ottesen, B., Wagner, G., Sundler, F., and Fahrenkrug, J. (1983). Vasoactive intestinal polypeptide (VIP) as a putative neurotransmitter in penile erection. *Life Sci. 33,* 383–391.

Yiangou, Y., Di Marzo, V., Spokes, R. A., Panico, M., Morris, H. R., and Bloom, S. R. (1987). Isolation, characterization, and pharmacological actions of peptide histidine valine 42, a novel prepro-vasoactive intestinal peptide derived peptide. *J. Biol. Chem. 262,* 14010–14013.

CHAPTER 11

Pharmacology of Acetylcholine and Vasoactive Intestinal Peptide in Perivascular Nerves

O. RORSTAD H. SHIRAHASE M. HUANG

I. Introduction

The traditional classification of vascular autonomic innervation by the sympathetic, parasympathetic, and sensory afferent systems continues to have heuristic utility notwithstanding the discovery of several neuropeptides that include the vasculature among their target organs of action. Although acetylcholine (ACh) is classically accepted as the principal parasympathetic neurotransmitter, experimental evidence has accumulated to suggest that it alone is not sufficient to account completely for parasympathetic effects in all vascular systems. As one example, administration of the muscarinic receptor antagonist, atropine, to cat cerebral arteries does not block neurogenic vasodilation induced by transmural nerve stimulation (Lee, 1980). Additionally, in this model neurogenic vasodilation does not require an intact endothelium whereas ACh-mediated vasodilation is endothelium dependent (Lee *et al.*, 1984). The neuropeptide most closely associated with the parasympathetic system is vasoactive intestinal peptide (VIP), consisting of 28 amino acids, which appears to colocalize with ACh in selected cholinergic neurons. VIP localization and actions, however, are not confined to the parasympathetic system nor is VIP the only neuropeptide reported to be associated with the parasympathetic system. Neuropeptide Y (NPY), which is usually grouped with the sympathetic neurotransmitters, has been localized to neurons of the parasympathetic sphenopalatine and otic ganglia of the rat (Leblanc *et al.*, 1987; N. Suzuki *et al.*, 1990b). The vascular actions of NPY are discussed in this volume under the section dealing with the sympathetic system. The literature on vascular actions of ACh has been critically reviewed by Duckles (1988) and Bevan and Brayden (1987) and, therefore, only selected recent advances in the study of the cholinergic vascular system will be presented with a particular focus on vascular cholinergic receptors. The vascular actions and receptor for VIP will also be emphasized, recognizing that vascular VIP innervation probably does not completely originate from the parasympathetic nervous system.

II. The Vascular Cholinergic System

A. *Cholinergic Vascular Innervation*

Innervation of blood vessels by ACh-containing nerves has been proposed for several years based predominately on the histochemical localization of the inactivating enzyme, acetylcholinesterase, and the biosynthetic enzyme, choline acetyltransferase (ChAT). Recent critical reviews point out the greater reliability of ChAT as an immunohistochemical marker for cholinergic nerves (Duckles, 1988; Bevan and Brayden, 1987). The cholinergic innervation of the cerebral circulation in particular has been studied in detail by this methodology perhaps in part because of the rich innervation of cerebral blood vessels. Saito *et al.* (1985) localized ChAT positive nerve fibers to the circle of Willis, the anterior and middle cerebral arteries, and the basilar artery of rats, cats, dogs, and pigs. This study identified a similar distribution of ChAT-containing nerves to rabbit brain arteries with the exception of few fibers to the vertebralbasilar vessels. The presence of ChAT in dog and human cerebral arteries has been confirmed by a biochemical assay for this enzyme (Tsukahara *et al.*, 1986). Retrograde tracing studies have found ChAT-containing nerve fibers in the rat middle cerebral artery that originate from the sphenopalatine, otic, and internal carotid ganglia (N. Suzuki *et al.*, 1990a). Parallel localization of ChAT with VIP was observed in this and other studies that will be discussed in more detail below.

Cholinergic nerves have been identified in the tunica adventitia (Saito *et al.*, 1985) and also in direct association with the basal lamina of capillary endothelial cells in the microvasculature (Armstrong, 1986; Arneric *et al.*, 1988). Rat cerebral cortical microvessels, consisting predominately of capillaries, possess a high level of ChAT activity measured by chemical assays (Estrada and Krause, 1982; Arneric *et al.*, 1988). Because of the heterogeneous cellular nature of microvessel preparations, the ChAT activity may be localized to residual neuronal elements or endothelial cells (Milner *et al.*, 1989, 1990).

B. *Vascular Actions of Acetylcholine*

Since it is not possible to do credit to the intricate and vast body of literature on cholinergic vascular actions in the present chapter, we will attempt to highlight some of the major observations in this area of study. Recent detailed reviews have been published by Bevan and Brayden (1987), Duckles (1988), and Kalsner (1989). The vascular actions of ACh are complex, including both vasodilation and vasoconstriction with levels of action at the endothelium, vascular smooth muscle cell, and adrenergic nerve terminals by inhibiting release of norepinephrine. An early experimental paradox that ACh produced vasodilation when administered *in vivo* whereas it caused vasoconstriction or no effect when studied *in vitro* was resolved in large part by the discovery of Furchgott and Zawadzki (1980) that the vasodilatory effect of ACh can be mediated via an endothelium-derived relaxing factor (EDRF). Although endothelium-dependent vasorelaxation (with or without endothelium-independent vasoconstriction) may predominate in many blood vessels (Furchgott and Zawadzki, 1980; Duckles, 1988; Chen and Suzuki, 1989), recent evidence has emerged to support additional mechanisms of ACh-mediated endothelium-dependent vasoconstriction in selected systems, such as canine cerebral arteries (Shirahase *et al.*, 1987) and rabbit pulmonary artery (Altiere *et al.*, 1986). Both endothelium-dependent vasorelaxation, as in the rat gastric vasculature (Kitagawa *et al.*, 1987), and endothelium-dependent vasoconstriction, as in isolated canine cerebral arteries (Usui *et al.*, 1986), may be elicited by nerve stimulation. To add another facet of complexity, neurogenic cholinergic vasodilation has been reported to occur in the cat auricular artery independently of the endothelium (Brayden and Bevan, 1985). In an analogous manner, neurogenic cholinergic vasoconstriction, which occurs in the coronary arteries of several species including humans, probably arises from ACh released from vascular nerves acting on medial smooth muscle cells (Kalsner, 1985; Kalsner, 1989; Kalsner and Quillan, 1989). This phenomenon also highlights the considerable differences in cholinergic vascular responses among different animal species and regional vascular beds (Kalsner, 1989). It is debatable whether neuronally released ACh could diffuse to act directly on an endothelial cell since cholinergic nerves have not been reported to our knowledge to innervate the endothelium of larger vessels (Duckles, 1988). However, the recent demonstration of ACh release from cultured human endothelial cells and the presence of ChAT in endothelial cells raises the possibility of an alternative, nonneuronal local mechanism of vascular regulation involving ACh (Milner *et al.*, 1989, 1990).

C. Vascular Cholinergic Receptors

Three subclasses of muscarinic cholinergic receptors have been proposed on the basis of pharmacological criteria and designated M_1 (predominately nervous system), M_2 (predominately heart), and M_3 (predominately glandular and smooth muscle) (Goyal, 1989; Watson and Abbott, 1991; Hulme *et al.*, 1990). The structure of the genes (m1, m2, m3) coding for these three pharmacologically defined subtypes and two other muscarinic receptor genes of uncertain pharmacological function (m4, m5) has been determined. The muscarinic receptor polypeptides range in size from 460 to 590 amino acids in the human and possess seven transmembrane domains that are characteristic of a receptor coupled to a guanine nucleotide regulatory protein. Doods *et al.* (1989) extensively reviewed muscarinic receptors in the heart and vascular system at a time when there was less agreement on the nature of muscarinic subtypes than exists at present. Accordingly, the present review will attempt predominately to place recent studies of vascular cholinergic subtypes into the context of the M_1, M_2, and M_3 pharmacological subclassification, as recommended by the Nomenclature Committee of the Fourth International Symposium on Subtypes of Muscarinic Receptors (1989).[1]

A number of selective muscarinic antagonists have been developed that have permitted classical functional analyses of vascular biological responses to be undertaken. Based on this approach, cholinergic vasoconstriction in coronary arteries has been attributed to the M_3 muscarinic subtype by Duckles and Garcia-Villalon (1990) using isolated bovine coronary arteries, van Chaldorp and van Zwieten (1989) using isolated porcine coronary arteries, and Bognar *et al.*, (1990) utilizing the perfused rat heart preparation. Similarly, cholinergic endothelial-dependent relaxation of the isolated rabbit ear artery (Duckles and Garcia-Villalon, 1990), the rat pulmonary artery (McCormack *et al.*, 1988), and the rat middle cerebral artery (Dauphin and Hamel, 1990), has been associated with the M_3 receptor subtype. In contrast, van Chaldorp and van Zwieten (1989) found that cholinergic contraction of the isolated porcine basilar artery possessed functional M_2 characteristics.

A large number of radioligand-binding studies have identified muscarinic receptor sites on both the smooth muscle and endothelial vascular components although the characterization of vascular muscarinic subtypes by radioligand-binding techniques is at an early stage at present. Vascular muscarinic receptors have been detected by the use of the nonselective muscarinic antagonist [^{3}H]quinuclidinyl benzilate (QNB) in bovine and human cerebral blood vessels, including microvessels (Estrada and Krause, 1982; Tsukahara *et al.*, 1986), bovine, dog, and rabbit aorta (Yamanaka *et al.*, 1986; Tsukahara *et al.*, 1989a), dog portal vein (Taniguchi *et al.*, 1983), porcine coronary artery (Rinner *et al.*, 1988), bovine splenic vein (Brunner and Kukovetz, 1986), and rabbit ear artery (Hynes *et al.*, 1986). Scatchard analyses of binding parameters in these studies have been most consistent with a single receptor site for muscarinic antagonists. The muscarinic receptor in both smooth muscle and endothelial layers of rabbit aorta and ear artery has been found to differ from the M_1 (nervous system) muscarinic subtype based on low affinity for binding of the M_1-selective antagonist, pirenzepine (Tsukahara *et al.*, 1989a; Hynes *et al.*, 1986). Entzeroth *et al.* (1990) concluded that membranes prepared from porcine coronary arteries following removal of the adventitial and endothelial layers contained a mixed population of M_2 and M_3 receptor subtypes based on competitive binding of selective muscarinic antagonists with [^{3}H]*N*-methyl-scopolamine. Other investigators have confirmed muscarinic receptor binding to smooth muscle using membrane preparations of bovine coronary artery (Brunner *et al.*, 1990) and by autoradiography of arterial tissue sections (Stephenson *et al.*, 1988). Muscarinic receptors have been localized to the endothelium by autoradiographic or binding studies of rabbit aorta and human cerebral arteries (Sim and Manjeet, 1989; Tsukahara *et al.*, 1989b). Brunner and Kukovetz (1991) found that freshly isolated

1. Subtypes of Muscarinic Receptors IV. Proceedings of the Fourth International Symposium on Subtypes of Muscarinic Receptors, Weisbaden, Germany, 1989. *In* "Trends in Pharmacological Sciences" (Suppl. Dec. 1989) (R. R. Levine and N. J. M. Birdsall, eds.), Elsevier Trends Journal, Cambridge, U.K.

bovine aortic endothelial cells possessed a receptor most closely resembling the M_1 subtype. Interestingly, Yamanaka *et al.* (1986) had previously assigned the bovine aortic muscarinic receptor as an M_1 subtype. The apparent discrepancy in the endothelial muscarinic receptor subtypes identified by functional assays (M_3) and radioligand binding (M_1) remains to be resolved.

D. Cholinergic Effector Mechanisms

The muscarinic receptor in numerous model systems has shown coupling via guanine nucleotide regulatory proteins to a variety of effector mechanisms, including stimulation of phosphoinositide metabolism, inhibition of adenylate cyclase, and activation of a membrane potassium channel (Hulme *et al.*, 1990). Studies of cholinergic effector systems in isolated endothelial cells can be summarized as having provided a glimpse of a complex and intriguing array of interacting regulatory mechanisms. For example, ACh or carbachol has been demonstrated to (i) stimulate inositol phosphate synthesis (Grigorian *et al.*, 1989), (ii) activate a membrane potassium channel (Olesen *et al.*, 1988), (iii) increase intracellular calcium concentration (Danthuluri *et al.*, 1988), and (iv) inhibit basal production of cyclic AMP (Sim and Manjeet, 1990). The picture may be more straightforward in the case of studies of isolated or skinned coronary smooth muscle whereby ACh stimulates phosphoinositol metabolism, increases intracellular calcium concentration, and generates contraction (T. Itoh *et al.*, 1988; Sim and Manjeet, 1990).

III. The Vascular VIP System

A. Introduction

VIP was originally isolated by Said and Mutt (1970) from an intestinal extract as a consequence of its potent hypotensive property. VIP is synthesized and distributed predominately in neurons localized to both the central and peripheral nervous systems (Fahrenkrug, 1989). Consistent with VIP's importance as a biological regulator, its primary structure is highly conserved in the animal kingdom, being identical in several mammals and differing slightly in the guinea pig, chicken, dogfish, and codfish (cited by Thwaites *et al.*, 1989). Structurally, VIP is homologous to the secretin-glucagon family of peptides that includes among its members the growth hormone–releasing factor (GRF), the PHI (peptide histidine isoleucine) (Fahrenkrug, 1989), and the newly discovered PACAP (pituitary adenylate cyclase activating peptide) groups (Miyata *et al.*, 1989) (Fig. 1). The PHI peptides contain approximately 50% homology with VIP and display considerable conservation of sequence between PHI members from different animal species (cited in Huang *et al.*, 1989). The human gene for VIP has been characterized structurally and contains the coding sequence for human PHI upstream of the VIP-coding region (N. Itoh *et al.*, 1983). VIP and PHI are synthesized from the same polypeptide precursor and are colocalized in and coreleased from selected cells (Fahrenkrug, 1987). The biological actions of VIP can be broadly summarized as relaxation of smooth

	1	2	3	4	5	6	7	8	9	10	11	12	13	14	15	16	17	18	19	20	21	22	23	24	25	26	27	28	29
VIP	H	S	D	A	V	F	T	D	N	Y	T	R	L	R	K	Q	M	A	V	K	K	Y	L	N	S	I	L	N*	
PHI	H	A	D	G	V	F	T	S	D	F	S	R	L	L	G	Q	L	S	A	K	K	Y	L	E	S	L	I*		
PHM	H	A	D	G	V	F	T	S	D	F	S	K	L	L	G	Q	L	S	A	K	K	Y	L	E	S	L	M*		
PACAP-27	H	S	D	G	I	F	T	D	S	Y	S	R	Y	R	K	Q	M	A	V	K	K	Y	L	A	A	V	L*		
rGRF	H	A	D	A	I	F	T	S	S	Y	R	R	I	L	G	Q	L	Y	A	R	K	L	L	H	E	I	M	N	R...
SECRETIN	H	S	D	G	T	F	T	S	E	L	S	R	L	R	D	S	A	R	L	Q	R	L	L	Q	G	L	V*		
GLUCAGON	H	S	Q	G	T	F	T	S	D	Y	S	K	Y	L	D	S	R	R	A	Q	D	F	V	Q	W	L	M	N	T

Figure 1 Amino acid sequences common to those in VIP are underlined. For rat (r) GRF, only the N-terminal 29 amino acid sequence is shown. The standard single amino acid code is used: A, Ala; D, Asp; E, Glu; F, Phe; G, Gly; H, His; I, Ile; K, Lys; L, Leu; M, Met; N, Asn; Q, Gln; R, Arg; S, Ser; T, Thr; V, Val; W, Trp; Y, Tyr. * indicates amidation.

muscle, including blood vessels, and regulation of secretion of a variety of molecules ranging from water and ions to enzymes and hormones (Fahrenkrug, 1989).

B. Vascular Innervation by VIP

The potential role for VIP as a vasodilatory neurotransmitter has been discussed in recent reviews by Bevan and Brayden (1987), Edvinsson and Uddman (1988), Edvinsson *et al.* (1988), Forssmann *et al.* (1988), Smitherman *et al.* (1988) and Mione *et al.* (1990). Additionally, vascular innervation by VIP-immunoreactive (VIP-IR) nerves is described in detail in Chapter 10 by Uddman *et al.* Immunohistochemical studies have demonstrated a widespread distribution of VIP-IR nerves throughout the cardiovascular system, including blood vessels to the brain, heart, respiratory, gastrointestinal, and genitourinary tracts (Said, 1982; Larsson *et al.*, 1976; Edvinsson *et al.*, 1980; Uddman *et al.*, 1981; Furness *et al.*, 1984). The VIP-IR innervation appears particularly prominent to the cerebral, mesenteric, and genitourinary vascular beds. The presence of VIP in a variety of blood vessels from different animal species has been confirmed by radioimmunoassay of vascular extracts (Duckles and Said, 1982; Edvinsson *et al.*, 1980; Brum *et al.*, 1986; Sidawy *et al.*, 1989). Perivascular VIP-IR nerves are typically localized in the adventitial layer and at the adventitial-medial junction and terminate in the proximity of vascular smooth muscle by light and electron microscopic immunocytochemistry (Larsson *et al.*, 1976; Edvinsson *et al.*, 1980, 1989b; Nakakita, 1990). In addition, a close association of VIP-IR nerves with capillaries has been observed in the coronary microvasculature (Weihe *et al.*, 1984). Several studies of the pathways and origins of VIP-IR vascular nerves have been conducted utilizing histochemical tracing and lesioning methods. A significant innervation of cerebral arteries appears to arise from extracranial parasympathetic sources such as the sphenopalatine and otic ganglia (Hara and Weir, 1988; Gibbins *et al.*, 1984; Hara *et al.*, 1985; N. Suzuki *et al.*, 1988; Edvinsson *et al.*, 1989a). In contrast, a sympathetic cerebral vascular innervation by VIP-IR nerves arising from the superior cervical ganglia has been proposed by Uemura *et al.* (1987) based on observation of decreased cerebral vascular VIP-IR innervation following extirpation of the superior cervical ganglia in dogs. However, Larsson *et al.* (1976) and Edvinsson *et al.* (1980) reported that the similar procedure in rats and cats did not affect VIP-IR nerves in cerebral blood vessels. Uemura *et al.* (1987) have suggested that differences in methodology and animal species may explain the apparent discrepancy. VIP-IR innervation of cerebral blood vessels may also arise directly from cerebral cortical neurons (Eckenstein and Baughman, 1984). A close association of cholinergic and VIP-IR neurons and colocalization of cholinergic markers and VIP-IR have been observed in cerebral cortical neurons, cerebral and peripheral vascular nerves, and nerves to exocrine glands (Eckenstein and Baughman, 1984; Miao and Lee, 1990; Lundberg *et al.*, 1980; Hara *et al.*, 1985; Kaji *et al.*, 1988). A recent study using sequential double-labeling methodology quantified less than 5% of cerebral vascular nerves as colocalizing ChAT and VIP-IR (Miao and Lee, 1990). Nevertheless the close association of cholinergic and VIP-IR innervation raises the possibility of functional interactions between these neurotransmitter systems. Colocalization of VIP-IR with other neuropeptides has been demonstrated in the specific instances of (i) PHI in perivascular nerves (Fahrenkrug, 1987; Edvinsson and McCulloch, 1985), (ii) NPY in parasympathetic cerebral vascular nerves and cardiac nerves (Forsgren, 1989; Leblanc *et al.*, 1987; N. Suzuki *et al.*, 1990b), and (iii) calcitonin gene-related peptide (CGRP) in the sympathetic stellate ganglion, a source of cholinergic innervation to sweat glands (Landis and Fredieu, 1986; Lindh *et al.*, 1987).

C. Vascular Actions of VIP

VIP has been observed to exert a vasodilatory action in several mammalian species, including humans, and involving a variety of vascular beds that generally correspond to the distribution of VIP-IR nerves. The cerebral and cardiac vasculature has been most extensively studied utilizing isolated arterial strips and selected *in vivo* models of regional blood flow or vascular caliber (citations are to be found in the review articles listed above and Duckles and Said, 1982; Larsson *et al.*, 1976; Wei *et al.*, 1980; Uddman *et al.*, 1981; Lee *et al.*, 1984; Y. Suzuki *et al.*, 1984, 1985; H. Itoh *et al.*, 1990). As an example Y. Suzuki *et al.* (1985) observed that VIP was typically a more potent vasorelaxant agent on isolated bovine and porcine blood vessels than the homologous PHI or GRF peptides whereas secretin and glucagon were inactive (Fig. 2). With selected exceptions, the PHI peptides have proven to be less potent than VIP as vasodilators in several studies utilizing isolated arterial segments or organ-specific blood flow models *in vivo* (reviewed in Huang *et al.*, 1989; H. Itoh *et al.*, 1990; and Y. Suzuki *et al.*, 1991). Full vasodilatory action of VIP requires the integrity of the complete polypeptide chain. Deletion of one or more amino acid residues from the amino terminus leads to a great reduction of vascular activity and smaller amino or carboxyl terminal fragments such as VIP(1–12), VIP(10–28), and several others are devoid of activity (Bodanszky *et al.*, 1973; Fournier *et al.*, 1984; Y. Suzuki *et al.*, 1984). Minor modifications to the interior of the VIP molecule have either a moderate or no effect on vasodilatory potency. Analogs with a single modification, such as the iodinated derivatives [I-Tyr^{10}]VIP and [I-Tyr^{22}]VIP and the oxidized derivatives [O-Met^{17}]VIP and [O_2-Met^{17}]VIP, retain full vasorelaxant activity *in vitro* (Y. Suzuki *et al.*, 1984; McMaster *et al.*,

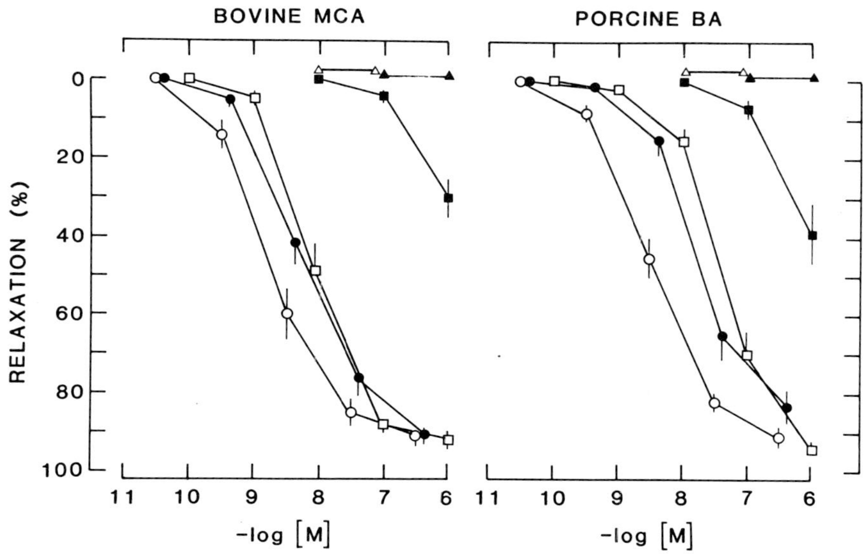

Figure 2 Relaxation of helical strips of bovine middle cerebral artery (MCA) and porcine basilar artery (BA) by VIP (○), PHM (●), PHI (□), rat GRF (■), human pancreatic GRF(1–40) amide (△), and secretin or glucagon (▲). (Reproduced with permission from Y. Suzuki *et al.*, 1985.)

1987). However, VIP derivatives that are both iodinated and oxidized, such as [I-Tyr^{10}, O-Met^{17}]VIP and [I-Tyr^{22}, O-Met^{17}]VIP, have 77% and 42%, respectively, of the vascular effect of VIP (McMaster *et al.,* 1987).

Mention was made in the introduction of the existence in cat cerebral arteries of atropine-resistant, endothelium-independent vasodilation that can be induced by transmural nerve stimulation and is hypothesized to be mediated by VIP release from vascular nerves (Lee, 1980; Lee *et al.,* 1984). Further evidence that VIP may represent the vasodilatory neurotransmitter in this particular model includes (i) correlation of the existence and magnitude of vasodilation after transmural nerve stimulation with the extent of innervation of the artery by VIP-IR nerves, (ii) inhibition of neurogenic vasodilation by administration of VIP antibody, which presumably neutralizes locally released VIP, and (iii) an increase in the release of VIP, measured by radioimmunoassay, after transmural stimulation of arteries that exhibit neurogenic vasodilation (Brayden and Bevan, 1986; Duckles and Said, 1982; Bevan *et al.,* 1984). Release of radioimmunoassayable VIP has also been documented in association with relaxation of the portal vein and penile erection in response to nerve stimulation (Hellstrand *et al.,* 1985; Anderson *et al.,* 1987). Administration of VIP antiserum to cats *in vivo* inhibits the increase in carotid blood flow resulting from stimulation of the brain stem locus ceruleus and the sphenopalatine ganglion whose efferent pathways correspond to the parasympathetic system (Goadsby and MacDonald, 1985). VIP-induced vasodilation was originally found to be independent of the endothelium and this observation was confirmed in a variety of blood vessels, such as cat cerebral artery (Duckles and Said, 1982; Lee *et al.*, 1984; Edvinsson *et al.*, 1985), porcine and bovine coronary arteries (Beny *et al.*, 1986; Forssmann *et al.*, 1988; H. Itoh *et al.*, 1990), rat aorta (Schoeffter and Stoclet, 1985) and bovine, human, rat, rabbit, and guinea pig pulmonary arteries (Barnes *et al.*, 1986; Sata *et al.*, 1986; Greenberg *et al.*, 1987). However, the existence of an alternative endothelial-dependent mechanism of vasodilation to VIP has gained credibility based on independent studies of rat aorta (Davies and Williams, 1984; Sata *et al.*, 1988), human celiac artery (Thom *et al.*, 1987), and bovine pulmonary artery (Ignarro *et al.*, 1987), notwithstanding the discrepant findings with respect to the rat aorta and bovine pulmonary artery. The involvement of the endothelium in selected instances of VIP vasodilation suggests the existence of more complex and interesting regulatory mechanisms for VIP vascular actions than was thought previously.

D. The Vascular Receptor for VIP

A VIP binding site in membranes prepared from bovine and rat arteries has been identified that satisfies the classical criteria for receptor–radioligand binding, including saturability, dependence on time and temperature, reversibility, high affinity (equilibrium dissociation constant approximately 0.1–0.2 nM), and specificity for VIP and related peptides (Huang and Rorstad, 1987; Y. Suzuki *et al.*, 1985). Additional studies of vascular VIP receptors have utilized isolated intact rat aortic smooth muscle cells (Hirata *et al.*, 1985) or sections of arteries in either radioligand binding (F. Amenta *et al.*, 1989) or autoradiographic analyses (Barnes *et al.*, 1986; F. Amenta *et al.*, 1989; Sidawy *et al.*, 1989). The use of a monoiodinated, homogeneous, and biologically active radioligand, $[Tyr(^{125}I)^{10}]$VIP, which is generated by oxidative iodination and purification by high-pressure liquid chromatography, has been important for the demonstration of vascular VIP receptors in membrane-binding studies (McMaster *et al.*, 1987). Likewise, in such experiments the method of separation of membrane-bound and free radioligand has been a key technological hurdle (Huang and Rorstad, 1987; Huang *et al.*, 1988). Scatchard analyses have identified two vascular VIP binding sites of high and low affinity in bovine cerebral and coronary arterial membranes, rat mesenteric arterial membranes, and rat isolated aortic smooth muscle cells (Huang and Rorstad, 1987; Hirata *et al.*, 1985; Y. Suzuki *et al.*, 1985) or one high-affinity binding site in sections of rat mesenteric arteries (F. Amenta *et al.*, 1989). The binding properties of the vascular VIP receptor are perturbed by reduction of disulfide bonds, as is the case for VIP receptors in nonvascular tissue, and incubation with guanine nucleotide analogs results in a reduction of high-affinity binding as is characteristic of receptors associated with a guanine nucleotide regulatory protein (Huang and Rorstad, 1989).

The pharmacological and biochemical properties of the VIP receptor have been extensively studied in nonvascular tissues (Luis *et al.*, 1988; Robberecht *et al.*, 1990). Covalent affinity labeling and gel electrophoresis of the high-affinity VIP receptor from rat sources have shown an apparent molecular weight between 46,000 to 73,000 (Luis *et al.*, 1988). The high-affinity VIP receptor is associated with a guanine nucleotide binding protein and appears to be a monomeric glycoprotein (Luis *et al.*, 1988). Recently, Couvineau *et al.* (1990) purified the rat liver receptor to homogeneity by affinity chromatography and assigned it a molecular weight of 53,000. Very recently the cloning and expression of the human VIP receptor were reported utilizing cDNA clones from Nalm 6 lymphoblasts and the HT-29 line of colonic carcinoma cells (Sreedharan *et al.*, 1991). The deduced receptor polypeptide contained 362 amino acids and shared a seven transmembrane segment hydrophobic profile with other G protein coupled receptors. Transfection of the receptor cDNA to other cell lines resulted in receptor binding of ^{125}I-labeled VIP and VIP stimulation of cAMP accumulation. Subsequent to this discovery, the question of structural heterogeneity of VIP receptors in different tissues will likely be resolved in due course by application of molecular biology technology, as has been the case for other neurotransmitter receptors.

The vascular VIP receptor is functionally coupled to vasorelaxation as reflected in the good correlation between the rank orders of potency of VIP and related peptides in assays of competitive receptor binding and relaxation of arterial strips (Figs. 2 and 3). The PHI and GRF families of peptides are less potent than VIP, and secretin and glucagon are inactive in receptor binding assays in agreement with the bioassay findings (Huang and Rorstad, 1987; Y. Suzuki *et al.*, 1985). The existence of a distinct VIP-preferring receptor, as opposed to the receptor for the homologous hormone, secretin, is widely accepted (Robberecht *et al.*, 1982). However, the delineation of subclasses of the VIP receptor has proven difficult and incomplete as reviewed in detail by Robberecht *et al.* (1990). Pharmacological studies have indicated the existence of heterogeneity of VIP receptors between different animal species, such as rat, human, and guinea pig. In addition a novel VIP receptor has been identified in human clonal cell lines, such as SUP-T1 lymphoma cells based on high affinity for helodermin, a peptide from lizard venom, compared to untransformed tissues and other cell lines (Robberecht *et al.*, 1988). With respect to the vascular VIP receptor, rat and bovine arteries are distinguishable in radioligand-binding studies by the relative potency of rat GRF, whereas the relative potencies of several other peptides were similar in the two species (Fig. 3). Heterogeneity of VIP receptors within a single species, the rat, has been suggested by the rank order of potency of VIP analogs in competitive binding studies (Fig. 4). The VIP receptor in rat arteries differs from that in liver, pituitary, or brain, particularly in its affinity for [D-Ala^4]VIP (Rorstad *et al.*, 1990). Unfortunately, the VIP antagonists developed to date, VIP(10–28), [4-Cl-D-Phe^6,Leu^{17}]VIP, and [*N*-Ac-Tyr^1, D-Phe^2]GRF(1–29)-NH$_2$, have had low affinity for the vasculature and have not differentiated between VIP receptors in different rat tissues (Rorstad *et al.*, 1990). Morris and Murphy (1989) have proposed a second VIP receptor in the guinea pig uterine artery that mediates vasoconstriction at high concentration of VIP or VIP(10–28). The functional significance of this observation vis-a-vis the more widely recognized vasorelaxant function of VIP remains to be determined.

VIP receptors have been localized by autoradiographic techniques to the walls of bovine cerebral (Poulin *et al.*, 1986) and pulmonary arteries (Barnes *et al.*, 1986),

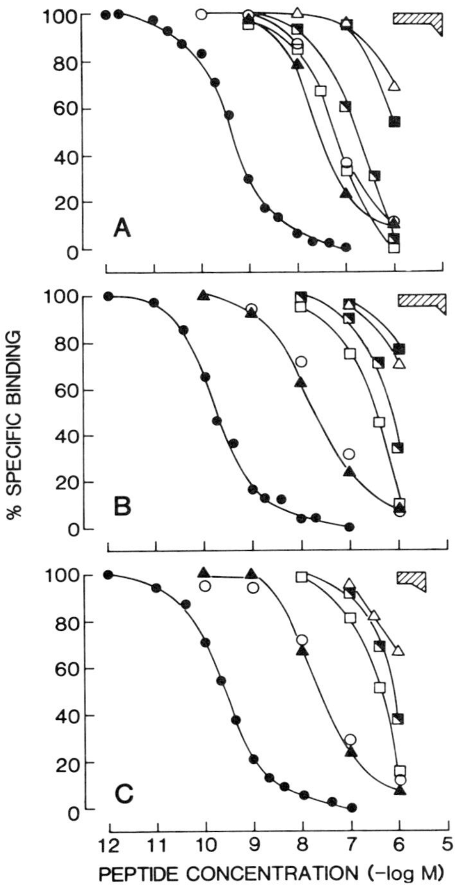

Figure 3 Competition by unlabeled VIP and related peptides for specific binding of [Tyr(^{125}I)^{10}]VIP to arterial membranes from (A) rat superior mesenteric artery, (B) bovine circumflex coronary artery, and (C) bovine left descending and right coronary arteries. The competing peptides were VIP (●), PHI (▲), PHM (○), rat GRF(1–40) amide (□), human pancreatic GRF(1–40) amide (■), bovine GRF(1–44) amide (◪), and VIP(10–28) (△). The shaded area encompasses responses to secretin and glucagon. Results of specific binding of radiolabeled VIP in the presence of a competing peptide are expressed as a percentage of maximal specific binding in the absence of the peptide. (Reproduced with permission from Huang and Rorstad, 1987.)

rabbit mesenteric, renal and iliac arteries (Sidawy *et al.*, 1989), and the rat mesenteric artery (F. Amenta *et al.*, 1989). The VIP binding has been most prominent over the smooth muscle-containing medial layer (Poulin *et al.*, 1986; F. Amenta *et al.*, 1989; Sidawy *et al.*, 1989). A localization of VIP receptors on smooth muscle cells was also confirmed by Hirata *et al.* (1985) utilizing isolated cultured rat aortic

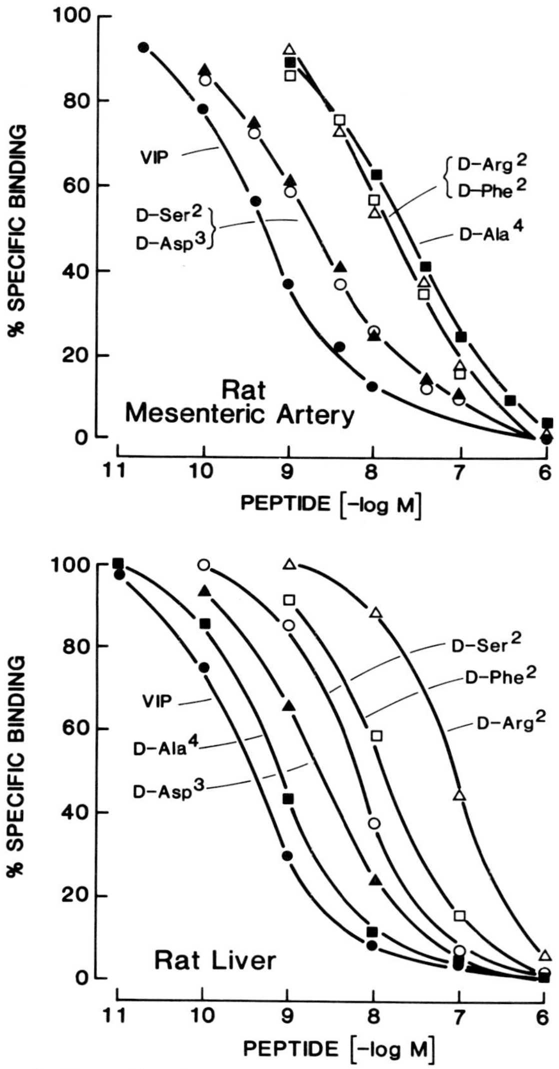

Figure 4 Competition for specific binding of [Tyr(^{125}I)^{10}]VIP to membrane preparations of rat mesenteric artery and rat liver by VIP (●) and its analogs: [D-Ser^2]VIP (○), [D-Arg^2]VIP (△), [D-Phe^2]VIP (□), [D-Asp^3]VIP (▲), and [D-Ala^4]VIP (■). Results are expressed as described in the legend to Fig. 3. (Reproduced with permission from Rorstad *et al.*, 1990.)

smooth muscle cells. VIP receptors have not been discernable in the intimal layer in autoradiography studies (Barnes *et al.*, 1986; Sidawy *et al.*, 1989).

Since the PHI peptides are coreleased and share vasorelaxant actions with VIP, the possibility arises that they may interact with a common vascular receptor (Fahrenkrug, 1987). Although selected examples exist of binding sites in nonvascular tissues with greater or equal affinity for PHI compared to VIP, the prevailing evidence favors an interpretation that PHI acts on the vasculature via binding to a VIP-preferring receptor (reviewed in Huang *et al.*, 1989). Three types of experiments support this opinion: (i) VIP and PHI do not have an additive effect on vascular relaxation (Edvinsson and McCulloch, 1985), the activation of vascular adenylate cyclase in cerebral microvessels or arteries (Huang and Rorstad, 1984), nor the accumulation of cAMP in choroid plexus epithelial cells (Crook and Pruisiner, 1986); (ii) porcine coronary artery segments show cross-desensitization to the relaxant effect of VIP and PHI (Fig. 5); and (iii) radioligand-binding studies

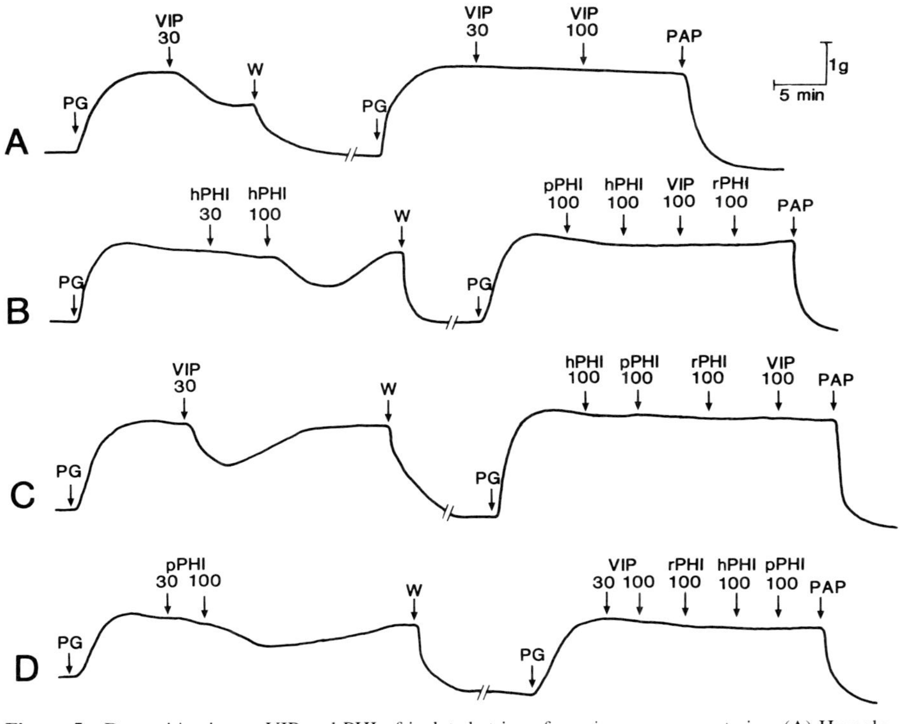

Figure 5 Desensitization to VIP and PHI of isolated strips of porcine coronary arteries. (A) Homologous desensitization to VIP; (B–D) cross-desensitization between VIP and three forms of PHI from rat (r), porcine (p), and human (h) species. Abbreviations: PG, addition of $PGF_{2\alpha}$ to increase arterial tension; W, wash with buffer; PAP, addition of papaverine to determine relaxation responses; 30 and 100, nanomolar concentrations of peptides. (Reproduced with permission from Huang *et al.*, 1989.)

using either radiolabeled VIP or PHI consistently identify a receptor with higher affinity for VIP than PHI in bovine, porcine, and rat arteries (Huang *et al.*, 1989). Likewise, other normal rat tissues, such as brain, anterior pituitary, liver, and uterus, contain a VIP-preferring receptor with lower affinity for PHI irrespective of whether radiolabeled VIP or PHI is utilized in radioligand-binding studies (Huang and Rorstad, 1990).

E. Effector Mechanisms for VIP

The postreceptor effector mechanisms responsible for VIP-induced vasodilation appear to involve activation of adenylate cyclase and generation of cyclic adenosine monophosphate (cAMP) as key regulatory steps. Abundant studies have demonstrated VIP stimulation in assays of (i) adenylate cyclase activity using cerebral and renal microvessels (Griffith and Simmons, 1990; Huang and Rorstad, 1983), cerebral arteries (Huang and Rorstad, 1984), and isolated aortic smooth muscle cells (Hirata *et al.*, 1985; Ganz *et al.*, 1986); and (ii) cAMP accumulation in cerebral arteries (Edvinsson *et al.*, 1985), rat aorta (Schoeffter and Stoclet, 1985; Sata *et al.*, 1988), mesenteric artery (T. Itoh *et al.*, 1985; Ganz *et al.*, 1986), pulmonary artery (Ignarro *et al.*, 1987), portal vein (D. Amenta *et al.*, 1988), and choroid plexus (Crook and Prusiner, 1986; Lindvall *et al.*, 1985). An increase in the concentration of cAMP appears to be a common mediator underlying the cerebral vasodilatory action of beta-adrenergic receptor agonists, adenosine, VIP, and CGRP (Edvinsson *et al.*, 1985; Y. Suzuki *et al.*, 1988). This contention is supported by the observation that forskolin, a receptor-independent activator of adenylate cyclase, both relaxes cerebral arteries and stimulates cerebral vascular adenylate cyclase with quantitatively comparable concentration-response curves (Y. Suzuki *et al.*, 1988). Removal of the endothelium was found to inhibit VIP stimulation of cAMP accumulation in rat aorta and bovine pulmonary artery (Ignarro *et al.*, 1987; Sata *et al.*, 1988), two vessels in which VIP-induced relaxation was also endothelium dependent. However, D. Amenta *et al.* (1988) observed that VIP stimulation of cAMP accumulation in the rat portal vein was independent of the endothelium. The role of cyclic guanosine monophosphate (cGMP) is uncertain since VIP has been found to stimulate endothelium-dependent cGMP accumulation in the bovine pulmonary artery (Ignarro *et al.*, 1987) but to have no effect on cGMP concentration in rat aorta (Schoeffter and Stoclet, 1985).

Prostanoids may participate in regulating VIP vasodilation although published findings have been contradictory and complex. For instance, VIP relaxation of the bovine pulmonary artery was observed to be endothelium dependent and indomethacin sensitive by Ignarro *et al.* (1987) but conversely, endothelium independent and indomethacin insensitive by Barnes *et al.* (1986) and Greenberg *et al.* (1987). Indomethacin has been reported to inhibit dilation of cerebral arterioles and pulmonary vasculature *in vivo* (Wei *et al.*, 1980; Kulik *et al.*, 1984). In contrast, inhibitors of arachidonate metabolism have had no effect on VIP-induced vasodilation or cAMP accumulation in studies of isolated cerebral arteries (Edvinsson *et al.*, 1985), rat aorta (Sata *et al.*, 1988), human myometrial and stem villous arteries (Hansen *et al.*, 1988), and human coronaries *in vivo* (Smitherman *et al.*, 1989). Uniquely, Davies and Williams (1984) observed that the lipoxygenase inhibitor eicosatetraynoic acid (ETYA), but not the cyclooxygenase inhibitor, indomethacin, antagonized endothelium-dependent VIP relaxation of rat aortic strips. Finally, VIP-induced dilation of cerebral arterioles does not appear to require the superoxide anion radical, an entity capable of being generated by arachidonate metabolism (Ballon *et al.*, 1986).

The involvement of intracellular calcium-mobilizing mechanisms in VIP-induced vasodilation has been studied little although observations from other model systems suggest that such perturbations to calcium concentrations could potentially contribute to the vascular actions of VIP. The membrane potential and resistance of vascular smooth muscle cells do not change in response to VIP, but T. Itoh *et al.* (1985) have proposed on the basis of indirect experiments that VIP may inhibit norepinephrine-induced intracellular calcium-releasing mechanisms thereby antagonizing norepinephrine-induced vasoconstriction. Alternative suggestions for mechanisms of VIP-induced relaxation have included VIP-induced calcium extrusion from the cell or increased sequestration of calcium in intracellular storage sites (Bolton *et al.*, 1981; T. Itoh *et al.*, 1985). VIP has been reported to inhibit the synthesis of inositol trisphosphate in the lower esophageal sphincter, which would tend to inhibit the release of calcium from intracellular stores (Szewczak *et al.*, 1990). Transduction mechanisms for VIP probably differ among tissues since VIP stimulates inositol phosphate metabolism in nervous tissue (Audigier *et al.*, 1986; Malhotra *et al.*, 1988). Finally, phorbol esters, which activate protein kinase C, cause enhancement of VIP-induced cAMP accumulation in cultured vascular smooth muscle cells (Nabika *et al.*, 1985).

IV. Conclusion

Significant parallels exist between the cholinergic and VIP systems, such as their common parasympathetic innervation, including neuronal colocalization, and their vasodilatory actions, which appear to be complementary as opposed to redundant. For instance, the two systems differ fundamentally in terms of their predominant effector mechanisms for vasodilation: generation of EDRF in the case of ACh and increased accumulation of cAMP by VIP. However, notable functional interactions between the cholinergic and VIP systems are exemplified by VIP facilitation of muscarinic neurotransmission in the superior cervical ganglion (Kawatani *et al.*, 1985) perhaps by potentiating muscarinic stimulation of inositol phosphate metabolism (Raiteri *et al.*, 1987). Considering actions at the level of receptor function, chronic atropine treatment of rats increases the number of VIP receptors in the cerebral cortex and salivary gland (Hedlund *et al.*, 1983; Abens *et al.*, 1984), whereas carbachol treatment of guinea pigs decreases the number of VIP receptors in pancreas (Murakami *et al.*, 1989). At present it is not known whether such interactions occur in the vasculature, or, if so, whether they would have physiological significance of the overall regulation of blood flow.

Acknowledgments

We thank Ms. Eunice Jacobs for preparing the manuscript. Portions of the studies discussed in this chapter were supported by the Alberta Heart and Stroke Foundation and the Medical Research Council of Canada.

References

Abens, J., Westlind, A., and Bartfai, T. (1984). Chronic atropine treatment causes increase in VIP receptors in rat cerebral cortex. *Peptides 5,* 375–377.

Altiere, R. J., Kiritsy-Roy, J. A., and Catravas, J. D. (1986). Acetylcholine-induced contractions in isolated rabbit pulmonary arteries: Role of thromboxane A_2. *J. Pharmacol. Exp. Ther. 236,* 535–541.

Amenta, D., Iacopino, L., and Amenta, F. (1988). Vasoactive intestinal peptide-sensitive cyclic adenosine monophosphate generating system in the rat portal vein. *Arch. Int. Pharmacodyn. 291,* 88–95.

Amenta, F., Cavallotti, C., Collier, W. L., Ferrante, F., Geppetti, P., and Ricci, A. (1989). Autoradiographic localization of vasoactive intestinal polypeptide receptors in the rat mesenteric vascular tree. *Regul. Pept. 26,* 9–17.

Andersson, P.-O., Bjornberg, J., Bloom, S. R., and Mellander, S. (1987). Vasoactive intestinal polypeptide in relation to penile erection in the cat evoked by pelvic and by hypogastric nerve stimulation. *J. Urol. 138,* 419–422.

Armstrong, D. M. (1986). Ultrastructural characterization of choline acetyltransferase-containing neurons in the basal forebrain of rat: Evidence for a cholinergic innervation of intracerebral blood vessels. *J. Comp. Neurol. 250,* 81–92.

Arneric, S. P., Honig, M. A., Milner, T. A., Greco, S., Iadecola, C., and Reis, D. J. (1988). Neuronal and endothelial sites of acetylcholine synthesis and release associated with microvessels in rat cerebral cortex: Ultrastructural and neurochemical studies. *Brain Res. 454,* 11–30.

Audigier, S., Barberis, C., and Jard, S. (1986). Vasoactive intestinal polypeptide increases inositol phospholipid breakdown in the rat superior cervical ganglion. *Brain Res. 376,* 363–367.

Ballon, B. J., Wei, E. P., and Kontos, H. A. (1986). Superoxide anion radical does not mediate vasodilation of cerebral arterioles by vasoactive intestinal polypeptide. *Stroke 17,* 1287–1290.

Barnes, P. J., Cadieux, A., Carstairs, J. R., Greenberg, B., Polak, J. M., and Rhoden, K. (1986). Vasoactive intestinal peptide in bovine pulmonary artery: Localization, function and receptor autoradiography. *Br. J. Pharmacol. 89,* 157–162.

Beny, J. L., Brunet, P. C., and Huggel, H. (1986). Effect of mechanical stimulation, substance P and vasoactive intestinal polypeptide on the electrical and mechanical activities of circular smooth muscles from pig coronary arteries contracted with acetylcholine: Role of endothelium. *Pharmacology 33,* 61–68.

Bevan, J. A., and Brayden, J. E. (1987). Nonadrenergic neural vasodilator mechanisms. *Circ. Res. 60,* 309–326.

Bevan, J. A., Moscowitz, M., Said, S. I., and Buga, G. (1984). Evidence that vasoactive intestinal polypeptide is a dilator transmitter to some cerebral and extracerebral cranial arteries. *Peptides 5,* 385–388.

Bodanszky, M., Klausner, Y. S., and Said, S. I. (1973). Biological activities of synthetic peptides corresponding to fragments of and to the entire sequence of the vasoactive intestinal peptide. *Proc. Natl. Acad. Sci. U.S.A. 70,* 382–384.

Bognar, I. T., Beinhauer, B., Kann, P., and Fuder, H. (1990). Different muscarinic receptors mediate autoinhibition of acetylcholine release and vagally-induced vasoconstriction in the rat isolated perfused heart. *Naunyn-Schmiedeberg's Arch. Pharmacol. 341,* 279–287.

Bolton, T. B., Lang, R. J., and Ottesen, B. (1981). Mechanism of action of vasoactive intestinal polypeptide on myometrial smooth muscle of rabbit and guinea-pig. *J. Physiol. 318,* 41–55.

Brayden, J. E., and Bevan, J. A. (1985). Neurogenic muscarinic vasodilation in the cat. An example of endothelial cell-independent cholinergic relaxation. *Circ. Res. 56,* 205–211.

Brayden, J. E., and Bevan, J. A. (1986). Evidence that vasoactive intestinal polypeptide (VIP) mediates neurogenic vasodilation of feline cerebral arteries. *Stroke 17,* 1189–1192.

Brum, J. M., Bove, A. A. Sufan, Q., Reilly, W., and Go, V. L. W. (1986). Action and localization of vasoactive intestinal peptide in the coronary circulation: Evidence for nonadrenergic, noncholinergic coronary regulation. *J. Am. Coll. Cardiol. 7,* 406–413.

Brunner, F., and Kukovetz, W. R. (1986). Muscarinic receptors of the vascular bed: Radioligand binding studies on bovine splenic veins. *J. Cardiovasc. Pharmacol. 8,* 712–721.

Brunner, F., and Kukovetz, W. R. (1991). Radioligand binding to muscarinic receptors of bovine aortic endothelial cells. *Br. J. Pharmacol. 102,* 373–380.

Brunner, F., Kuhberger, E., Brockmeier, D., and Kukovetz, W. R. (1990). Evidence for muscarinic receptors in endothelial cells from combined functional and binding studies. *Eur. J. Pharmacol. 187,* 145–154.

Chen, G., and Suzuki, H. (1989). Direct and indirect actions of acetylcholine and histamine on intrapulmonary artery and vein muscles of the rat. *Jpn. J. Physiol. 39,* 51–65.

Couvineau, A., Voisin, T., Guijarro, L., and Laburthe, M. (1990). Purification of vasoactive intestinal peptide receptor from porcine liver by a newly designed one-step affinity chromatography. *J. Biol. Chem. 265,* 13386–13390.

Crook, R. B., and Prusiner, S. B. (1986). Vasoactive intestinal peptide stimulates cyclic AMP metabolism in choroid plexus epithelial cells. *Brain Res. 384,* 138–144.

Danthuluri, N. R., Cybulsky, M. I., and Brock, T. A. (1988). ACh-induced calcium transients in primary cultures of rabbit aortic endothelial cells. *Am. J. Physiol. 255,* H1549–H1553.

Dauphin F., and Hamel, E. (1990). Muscarinic receptor subtype mediating vasodilation in feline middle cerebral artery exhibits M_3 pharmacology. *Eur. J. Pharmacol. 178,* 203–213.

Davies, J. M., and Williams, K. I. (1984). Endothelial-dependent relaxant effects of vasoactive intestinal polypeptide and arachidonic acid in rat aortic strips. *Prostaglandins 27,* 195–202.

Doods, H. N., Dammgen, J., Mayer, N., Rinner, I., and Trach, V. (1989). Muscarinic receptors in the heart and vascular system. *In* "Progress in Pharmacology and Clinical Pharmacology." (P. A. van Zwieten and E. Schonbaum, eds.), Vol. 7, No.1, pp. 47–72. Gustav Fischer Verlag, Stuttgart, New York.

Duckles, S. P. (1988). Acetylcholine. *In* "Nonadrenergic Innervation of Blood Vessels". (G. Burnstock and S. G. Griffith, eds.), Vol. 1, pp. 15–26. CRC Press Inc., Boca Raton, Florida.

Duckles, S. P., and Garcia-Villalon, A. L. (1990). Characterization of vascular muscarinic receptors: Rabbit ear artery and bovine coronary artery. *J. Pharmacol. Exp. Ther. 253,* 608–613.

Duckles, S. P., and Said, S. I. (1982). Vasoactive intestinal peptide as a neurotransmitter in the cerebral circulation. *Eur. J. Pharmacol. 78,* 371–374.

Eckenstein, F., and Baughman, R. W. (1984). Two types of cholinergic innervation in cortex, one colocalized with vasoactive intestinal polypeptide. *Nature* (London) *309,* 153–155.

Edvinsson, L., and McCulloch, J. (1985). Distribution and vasomotor effects of peptide HI (PHI) in feline cerebral blood vessels *in vitro* and *in situ. Regul. Pept. 10,* 345–356.

Edvinsson, L., and Uddman, R. (1988). Vasoactive intestinal polypeptide (VIP): A putative neurotransmitter in the cardiovascular system. *In* "Nonadrenergic Innervation of Blood Vessels." (G. Burnstock and S. G. Griffith, eds.), Vol. 1, pp. 101–125. CRC Press Inc., Boca Raton, Florida.

Edvinsson, L., Fahrenkrug, J., Hanko, J., Owman, C., Sundler, F., and Uddman, R. (1980). VIP (vasoactive intestinal polypeptide)-containing nerves of intracranial arteries in mammals. *Cell Tissue Res. 208,* 135–142

Edvinsson, L., Fredholm, B. B. Hamel, E., Jansen, I., and Verrecchia, C. (1985). Perivascular peptides relax cerebral arteries concomitant with stimulation of cyclic adenosine monophosphate accumulation or release of an endothelium-derived relaxing factor in the cat. *Neurosci. Lett. 58,* 213–217.

Edvinsson, L., McCulloch, J., Kelly, P. A. T., and Tuor, U. I. (1988). Role of vasoactive intestinal peptide and peptide histidine isoleucine in the cerebral circulation. *Ann. N.Y. Acad. Sci. 527,* 378–392.

Edvinsson, L., Hara, H., and Uddman, R. (1989a). Retrograde tracing of nerve fibers to the rat middle cerebral artery with true blue: Colocalization with different peptides. *J. Cereb. Blood Flow Metab. 9,* 212–218.

Edvinsson, L., Gulbenkian, S., Jansen, I., Wharton, J., Cervantes, C., and Polak, J. M. (1989b). Comparison of peptidergic mechanisms in different parts of the guinea pig superior mesenteric artery: Immunocytochemistry at the light and ultrastructural levels and responses *in vitro* of large and small arteries. *J. Auton. Nerv. Syst. 28,* 141–154.

Entzeroth, M., Doods, H. N., and Mayer, N. (1990). Characterization of porcine coronary muscarinic receptors. *Naunyn-Schmiedeberg's Arch. Pharmacol. 341,* 432–438.

Estrada, C., and Krause, D. N. (1982). Muscarinic cholinergic receptor sites in cerebral blood vessels. *J. Pharmacol. Exp. Ther. 221,* 85–90.

Fahrenkrug, J. (1987). Co-existence and co-secretion of the structurally related peptides VIP and PHI. *Scand. J. Clin. Lab. Invest. 47* (Suppl. 186), 43–50.

Fahrenkrug, J. (1989). VIP and autonomic neurotransmission. *Pharmacol. Ther. 41,* 515–534.

Forsgren, S. (1989). Vasoactive intestinal polypeptide-like immunoreactivity in the bovine heart: High degree of coexistence with neuropeptide Y-like immunoreactivity. *Cell Tissue Res. 256,* 125–135.

Forssmann, W. G., Triepel, J., Daffner, C., Heym, Ch., Cuevas, P., Noble, M. I. M., and Yanaihara, N. (1988). Vasoactive intestinal peptide in the heart. *Ann. N. Y. Acad. Sci. 527,* 405–420.

Fournier, A., Saunders, J. K., and St-Pierre, S. (1984). Synthesis, conformational studies and biological activities of VIP and related fragments. *Peptides 5,* 169–177.

Furchgott, R. F., and Zawadzki, J. V. (1980). The obligatory role of endothelial cells in the relaxation of arterial smooth muscle by acetylcholine. *Nature* (London) *288,* 373–376.

Furness, J. B., Costa, M., Papka, R. E., Della, N. G., and Murphy, R. (1984). Neuropeptides contained in peripheral cardiovascular nerves. *Clin. Exp. Hypertens. A6,* 91–106.

Ganz, P., Sandrock, A. W., Landis, S. C., Leopold, J., Gimbrone, M. A., Jr., and Alexander, R. W. (1986). Vasoactive intestinal peptide: Vasodilation and cyclic AMP generation. *Am. J. Physiol. 250,* H755–H760.

Gibbins, I. L., Brayden, J. E., and Bevan, J. A. (1984). Distribution and origins of VIP-immunoreactive nerves in the cephalic circulation of the cat. *Peptides 5,* 209–212.

Goadsby, P. J., and MacDonald, G. J. (1985). Extracranial vasodilation mediated by vasoactive intestinal polypeptide (VIP). *Brain Res. 329,* 285–288.

Goyal, R. K. (1989). Muscarinic receptor subtypes. Physiology and clinical implications. *N. Engl. J. Med. 321,* 1022–1029.

Greenberg, B., Rhoden, K., and Barnes, P. J. (1987). Relaxant effects of vasoactive intestinal peptide and peptide histidine isoleucine in human and bovine pulmonary arteries. *Blood Vessels 24,* 45–50.

Griffiths, N. M., and Simmons, N. L. (1990). Vasoactive intestinal peptide stimulation of adenylate cyclase activity in isolated rabbit renal arterial microvessels. *Exp. Physiol. 75,* 601–604.

Grigorian, G. Y., Mirzapoyazova, T. Y., Resink, T. J., Danilov, S. M., and Tkachuk, V. A. (1989). Regulation of phosphoinositide turnover in endothelium from human pulmonary artery, aorta and umbilical vein. Antagonistic action of the β-adrenoceptor coupled adenylate cyclase system. *J. Mol. Cell. Cardiol. 21* (Suppl. I), 119–123.

Hansen, V., Maigaard, S., Allen, J., and Forman, A. (1988). Effects of vasoactive intestinal polypeptide and substance P on human intramyometrial arteries and stem villous arteries in term pregnancy. *Placenta 9,* 501–506.

Hara, H., and Weir, B. (1988). Pathway of nerves with vasoactive intestinal polypeptide-like immunoreactivity to the major cerebral arteries of the rat. *Cell Tissue Res. 251,* 275–280.

Hara, H., Hamill, G. S., and Jacobowitz, D. M. (1985). Origin of cholinergic nerves to the rat major cerebral arteries: Coexistence with vasoactive intestinal polypeptide. *Brain Res. Bull. 14,* 179–188.

Hedlund, B., Abens, J., and Bartfai, T. (1983). Vasoactive intestinal polypeptide and muscarinic receptors: Supersensitivity induced by long-term atropine treatment. *Science 220,* 519–521.

Hellstrand, P., Fahrenkrug, J., Uddman, R., and Jarhult, J. (1985). Role of vasoactive intestinal polypeptide (VIP) in the neurogenic vasodilation of the portal vein in the rabbit. *Regul. Pept. 12,* 309–316.

Hirata, Y., Tomita, M., Takata, S., and Fujita, T. (1985). Functional receptors for vasoactive intestinal peptide in cultured vascular smooth muscle cells from rat aorta. *Biochem. Biophys. Res. Commun. 132*. 1079–1087.

Huang, M., and Rorstad, O. P. (1983). Effects of vasoactive intestinal polypeptide, monoamines, prostaglandins, and 2-chloroadenosine on adenylate cyclase in rat cerebral microvessels. *J. Neurochem. 40,* 719–726.

Huang, M., and Rorstad, O. P. (1984). Cerebral vascular adenylate cyclase: Evidence for coupling to receptors for vasoactive intestinal peptide and parathyroid hormone. *J. Neurochem. 43,* 849–856.

Huang, M., and Rorstad, O. P. (1987). VIP receptors in mesenteric and coronary arteries: A radioligand binding study. *Peptides 8,* 477–485.

Huang, M., and Rorstad, O. P. (1989). Effects of GTP analogs and dithiothreitol on the binding properties of the vascular vasoactive intestinal peptide receptor. *Can. J. Physiol. Pharmacol. 67,* 851–856.

Huang, M., and Rorstad, O. P. (1990). PHI preferentially binds to VIP receptors in normal rat tissues. *Peptides 11,* 1015–1020.

Huang, M., Metherell, L., and Rorstad, O. P. (1988). Utility of polycation-treated filters for the assay of receptors for VIP. *J. Receptor Res. 8,* 831–838.

Huang, M., Itoh, H., Lederis, K., and Rorstad, O. P. (1989). Evidence that vascular actions of PHI are mediated by a VIP-preferring receptor. *Peptides 10,* 993–1001.

Hulme, E. C., Birdsall, N. J. M., and Buckley, N. J. (1990). Muscarinic receptor subtypes. *Annu. Rev. Pharmacol. Toxicol. 30,* 633–673.

Hynes, M. R., Banner, W., Jr., Yamamura, H. I., and Duckles, S. P. (1986). Characterization of muscarinic receptors of the rabbit ear artery smooth muscle and endothelium. *J. Pharmacol. Exp. Ther. 238,* 100–105.

Ignarro, L. J., Byrns, R. E., Buga, G. M., and Wood, K. S. (1987). Mechanisms of endothelium-dependent vascular smooth muscle relaxation elicited by bradykinin and VIP. *Am. J. Physiol. 253,* H1074–H1082.

Itoh, H., Lederis, K. P., and Rorstad, O. P. (1990). Relaxation of isolated bovine coronary arteries by vasoactive intestinal peptide. *Eur. J. Pharmacol. 181,* 199–205.

Itoh, N., Obata, K., Yanaihara, N., and Okamoto, H. (1983). Human preprovasoactive intestinal polypeptide contains a novel PHI-27-like peptide, PHM-27. *Nature* (London) *304,* 547–549.

Itoh, T., Sasaguri, T., Makita, Y., Kanmura, Y., and Kuriyama, H. (1985). Mechanisms of vasodilation induced by vasoactive intestinal polypeptide in rabbit mesenteric artery. *Am. J. Physiol. 249,* H231–H240.

Itoh, T., Kubota, Y., and Kuriyama, H. (1988). Effects of phorbol ester on acetylcholine-induced Ca^{2+} mobilization and contraction in the porcine coronary artery. *J. Physiol.* (London) *397,* 401–419.

Kaji, A., Shigematsu, H., Fujita, K., Maeda, T., and Watanabe, S. (1988). Parasympathetic innervation of cutaneous blood vessels by vasoactive intestinal polypeptide-immunoreactive and acetylcholinesterase-positive nerves: Histochemical and experimental study on rat lower lip. *Neuroscience 25,* 353–362.

Kalsner, S. (1985). Cholinergic mechanisms in human coronary artery preparations: Implications of species differences. *J. Physiol.* (London) *358:* 509–526.

Kalsner, S. (1989). Cholinergic constriction in the general circulation and its role in coronary artery spasm. *Circ. Res. 65,* 237–257.

Kalsner, S., and Quillan, M. (1989). Cholinergic contraction to field stimulation in coronary arteries of cattle. *J. Pharmacol. Exp. Ther. 249,* 785–789.

Kawatani, M., Rutigliano, M., and de Groat, W. C. (1985). Depolarization and muscarinic excitation induced in a sympathetic ganglion by vasoactive intestinal polypeptide. *Science 229,* 879–881.

Kitagawa, H., Takeda, F., and Kohei, H. (1987). Endothelium-dependent increases in rat gastric mucosal hemodynamics induced by acetylcholine and vagal stimulation. *Eur. J. Pharmacol. 133,* 57–63.

Kulik, T. J., Johnson, D. E., Elde, R. P., and Lock, J. E. (1984). Pulmonary vascular effects of vasoactive intestinal peptide in conscious newborn lambs. *Am. J. Physiol. 246,* H716–H719.

Landis, S. C., and Fredieu, J. R. (1986). Coexistence of calcitonin gene-related peptide and vasoactive intestinal peptide in cholinergic sympathetic innervation of rat sweat glands. *Brain Res. 377,* 177–181.

Larsson, L.-I., Edvinsson, L., Fahrenkrug, J., Hakanson, R., Owman, Ch., Schaffalitzky de Muckadell, O., and Sundler, F. (1976). Immunohistochemical localization of a vasodilatory polypeptide (VIP) in cerebrovascular nerves. *Brain Res. 113,* 400–404.

Leblanc, G. G., Trimmer, B. A., and Landis, S. C. (1987). Neuropeptide Y-like immunoreactivity in rat cranial parasympathetic neurons: Coexistence with vasoactive intestinal peptide and choline acetyltransferase. *Proc. Nat. Acad. Sci. U.S.A. 84,* 3511–3515.

Lee, T. J.-F. (1980). Direct evidence against acetylcholine as the dilator transmitter in the cat cerebral artery. *Eur. J. Pharmacol. 68,* 393–394.

Lee, T. J.-F., Saito, A., and Berezin, I. (1984). Vasoactive intestinal polypeptide-like substance: The potential transmitter for cerebral vasodilation. *Science 224,* 898–901.

Lindh, B., Lundberg, J. M., Hokfelt, T., Elfvin, L. G., Fahrenkrug, J., and Fischer, J. (1987). Coexistence of CGRP- and VIP-like immunoreactivities in a population of neurons in the cat stellate ganglia. *Acta Physiol. Scand. 131,* 475–476.

Lindvall, M., Gustafson, A., Hedner, P., and Owman, C. (1985). Stimulation of cyclic adenosine 3′, 5′-monophosphate formation in rabbit choroid plexus by β-receptor agonists and vasoactive intestinal polypeptide. *Neurosci. Lett. 54,* 153–157.

Luis, J., Martin, J.-M., El Battari, A., Marvaldi, J., and Pichon, J. (1988). The vasoactive intestinal peptide (VIP) receptor: Recent data and hypothesis. *Biochimie 70,* 1311–1322.

Lundberg, J. M., Anggard, A., Fahrenkrug, J., Hokfelt, T., and Mutt, V. (1980). Vasoactive intestinal polypeptide in cholinergic neurons of exocrine glands: Functional significance of coexisting transmitters for vasodilation and secretion. *Proc. Natl. Acad. Sci. U.S.A. 77,* 1651–1655.

Malhotra, R. K., Wakade, T. D., and Wakade, A. R. (1988). Vasoactive intestinal polypeptide and muscarine mobilize intracellular Ca^{2+} through breakdown of phosphoinositides to induce catecholamine secretion. *J. Biol. Chem. 263,* 2123–2126.

McCormack, D. G., Mak, J. C., Minette, P., and Barnes, P. J. (1988). Muscarinic receptor subtypes mediating vasodilation in the pulmonary artery. *Eur. J. Pharmacol. 158,* 293–297.

McMaster, D., Suzuki, Y., Rorstad, O. P., and Lederis, K. (1987). Iodinated derivatives of vasoactive intestinal peptide (VIP), PHI and PHM: Purification, chemical characterization and biological activity. *Peptides 8,* 663–676.

Miao, F. J.-P., and Lee, T. J.-F. (1990). Cholinergic and VIPergic innervation in cerebral arteries: A sequential double-labeling immunohistochemical study. *J. Cereb. Blood Flow Metab. 10,* 32–37.

Milner, P., Ralevic, V., Hopwood, A. M., Feher, E., Lincoln, J., Kirkpatrick, K. A., and Burnstock, G. (1989). Ultrastructural localization of substance P and choline acetyltransferase in endothelial cells of rat coronary artery and release of substance P and acetylcholine during hypoxia. *Experientia 45,* 121–125.

Milner, P., Kirkpatrick, K. A., Ralevic, V., Toothill, V., Pearson, J., and Burnstock, G. (1990). Endothelial cells cultured from human umbilical vein release ATP, substance P and acetylcholine in response to increased flow. *Proc. R. Soc. London B 241,* 245–248.

Mione, M. C., Ralevic, V., and Burnstock, G. (1990). Peptides and vasomotor mechanisms. *Pharmacol. Ther. 46,* 429–468.

Miyata, A., Arimura, A., Dahl, R. R., Minamino, N., Uehara, A., Jiang, L., Culler, M. D., and Coy, D. H. (1989). Isolation of a novel 38 residue-hypothalamic polypeptide which stimulates adenylate cyclase in pituitary cells. *Biochem. Biophys. Res. Commun. 164,* 567–574.

Morris, J. L., and Murphy, R. (1989). Analogues of vasoactive intestinal peptide (VIP) contract the guinea-pig uterine artery but do not antagonize VIP-induced relaxations. *Eur. J. Pharmacol. 162,* 375–379.

Murakami, M., Vinayek, R., Jensen, R. T., and Gardner, J. D. (1989). Carbachol-induced down-regulation of high affinity receptors for vasoactive intestinal peptide. *Am. J. Physiol. 257,* G402–G408.

Nabika, T., Nara, Y., Yamori, Y., Lovenberg, W., and Endo, J. (1985). Angiotension II and phorbol ester enhance isoproterenol- and vasoactive intestinal peptide (VIP)-induced cyclic AMP accumulation in vascular smooth muscle cells. *Biochem. Biophys. Res. Commun. 131,* 30–36.

Nakakita, K. (1990). Peptidergic innervation in the cerebral blood vessels of the guinea pig: An immunohistochemical study. *J. Cereb. Blood Flow Metab. 10,* 819–826.

Olesen, S.-P., Davies, P. F., and Clapham, D. E. (1988). Muscarinic-activated K^+ current in bovine aortic endothelial cells. *Circ. Res. 62,* 1059–1064.

Poulin, P., Suzuki, Y., Lederis, K., and Rorstad, O. P. (1986). Autoradiographic localization of binding sites for vasoactive intestinal peptide (VIP) in bovine cerebral arteries. *Brain Res. 381,* 382–384.

Raiteri, M., Marchi, M., and Paudice, P. (1987). Vasoactive intestinal polypeptide (VIP) potentiates the muscarinic stimulation of phosphoinositide turnover in rat cerebral cortex. *Eur. J. Pharmacol. 133,* 127–128.

Rinner, I., Doods, H. N., van Charldorp, K. J., Davidesko, D., and van Zwieten, P. A. (1988). Binding of muscarinic receptor antagonists to pig coronary smooth muscle. *Naunyn-Schmeideberg's Arch. Pharmacol. 337,* 146–150.

Robberecht, P., Chatelain, P., Waelbroeck, M., and Christophe, J. (1982). Heterogeneity of VIP-recognizing binding sites in rat tissues. *In* "Vasoactive Intestinal Peptide." (S. I. Said, ed.), pp. 323–332. Raven Press, New York.

Robberecht, P., Waelbroeck, M., De Neef, P., Tastenoy, M., Gourlet, P., Cogniaux, J., and Christophe, J. (1988). A new type of functional VIP receptor has an affinity for helodermin in human SUP-T1 lymphoblasts. *FEBS Lett. 228,* 351–355.

Robberecht, P., Cauvin, A., Gourlet, P., and Christophe, J. (1990). Heterogeneity of VIP receptors. *Arch. Int. Pharmacodyn. 303,* 51–66.

Rorstad, O., P., Wanke, I., Coy, D. H., Fournier, A., and Huang, M. (1990). Selectivity for binding of peptide analogs to vascular receptors for vasoactive intestinal peptide. *Mol. Pharmacol. 37,* 971–977.

Said, S. I. (1982). "Vasoactive Intestinal Peptide." Raven Press, New York.

Said, S. I., and Mutt, V. (1970). Polypeptide with broad biological activity: Isolation from small intestine. *Science 169,* 1217–1218.

Saito, A., Wu, J.-Y., and Lee, T. J.-F. (1985). Evidence for the presence of cholinergic nerves in cerebral arteries: An immunohistochemical demonstration of choline acetyltransferase. *J. Cereb. Blood Flow Metab. 5,* 327–334.

Sata, T., Misra, H. P., Kubota, E., and Said, S. I. (1986). Vasoactive intestinal polypeptide relaxes pulmonary artery by an endothelium-independent mechanism. *Peptides 7* (Suppl. 1), 225–227.

Sata, T., Linden, J., Liu, L-W., Kubota, E., and Said, S. I. (1988). Vasoactive intestinal peptide evokes endothelium-dependent relaxation and cyclic AMP accumulation in rat aorta. *Peptides 9,* 853–858.

Schoeffter, P., and Stoclet, J. C. (1985) Effect of vasoactive intestinal polypeptide (VIP) on cyclic AMP level and relaxation in rat isolated aorta. *Eur. J. Pharmacol. 109,* 275–279.

Shirahase, H., Usui, H., Kurahashi, K., Fujiwara, M., and Fukui, K. (1987). Possible role of endothelial thromboxane A_2 in the resting tone and contractile responses to acetylcholine and arachidonic acid in canine cerebral arteries. *J. Cardiovasc. Pharmacol. 10,* 517–522.

Sidawy, A. N., Sayadi, H., Harmon, J. W., Termanini, B., Andrews, B., De Palma, R. G., and Korman, L. Y. (1989). Distribution of vasoactive intestinal peptide and its receptors in the arteries of the rabbit. *J. Surg. Res. 47,* 105–111.

Sim, M. K., and Manjeet, S. (1989). Endothelial muscarinic receptors in the rabbit aorta. *Eur. J. Pharmacol. 163,* 399–400.

Sim, M. K., and Manjeet, S. (1990). Properties of muscarinic receptors mediating second messenger responses in the rabbit aorta. *Eur. J. Pharmacol. 189,* 399–404.

Smitherman, T. C., Dehmer, G. J., and Said, S. I. (1988). Vasoactive intestinal peptide as a coronary vasodilator. *Ann. N. Y. Acad. Sci. 527,* 421–430.

Smitherman, T. C., Popma, J. J., Said, S. I., Krejs, G. J., and Dehmer, G. J. (1989). Coronary hemodynamic effects of intravenous vasoactive intestinal peptide in humans. *Am. J. Physiol. 257,* H1254–H1262.

Sreedharan, S. P., Robichon, A., Peterson, K. E., and Goetzl, E. J. (1991). Cloning and expression of the human vasoactive intestinal peptide receptor. *Proc. Natl. Acad. Sci. U.S.A. 88,* 4986–4990.

Stephenson, J. A., Gibson, R. E., and Summers, R. J. (1988). An autoradiographic study of muscarinic cholinoceptors in blood vessels: No localization on vascular endothelium. *Eur. J. Pharmacol. 153,* 271–283.

Suzuki, N., Hardebo, J. E., and Owman, C. (1988). Origins and pathways of cerebrovascular vasoactive intestinal polypeptide-positive nerves in rat. *J. Cereb. Blood Flow Metab. 8,* 697–712.

Suzuki, N., Hardebo, J. E., and Owman, C. (1990a). Origins and pathways of choline acetyltransferase-positive parasympathetic nerve fibers to cerebral vessels in rat. *J. Cereb. Blood Flow Metab. 10,* 399–408.

Suzuki, N., Hardebo, J. E., Kahrstrom, J., and Owman, C. (1990b). Neuropeptide Y co-exists with vasoactive intestinal polypeptide and acetylcholine in parasympathetic cerebrovascular nerves originating in the sphenopalatine, otic, and internal carotid ganglia of the rat. *Neuroscience 36,* 507–519.

Suzuki, Y., McMaster, D., Lederis, K., and Rorstad, O. P. (1984). Characterization of the relaxant effects of vasoactive intestinal peptide (VIP) and PHI on isolated brain arteries. *Brain Res. 322,* 9–16.

Suzuki, Y., McMaster, D., Huang, M., Lederis, K., and Rorstad, O. P. (1985). Characterization of functional receptors for vasoactive intestinal peptide in bovine cerebral arteries. *J. Neurochem. 45,* 890–899.

Suzuki, Y., Huang, M., Lederis, K., and Rorstad, O. P. (1988). The role of adenylate cyclase in relaxation of brain arteries: Studies with forskolin. *Brain Res. 457,* 241–245.

Suzuki, Y., Satoh, S., Lederis, K., and Rorstad, O. P. (1991). Cerebral vascular effects of peptide histidine methionine in canine and bovine species. *Pharmacology 42,* 241–245.

Szewczak, S. M., Behar, J., Billett, G., Hillemeier, C., Rhim, B. Y., and Biancani, P. (1990). VIP-induced alterations in cAMP and inositol phosphates in the lower esophageal sphincter. *Am. J. Physiol. 259,* G239–G244.

Taniguchi, T., Tsukahara, T., and Fujiwara, M. (1983). Evidence for muscarinic cholinergic receptors in dog portal vein: Binding of [^{3}H]quinuclidinyl benzilate. *Life Sci. 32,* 1757–1763.

Thom, S., Hughes, A., Martin, G., and Sever, P. S. (1987). Endothelium-dependent relaxation in isolated human arteries and veins. *Clin. Sci. 73,* 547–552.

Thwaites, D. T., Young, J., Thorndyke, M. C., and Dimaline, R. (1989). The isolation and chemical characterization of a novel vasoactive intestinal peptide-related peptide from a teleost fish, the cod, Gadus morhua. *Biochim. Biophys. Acta 999,* 217–220.

Tsukahara, T., Usui, H., Taniguchi, T., Shimohama, S., Fujiwara, M., and Handa, H. (1986). Characterization of muscarinic cholinergic receptors in human and dog cerebral arteries. *Stroke 17,* 300–305.

Tsukahara, T., Hongo, K., Kassell, N. F., and Ogawa, H. (1989a). Characterization of muscarinic cholinergic receptors on the endothelium and the smooth muscle of the rabbit thoracic aorta. *J. Cardiovasc. Pharmacol. 13,* 870–878.

Tsukahara, T., Kassell, N. F., Hongo, K., Vollmer, D. G., and Ogawa, H. (1989b). Muscarinic cholinergic receptors on the endothelium of human cerebral arteries. *J. Cereb. Blood Flow Metab. 9,* 748–753.

Uddman, R., Alumets, J., Edvinsson, L., Hakanson, R., and Sundler, F. (1981). VIP nerve fibres around peripheral blood vessels. *Acta Physiol. Scand. 112,* 65–70.

Uemura, Y., Sugimoto, T., Kikuchi, H., and Mizuno, N. (1987). Changes of vasoactive intestinal polypeptide-like immunoreactivity in cerebrovascular nerve fibers after cervical sympathectomy in the dog. *Neurosci. Lett. 82,* 6–10.

Usui, H., Fujiwara, M., Tsubomura, T., Kurahashi, K., Nomura, S., and Mizuno, N. (1986). Possible involvement of prostanoids in cholinergic contractile response of canine cerebral artery. *In* "Neural Regulation of Brain Circulation." (C. Owman and J. E. Hardebo, eds.), pp. 261–272. Elsevier Science Publishers, Amsterdam.

van Charldorp, K. I., and van Zwieten, P. A. (1989). Comparison of the muscarinic receptors in the coronary artery, cerebral artery and atrium of the pig. *Naunyn-Schmiedeberg's Arch Pharmacol. 339,* 403–408.

Watson, S., and Abbott, A. (1991). TiPS Receptor Nomenclature Supplement 1991. Trends in Pharmacological Science, Vol. 12, Elsevier Science Publishers, Cambridge, U.K.

Wei, E. P., Kontos, H. A., and Said, S. I. (1980). Mechanism of action of vasoactive intestinal polypeptide on cerebral arterioles. *Am. J. Physiol. 239,* H765–H768.

Weihe, E., Reinecke, M., and Forssmann, W. G. (1984). Distribution of vasoactive intestinal polypeptide-like immunoreactivity in the mammalian heart. Interrelation with neurotensin- and substance P-like immunoreactive nerves. *Cell Tissue Res. 236,* 527–540.

Yamanaka, K., Hashimoto, S., and Muramatsu, I. (1986). Species difference in the muscarinic receptors of thoracic aorta. *J. Pharmacol. Exp. Ther. 237,* 980–986.

Part IV

Sensory Afferent System

CHAPTER 12

Pattern of Innervation of Sensory Afferent Systems in the Circulation: Immunocytochemistry and Ultrastructure

SÉRGIO GULBENKIAN JOHN WHARTON

I. Introduction

The sensory nervous system is involved in the evaluation of environmental changes related to bodily functions and interacts closely with components of the autonomic nervous system. However, relatively little anatomical data are available on the sensory innervation of the vascular system. In fact, the exact localization of sensory neurons as well as the structure of their nerve endings and the type of neurotransmitter(s) used are often a matter of speculation. Sensory cell bodies are situated in cranial and spinal ganglia and have two long neuronal processes, one directed peripherally to its target organ and one centrally, towards higher integrative levels. It is assumed that the peripheral projections of sensory neurons collect the information whereas the central branches convey it to the central nervous system. This simplistic view was questioned following the demonstration that arteriolar dilatation and plasma protein extravasation were induced by stimulating the peripheral part of transected dorsal roots or sensory nerves (Chapman and Goodell, 1964; Jancsó *et al.*, 1967; Pernow, 1983). These observations led to the notion that the peripheral branches of sensory neurons might be a part of local regulator systems as well as conveyors of sensory information to central centers (see Holzer, 1988).

For many years, morphological studies on the sensory innervation of the vascular system were based on methylene blue staining and silver impregnation techniques (Nonidez, 1937, 1941; Grigor'eva, 1962; Abraham, 1969). However, the results obtained were limited due to the fact that staining is inconsistent and does not allow the discrimination of nerve subpopulations. Sensory nerve terminals have also been shown to be positive for the histochemical reaction for acetylcholinesterase (AChE), the degradative enzyme for acetylcholine (Cauna, 1969; Cauna *et al.*, 1969; Tranum-Jensen, 1979), thus implying that the AChE stain could not be used as a specific marker for afferent nerve fibers.

Significant advances have been made in the morphological analysis of presumptive afferent nerve endings in specific arterial baroreceptor and chemoreceptor areas by the introduction of electron-microscopical methods (Rees, 1967; Kondo, 1971; Chiba, 1972; Kraus, 1979; Pallot, 1983).

However, only with the recent use of immunocytochemical techniques has our knowledge on the distribution of sensory afferent nerve fibers supplying the vasculature expanded considerably. These techniques have allowed the demonstration of a variety of peptides within these nerves (Wharton and Gulbenkian, 1987; Owman, 1988).

In this paper, the main ultrastructural features of sensory nerve terminals innervating blood vessels are described. Then, the immunocytochemical evidence suggesting that several neuropeptides may be transmitters and/or modulator substances in the afferent innervation of the mammalian vascular system is focused. Specific arterial baroreceptor (e.g., aortic and carotid sinuses) and chemoreceptor areas (e.g., aortic, pulmonary, and carotid bodies) are not discussed.

II. General Model of Innervation of Blood Vessels

Nerve fibers supplying blood vessels are generally arranged in two plexuses: an outer plexus usually located in the middle of the adventitia and an inner plexus, or terminal ground plexus, which derives from the former and is typically restricted to the adventitial-medial border (Burnstock *et al.*, 1980). However, in some vessels, such as the sheep carotid artery (Keatinge, 1966), the rabbit saphenous artery (Bevan and Purdy, 1973) and cutaneous veins (Ehinger *et al.*, 1966), nerves are found within the smooth muscle media layer. Generally, terminal parts of the nerves innervating blood vessels generally consist of nonmyelinated axons surrounded by a Schwann cell sheath and exhibit, in the vicinity of the adventitial-medial border, periodic swellings or varicosities. These nerve varicosities are partly free of a Schwann cell sheath in the area directed towards the adjacent smooth muscle layer and are generally filled with secretory vesicles.

III. Ultrastructure of Vascular Sensory Nerve Terminals

The presence of an unusual abundance of mitochondria within nerve varicosities has been repeatedly shown to be one of the most distinguishing features of the sensory nerve terminals (Fig. 1). Nerve endings showing these ultrastructural characteristics have been observed in cerebral (Iwayama *et al.*, 1970), coronary (Gulbenkian *et al.*, 1990), and mesenteric arteries (Fig. 2). In rat cerebral arteries, mitochondria-rich nerve varicosities were interpreted as sensory in nature because they could be traced back to myelinated fibers (Hagen and Wittkowski, 1969; Burnstock *et al.*, 1970) and did not degenerate after sympathetic denervation by removal of the superior cervical ganglion (Iwayama *et al.*, 1970).

The presumptive sensory nerve terminals supplying the vasculature may contain, beside a large number of mitochondria, small clear and/or dense-cored vesicles

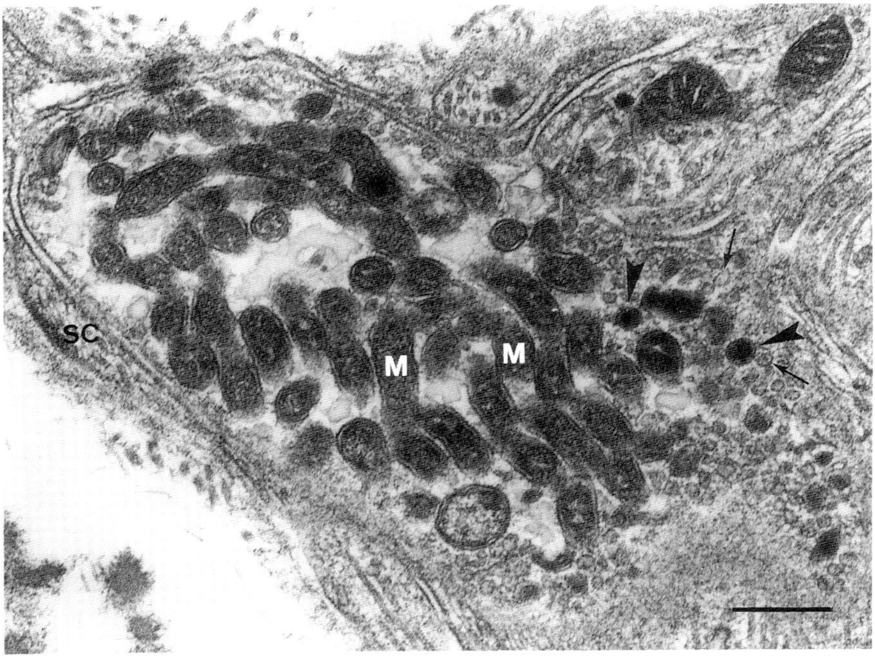

Figure 1 Electron micrograph of an axon varicosity in the adventitia of a human coronary artery. The varicosity contains numerous elongated mitochondria (M) together with large granular vesicles *(arrowheads)* and small granular and agranular vesicles *(arrows)*. The varicosity is partly ensheathed by a Schwann cell (SC). Bar = 400 nm.

(30–60 nm in diameter) and large dense-cored vesicles (80–150 nm in diameter) (Figs. 1 and 3B), autophagic vacuoles and pleomorphic dense bodies (Fig. 2), large amounts of glycogen-like granules (Fig. 3A), and myelin figures (Fig. 3B). These types of varicosities have been postulated to represent nerve specializations for pressure or tension reception based on the structural analogy they share with nerve endings in several specialized mechanoreceptors, such as muscle spindles (Merrillees, 1960; Landon, 1972), Golgi tendon organs (Schoultz and Swett, 1972), Pacini bodies (Nishi *et al.*, 1969; Spencer and Shaumburg, 1973), Meissner bodies (Cauna and Ross, 1960), and those in the wall of the carotid sinus (Chiba, 1972; Bock and Gorgas, 1976; Shin *et al.*, 1987).

The large number of mitochondria within sensory nerve endings is generally interpreted as indicating a high metabolic rate of mechanoreceptors (Bock and Gorgas, 1976; Krauhs, 1979). Further support for this hypothesis comes from a study by Hanker *et al.* (1973) demonstrating that mitochondria in sensory nerve endings expressed a high level of cytochrome oxidase activity.

According to recent theories concerning mechanoreceptor function (see Shepherd, 1988), mechanical stress exerted on proteins within the nerve membrane leads

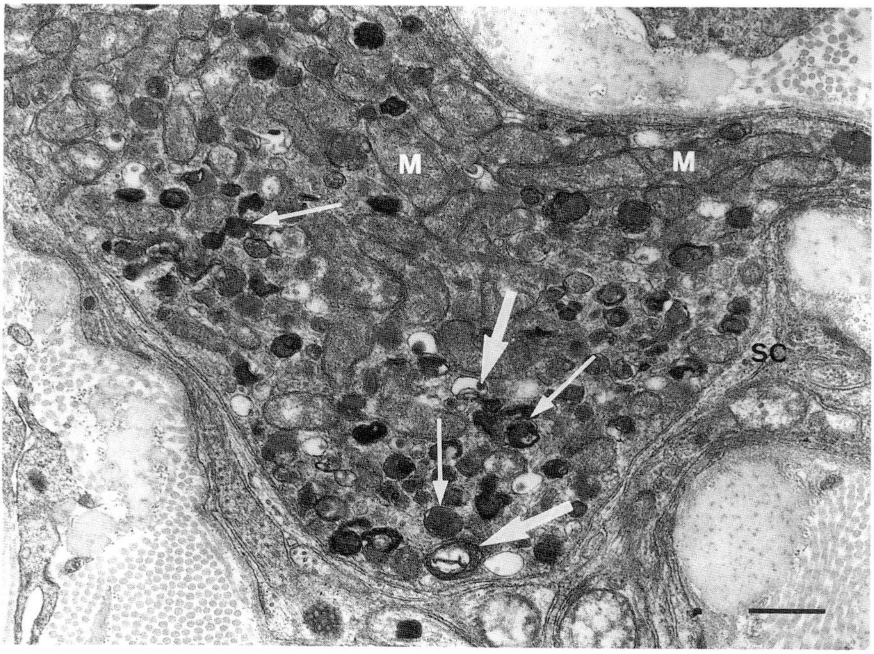

Figure 2 Guinea pig superior mesenteric artery. Electron micrograph of an axon varicosity filled with mitochondria (M), pleomorphic dense bodies *(small arrows)*, and autophagic vacuoles *(large arrows)*. SC, Schwann cell. Bar = 400 nm.

to changes in their molecular conformation and in conductivity of channel proteins. The inflow of ions produces, thereby, a receptor potential. The process of restoring a receptor potential may require large amounts of energy and be associated with a rapid turnover of mitochondria.

The myelin figures, autophagic vacuoles, broken mitochondria, and pleomorphic dense bodies observed in presumptive sensory nerve terminals are thought to indicate processes of degeneration. It has been suggested that peripheral receptors are subjected to a continuous process of replacement and reorganization of neural material (Cauna and Ross, 1960; Tranum-Jensen, 1975; Bock and Gorgas, 1976; Krauhs, 1979). Although little is known about the meaning of the presence of glycogen-like granules within mitochondria-rich varicosities, this feature is likely to be related to metabolism of the sensory nerve ending.

IV. Neuropeptides in Sensory Afferent Systems

To date, of the neuropeptides identified in afferent perivascular nerve fibers, the most widely distributed are calcitonin gene-related peptide (CGRP) and the tachykinins; from functional evidence it appears that at least some of these substances play a transmitter and/or modulator role (Franco-Cereceda, 1988; Holzer, 1988).

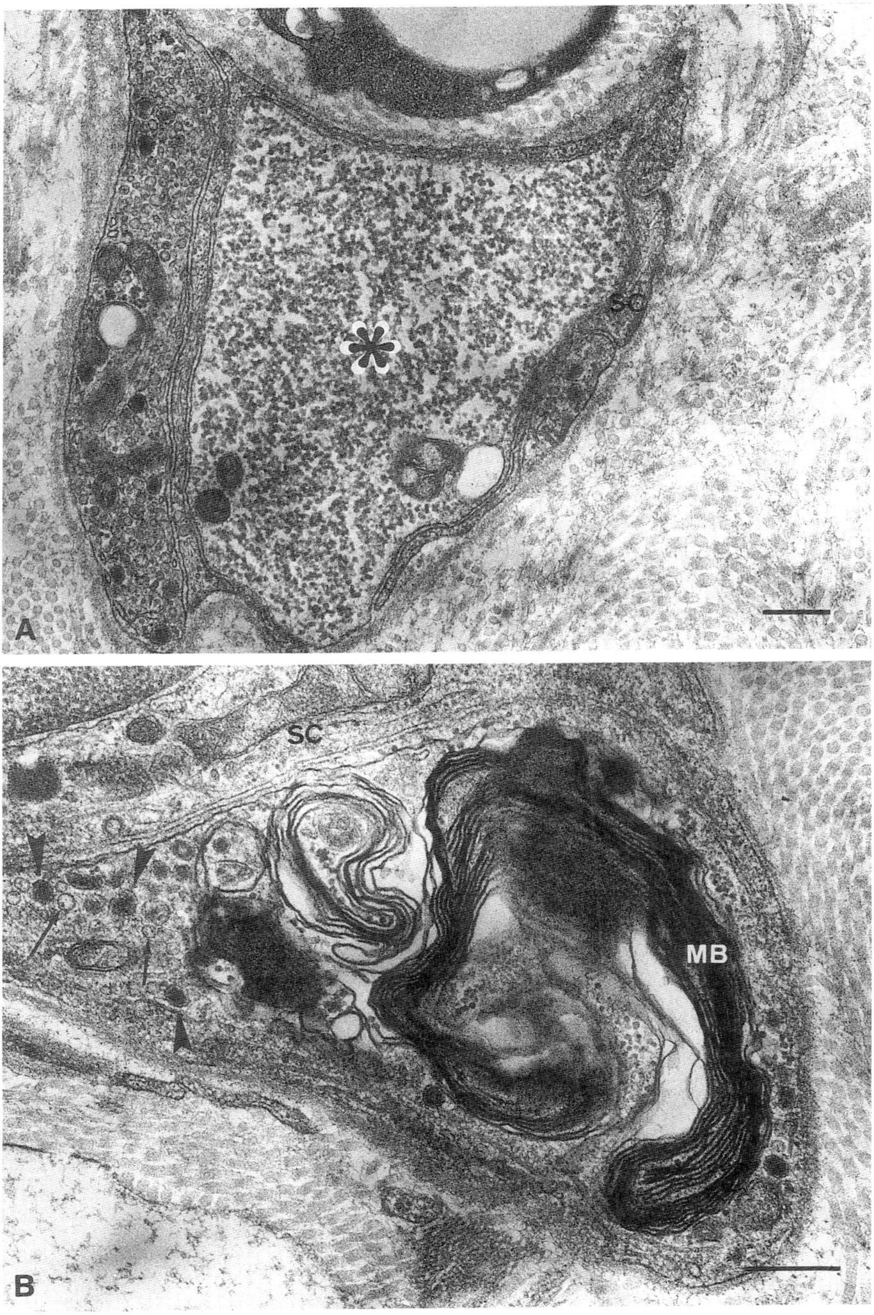

Figure 3 Electron micrographs of axon varicosities in the adventitia of a human temporal artery. These axon varicosities contain large amounts of glycogen-like granules (*asterisk;* Fig. 3A) or large myelin bodies (MB; Fig. 3B) together with some large granular vesicles *(arrowheads)* and small clear vesicles *(arrows)*. SC, Schwann cell. Bar = 400 nm.

A. Tachykinins

The tachykinins are a family of peptides that share a common C-terminal amino acid sequence, Phe-X-Gly-Leu-Met-NH_2 (Erspamer, 1981). Substance P was thought to be the only mammalian tachykinin until three new tachykinins have been recently isolated from porcine spinal cord and brain and named neurokinin A, neurokinin B, and neuropeptide K (Kimura *et al.*, 1983; Tatemoto *et al.*, 1985). The primary structure of two preprotachykinins has also been determined, one containing the substance P sequence (α-preprotachykinin) and the other containing the substance P sequence as well as that for neurokinin A and neuropeptide K (β-preprotachykinin) (Nawa *et al.*, 1984; Tatemoto *et al.*, 1985).

1. Immunocytochemical Distribution

Substance P is an 11 amino acid peptide that was originally isolated from equine brain and intestine by Von Euler and Gaddum (1931). Knowledge of its sequence and its subsequent synthesis enabled specific antisera to be raised and led to a considerable amount of data on the localization and potential functions of substance P in specific neuronal populations.

The two most striking features of the immunocytochemical localization of this neuropeptide in the periphery are that practically all blood vessels of the body are supplied by nerve fibers containing substance P and those nerve fibers seem to be peripheral branches of primary sensory neurons. The morphological evidence that substance P-containing perivascular nerve fibers are sensory in origin comes from immunocytochemical studies showing that this neuropeptide occurs in known sensory neurons and their peripheral projections (Harmer *et al.*, 1980; Matsuyama *et al.*, 1984; Urban and Papka, 1985).

The sensory origin of substance P-immunoreactive nerve fibers has also been substantiated by using the selective sensory neurotoxin capsaicin. Systemic capsaicin treatment of guinea pigs and rats leads to a marked loss of substance P immunoreactivity in the vasculature (Furness *et al.*, 1982; Barja *et al.*, 1983; Wharton *et al.*, 1986; Gulbenkian *et al.*, 1987). The substance P immunoreactivity is, however, unaffected by 6-hydroxydopamine, which degenerates the noradrenergic nerves to the heart and blood vessels (Furness *et al.*, 1982).

In some species, such as the guinea pig, practically every vascular bed is supplied with capsaicin-sensitive substance P-immunoreactive nerve fibers (Furness *et al.*, 1982). The highest density of substance P-immunoreactive nerve networks is found in large vessels near the heart (pulmonary trunk, aortic arch, thoracic aorta, common carotids, vena cava) and, in general, the density of the innervation decreases as conducting vessels are followed away from the heart. This pattern of reduction of nerve density with declining vessel diameter is less apparent in mesenteric arteries and their branches to the viscera (Edvinsson *et al.*, 1989), which are more densely supplied than are other arteries of comparable size, for example, the subclavian and iliac arteries. Vessels supplying the reproductive and urinary organs also possess a moderately dense nerve supply, as do the cerebral (Yamamoto *et al.*, 1983;

Nakakita, 1990) and coronary arteries (Gulbenkian *et al.*, 1990). On the venous side, there is a rich supply of substance P-containing nerves to the vena cava near the heart, but other veins are innervated very sparsely, or not at all.

As with other types of peptide-containing perivascular nerves, there are species variations in the distribution of substance P-immunoreactivity. In contrast to the guinea pig, the rat vascular system contains significantly fewer substance P-immunoreactive nerve fibers. In the rat, substance P-immunoreactive perivascular nerve fibers also show marked variations in their sensitivity to capsaicin (Matsuyama *et al.*, 1986; Wharton *et al.*, 1986). In man, nerves containing substance P immunoreactivity have been detected in several vascular beds (Weihe *et al.*, 1981; Edvinsson *et al.*, 1987; Franco-Cereceda *et al.*, 1987; Allen *et al.*, 1989; Wharton *et al.*, 1990), but, in general, they appear to have a relatively sparse distribution. In the cat, however, a relatively dense nerve supply of substance P-immunoreactive nerve fibers has been shown to occur around the cerebral vasculature (Fig. 4), with anterior vessels in the Circle of Willis receiving a more dense supply than those in the posterior circulation (Edvinsson *et al.*, 1981; Liu-Chen *et al.*, 1983).

Recent immunocytochemical studies have indicated that several tachykinins (i.e., substance P, neurokinin A, and neuropeptide K) coexist in capsaicin-sensitive neurons (Fig. 5). Peripheral projections of those sensory neurons have been found associated with blood vessels supplying various organs, such as the brain, heart,

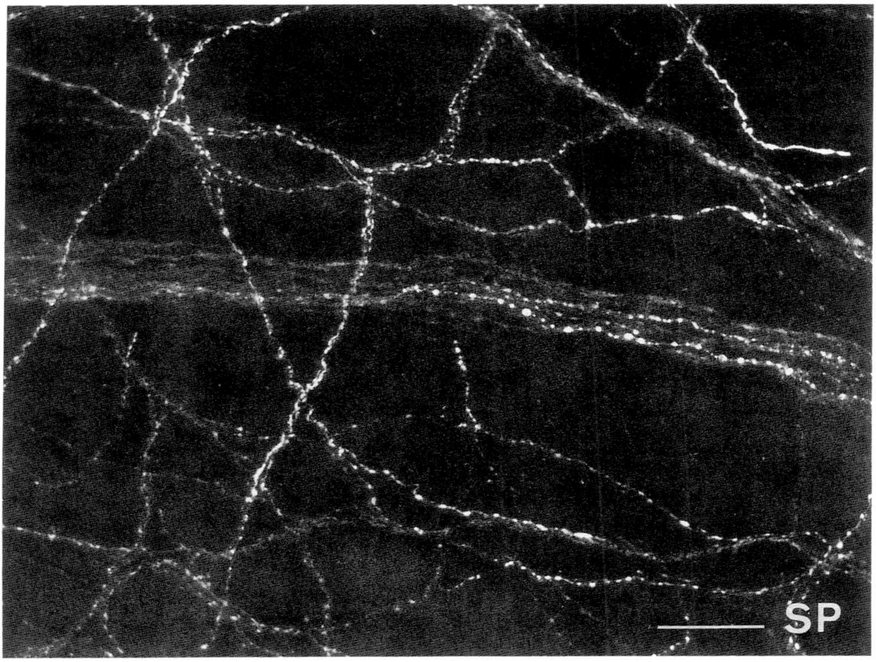

Figure 4 Whole mount preparation of a cat middle cerebral artery that contains a network of varicose nerve fibers displaying substance P (SP) immunoreactivity. Bar = 50 μm.

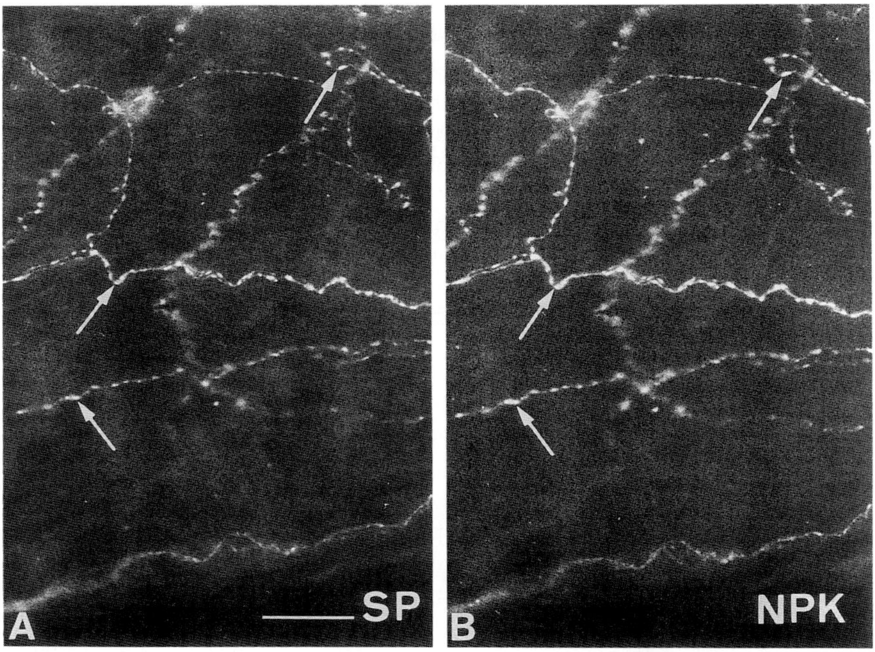

Figure 5 Whole mount preparation of a guinea pig left anterior descending coronary artery immunostained for (A) substance P (SP) and (B) neuropeptide K (NPK). The colocalization of SP with NPK was determined by double immunostaining the same preparation. Varicose nerve fibers *(arrows)* displaying SP-immunofluorescence staining also possess NPK immunoreactivity. Bar = 50 μm.

lungs, and gut (Hua *et al.*, 1985; Edvinsson *et al.*, 1988, 1989; Gulbenkian *et al.*, 1990).

B. Calcitonin Gene-Related Peptide (CGRP)

CGRP is a 37 amino acid peptide encoded by the calcitonin gene and its expression is dependent on tissue-specific alternative RNA processing (Amara *et al.*, 1982). Two forms of CGRP that are derived from related α- and β-genes have been identified in both the rat (Amara *et al.*, 1985) and man (Steenbergh *et al.*, 1985).

1. Immunocytochemical Distribution

From immunocytochemical studies on the distribution of CGRP in the central and peripheral nervous systems, it has been suggested that this peptide might subserve a sensory function. CGRP immunostaining has been localized in a population of primary sensory neurons in spinal and cranial ganglia as well as in sensory-receiving areas of the spinal cord (Rosenfeld *et al.*, 1983; Gibson *et al.*, 1984). It has also been shown that CGRP immunoreactivity often coexists with substance P immunoreactivity in a population of sensory ganglion cells in the rat and guinea pig (Gibson *et al.*, 1984; Wiesenfeld-Hallin *et al.*, 1984; Gibbins *et al.*, 1985).

After capsaicin treatment both CGRP and substance P immunoreactivity are depleted from peripheral projections of sensory neurons supplying the vasculature (Gibbins *et al.*, 1985; Wharton *et al.*, 1986; Uddman *et al.*, 1986). The degree of coexistence of the two peptides appears to be very high in the guinea pig, with some 90% of nerve cell bodies in the dorsal root ganglia displaying both CGRP and substance P immunoreactivity (Gibbins *et al.*, 1985). In the rat, CGRP immunoreactivity appears to occur in two different populations of sensory neurons that vary in their soma size (Gibson *et al.*, 1984; Lee *et al.*, 1985). One population of sensory neurons (small to medium-sized cells) contains both CGRP and substance P and is sensitive to capsaicin, whereas the other (large cells) possesses only CGRP and does not appear to be affected by the sensory neurotoxin (Matsuyama *et al.*, 1986).

At the ultrastructural level, the use of double immunogold staining procedures has allowed to demonstrate that CGRP and substance P immunoreactivities are consistently found colocalized in the same large granular vesicles (70–150 nm in diameter) in both sensory ganglion cells and varicosities of perivascular nerve fibers (Fig. 6) of the guinea pig (Gulbenkian *et al.*, 1986; Edvinsson *et al.*, 1989). These

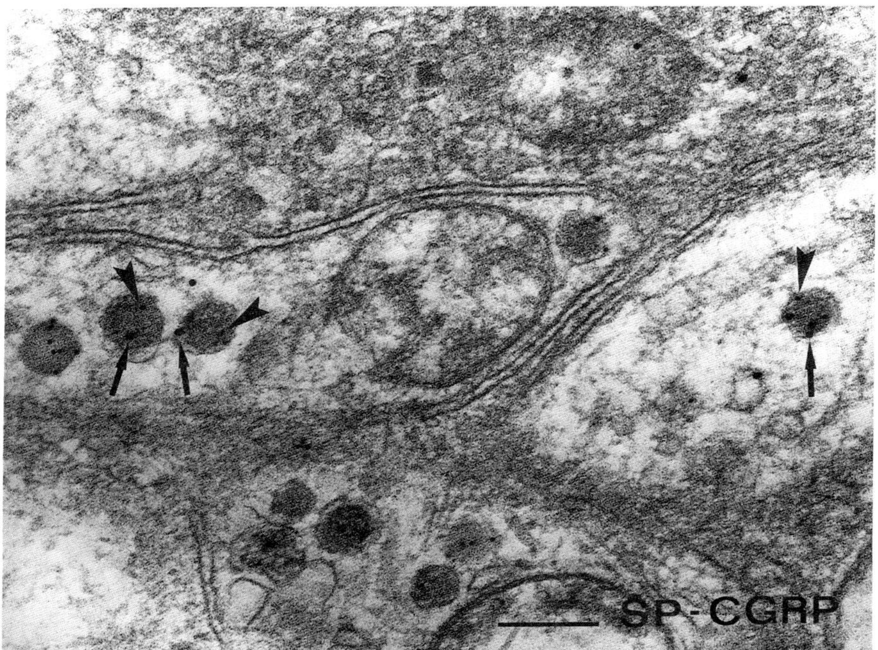

Figure 6 Electron micrograph of axon varicosities in the adventitia of a guinea pig superior mesenteric artery. Colocalization of substance P (SP) and CGRP immunoreactivity in the same large granular vesicles is demonstrated by double immunogold staining. SP, 10-nm gold particles *(arrowheads);* CGRP, 15-nm gold particles *(arrows).* Bar = 200 nm.

vesicles are ultrastructurally different from both the small, granular, adrenergic vesicles and the small, agranular, cholinergic, vesicles (30–60 nm in diameter) and are characterized as p-type (peptidergic-type) vesicles (Baumgarten *et al.*, 1970). Furthermore, the demonstration that substance P, neurokinin A, and neuropeptide K are derived from a common precursor molecule (e.g., Section IV.A) suggests that several peptides of the tachykinin family may be costored with CGRP in the same p-type vesicles.

The evidence that perivascular CGRP-containing nerve fibers are sensory in origin is also based on the use of selective surgical lesions to denervate peripheral targets of their sensory innervation. Thus, lesions to the trigeminal ganglia produces the depletion of both CGRP and substance P immunoreactivity from nerve fibers supplying cerebral blood vessels of the rat (Matsuyama *et al.*, 1986), guinea pig (Wanaka *et al.*, 1986), and cat (Uddman *et al.*, 1985).

In the guinea pig and rat, where an extensive survey has been made (Mulderry *et al.*, 1985; Wharton *et al.*, 1986; Uddman *et al.* 1986), and in other species, where less systematic studies have been undertaken, CGRP-immunoreactive nerves have a very widespread distribution throughout the vascular system. Like for substance P (e.g., Section IV.A), CGRP-immunoreactive nerve fibers are more abundant in the guinea pig than in the rat. Furthermore, marked regional variations are observed in the density of the CGRP-immunoreactive innervation in both species (Table 1). In the guinea pig, the number and distribution of CGRP-immunoreactive nerve fibers closely parallel that of fibers containing substance P, the two immunoreactivities being found invariably to coexist in the same perivascular nerve networks. In the rat, CGRP-immunoreactive nerve fibers have also a similar distribution to those containing substance P, however, in contrast to the guinea pig, the former are far more numerous. Human blood vessels generally display a sparse (CGRP-immunoreactive innervation (Fig. 7).

V. Conclusions

The use of immunocytochemical techniques at both the light and electron microscopical levels has allowed to demonstrate that CGRP and substance P occur in specific subpopulations of afferent nerves supplying the mammalian vascular system. These neuropeptides arise from cell bodies located in sensory ganglia and often coexist in capsaicin-sensitive afferent nerve terminals. CGRP- and substance P-immunoreactive nerve fibers are widespread throughout vascular tissues and exhibit both regional and species variations in their distribution. CGRP and substance P immunoreactivities occur extensively in the vascular innervation of the guinea pig, rat, and cat. However, in man, relatively few nerve fibers appear to display immunoreactivity for those neuropeptides.

There is also evidence that in addition to substance P, other tachykinins (neurokinin A and neuropeptide K) may also be present in capsaicin-sensitive nerves and it is conceivable that more than one tachykinin may be synthesized and released together with CGRP from the same sensory neuron. This is further supported by the

Table 1 Relative Densities of Supply of CGRP-Immunoreactive Nerve Fibers in Arteries and Veins of the Guinea Pig and Rat[a]

Artery	Guinea pig	Rat
Cerebral	++	++
Carotid	+++	+
Subclavian	+	+
Coronary	++	+
Pulmonary trunk	+++	++
Intrapulmonary	++	+
Aortic arch	+++	+
Thoracic aorta	+++	+
Abdominal aorta	++	+
Superior mesenteric	++++	+++
Mesenteric (near gut)	+++	+++
Femoral	+	+
Renal	++	++
Uterine	++	++
Testicular	++	++
Vein	**Guinea pig**	**Rat**
Jugular	+	0
Superior vena cava	++	+
Inferior vena cava	++++	+
Abdominal vena cava	+	+
Mesenteric	++	+
Iliac	+	+
Hepatic portal	+	++

[a] *Abbreviations:* 0, no immunostained fibers detected; +, nerve fibers rarely detected; ++, scattered nerve fibers; +++, moderate number of fibers; ++++, numerous nerve fibers.

electron microscopical immunocytochemical observation that CGRP, substance P, and neuropeptide K are co-stored in the same p-type vesicles in both sensory ganglia cells and perivascular nerve terminals of the guinea pig. A few p-type vesicles are also observed within the mitochondria-rich, presumptive mechanoreceptor nerve terminals supplying a variety of blood vessels (e.g., Section III) and it is possible that they contain the putative sensory transmitters CGRP and tachykinins.

The functional significance of the presence of multiple putative peptide transmitters within afferent nerve fibers supplying the vasculature has yet to be determined. However, if CGRP and several tachykinins coexist in the same secretory vesicles and are coreleased, it is likely that interactions between them will occur. Support for this hypothesis comes from pharmacological experiments showing that, for instance, CGRP can potentiate the plasma protein extravasation induced by tachykinins in neurogenic inflammation (Gamse and Saria, 1985). It is suggested that

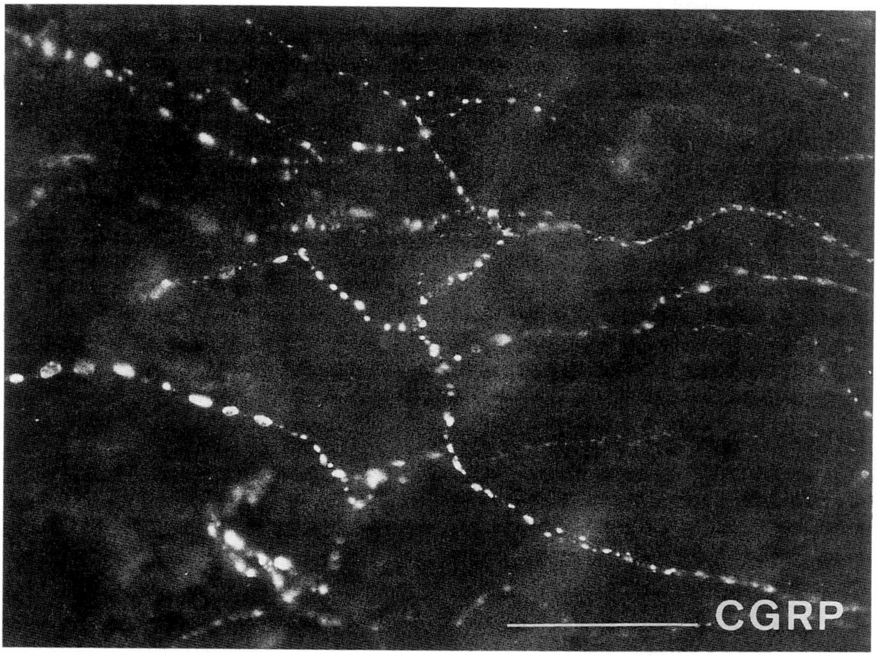

Figure 7 CGRP immunoreactivity localized to varicose nerve fibers in a whole mount preparation of a human left circumflex coronary artery. Bar = 50 μm.

vasodilatation (Brain and Williams, 1985), inhibition of substance P degradation (Le Greves *et al.*, 1985), and potentiation of the substance P release (Oku *et al.*, 1987) could be part of the mechanisms by which CGRP potentiates the response to tachykinins.

Finally, it seems that afferent neurons, in addition to transmitting information for centrally mediated reflexes, may also be involved in the regulation of vascular tone by local release of tachykinins and/or CGRP.

References

Abraham, A. (1969). "Microscopic innervation of the heart and blood vessels in vertebrates including man." Pergamon Press, Oxford.

Allen, K., Wharton, J., Polak, J. M., and Haworth, S. (1989). A study of nerves containing peptides in the pulmonary vasculature of healthy infants and children and of those with pulmonary hypertension. *Br. Heart J. 62,* 353–360.

Amara, S. G., Jonas, V., Rosenfeld, M. G., Ong, E. S., and Evans, R. M. (1982). Alternative RNA processing in calcitonin gene expression generates mRNAs encoding different polypeptide products. *Nature 298,* 240–244.

Amara, S. G., Arriza, J. L., Leff, S. E., Swanson, L. W., Evans, R. M., and Rosenfeld, M. G. (1985). Expression in brain of a messenger RNA encoding a novel neuropeptide homologous to calcitonin gene-related peptide. *Science 229,* 1094–1097.

Barja, F., Mathison, R., and Huggel, H. (1983). Substance P-containing nerve fibres in large peripheral blood vessels of the rat. *Cell Tiss. Res. 229,* 411–422.

Baumgarten, H. G., Holstein, A. F., and Owman, C. H. (1970). Auerbach's plexus of mammals and man: Electron microscopical identification of three different types of neuronal processes in myenteric ganglia of the large intestine from rhesus monkeys, guinea pigs and man. *Z. Zellforsch 106,* 376–397.

Bevan, J. A. and Purdy, R. E. (1973). Variations in adrenergic innervation and contractile responses of the rabbit saphenous artery. *Circ. Res. 32,* 746–751.

Bock, P., and Gorgas, K. (1976). Fine structure of baroreceptor terminals in the carotid sinus of guinea pigs and mice. *Cell Tiss. Res. 170,* 95–112.

Brain, S. D., and Williams, T. J. (1985). Inflammatory oedema induced by synergism between calcitonin gene-related peptide (CGRP) and mediators of increased vascular permeability. *Br. J. Pharmacol. 86,* 855–860.

Burnstock, G., Gannon, B., and Iwayama, T. (1970). Sympathetic innervation of vascular smooth muscle in normal and hypertensive animals. *Circ. Res. 26,27* (Suppl. 11), 5–24.

Burnstock, G., Chamley, J. H., and Campbell, G. R. (1980). The innervation of arteries. *In* "Structure and function of the circulation." (J. Schwartz, N. T. Werthessen, and S. Wolf, eds.), Vol. 1, pp. 729–767. Plenum Publishing Corp., New York.

Cauna, N. (1969). The fine morphology of the sensory receptor organs in the auricle of the rat. *J. Comp. Neurol. 136,* 81–98.

Cauna, N., and Ross, L. L. (1960). The fine structure of Meissner's touch corpuscules of human fingers. *J. Biophys. Biochem. Cytol. 8,* 467–482.

Cauna, N., Hinderer, K. H., and Wentges, R. T. (1969). Sensory receptor organs of the human nasal respiratory mucosa. *Am. J. Anat. 124,* 187–209.

Chapman, L. F., and Goodell, H. (1964). The participation of the nervous system in the inflammatory reaction. *Ann. N.Y. Acad. Sci. 116,* 990–1017.

Chiba, T. (1972). Fine structure of the baroreceptor nerve terminals in the carotid sinus of the dog. *J. Electron. Microsc. 21,* 139–148.

Edvinsson, L., McCulloch, H., and Uddman, R. (1981). Substance P: Immunohistochemical localization and effect upon cat pial arteries *in vitro* and *in situ. J. Physiol. 318,* 251–258.

Edvinsson, L., Ekman, R., Jansen, I., Ottosson, A., and Uddman, R. (1987). Peptide-containing nerve fibres in human cerebral arteries: Immunocytochemistry, radioimmunoassay and *in vitro* pharmacology. *Ann. Neurol. 21,* 431–437.

Edvinsson, L., Brodin, E., Jansen, I., and Uddman, R. (1988). Neurokinin A in cerebral vessels: Characterization, localization and effects *in vitro. Reg. Peptides 20,* 181–197.

Edvinsson, L., Gulbenkian, S., Jansen, I., Wharton, J., Cervantes, C., and Polak, J. M. (1989). Comparison of peptidergic mechanisms in different parts of the guinea pig superior mesenteric artery: Immunocytochemistry at the light and ultrastructural levels and responses *in vitro* of large and small arteries. *J. Auton. Nerv. Syst. 28,* 141–154.

Ehinger, G., Falck, B., and Sporrong, B. (1966). Adrenergic fibres to the heart and peripheral vessels. *Bibl. Anat. 8,* 35–45.

Erspamer, V. (1981). The tachykinin peptide family. *Trends Neurosci. 4,* 267–269.

Franco-Cereceda, A. (1988). Calcitonin gene-related peptide and tachykinins in relation to local sensory control of cardiac contractility and coronary vascular tone. *Acta Physiol. Scand. 113* (Suppl. 569), 3–63.

Franco-Cereceda, A., Henke, H., Lundberg, J. M., Petermann, J. B., Hökfelt, T., and Fischer, J. A. (1987). Calcitonin gene-related peptide (CGRP) in capsaicin-sensitive substance P-immunoreactive sensory neurons in animals and man: Distribution and release by capsaicin. *Peptides 8,* 399–410.

Furness, J. B., Papka, R. E., Della, N. G., Costa, M., and Eskay, R. L. (1982). Substance P-like immunoreactivity in nerves associated with the vascular system of guinea pigs. *Neuroscience 7,* 447–459.

Gamse, R., and Saria, A. (1985). Potentiation of tachykinin-induced plasma protein extravasation by calcitonin gene-related peptide. *Eur. J. Pharmacol 114,* 61–6.

Gibbins, I. L., Furness, J. B., Costa, M., MacIntyre, I., Hillyard, C. J., and Girgis, S. (1985). Co-localisation of calcitonin gene-related peptide-like immunoreactivity with substance P in cutaneous, vascular and visceral sensory neurons of guinea pigs. *Neurosci. Lett 57,* 125–130.

Gibson, S. J., Polak, J. M., Bloom, S. R., Sabate, I. M., Mulderry, P. M., Ghatei, M. A., McGregor, G. P., Morrison, J. F. B., Kelly, J. S., Evans, R. M., and Rosenfeld, M. G. (1984). Calcitonin gene-related peptide immunoreactivity in the spinal cord of man and of eight other species. *J. Neurosci. 4,* 3101–3111.

Grigor'eva, T. A. (1962). "The innervation of blood vessels." Pergamon Press, Oxford.

Gulbenkian, S., Merighi, A., Wharton, J., Varndell, I. M., and Polak, J. M. (1986). Ultrastructural evidence for the coexistence of calcitonin gene-related peptide and substance P in secretory vesicles of peripheral nerves in the guinea pig. *J. Neurocytol. 15,* 535–542.

Gulbenkian, S., Wharton, J., and Polak, J. M. (1987). The visualisation of cardiovascular innervation in the guinea pig using an antiserum to protein gene product 9.5 (PGP 9.5). *J. Auton. Nerv. Syst. 18,* 235–247.

Gulbenkian, S., Edvinsson, L., Saetrum Opgaard, O., Wharton, J., Polak, J. M., and David-Ferreira, J. F. (1990). Peptide-containing nerve fibres in guinea pig coronary arteries: Immunohistochemistry, utrastructure and vasomotility. *J. Auton. Nerv. Syst. 31,* 153–168.

Hagen, E., and Wittkowski, W. (1969). Light und electronenmikroskopishe untersuchung zur innervation der piagefasse. *Z. Zellforch 95,* 429–444.

Hanker, J. S., Dixon, A. D., and Moore, H. G. (1973). Cytochrome oxidase activity of mitochondria in sensory nerve endings of mouse palatal rugae. *J. Anat. 116,* 93–102.

Harmer, A., Schofield, J. G., and Keen, P. (1980). Cycloheximide-synthesis of substance P by isolated dorsal root ganglia. *Nature 284,* 267–269.

Holzer, P. (1988). Local effector functions of capsaicin-sensitive sensory nerve endings: Involvement of tachykinins, calcitonin gene-related peptide and other neuropeptides. *Neuroscience 24,* 739–768.

Hua, X-Y., Theodorsson-Norheim, E., Brodin, E., Lundberg, J. M., and Hökfelt, T. (1985). Multiple tachykinins (neurokinin A, neuropeptide K and substance P) in capsaicin-sensitive sensory neurons in the guinea pig. *Reg. Peptides 13,* 1–19.

Iwayama, T., Furness, J. B., and Burnstock, G. (1970). Dual adrenergic and cholinergic innervation of the cerebral arteries of the rat. An ultrastructural study. *Circ. Res. 26,* 635–646.

Jancsó, N., Jancsó-Gäbor, A., and Szolcsányi, J. (1967). Direct evidence for neurogenic inflammation and its prevention by denervation and by pretreatment with capsaicin. *Br. J. Pharmacol. 31,* 138–151.

Keatinge, W. R. (1966). Electrical and mechanical responses of arteries to stimulation of sympathetic nerves. *J. Physiol.* (London) *185,* 701–715.

Kimura, S., Okada, M., Sugita, Y., Kanazawa, T., and Munekata, E. (1983). Novel neuropeptides, neurokinin α and β isolated from porcine spinal cord. *Proc. Jpn. Acad. 59B,* 101–104.

Kondo, H. (1971). An electron microscopic study on innervation of the carotid body of guinea pig. *J. Ultrastruct. Res. 37,* 544–562.

Krauhs, J. M. (1979). Structure of rat aortic baroreceptors and their relationship to connective tissue. *J. Neurocytol. 8,* 401–414.

Landon, D. N. (1972). The fine structure of the equatorial regions of developing muscle spindles in the rat. *J. Neurocytol. 1,* 189–210.

Lee, Y., Takami, K., Kawai, Y., Girgis, S., Hillyard, J., MacIntyre, I., Emson, P. C., and Tohyama, M. (1985). Distribution of calcitonin gene-related peptide in the rat peripheral nervous system with reference to its coexistence with substance P. *Neuroscience 15,* 1227–1237.

Le Greves, P., Nyberg, F., Terenius, L., and Hökfelt, T. (1985). Calcitonin gene-related peptide is a potent inhibitor of substance P degradation. *Eur. J. Pharmacol. 115,* 309–311.

Liu-Chen, L. Y., Mayberg, M. R., and Moskowitz, M. A. (1983). Immunohistochemical evidence for a substance P-containing trigeminovascular pathway to pial arteries in cats. *Brain Res. 268,* 162–166.

Matsuyama, T., Matsumoto, M., Shiosaka, S., Hayakawa, T., Yoneda, S., Kimura, K., Abe, H., and

Tohyama, M. (1984). Dual innervation of substance P-containing neuron system in the wall of the cerebral arteries. *Brain Res. 322,* 144–147.

Matsuyama, T., Wanaka, A., Yoneda, S., Kimura, K., Kamada, T., Girgis, S., MacIntyre, I., Emson, P. C., and Tohyama, M. (1986). Two distinct calcitonin gene-related peptide-containing peripheral nervous systems: Distribution and quantitative differences between the iris and cerebral artery with reference to substance P. *Brain Res. 373,* 205–212.

Merrillees, N. C. R. (1960). The fine structure of muscle spindles in the lumbrical muscles of the rat. *J. Biophys. Biochem. Cytol. 7,* 725–740.

Mulderry, P. K., Ghatei, M. A., Rodrigo, J., Allen, J. M., Rosenfeld, M. G., Polak, J. M., and Bloom, S. R. (1985). Calcitonin gene-related peptide in cardiovascular tissues of the rat. *Neuroscience 14,* 947–954.

Nakakita, K. (1990). Peptidergic innervation in the cerebral blood vessels of the guinea pig: An immunohistochemical study. *J. Cereb. Blood Flow Metabol. 10,* 819–826.

Nawa, H., Kotani, H., and Nakanishi, S. (1984). Tissue-specific generation of two preprotachykinin mRNAs from one gene by alternative RNA splicing. *Nature 312,* 729–734.

Nishi, K., Oura, C., and Pallie, W. (1969). Fine structure of Pacinian corpuscules in the mesentery of the cat. *J. Cell Biol. 43,* 539–552.

Nonidez, J. F. (1937). Identification of the receptor areas in the venae cavae and pulmonary veins which initiate reflex cardiac acceleration (Brainbrige's reflex). *Am. J. Anat. 61,* 203–231.

Nonidez, J. F. (1941). Studies on the innervation of the heart. II. Afferent nerve endings in the large arteries and veins. *Am. J. Anat. 68,* 151–189.

Oku, R., Satoh, M., Fujii, N., Otaka, A., Yajima, H., and Takagi, H. (1987). Calcitonin gene-related peptide promotes mechanical nociception by potentiating release of substance P from the spinal dorsal horn in rats. *Brain Res. 403,* 350–354.

Owman, C. (1988). Autonomic innervation of the cardiovascular system. *In* "Handbook of chemical neuroanatomy—The peripheral nervous system." (A. Bjorklund, T. Hökfelt, and C. Owman, eds.), Vol. 6, pp. 327–389. Elsevier Science Publishers B. V., Amsterdam.

Pallot, D. J. (1983). Relationship between ultrastructure and function in peripheral arterial chemoreceptors. *In* "Physiology of the peripheral arterial chemoreceptors." (H. Acker and R. G. O'Regan, eds.), pp. 1–19. Elsevier Science Publishers B. V., Amsterdam.

Pernow, B. (1983). Substance P—a putative mediator of antidromic vasodilatation. *Gen. Pharmacol. 14,* 13–16.

Rees, P. M. (1967). Observations on the fine structure and distribution of presumptive baroreceptor nerves at the carotid sinus. *J. Comp. Neurol. 131,* 517–548.

Rosenfeld, M. G., Mermod, J. J., Amara, S. G., Swanson, L. W., Rivier, J., Vale, W. W., and Evans, R. M. (1983). Production of a novel neuropeptide encoded by the calcitonin gene via tissue-specific RNA processing. *Nature 304,* 129–135.

Schoultz, T., and Swett, J. E. (1972). The fine structure of the Golgi tendon organ. *J. Neurocytol. 1,* 1–26.

Shepherd, G. M. (1988). "Neurobiology." Oxford University Press, Oxford.

Shin, H. S., Hulbert, W. C., and Biggs, D. F. (1987). Observations on the fine structure of the baroreceptors and adrenergic innervation of the guinea-pig carotid sinus. *J. Morphol. 194,* 65–74.

Spencer, P. S., and Schaumburg, H. H. (1973). An ultrastructural study of the inner core of the Pacinian corpuscule. *J. Neurocytol. 2,* 217–235.

Steenbergh, P. H., Höppener, J. W. M., Zandberg, J., Lips, C. J. M., and Jansz, H. S. (1985). A second human calcitonin/CGRP gene. *FEBS Lett. 183,* 403–407.

Tatemoto, K., Lundberg, J. M., Jörnvall, H., and Mutt, V. (1985). Neuropeptide K: Isolation, structure and biological activities of a novel brain tachykinin. *Biochem. Biophys. Res. Commun. 128,* 947–953.

Tranum-Jensen, J. (1975). The ultrastructure of the sensory end-organs (baroreceptors) in the atrial endocardium of young mini-pigs. *J. Anat. 119,* 255–275.

Tranum-Jensen, J. (1979). Ultrastructural studies on atrial nerve-end formations in mini-pigs. *In* "Cardiac receptors". (R. Mainsworth, C. Kidd, and R. J. Linden, eds.), pp. 27–50. Cambridge University Press, Cambridge.

Uddman, R., Edvinsson, L., Ekman, R., Kingman, T., and McCulloch, J. (1985). Innervation of the feline cerebral vasculature by nerve fibres containing calcitonin gene-related peptide: Trigeminal origin and coexistence with substance P. *Neurosci. Lett. 62,* 131–136.

Uddman, R., Edvinsson, L., Ekblad, E., Håkanson, R., and Sundler, F. (1986). Calcitonin gene-related peptide (CGRP): Perivascular distribution and vasodilatory effects. *Reg. Peptides 15,* 1–23.

Urban, L., and Papka, R. E. (1985). Origin of small primary afferent substance P-immunoreactive nerve fibres in the guinea pig heart. *J. Auton. Nerv. Syst. 12,* 321–331.

Von Euler, U. S., and Gaddum, J. H. (1931). An unidentified depressor substance in certain tissue extracts. *J. Physiol. 192,* 74–87.

Wanaka, A., Matsuyama, T., Yoneda, S., Kimura, K., Kamada, T., Girgis, S., MacIntyre, I., Emson, P. C., and Tohyama, M. (1986). Origins and distribution of calcitonin gene-related peptide-containing nerves in the wall of the cerebral arteries of the guinea pig with special reference to the coexistence with substance P. *Brain Res. 369,* 185–192.

Weihe, E., Reinecke, M., Opherk, D., and Forssmann, W. G. (1981). Peptidergic innervation (substance P) in the human heart. Interrelation with neurotensin- and substance P-like immunoreactive nerves. *J. Mol. Cell Cardiol. 13,* 331–333.

Wharton, J., and Gulbenkian, S. (1987). Peptides in the mammalian cardiovascular system. *Experientia 43,* 821–832.

Wharton, J., Gulbenkian, S., Mulderry, P. K., Ghatei, M. A., McGregor, G. P., Bloom, S. R., and Polak, J. M. (1986). Capsaicin induces a depletion of calcitonin gene-related peptide (CGRP)-immunoreactive nerves in the cardiovascular system of the guinea pig and rat. *J. Auton. Nerv. Syst. 16,* 289–309.

Wharton, J., Polak, J. M., Gordon, L., Banner, N. R., and Yacoub, M. H. (1990). Immunohistochemical demonstration of human cardiac innervation before and after transplantation. *Circ. Res. 66,* 900–912.

Wiesenfeld-Hallin, Z., Hökfelt, T., Lundberg, J. M., Forsmann, W. G., Reinecke, M., Tschopp, F. A., and Fisher, J. A. (1984). Immunoreactive calcitonin gene-related peptide and substance P coexist in sensory neurons in spinal cord and interact in spinal behavioural responses of the rat. *Neurosci. Lett. 52,* 199–204.

Yamamoto, K., Matsuyama, T., Shiosaka, S., Inagaki, S., Senba, E., Shimizu, Y., Ishimoto, I., Hayakawa, T., Matsumoto, M., and Tohyama, M. (1983). Overall distribution of substance P-containing nerves in the wall of the cerebral arteries of the guinea pig and its origin. *J. Comp. Neurol. 215,* 421–426.

CHAPTER 13

Calcitonin Gene-Related Peptide and Blood Vessels

INGER JANSEN LEIF NILSSON LARS EDVINSSON

I. Introduction

Blood vessels are innervated by unmyelinated and thinly myelinated primary sensory neurons containing substance P (SP), neurokinin A (NKA), and calcitonin gene-related peptide (CGRP). This class of neurons exerts a twofold control over physiological and pathophysiological functions (Szolcsányi, 1984; Lembeck, 1988) comprising transmission of sensory (afferent) signals to the CNS and release of neurotransmitters to regulate tissue functions locally (efferent) (Holzer, 1988; Maggi and Meli, 1988). The efferent function of small diameter primary sensory neurons is to participate in the regulation of local blood flow, inflammatory processes, and smooth muscle cell functions—all with or without the additional involvement of the large reflex pathways.

Each of the sensory neuropeptides *per se* induces relaxation of blood vessels. The effect is mediated through receptors coupled to different second-messenger systems, such as cyclic AMP and cyclic GMP and/or the release of endothelium-derived relaxing factors (EDRF) (Fig. 1).

There is growing evidence for the existence of different types of receptors for CGRP. Most studies of CGRP receptors have centered on the effect of CGRP analogs, using *in vitro* methods and receptor binding studies. Recent investigations have identified fragments of CGRP that exert antagonistic effects (Chiba *et al.*, 1989; Dennis *et al.*, 1989, 1990; Donoso *et al.*, 1990). These fragments have been examined in many tissues using a variety of methods. The present review examines current experience with CGRP analogs and in particular the antagonist, human α-CGRP(8–37), on vasomotor responses in different parts of the circulation.

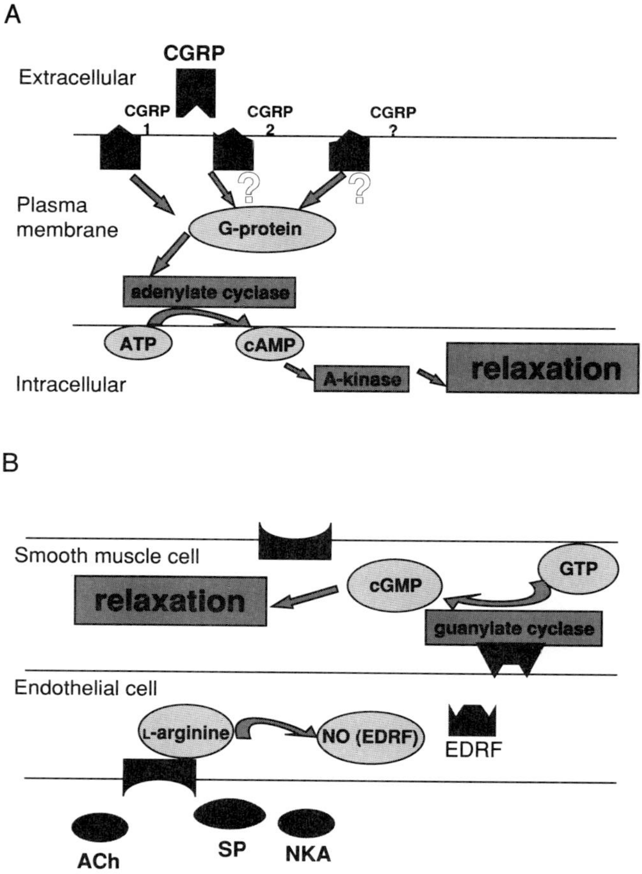

Figure 1 The effect of sensory neuropeptides on the vascular bed is mediated via different mechanisms of action. (A) CGRP receptors are coupled to a stimulatory G-protein that activates adenylate cyclase. This triggers the formation of cyclic AMP that diffuses through the cytoplasm and activates cyclic AMP-dependent protein kinase. The subsequent decrease in free intracellular calcium causes relaxation of blood vessels. (B) Tachykinin receptors are probably located at the membrane of endothelial cells. Binding to these receptors results in the formation of nitric oxide (NO) from L-arginine within the endothelial cell. NO diffuses rapidly into smooth muscle cells where it binds to guanylate cyclase that is activated. Cyclic GMP is formed from GTP, causing relaxation by decreasing the intracellular levels of free calcium.

II. Responses to CGRP in Different Vascular Beds

A. Coronary Circulation and Aorta

Substance P and CGRP coexist in a population of sensory nerves, innervating the coronary epicardial, and intramyocardial arteries (Gulbenkian *et al.*, 1990). Low concentrations of rat α-CGRP cause relaxation of rat thoracic aorta; the effect requires an intact endothelium (Brain *et al.*, 1985; Kubota *et al.*, 1985; Grace *et al.*, 1987). This relaxation is considerably less than that observed with another endothelium-dependent dilator, acetylcholine. It is well known that acetylcholine causes the release of EDRF, mostly believed to be nitric oxide (NO), which stimulates guanylate cyclase and which in turn induces relaxation of vascular smooth muscle cells (Fig. 1). Although CGRP is dependent on an intact endothelium to cause relaxation of the thoracic aorta, no parallel increase in cyclic GMP levels is observed (Grace *et al.*, 1987). It has been shown in cultured rat thoracic aorta smooth muscle cells and bovine endothelial cells that CGRP activates adenylate cyclase in concentrations comparable to the apparent K_d value from binding studies (Hirata *et al.*, 1988). Furthermore, in rings of rat proximal epicardial coronary arteries the relaxations induced by rat α-CGRP are abolished by removal of the endothelium or treatment with indomethacin but not by administration of methylene blue. Thus, the endothelial factor released by CGRP in rat proximal epicardial coronary arteries appears to be a cyclooxygenase product. In distal intramyocardial coronary arteries the relaxation is not affected by any of these treatments (Prieto *et al.*, 1991). In contrast, in the rat thoracic aorta it has been shown that CGRP induces elevation of cyclic GMP and cyclic AMP (Marshall, 1991). Nitro-L-arginine, an inhibitor of NO synthase, inhibits the production of cyclic GMP but not cyclic AMP. Thus, the activation of guanylate cyclase seems to depend on an increase in cyclic AMP. In precontracted human (Franco-Cereceda and Rudehill, 1989), bovine (Greenberg *et al.*, 1987), and guinea pig coronary arteries (Sätrum *et al.*, unpublished) α-CGRP acts as a strong vasodilator; the effect is not dependent on the presence of endothelium. Infusion of CGRP increases coronary blood flow in the isolated rabbit heart (Marshall *et al.*, 1986c) and dilates coronary arteries in man during coronary angiography (McEwan *et al.*, 1985). The available data clearly demonstrates the need for further careful experimentation of the coronary circulation—many questions prevail.

B. Pulmonary Circulation

In precontracted isolated rings of guinea pig pulmonary artery SP induces a prompt and transient relaxation. In rings, in which the endothelium is removed beforehand, SP evokes a transient small contraction. In the same preparation, CGRP causes endothelium-independent relaxation (Maggi *et al.*, 1990; Cardell *et al.*, unpublished). Capsaicin, which releases CGRP and SP, causes an endothelium-independent relaxation of the guinea pig pulmonary artery, which suggests that the main effect is caused by CGRP or a similar agent. However, the contribution of a small tachykinin-mediated and endothelium-dependent component in the overall relaxation cannot be ruled out (Maggi *et al.*, 1990). In human pulmonary arteries and veins, CGRP induces relaxation; the response is somewhat more potent in veins than in arteries. Removal of the endothelium does not diminish the vasodilator potency of CGRP in either vessel type (McCormack *et al.*, 1989).

C. Mesenteric Circulation

In guinea pig and rat, the superior mesenteric artery seems to contain the highest level of CGRP as compared to other blood vessels in these species (Mulderry *et al.*, 1985; Wharton *et al.*, 1986). SP and CGRP immunoreactive nerves have also been demonstrated in the mesenteric artery and vein of rat (Edvinsson *et al.*, 1989a). The sensory nerve terminals are located more distally in the adventitia as compared to the "classical" autonomic nerve terminals (Edvinsson *et al.*, 1989a; Gulbenkian *et al.*, 1986). The density of the CGRP, SP, and NKA containing nerve fibers appears to be similar in both proximal and distal regions of the superior mesenteric artery. However, in the proximal part the nerves have a circular orientation around the vessel circumference, whereas in the distal superior mesenteric artery they run parallel to the long axis of the vessel. Despite this difference in anatomical arrangement, there is no difference in relaxant responses neither in potency nor in maximum efficacy to CGRP, SP, and NKA (Edvinsson *et al.*, 1989a). Human α-CGRP has been shown to be a potent dilator in human, rat, and rabbit perfused mesenteric vasculature (Marshall *et al.*, 1986b; Marshall *et al.*, 1988). The response to human α-CGRP appears to be more potent, relative to nitroprusside, on the mesenteric resistance arterioles than on the large conduit arteries in both man and rat (Marshall *et al.*, 1986a,b; 1988). The effect of human α-CGRP in the human perfused mesenteric vasculature has a longer duration as compared to the effect of nitroprusside (Marshall *et al.*, 1986b; Marshall *et al.*, 1988).

Stimulation of nerves to resistance blood vessels of the rat mesenterium produces a neurogenic vasodilatation, probably via the release of CGRP (Kawasaki *et al.*, 1988). Rat femoral artery and vein contain relatively few CGRP and SP nerve fibers and the peptides only produce weak (10–25%) relaxation (Edvinsson *et al.*, 1989b). Neuropeptide Y (NPY), a vasoconstrictor peptide, present in sympathetic perivascular nerves, is colocalized with noradrenaline. Five and 10 nM of NPY inhibit the dilatation of the mesenteric resistance vessels induced by perivascular nerve stimulation. NPY has no effect by itself on vasodilatation in response to bolus infusion of rat CGRP. Thus, NPY was thought to modulate the release of CGRP from CGRP-containing nerves in mesenteric resistance vessels (Kawasaki *et al.*, 1991).

CGRP acts as a local factor stimulating proliferation in cultured endothelial cells from human umbilical vein. It has been suggested that this mechanism of action is associated with cAMP formation. This effect of CGRP may be important for the formation of new vessels during physiological and pathophysiological events, such as ischemia, inflammation, and wound healing (Hægerstrand *et al.*, 1990).

D. *Cerebral Circulation*

CGRP-containing fibers are present in perikarya in the trigeminal ganglion of all species examined, including man. In these perikarya, CGRP is frequently colocalized with SP although the number of CGRP-containing cells exceeds the number of SP-containing cells (Rosenfeld *et al.*, 1983; Uddman *et al.*, 1985; Lee *et al.*, 1985; McCulloch *et al.*, 1986). The major cerebral arteries (anterior, middle, and posterior cerebral arteries, vertebral and basilar arteries), and pial arterioles on the cortical surface, are invested with fine varicosed nerve fibers that contain CGRP.

In animals in which the trigeminal nerve has been surgically divided 14 days before sacrifice, there is a complete absence of CGRP and SP fibers ipsilateral to the lesion (Uddman *et al.*, 1985; McCulloch *et al.*, 1986). Interestingly, cerebrovascular levels of CGRP are markedly elevated (by 70%) after cervical sympathectomy, putatively because of diminished competition from sympathetic fibers for nerve growth factor (Schon *et al.*, 1985).

The sensory neuropeptides CGRP and SP are cerebral vasodilators in all species tested (Edvinsson *et al.*, 1981, 1985, 1987; Hanko *et al.*, 1985; McCulloch *et al.*, 1986; Edvinsson and Jansen, 1987; Saito *et al.*, 1989; Jansen *et al.*, 1991; Jansen, 1992; Jansen *et al.*, 1992). The relative potency varies considerably among species; in the cat middle cerebral artery CGRP induces a greater relaxation of cerebrovascular smooth muscle than does SP although the EC_{50} (the concentration effecting half maximal response) values are similar for the two peptides; in human pial arteries, the maximum responses to CGRP and SP are similar, but CGRP induces relaxation at considerably lower concentrations. The relaxation induced by CGRP in cerebral vessels from man, guinea pig, and cat is endothelium independent and paralleled with an increase in cAMP levels (Edvinsson *et al.*, 1985; Saito *et al.*, 1989; Jansen *et al.*, 1992). CGRP is a potent vasodilator of arterioles while it has virtually no effect on pial veins *in vivo* (Fig. 2).

III. Vasomotor Responses *in Vitro*

A. *Agonist Responses*

In man, all CGRP agonists induce strong and potent concentration-dependent relaxations of precontracted cerebral (pial) arteries. Human calcitonin only induces weak relaxation. The order of potency is rat β-CGRP = human α-CGRP > rat α-CGRP > human β-CGRP > calcitonin (Table 1A).

In cat middle cerebral artery segments, the CGRP agonists induce relaxation with maximum responses between 88 and 97% of precontraction. The order of potency is rat β-CGRP = rat α-CGRP ≥ human β-CGRP > human α-CGRP (Table 1B). In the guinea pig basilar artery, the order of potency is rat α-CGRP ≥ rat β-CGRP ≥ human α-CGRP ≥ human β-CGRP. All CGRP analogs dilate with a maximum response between 80 and 87% of the $PGF_{2\alpha}$-induced precontraction. As in human vessels, the response to calcitonin of the guinea pig basilar artery is low (12%), indicative of partial agonism as the potency is comparable to that of the CGRP

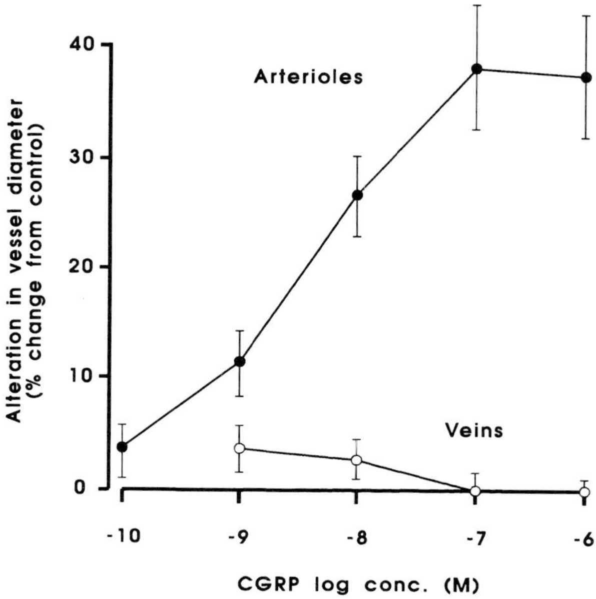

Figure 2 Alterations in feline cerebral vascular diameter (expressed as percentage change from preinjected level) resulting from perivascular microinjection of calcitonin gene-related peptide. Data are presented as mean ± SEM; ● arteries; ○ veins. Veins did not reveal any significant caliber alteration following application of calcitonin gene-related peptide.

analogs (Table 1C). With data as these, the CGRP responses were initially characterized as indicative of a CGRP receptor, while the second type of receptors in this group showed full agonism to calcitonin. The receptors have been named CGRP and calcitonin receptors, respectively.

B. Antagonistic Experiments

C-terminal fragments of human α-CGRP, human α-CGRP(8–37), and [Tyr°]CGRP(28–37) have been reported as putative receptor antagonists (Chiba *et al.*, 1989; Chakder and Rattan, 1990; Dennis *et al.*, 1990; Donoso *et al.*, 1990; Maton *et al.*, 1988). This has provided us with new information and a more specific pharmacological characterization of CGRP receptors. Human α-CGRP(8–37) has antagonistic ef- on CGRP-induced adenylate cyclase activity in liver cell membranes and acts as a competitive human α-CGRP receptor antagonist in the cardiovascular system of the rat and in the guinea pig atrium and ileum (Dennis *et al.*, 1990; Donoso *et al.*, 1990). Human α-CGRP(8–37) shows a weak antagonistic effect on CGRP-mediated inhibition of the electrically induced twitch response of the rat vas deferens and is ineffective against human α-CGRP-induced hyperthermia. The existence of at least two classes of CGRP receptors has thus been proposed, one being sensitive and the other being insensitive to human α-CGRP(8–37) (Dennis *et al.*, 1990). [Tyr°]CGRP (28–37) has been found to block the CGRP-induced fall in resting tension of the internal anal sphincter muscle and of CGRP-stimulated increase in amylase secretion from pancreatic acini (Chakder and Rattan, 1990; Maton *et al.*, 1988).

In cerebral blood vessels, a single dose of human α-CGRP(8–37) (10^{-6} M) does not induce relaxation of the precontracted vessel segments (Jansen, 1991; Jansen *et al.*, 1992). In man, cat, and guinea pig, human α-CGRP(8–37) induces a parallel shift of the concentration-response curves induced by either human α-CGRP or rat α-CGRP without any change in maximum effect (Fig. 3). The calculated pA_2 values (dissociation constant) on human α-CGRP-induced relaxation are in man 6.70. In cat, relaxation induced by human α-CGRP and rat α-CGRP has pA_2 values of 7.14 and 6.10, and in guinea pig 7.12 and 6.93, respectively. In contrast, human α-CGRP(8–37) does not antagonize the dilator response to human β-CGRP in cerebral arteries from man, cat, and guinea pig (Fig. 4). Similar findings have recently been reported in the microvascular circulation of rabbit dorsal skin where human α-CGRP(8–37) is a more potent antagonist of vasodilatation induced by human α-CGRP than on responses induced by human β-CGRP (Hughes and Brain, 1991). On the other hand, [Tyr°]CGRP(28–37) does not induce relaxation of the precontracted

Table 1 Relaxant Responses of (A) Human (B) Cat, and (C) Guinea Pig Cerebral Arteries to Different CGRP Agonists[a]

	pD_2	I_{max}
A		
rat α-CGRP	9.6 ± 0.2	80 ± 7
human α-CGRP	9.9 ± 0.3	91 ± 3
rat β-CGRP	10.3 ± 0.3	81 ± 8
human β-CGRP	9.3 ± 0.2	71 ± 10
human calcitonin	7.7 ± 0.1	19 ± 6
B		
rat α-CGRP	9.1 ± 0.2	88 ± 4
human α-CGRP	8.6 ± 0.1	97 ± 1
rat β-CGRP	9.2 ± 0.2	96 ± 3
human β-CGRP	9.0 ± 0.3	89 ± 5
C		
rat α-CGRP	8.9 ± 0.2	87 ± 8
human α-CGRP	8.5 ± 0.3	87 ± 8
rat β-CGRP	8.6 ± 0.2	87 ± 8
human β-CGRP	8.2 ± 0.2	80 ± 9
human calcitonin	8.6 ± 0.4	12 ± 3
salmon calcitonin	8.1 ± 0.4	12 ± 4

[a] *Abbreviations:* pD_2, log concentration of agonist eliciting half maximum relaxation; I_{max}, maximum relaxant response. Data are presented as mean relaxant responses ± SEM

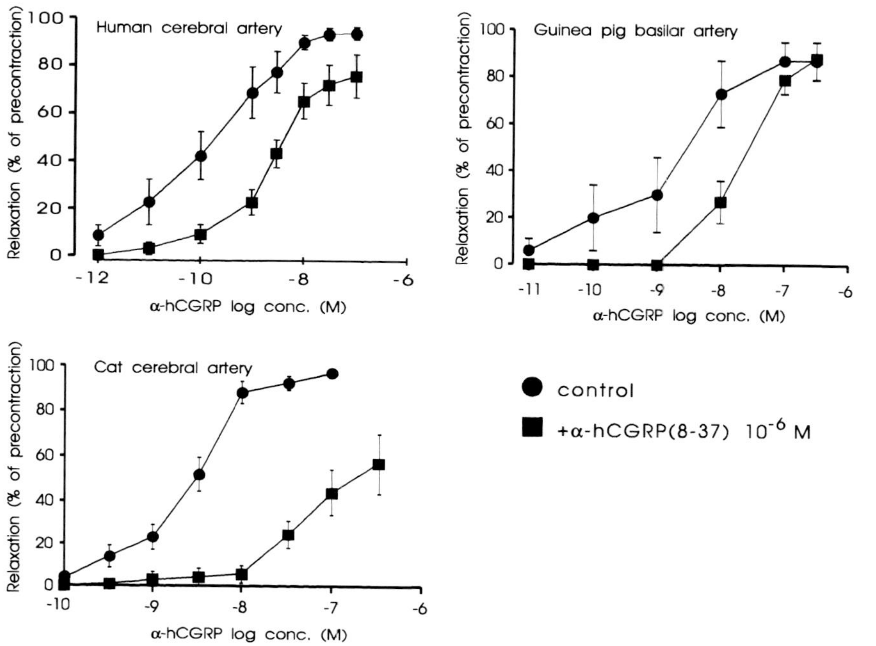

Figure 3 The CGRP antagonist human α-CGRP(8-37) (10^{-6} M) significantly blocks the relaxation induced by human α-CGRP in cerebral arteries from man, cat, and guinea pig. Values given represent the mean ± SEM; n = 6–10.

vessel segments or antagonize CGRP-induced relaxation in the guinea pig basilar artery (Jansen, 1992).

The potent antagonistic effect of human α-CGRP(8–37) on cerebral vessels from man, cat, and guinea pig has been taken as evidence for the presence of two types of CGRP receptors; CGRP-1 receptors being sensitive to human CGRP(8–37) while the lack of antagonistic effect of β-CGRP-induced relaxation indicates a response mediated via a second type of CGRP receptors, the CGRP-2 subtype. In analogy with this, α-CGRP-induced hyperthermia is not affected by human α-CGRP(8–37) (Dennis *et al.*, 1990).

IV. Second-Messenger Studies

A. Cyclic AMP

In cultured rat smooth muscle cells, CGRP stimulates the formation of cAMP in a concentration-dependent manner (Kubota *et al.*, 1985). In cat and human cerebral arteries, CGRP induces an increase of cAMP levels in parallel with relaxation (Edvinsson *et al.*, 1985; Jansen *et al.*, 1992) (Fig. 5A). In human cerebral arteries, the antagonist human α-CGRP(8–37) has virtually no effect on cAMP formation (Fig. 5B). Human α-CGRP(8–37) (10^{-6} M) gives a significant decrease in max-

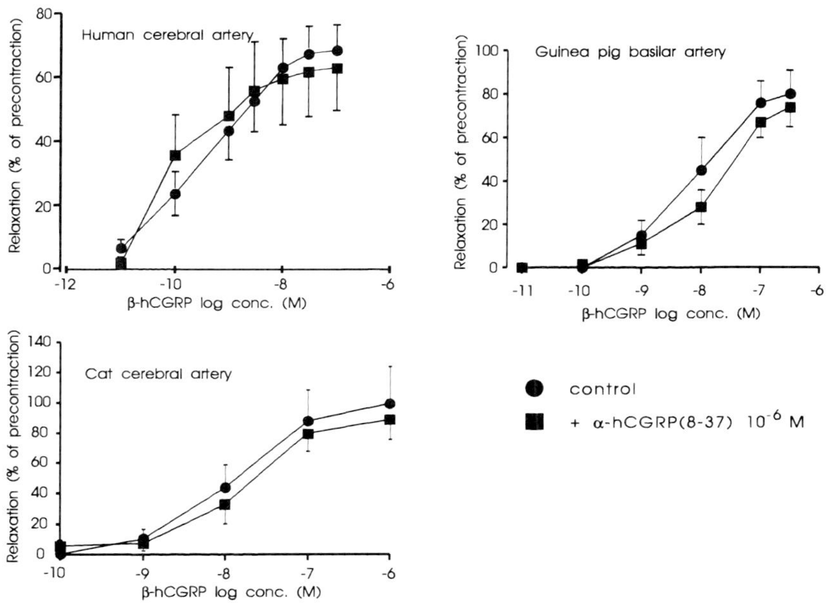

Figure 4 The relaxation induced by human β-CGRP in cerebral arteries from man, cat, and guinea pig is not blocked by the CGRP antagonist human α-CGRP(8–37) (10^{-6} M). Values given represent the mean ± SEM; n = 6–10.

imum amount of cAMP formed by human α-CGRP (Fig. 5B). These findings show that the CGRP-1 receptor in human cerebral vessels is coupled to adenylate cyclase. Incubation of guinea pig cerebral arteries with SP or NKA does not result in any significant elevation of cyclic AMP levels (Jansen, unpublished; Edvinsson *et al.*, 1985).

B. Cyclic GMP

In some vascular preparations, CGRP is dependent on an intact endothelium to induce relaxation (Table 2). Binding to an endothelial receptor is generally thought to trigger the synthesis of nitric oxide (NO) from L-arginine (Fig. 1). The unique chemistry of NO allows this molecule, with the capacity to diffuse rapidly into nearby cells, to stimulate cyclic GMP formation (Ignarro, 1991). However, controversy exists regarding the involvement of guanylate cyclase in vessels in which CGRP induces endothelium-dependent relaxation (Table 2). Grace and co-workers (1987) have reported that CGRP induces endothelium-dependent relaxation of the thoracic aorta. This relaxation does not occur in parallel with an increase in cyclic GMP formation. Prieto *et al.* (1991) showed that the endothelium-dependent relaxation of the rat thoracic aorta was antagonized by the cyclooxygenase inhibitor indomethacin but not by methylene blue, an inhibitor of guanylate cyclase. Furthermore, in culture of rat aorta smooth muscle cells CGRP has no effect on formation

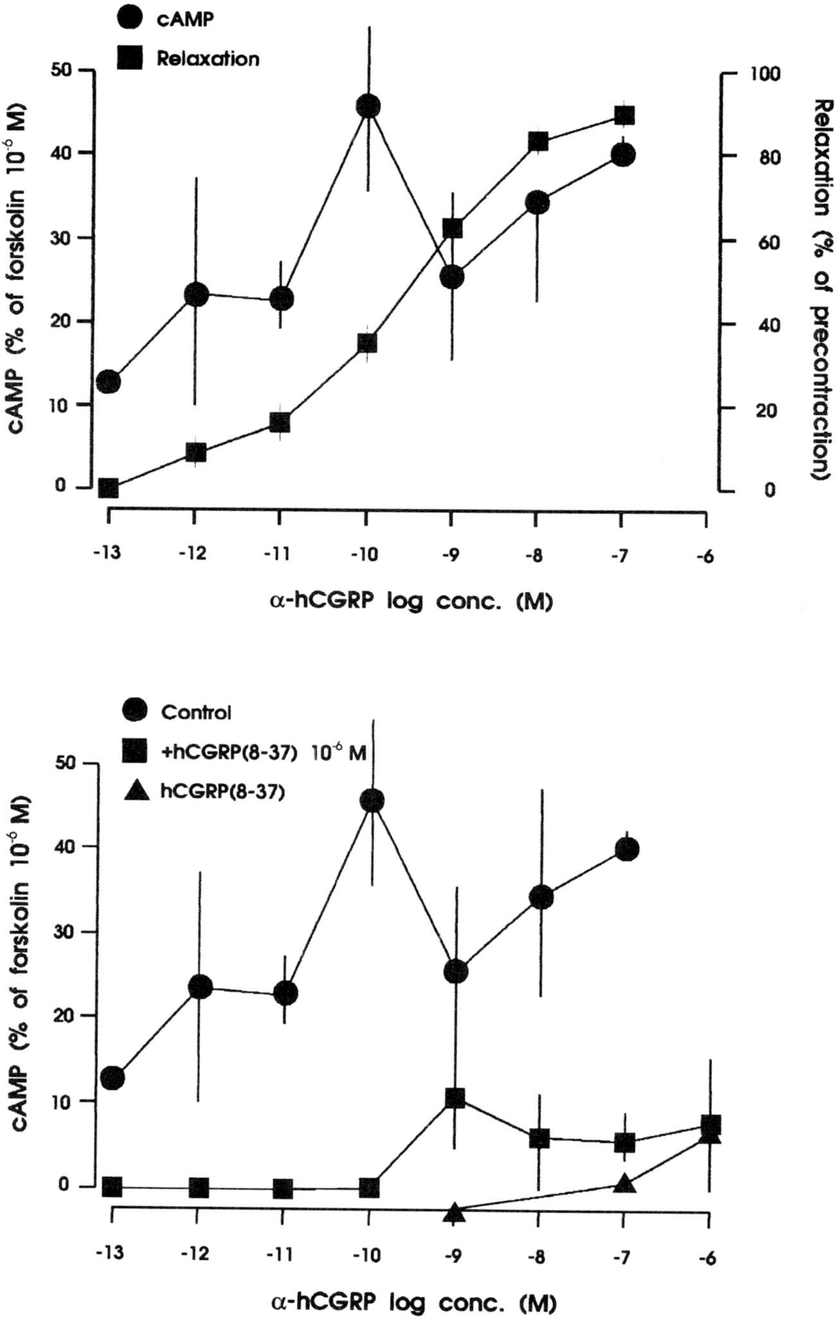

Figure 5 Human α-CGRP increases the formation of cyclic AMP in human cerebral arteries. *Upper panel:* A parallel response of cyclic AMP formation and relaxation. *Lower panel:* The antagonist human α-CGRP(8–37) (10^{-9}–10^{-6} M) only induces a slight increase in cyclic AMP formation by itself. When the vessels are incubated in the presence of human α-CGRP(8–37) (10^{-6} M), the formation of cyclic AMP by human α-CGRP is markedly reduced. Values given represent the mean ± SEM; n = 3–8.

Table 2 An Overview of Vascular Preparations with Endothelium-Dependent Relaxant Responses to CGRP and Available Data on Second-Messenger Systems Activated[a]

Tissue	Adenylate cyclase	Guanylate cyclase	References
Human radial artery	N.d.	N.d.	Thom *et al.*, 1987
Human coronary artery	N.d.	N.d.	Thom *et al.*, 1987
Human gastric arteries	N.d.	N.d.	Thom *et al.*, 1987
Rat thoracic aorta	N.d.	N.d.	Brain *et al.*, 1985
Rat thoracic aorta	Yes	No	Kubota *et al.*, 1985
Rat thoracic aorta	N.d.	No	Grace *et al.*, 1987
Rat thoracic aorta	Yes	Yes	Marshall, 1991
Cultured rat aorta smooth muscle cells	Yes	No	Hirata *et al.*, 1988

[a]N.d., not determined.

of cyclic GMP (Kubota *et al.*, 1985). These findings indicate that the effect could be mediated by release of endothelium-produced prostacyclin. In contrast, Marshall (1991) has demonstrated a CGRP-induced elevation of cyclic GMP in the rat thoracic aorta. It was suggested that the activation of guanylate cyclase was triggered by CGRP-induced increase in cyclic AMP. In our studies on cerebral arteries, CGRP induced an endothelium-independent relaxation, occuring in parallel with activation of adenylate cyclase (Edvinsson *et al.*, 1985; Jansen *et al.*, 1992). In preliminary studies on the same tissue preparation we found no stimulating effect of CGRP on the formation of cyclic GMP (Jansen *et al.*, unpublished). This is supported by the lack of antagonistic effect of methylene blue on CGRP-induced relaxation (Fig. 6).

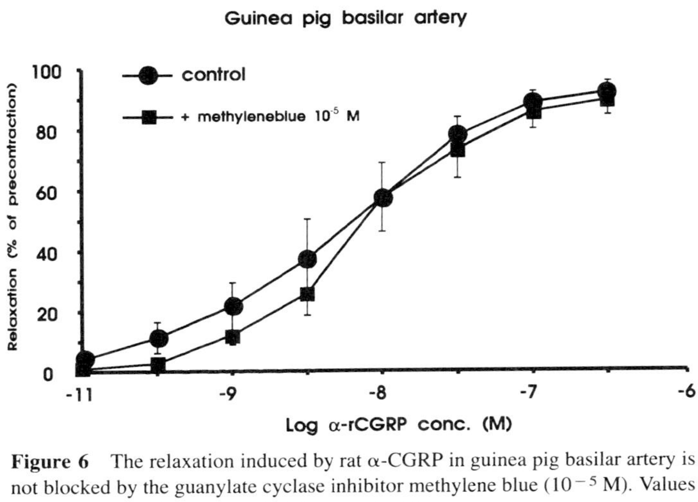

Figure 6 The relaxation induced by rat α-CGRP in guinea pig basilar artery is not blocked by the guanylate cyclase inhibitor methylene blue (10^{-5} M). Values given represent the mean ± SEM; n = 10–12.

C. Intracellular Ca $^{2+}$

Noradrenaline-induced contraction of the rat tail artery consists of two components; one being due to the entry of extracellular calcium via receptor-operated calcium channels, and the other on mobilization of intracellular calcium. CGRP does not inhibit the contraction induced by KCl or noradrenaline when calcium is added in increasing doses to a calcium-deficient buffer solution. This has been taken as evidence that CGRP has no effect on extracellular calcium entry. On the other hand, CGRP produces a concentration-dependent inhibition of noradrenaline-induced contraction in calcium free buffer solution. Thus, it has been suggested that CGRP exerts relaxation by inhibiting the mobilization of intracellular calcium in this system (Kline and Pang, 1988).

D. K^+*-Channels*

In a recent study, it was found that CGRP causes hyperpolarization of the smooth muscle of rabbit mesenteric artery by activating single K^+-channels (Nelson *et al.*, 1990). This hyperpolarization is reversed by glibenclamide, a blocker of ATP-sensitive potassium channels (K_{ATP}). Also the relaxant response to CGRP is partially reversed by glibenclamide. Thus, in rabbit mesenteric artery hyperpolarization by activating K_{ATP} seems to be responsible for part of the observed relaxation induced by CGRP (Nelson *et al.*, 1990). Similar findings have been reported for the rabbit ophthalmic artery; CGRP-induced relaxation and hyperpolarization were partly inhibited by glibenclamide (Zschauer *et al.*, 1991). In contrast, glibenclamide has no effect on the relaxation induced by CGRP in pig vesical (Persson *et al.*, 1991), rat coronary (Prieto *et al.*, 1991), and guinea pig basilar arteries (Nilsson *et al.*, 1991). The reason for this dicotomy is not clear. A difference in receptor population may be the underlying explanation.

V. Ligand Binding Studies

The relative affinity of fragments of human α-CGRP for the [^{125}I]human α-CGRP binding sites has been examined in rat brain, guinea pig atria, and guinea pig vas deferens. Human α-CGRP(8–37) and human α-CGRP(9–37) show higher affinities for [^{125}I] human α-CGRP in the brain and vas deferens preparations than unlabeled human α-CGRP. In the atrial preparation, human α-CGRP, human α-CGRP(8–37), and human α-CGRP(9–37) possess similar affinities for [^{125}I] human α-CGRP binding. The two shorter fragments, human α-CGRP(10–37) and human α-CGRP(11–37) are less potent than the complete peptide in competing for [^{125}I] human α-CGRP binding sites in all preparations (Mimeault *et al.*, 1991). Specific binding sites for CGRP have also been found in cultures of rat vascular smooth muscle cells and bovine endothelial cells. Rat α-CGRP and human α-CGRP appear to have a similar binding affinity, while human calcitonin is less potent, in inhibiting [^{125}I] rat α-CGRP binding to its receptors in both cell types (Hirata *et al.*, 1988).

Autoradiographic binding studies of [^{125}I] human α-CGRP in human pulmonary arteries and veins have demonstrated a high density of CGRP receptors, particularly located in the smooth muscle cell layer. The density of the autoradiographic grains (specific binding) is similar in both arteries and veins (McCormack *et al.*, 1989). Similar results have been obtained in rat aorta and coronary arteries (Sigrist *et al.*, 1986).

There is a difference in the distribution of α-CGRP binding sites in the gastrointestinal tract. Small arteries in the proximal (from the esophagus to the duodenum) but not the distal (jejunum, ileum, and colon) gastrointestinal tract express α-CGRP binding sites (Gates *et al.*, 1989). Veins and venules do not. This correlates well with previous studies that show that CGRP dilates resistance vessels and not capacitance vessels (Benjamin *et al.*, 1987).

VI. *In Vivo* Experiments

Infusion of CGRP in man evokes a decrease in diastolic blood pressure and an increase in heart rate and skin temperature (Struthers *et al.*, 1986). Regions of the circulation that respond particularly well with dilatation after *in vivo* administration of CGRP are the superficial subcutaneous vasculature (Williams *et al.*, 1988) and the superior mesenteric and carotid arteries (Mulholland *et al.*, 1991). In the rat mesenteric circulation human α-CGRP but not calcitonin gives a reduction in blood pressure (Marshall *et al.*, 1986b). Gardiner and colleagues have undertaken a series of studies aimed at characterizing the regional hemodynamic effects of the rat and human variants of α- and β-CGRP. Human α-CGRP-induced tachycardia is accompanied by a fall in mean arterial blood pressure, while human β-CGRP is without significant effects on mean arterial blood pressure at a time when it causes an increase in heart rate (Gardiner *et al.*, 1989a,b). However, intravenous infusion with β-CGRP in conscious, unrestrained rats, causes dose-dependent renal and hindquarters vasodilatation not different from those seen with α-CGRP. In the highest doses tested, α- and β-CGRP induce mesenteric vasoconstriction and renal vasodilatation. At intermediate doses α-CGRP caused more significant hemodynamic changes than β-CGRP. Thus, the experiments suggested that the α-form had more potent hemodynamic effects, and furthermore, the human form was more potent than the rat form. In general, the potency data seen in isolated vessel segments were similar to those seen *in vivo*.

Human α-CGRP has been tested for a possible clinical therapeutic role in vasospastic disorders. A combined treatment of low doses of human α-CGRP and the calcium entry blocker nimodipine improves internal carotid blood flow without compromising blood flow to the kidney (Gardiner *et al.*, 1990a). It has been suggested that this administration might be a helpful treatment of patients after cerebral vasopasm and in subjects with impaired renal perfusion. To be clinically effective the infusion has to occur for several days and hence the prolonged infusion of human α-CGRP (Gardiner *et al.*, 1991a). Thus, in the clinical setting, maintained increases in internal carotid blood flow can be achieved by individually adjusted infusions of human α-CGRP.

Using the regional hemodynamic approach, Gardiner and colleagues also observed that human α-CGRP(8–37) was an effective antagonist of the cardiovascular actions of human α-CGRP *in vivo* (Gardiner *et al.*, 1990b). This antagonist had no significant effect on blood pressure, heart rate, or common or internal carotid haemodynamics although it caused significant, reversible, inhibition of the vasodilator effects of human α-CGRP in conscious Long Evans rats (Gardiner *et al.*, 1991b). Human α-CGRP(28–37) had no cardiovascular effects itself and did not influence responses to human α-CGRP. These results suggested that the carotid hemodynamic effects of human α-CGRP were largely due to activation of the CGRP-1 receptor subtype. Future experimentation and biochemistry will provide additional data to more firmly differentiate possible subtypes of CGRP receptors, particularly molecular biology advancements and cloning of CGRP receptors will ultimately demonstrate the presence and location of different CGRP receptor subtypes.

References

Benjamin, N., Dollery, C. T., Fuller, R. W., Larkin, S., and McEwan, S. E. (1987). The effects of calcitonin gene-related peptide and substance P on resistance and capacitance vessels. *Br. J. Pharmac. 90,* 39.

Brain, S. D., Williams, T. J., Tippins, J. R., Morris, H. R., and MacIntyre, I. (1985). Calcitonin gene-related peptide is a potent vasodilator. *Nature 313,* 54–56.

Chakder, S., and Rattan, S. (1990). [Tyr°]-calcitonin gene-related peptide 28–37 (rat) as a putative antagonist of calcitonin gene-related peptide response on opossum internal anal sphincter smooth muscle. *J. Pharmacol. Exp. Therap. 253,* 200–206.

Chiba, T., Yamaguchi, A., Yamatani, T., Nakamura, A., Morishita, T., Inui, T., Fukase, M., Noda, T., and Fujita, T. (1989). Calcitonin gene-related peptide receptor antagonist human CGRP-(8–37). *Am. J. Physiol. 256,* E331–E335.

Dennis, T., Fournier, A., St. Pierre, S., and Quirion, R. (1989). Structure–activity profile of calcitonin gene-related peptide in peripheral and brain tissues. Evidence for receptor multiplicity. *J. Pharmacol. Exp. Ther. 251,* 718–725.

Dennis, T., Fournier, A., Cadieux, A., Pomerleau, Jolicoeur, F. B., St. Pierre, S., and Quirion, R. (1990). Human CGRP(8–37) a calcitonin gene-related peptide antagonist revealing calcitonin gene-related peptide receptor heterogeneity in brain and periphery. *J. Pharmacol. Exp. Ther. 254,* 123–128.

Donoso, V. M., Fournier, A., St. Pierre, S., and Huidobro-Toro, J. P. (1990). Pharmacological characterization of CGRP1 receptor subtype in the vascular system in the rat: Studies with human CGRP fragments and analogs. *Peptides 11,* 885–889.

Edvinsson, L. (1985). Functional role of perivascular peptides in the control of the cerebral circulation. *Trends Neurosci. 8,* 126–131.

Edvinsson, L., and Jansen, I. (1987). Characterization of tachykinin receptors in isolated cerebral arteries of guinea pig. *Br. J. Pharmacol. 90,* 553–559.

Edvinsson, L., McCulloch, J., and Uddman, R. (1981). Substance P: Immunohistochemical localization and effects upon cat pial arteries *in vitro* and *in situ*. *J. Physiol. 318,* 251–258.

Edvinsson, L., Fredholm, B. B., Hamel, E., Jansen, I., and Verrecchia, C. (1985). Perivascular peptides relax cerebral arteries concomitant with stimulation of cyclic adenosine monophosphate accumulation or release of an endothelium derived relaxing factor in the cat. *Neurosci. Lett. 58,* 213–217.

Edvinsson, L., Ekman, R., Jansen, I., Ottosson, A., and Uddman, R.(1987). Peptide containing nerve fibers in human cerebral arteries: Immunocytochemistry, radioimmunoassay and *in vitro* pharmacology. *Ann. Neurol. 21,* 431–437.

Edvinsson, L., Gulbenkian, S., Jansen, I., Wharton, J., Cervantes, C., and Polak, J. (1989a). Comparison of peptidergic mechanisms in different parts of the guinea pig superior mesenteric artery: Immunocytochemistry at the light and ultrastructural levels and responses *in vitro* of large and small arteries. *J. Auton. Nerv. Syst. 28,* 141–154.

Edvinsson, L., Gulbenkian, S., Wharton, J., Jansen, I., and Polak, J. (1989b). Peptide-containing nerves in the rat femoral artery and vein. An immunocytochemical and vasomotor study. *Blood Vessels. 26,* 254–271.

Franco-Cereceda, A., and Rudehill, A. (1989). Capsaicin-induced vasodilatation of human coronary arteries *in vitro* is mediated by calcitonin gene-related peptide rather than substance P or neurokinin A. *Acta Physiol. Scand. 136,* 575–580.

Gardiner, S. M., Compton, A. M., and Bennett, T.(1989a). Regional haemodynamic effects of human α- and β-calcitonin gene-related peptide in conscious Wistar rats. *Br. J. Pharmacol. 98,* 1225–1232.

Gardiner, S. M., Compton, A. M., and Bennett, T. (1989b). Regional haemodynamic effects of calcitonin gene-related peptide. *Am. J. Physiol. 256,* 332–338.

Gardiner, S. M., Compton, A. M., Bennett, T., Kemp, P. A., and Ney, U. (1990a). Synergistic internal carotid vasodilator effects of human α-calcitonin gene-related peptide and nimodipine in conscious rats. *Br. J. Pharmacol. 99,* 830–834.

Gardiner, S. M., Compton, A. M., Kemp, P. A., Bennett, T., Bose, C., Foulkes, R., and Hughes, B. (1990b). Antagonistic effect of human α-CGRP (8–37) on the *in vivo* regional haemodynamic actions of human α-CGRP. *Biochem. Biophys. Res. Comm. 171,* 938–943.

Gardiner, S. M., Compton, A. M., Kemp, P. A., Bennett, T., Foulkes, R., and Hughes, B.(1991a). Regional haemodynamic effects of prolonged infusions of human α-calcitonin gene-related peptide in conscious Long Evans rats. *Br. J. Pharmacol. 103,* 1509–1514.

Gardiner, S. M., Compton, A. M., Kemp, P. A., Bennett, T., Bose, C., Foulkes, R., and Hughes, B. (1991b). Human α-calcitonin gene-related peptide (CGRP)-(8–37), but not -(28–37), inhibits carotid vasodilator effects of human α-CGRP *in vivo. Eur. J. Pharmacol. 199,* 375–378.

Gates, T. S., Zimmerman, R. P., Mantyh, C. R., Vigna, S. R., and Mantyh, P. W. (1989). Calcitonin gene-related peptide-α receptor binding sites in the gastrointestinal tract. *Neuroscience 31,* 757–770.

Grace, C. G., Dusting, G. J., Kemp, B. E., and Martin, T. J. (1987). Endothelium and the vasodilator action of rat calcitonin gene-related peptide. *Br. J. Pharmac. 91,* 729–733.

Greenberg, B., Rhoden, K., and Barnes, P. (1987). Calcitonin gene-related peptide (CGRP) is a potent non-endothelium-dependent inhibitor of coronary vasomotor tone. *Br. J. Pharmacol. 92,* 789–794.

Gulbenkian, S., Merighi, A., Wharton, J., Varndell, I. M., and Polak, J. M. (1986). Ultrastructural evidence for the coexistence of calcitonin gene-related peptide and substance P in sensory vesicles of peripheral nerves in the guinea pig. *J. Neurocytol. 15,* 535–542.

Gulbenkian, S., Edvinsson, L., Saetrum-Opgaard, O., Wharton, J., Polak, J. M., and David-Ferreira, J. F. (1990). Peptide-containing nerve fibres in guinea pig coronary arteries: Immunohistochemistry, ultrastructure and vasomotility. *J. Auton. Nerv. Syst. 31,* 153–168.

Hanko, J., Hardebo, J. E., Kåhrström, J., Owman, C., and Sundler, F. (1985). Calcitonin gene-related peptide is present in mammalian cerebrovascular nerve fibers and dilates pial and peripheral arteries. *Neurosci. Lett. 57,* 91–95.

Hirata, Y., Takagi, Y., Takata, S., Fukuda, Y., Yoshimi, H., and Fujita, T.(1988). Calcitonin gene-related peptide receptor in cultured vascular smooth muscle and endothelial cells. *Biochem. Biophys. Res. Commun. 151,* 1113–1121.

Holzer, P. (1988). Local effector functions of capsaicin-sensitive sensory nerve endings: Involvement of tachykinins, calcitonin gene-related peptide and other neuropeptides. *Neuroscience 24,* 739–768.

Hughes, S. R., and Brain, S. D. (1991). A calcitonin gene-related peptide (CGRP) antagonist ($CGRP_{8-37}$) inhibits microvascular responses induced by CGRP and capsaicin in skin. *Br. J. Pharmac. 104,* 738–742.

Hægerstrand, A., Dalsgaard, C.-J., Jonzon, B., Larsson, O., and Nilsson, J. (1990). Calcitonin gene-related peptide stimulates proliferation of human endothelial cells. *Proc. Natl. Acad. Sci. U.S.A. 87,* 3299–3303.

Ignarro, L. J. (1991). Heme-dependent activation of guanylate cyclase by nitric oxide: A novel signal transduction mechanism. *Blood Vessels 28,* 67–73.

Jansen, I. (1992). Characterization of calcitonin gene-related peptide (CGRP) receptors in guinea pig basilar artery. *Neuropeptides 21,* 73–79.

Jansen, I., Alafaci, C., McCulloch, J., Uddman, R., and Edvinsson, L. (1991). Tachykinins (substance P, neurokinin A, neuropeptide K, neurokinin B) in the cerebral circulation: Vasomotor responses *in vitro* and *in situ*. *J. Cereb. Blood Flow Metab. 11,* 567–575.

Jansen, I., Mortensen, A., and Edvinsson, L. (1992). Characterization of calcitonin gene-related peptide (CGRP) receptors in human cerebral vessels: Vasomotor responses and cAMP accumulation. *Ann. N.Y. Acad. Sci. 657,* 435–440.

Kawasaki, H., Takasaki, K., Saito, A., and Goto, K. (1988). Calcitonin gene-related peptide acts as a novel vasodilator neurotransmitter in mesenteric resistance vessels of the rat. *Nature* (Lond.) *335,* 164–167.

Kawasaki, H., Nuki, C., Saito, A., and Takasaki, K. (1991). NPY modulates neurotransmission of CGRP-containing vasodilator nerves in rat mesenteric arteries. *Am. J. Physiol. 261,* 683–690.

Kline, L., and Pang, P.(1988). Calcitonin gene-related peptide relaxed rat tail artery helical strips *in vitro* in an intracellular calcium-dependent manner. *Eur. J. Pharmacol. 150,* 233–238.

Kubota, M., Moseley, J. M., Butera, L., Dusting, G. J., and Martin, T. J. (1985). Calcitonin gene-related peptide stimulates cyclic AMP formation in rat aortic smooth muscle cells. *Biochem. Biophys. Res. Comm. 132,* 88–94.

Lee, T., Kawai, Y., Shiosaka, S., Takami, K., Kiyama, H., Hillyard, C. J., Girgis, S., MacIntyre, I., Emson, P. C., and Tohyama, M.(1985). Coexistence of calcitonin gene-related peptide and substance P-like peptide in single cells of the trigeminal ganglion of the rat: Immunohistochemical analysis. *Brain Res. 330,* 194–196.

Lembeck, F.(1988). The 1988 Ulf von Euler lecture. Substance P: From extract to excitement. *Acta Physiol. Scand. 133,* 434–454.

Maggi, C. A., and Meli, A.(1988). The sensory-efferent function of capsaicin-sensitive sensory neurones. *Gen. Pharmacol. 19,* 1–43.

Maggi, C. A., Patacchini, R., Perretti, F., Tramontana, M., Manzini, S., Geppetti, P., and Santicioli, P. (1990). Sensory nerves, vascular endothelium and neurogenic relaxation of the guinea-pig isolated pulmonary artery. *Naunyn-Schmiedberg's Arch. Pharmacol. 342,* 78–84.

Marshall, I. (1991). Vasodilator action of CGRP. *Regulatory Peptides 34,* 82.

Marshall, I., Al-Kazwini, S. J., Holman, J. J., Broad, P. M., Edbrooke, M. R., and Craig, R. K. (1986a). Structure, function and expression of the human calcitonin/alpha-CGRP gene. *In* "Neuroendocrine Molecular Biology." (G. Fink, A. J. Harmar, and K. W. McKerns, eds.), pp. 1–16. Plenum Press, New York.

Marshall, I., Al-Kazwini, S. J., Holman, J. J., and Craig, R. K. (1986b). Human and rat alpha-CGRP but not calcitonin cause mesenteric vasodilatation in rats. *Eur. J. Pharmac. 123,* 217–222.

Marshall, I., Al-Kazwini, S. J., Roberts, P. M. Shepperson, N. B., Adams, M., and Craig, R. K. (1986c). Cardiovascular effects of human and rat CGRP compared in the rat and other species. *Eur. J. Pharmac. 123,* 207–216.

Marshall, I., Al-Kazwini, S. J., Holman, J. J., and Craig, R. K. (1988). Human alpha-calcitonin gene-related peptide (CGRP) is a potent vasodilator in human mesenteric vasculature. *Br. J. Clin. Pharmac. 26,* 691–695.

Maton, N., Webber, P., Zhou, Z. C., Gardner, J. D., and Jensen, R. T. (1988). Discovery of a peptide that acts as an antagonist of calcitonin gene-related peptide (CGRP) *Gastroenterology 94,* A289.

McCormack, D. G., Mak, J. C. W., Coupe, M. O., and Barnes, P. J. (1989). Calcitonin gene-related peptide vasodilatation of human pulmonary vessels. *J. Appl. Physiol. 67,* 1265–1270.

McCulloch, J., Uddman, R., Kingman, T. A., and Edvinsson, L. (1986). Calcitonin gene-related peptide: Functional role in cerebrovascular regulation. *Proc. Natl. Acad. Sci. U.S.A. 83,* 5741–5745.

McEwan, J., Larkin, S., Davies, G., Chierchia, S., Brown, M., Stevenson, J., McIntyre, I., and

Maseri, A. (1985). Calcitonin gene-related peptide: A potent dilator of epicardial coronary arteries. *Circulation 74,* 1243–1247.

Mimeault, M., Fournier, A., Dumont, Y., St-Pierre, S., and Quirion, R. (1991). Comparative affinities and antagonistic potencies of various human calcitonin gene-related peptide fragments on calcitonin gene-related peptide receptors in brain and periphery. *J. Pharmacol. Exp. Ther. 258,* 1084–1090.

Mulderry, P. K., Ghatei, M. A., Rodrigo, J., Allen, J., Rosenfeld, M. G., Polak, J. M., and Bloom, S. R.(1985). Calcitonin gene-related peptide in cardiovascular tissues of the rat. *Neuroscience 14,* 947–954.

Mulholland, M. W., Sarpa, M. S., Delvalle, J., and Messina, L. M. (1991). Splanchnic and cerebral vasodilatory effects of calcitonin gene-related peptide I in humans. *Ann. Surg. 214,* 440–447.

Nelson, M. T., Huang, Y., Brayden, J. E., Hescheler, J., and Standen, N. B. (1990). Arterial dilations in response to calcitonin gene-related peptide involve activation of K^+ channels. *Nature 344,* 770–773.

Nilsson, L., Edvinsson, L., and Jansen, I. (1991). Mechanisms of action of the dilatory response to calcitonin gene-related peptide in guinea pig basilar artery. *Ann. N.Y. Acad. Sci. 657,* 510–512.

Persson, K., Garcia-Pascual, A., and Anderson, K.-E. (1991). Differences in the actions of calcitonin gene-related peptide on pig detrusor and vesical arterial smooth muscle. *Acta Physiol. Scand. 143,* 45–53.

Prieto, D., Benedito, S., and Berg Nyborg, N. C. (1991). Heterogeneous involvement of endothelium in calcitonin gene-related peptide-induced relaxation in coronary arteries from rat. *Br. J. Pharmacol. 103,* 1764–1768.

Rosenfeld, M. G., Mermod, J. J., Amara, S. G., Swanson, L. W., Sawchenko, P. E., Rivier, J., Vale, J. E., and Evans, R. M. (1983). Production of a novel neuropeptide encoded by the calcitonin gene via tissue-specific RNA processing. *Nature* (Lond.) *304,* 129–135.

Saito, A., Masaki, T., Uchiyama, Y., Lee, T. J. F., and Goto, K. (1989). Calcitonin gene-related peptide and vasodilator nerves in large cerebral arteries of cats. *J. Pharmacol. Exp. Ther. 248,* 455–462.

Schon, F., Ghatei, M., Allen, J. M., Mulderry, P. K., Kelly, J. S., and Bloom, S. R. (1985). The effect of sympathectomy on calcitonin gene-related peptide levels in the rat trigeminovascular system. *Brain Research 384,* 197–200.

Sigrist, S., Franco-Cereceda, A., Muff, R., Henke, H., Lundberg, J. M., and Fischer, J. A. (1986). A specific receptor and cardiovascular effects of calcitonin gene-related peptide. *Endocrinology 119,* 381–389.

Struthers, A. D., Brown, M. J., MacDonald, D. W. R., Beacham, J. L., Stevenson, J. C., Morris, H. R., and MacIntyre, I. (1986). Human calcitonin gene-related peptide: A potent endogeneous vasodilator in man. *Clin. Sci. 70,* 389–393.

Szolcsányi, J. (1984). Capsaicin-sensitive chemoceptive neuronal system with dual sensory-efferent function. *In* "Antidromic Vasodilatation and Neurogenic Inflammation." (L. A. Chahl, J. Szolcsányi, and F. Lembeck, eds.), pp. 27–52. Akademia Kiado, Budapest.

Thom, S. McG., Hughes, A. D., Goldberg, P., Martin, G., Schachter, M., and Sever, P. S. (1987). The actions of calcitonin gene-related peptide and vasoactive intestinal peptide as vasodilators in man *in vivo* and *in vitro*. *Br. J. Clin. Pharmacol. 24,* 139–144.

Uddman, R., Edvinsson, L., Ekman, R., McCulloch, J., and Kingman, T. A. (1985). Innervation of the feline cerebral vasculature by nerve fibers containing calcitonin gene-related peptide: Trigeminal origin and co-existence with substance P. *Neurosci. Lett. 62,* 131–136.

Wharton, J., Gulbenkian, S., Mulderry, P. K., Ghatei, M. A., McGregor, G. P., Bloom, S. R., and Polak, J. M. (1986). Capsaicin induces a depletion of calcitonin gene-related peptide (CGRP)-immunoreactive nerves in the cardiovascular system of guinea pig and rat. *J. Auton. Nerv. Syst. 16,* 289–309.

Williams, G., Cardoso, H., Ball, J. A., Mulderry, P. K., Cooke, E., and Bloom, S. R. (1988). Potent and comparable vasodilator actions of α- and β-calcitonin gene-related peptides on the superficial subcutaneous vasculature of man. *Clin. Sci. 75,* 309–313.

Zschauer, A., Uusitalo, H., and Brayden, J. E. (1991). Characteristics of CGRP-induced relaxation in rabbit ophtalmic artery. *Regul. Peptides 34,* 116.

CHAPTER 14

Characterization of Neurokinin Receptors in the Circulation

DOMENICO REGOLI NOURREDINE ROUISSI
PEDRO D'ORLÉANS-JUSTE

I. Introduction

A. *Cardiovascular Effects of Substance P and Related Neurokinins*

A potent hypotensive effect of substance P (SP) in the rabbit was described sixty years ago by von Euler and Gaddum [1] who also observed that it was not blocked by atropine. Euler and Gaddum suggested that the SP-induced hypotension is due to peripheral vasodilation since it occurs rapidly (in 1.1 to 2.5 s) and equally well after intravenous or intraarterial application of SP. This suggests that SP is not inactivated in the lung, in accord with Vane [2]. Later studies showed that substance P is inactive on heart rate and contractility but is able to reduce peripheral vascular resistance in several vascular beds of various species, including the coronary, cerebral, mesenteric, renal, and cutaneous circulation of the dog [3,4,5,6]. In man, substance P increases the forearm blood flow when infused at doses from 1 to 6.5 ng/min and the effect is dose dependent [7]. Recently, the group of Maseri [8] has performed a systematic investigation of the coronary vasodilator effect of substance P in humans. It has been shown that substance P at very small doses (2.8–11.2 pmol/min) induces epicardial coronary artery dilatation without exerting any effect on other peripheral vascular beds. SP reverses the effect of the constrictor ergonovine [8]. A summary of relevant work performed *in vivo* and in isolated vascular beds is presented in Tables 1A and 1B.

Peripheral vasodilatation induced by substance P in man [7] and in the dog [5] is not modified either by atropine, propranolol, or indomethacin and was attributed by Eklund *et al.* [7] to a direct relaxing effect of the peptide on the arterial smooth muscles. A few years later, Furchgott [11] and other investigations [12,13,14] observed that the presence of endothelium is required *in vitro* for allowing the

Table 1A Vascular Responses to Neurokinins *in Vivo* and in Perfused Organs

	Effect	Agonist	Threshold	Reference
Human arterial pressure	R	SP	44.8 pmol	*a*
Human coronary vessels	R	SP	2.8 pmol	*a*
Human hand dorsal vein	R	SP	17 pmol	*b*
Anesthetized dog	R	SP	100 nmol/kg	*c*
Anesthetized rat	R	SP	20 nmol/kg	*d*
Isolated dog heart	R	SP	40 pmol/100 g	*e*
Canine portal circuit	R	SP	1 fmol	*f*
Dog coronary bed	R	SP	100 nmol/kg	*c*
Rat venous mesenteric bed	C	NKB	100 pmol	*g*
Rat arterial mesenteric bed	Inactive	SP	10 nmol	*g*
Rat plasma extravasation	R	SP	50 pmol	*h*

Table 1B Vascular Responses to Neurokinins in Isolated Vascular Preparations

	Endothelium	Effect	Agonist	pD_2	Reference
Human epicardial coronary artery	Yes	R	SP	9.0	*i*
Dog carotid artery	Yes	R	SP	10.0	*j*
Dog renal artery	Yes	R	SP	10.5	*k*
Rabbit pulmonary artery	Yes	R	SP	9.6	*l*
Rabbit pulmonary artery	No	C	NKA	8.2	*l*
Rabbit aorta	No	C	NKA	8.06	*m*
Pig coronary artery	Yes	R	SP	9.4	*n*
Rabbit jugular vein	No	C	SP	8.8	*o*
Rabbit vena cava	No	C	SP	7.9	*p*
Rat portal vein	No	C	NKB	7.7	*q*

Abbreviations: R, relaxation; C, constriction in A, contraction in B; pD_2, the log of the molar concentration of agonist that produces 50% of the maximum effect.

[a] Crossman, D. C., Larkin, S. W., Fuller, R. W., Davies, G. J., and Maseri, A. (1989). *Circulation 80,* 475.

[b] Mattuci-Cerinic, M., Pietrini, V., and Marabini, S. (1990). *Clin. Exp. Rheumatol. 8,* 565.

[c] Maxwell, G. M. (1968). *Br. J. Pharmacol. Chemother. 32,* 514.

[d] Couture, R., Laneuville, O., Guimond, C., Drapeau, G., and Regoli, D. (1989). *Naunyn-Schmiedeberg's Arch. Pharmacol. 340,* 547.

[e] Losay, S., Mroz, E. A., Tregear, G. W., Leeman, S. E., and Gamble, W. I. (1977). "Substance P," V. S. Von Euler and B. Pernow, ed. (p. 287), Raven Press, New York.

[f] Withrington, P. (1987). *J. Hepatol, 4,* 16.

[g] D'Orléans-Juste, P., Claing, A., Telemaque, S., Warner, T. D., and Regoli, D. (1991). *Eur. J. Pharmacol. 204,* 329.

[h] Devor, M., Papir-Kricheli, D., Nachmias, E., Rosenthal, F., Gilon, F. C., Chorev, M., and Selinger, S. (1989). *Neurosci. Letters 103,* 203.

[i] Chester, A. H., O'Neil, G. S., Tadjkarimi, S., Palmer, R. N. J., Moncada, S., and Yacoub, M. H. (1990). *Int. J. Cardiol. 29,* 305.

[j] D'Orléans-Juste, P., Dion, S., Mizrahi, J., and Regoli, D. (1985). *Eur. J. Pharmacol. 114,* 9.

(*continued*)

Table 1B (*Continued*)

[k] Regoli, D. (personal communication).
[l] D'Orléans-Juste, P., Dion, S., Drapeau, G., and Regoli, D. (1986). *Eur. J. Pharmacol. 125,* 37.
[m] Constantine, J. W., Lebel, W. S., and Woody, H. A. (1990). *Naunyn-Schmiedeberg's Arch. Pharmacol. 342,* 722.
[n] Gulati, N., Mathison, R., Huggel, H., Regoli, D., and Beny, J. L. (1987). *Eur. J. Pharmacol. 136,* 149.
[o] Nantel, F., Rouissi, N. E., Rhaleb, N.-E., Dion, S., Drapeau, G., and Regoli, D. (1990). *Eur. J. Pharmacol. 179,* 457.
[p] Nantel, F. (1991). Master Thesis, University of Sherbrooke.
[q] Mastrangelo, D., Mathison, R., Huggel, H. J., Dion, S., D'Orleans-Juste, P., Rhaleb, N. E., Drapeau, G., Rovero, P., and Regoli, D. (1987). *Eur. J. Pharmacol. 134,* 321.

relaxant effect of acetylcholine, bradykinin, and substance P in isolated arterial strips. More recently, substance P and bradykinin have been shown to promote the release of prostacyclin [15,16]. As a peripheral vasodilator, substance P is as active as prostaglandins and, in some species, (e.g., the dog) it is 30 to 100 times more potent than bradykinin [6]. In other sections of the vasculature, substance P has been shown to increase capillary permeability [9] and contract the veins in man [7] and in animals either by a direct activation of the smooth muscle fiber [10,17] or indirectly, by promoting the release of catecholamines, since the effect can be prevented by phentolamine [7]. These data indicate that substance P and the newly discovered neurokinin A (NKA) [18,20] and neurokinin B (NKB) [19] act on several cell types in peripheral vessels, in particular, the arterial and capillary endothelia, the arterial and venous smooth muscle fibers, and possibly other cell types. The multiple biological effects of substance P and related neurokinins in the cardiovascular system are mediated by several receptor types that have been classified and named at the Montreal International Symposium on Neurokinins in 1986 [21]. Primary structures of neurokinins as well as the three receptor types and their localization in the cardiovascular system are schematically presented in Table 2.

Table 2 Neurokinins, Receptors, and Receptor Localization in the Cardiovascular System

Neurokinin	Receptor	Receptor localization	Reference
ubstance P			
H. Arg-Pro-Lys-Pro-Gln-Gln-*Phe*-Phe-*Gly-Leu-Met*.NH_2	NK-1	Endothelium	[36]
		Smooth muscles	[17]
eurokinin A			
H.His-Lys-Thr-Asp-Ser-*Phe*-Val-*Gly-Leu-Met*.NH_2	NK-2	Smooth muscles	[58]
		Sympathetic nerves	[52]
eurokinin B			
H.Asp-Met-His-Asp-Phe-*Phe*-Val-*Gly-Leu-Met*.NH_2	NK-3	Smooth muscles	[36]
		Parasympathetic nerves	[54]

II. Neurokinins in Peripheral Vessels: Localization and Functions

Substance P–like immunoreactivity has been demonstrated in vessels of various species, such as the rat [22], the guinea pig [23], cat [24], and man [25–26] among others. SP-like immunoreactivities are localized in nerve fibers (probably sensory nerve fibers), the majority of which can be eliminated by pretreatment with capsaicin [25]: SP-like immunoreactivity may consist not only of SP but also of NKA and NKB since the three neuropeptides have been identified in two different genes and are expressed and possibly released by the same neurons [18,19,20,26]. Moreover, SP and related neurokinins coexist with CGRP [24]. When released by chemical (e.g., capsaicin) or physical (e.g., electrical impulses) stimulations, the sensory neuropeptides produce vasodilatation and increase blood flow in the nasal mucosa of the cat [24,26] and these changes may be associated with local edema. Intraarterial infusions of sensory neuropeptides in the nasal mucosa of the cat [24] induce similar changes of mucosal blood flow as capsaicin or electrical stimulation. SP is the most potent agent, NKA and NKB being very weak; the effect of SP is however of shorter duration than that of CGRP [25]. These data suggest that SP may play a major role in the tissue reactions to neurogenic inflammation, and the receptor involved in these reactions may be of the NK-1 type since this is the receptor that mediates the release of endothelium-derived relaxing factor (EDRF) from the endothelium, the liberation of histamine, and other phenomena related to the inflammatory reaction [27,28]. Such an interpretation should be confirmed by the use of selective NK-1 receptor antagonists. It thus appears that the widespread presence of neurokinins in various vessels described [23] as well as the intense network of sensory fibers detected in cerebral blood vessels of various mammals, including man [25], and other recent findings reviewed by Stjarne *et al.* [24,26] support the interpretation that antidromic stimulation of sensory fibers promotes the release of neuropeptides in the musculature or at the adventitia-medial junction of arteries and veins, and thus allows neuropeptides to act on endothelia or smooth muscle fibers.

As pointed out before [29] by two of us, substance P and related neurokinins may act near the site of their release, which may be far away from the endothelium in large vessels but, perhaps, not as far in small vessels, which are instrumental for the regulation of peripheral vascular resistance.

III. Pharmacological Characterization of Neurokinin Receptors in Isolated Vessels

A variety of isolated organs has been used since 1931 to evaluate the myotropic (relaxant or contractile) effects of substance P and related neurokinins. Isolated vessels of various species have been extensively studied in our and other laboratories [28,30,33,35,48] and the most useful preparations are mentioned in Table 1B. It has been shown that SP, NKA, and NKB induce endothelium-dependent relaxation of the isolated arteries and endothelium-independent contraction of the veins. In some arteries of the rabbit (the rabbit aorta and pulmonary artery), a biphasic effect has been described [33,58], and removal of endothelium leaves a pure contractile response particularly sensitive to NKA. Contractile response of the rat portal-mesenteric vein is sensitive to NKB. Worthy of notice is the fact that the preparations in which SP is the most potent agonist show high sensitivity for this peptide (pD_2 8.8–10), while the rabbit arteries and the rat portal vein show lower sensitivity to their preferred stimulants, NKA and NKB. Among the preparations analyzed in Table 1B, those that have been underlined are monoreceptor systems that have been extensively used for measuring the biological activities of a large number of analogs and derivatives of neurokinins during the search for selective agonists and antagonists [37,31,36]. At the same time, these vascular preparations provide the best bioassays for neurokinins [31,36] and for studying the three functional neurokinin receptor types that are present in the cardiovascular system. Two of these isolated preparations, the dog carotid artery that responds to substance P by a relaxation mediated through EDRF and the rabbit jugular vein that contracts to substance P, have been used for the NK-1 receptor in order to investigate the two basic peripheral vascular effects of neurokinins, arterial vasodilatation, and venoconstriction.

The three neurokinin receptors have been characterized by using (a) the peptides identified in mammals (SP, NKA, NKB); (b) physalaemin, eledoisin, and kassinin, three of the tachykinins occurring in lower species; and (c) the C-terminal octapeptide (for SP) or the heptapeptide (for NKA and NKB) fragments (Table 3). Selective agonists for each of the neurokinin receptors that have been identified in our laboratories [36,38,39] have also been used (Table 4) to further characterize the neurokinin receptors by agonists and establish the orders of potency of agonists in Schema 1.

The same compounds have been used to compare vascular and nonvascular preparations (Table 5). The nonvascular preparations, the guinea pig ileum, the rat duodenum, and the hamster urinary bladder are those most frequently used by other investigators in structure–activity studies of neurokinins [40,41,42]. The data presented in Table 5 indicate that the NK-1 system of the guinea pig ileum is not a pure NK-1 because it shows high sensitivity to [$MePhe^7$]-NKB. The rat duodenum is very sensitive to the NK-2 and NK-3 selective agonists but respond also to [Sar^9,$Met(O_2)^{11}$]SP. The hamster urinary bladder responds to both [βAla^8]NKA [4–10] and [$MePhe^7$]NKB [43]. Because of the presence of more than one type of neurokinin receptors, these organs are not to be recommended for identification and characterization of neurokinin antagonists [36].

Neurokinin receptors have been further characterized by using antagonists, and the results are illustrated in Table 6. A series of compounds of different molecular sizes were investigated starting with the undecapeptide Spantide that was discovered in the early 1980s by Folkers *et al.* [40]. C-terminal octapeptide sequences containing the same substitution as in Spantide were prepared in the early 1980s [44] and a hexapeptide (MM5-150) was also developed [45]. These three types of peptides contain two D-Trp residues in their sequence and show very similar antag-

Table 3 Biological Activities of Substance P, Related Neurokinins, and Tachykinins and Neurokinin Fragments in Isolated Vascular Preparations That Possess a Single Receptor Type (NK-1, NK-2, or NK-3)[a]

	NK-1				NK-2		NK-3	
	Dog carotid artery		Rabbit jugular vein		Rabbit pulmonary artery		Rat portal vein	
Peptide	pD_2	R.A.	pD_2	R.A.	pD_2	R.A.	pD_2	R.A.
Substance P	10.00	100	8.83	100	6.13	100	5.82	100
Neurokinin A	9.40	25	7.65	7	8.22	12303	6.45	427
Neurokinin B	8.90	8	7.84	10	7.45	2089	7.68	7244
Physalaemin	10.00	100	8.88	112	6.02	78	6.18	229
Eledoisin	9.75	32	8.43	43	8.22	12303	7.11	1950
Kassinin	9.20	16	8.48	45	7.66	3388	7.20	2399
SP (4–11)	9.67	47	8.62	61	6.23	126	5.87	112
NKA (4–10)	8.62	4	7.40	4	8.52	24547	6.79	933
NKB (4–10)	8.70	5	6.07	0.2	7.95	6607	6.67	708

[a]NK-1, NK-2, NK-3, neurokinin receptors according to the Montreal-nomenclature (Henry [21]); pD_2, the -log of the molar concentration of peptide producing 50% of the maximum effect; R.A., relative activity expressed in percent of that of SP (D.C.A., R.J.V.), NKA (R.P.A.), or NKB (R.P.V.).

Table 4 Biological Activities of Selective Neurokinin Agonists in Four Isolated Vessels Possessing a Single Neurokinin Receptor Type [a]

	NK-1				NK-2		NK-3	
	Dog carotid artery		Rabbit jugular vein		Rabbit pulmonary artery		Rat portal vein	
Peptide	pD_2	R.A.	pD_2	R.A.	pD_2	R.A.	pD_2	R.A.
[Sar^9,$Met(O_2)^{11}$]SP	10.45	282	8.86	107	Inactive		Inactive	
[Arg^6,Sar^9,$Met(O_2)^{11}$]SP (6–11)	10.24	174	8.99	144.5	Inactive		Inactive	
[Nle^{10}]NKA (4–10)	7.64	4.4	n.d.		8.01	62	6.49	110
[βAla^8]NKA (4–10)	6.70	0.05	6.23	0.3	8.60	240	6.13	48
[$MePhe^7$]NKB (4–10)	7.09	2	6.07	1	Inactive		7.58	79
[$MePhe^7$]NKB	7.15	0.1	6.23	0.3	5.24	1	8.30	417

[a] Abbreviations and captures as in Table 3; n.d.: not determined.

Schema 1

		Neurokinins	Tachykinins	Fragments	Selective agonists[a]
NK-1	Dog carotid artery	SP > NKA > NKB	PHY > ELE > KAS	SP (4–11) > NKA (4–10) > NKB (4–10)	A > B and C
	Rabbit jugular vein	SP > NKA > NKB	PHY > ELE > KAS	SP (4–11) > NKA (4–10) > NKB (4–10)	A > B and C
NK-2	Rabbit pulmonary artery	NKA > NKB > SP	ELE > KAS > PHY	NKA (4–10) > NKB (4–10) > SP (4–11)	B > C > A
NK-3	Rat portal vein	NKB > NKA > SP	KAS > ELE > PHY	NKA (4–10) = NKB (4–10) > SP (4–11)	B > B > A

[a] A, [Sar^9,Met$(O_2)^{11}$]SP; B, [βAla^8]NKA (4–10); C, [MePhe^7]NKB.

Table 5 Activities of Neurokinin and Selective Agonists on Vascular and Nonvascular Preparations[a]

	NK-1			NK-2		NK-3	
Peptide	D.C.A.	R.J.V.	G.P.I.	R.P.A.	R.D.	R.P.V.	H.U.B.
Substance P	10.00	8.83	8.78	613	6.50	5.82	5.57
Neurokinin A	9.40	7.65	8.40	822	8.22	6.45	7.40
Neurokinin B	8.90	7.84	8.64	7.45	8.15	7.68	7.20
[Sar^9,Met(O$_2$)^{11}]SP	10.45	8.86	8.91	Inactive	6.67	Inactive	Inactive
[β-Ala^8]NKA (4–10)	6.70	6.23	7.33	8.60	8.04	6.13	7.16
[MePhe^7]NKB	7.15	6.23	8.76	5.24	8.66	8.30	6.14
[Arg^6,Sar^9,Met(O$_2$)^{11}]SP (6–11)	10.24	8.99	9.2	Inactive	6.87	Inactive	Inactive

[a]Values are pD$_2$ (see definition in Table 3). D.C.A., dog carotid artery; R.J.V., rabbit jugular vein; G.P.I., guinea pig ileum; R.P.A., rabbit pulmonary artery; R.D., rat duodenum; R.P.V., rat pulmonary vein; H.U.B., hamster urinary bladder.

onistic activities on the NK-1 system of the dog carotid artery and rabbit jugular vein. They are weaker by 1 to 1.5 log unit on the NK-2 receptor, but Spantide is fairly active (pA$_2$ 6.2) in the rat portal vein (NK-3). The octapeptide sequences are much weaker than Spantide on the NK-3 site, and the hexapeptide is inactive. The size of the antagonist was further reduced by Fujisawa [46] who patented a large number of tripeptides, including the compound reported in Table 6. Ac.Thr-D-Trp(For)-Phe-N.MeB$_z$ was found to be very active on the NK-1 receptor systems, particularly on the dog carotid artery (pA$_2$ 7.8) and much weaker on the other two receptors. The compound shows therefore some selectivity for the NK-1 receptor. A nonpeptidic compound, SN 289.428 ((±)CP 96345), prepared by the Eli Lilly investigators following the indication of a Pfizer patent [47] was tested and found to be extremely potent on the NK-1 and very weak on the other two systems. This compound is therefore selective for the NK-1 receptor (Table 6) and appears to be promising for therapeutic applications.

IV. Mediators and Second Messengers Involved in the Cardiovascular Effects of Neurokinins

By activating at least three receptor types, substance P and related neurokinins exert a large variety of biological effects in the cardiovascular and in other systems: these include smooth muscle contraction, exocrine and endocrine secretions, neurons activation, and local release of various autacoids (histamine, prostanoids) and neurotransmitters (noradrenaline, acetylcholine). Receptors for neurokinins are present in various cell types, such as smooth muscles, endothelia, epithelia, blood and tissue cells, neurons, etc. Neurokinins exert direct effects when the response occurs in the same cell as the receptor target cell: the effect is considered indirect when it occurs in another cell than that in the target. In this case, a mediator is needed, a

Table 6 Antagonists Activities of Several Compounds on Vascular Neurokinin Receptors[a]

Peptide	NK-1		NK-2	NK-3
	Dog carotid artery pA_2	Rabbit jugular vein pA_2	Rabbit pulmonary artery pA_2	Rat portal vein pA_2
D-Arg[D-Trp^{7,9},Nle^{11}]SP (Spantide)	n.d.	6.78	5.78	6.18
[D-Pro^4,D-Trp^{7,9}Nle1^{M}]SP (4–11)	6.8	6.29	5.1	5.4
[D-Tyr^4,D-Trp^{7,9},NleA^{11}]SP (4–11)	7.0	6.07	5.4	5.8
MM5-150[b]	n.d.	n.d.	4.5	Inact.
Ac.Thr-D-Trp(For)-Phe-N.MeB$_z$	7.79	7.0	5.4	5.8
SN 289.428[c]	9.52	9.50	4.0	5.6

[a] pA_2 pA_{10}, the -log of the molar concentration of antagonist that reduces the effect of a double dose of agonist to that of a single dose (Schild, 1943); n.d., not determined.
[b] MM5-150, PHPA-D-Trp-Phe-D-Trp-Leu-Nle-NH_2, PHPA, parahyohoxyphenylacetyl-.
[c] Jukic, D., Rouissi, N., Laprise, R., and Regoli, D. (1991). *Life Science 49,* 1463.

substance (e.g., EDRF, acetylcholine, noradrenaline) that is released from the target cell and act on the effector cell. Examples of such effects are given in Table 7, where mediators and second messengers of the three neurokinin receptors are analyzed. Thus, relaxing (vasodilator) effects of substance P occurring in isolated arteries of the dog and the rabbit (Table 7) are due to the activation of a NK-1 receptor present in endothelial cells, which subsequently stimulates a Ca^{2+} and calmodulin system that leads to synthesis and release of EDRF (nitric oxide) [37]. EDRF penetrates the smooth muscle and activates the soluble guanylate cyclase and the production of cGMP [61] that relaxes the smooth muscle fibers. Relaxation of isolated arteries in response to neurokinins occurs only when the endothelial layer is intact [11,28]. Conversely, contractions of some rabbit veins (the rabbit jugular or cava veins) occur in the absence of endothelium and are expected to be due to activation of receptors in the smooth muscle fibers. Indirect evidence obtained with a systematic pharmacological approach by Nantel *et al.* [48] lead to the conclusion that contraction of venous smooth muscle is mediated by calcium and by the activation of phosphoinositides. Indeed, this mechanism has been shown to be present in the guinea pig ileum where NK-1 receptors coupled to G-proteins are linked to the phosphoinositide transmembrane-signaling pathway [49,50]. Activation of NK-1 receptors in primary cultures of mouse cortical astrocytes lead also to stimulation of phosphoinositide hydrolysis [51].

NK-2 receptors that have been found in some rabbit arteries (rabbit aorta and pulmonary artery) (Table 7) have been shown to mediate either contraction or cell multiplication (mitogenesis) in smooth muscles. Indirect evidence [17] suggests that calcium is essential for contraction of the rabbit pulmonary artery since contraction is inhibited by calcium and calmodulin antagonists. As for the rabbit aorta, the effect of SP and NKA to stimulate DNA synthesis in connective tissue cells was associated to activation of inositol phosphate and it was blocked by pertussis toxin, as well as by Ca^{2+} and Ca^{2+}-calmodulin antagonists [34]. In addition to these two direct effects, NK-2 receptors act indirectly to increase heart rate in rats [36] or potentiate the effect of electrical stimulation in vasa deferentia [52]. These indirect effects have been attributed to the ability of NKA and related neurokinins to promote the release of noradrenaline from the sympathetic nerve endings. It remains to be established if calcium and phosphoinositides are the mediators for this effect. The same is true for NK-3 receptors that have been shown to mediate contractions of the rat portal vein [35] and constriction of the rat venous mesenteric vascular bed [53] by acting directly on the venous smooth muscle. NK-3 receptors have also been demonstrated in parasympathetic nerve terminals [54] where they promote the release of acetylcholine. Indeed, NK-3 receptor activation by the selective agonist senktide has been linked to the phosphoinositide pathway both in the myenteric plexus of the guinea pig ileum [49] and in the dorsal horn of the rat spinal cord [55].

It therefore appears that activation of the Ca^{2+}-calmodulin system and of the phosphoinositides pathway is involved in all biological effects of neurokinins and represents the second messengers of the three neurokinin receptor types. This is consistent with the fact that the three functional sites belong to a superfamily of G-

Table 7 Cellular Mediators and Intracellular Messengers Involved in the Myotropic Responses in Isolated Organs to Neurokinins

Preparation	Receptor	Target cell	Second messenger	Mediator	Effector cell	Second messenger	Effect	Reference
Dog carotid artery	NK-1	Endothelial	Ca^{2+}-Calmodulin[a]	EDRF	Vascular	cGMP[a]	R[j]	*b*
Dog renal artery	NK-1	Endothelial	Ca^{2+}-Calmodulin[a]	EDRF	Vascular	cGMP[a]	R	*c*
Rabbit jugular vein	NK-1	Smooth muscle	Ca^{2+}		Vascular		C[j]	*d*
Rabbit vena cava	NK-1	Smooth muscle	Ca^{2+}		Vascular		C	*d*
Rabbit pulmonary artery*	NK-1	Endothelial	Ca^{2+}-Calmodulin[a]	EDRF	Vascular	cGMP	R	*j*
Rabbit pulmonary artery†	NK-2	Smooth muscle	Ca^{2+}		Vascular		C	*e*
Rat portal vein	NK-3	Smooth muscle	Ca^{2+}		Vascular		C	*f*
Guinea pig ileum‡	NK-1	Smooth muscle	IP_3		Intestinal		C	*g*
Guinea pig ileum	NK-3	Parasymp. nerve	IP_3	Ach	Intestinal	Ca^{+2}	C	*h*
Rat aorta	NK-2	Smooth muscle	IP_3		Vascular		Mitogenic	*i*

Abbreviations: IP_3; Inositol triphosphate; EDRF: endothelium-derived relaxing factor; Ach: acetylcholine; cGMP guanosine 3′:5′-cyclic monophosphate-; R: relaxation; C: contraction.

*With endothelium. †Without endothelium. ‡In presence of atropine.

[a] Busse, R., and Mulsch, A. (1990). *Febs Lett. 265,* 133.

[b] D'Orléans-Juste, P., Dion, S., Mizrahi, S., and Regoli, D. (1985). *Eur. J. Pharmacol. 114,* 9.

[c] Rhaleb, N. (personal communication).

[d] Nantel, F. (personal communication).

[e] Nantel, F., Rouissi, N., Rhaleb, N.-E., Jukic, D., and Regoli, D. (1991). *J. Cardiov. Pharmacol. 18,* 348.

[f] Sutter, M. (1990). *Pharmacol. Review 42,* 287.

[g] Watson, S., and Downes, C. P. (1983). *Eur. J. Pharmacol. 93,* 245–253.

[h] Gard, S., Watson, S. P., Maggio, E., To, H. P., and Watling, K. J. (1990). *Br. J. Pharmacol. 99,* 767.

[i] Hultgardh-Nilsson, A., Nilsson, J., Jonzon, B., and Dalsgaard, C. J. (1988). *J. Cell Physiol. 137,* 141.

[j] D'Orléans-Juste, P., Dion, S., Drapeau, G., and Regoli, D. (1986). *Eur. J. Pharmacol. 125,* 37.

protein-coupled receptors that have been identified by Nakanishi and his co-workers (20,56).

V. Biochemical Characterization of Neurokinin Receptors: Binding Assays and Autoradiography

Binding assays on blood vessels have been successfully used for a variety of peptides, including ANF [61,62] but not for the neurokinins. Binding assays for the peptide have been extensively utilized to characterize neurokinin-binding sites in the brain and some peripheral organs (e.g., the guinea pig ileum) [49,57] but not in arteries, veins, or the heart. The only available data for neurokinins receptors in the vessels are historadiographic demonstrations of substance P receptors in arterial endothelium and in the muscularis and adventitia of veins [60] obtained with Bolton-Hunter-SP (BH-SP). This is illustrated in Fig. 1, which is taken from Ste-

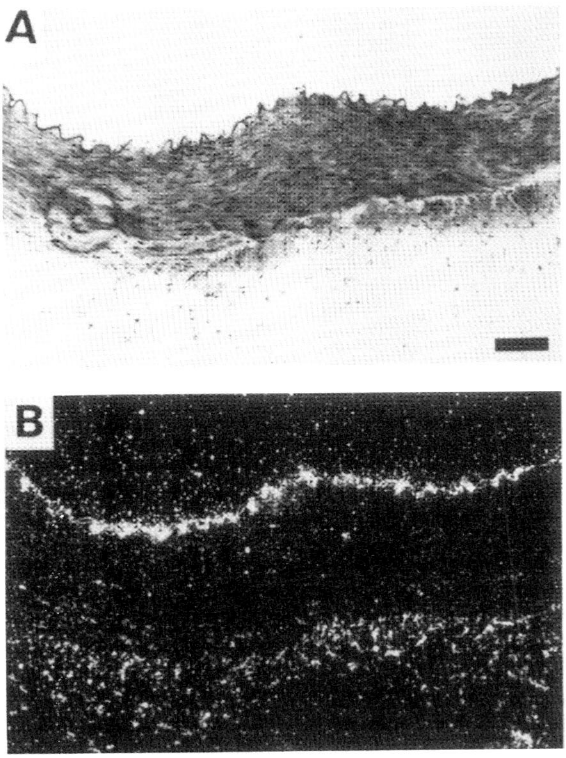

Figure 1 Autoradiographic localization of BHSP binding to sections of dog splenic artery. (A) Lightfield micrograph and (B) darkfield micrograph (silver grain localization) showing the intact endothelium of the artery. BHSP binding is associated with the endothelium of the intact artery. (Bar = 100 μM.) (From Stephenson *et al.* [60].)

phenson *et al.* [60] at the Montreal International Symposium on Neurokinins. It is evident that BH-SP labeled with ^{125}Iodine binds to the endothelia and to scattered cells in the adventitia but not at all to the smooth muscle of the dog splenic artery. No indication of SP-receptor binding to nervous sensory fibers (within the vessels) has been obtained.

VI. Conclusions

Sensory neuropeptides, such as substance P, and related neurokinins have been found in peripheral vessels and in the heart. Neurokinins are potent vasodilators that induce hypotension and changes of heart rate. These effects, as well as many other investigated organs and systems (plasma extravasation, venoconstriction, bronchoconstriction, exocrine secretions, motility activation in the gut and the urinary system, etc.), appear to be mediated by three functional sites, NK-1, NK-2, and NK-3, which belong to a superfamily of G-protein-coupled receptors that are linked to phosphoinositide signaling pathways. NK-1 receptors are present in endothelia, smooth muscles, and other cells (mast cells, lymphocytes, etc.), while NK-2 and NK-3 receptors are expected to be found in smooth muscles and the nerve endings of the autonomic nervous systems, the NK-2 in the sympathetic, the NK-3 in the parasympathetic.

It thus appears that neurokinins may modulate peripheral vascular resistance and cause plasma extravasation by acting on various cells within the vessels, namely the endothelia from which they promote the release of vasodilatory agents, an arterial and venous smooth muscles on which they act directly, and on autonomic nervous from which they promote the release of neurotransmitters. Neurokinins are contained in sensory nerves and may be released locally to produce the above-mentioned effects. In this way, SP and related neurokinins, as well as their receptors, play an important role in cardiovascular homeostasis and the regulation of local blood flows.

Acknowledgments

The authors acknowledge the secretarial help of C. Théberge and L. Bérard for typing the manuscript and the tables, as well as the technical work of M. Boussougou and R. Laprise. We thank D. Jukic and G. Drapeau for the synthesis of some of the peptides, and S. Dion, N. E. Rhaleb, F. Nantel and A. Claing, for the biological assays. The experimental work described in this paper has been performed with the financial support of the Medical Research Council of Canada (M.R.C.C.). D.R. is a career investigator of the M.R.C.C., P.D.J. is a scholar of the Canadian Heart Foundation, and N.R. is a student of the University of Sherbrooke.

References

1. Euler, U. S. V., and Gaddum, J. H. (1931). An unidentified depressor substance in certain tissue extracts. *J. Physiol.* *72*, 74–84.

2. Vane, J. R. (1969). The release and fate of vasoactive hormones in the circulation. *Br. J. Pharmac.* *35*, 209–242.

3. Maxwell, G. M. (1968). Actions of substance P on the general, pulmonary and coronary hemodynamics and metabolism in intact dogs. *Brit. J. Pharmacol. 32,* 514–522.

4. Hallberg, D., and Pernow, B. (1975). Effect of substance P on various vascular beds in the dog. *Acta Physiol. Scand. 63,* 277–285.

5. Burcher, E., Atterhog, J. H., Pernow, B., and Rosell, S. (1977). Cardiovascular effects of substance P: Effects on the heart and regional blood flow in the dog. *In* "Substance P." (U. von Euler and B. Pernow eds.), pp. 261–268. Raven Press, N.Y.

6. Lortie, M., Plante, G., Rouissi, N., and Regoli, D. unpublished.

7. Eklund, B., Jogestrand, T., and Pernow, B. (1977). Effect of substance P on resistance and capacitance vessels in the human forearm. *In* "Substance P." (U. von Euler and B. Pernow eds.), pp. 275–285. Raven Press, New York.

8. Crossman, D. C., Larkin, S. W., Fuller, R. W., Davies, G. J., and Maseri, A. (1989). Substance P dilates epicardial coronary arteries and increases coronary blood flow in humans. *Circulation 80,* 475–484.

9. Devor, M., Papir-Kricheli, D., Nachmias, E., Rosenthal, F., Gilon, C., Chorev, H., and Selinger, Z. (1988). Substance P induced cutaneous plasma extravasation in rats is mediated by NK-1 tachykinin receptors. *Neurosci. Lett. 103,* 203–208.

10. Nantel, F., Rouissi, N.-E., Rhaleb, N.-E., Dion, S., Drapeau, G., and Regoli, D. (1990). The rabbit jugular vein is a contractile NK-1 receptor system. *Eur. J. Pharmacol. 179,* 457–462.

11. Furchgott, R. F.(1981). The requirements for endothelial cells in the relaxation of arteries by acetylcholine and some other vasodilators. *TIPS 2,* 173–176.

12. Chand, N., and Altura, B. M. (1981). Acetylcholine and bradykinin relax intrapulmonary arteries by acting on endothelial cells. Role in lung vascular diseases. *Science 213,* 1376–1379.

13. DeMey, J. C., Claeys, M., and Vanhoutte, P. M. (1982). Endothelium dependent inhibitory effects of acetylcholine, adenosine triphosphate, thrombin and arachidonic acid in the canine femoral artery. *J. Pharmacol. Exp. Ther. 222,* 166–173.

14. Regoli, D. (1982). Vascular receptors for polypeptides. *TIPS 3,* 286–288.

15. Marceau, F., and Tremblay, B. (1989). Prostacyclin release induced by neurokinins in cultured human endothelial cells. *Can. J. Physiol. Pharmacol. 67,* 159–162.

16. D'Orléans-Juste, P., de Nucci, G., and Vane, J. R. (1989). Kinins act on B_1 or B_2 receptors to release conjointly endothelium-derived relaxing factor and prostacyclin from bovine aortic endothelial cells. *Brit. J. Pharmacol. 96,* 920–926.

17. Nantel, F. (Personal communication).

18. Nawa, H., Hirose, T., Takashima, H., Inayama, S., and Nakanishi, S. (1983). Nucleotide sequences of clone cDNAs for two types of bovine brain substance P precursors. *Nature 306,* 32–36.

19. Kimura, S., Osakada, M., Sugita, Y., Kangawa, I., and Munekata, E. (1983). Novel neuropeptides, neurokinin A and B isolated from porcine spinal cord. *Proc. Jap. Acad. Sci. 59,* 101–104.

20. Nakanishi, S. (1986). Structure and regulation of the preprotachykinin gene. *TINS 9,* 41–44.

21. Henry, J. L. (1987). Discussion on nomenclature for tachykinins and tachykinin receptors. *In* "Substance P and Neurokinins." (J. L. Henry, R. Couture, A. C. Cuello, G. Pelletier, R. Quirion, and D. Regoli eds.), pp. 17–18. Springer-Verlag, New York.

22. Scott, T. M., Robinson, J., and Foote, J. (1987). Development of the substance P like immunoreactive innervation of the mesenteric vascular bed in normotensive and hypertensive rats. *In* "Substance P and Neurokinins." (J. L. Henry, R. Couture, A. C. Cuello, G. Pelletier, R. Quirion, and D. Regoli, eds.), pp. 217–219. Springer-Verlag, New York.

23. Furness, J. B., and Costa, M. (1982). Substance P-like immunoreactivity in nerves associated with the vascular system of guinea pigs. *Neuroscience 7,* 447–459.

24. Stjarne, P., Lundbald, L., Anggard, A., Hokfelt, T., and Lundberg, M. (1989). Tachykinins and calcitonin gene-related peptide: Co-existence in sensory nerves of the nasal mucosa and effects on blood flow. *Cell. Tissue Res. 256,* 439–446.

25. Ullman, R., Malm, L., and Sundler, F. (1983). Substance P containing nerve fibers in the nasal mucosa. *Arch. Oto Rhino Laryngol. 238,* 9–16.

26. Stjarne, P., Lundbald, L., Lundberg, J. M., and Anggard, A. (in press). Capsaicine and nicotine sensitive afferent neurons and nasal secretion in healthy human volunteers and in patients with vasomotor rhinithis. *Br. J. Pharmacol.*

27. Lundberg, J. M., Franco-Cereceda, A., Xua, X., Hokfelt, T., and Fisher, J. A. (1985). Coexistence of substance P and calcitonin gene-related peptide-like immunoreactivity in sensory nerves in relation to cardiovascular and bronchoconstrictor effects of capsaicin. *Eur. J. Pharmacol. 108,* 315–319.

28. D'Orléans-Juste, P., Dion, S., Drapeau, G., and Regoli, D. (1985). Effects of peptides and non-peptides on isolated arterial smooth muscle, role of endothelium. *Eur. J. Pharmacol. 114,* 9–21.

29. Regoli, D, and D'Orléans-Juste, P. (1987). Substance P and neurokinins in vascular neuroeffector transmission. *In* "VIth International Symposium on Vascular Neuroeffector Mechanism." (J. A. Bevan, J. Majowski, R. A. Maxwell, and D. F. Story, eds.); pp. 131–139. ICSW Press, Paris.

30. Couture, R., and Regoli, D. (1982). Mini Review. Smooth muscle pharmacology of substance P. *Pharmacol. 24,* 1–25.

31. Regoli, D., Drapeau, G., Dion, S., and D'Orléans-Juste, P. (1987). Pharmacological receptors for substance P and neurokinins. *Life Sci. 40,* 109–117.

32. Regoli, D., Drapeau, G., Dion, S., and D'Orléans-Juste, P. (1987). Receptors for neurokinins in peripheral organs. *In* "Substance P and Neurokinins." (J. L. Henry, R. Couture, A. C. Cuello, G. Pelletier, R. Quirion, and D. Regoli, eds.), pp. 99–107. Springer-Verlag, New York.

33. Constantine, J. W., Lebel, W. S., and Woody, H. A. (1990). Smooth muscle of rabbit isolated aorta contains the NK-2 tachykinin receptor. *Naunyn-Schmiedeberg's Arch. Pharmacol. 342.* 722–724.

34. Holtgardh-Nilsson, A., Nilson, J., Jouzon, B, and C. J. Dalsgaard (1988). Coupling between inositol phosphate formation and DNA synthesis in smooth muscle cells stimulated with neurokinin A. *J. Cell. Physiol. 137,* 141–145.

35. Mastrangello, P., Mathison, R., Huggel, H. J., Dion, S., D'Orléans-Juste, P., Rhaleb, M.-E., Drapeau, G., Rovero, P., and Regoli, D. (1987). The rat isolated portal vein: A preparation sensitive to neurokinins, particularly neurokinin B. *Eur. J. Pharmacol. 134,* 321–325.

36. Regoli, D., Drapeau, G., Dion, S., and Couture, R. (1988). New selective agonists for meurokinins receptors: Pharmacological tools for receptor characterization. *Trends Pharmacol. Sci. 9,* 290–295.

37. Busse, R., and Mulsch, A. (1990). Calcium-dependent nitric oxide synthesis in endothelial cytosol is mediated by calmodulin, *FEBS Lett. 263,* 133.

38. Drapeau, G., D'Orléans-Juste, P., Dion, S., Rhaleb, N.-E., Rouissi, N.-E., and Regoli, D. (1987). Selective agonists for substance P and neurokinins. *Neuropeptides 10,* 43–45.

39. Rovero, P., Pestellini, V., Rhaleb, N.-E., Dion, S., Rouissi, N., Tousignant, C., Telemaque, S., Drapeau, G., and Regoli, D. (1989). Structure activity studies of neurokinin A. *Neuropeptides 13,* 263–270.

40. Folkers, K., Rosell, S., Hakanson, R., Chu, J. Y., Lu, L. A., Leander, S. Tang, E. F. L., and Ljungqvist, A.(1985). Chemical design of antagonists of substance P. *In* "Tachykinin Antagonists." (R. Hakanson and F. Sundler eds.), pp. 259–266. Elsevier Scientific, Amsterdam.

41. Minamino, N., Masuda, H., Kangawa, R., and H. Matsuo. (1984). Regional distribution of neuromedin K and neuromedin L in rat brain and spinal cord. *Biochem. Biophys. Res. Commun. 124,* 731–738.

42. Buck, S. H., Burcher, E., Schults, C. W., Lovenberg, W., and O'Donahue, T. L. Novel pharmacology of substance K binding sites: A third type of neurokinin receptor. *Science 226,* 887–989.

43. Dion, S., D'Orléans-Juste, P., Drapeau, G., Rhaleb, N.-E., Rouissi, N., Tousignant, C., and Regoli, D.(1987). Characterization of neurokinin receptors in various isolated organs by the use of selective agonists. *Life Sci. 41,* 2269–2278.

44. Escher, E., Parent, P., Mizrahi, J., D'Orléans-Juste, P., Dion, S., Drapeau, G., and Regoli, D. (1985). *In* "Tachykinin Antagonists." (R. Hakanson and F. Sundler eds.), pp. 267–276. Elsevier Scientific, Amsterdam.

45. Schmitt, P., Mayer, M., Magency, S., Michelot, R., and Potier, P. (1988). N-acylated pentapeptides antagonists of substance P on guinea pig ileum. *Biochem. Biophys. Res. Commun. 156,* 323–327.

46. Fujisawa's European Patent no. 0333174 A_2.
47. Pfizer patent application PCT/US89/05338.
48. Nantel, F., Rouissi, N., Rhaleb, N., Jukic, D., and Regoli, D. (1991). Pharmacological evaluation of the cellular mechanism of vasoactive peptides on venous muscle. *J. Cardiovasc. Pharmacol. 18,* 348–405.
49. Guard, S., Watling, R. J., and Watson, S. P. (1989). NK-1 tachykinins receptors are linked to inositol phospholipid hydrolysis in the guinea pig ileum longitudinal muscle-myenteric plexus preparation. *Br. J. Pharmacol. 94,* 148–154.
50. Guard, S., and Watson, S. P.(1991). Tachykinin receptor types: Classification and membrane signalling mechanisms. *Neurochem. Int. 18,* 149–165.
51. Torrens, Y., Daguet De Montety, M. C., El Etr, M., Beaujouan, J. C., and Glowinski, J. (1989). Tachykinin receptors of NK-1 type (substance P) coupled positively to phospholipase C on cortical astrocytes from the new born mouse in primary culture. *J. Neurochem. 52,* 1913–1918.
52. Tousignant, C., Dion, S., Drapeau, G., and Regoli, D. (1987). Characterization of pre- and postjunctional receptors for kinins and neurokinins in the rat vas deferens. *Neuropeptides 9,* 333–343.
53. D'Orléans-Juste, P., Claing, A., Telemaque, S., Warner, T. D., and Regoli, D. (1991). Neurokinins produce selective venoconstriction via NK-3 receptors in the rat mesenteric vascular bed. *Eur. J. Pharmacol. 204,* 329–334.
54. Laufer, R., Wormser, V., Friedman, Z. Y., Gilon, C., Chorev, M., and Selinger, Z. (1985). Neurokinin B is a preferred agonist for a neuronal substance P receptor and its action is antagonized by enkephalin. *Proc. Natl. Acad. Sci. U.S.A. 82,* 7444–7448.
55. Ninkovic, M., Beaujouan, J. C., Torrens, Y., Saffroy, M., Hall, M. D., and Glowinski, J. (1985). Differential localization of tachykinin receptors in rat spinal cord. *Eur. J. Pharmacol. 106,* 463–464.
56. Nakanishi, S. (1987). Substance P precursor and kininogen: Their structure, gene organizations and regulation. *Physiol. Rev. 67,* 1117–1142.
57. Watson, S. P., and Iversen, L. L. (1984). ^{3}H-substance P binding to guinea-pig ileum longitudinal smooth muscle membranes. *Regul. Pept. 8,* 273–281.
58. D'Orléans-Juste, P., Dion, S., Drapeau, G., and Regoli, D. (1986). Different receptors are involved in the endothelium-mediated relaxation and the smooth muscle contraction of the rabbit pulmonary artery in response to substance P and related neurokinins. *Eur. J. Pharmacol. 125,* 37–44.
59. Palmer, R. J. M., Ferrigo, A. G., and Moncada, S. (1988). Nitric oxide release accounts for the biological activity of endothelium derived relaxing factor. *Nature 327,* 524–525.
60. Stephenson, J. A., Summers, R. J., and Burcher, E. (1987). Endothelial-dependent binding of [^{125}I]-Bolton-Hunter substance P in blood vessels. *In* "Substance P and Neurokinins." (J. L. Henry, R. Couture, A. C. Cuelllo, G. Pelletier, R. Quirion, and D. Regoli, eds.) pp. 53–55. Springer-Verlag, New York, U.S.A.
61. Ignarro, L. J., Harbison, R. G., Wood, K. S., and Kadowitz, P. J. (1986). Activation of purified soluble guanylate cyclase by endothelium derived relaxing factor from artery and vein. Stimulation by acetylcholine, bradykinin and arachidonic acid. *J. Pharmacol. Exp. Ther. 237,* 893–900.
62. Hirata, Y., Tomita, M., Yoshimi, H., and Ikeda, M. (1984). Specific receptors for atrial natriuretic factor (ANF) in cultured vascular smooth muscle cells of rat aorta. *Biochem. Biophys. Res. Commun. 125,* 562–568.
63. Napier, M. A., Vandlen, R. L., Albers-Schonberg, G., Nutt, R. F., Brady, S., Lyle, T., Winquist, R., Faison, E. P., Heinel, L. A., and Blaine, E. H. (1984). Specific membrane receptors for atrial natriuretic factor in renal and vascular tissues. *Proc. Natl. Acad. Sci. U.S.A. 81,* 5946–5950.

CHAPTER 15

Peptidergic Sensory Neurons in the Local Regulation of Splanchnic Blood Flow

PETER HOLZER

I. Introduction

The cell bodies of primary afferent (sensory) neurons are located in the spinal (dorsal root) and cranial sensory ganglia and send fibers in both the central and peripheral direction. Apart from perceiving and transmitting information to the central nervous system, unmyelinated sensory neurons have been recognized to act as "effector" neurons to control blood flow and other vascular functions (for reviews see Lundberg and Saria, 1987; Chahl, 1988; Holzer, 1988; Lembeck and Holzbauer, 1988; Maggi and Meli, 1988). This control function is exerted by release of peptide transmitters from the peripheral sensory nerve terminals and is referred to as "local effector" (Holzer, 1988) or "local efferent" (Szolcsányi, 1984; Maggi and Meli, 1988) role of afferent neurons. Historically, this role was first portrayed by the finding that stimulation of the peripheral ends of cut dorsal roots induced vasodilatation in the skin area supplied by the respective neurons (Stricker, 1876; Bayliss, 1901). "Antidromic vasodilatation," as this response commonly is called, can be observed in many tissues, including the gastrointestinal tract (see Holzer, 1988). This chapter reviews the role of sensory neurons in the local control of blood flow in the splanchnic vascular bed.

II. Identity of Sensory Neurons Regulating Blood Flow

A. *Sensitivity to Capsaicin*

A property of the sensory neurons governing the circulatory system is their sensitivity to capsaicin that has been established as an important probe for sensory neuron mechanisms in mammals. Chemically related to vanillyl amide (8-methyl-*N*-vanillyl-6-nonenamide), capsaicin is the pungent ingredient of red peppers. At low doses (in the μg/kg range) this drug excites afferent neurons having small cell bodies and unmyelinated (C-fiber) or thinly myelinated (Aδ-fiber) axons, an action that is soon followed by desensitization to the drug and block of nerve conduction (see Kaufman *et al.*, 1982; Longhurst *et al.*, 1984; Szolcsányi, 1984, 1990; Holzer, 1991). Systemic administration of high doses of capsaicin (in the mg/kg range) damages these neurons, and the extent of damage (ultrastructural changes or degeneration of C-fiber and Aδ-fiber afferent neurons) depends on the dosage, route of administration, animal species, and age of the animals (see Jancsó *et al.*, 1987; Szolcsányi, 1990; Holzer, 1991). In isolated preparations or tissue cultures of rat sensory neurons capsaicin produces half-maximal excitation at 0.1–0.5 μM, and concentrations of >1 μM exert a definitely neurotoxic action within 5–20 min (Holzer, 1991). These effects are mediated by specific receptors for capsaicin-like substances, which are coupled to nonselective cation channels in the cell membrane. In line with this mechanism, intracellular accumulation of Ca^{2+} and NaCl is the major factor that determines capsaicin's neurotoxicity (Bevan and Szolcsányi, 1990). Importantly, the "sensory neuron-selective" effects of capsaicin appear to be confined to mammals and show marked species differences even within this class of vertebrates (Holzer, 1991). Capsaicin's specificity for sensory neurons is not absolute, and particularly the acute actions of capsaicin are not restricted to afferent neurons (see Holzer, 1991). Thus, some of the acute effects of the drug on blood vessels appear to result from a direct action on the vascular muscle (Donnerer and Lembeck, 1982; Duckles, 1986; Bény *et al.*, 1989; Edvinsson *et al.*, 1990; Moritoki *et al.*, 1990) and endothelium (Kenins *et al.*, 1984). The label "*capsaicin-sensitive*" is applied to those afferent neurons that acutely are *excited and* in the long term are *damaged* by the drug. This population of neurons is heterogeneous and comprises most *but not all* C-fiber and some Aδ-fiber primary afferent neurons, which are connected to chemonociceptors, polymodal nociceptors, or some warmth receptors (Szolcsányi, 1990; Holzer, 1991).

B. Peptides as Transmitters

The local effector functions of sensory neurons are mediated by transmitter substances released from their peripheral terminals (Holzer, 1988). Capsaicin-sensitive and fine afferent neurons in general contain a number of bioactive peptides that are localized within synaptic vesicles. These peptides (see Table 1) comprise substance P (SP), neurokinin A (NKA), and calcitonin gene-related peptide (CGRP) and form different pathway-specific patterns of coexistence in different sensory neurons (Gibbins *et al.*, 1987; O'Brien *et al.*, 1989). Criteria to firmly establish these peptides as sensory neurotransmitters have thus far only been met for SP/NKA and CGRP. The peptides are synthesized in the somata of sensory neurons (Keen *et al.*, 1982; Rosenfeld *et al.*, 1983; Nakanishi, 1987) and brought to the nerve endings by way of axonal transport. Importantly, the bulk of peptide is transported to the peripheral, but not central, endings of sensory neurons (Brimijoin *et al.*, 1980; Keen *et al.*, 1982). Stimulus-induced exocytotic release of the peptides from synaptic vesicles in the peripheral terminals depends on the presence of extracellular Ca^{2+} (Maggi *et al.*, 1989b; Amann, 1990). Due to their high activity in dilating certain blood vessels, CGRP and SP/NKA are foremost candidate mediators of the vascular

Table 1 Some Bioactive Peptides Present in Capsaicin-Sensitive Primary Afferent Neurons

Peptide	Selected references
Bombesin/gastrin-releasing peptide	Decker *et al.*, 1985
Calcitonin gene-related peptide	Rosenfeld *et al.*, 1983; Gibbins *et al.*, 1985; Skofitsch and Jacobowitz, 1985b; Wharton *et al.*, 1986; Franco-Cereceda *et al.*, 1987
Cholecystokinin	Jancsó *et al.*, 1981; Gibbins *et al.*, 1987
Corticotropin-releasing factor	Skofitsch *et al.*, 1985
Dynophin	Gibbins *et al.*, 1987; Weihe, 1990
Galanin	Skofitsch and Jacobowitz, 1985a
Leucine enkephalin	Weihe, 1990
Neurokinin A	Maggio and Hunter, 1984; Hua *et al.*, 1985
Peptide histidine methionine	Chéry-Croze *et al.*, 1989
Somatostatin	Jancsó *et al.*, 1981; Gamse *et al.*, 1981; Nagy *et al.*, 1981
Substance P	Jessell *et al.*, 1978; Gamse *et al.*, 1980; Nagy *et al.*, 1980; Janscó *et al.*, 1981
Vasoactive intestinal polypeptide	Jancsó et al., 1981; Skofitsch *et al.*, 1985

effects of sensory nerve stimulation (see Holzer, 1988). Pharmacological identity of the vascular actions of exogenously administered and endogenously released CGRP, SP, and NKA has been proven by the use of immunoblockade or specific antagonists. The simultaneous release of coexisting sensory neuropeptides (Saria *et al.*, 1986) is likely to give rise to an intricate control of vascular functions (see Brain and Williams, 1989).

C. Innervation of Splanchnic Vascular Bed by Peptidergic Sensory Neurons

SP- and CGRP-containing varicose and smooth nerve fibers of sensory origin innervate blood vessels throughout the body, but the density of fibers around arteries is in general higher than it is around veins (Furness *et al.*, 1982; Barja *et al.*, 1983; Uddman *et al.*, 1986; Gibbins *et al.*, 1988). The mesenteric arteries, especially the superior mesenteric artery of the rat and guinea pig, receive a very rich innervation by peptidergic sensory nerve fibers sensitive to the neurotoxic action of capsaicin (Furness *et al.*, 1982; Barja *et al.*, 1983; Mulderry *et al.*, 1985; Uddman *et al.*, 1986; Wharton *et al.*, 1986; Kawasaki *et al.*, 1988). The network of fibers is very dense in the main vessels and becomes loose towards the precapillary arterioles (Furness *et al.*, 1982). The veins of peripheral vascular beds contain only a few or even no discernible peptidergic nerve fibers. The axons run primarily in the connective tissue (tunica adventitia) surrounding the vessels and at the border between adventitia and media (muscle layer) but also in the media itself.

In the alimentary canal it is predominantly submucosal and mucosal blood vessels, especially arterioles, that are innervated by capsaicin-sensitive peptidergic afferent fibers (Furness *et al.*, 1982; Minagawa *et al.*, 1984; Papka *et al.*, 1984;

Sharkey *et al.*, 1984; Gibbins *et al.*, 1985, 1988; Rodrigo *et al.*, 1985; Uddman *et al.*, 1986; Lee *et al.*, 1987; Sternini *et al.*, 1987; Su *et al.*, 1987; Galligan *et al.*, 1988; Green and Dockray, 1988; Varro *et al.*, 1988; Chéry-Croze *et al.*, 1989). There are species differences in the neuropeptide expression of nerve fibers in the gastrointestinal tract (Brodin *et al.*, 1983b; Keast *et al.*, 1985; Sundler *et al.*, 1991), and this is particularly well exemplified by an abundance of CGRP-containing nerve fibers in the rat stomach (Green and Dockray, 1988; Varro *et al.*, 1988; Sundler *et al.*, 1991) compared with a scarcity or absence of CGRP-containing nerve fibers in the porcine and human stomach (Tsutsumi and Hara, 1989; Sundler *et al.*, 1991). Most, if not all, CGRP-containing nerve fibers in the rat stomach represent extrinsic sensory neurons that originate from dorsal root ganglia (Sternini *et al.*, 1987; Su *et al.*, 1987; Green and Dockray, 1988; Varro *et al.*, 1988) while vagal sensory fibers contribute little to the CGRP content of the rat stomach (Green and Dockray, 1988; Varro *et al.*, 1988). Vagal CGRP-containing sensory neurons, however, innervate the esophagus (Rodrigo *et al.*, 1985). SP and CGRP coexist in the perivascular sensory axons, but the overlap is not complete inasmuch as more fibers contain CGRP than SP (Gibbins *et al.*, 1985; Rodrigo *et al.*, 1985; Uddman *et al.*, 1986; Lee *et al.*, 1987; Su *et al.*, 1987; Green and Dockray, 1988; Lindh *et al.*, 1988; Sundler *et al.*, 1991). Most of the CGRP present in sensory neurons is CGRP-α (or CGRP-I) (Mulderry *et al.*, 1988; Varro *et al.*, 1988; Sternini and Andersson, 1990) although some CGRP-β (or CGRP-II) may also be expressed (Noguchi *et al.*, 1990; Sternini and Andersson, 1990). In contrast, the only form of CGRP present in enteric neurons is CGRP-β (Mulderry *et al.*, 1988; Sternini and Andersson, 1990).

III. Splanchnic Circulation

A. *Blood Flow in Mesenteric Arteries*

1. Local Regulation of Mesenteric Arterial Tone

a. NONADRENERGIC NONCHOLINERGIC VASODILATATION MEDIATED BY CAPSAICIN-SENSITIVE NEURONS. The mesenteric arteries are innervated not only by sympathetic vasoconstrictor fibers but also by nonadrenergic noncholinergic (NANC) vasodilator fibers. Thus, when sympathetic nerve activity and noradrenergic transmission are blocked, electrical nerve stimulation of the nerve plexus around isolated mesenteric arteries of the rat causes dilatation of the vessels precontracted by α-adrenoceptor agonists (Kawasaki *et al.*, 1988, 1990a; Axelsson *et al.*, 1989; Han *et al.*, 1990b) or endothelin (Kawasaki *et al.*, 1990b). The vasodilator response is independent of cholinergic transmission (Kawasaki *et al.*, 1988; Axelsson *et al.*, 1989; Han *et al.*, 1990b), endothelium-derived factors (Axelsson *et al.*, 1989; Han *et al.*, 1990b), histamine, and prostaglandins (Kawasaki *et al.*, 1988; Axelsson *et al.*, 1989). However, the NANC vasodilator fibers in the rat mesenteric arteries are sensitive to the stimulant and neurotoxic actions of capsaicin, which indicates their primary afferent nature. Intravascular administration of capsaicin, to stimulate sensory neurons, causes NANC dilatation of rat mesenteric arteries *in vitro* (Manzini and Perretti, 1988; Fujimori *et al.*, 1990) and *in vivo* (Hottenstein *et al.*, 1991). Blood flow also rises after periarterial administration of capsaicin to the superior mesenteric artery of rat (Hottenstein *et al.*, 1991) and dog (Rózsa *et al.*, 1984) *in vivo*. Capsaicin pretreatment of the rat isolated mesenteric arterial bed, to defunctionalize sensory nerve fibers, inhibits the vasodilator response to both acute capsaicin application (Manzini and Perretti, 1988) and electrical nerve stimulation (Fujimori *et al.*, 1990; Han *et al.*, 1990b; Kawasaki *et al.*, 1990a). In contrast, bovine mesenteric arteries do not relax in response to capsaicin, and NANC vasodilatation remains unaffected by pretreatment of the tissue with capsaicin (Axelsson *et al.*, 1989).

b. Release and Effects of Sensory Neuropeptides on Mesenteric Blood Flow. The peptidergic nature of the capsaicin-sensitive nerve fibers producing NANC dilatation of rat mesenteric arteries is shown by the finding that electrical stimulation of the periarterial nerve plexus leads to a Ca^{2+}-dependent release of CGRP (Fujimori *et al.*, 1989, 1990). Sensory nerve stimulation by acute capsaicin also releases CGRP, and peptide release evoked by both capsaicin and electrical stimulation is inhibited after capsaicin-induced defunctionalization of sensory nerve fibers (Fujimori *et al.*, 1990). CGRP is very potent in dilating isolated mesenteric arteries from rat (Marshall *et al.*, 1986; Uddman *et al.*, 1986; Kawasaki *et al.*, 1988, 1990a; Manzini and Perretti, 1988; Han *et al.*, 1990b), rabbit (Nelson *et al.*, 1990), and man (Törnebrandt *et al.*, 1987), and this high activity of CGRP is matched by an abundance of CGRP fibers and CGRP binding sites in the superior mesenteric artery of the rat (Wimalawansa and MacIntyre, 1988). Both in the rat and rabbit mesenteric vascular bed rat CGRP-α is more potent as a dilator than human CGRP-α (Marshall *et al.*, 1986). Isolated bovine mesenteric arteries, though, appear to be insensitive to CGRP (Axelsson *et al.*, 1989). The relaxation of rabbit mesenteric arteries induced by rat CGRP-α appears to arise primarily from an activation of K_{ATP} channels, which leads to hyperpolarization of arterial smooth muscle (Nelson *et al.*, 1990).

The effect of SP is variable depending on species and extra-/intravascular route of administration. SP is very weak in relaxing isolated bovine mesenteric arteries (Axelsson *et al.*, 1989). In the rat isolated mesenteric vascular bed, SP, NKA, and NKB fail to cause any effect (Barja *et al.*, 1983; Kawasaki *et al.*, 1988, 1990a) whereas intravenous administration of SP to the rat *in vivo* increases mesenteric blood flow (Rózsa and Jacobson, 1989). Isolated mesenteric arteries from guinea pig (Bolton and Clapp, 1986), rabbit (Stewart-Lee and Burnstock, 1989), and man (Törnebrandt *et al.*, 1987) are relaxed by SP. This effect of the peptide is mediated by NK-1 receptors located on the endothelium (Bolton and Clapp, 1986; Stewart-Lee and Burnstock, 1989). Intravascular administration of SP to the dog (Hallberg and Pernow, 1975; Rózsa *et al.*, 1984) and pig (Schrauwen and Houvenaghel, 1980) *in vivo* evokes hyperemia in the superior mesenteric artery.

Vasoactive intestinal polypeptide (VIP) dilates isolated mesenteric arteries from rat (Kawasaki *et al.*, 1988) and man (Törnebrandt *et al.*, 1987), and this activity is consistent with the presence of many VIP receptors in the mesenteric arteries of the rabbit (Huang and Rorstad, 1987; Sidawy *et al.*, 1989). VIP, however, is only weakly active in relaxing bovine mesenteric arteries *in vitro* (Axelsson *et al.*, 1989). In the rat (Holliger *et al.*, 1983; Rózsa and Jacobson, 1989) and dog (Rózsa *et al.*, 1985) *in vivo,* mesenteric blood flow is enhanced following intravascular administration of VIP. While cholecystokinin-octapeptide likewise increases mesenteric blood flow in the dog (Rózsa *et al.*, 1985) and rat (Rózsa and Jacobson, 1989), somatostatin reduces mesenteric blood flow in the dog and diminishes the vasodilator effect of intraarterial capsaicin (Rózsa *et al.*, 1984, 1985). In contrast, human mesenteric artery preparations respond to somatostatin and methionine-enkephalin with relaxation (Törnebrandt *et al.*, 1987).

c. MEDIATOR ROLE OF SENSORY NEUROPEPTIDES. A mediator role of peptides in NANC dilatation of mesenteric arteries was first shown by the inhibitory effect of α-chymotrypsin, a proteolytic enzyme (Manzini and Perretti, 1988). Specifically, CGRP seems to be a major NANC vasodilator transmitter in the mesenteric vasculature of the rat because desensitization to CGRP (Han *et al.*, 1990a), a CGRP antagonist (Han *et al.*, 1990a), and a CGRP antibody (Han *et al.*, 1990b) are able to reduce vasodilatation evoked by periarterial nerve stimulation. In contrast, tachykinins do not influence mesenteric blood flow in the rat (Kawasaki *et al.*, 1988, 1990a). However, in the dog *in vivo* combined administration of antibodies to SP, VIP and cholecystokinin-octapeptide have been reported to reduce the mesenteric vasodilator effect of capsaicin by 80%, the remaining vasodilatation being abolished by atropine (Rózsa *et al.*, 1985). Conversely, an antibody to somatostatin facilitates mesenteric blood flow by itself and further enhances the vasodilatation induced by capsaicin in the dog *in vivo* (Rózsa *et al.*, 1985).

d. INTERACTION OF SENSORY AND SYMPATHETIC NEURONS. On the one hand, sympathetic nerve activity and noradrenaline are able to inhibit vasodilatation induced by both CGRP and electrical stimulation of isolated preparations (Kawasaki *et al.*, 1988, 1990b). The available evidence indicates that endogenous noradrenaline released from sympathetic nerve endings inhibits NANC vasodilatation by a presynaptic site of action, probably by suppressing the release of the NANC vasodilator transmitters via α_2-adrenoceptors (Kawasaki *et al.*, 1990b). Vice versa, the NANC vasodilator nerves control the activity of sympathetic vasoconstrictor nerves since defunctionalization of the vasodilator nerves augments the vasoconstrictor effect of exogenous noradrenaline (Kawasaki *et al.*, 1990a) and electrical stimulation of periarterial sympthetic neurons (Kawasaki *et al.*, 1990a; Remak *et al.*, 1990) on isolated mesenteric arteries from the rat. Because the release of endogenous noradrenaline is not altered by CGRP or defunctionalization of sensory neurons, while CGRP inhibits the vasoconstrictor response to sympathetic nerve stimulation or exogenous noradrenaline, it can be inferred that the NANC vasodilator nerves inhibit noradrenaline-evoked vasoconstriction by a postjunctional mechanism of action (Han *et al.*, 1990b; Kawasaki *et al.*, 1990a).

These interactions between capsaicin-sensitive vasodilator nerves and sympathetic neurons are also seen in the rat *in vivo*. Periarterial nerve stimulation reduces blood flow in the superior mesenteric artery, but peak vasoconstriction is soon followed by a recovery of blood flow (Remak *et al.*, 1990) and hyperemia after cessation of nerve stimulation (Hottenstein *et al.*, 1990). Both the initial autoregulatory escape from sympathetic vasoconstriction (Remak *et al.*, 1990) and poststimulation hyperemia (Hottenstein *et al.*, 1990) are inhibited by ablation of capsaicin-sensitive afferent neurons. Following defunctionalization of sympathetic neurons with reserpine, periarterial nerve stimulation causes vasodilatation that also is blocked by ablation of sensory neurons (Remak *et al.*, 1990). Experiments involving surgery indicate that the sensory nerve-mediated autoregulatory escape response seems to result from a peripheral interaction between vasodilator and vasoconstrictor nerve fibers (Remak *et al.*, 1990). An additional interaction between sensory and sympathetic neurons is reflected by the observation that ablation of capsaicin-sensitive afferent neurons can lead to an increase in the transmitter content and innervation density of sympathetic nerve endings in certain tissues whereas long-term ablation of sympathetic neurons is followed by an increase in the sensory innervation (Terenghi *et al.*, 1986; Nielsch and Keen, 1987; Luthman *et al.*, 1989; Aberdeen *et al.*, 1990). These reciprocal alterations are seen in newborn and young rats in which the development and maturation of both sensory and sympathetic neurons depend on nerve growth factor, and the promotion of one of the two populations of neurons after ablation of the other may arise from the availability of additional nerve growth factor for the surviving population of neurons (Terenghi *et al.*, 1986; Nielsch and Keen, 1987; Luthman *et al.*, 1989; Aberdeen *et al.*, 1990).

2. Reflex Regulation of Mesenteric Arterial Tone

Mesenteric blood flow can be influenced by chemical or thermal stimuli applied to the mucosal or serosal surface of the intestine. Thus, blood flow rises when capsaicin is administered into the small intestinal lumen of the dog (Rózsa *et al.*, 1986) and rat (Rózsa and Jacobson, 1989; Hottenstein *et al.*, 1991) or when bile-oleate is placed into the rat jejunum (Rózsa and Jacobson, 1989). The responses to both capsaicin and bile-oleate result from a neural reflex in which capsaicin-sensitive afferent neurons participate. The transmitter mediating the vasodilator response to intraluminal bile-oleate is NANC in nature and not antagonized by antibodies to cholecystokinin-octapeptide or SP (Rózsa and Jacobson, 1989). Any involvement of CGRP has not yet been explored, but an antibody to VIP reduces bile-oleate-evoked mesenteric hyperemia (Rózsa and Jacobson, 1989). The origin of VIP, though, is not clear. In the guinea pig, mesenteric arteries are densely supplied by *capsaicin-insensitive* nerve fibers containing VIP (Della *et al.*, 1983) and originating from enteric neurons that send intestinofugal axons along the vessels (Lindh *et al.*, 1988; Vickers *et al.*, 1990). It appears, therefore, that bile-oleate-induced mesenteric hyperemia is due to a reflex involving both primary afferent (Rózsa and Jacobson, 1989) and enteric neurons, a possibility that is quite conceivable in view of the reported synaptic contacts between sensory nerve fibers and enteric neurons (Neuhuber, 1987). At any rate, it seems unlikely that mesenteric hyperemia is the result of "simple" axon reflexes between intestinal and mesenteric collaterals of sensory neurons (Rózsa *et al.*, 1986; Rózsa and Jacobson, 1989).

Intestinal warming to 45°C gives rise to both constrictor and dilator reflex responses in the superior mesenteric artery of the rat (Rózsa *et al.*, 1988). The reflex pathways run in the splanchnic but not vagal nerves and involve capsaicin-sensitive neurons as afferent reflex arc (Rózsa *et al.*, 1988). Vasoconstriction is mediated by sympathetic efferent neurons. The mesenteric vasodilator response to intestinal warming, which is seen after defunctionalization of sympathetic neurons, is relayed by pathways distal to the celiac/superior mesenteric ganglion complex and mediated by a NANC transmitter (Rózsa *et al.*, 1988). Both vasoconstriction and vasodilatation are inhibited by a SP antibody (Rózsa *et al.*, 1988), indicating that SP may serve as a transmitter released from sensory neurons. Since, however, SP is unable to dilate mesenteric arteries of the rat (Barja *et al.*, 1983; Kawasaki *et al.*, 1988, 1990a), this peptide is unlike to be the vasodilator transmitter. The possible participation of CGRP and VIP has not yet been tested.

B. Blood Flow in the Gastrointestinal Mucosa

1. Hyperemia in the Gastrointestinal Mucosa Mediated by Capsaicin-Sensitive Neurons

The exploration of a possible role of peptidergic sensory neurons in gastrointestinal functions is hampered by the fact that both extrinsic afferent and intrinsic enteric neurons have many neuropeptide transmitters in common as is the case for SP, NKA, CGRP, VIP, somatostatin, and opioid peptides (see Ekblad *et al.*, 1985; Costa *et al.*, 1986; Kirchgessner *et al.*, 1988). In addition, some peptides occur not only in enteric and primary afferent neurons but also in parasympathetic (e.g., VIP, see Lundberg, 1981) and sympathetic (e.g., somatostatin, see Costa and Furness, 1984) efferent neurons. Sensory neuron mechanisms, though, can selectively be probed by capsaicin because this drug does not directly stimulate enteric neurons (Barthó and Szolcsányi, 1978; Holzer, 1984) and appears to lack neurotoxicity on enteric neurons (Holzer *et al.*, 1980; Barthó *et al.*, 1982; Furness *et al.*, 1982; Donnerer *et al.*, 1984; Holzer, 1984; Papka *et al.*, 1984; Sharkey *et al.*, 1984; Kirchgessner *et al.*, 1988; Geppetti *et al.*, 1988; Green and Dockray, 1988).

Electrical stimulation of the peripheral vagus nerve increases gastric mucosal blood flow, which in the rat is in part mediated by cholinergic neurons and in part by noncholinergic neurons (Morishita and Guth, 1986) that are sensitive to the neurotoxic action of capsaicin and hence most probably of primary afferent origin (Thiefin *et al.*, 1990). Intragastric or submucosal application of capsaicin, to stimulate sensory nerve endings in the rat stomach, dilates submucosal arterioles (Chen *et al.*, 1991) and enhances blood flow through the mucosa (Lippe *et al.*, 1989b; Holzer *et al.*, 1990a, 1991b; Leung, 1990a; Li *et al.*, 1991), but this effect is mediated by afferent nerve fibers originating from dorsal root ganglia and passing through the celiac/superior mesenteric ganglion complex (Li *et al.*, 1991). Hyperemic responses to intraluminal capsaicin are also seen in the mucosa and muscle of the rabbit esophagus (Trad *et al.*, 1990) and in the mucosa of the rat colon (Leung, 1990b). Capsaicin-evoked hyperemia in the rat gastric mucosa involves a neural reflex (Holzer *et al.*, 1991b) and is abolished by chemical ablation of afferent nerve fibers (Lippe *et al.*, 1989b; Leung, 1990a; Holzer *et al.*, 1991b). Capsaicin-sensitive afferent neurons are furthermore engaged in the autoregulatory escape reaction from adrenaline-induced vasoconstriction in the gastric mucosa (Leung, 1990a). Of considerable pathophysiological significance is the finding that the hyperemic response of the gastric mucosa to acid back diffusion (Holzer *et al.*, 1991a; Raybould *et al.*, 1990) and of the colonic mucosa to acetic acid (Leung, 1990b) depends on the integrity of the sensory innervation of the tissue. Gastric mucosal hyperemia due to acid back diffusion is mediated by sensory nerve fibers, which originate from dorsal root ganglia (Raybould *et al.*, 1990), and in addition relies on functional pathways through the celiac/superior mesenteric ganglion (Lippe and Holzer, 1991). Gastric mucosal vasodilatation elicited by pentagastrin is not mediated by capsaicin-sensitive neurons (Livingston *et al.*, 1990).

2. Release and Effects of Sensory Neuropeptides on Blood Flow in the Gastrointestinal Mucosa

Stimulation of the vagus nerve in the cat results in the appearance of SP in the lumen of the stomach (Uvnäs-Wallensten, 1978) and upper small intestine (Grönstad *et al.*, 1985) and in an increase in the plasma levels of SP in the portal vein blood (Grönstad *et al.*, 1983), but the origin of the peptide is not known. Peptide release from sensory nerve fibers can be explored by the use of capsaicin because this drug does not seem to influence the release of SP from enteric neurons (Donnerer *et al.*, 1984; Holzer, 1984). Capsaicin releases CGRP and SP from isolated tissues of the guinea pig gallbladder (Maggi *et al.*, 1989c) and stomach (Renzi *et al.*, 1988). In the rat isolated stomach, capsaicin is very effective in releasing CGRP but only weakly active in releasing SP (Renzi *et al.*, 1990; Geppetti *et al.*, 1991). The acute effect of capsaicin to induce peptide release is abolished after defunctionalization of capsaicin-sensitive neurons (Renzi *et al.*, 1988; Maggi *et al.*,1989a,c; Geppetti *et al.*, 1991). Intravascular capsaicin facilitates release of SP into the vascular bed of the guinea pig small intestine *in vitro* (Donnerer *et al.*, 1984) and into the gastric lumen of the guinea pig *in vivo* (Renzi *et al.*, 1988). Likewise, intraarterial administration of capsaicin to the vascular bed of the rat stomach releases both SP (Kwok and McIntosh, 1990) and CGRP (Gray *et al.*, 1989; Holzer *et al.*, 1990b). Since the major source of CGRP in the rat stomach is primary afferent neurons (Green and Dockray, 1988), it would seem that the CGRP release induced by dibutyryl cyclic AMP or theophylline in this organ (Inui *et al.*, 1989) is also derived from sensory nerve endings. In contrast, capsaicin fails to augment the release of CGRP in the isolated small intestine of humans but is able to evoke a release of VIP (Maggi *et al.*, 1989a).

The effects of SP on gastrointestinal blood flow exhibit regional and species differences. Intravenous SP increases blood flow in the stomach and small intestine of the dog (Yeo *et al.*, 1982b). Similarly, intraluminal application of SP stimulates mucosal but not muscular blood flow in the feline jejunum (Yeo *et al.*, 1982a), an effect that does not involve histamine, prostaglandins, acetylcholine, or noradrenaline (Grönstad *et al.*, 1986). SP-induced arteriolar dilatation is also seen in the guinea pig isolated small intestine (Galligan *et al.*, 1990). In contrast, SP and NKA fail to alter mucosal blood flow in the rat stomach (Yokotani and Fujiwara, 1985; Holzer and Guth, 1991) while SP can reduce gastric mucosal blood flow stimulated either by bethanechol or vagal nerve stimulation (Yokotani and Fujiwara, 1985). The inconsistent activity of SP may be related to the scarcity of SP receptors on vessels of the digestive tract. Submucosal arterioles and venules in the human and canine bowel express only a low concentration of NK-1 but not NK-2 or NK-3 receptor sites (Gates *et al.*, 1988; Mantyh *et al.*, 1988a,b, 1989), and in the guinea pig ileum only large but not small blood vessels display a small number of NK-1 binding sites (Burcher and Bornstein, 1988).

Intravenous administration of human CGRP-α and CGRP-β to rabbits potently stimulates mucosal blood flow in the stomach (Bauerfeind *et al.*, 1989). The vasodilator activity of CGRP in the stomach is considerably more pronounced than in duodenum, in which only CGRP-α but not CGRP-β produces mucosal hyperemia (Bauerfeind *et al.*, 1989). A similar regional heterogeneity is seen in the guinea pig, in which rat CGRP-α is a potent dilator of the isolated gastroepiploic artery (Uddman *et al.*, 1986), but fails to dilate submucosal arterioles in the small intestine (Galligan *et al.*, 1990). In the rat stomach, topical application of the peptide (0.1–10 nM) dilates submucosal arterioles (Chen *et al.*, 1991), and intravenous (Lippe *et al.*, 1989a) and close arterial (Holzer and Guth, 1991; Li *et al.*, 1991) administration of rat CGRP-α potently augments mucosal blood flow. The relaxant action of CGRP on precontracted arteries from the human stomach depends on the integrity of the vascular endothelium (Thom *et al.*, 1987). The particularly high vasodilator activity of CGRP in the stomach is in keeping with the particularly high number of vascular CGRP receptors in this region. CGRP-α binding sites are present on medium-sized and small arteries and on arterioles of the canine esophagus, stomach, and duodenum, where they are expressed on both smooth muscle (tunica media) and endothelium. In contrast, CGRP-α binding sites are absent from vessels in the jejunum and more distal digestive system (Gates *et al.*, 1989). Veins and venules do not express binding sites for CGRP throughout the gastrointestinal tract of the dog (Gates *et al.*, 1989).

VIP, as its name implies, produces vasodilatation in the digestive system (Said and Mutt, 1970). VIP-evoked relaxation of precontracted arteries from the human stomach requires an intact endothelium (Thom *et al.*, 1987), and the dilator effect of VIP on submucosal arterioles in the guinea pig small intestine is attributed to the activation of enteric cholinergic vasodilator neurons (Galligan *et al.*, 1990). Intravascular administration of VIP enhances mucosal blood flow in the stomach of dog (Ito *et al.*, 1988) and rat (Holzer and Guth, 1991) and in the canine small intestine (Kachelhoffer *et al.*, 1974). The vasodilator activity of VIP is consistent with the presence of many VIP receptors in the small intestinal mucosa of the rat and rabbit (Sayadi *et al.*, 1988) and on the smooth muscle of submucosal arterioles throughout the human gastrointestinal tract (Zimmerman *et al.*, 1989). Somatostatin is known to reduce gastrointestinal blood flow (see Jodal and Lundgren, 1989) although the pertinent reports are not consistent with each other (see Leung and Guth, 1985).

3. Mediator Role of Sensory Neuropeptides

The vasodilator response of the rat gastric mucosa to sensory nerve stimulation by capsaicin is mediated by a NANC transmitter (Lippe *et al.*, 1989b). Prostaglandins, histamine, and acetylcholine acting via muscarinic receptors have been ruled out as mediators of gastric mucosal hyperemia due to acid back diffusion (Holzer *et al.*, 1991a; Lippe and Holzer, 1991). However, the dilator response of submucosal arterioles to topical capsaicin is inhibited by topical administration of a CGRP antagonist, $CGRP_{8-37}$, which indicates that the response is, at least in part, mediated by local release of the sensory vasodilator peptide CGRP (Chen *et al.*, 1991). This inference is strengthened by the observation that close arterial administration of the CGRP antagonist to the rat stomach also attenuates the mucosal hyperemia caused by intragastric capsaicin (Li *et al.*, 1991).

4. Role of Gastrointestinal Blood Flow in Mucosal Protection

Maintaining or increasing mucosal blood flow is thought to be a central element in the protection of the gastrointestinal mucosa from endogenous and exogenous injurious factors (see Guth and Leung, 1987; Oates, 1990). Sensory nerve-mediated mucosal hyperemia in the gastrointestinal tract is likely to play a similar role for two reasons. On the one hand, ablation of capsaicin-sensitive afferent neurons causes aggravation of experimentally induced lesion formation in the gastric (Szolcsányi and Barthó, 1981; Evangelista *et al.*, 1986; Holzer and Sametz, 1986; Esplugues and Whittle, 1990), duodenal (Maggi *et al.*, 1987), and colonic (Evangelista and Meli, 1989) mucosa. On the other hand, intragastric administration of capsaicin, to stimulate sensory nerve fibers, is able to protect against experimentally imposed lesions in the rat gastric (Szolcsányi and Barthó, 1981; Holzer and Lippe, 1988; Holzer *et al.*, 1989, 1990a) and distal colonic (Endoh and Leung, 1990) mucosa. Correlative evidence indeed suggests that the protective effect of sensory nerve stimulation is due to increased blood flow in the gastric mucosa (Holzer *et al.*, 1991b). In keeping with this inference is the finding that intraarterial CGRP that potently stimulates blood flow in the rat gastric mucosa (Holzer and Guth, 1991; Li *et al.*, 1991) also potently prevents experimental lesion formation in the stomach (Lippe *et al.*, 1989a; Whittle, 1990). In contrast, SP and NKA fail to alter mucosal blood flow in the rat stomach (Yokotani and Fujiwara, 1985; Holzer and Guth, 1991), and subcutaneously administered SP is unable to prevent gastric damage (Evangelista *et al.*, 1989). There is indirect evidence that different vasodilator and vasoconstrictor systems interact in the maintenance of mucosal integrity in the gastrointestinal mucosa. Platelet-activating factor (PAF) is a potent endogenous vasoconstrictor, and its vasoconstrictor effect on the rat gastric mucosa is amplified by prior defunctionalization of capsaicin-sensitive afferent neurons (Pique *et al.*, 1990). Endothelium-derived nitric oxide represents another important vasodilator system in the rat gastric mucosa, and inhibition of nitric oxide biosynthesis is followed by gastric mucosal vasoconstriction (Pique *et al.*, 1989; Walder *et al.*, 1990). Neither ablation of sensory neurons (Holzer and Sametz, 1986; Whittle *et al.*, 1990) nor inhibition of nitric oxide synthesis (Whittle *et al.*, 1990) alone causes gastric mucosal damage. When, however, both vasodilator systems are eliminated extensive damage develops in response to challenge of the gastric mucosa with acid (Whittle *et al.*, 1990). A similar interaction may exist between sensory neurons in the rat gastric mucosa and the vasoconstrictor endothelin-1 (Whittle, 1990).

C. Cardiovascular Reflexes Arising from Stimulation of Visceral Afferent Neurons

Capsaicin-sensitive afferent nerve fibers in the gastrointestinal tract participate in the reflex regulation of the cardiovascular system. Activation of these afferent fibers gives rise to sympathetically mediated hypertension and tachycardia in the dog (Longhurst *et al.*, 1980) and cat (Ordway and Longhurst, 1983). In the rat, activation of capsaicin-sensitive afferent nerve fibers by distension of the small intestine is followed by both hypertensive and hypotensive responses (Lembeck and Skofitsch, 1982). The extent to which mesenteric and gastrointestinal circulation are affected by these cardiovascular reactions is not known.

IV. Perspectives for Physiology and Pathophysiology

The experimental data surveyed here identify a group of fine primary afferent neurons to be implicated in the neural control of splanchnic circulation and provide completely new insights into the neural control of tissue blood flow. Apart from reflexogenic changes in the cardiovascular system, it is also by a local control of the vascular effector systems that these neurons regulate the diameter of small blood vessels. This dual function implies that, in response to a stimulus, afferent neurons not only convey messages to the central nervous system but also release mediator substances from their peripheral endings. Perivascular sensory nerve endings thus serve both a receptive and an effector control function. Since there is no proof for the existence of specialized neuroeffector junctions, the mediators probably have to diffuse some distance to make contact with the effector cells. Regional and species differences in the vasodilator consequences of sensory nerve stimulation are likely to arise from differences in the pathway-specific patterns of coexisting peptides in sensory neurons (Gibbins *et al.*, 1987; O'Brien *et al.*, 1989) and differences in the expression of peptide receptors on blood vessels (see Gates *et al.*, 1989). Functionally, therefore, different multimessenger systems will give rise to diverse vascular actions depending on the mixture of vasoactive peptides released from the nerve endings and the effect produced by the peptides in combination.

It is important to recognize that the local effector functions of sensory nerve endings do not operate in isolation but are in mutual interaction with other vascular control systems, such as the sympathetic nervous system and vasoactive factors derived, e.g., from the endothelium. The vascular system thus is under a complex multifactorial control, and the question presents itself whether the component contributed by afferent neurons is of physiological or pathophysiological significance. Most of the afferent neurons, which when stimulated have been found to produce vasodilatation, are sensitive to the excitatory and neurotoxic actions of capsaicin and thus fall within a group of neurons with chemonociceptors and polymodal nociceptors (Szolcsányi, 1990; Holzer, 1991). Noxious stimuli are therefore to be expected to activate these neurons and to give rise to both nociception and local vascular changes although it is not yet known whether the two functions are activated in parallel. In addition to potentially injurious stimuli such as hydrochloric acid (Holzer *et al.*, 1991a), acetic acid (Leung, 1990b), and platelet-activating factor (Pique *et al.*, 1990), however, certain physiological stimuli, such as the "postprandial" presence of bile-oleate (Rózsa and Jacobson, 1989), are also able to stimulate sensory neurons and to produce hyperemia in the splanchnic vascular bed. Related to the conditions under which sensory neurons are activated is the question whether the vascular actions exerted by these neurons have a bearing for the moment-to-moment control of vascular diameter. There is at present only circumstantial evidence that sensory neurons control basal vascular tone (Kawasaki *et al.*, 1990c,d) and basal mesenteric blood flow (Hottenstein *et al.*, 1991). Sensory nerve-mediated vasodilatation, though, is likely to reflect an emergency reaction that is called into operation when injurious events challenge homeostasis. It is an ingenious design that those neurons that detect potential threats to the integrity of the tissue are themselves able to initiate protective hyperemia and other vascular reactions. The sensitivity of sensory neurons to hydrogen ions (Bevan and Yeats, 1989; Geppetti *et al.*, 1991) is of particular significance in the upper gastrointestinal tract in which acid is a constant hazard. In this region sensory neurons subserve an emergency role in that they signal for a protective increase in mucosal blood flow in the face of pending acid injury (Holzer *et al.*, 1991a).

Dysfunction of this neurogenic vasodilator system is expected to give rise to homeostatic disorders and inadequate reactions to challenges of homeostasis, but the pathophysiological potential of the system has not yet been explored in full detail. Vascular tone in the splanchnic bed is of significance for the whole cardiovascular system (see Greenway, 1983), and there is circumstantial evidence that malfunction of the sensory vasodilator system in the mesenteric arteries is a factor in the etiology of hypertension (Kawasaki *et al.*, 1990c,d). Thus, spontaneously hypertensive rats display a progressive decrease in the density of CGRP-containing nerve fibers around mesenteric arteries and in the stimulus-induced release of CGRP from periarterial neurons, an impairment of NANC vasodilatation and accentuation of sympathetic vasoconstrictor responses (Kawasaki *et al.*, 1990c,d). Inflammatory bowel disease in man is associated with a dramatic upregulation of SP receptors on submucosal arterioles and venules that normally express only a low concentration of SP receptor sites (Mantyh *et al.*, 1988a, 1989); the precise functional implications of these changes have not yet been determined. In conclusion, the data reviewed here provide novel perspectives in the physiology and pathophysiology of the neural regulation of the vascular system and identify an important target for the development of drugs and therapeutic strategies in the treatment of vascular dysfunction.

Acknowledgments

I am very grateful to Drs. Irmgard Th. Lippe and Ulrike Holzer-Petsche for their helpful comments on the manuscript. Research done in the author's laboratory was supported by the Austrian Scientific Research Funds (Grant Nos. 4641, 5552, and 7845), the Austrian National Bank (Grant 4207), and by the Franz Lanyar Foundation at the Medical Faculty of the University of Graz.

References

Aberdeen, J., Corr, L., Milner, P., Lincoln, J., and Burnstock, G. (1990). Marked increases in calcitonin gene-related peptide-containing nerves in the developing rat following long-term sympathectomy with guanethidine. *Neuroscience 35,* 175–184.

Amann, R.(1990). Desensitization of capsaicin-evoked neuropeptide release—influence of Ca^{2+} and temperature. *Naunyn-Schmiedeberg's Arch. Pharmacol. 342,* 671–676.

Axelsson, K. L., Ljusegren, M. E., Ahlner, J., and Grundström, N. (1989). A novel neurogenic vasodilator mechanism in bovine mesenteric artery. *Circ. Res. 65,* 903–908.

Barja, F., Mathison, R., and Huggel, H. (1983). Substance P-containing nerve fibers in large peripheral blood vessels of the rat. *Cell Tiss. Res. 229,* 411–422.

Barthó, L., and Szolcsányi, J. (1978). The site of action of capsaicin on the guinea-pig isolated ileum. *Naunyn-Schmiedeberg's Arch. Pharmacol. 305,* 75–81.

Barthó, L., Sebök, B., and Szolcsányi, J. (1982). Indirect evidence for the inhibition of enteric substance P neurones by opiate agonists but not by capsaicin. *Eur. J. Pharmacol. 77,* 273–279.

Bayliss, W. M. (1901). On the origin from the spinal cord of the vaso-dilator fibres of the hind-limb, and on the nature of these fibres. *J. Physiol.* (London) *26,* 173–209.

Bauerfeind, P., Hof, R., Cucala, M., Siegrist, S., von Ritter, Ch., Fischer, J. A., and Blum, A. L. (1989). Effects of hCGRP I and II on gastric blood flow and acid secretion in anesthetized rabbits. *Am. J. Physiol. 256,* G145–G149.

Bény, J.-L., Brunet, P. C., and Huggel, H. (1989). Effects of substance P, calcitonin gene-related peptide and capsaicin on tension and membrane potential of pig coronary artery *in vitro. Regul. Pept. 25,* 25–36.

Bevan, S., and Szolcsányi, J. (1990). Sensory neuron-specific actions of capsaicin—mechanisms and applications. *Trends Pharmacol. Sci. 11,* 330–333.

Bevan, S., and Yeats, J. C. (1989). Protons activate a sustained inward current in a sub-population of rat isolated dorsal root ganglion (DRG) neurones. *J. Physiol.* (London) *417,* 81P.

Bolton, T. B., and Clapp, L. H. (1986). Endothelial-dependent relaxant actions of carbachol and substance P in arterial smooth muscle. *Br. J. Pharmacol. 87,* 713–723.

Brain, S. D., and Williams, T. J. (1989). Interactions between the tachykinins and calcitonin gene-related peptide lead to the modulation of oedema formation and blood flow in rat skin. *Br. J. Pharmacol. 97,* 77–82.

Brimijoin, S., Lundberg, J. M., Brodin, E., Hökfelt, T., and Nilsson, G. (1980). Axonal transport of substance P in the vagus and sciatic nerves of the guinea pig. *Brain Res. 191,* 443–457.

Brodin, E., Sjölund, K., Håkanson, R., and Sundler, F. (1983b). Substance P-containing nerve fibers are numerous in human but not in feline intestinal mucosa. *Gastroenterology 85,* 557–64.

Burcher, E., and Bornstein, J. C. (1988). Localization of substance P binding sites in submucous plexus of guinea pig ileum, using whole-mount autoradiography. *Synapse 2,* 232–239.

Chahl, L. A. (1988). Antidromic vasodilatation and neurogenic inflammation. *Pharmacol. Ther. 37,* 275–300.

Chen, R., Li, D. S., and Guth, P. H.(1991). Capsaicin-sensitive afferent nerve induced gastric arteriolar dilatation is mediated, in part, by calcitonin gene-related peptide (CGRP). *Gastroenterology 100,* A42.

Chéry-Croze, S., Bosshard, A., Martin, H., Cuber, J. C., Charnay, Y., and Chayvialle, J. A. (1989). Peptide immunocytochemistry in afferent neurons from lower gut in rats. *Peptides 9*, 873–881.

Costa, M., and Furness, J. B.(1984). Somatostatin is present in a subpopulation of noradrenergic nerve fibres supplying the intestine. *Neuroscience 13*, 911–919.

Costa, M., Furness, J. B., and Gibbins, I. L. (1986). Chemical coding of enteric neurons. *Prog. Brain Res. 68*, 217–239.

Decker, M. W., Towle, A. C., Bissette, G., Mueller, R. A., Lauder, J. M., and Nemeroff, C. B.(1985). Bombesin-like immunoreactivity in the central nervous system of capsaicin-treated rats: A radioimmunoassay and immunohistochemical study. *Brain Res. 342*, 1–8.

Della, N. G., Papka, R. E., Furness, J. B., and Costa, M. (1983). Vasoactive intestinal peptide-like immunoreactivity in nerves associated with the cardiovascular system of guinea pigs. *Neuroscience 9*, 605–619.

Donnerer, J., and Lembeck, F. (1982). Analysis of the effects of intravenously injected capsaicin in the rat. *Naunyn-Schmiedeberg's Arch. Pharmacol. 320*, 54–57.

Donnerer, J., Barthó, L., Holzer, P., and Lembeck, F. (1984). Intestinal peristalsis associated with release of immunoreactive substance P. *Neuroscience 11*, 913–918.

Duckles, S. P. (1986). Effects of capsaicin on vascular smooth muscle. *Naunyn-Schmiedeberg's Arch. Pharmacol. 333*, 59–64.

Edvinsson, L., Jansen, I., Kingman, T. A., and McCulloch, J. (1990). Cerebrovascular responses to capsaicin *in vitro* and *in situ*. *Br. J. Pharmacol. 100*, 312–318.

Ekblad, E., Ekelund, M., Graffner, H., Håkanson, R., and Sundler, F. (1985). Peptide-containing nerve fibers in the stomach wall of rat and mouse. *Gastroenterology 89*, 73–85.

Endoh, K., and Leung, F. W. (1990). Topical capsaicin protects the distal but not the proximal colon against acetic acid injury. *Gastroenterology 98*, A446.

Esplugues, J. V., and Whittle, B. J. R. (1990). Morphine potentiation of ethanol-induced gastric mucosal damage in the rat. Role of local sensory afferent neurons. *Gastroenterology 98*, 82–89.

Evangelista, S., and Meli, A. (1989). Influence of capsaicin-sensitive fibers on experimentally-induced colitis in rats. *J. Pharmac. Pharmacol. 41*, 574–575.

Evangelista, S., Maggi, C. A., and Meli, A. (1986). Evidence for a role of adrenals in the capsaicin-sensitive "gastric defence mechanism" in rats. *Proc. Soc. Exp. Biol. Med. 182*, 568–569.

Evangelista, S., Lippe, I. Th., Rovero, P., Maggi, C. A., and Meli, A. (1989). Tachykinins protect against ethanol-induced gastric lesions in rats. *Peptides 10*, 79–81.

Franco-Cereceda, A., Henke, H., Lundberg, J. M., Petermann, J. B., Hökfelt, T., and Fischer, J. A. (1987). Calcitonin gene-related peptide (CGRP) in capsaicin-sensitive substance P-immunoreactive sensory neurons in animals and man: Distribution and release by capsaicin. *Peptides 8*, 399–410.

Fujimori, A., Saito, A., Kimura, S., Watanabe, T., Uchiyama, Y., Kawasaki, H., and Goto, K. (1989). Neurogenic vasodilation and release of calcitonin gene-related peptide (CGRP) from perivascular nerves in the rat mesenteric artery. *Biochem. Biophys. Res. Commun. 165*, 1391–1398.

Fujimori, A., Saito, A, Kimura, S., and Goto, K. (1990). Release of calcitonin gene-related peptide (CGRP) from capsaicin-sensitive vasodilator nerves in the rat mesenteric artery. *Neurosci. Lett. 112*, 173–178.

Furness, J. B., Papaka, R. E., Della, N. G., Costa, M., and Eskay, R. L. (1982). Substance P-like immunoreactivity in nerves associated with the vascular system of guinea-pigs. *Neuroscience 7*, 447–459.

Galligan, J. J., Costa, M., and Furness, J. B. (1988). Changes in surviving nerve fibers associated with submucosal arteries following extrinsic denervation of the small intestine. *Cell Tiss. Res. 253*, 647–656.

Galligan, J. J., Jiang, M.-M., Shen, K.-Z., and Surprenant, A. (1990). Substance P mediates neurogenic vasodilatation in extrinsically denervated guinea-pig submucosal arterioles. *J. Physiol.* (London) *420*, 267–280.

Gamse, R., Holzer, P., and Lembeck, F. (1980). Decrease of substance P in primary sensory neurones and impairment of neurogenic plasma extravasation by capsaicin. *Br. J. Pharmacol. 68*, 207–213.

Gamse, R., Leeman, S. E., Holzer, P., and Lembeck, F. (1981). Differential effects of capsaicin on the content of somatostatin, substance P, and neurotensin in the nervous system of the rat. *Naunyn-Schmiedeberg's Arch. Pharmacol. 317,* 140–148.

Gates, T. S., Zimmerman, R. P., Mantyh, C. R., Vigna, R., Maggio, J. E., Welton, M. L., Passaro, E. P., and Mantyh, P. W. (1988). Substance P and substance K receptor binding sites in the human gastrointestinal tract: Localization by autoradiography. *Peptides 9,* 1207–1219.

Gates, T. S., Zimmerman, R. P., Mantyh, C. R., Vigna, S. R., and Mantyh, P. W. (1989). Calcitonin gene-related peptide-α receptor binding sites in the gastrointestinal tract. *Neuroscience 31,* 757–770.

Geppetti, P., Frilli, S., Renzi, D., Santicioli, P., Maggi, C. A., Theodorsson, E., and Fanciullacci, M. (1988). Distribution of calcitonin gene-related peptide-like immunoreactivity in various rat tissues: Correlation with substance P and other tachykinins and sensitivity to capsaicin. *Regul. Pept. 23,* 289–298.

Geppetti, P., Tramontana, M., Evangelista, S., Renzi, D., Maggi, C. A., Fusco, B. M., and Del Bianco, E. (1991). Differential effect on neuropeptide release of different concentrations of hydrogen ions on afferent and intrinsic neurons of the rat stomach. *Gastroenterology 101,* 1505–1511.

Gibbins, I. L., Furness, J. B., Costa, M., MacIntyre, I., Hillyard, C. J., and Girgis, S. (1985). Co-localization of calcitonin gene-related peptide-like immunoreactivity with substance P in cutaneous, vascular and visceral sensory neurons of guinea pigs. *Neurosci. Lett. 57,* 125–130.

Gibbins, I. L., Furness, J. B., and Costa, M. (1987). Pathway-specific patterns of the co-existence of substance P, calcitonin gene-related peptide, cholecystokinin and dynorphin in neurons of the dorsal root ganglia of the guinea-pig. *Cell Tiss. Res. 248,* 417–437.

Gibbins, I. L., Morris, J. L., Furness, J. B., and Costa, M. (1988). Innervation of systemic blood vessels. *In* "Nonadrenergic Innervation of Blood Vessels, Volume II: Regional Innervation." (G. Burnstock and S. G. Griffith, eds.), pp. 1–36. CRC Press, Boca Raton.

Gray, J. L., Bunnett, N. W., Mulvihill, S. J., and Debas, H. T. (1989). Capsaicin stimulates release of calcitonin gene-related peptide (CGRP) from the isolated perfused rat stomach. *Gastroenterology 96,* A181.

Green, T., and Dockray, G. J. (1988). Characterization of the peptidergic afferent innervation of the stomach in the rat, mouse, and guinea-pig. *Neuroscience 25,* 181–193.

Greenway, C. V. (1983). Role of splanchnic venous system in overall cardiovascular homeostasis. *Fed. Proc. 42,* 1678–1684.

Grönstad, K. O., Ahlman, H., Zinner, M. J., and Jaffe, B. M. (1983). The effect of vagal nerve stimulation on feline portal venous levels of substance P. *In* "Substance P Dublin 1983." (P. Skrabanek and D. Powell, eds.), pp. 153–154. Boole, Press, Dublin.

Grönstad, K. O., Nilsson, O., Dahlström, A., Price, B., Zinner, M. J., Jaffe, B. M., and Ahlman, H.(1985). The effects of vagal nerve stimulation on endoluminal release of serotonin and substance P into the feline small intestine. *Scand. J. Gastroenterol. 20,* 163–169.

Grönstad, K. O., Dahlström, A., Jaffe, B. M., Zinner, M. J., and Ahlman, H. (1986). Studies on the mucosal hyperaemia of the feline small intestine observed at endoluminal perfusion with substance P. *Acta Physiol. Scand. 128,* 97–108.

Guth, P. H., and Leung, F. W.(1987). Physiology of the gastric circulation. *In* "Physiology of the Gastrointestinal Tract." (L. R. Johnson, ed.), pp. 1031–1053. Raven Press, New York.

Hallberg, D., and Pernow, B. (1975). Effect of substance P on various vascular beds in the dog. *Acta Physiol. Scand. 93,* 277–285.

Han, S.-P., Naes, L., and Westfall, T. C. (1990a). Inhibition of periarterial nerve stimulation-induced vasodilation of the mesenteric arterial bed by CGRP (8-37) and CGRP receptor desensitization. *Biochem. Biophys. Res. Commun. 168,* 786–791.

Han, S.-P., Naes, L., and Westfall, T. C. (1990b). Calcitonin gene-related peptide is the endogenous mediator of nonadrenergic noncholinergic vasodilation in rat mesentery. *J. Pharmacol. Exp. Ther. 255,* 423–428.

Holliger, C., Radzyner, M., and Knoblauch, M. (1983). Effects of glucagon, vasoactive intestinal peptide, and vasopressin on villous microcirculation and superior mesenteric artery blood flow of the rat. *Gastroenterology 85,* 1036–1043.

Holzer, P. (1984). Characterization of the stimulus-induced release of immunoreactive substance P from the myenteric plexus of the guinea-pig small intestine. *Brain Res. 297,* 127–136.

Holzer, P. (1988). Local effector functions of capsaicin-sensitive sensory nerve endings: Involvement of tachykinins, calcitonin gene-related peptide and other neuropeptides. *Neuroscience 24,* 739–768.

Holzer, P. (1991). Capsaicin: Cellular targets, mechanisms of action and selectivity for thin sensory neurons. *Pharmacol. Rev. 43,* 143–201.

Holzer, P., and Guth, P. H. (1991). Neuropeptide control of rat gastric mucosal blood flow. Increase by calcitonin gene-related peptide and vasoactive intestinal polypeptide, but not substance P and neurokinin A. *Circ. Res. 68,* 100–105.

Holzer, P., and Lippe, I. Th. (1988). Stimulation of afferent nerve endings by intragastric capsaicin protects against ethanol-induced damage of gastric mucosa. *Neuroscience 27,* 981–987.

Holzer, P., and Sametz, W. (1986). Gastric mucosal protection against ulcerogenic factors in the rat mediated by capsaicin-sensitive afferent neurons. *Gastroenterology 91,* 975–981.

Holzer, P., Gamse, R., and Lembeck, F. (1980). Distribution of substance P in the rat gastrointestinal tract—lack of effect of capsaicin pretreatment. *Eur. J. Pharmacol. 61,* 303–307.

Holzer, P., Pabst, M. A., and Lippe, I. Th. (1989). Intragastric capsaicin protects against aspirin-induced lesion formation and bleeding in the rat gastric mucosa. *Gastroenterology 96,* 1425–1433.

Holzer, P., Pabst, M. A., Lippe, I. Th., Peskar, B. M., Peskar, B. A., Livingston, E. H., and Guth, P. H. (1990a). Afferent nerve-mediated protection against deep mucosal damage in the rat stomach. *Gastroenterology 98,* 838–848.

Holzer, P., Peskar, B. M., Peskar, B. A., and Amann, R. (1990b). Release of calcitonin gene-related peptide induced by capsaicin in the vascularly perfused rat stomach. *Neurosci. Lett. 108,* 195–200.

Holzer, P., Livingston, E. H., and Guth, P. H. (1991a). Sensory neurons signal for an increase in rat gastric mucosal blood flow in the face of pending acid injury. *Gastroenterology 101,* 416–423.

Holzer, P., Livingston, E. H., Saria, A., and Guth, P. H. (1991b). Sensory neurons mediate protective vasodilatation in rat gastric mucosa. *Am. J. Physiol. 260,* G363–G370.

Hottenstein, O. D., Remak, G., and Jacobson, E. D. (1990). Post-nerve stimulation hyperemia in rat intestinal circulation involves capsaicin-sensitive nerves. *Gastroenterology 98,* A175.

Hottenstein, O. D., Pawlik, W. W., Remak, G., and Jacobson, E. D. (1991). Capsaicin-sensitive nerves modulate resting blood flow and vascular tone in rat gut. *Naunyn-Schmiedeberg's Arch. Pharmacol. 343,* 179–184.

Hua, X.-Y., Theodorsson-Norheim, E., Brodin, E., Lundberg, J. M., and Hökfelt, T. (1985). Multiple tachykinins (neurokinin A, neuropeptide K and substance P) in capsaicin-sensitive sensory neurons in the guinea-pig. *Regul. Pept. 13,* 1–19.

Huang, M., and Rorstad, O. P. (1987). VIP receptors in mesenteric and coronary arteries: A radioligand study. *Peptides 8,* 477–485.

Inui, T., Chiba, T., Okimura, Y., Morishita, T., Nakamura, A., Yamaguchi, A., Yamatani, T., Kadowaki, S., Chihara, K., and Fujita, T. (1989). Presence and release of calcitonin gene-related peptide in rat stomach. *Life Sci. 45,* 1199–1206.

Ito, S., Ohga, A., and Ohta, T. (1988). Gastric vasodilatation and vasoactive intestinal peptide output in response to vagal stimulation in the dog. *J. Physiol.* (London) *404,* 669–682.

Jancsó, G., Hökfelt, T., Lundberg, J. M., Király, E., Halász, N., Nilsson, G., Terenius, L., Rehfeld, J., Steinbusch, H., Verhofstad, A. E. R., Said, S., and Brown, M. (1981). Immunohistochemical studies on the effect of capsaicin on spinal and medullary peptide and monoamine neurons using antisera to substance P, gastrin/CCK, somatostatin, VIP, enkephalin, neurotensin and 5-hydroxytryptamine. *J. Neurocytol. 10,* 963–980.

Jancsó, G., Király, E., Such, G., Joó, F., and Nagy, A. (1987). Neurotoxic effect of capsaicin in mammals. *Acta Physiol. Hung. 69,* 295–313.

Jessell, T. M., Iversen, L. L., and Cuello, A. C. (1978). Capsaicin-induced depletion of substance P from primary sensory neurons. *Brain Res. 152,* 183–188.

Jodal, M., and Lundgren, O. (1989). Neurohormonal control of gastrointestinal blood flow. *In* "The Gastrointestinal System." (S. G. Schultz, ed.), Handbook of Physiology, Section 6, Volume I, Part 2, pp. 1667–1707. American Physiological Society, Bethesda.

Kachelhoffer, J., Eloy, M. R., Pousse, A., Hohmatter, D., and Grenier, J. F. (1974). Mesenteric vasomotor effects of vasoactive intestinal polypeptide. Study on perfused isolated canine jejunal loops. *Pflügers Arch. 352,* 37–46.

Kaufman, M. P., Iwamoto, G. A., Longhurst, J. C., and Mitchell, J. H. (1982). Effects of capsaicin and bradykinin on afferent fibers with endings in skeletal muscle. *Circ. Res. 50,* 133–139.

Kawasaki, H., Takasaki, K., Saito, A., and Goto, K. (1988). Calcitonin gene-related peptide acts as a novel vasodilator neurotransmitter in mesenteric resistance vessels of the rat. *Nature 335,* 164–167.

Kawasaki, H., Nuki, C., Saito, A., and Takasaki, K. (1990a). Role of calcitonin gene-related peptide-containing nerves in the vascular adrenergic neurotransmission. *J. Pharmacol. Exp. Ther. 252,* 403–409.

Kawasaki, H., Nuki, C., Saito, A., and Takasaki, K. (1990b). Adrenergic modulation of calcitonin gene-related peptide (CGRP)-containing nerve-mediated vasodilation in the rat mesenteric resistance vessel. *Brain Res. 506,* 287–290.

Kawasaki, H., Saito, A., and Takasaki, K. (1990c). Age-related decrease of calcitonin gene-related peptide-containing vasodilator innervation in the mesenteric resistance vessel of the spontaneously hypertensive rat. *Circ. Res. 67,* 733–743.

Kawasaki, H., Saito, A., and Takasaki, K. (1990d). Changes in calcitonin gene-related peptide (CGRP)-containing vasodilator nerve activity in hypertension. *Brain Res. 518,* 303–307.

Keast, J. R., Furness, J. B., and Costa, M. (1985). Distribution of certain peptide-containing nerve fibers and endocrine cells in the gastrointestinal mucosa in five mammalian species. *J. Comp. Neurol. 236,* 403–422.

Keen, P., Harmar, A. J., Spears, F., and Winter, E. (1982). Biosynthesis, axonal transport and turnover of neuronal substance P. *In* "Substance P in the Nervous System." (R. Porter and M. O'Connor, eds.), pp. 145–160. Pitman, London.

Kenins, P., Hurley, J. V., and Bell, C. (1984). The role of substance P in the axon reflex in the rat. *Br. J. Dermatol. 111,* 551–559.

Kirchgessner, A. L., Dodd, J., and Gershon, M. D. (1988). Markers shared between dorsal root and enteric ganglia. *J. Comp. Neurol. 276,* 607–621.

Kwok, Y. N., and McIntosh, C. H. S. (1990). Release of substance P-like immunoreactivity from the vascularly perfused rat stomach. *Eur. J. Pharmacol. 180,* 201–207.

Lee, Y., Shiotani, Y., Hayashi, N., Kamada, T., Hillyard, C. J., Girgis, S. I., MacIntyre, I., and Tohyama, M. (1987). Distribution and origin of calcitonin gene-related peptide in the rat stomach and duodenum: An immunohistochemical study. *J. Neur. Transmiss. 68,* 1–14.

Lembeck, F., and Holzbauer, M. (1988). Neuronal mechanisms of cutaneous blood flow. *In* "Non-adrenergic Innervation of Blood Vessels, Volume II: Regional Innervation." (G. Burnstock and S. G. Griffith, eds.), pp. 119–132. CRC Press, Boca Raton.

Lembeck, F., and Skofitsch, G. (1982). Visceral pain reflex after pretreatment with capsaicin and morphine. *Naunyn-Schmiedeberg's Arch. Pharmacol. 321,* 116–122.

Leung, F. W. (1990a). Capsaicin-sensitive afferent sensory fibers mediate autoregulatory escape from adrenergic (epinephrine) vasoconstriction in the rat stomach. *Gastroenterology 98,* A187.

Leung, F. W. (1990b). Capsaicin-sensitive afferent sensory fibers mediate capsaicin-induced and injury-induced colonic hyperemia. *Gastroenterology 98,* A187.

Leung, F. W., and Guth, P. H. (1985). Dissociated effects of somatostatin on gastric acid secretion and mucosal blood flow. *Am. J. Physiol. 248,* G337–G341.

Li, D. S., Raybould, H. E., Quintero, E., and Guth, P. H. (1991). Role of calcitonin gene-related peptide in gastric hyperemic response to intragastric capsaicin. *Am. J. Physiol. 261,* G657–G667.

Lindh, B., Hökfelt, T., and Elfvin, L.-G. (1988). Distribution and origin of peptide-containing nerve fibers in the celiac superior mesenteric ganglion of the guinea-pig. *Neuroscience 26,* 1037–1071.

Lippe, I. Th., and Holzer, P. (1991). The rat gastric mucosal vasodilator response to acid back-diffusion is mediated by a neural reflex involving the celiac ganglion. *Gastroenterology 100,* A112.

Lippe, I. Th., Lorbach, M., and Holzer, P. (1989a). Close arterial infusion of calcitonin gene-related peptide into the rat stomach inhibits aspirin- and ethanol-induced hemorrhagic damage. *Regul. Pept. 26,* 35–46.

Lippe, I. Th., Pabst, M. A., and Holzer, P. (1989b). Intragastric capsaicin enhances rat gastric acid elimination and mucosal blood flow by afferent nerve stimulation. *Br. J. Pharmacol. 96,* 91–100.

Livingston, E. H., Holzer, P., Passaro, E. P., and Guth, P. H. (1990). Modulation of gastric acid secretion and blood flow by sensory nerves. *Gastroenterology 98,* A79.

Longhurst, J. C., Ashton, J. H., and Iwamoto, G. A.(1980). Cardiovascular reflexes resulting from capsaicin-stimulated gastric receptors in anesthetized dogs. *Circ. Res. 46,* 780–788.

Longhurst, J. C., Kaufman, M. P., Ordway, G. A., and Musch, T. I. (1984). Effects of bradykinin and capsaicin on endings of afferent fibers from abdominal visceral organs. *Am. J. Physiol. 247,* R552–R559.

Lundberg, J. M. (1981). Evidence for coexistence of vasoactive intestinal polypeptide (VIP) and acetylcholine in neurons of cat exocrine glands. *Acta Physiol. Scand. 112* (Suppl. 496, 1–57.

Lundberg, J. M., and Saria, A. (1987). Polypeptide-containing neurons in airway smooth muscle. *Ann. Rev. Physiol. 49,* 557–572.

Luthman, J., Stromberg, I., Brodin, E., and Jonsson, G. (1989). Capsaicin treatment to developing rats induces increase of noradrenaline levels in the iris without affecting the adrenergic terminal density. *Int. J. Develop. Neurosci. 7,* 613–620.

Maggi, C. A., and Meli, A. (1988). The sensory-efferent function of capsaicin-sensitive sensory neurons. *Gen. Pharmacol. 19,* 1–43.

Maggi, C. A., Evangelista, S., Abelli, L., Somma, V., and Meli, A. (1987). Capsaicin-sensitive mechanisms and experimentally induced duodenal ulcers in rats. *J. Pharmac. Pharmacol. 39,* 559–561.

Maggi, C. A., Santicioli, P., Del Bianco, E., Geppetti, P., Barbanti, G., Turini, D., and Meli, A. (1989a). Release of VIP- but not CGRP-like immunoreactivity by capsaicin from the human isolated small intestine. *Neurosci. Lett. 98,* 317–320.

Maggi, C. A., Santicioli, P., Geppetti, P., Parlani, M., Astolfi, M., Del Bianco, E., Patacchini, R., Giuliani, S., and Meli, A. (1989b). The effect of calcium free medium and nifedipine on the release of substance P-like immunoreactivity and contractions induced by capsaicin in the isolated guinea pig and rat bladder. *Gen. Pharmacol. 20,* 445–456.

Maggi, C. A., Santicioli, P., Renzi, D., Patacchini, R., Surrenti, C., and Meli, A. (1989c). Release of substance P- and calcitonin gene-related peptide-like immunoreactivity and motor response of the isolated guinea pig gallbladder to capsaicin. *Gastroenterology 96,* 1093–1101.

Maggio, J. E., and Hunter, J. C. (1984). Regional distribution of kassinin-like immunoreactivity in rat central and peripheral tissues and the effect of capsaicin. *Brain Res. 307,* 370–373.

Mantyh, C. R., Gates, T. S., Zimmerman, R. P., Welton, M. L., Passaro, E. P., Vigna, S. R., Maggio, J. E., Kruger, L., and Mantyh, P. W. (1988a). Receptor binding sites for substance P, but not substance K or neuromedin K, are expressed in high concentrations by arterioles, venules, and lymph nodules in surgical specimens obtained from patients with ulcerative colitis and Crohn disease. *Proc. Natl. Acad. Sci. U.S.A. 85,* 3235–3239.

Mantyh, P. W., Mantyh, C. R., Gates, T., Vigna, S. R., and Maggio, J. E. (1988b). Receptor binding sites for substance P and substance K in the canine gastrointestinal tract and their possible role in inflammatory bowel disease. *Neuroscience 25,* 817–837.

Mantyh, P. W., Catton, M. D., Boehmer, C. G., Welton, M. L., Passaro, E. P., Maggio, J. E., and Vigna, S. R. (1989). Receptors for sensory neuropeptides in human inflammatory diseases: implications for the effector role of sensory neurons. *Peptides 10,* 627–645.

Manzini, S., and Perretti, F. (1988). Vascular effects of capsaicin in isolated perfused rat mesenteric bed. *Eur. J. Pharmacol. 148,* 153–159.

Marshall, I., Al-Kazwini, S. J., Holman, J. J., and Craig, R. K. (1986). Human and rat alpha-CGRP but not calcitonin cause mesenteric vasodilatation in rats. *Eur. J. Pharmacol. 123,* 217–222.

Minagawa, H., Shiosaka, S., Inoue, H., Hayashi, N., Kasahara, A., Kamada, T., Tohyama, M., and Shitani, Y. (1984). Origins and three-dimensional distribution of substance P-containing structures in the rat stomach using whole-mount tissue. *Gastroenterology 86,* 51–59.

Morishita, T., and Guth, P. H. (1986). Vagal nerve stimulation causes noncholinergic dilatation of gastric arterioles. *Am. J. Physiol. 250,* G660–G664.

Moritoki, H., Takase, H., and Tanioka, A. (1990). Dual effects of capsaicin on responses of the rabbit ear artery to field stimulation. *Br. J. Pharmacol. 99,* 152–156.

Mulderry, P. K., Ghatei, M. A., Rodrigo, J., Allen, J. M., Rosenfeld, M. G., Polak, J. M., and Bloom, S. R. (1985). Calcitonin gene-related peptide in cardiovascular tissues of the rat. *Neuroscience 14,* 947–954.

Mulderry, P. K., Ghatei, M. A., Spokes, R. A., Jones, P. M., Pierson, A. M., Hamid, Q. A., Kanse, S., Amara, S. G., Burrin, J. M., Legon, S., Polak, J. M., and Bloom, S. R.(1988). Differential expression of α-CGRP and β-CGRP by primary sensory neurons and enteric autonomic neurons of the rat. *Neuroscience 25,* 195–205.

Nagy, J. I., Vincent, S. R., Staines, W. M. A., Fibiger, H. C., Reisine, T. D., and Yamamura, H. I. (1980). Neurotoxic action of capsaicin on spinal substance P neurons. *Brain Res. 186,* 435–444.

Nagy, J. I., Hunt, S. P., Iversen, L. L., and Emson, P. C. (1981). Biochemical and anatomical observations on the degeneration of peptide-containing primary afferent neurons after neonatal capsaicin. *Neuroscience 6,* 1923–1934.

Nakanishi, S. (1987). Substance P precursor and kininogen: Their structures, gene organizations, and regulation. *Physiol. Rev. 67,* 1117–1142.

Nelson, M. T., Huang, Y., Brayden, J. E., Hescheler, J., and Standen, N. B. (1990). Arterial dilations in response to calcitonin gene-related peptide involve activation of K^+ channels. *Nature 344,* 770–773.

Neuhuber, W. L. (1987). Sensory vagal innervation of the rat esophagus and cardia: A light and electron microscopic anterograde tracing study. *J. Auton. Nerv. Syst. 20,* 243–255.

Nielsch, U., and Keen, P. (1987). Effects of neonatal 6-hydroxydopamine administration on different substance P-containing sensory neurones. *Eur. J. Pharmacol. 138,* 193–197.

Noguchi, K., Senba, E., Morita, Y., Sato, M., and Tohyama, M. (1990). Co-expression of α-CGRP and β-CGRP mRNAs in the rat dorsal root ganglion cells. *Neurosci. Lett. 108,* 1–5.

Oates, P. J. (1990). Gastric blood flow and mucosal defense. *In* "Gastric Cytoprotection." (D. Hollander and A. S. Tarnawski, eds.), pp. 125–165. Plenum, New York.

O'Brien, C., Woolf, C. J., Fitzgerald, M., Lindsay, R. M., and Molander, C. (1989). Differences in the chemical expression of rat primary afferent neurons which innervate skin, muscle or joint. *Neuroscience 32,* 493–502.

Ordway, G. A., and Longhurst, J. C. (1983). Cardiovascular reflexes arising from the gallbladder of the cat. Effects of capsaicin, bradykinin, and distension. *Circ. Res. 52,* 26–35.

Papka, R. E., Furness, J. B., Della, N. G., Murphy, R., and Costa, M. (1984). Time course of effect of capsaicin on ultrastructure and histochemistry of substance P-immunoreactive nerves associated with the cardiovascular system of the guinea-pig. *Neuroscience 12,* 1277–1292.

Pique, J. M., Esplugues, J. V., and Whittle, B. J. R. (1989). The vasodilator role of endogenous nitric oxide in the rat gastric microcirculation. *Eur. J. Pharmacol. 174,* 293–296.

Pique, J. M., Esplugues, J. V., and Whittle, B. J. R. (1990). Influence of morphine or capsaicin pretreatment on rat gastric microcirculatory response to PAF. *Am. J. Physiol. 258,* G352–G357.

Raybould, H. E., Holzer, P., Sternini, C., and Eysselein, V. E. (1990). Selective ablation of spinal sensory neurons containing CGRP inhibits the increase in rat gastric mucosal blood flow due to acid back-diffusion. *Gastroenterology 98,* A198.

Remak, G., Hottenstein, O. D., and Jacobson, E. D. (1990). Sensory nerves mediate neurogenic escape in rat gut. *Am. J. Physiol. 258,* H778–H786.

Renzi, D., Santicioli, P., Maggi, C. A., Surrenti, C., Pradelles, P., and Meli, A. (1988). Capsaicin-induced release of substance P-like immunoreactivity from the guinea pig stomach *in vitro* and *in vivo. Neurosci. Lett. 92,* 254–258.

Renzi, D., Evangelista, S., Mantellini, P., Santicioli, P., Maggi, C. A., and Surrenti, C. (1990). Simultaneous release of calcitonin gene-related peptide- and tachykinin-like immunoreactivity by capsaicin from rat stomach. *Gastroenterology 98,* A384.

Rodrigo, J., Polak, J. M., Fernandez, L., Ghatei, M. A., Mulderry, P., and Bloom, S. R. (1985). Calcitonin gene-related peptide immunoreactive sensory and motor nerves of the rat, cat, and monkey esophagus. *Gastroenterology 88,* 444–451.

Rosenfeld, M. G., Mermod, J.-J., Amara, S. G., Swanson, L. W., Sawchenko, P. E., Rivier, J., Vale, W. W., and Evans, R. M. (1983). Production of a novel neuropeptide encoded by the calcitonin gene via tissue-specific RNA processing. *Nature 304,* 129–135.

Rózsa, Z., and Jacobson, E. D.(1989). Capsaicin-sensitive nerves are involved in bile-oleate-induced intestinal hyperemia. *Am. J. Physiol. 256,* G476–G481.

Rózsa, Z., Jancsó, G., and Varró, V. (1984). Possible involvement of capsaicin-sensitive sensory nerves in the regulation of intestinal blood flow in the dog. *Naunyn-Schmiedeberg's Arch. Pharmacol. 326,* 352–356.

Rózsa, Z., Varró, V., and Jancsó, G. (1985). Use of immunoblockade to study the involvement of peptidergic afferent nerves in the intestinal vasodilatory response to capsaicin in the dog. *Eur. J. Pharmacol. 115,* 59–64.

Rózsa, Z., Sharkey, K. A., Jancsó, G., and Varró, V. (1986). Evidence for a role of capsaicin-sensitive mucosal afferent nerves in the regulation of mesenteric blood flow in the dog. *Gastroenterology 90,* 906–910.

Rózsa, Z., Mattila, J., and Jacobson, E. D.(1988). Substance P mediates a gastrointestinal thermoreflex in rats. *Gastroenterology 95,* 265–276.

Said, S. I., and Mutt, V. (1970). Potent peripheral and splanchnic vasodilator peptide from normal gut. *Nature 225,* 863–864.

Saria, A., Gamse, R., Petermann, J., Fischer, J. A., Theodorsson-Norheim, E., and Lundberg, J. M. (1986). Simultaneous release of several tachykinins and calcitonin gene-related peptide from rat spinal cord slices. *Neurosci. Lett. 63,* 310–314.

Sayadi, H., Harmon, J. W., Moody, T. W., and Korman, L. Y. (1988). Autoradiographic distribution of vasoactive intestinal polypeptide receptors in rabbit and rat small intestine. *Peptides 9,* 23–30.

Schrauwen, E., and Houvenaghel, A. (1980). Substance P: A powerful intestinal vasodilator in the pig. *Pflügers Arch. 386,* 281–284.

Sharkey, K. A., Williams, R. G., and Dockray, G. J. (1984). Sensory substance P innervation of the stomach and pancreas. Demonstration of capsaicin-sensitive sensory neurons in the rat by combined immunohistochemistry and retrograde tracing. *Gastroenterology 87,* 914–921.

Sidawy, A. N., Sayadi, H., Harmon, J. W., Termanini, B., Andrews, B., DePalma, R. G., and Korman, L. Y. (1989). Distribution of vasoactive intestinal peptide and its receptors in the arteries of the rabbit. *J. Surg. Res. 47,* 105–111.

Skofitsch, G., and Jacobowitz, D. M. (1985a). Galanin-like immunoreactivity in capsaicin-sensitive sensory neurons and ganglia. *Brain Res. Bull. 15,* 191–195.

Skofitsch, G., and Jacobowitz, D. M. (1985b). Calcitonin gene-related peptide coexists with substance P in capsaicin sensitive neurons and sensory ganglia of the rat. *Peptides 6,* 747–754.

Skofitsch, G., Zamir, N., Helke, C. J., Savitt, J. M., and Jacobowitz, D. M. (1985). Corticotropin releasing factor-like immunoreactivity in sensory ganglia and capsaicin sensitive neurons of the rat central nervous system: Colocalization with other neuropeptides. *Peptides 6,* 307–318.

Sternini, C., and Andersson, K. (1990). Differential distribution of α- and β-calcitonin gene-related peptide (CGRP) mRNAs in enteric and sensory neurons. *Gastroenterology 98,* A526.

Sternini, C., Reeve, J. R., and Brecha, N. (1987). Distribution and characterization of calcitonin gene-related peptide immunoreactivity in the digestive system of normal and capsaicin-treated rats. *Gastroenterology 93,* 852–862.

Stewart-Lee, A., and Burnstock, G. (1989). Actions of tachykinins on the rabbit mesenteric artery: Substance P and [Glp^6,L-Pro^9]SP$_{6-11}$ are potent agonists for endothelial neurokinin-1 receptors. *Br. J. Pharmacol. 97,* 1218–1224.

Stricker, S. (1876). Untersuchungen über die Gefässwurzeln des Ischiadicus. *Sitz.-Ber. Kaiserl. Akad. Wiss.* (Wien) *3,* 173–185.

Su, H. C., Bishop, A. E., Power, R. F., Hamada, Y., and Polak, J. M. (1987). Dual intrinsic and extrinsic origins of CGRP- and NPY-immunoreactive nerves of rat gut and pancreas. *J. Neurosci. 7,* 2674–2687.

Sundler, F., Ekblad, E., and Håkanson, R. (1991). Occurrence and distribution of SP- and CGRP-containing nerve fibers in the gastric mucosa. Species differences. *In* "Sensory Nerves and Neuropeptides in Gastroenterology: From Basic Science to Clinical Perspectives." (M. Costa, C. Surrenti, S., Gorini, C. A., Maggi, and A. Meli, eds.), pp. 29–37. Plenum Press, London.

Szolcsányi, J. (1984). Capsaicin-sensitive chemoceptive neural system with dual sensory-efferent function. *In* "Antidromic Vasodilatation and Neurogenic Inflammation." (L. A. Chahl, J., Szolcsányi, and F. Lembeck, eds.), pp. 27–52. Akadémiai Kiadó, Budapest.

Szolcsányi, J. (1990). Capsaicin, irritation, and desensitization. Neurophysiological basis and future perspectives. *In* "Irritation." (B. G. Green, J. R. Mason, and M. R. Kare, eds.), Chemical Senses, Vol. 2, pp. 141–168. Marcel Dekker, New York and Basel.

Szolcsányi, J., and Barthó, L. (1981). Impaired defense mechanism to peptic ulcer in the capsaicin-desensitized rat. *In* "Gastrointestinal Defense Mechanisms." (G. Mózsik, O. Hänninen, and T. Jávor, eds.), Advances in Physiological Sciences, Vol. 15, pp. 39–51. Pergamon Press and Akadémiai Kiadó, Oxford and Budapest.

Terenghi, G., Zhang, S.-Q., Unger, W. G., and Polak, J. M. (1986). Morphological changes of sensory CGRP-immunoreactive and sympathetic nerves in peripheral tissues following chronic denervation. *Histochemistry 86,* 89–95.

Thiefin, G., Raybould, H. E., Leung, F. W., Taché, Y., and Guth, P. H. (1990). Capsaicin-sensitive afferent fibers contribute to gastric mucosal blood flow response to electrical vagal stimulation. *Am. J. Physiol. 259,* G1037–G1043.

Thom, S. M., Hughes, A. D., Goldberg, P., Martin, G., Schachter, M., and Sever, P. S. (1987). The actions of calcitonin gene-related peptide and vasoactive intestinal peptide as vasodilators in man *in vivo* and *in vitro. Br. J. Clin. Pharmacol. 24,* 139–144.

Törnebrandt, K., Nobin, A., and Owman, Ch. (1987). Contractile and dilatory action of neuropeptides on isolated human mesenteric blood vessels. *Peptides 8,* 251–256.

Trad, K. S., Harmon, J. W., Fernicola, M. T., Hakki, F. Z., Rao, J. S., Dziki, A. J., and Bass, B. L. (1990). Evidence for a role of capsaicin-sensitive mucosal afferent nerves in the regulation of esophageal blood flow in rabbits. *Gastroenterology 98,* A139.

Tsutsumi, Y., and Hara, M. (1989). Peptidergic nerve fibers in human stomach and duodenal bulb: Immunohistochemical demonstration of changes in fiber distribution in chronic gastritis. *Biomed. Res. 10,* 209–216.

Uddman, R., Edvinsson, L., Ekblad, E., Håkanson, R., and Sundler, F. (1986). Calcitonin gene-related peptide (CGRP): Perivascular distribution and vasodilatory effects. *Regul. Pept. 15,* 1–23.

Uvnäs-Wallensten, K. (1978). Release of substance P-like immunoreactivity into the antral lumen of cats. *Acta Physiol. Scand. 104,* 464–468.

Varro, A., Green, T., Holmes, S., and Dockray, G. J. (1988). Calcitonin gene-related peptide in visceral afferent nerve fibres: Quantification by radioimmunoassay and determination of axonal transport rates. *Neuroscience 26,* 927–932.

Vickers, J. C., Costa, M., Vitadello, M., Dahl, D., and Marotta, C. A. (1990). Neurofilament protein-triplet immunoreactivity in distinct subpopulations of peptide-containing neurons in the guinea-pig coeliac ganglion. *Neuroscience 39,* 743–759.

Walder, C. E., Thiemermann, C., and Vane, J. R. (1990). Endothelium-derived relaxing factor participates in the increased blood flow in response to pentagastrin in the rat stomach mucosa. *Proc. Royal Soc. London B 241,* 195–200.

Weihe, E. (1990). Neuropeptides in primary afferent neurons. *In* "The Primary Afferent Neuron." (W. Zenker, and W. L. Neuhuber, eds.), pp. 127–159. Plenum Press, New York.

Wharton, J., Gulbenkian, S., Mulderry, P. K., Ghatei, M. A., McGregor, G. P., Bloom, S. R., and Polak, J. M. (1986). Capsaicin induces a depletion of calcitonin gene-related peptide (CGRP)-immunoreactive nerves in the cardiovascular system of the guinea-pig and rat. *J. Auton. Nerv. Syst. 16,* 289–309.

Whittle, B. J. R. (1990). Pharmacological manipulation of vasoactive mediators in gastric damage and protection. *Eur. J. Pharmacol. 183,* 127–128.

Whittle, B. J. R., Lopez-Belmonte, J., and Moncada, S. (1990). Regulation of gastric mucosal integrity by endogenous nitric oxide: Interactions with prostanoids and sensory neuropeptides in the rat. *Br. J. Pharmacol. 99,* 607–611.

Wimalawansa, S. J., and MacIntyre, I. (1988). Calcitonin gene-related peptide and its specific binding sites in the cardiovascular system. *Int. J. Cardiol. 20,* 29–37.

Yeo, C. J., Jaffe, B. M., and Zinner, M. J. (1982a). Local regulation of blood flow in the feline jejunum. A possible role for endoluminally released substance P. *J. Clin. Invest. 70,* 1329–1333.

Yeo, C. J., Zinner, J. M., Denoy, D., and Jaffe, B. M. (1982b). Intravenous substance P infusion in conscious dogs. Its effect on hemodynamics, regional blood flow, glucose, insulin, and cortisol. *Gastroenterology 82,* 1216.

Yokotani, K., and Fujiwara, M. (1985). Effects of substance P on cholinergically stimulated gastric acid secretion and mucosal blood flow in rats. *J. Pharmacol. Exp. Ther. 232,* 826–830.

Zimmerman, R. P., Gates, T. S., Mantyh, C. R., Vigna, S. R., Welton, M. L., Passaro, E. P., and Mantyh, P. W. (1989). Vasoactive intestinal polypeptide receptor binding sites in the human gastrointestinal tract: Localization by autoradiography. *Neuroscience 31,* 771–783.

PART V

Clinical Application

CHAPTER 16

Neuropeptides in Human Cardiovascular Regulation

JOHN PERNOW

I. Introduction

It has long been known that the principal postganglionic transmitter substances in the autonomic nervous system regulating cardiovascular function are noradrenaline in sympathetic nerves and acetylcholine in parasympathetic nerves. Furthermore, a large number of therapeutic agents used in the treatment of cardiovascular disorders act by interfering with the mechanisms of these neurotransmitters. Over the past few years, however, it has become increasingly evident that a number of different peptides coexist with the classical neurotransmitters in nerves involved in cardiovascular regulation. The list of these peptides is steadily growing and their coexistence with classical neurotransmitters has been described in several reviews (Hökfelt *et al.*, 1980; Lundberg and Hökfelt, 1983; Owman, 1988). Studies of the distribution and function of neuropeptides as possible transmitters or modulators have mainly been performed in experimental animals. However, although the combination of peptide/classical transmitter varies from one species to another, both the presence and the functional effects of the peptides appear to be highly comparable between most experimental animals and humans.

This overview will focus on the distribution and pharmacological actions of peptides present in nerves that are involved in cardiovascular regulation in humans. It deals with peptides present in the peripheral nervous system (i.e., sympathetic, parasympathetic, and sensory nerves). Examples will also be given of some pathophysiological conditions in which neuropeptides may be of importance. Although many studies describe the actions of neuropeptides possibly involved in central cardiovascular regulation in experimental animals (for review see Reid and Rubin, 1987), data from human studies are lacking. Additional peptides located in peripheral structures other than nerves, and hence at the present stage not considered to exert any major function as peripheral neuropeptides (e.g., angiotensin II, atrial natriuretic peptide, and endothelin), may also play important roles in cardiovascular regulation but falls outside the topic of this overview.

II. Sympathetic Nerves

A. *Neuropeptide Y*

1. Distribution

Neuropeptide Y (NPY) is a 36 amino acid peptide originally isolated and sequenced by Tatemoto (1982). The structure of NPY is strikingly similar to that of peptide YY, which is present in endocrine cells of the intestine and pancreatic polypeptide. NPY is widely distributed in sympathetic nerves supplying the cardiovascular system. Thus, sympathetic ganglia and nerve fibers around blood vessels and in the heart have been shown to be densely immunoreactive to antisera directed against NPY (Lundberg *et al.,* 1983; Gu *et al.,* 1983; Edvinsson *et al.,* 1985; J. Pernow *et al.,* 1987c). Simultaneous staining with antisera to noradrenaline-synthetizing enzymes has revealed that NPY and noradrenaline most likely coexist in the same nerves. The perivascular innervation of NPY-containing fibers is generally richer around arteries than veins, and more dense around smaller blood vessels than large ones (Lundberg *et al.,* 1983; J. Pernow *et al.,* 1987c). In the heart, many NPY-immunoreactive fibers are seen in the atria, particularly around the S-A and A-V nodes, whereas they are less abundant in the ventricles. The coronary arteries are also richly innervated with NPY-immunoreactive nerves (Lundberg *et al.,* 1983; Gu *et al.,* 1983). In addition, NPY is present in the adrenal medulla, where it seems to be confined mainly to adrenaline-containing cells (Lundberg *et al.,* 1986a). However, in comparison with other species the human adrenal contains relatively small amounts of NPY.

2. Cardiovascular Effects

Administration of NPY either *in vitro* or *in vivo* causes dose-related vasoconstriction. On isolated vascular segments *in vitro,* NPY induces slowly developing, long-lasting, and potent contraction of cerebral arteries (Mejia *et al.,* 1988; Uddman and Edvinsson, 1989), gastrointestinal arteries (Hanko *et al.,* 1986), renal arteries, skeletal muscle arteries (J. Pernow *et al.,* 1987b), and coronary arteries (Franco-Cereceda and Lundberg, 1987; Tseng *et al.,* 1988). The potency of NPY is generally 10–100 times higher than that of noradrenaline on a molar basis, and a threshold response is often observed at a concentration of about 1 nM (Fig. 1). In skeletal muscle, NPY is more active on arteries than on veins (J. Pernow *et al.,* 1987c). In addition to having direct vasoconstrictor effects, low concentrations of NPY have been shown to potentiate the vasoconstrictor response to noradrenaline in coronary (Tseng *et al.,* 1988) and omental arteries (Edvinsson *et al.,* 1985), which may indicate a cooperative mechanism between the two agents when released from sympathetic nerves. This potentiation does not seem to be a general phenomenon, however, as it has not been observed in cerebral, mesenteric, or skeletal muscle arteries (Hanko *et al.,* 1986; J. Pernow *et al.,* 1987b; Uddman and Edvinsson, 1989).

Under *in vivo* conditions, local infusion of NPY into the brachial artery of healthy

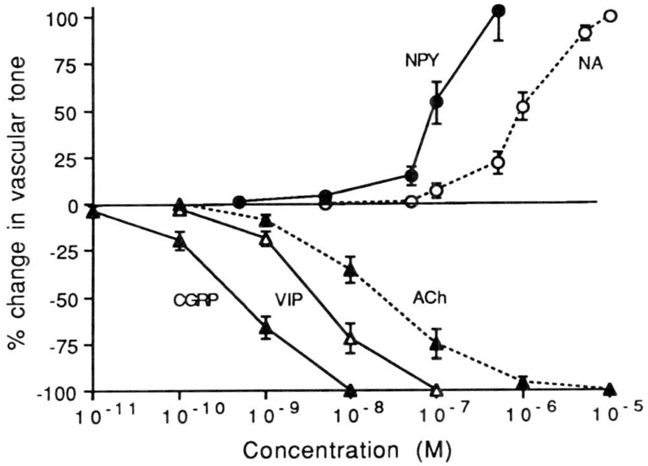

Figure 1 Concentration-response curves of the vasoconstrictor and vasodilator effects of neuropeptide Y (NPY), noradrenaline (NA), calcitonin gene-related peptide (CGRP), vasoactive intestinal polypeptide (VIP), and acetylcholine (ACh) on human skeletal muscle arteries *in vitro*. Values are means and SEM. (Modified from J. Pernow *et al.*, 1987b, and J. Pernow, 1989.)

volunteers evoked a dose-dependent reduction in forearm blood flow by 55% and an increase in venous tone by 135%, indicating constriction of both arterial resistance vessels and veins (J. Pernow *et al.*, 1987a). A threshold response was observed at a plasma concentration of 3 mM. The vasoconstrictor response to the highest dose of NPY was still present 25 min after the infusion, showing a very long-lasting effect compared to that of noradrenaline, which lasted less than 5 min. Intravenous infusion of NPY at a dose that increased plasma NPY levels from 12 to 356 pM did not affect systemic blood pressure, heart rate, or forearm blood flow (J. Pernow *et al.*, 1987a). However, NPY evoked reduction in renal and splanchnic blood flow by 13 and 18%, respectively, at 261 pM during prolonged (40 min) intravenous administration (Ahlborg *et al.*, 1989, 1992). Infusion of NPY into the coronary artery of patients with angina pectoris induced chest pain, ECG changes indicating ischemia, and coronary vasoconstriction as shown by arteriography (Clarke *et al.*, 1987). The ischemia was completely reversed by administration of isosorbide dinitrate. Following studies on isolated specimens from the human heart, it has been concluded that NPY does not exert any direct inotropic or chronotropic effects, however (Franco-Cereceda *et al.*, 1987a; Michel *et al.*, 1989).

The vasoconstrictor response to NPY is not affected by α- or β-adrenoceptor blockade, indicating that it acts via specific receptors (Edvinsson *et al.*, 1985; J. Pernow *et al.*, 1987b). Furthermore, the effect is not dependent on endothelium-derived mediators (J. Pernow and Lundberg, 1988). On the other hand, the vasoconstrictor effect of NPY on both cerebral and peripheral arteries is greatly reduced or abolished in the presence of calcium antagonists, such as nifedipine or diltiazem

(J. Pernow *et al.*, 1987b; Uddman and Edvinsson, 1989). Moreover, the NPY-induced reduction in forearm blood flow observed during intraarterial infusion is significantly attenuated and of a shorter duration after oral administration of nifedipine (J. Pernow *et al.*, 1989). These findings would suggest that the vasoconstrictor effect of NPY is dependent on influx of extracellular calcium. However, removal of extracellular calcium does not inhibit the NPY-induced vasoconstriction in skeletal muscle arteries, indicating that also other actions of the calcium antagonist may underly the inhibition of NPY-induced vasoconstrictor responses (J. Pernow, 1988).

In addition to having direct cardiovascular effects, NPY has been shown to inhibit nerve stimulation-evoked release of noradrenaline from perivascular nerves (J. Pernow *et al.*, 1987b), indicating a prejunctional modulatory action on transmitter release. This effect is observed at concentrations similar to those at which NPY evokes vasoconstriction, and it is not affected by calcium antagonists or α-adrenoceptor antagonists.

3. Plasma Levels and Release of NPY

The levels of NPY-like immunoreactivity observed in plasma under basal conditions in healthy subjects vary depending on the assay used. In a series of investigations that includes a large number of observations using a well-characterized antiserum (N1), the plasma levels of NPY-like immunoreactivity in healthy volunteers fall within the range of 12–30 pM (see Lundberg *et al.*, 1987; J. Pernow, 1988). Similar levels are reported in most experimental animals (e.g., pig, dog, cat, and guinea pig). Other groups have reported basal plasma levels of NPY in humans that are in the range of 35–106 pM (Adrian *et al.*, 1983; Morris *et al.*, 1986; Takahashi *et al.*, 1987; Maisel *et al.*, 1989; Edvinsson *et al.*, 1991). The variation in plasma NPY probably relates to the fact that different extraction procedures are used and that the antisera show different cross-reactivity patterns. The reported basal concentrations are all lower than those required to evoke vasoconstriction and therefore probably reflect spillover from release. The disappearance of NPY from plasma has been determined to be biphasic with half-lives of 4.1 and 20 min for the two slopes, respectively, (J. Pernow *et al.*, 1987a). The splanchnic region seems to be important for clearance of circulating NPY since a 59% extraction was observed over this region during intravenous infusion of NPY (Ahlborg *et al.*, 1989). There is, on the other hand, no extraction over the kidney (Ahlborg *et al.*, 1989; 1992) or forearm (J. Pernow *et al.*, 1987a).

Increased plasma levels as an indication of release of NPY have been observed in a number of different physiological and pathophysiological situations (see Table 1). During activation of the sympathetic nervous system by physical exercise there is an increase in peripheral plasma levels of NPY as well as of noradrenaline and adrenaline (Fig. 2). During both graded (Lundberg *et al.*, 1985a) and prolonged (J. Pernow *et al.*, 1986) bicycle exercise plasma levels of NPY increase 3- to 4-fold. The maximal systemic plasma levels attained during exercise (80–100 pM) are not high enough to exert vasoconstrictor effects. The concentration close to the release site is

Table 1 Situations in Which Increased Plasma Levels of NPY-Like Immunoreactivity Are Observed

Situation	% Increase[a] NPY	NA	Reference
Bicycle exercise			
graded	183	2130	Lundberg *et al.*, 1985a
	57	—	Morris *et al.*, 1986
	190	1577	J. Pernow *et al.*, 1988
graded + α-blockade	779	2877	J. Pernow *et al.*, 1988
prolonged	326	872	J. Pernow *et al.*, 1986
Cold pressor test	—	—	Morris *et al.*, 1986
Blood pressure reduction	105	160	Eckberg *et al.*, 1988
Infants after delivery	—	—	Lundberg *et al.*, 1986b
Thoracic surgery	200	223	Lundberg *et al.*, 1985b
Exposure to cigarette smoke	53	150	Rudehill *et al.*, 1989
Hypoglycemia	48	nd	Takahashi *et al.*, 1988
Congestive heart failure	77	147	Maisel *et al.*, 1989
	907	—	Hulting *et al.*, 1990
Hypertension	37	—	Edvinsson *et al.*, 1991
Septic shock	—	nd	Watson *et al.*, 1988
Renal failure	148	nd	Takahashi *et al.*, 1987
Phaeochromocytoma patients	736	nd	Adrian *et al.*, 1983
	556	—	Lundberg *et al.*, 1986a
	282	nd	Takahashi *et al.*, 1987

[a]The increase is expressed in percent above basal level or a control group. The associated increase in plasma noradrenaline is also given for comparison. — indicates that the exact values or that control values are not given; nd, not determined.

likely to be considerably higher, however, and this indicates that NPY does not act as a circulating hormone but rather as a locally acting vasoconstrictor (J. Pernow, 1988). The plasma levels of NPY during exercise correlate significantly better to those of noradrenaline than to those of adrenaline, suggesting that the peptide is released from the sympathetic nerves rather than the adrenal medulla. The increase in plasma NPY levels generally occurs later and at a higher work load than the increase in noradrenaline concentrations (Fig. 2), which may reflect that NPY is released mainly at a high degree of sympathetic activity. Furthermore, the correlation between plasma NPY and noradrenaline represents a sigmoidal curve rather than a straight line, supporting the view that NPY is released mainly when noradrenaline release is large (J. Pernow, 1988). Moreover, during isometric handgrip and orthostatic tests, when the sympathetic nervous system is activated less than during bicycle exercise, no release of NPY can be detected whereas plasma noradrenaline levels increase 2-fold (J. Pernow *et al.*, 1986). In accordance with this observation, animal experiments have shown that NPY is released mainly upon high-frequency sympathetic nerve stimulation (Lundberg *et al.*, 1987; J. Pernow, 1988). The release of NPY, like the release of noradrenaline, seems to be regulated

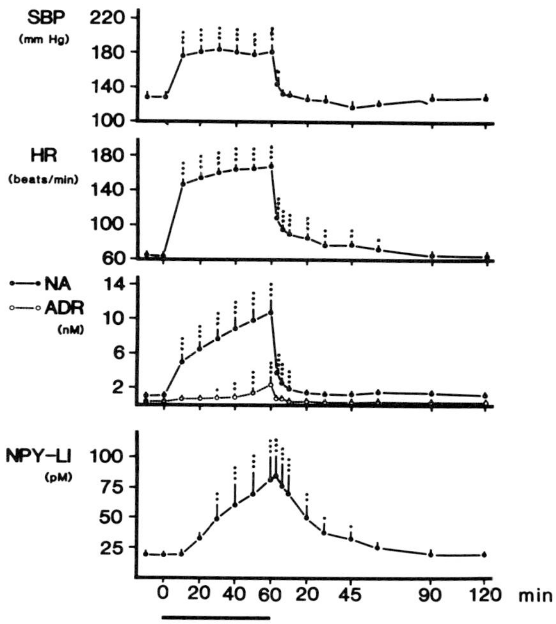

Figure 2 Effects of bicycle exercise during 60 min at a constant work load (70% of maximum oxygen uptake) on systolic blood pressure (SBP), heart rate (HR), arterial plasma levels of adrenaline (ADR), and cubital venous plasma levels of noradrenaline (NA) and neuropeptide Y-like immunoreactivity (NPY-LI). Values are means and SEM. Significant differences from basal levels are indicated; $*P<0.05$, $**P < 0.01$, $***P < 0.001$. (From J. Pernow *et al.*, 1986, with permission.)

via α-adrenoceptors since pretreatment with the α_2-adrenoceptor agonist clonidine attenuated, and the α-adrenoceptor antagonist phentolamine enhanced the exercise-induced increase in plasma levels of both NPY and noradrenaline (J. Pernow *et al.*, 1988). However, whereas resting NPY levels were unaffected by the drugs, resting noradrenaline levels were altered in a manner similar to that observed during exercise, again indicating that the release of NPY is relatively small under basal conditions with a low degree of sympathetic activity.

Determinations of local NPY release during physical exercise have been performed by measuring arteriovenous gradients over various vascular beds. A significant net corelease of NPY and noradrenaline was observed from the heart (Kaijser *et al.*, 1990) and the kidney (Tidgren, 1989) but not from the forearm (J. Pernow *et al.*, 1986). The release of NPY from the heart during exercise was enhanced when the subjects breathed hypoxic gas resulting in a marked decrease in arterial pO_2 (Fig. 3). The enhancement of NPY release was larger than that of noradrenaline, indicating that hypoxia may preferentially facilitate NPY release either via a direct action or secondary to increased sympathetic activity. This finding and the observation that NPY is a potent coronary vasoconstrictor open up the possibility that NPY may be a factor involved in the events occurring during ischemic heart disease.

Increased circulating plasma levels of NPY are also observed during situations of reflex sympathoadrenal activation other than physical exercise (Table 1). High levels are observed in infants immediately after delivery, indicating that NPY may play a role in the stress reaction in association with birth (Lundberg *et al.*, 1986b). Insulin-induced hypoglycemia, during which there is marked activation of the adrenals, has been reported to be associated with only a moderate increase (less than 2-fold) (Takahashi *et al.*, 1988) or no change (Russel *et al.*, 1989) in systemic plasma NPY levels even though the adrenaline levels increased 15-fold. This suggests that the adrenal is not a major source of circulating NPY during sympathoadrenal activation.

A number of different pathophysiological conditions are also known to be associated with elevated plasma NPY levels (Table 1). As mentioned earlier, the observation that NPY is a potent coronary vasoconstrictor, and that its release from the heart is enhanced during hypoxic conditions, indicates that it may be of importance in ischemic heart disease. Accordingly, a study performed on patients on admission to

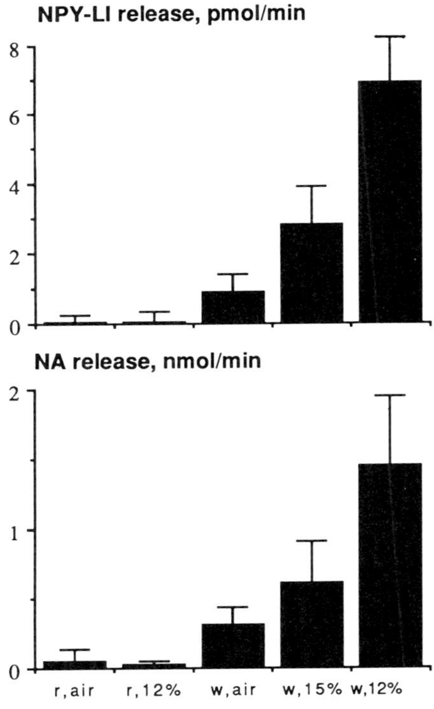

Figure 3 Net release of neuropeptide Y-like immunoreactivity (NPY-LI) and noradrenaline (NA) from the heart at rest (r) breathing air and 12% oxygen and during submaximal work (w) breathing air, 15% and 12% oxygen. Values are means and SEM. (From Kaijser *et al.*, 1990, with permission.)

a coronary care unit showed that 37 of 113 patients (33%) with acute myocardial infarction had increased NPY plasma levels (Hulting *et al.*, 1990). Elevated plasma NPY levels have been observed in patients with congestive heart failure (Maisel *et al.*, 1989; Hulting *et al.*, 1990). Hulting *et al.* (1990) found that 90% of patients with pulmonary edema and 100% of patients in cardiogenic shock had elevated plasma NPY levels, with an average magnitude of 6- and 11-fold, respectively. Furthermore, Edvinsson *et al.* (1991) recently reported that patients with severe hypertension (>120 mmHg diastolic pressure) had 40% higher circulating plasma levels of NPY-like immunoreactivity than a control group. However, the levels were still significantly elevated 2–3 weeks after treatment and reduction of the blood pressure, indicating that the levels of NPY-like immunoreactivity were not directly related to the blood pressure. In addition, in that study (Edvinsson *et al.*, 1991), only a minor portion of the immunoreactivity detected by the assay corresponded to human NPY, suggesting that the role of NPY in hypertension still remains unclear. Circulating plasma NPY levels are also increased in patients in septic shock (Watson *et al.*, 1988), a condition characterized by increased sympathetic activation and peripheral vasoconstriction. Moreover, phaeochromocytoma tumors frequently contain large amounts of NPY (Adrian *et al.*, 1983; Lundberg *et al.*, 1986a). Plasma from patients with these tumors has about 8 times higher concentration of NPY during basal conditions than plasma from normal subjects. In connection with surgery and manipulation of the tumor there is a marked additional increase in plasma NPY (up to 2000 pM). Interestingly, this increase is associated with an elevation in blood pressure despite pretreatment with an α-adrenoceptor antagonist (Lundberg *et al.*, 1986a). These observations suggest a possible role of NPY in various cardiovascular disorders and this issue is further described in Chapters 17 and 18.

The cardiovascular actions and release of NPY in humans indicate that it may be involved as a sympathetic cotransmitter together with noradrenaline in the regulation of the cardiovascular system both under physiological and pathophysiological conditions. The results described suggest that its importance may increase with the degree of the sympathetic activity. Furthermore, interference with NPY mechanisms (e.g., release and vasoconstriction) should be taken into consideration when evaluating the effects of various drugs used in the treatment of cardiovascular disorders (see Lundberg *et al.*, 1987, 1990).

B. Somatostatin

Somatostatin (SOM) was first isolated as a 14 amino acid peptide and was subsequently found to exist also in a 28 amino acid form (Pradayrol *et al.*, 1980). The distribution pattern of SOM in the peripheral nervous system is complex. Studies using immunohistochemical techniques have shown that SOM is present in the sympathetic and sensory nervous system (Hökfelt *et al.*, 1980) as well as in cardiac vagal nerves (see Section III.B; Day *et al.*, 1985; Franco-Cereceda *et al.*, 1986) besides being widely distributed in intestinal nerve fibers (Costa and Furness, 1984).

In sympathetic ganglion cells SOM is colocalized with noradrenaline, and in sensory nerves it is present together with substance P (Hökfelt *et al.*, 1980).

Apart from the effects of SOM in the gastrointestinal tract, where it acts as a potent inhibitor of secretion of other peptides (Bloom, 1978), early studies of the cardiovascular actions of SOM in humans revealed that it causes bradycardia and reduction in cardiac output (Rosenthal *et al.*, 1978; Chernish *et al.*, 1979; Hallemans *et al.*, 1981) but only a small transient increase (Chernish *et al.*, 1979; Hallemans *et al.*, 1981) or no change in systemic blood pressure (Rosenthal *et al.*, 1978). It has also been demonstrated that SOM inhibits renin release in patients with high renin essential hypertension (Rosenthal *et al.*, 1978). In normotensive subjects, SOM attenuates renin release when the renin production is stimulated by β-adrenoceptor agonists (Rosenthal *et al.*, 1977) or furosemide (Rosenthal *et al.*, 1978).

The mechanism of the cardioinhibitory effects of SOM has been further evaluated using isolated preparations of the human heart. SOM evoked a dose-dependent inhibition of noradrenaline-induced stimulation of atrial contractility, the inhibition being approximately 50% at 1 μM (Franco-Cereceda *et al.*, 1986). This inhibitory effect was not affected by atropine, indicating that SOM exerts a direct effect on the myocardium independent of muscarinic receptors. Furthermore, SOM suppresses the automaticity and the triggered activity of isolated human atrial fibers (Hou *et al.*, 1987). In patients with paroxysmal supraventricular and junctional tachycardia, SOM has been shown to restore sinus rhythm as effectively as the calcium antagonist verapamil (Greco *et al.*, 1984; Webb *et al.*, 1986). During sinus rhythm, the heart rate is reduced and the atrioventricular conductance is prolonged by SOM. The mode of action of SOM in these situations has been suggested to involve a decrease in calcium influx, resembling the action of calcium antagonists. In the guinea pig, the negative inotropic effect of SOM can be potentiated by reducing the extracellular calcium level and antagonized by increasing it (Quirion *et al.*, 1979). The similarity in effects on the human heart by SOM and verapamil observed by Webb *et al.* (1986) supports this view.

Studies on the effects of SOM on the peripheral circulation have revealed that it effectively reduces splanchnic blood flow (Wahren and Felig, 1976). Subsequent investigations showed that SOM given as intravenous bolus injections (1 mg/kg) or infusions (7.5 mg/min) reduces portal venous pressure and hepatic blood flow by 9–28% (Bosch *et al.*, 1981; Eriksson *et al.*, 1984). These observations suggest that SOM could be a useful agent in the treatment of portal hypertension. The mechanism underlying this effect is unclear although suppression of the release of intestinal vasodilating hormones has been suggested (Bosch *et al.*, 1988). One disadvantage with using SOM as a therapeutic agent is its short half-life of only 2–3 min (Webb *et al.*, 1986). Recently, a SOM analog, SMS 201–995, with a half-life of approximately 1 h was reported to have effects on splanchnic blood flow that are similar to but more prolonged than those of SOM (Eriksson and Wahren, 1989).

These findings indicate possible and complex actions of SOM via different branches of the autonomic nervous system on both the heart and peripheral circulation in addition to effects secondary to modulation of other vasoactive peptides.

C. Opioids

The opioid peptides are not only widely distributed in the central nervous system but are also present in the peripheral autonomic nervous system of both animals and man (see Hökfelt *et al.*, 1980; Lundberg and Hökfelt, 1983; Ganten *et al.*, 1984). In the sympathetic nervous system, enkephalins may be present as cotransmitters with catecholamines. Biochemical and immunohistochemical studies have shown the pentapeptides met-enkephalin and leu-enkephalin to be present in catecholamine-producing human adrenal medullary chromaffin cells (Lundberg *et al.*, 1979; Clement-Jones *et al.*, 1980b; Varndell *et al.*, 1982; Parmer and O'Connor, 1988). Met-enkephalin has also been demonstrated to coexist with NPY in a subpopulation of adrenal medullary cells (Osamura *et al.*, 1987). Furthermore, certain phaeochromocytoma contain large amounts of enkephalins (Sullivan *et al.*, 1978; Lundberg *et al.*, 1979). The predominant form of enkephalins present in phaeochromocytoma tumors seems to be met-enkephalin over leu-enkephalin by a factor of 10–80 to 1 (Clement-Jones *et al.*, 1980a; Parmer and O'Connor, 1988). In most studies, the presence and amount of enkephalin correlate with adrenaline rather than noradrenaline, indicating coexistence of enkephalin and adrenaline (Yoshimasa *et al.*, 1983; Parmer and O'Connor, 1988; Kodama *et al.*, 1990), whereas other studies have demonstrated met-enkephalin in both adrenaline- and noradrenaline-producing cells (Varndell *et al.*, 1982).

Enkephalins are present in human plasma as intact pentapeptides at a concentration of 14–140 pg/ml (25–250 pM; Clement-Jones *et al.*, 1982). Determination of plasma gradients of met-enkephalin over different vascular beds showed that the adrenal vein contained higher levels than peripheral plasma, indicating that the adrenal medulla is a major source of circulating enkephalin. However, the adrenal may not be the only source since plasma enkephalin could still be detected after bilateral adrenalectomy (Clement-Jones *et al.*, 1982), suggesting that spillover of release from postganglionic perivascular nerves is also possible (see Ganten *et al.*, 1984; Owman, 1988). Increased plasma levels of enkephalin have been reported in experimental animals following sympathetic nerve stimulation or hemorrhagic shock (Viveros *et al.*, 1979; Lang *et al.*, 1982) and in humans during physical exercise, indicating release in connection with increased sympathoadrenal activation (Ganten *et al.*, 1984). Elevated plasma enkephalin immunoreactivity has further been demonstrated in phaeochromocytoma patients with high adrenaline levels (Yoshimasa *et al.*, 1983).

The presence of enkephalins in the peripheral autonomic nervous system and their release into plasma may indicate a possible role of these peptides in peripheral cardiovascular regulation. Their actions may be either modulation of transmitter secretion or direct effects on blood vessels. In experimental animals, enkephalins are able to inhibit noradrenaline release from sympathetic nerves (Ganten *et al.*,

1984). Accordingly, met-enkephalin has been demonstrated to inhibit catecholamine secretion from human phaeochromocytoma cells (Yanase *et al.*, 1986).

The ability of opioid receptor-stimulating drugs to lower blood pressure in humans upon systemic administration is well known. Furthermore, the opioid receptor antagonist naloxone has been shown to prevent the fall in blood pressure during sleep (Rubin *et al.*, 1981) and may reverse the fall in blood pressure during septic shock (Peters *et al.*, 1981). However, the predominant site of action for opioids in blood pressure regulation seems to be the central nervous system rather than the periphery (Reid and Rubin, 1987). Zelis *et al.* (1974) reported that intravenous administration of morphine to normal volunteers resulted in a reduction in forearm vascular resistance and venous pressure of the hand. However, when administered locally into the brachial artery, morphine produced no effect on forearm vascular resistance. Furthermore, the dilator response to intravenous morphine was blocked by the α-adrenoceptor antagonist phentolamine, indicating that the dilator response was mediated via a central reflex action involving sympathetic vasomotor tone rather than a direct action on the blood vessels. Studies on isolated human mesenteric arteries and veins *in vitro* have not revealed any reproducible effects of met-enkephalin or leu-enkephalin on vascular tone (Törnebrandt *et al.*, 1987). Administration of the met-enkephalin analog D-Ala,MePhe,Met,enkephalin (DAMME) and metkephamid, which are specific agonists for μ and δ opioid receptors, produces orthostatic hypotension by a mechanism that appears to involve a reduction in the baroreflex sensitivity since heart rate and plasma noradrenaline did not increase appropriately in response to the reduction in systemic blood pressure, and supine blood pressure was unaffected (Rubin *et al.*, 1983; Pasanisi *et al.*, 1985).

These findings indicate that endogenous opioids may be involved in human blood pressure regulation although the exact site of action is still unclear and the peripheral actions may be of minor importance (see Ganten *et al.*, 1984; Reid and Rubin, 1987).

III. Parasympathetic Nerves

A. *Vasoactive Intestinal Polypeptide and Peptide Histidine-Methionine*

1. Distribution

Vasoactive intestinal polypeptide (VIP) was first isolated from the porcine intestine by Said and Mutt (1970). The human and porcine forms of the peptide have been shown to be identical (Carlquist *et al.*, 1982). VIP has a wide distribution in both the central and peripheral nervous system. In the periphery, VIP has been shown to be present in parasympathetic postganglionic nerves together with acetylcholine (Lundberg, 1981). An additional peptide called PHM (peptide with N-terminal histidine and C-terminal methionine amide), which is derived from the same precursor peptide, also coexists with VIP (Lundberg *et al.*, 1984). PHM represents the human form of PHI (peptide with N-terminal histidine and C-terminal isoleucine amide), which is present in the rat and pig. Data showing the distribution and function of VIP and PHM/PHI have recently been reviewed by Fahrenkrug (1989). In human tissue, immunoreactive staining with VIP antisera has revealed the presence of VIP in parasympathetic ganglia and nerves innervating blood vessels in the respiratory tract, the gastrointestinal tract, and cerebral arteries as well as in connection with myocardial cells and blood vessels in the heart (Lundberg *et al.*, 1984; Uddman and Edvinsson, 1989; see also Fahrenkrug, 1989).

2. Cardiovascular Effects

VIP is a strong and potent vasodilator and is likely to mediate the atropine-resistant vasodilatation evoked by parasympathetic nerve stimulation in experimental animals (Fahrenkrug, 1989). In humans, VIP evokes dose-dependent relaxation of isolated omental arteries and veins (Edvinsson *et al.*, 1985); arteries from exocrine glands (Larsson *et al.*, 1986); gastric, pulmonary, and transverse cervical arteries (Thom *et al.*, 1987); cerebral arteries (Uddman and Edvinsson, 1989); skeletal muscle arteries (J. Pernow, 1989); and coronary arteries (Franco-Cereceda and Rudehill, 1989) (Fig. 1). The vasodilator properties of PHM in submandibular arteries seem to be similar to those of VIP (Larsson *et al.*, 1986) whereas VIP was 10 times more potent than PHM as a dilator of cerebral arteries (Edvinsson *et al.*, 1987b). In accordance with the relaxation observed in isolated blood vessels *in vitro,* VIP produces reduction in systemic blood pressure and increase in heart rate upon intravenous infusion in humans (Table 2) (Domschke *et al.*, 1978; Thulin *et al.*, 1984; Frase *et al.*, 1987; Smitherman *et al.*, 1989). The reduction in blood pressure is produced by peripheral vasodilatation with a fall in peripheral resistance. During intravenous infusion of VIP at a dose that increased arterial plasma VIP levels from 45 to 180 pg/ml (13–54 pM), Smitherman *et al.* (1989) observed a decrease in mean arterial blood pressure, total peripheral resistance and coronary vascular resistance, and an increase in heart rate and cardiac output (Table 2). Interestingly, these plasma levels of VIP attained were lower than those reported in patients with VIP-secreting tumors (Bloom *et al.*, 1973). Local infusion of VIP into a coronary artery evoked a 46% decrease in coronary vascular resistance at a plasma level of 110 pg/ml (33 pM), indicating a direct coronary vasodilator effect *in vivo* (Smitherman *et al.*, 1989). In these experiments VIP did not increase coronary sinus blood levels of the prostaglandin I_2 metabolite 6-keto-prostaglandin $F_1\alpha$, and the coronary vasodilator effect of VIP was not affected by cyclooxygenase inhibitor, suggesting that its effect did not involve prostaglandin formation. It has further been reported that VIP induces vasodilatation in the forearm (Frase *et al.*, 1987; Thom *et al.*, 1987) and the splanchnic region (Thulin *et al.*, 1984) as well as cutaneous flushing (Domschke *et al.*, 1978). VIP most likely does not act as a circulating hormone but rather exerts local effects at the site of release from the parasympathetic nerves (Owman 1988; Fahrenkrug, 1989). Furthermore, circulating VIP is rapidly cleared from the circulation with a half-life of approximately 1 min (Domschke *et al.*, 1978). Intravenous infusion of PHM at a dose that increased

Table 2 Hemodynamic and Metabolic Effects of Intravenous VIP Administration[a]

	Baseline	Infusion 1	Infusion 2
Arterial VIP level (pg/ml)	45 ± 17	91 ± 24[b]	180 ± 35[b]
Coronary sinus VIP level (pg/ml)	45 ± 11	92 ± 23[b]	179 ± 48[b]
Heart rate (beats/min)	71 ± 12	77 ± 12[b]	81 ± 11[b]
Mean arterial pressure (mmHg)	106 ± 19	101 ± 18	94 ± 17[b]
Mean right atrial pressure (mmHg)	7 ± 4	7 ± 4	7 ± 4
Mean pulmonary arterial pressure (mmHg)	16 ± 6	15 ± 6	16 ± 7
Mean pulmonary capillary wedge pressure (mmHg)	9 ± 5	9 ± 4	9 ± 5
Cardiac output (l/min)	5.98 ± 1.44	7.20 ± 2.31[b]	7.59 ± 2.41[b]
Coronary blood flow (ml/min)	127 ± 66	150 ± 78	168 ± 76[b]
Systemic vascular resistance (dyn s cm^{-5})	1404 ± 487	1116 ± 353[b]	973 ± 285[b]
Coronary vascular resistance (mmHg ml^{-1} min)	0.94 ± 0.43	0.77 ± 0.35	0.63 ± 0.34[b]
Pulmonary vascular resistance (dyn s cm^{-5})	92 ± 32	75 ± 26[b]	70 ± 15[b]
Myocardial oxygen uptake (ml/min)	16 ± 9	19 ± 10	20 ± 11[b]
Arterial-coronary sinus O_2 (vol %)	12.0 ± 2.2	12.0 ± 2.3	11.7 ± 2.3
Coronary arterial oxygen extracted (%)	62 ± 10	62 ± 10	61 ± 10
Arterial 6-keto-prostaglandin $F_{1\alpha}$ level (pg/ml)	3.7 ± 2.9	2.4 ± 2.6	2.1 ± 1.9
Coronary sinus 6-keto-prostaglandin $F_{1\alpha}$ level (pg/ml)	2.7 ± 1.4	1.8 ± 1.1	2.3 ± 1.6

Source: Smitherman *et al.* (1989) with permission.
[a] Infusion 1 was 100 pmol kg^{-1} h^{-1} for 12 min followed by 50 pmol kg^{-1} h^{-1} during the hemodynamic measurements and blood sampling. Infusion 2 was 200 pmol kg^{-1} h^{-1} for 12 min followed by 100 pmol kg^{-1} h^{-1}. Significant differences from basal values are indicated.
[b] $P \leq 0.05$. Values are means ±SD.

plasma PHM levels to 1340 pM did not produce any affects on blood pressure or heart rate, indicating that it is less active *in vivo* than VIP (Fahrenkrug *et al.*, 1987).

The vasodilator effect of VIP does not involve cholinergic or adrenergic receptors, indicating activation via specific VIP receptors (Uddman and Edvinsson, 1989; Unwin *et al.*, 1987). In the heart, two classes of VIP receptors have been demonstrated based on the ability of VIP to stimulate cyclic AMP production (Christophe *et al.*, 1984). Removal of the vascular endothelium does not affect the vasodilator response to VIP in skeletal muscle arteries (J. Pernow, 1989), coronary arteries (Franco-Cereceda and Rudehill, 1989), or pulmonary arteries (Thom *et al.*, 1987), whereas the vasodilatation induced by the colocalized transmitter acetylcholine is endothelium dependent. The endothelium dependence of VIP seems to vary between different types of vessels, however, as the vasodilatation of VIP in gastric and transverse cervical arteries has been described to be endothelium dependent (Thom *et al.*, 1987).

As mentioned earlier, the fall in blood pressure observed during intravenous administration of VIP is accompanied by an increase in heart rate and cardiac output that may be explained by reflex sympathetic activation. Unwin *et al.* (1987), however, found that the percentual increase in heart rate induced by intravenous VIP was unchanged after β-adrenoceptor blockade. This finding would indicate that VIP exerts direct cardiostimulatory effects. Accordingly, VIP evokes positive inotropic effects on isolated human atria although the response is smaller than that evoked by noradrenaline (Franco-Cereceda *et al.*, 1987a).

B. Somatostatin

As previously mentioned (Section II.B), SOM has been demonstrated in the human heart. Immunohistochemical studies have revealed SOM-containing nerve fibers mainly in the myocardium of the right atrium, and single fibers in the left ventricle (Franco-Cereceda *et al.*, 1986). High concentrations of SOM have been detected in the A-V node and right atrium by biochemical methods (Day *et al.*, 1985). These SOM-containing nerve fibers are likely to be of vagal parasympathetic origin and SOM would thus coexist with acetylcholine. The cardiac effects of SOM (see Section II.B), which are similar to those of acetylcholine, as well as the distribution and the finding that SOM is released by vagal nerve stimulation in the heart of the toad (Campbell *et al.*, 1982) support this view. The demonstration of SOM in the human heart and its effects on cardiac contractility and conductance may thus suggest a physiological role in vagal cardiac control.

IV. Sensory Nerves

A. Tachykinins and Calcitonin Gene-Related Peptide

1. Distribution

The sensory nerves contain several peptides of separate origin. The group of peptides called tachykinins include, among others, substance P (SP), neurokinin A (NKA), and neuropeptide K (NPK), which are encoded from the same gene (for review see B. Pernow, 1983). Recently, an additional peptide, calcitonin gene-related peptide (CGRP), which is encoded from the same gene as calcitonin, has been shown to coexist with tachykinins in sensory C-fibers. In nervous tissue the main product is CGRP, whereas mainly calcitonin is formed in the thyroid. At least two forms of CGRP exist, α and β, which differ in three amino acids (see Tippins, 1986; Franco-Cereceda, 1988). Nerve terminals containing SP- and CGRP-like immunoreactivity have been observed in various areas of the human cardiovascular system, including the heart (Weihe *et al.*, 1981; Rechardt *et al.*, 1986; Franco-Cereceda *et al.*, 1987b), blood vessels in the skin (Dalsgaard, 1988), skeletal muscle (Uddman *et al.*, 1986), and cerebral arteries (Edvinsson *et al.*, 1987a). SP-like immunoreactivity has also been identified in arterial endothelial cells (Linnik and Moskowitz, 1989). In the heart, SP and CGRP immunoreactive varicose nerves are present both in association with myocardial cells and around blood vessels (Franco-Cereceda *et al.*, 1987b). By using biochemical methods, it was further shown that the CGRP levels were 3- to 4-fold higher in the atrium than in the ventricle.

2. Cardiovascular Effects

SP, CGRP, and NKA cause potent relaxation of isolated blood vessels from a variety of vascular regions, including coronary arteries (Franco-Cereceda and

Lundberg, 1987), skeletal muscle arteries (J. Pernow, 1989), omental arteries and veins (Edvinsson *et al.*, 1985), pulmonary arteries and veins (McCormack *et al.*, 1989), and cerebral arteries (Mejia *et al.*, 1988; Uddman and Edvinsson, 1989). CGRP and SP are considerably more potent than other vasodilators (e.g., VIP and acetylcholine) (Fig. 1).

Intravenous administration of SP and CGRP in humans evokes a fall in arterial blood pressure and an increase in heart rate (Eklund *et al.*, 1977; Gennari and Fischer, 1985; Franco-Cereceda *et al.*, 1987c). During intravenous infusion of SP there is an increase in skin and muscle blood flow, indicating vasodilatation, and a fall in forearm venous tone that is secondary to reflex mechanisms (Eklund *et al.*, 1977). Close arterial administration of SP in the forearm evokes an immediate and dose-related increase in forearm blood flow but does not affect venous tone. The arterial vasodilatation is not altered by α- or β-adrenoceptor antagonists, atropine, antihistamines, or indomethacin (Eklund *et al.*, 1977; B. Pernow, 1983). Intracoronary infusion of SP dilated the left anterior descending artery dose dependently, as revealed by angiography, with a maximal dilatation of 29% at a dose of 11.2 pmol/min (Crossman *et al.*, 1989). At this dose there was no change in systemic blood pressure. When injected intradermally, SP produced a flare, wheal, and itching (Hägermark *et al.*, 1978). This effect was inhibited by oral pretreatment with antihistamine or compound 48/80, an agent known to deplete mast cells of histamine, suggesting that it was indirect and caused by mast cell degranulation.

CGRP, like SP, produces a fall in blood pressure and increases in heart rate when administered intravenously to healthy volunteers (Gennari and Fischer, 1985; Franco-Cereceda *et al.*, 1987c). The cardiovascular effects of α- and β-CGRP are largely similar. In addition, intravenous CGRP causes increased inotropic effects (Franco-Cereceda *et al.*, 1987c), stimulation of renin release (Kurtz *et al.*, 1988), and vasodilatation in the forearm and skin (Jernbeck *et al.*, 1990). A fall in blood pressure is observed at a dose of approximately 500 pmol/min (Tippins, 1986; Jernbeck *et al.*, 1990) and at this dose forearm blood flow increased by 115% (Jernbeck *et al.*, 1990). Infusion of CGRP into the brachial artery resulted in a dose-dependent increase in forearm and cutaneous blood flow of the wrist that persisted for 90 min (Fig. 4). At the highest dose given (216 pmol/min) forearm blood flow increased by 177–352% and cutaneous blood flow increased 10-fold (Jernbeck *et al.*, 1990). In a study in which the vasodilator effects of CGRP and SP in the forearm were directly compared, McEwan *et al.* (1988) found that SP was 10 times more potent than CGRP on a molar basis. Furthermore, the duration of the effect produced by SP was considerably shorter than that produced by CGRP. Thus, the half-life of the biological effect of SP was 15 s while that of CGRP was 18 min. This difference parallels the observation that SP is rapidly broken down in human blood (B. Pernow, 1983), while CGRP has a half-life of 9.7 min (Struthers *et al.*, 1986). In addition, tachyphylaxis developed during prolonged infusion of SP but not during infusion of CGRP. Intracoronary infusion of CGRP caused an increase in diameter of various parts of the left anterior descending artery by up to 40% with the largest effect in the distal part of the artery (McEwan *et al.*, 1986). Intradermal injections of CGRP were found to induce cutaneous vasodilatation that lasted for 2 h after a 15

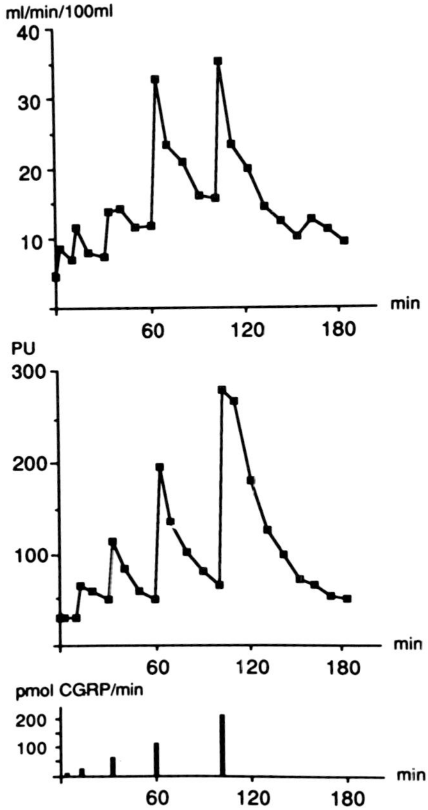

Figure 4 Effects of intraarterial infusion of calcitonin gene-related peptide (CGRP) on forearm blood flow (*top*) and cutaneous blood flow of the hand (*middle*) in one subject. The doses are indicated in bottom figure. (From Jernbeck *et al.*, 1990, with permission.)

fmol dose (Brain *et al.*, 1985). When CGRP and SP were injected simultaneously into the skin, the long-lasting vasodilatation by CGRP was converted into a transient response that lasted less than 1 h (Brain and Williams, 1988). The authors concluded that this phenomenon was due to proteases from mast cells that were released by SP, and that it may represent an interaction between the two peptides when coreleased from sensory nerves. Shawket *et al.* (1989) observed that intravenous infusion of CGRP caused flushing and increased blood flow in the skin of the face and hands in patients with Raynaud's phenomenon, whereas in controls CGRP caused flushing and increased blood flow only in the face. A similar difference was not observed with prostacyclin and it was suggested to be related to supersensitivity to CGRP due to deficiency of endogenous CGRP in Raynaud's phenomenon.

The vasodilatation induced by the sensory neuropeptides does not involve adrenergic, cholinergic, or histaminergic receptors, indicating existence of specific receptors. Furthermore, tachyphylaxis to SP does not influence the vasodilator response to CGRP, which suggests that these two peptides act via separate receptor mechanisms (Franco-Cereceda and Rudehill, 1989). This is further supported by the finding that SP causes relaxation of blood vessels by releasing endothelium-derived relaxing factor, whereas CGRP acts independently of the vascular endothelium (Mejia *et al.*, 1988; Franco-Cereceda and Rudehill, 1989; J. Pernow, 1989). Specific binding sites for SP and CGRP have been demonstrated on intestinal and pulmonary blood vessels, respectively (Mantyh *et al.*, 1988; McCormack *et al.*, 1989). On isolated specimens from the right atrium, CGRP stimulates contractility with a 10-fold-higher potency than noradrenaline, via a mechanism independent of adrenergic receptors (Franco-Cereceda *et al.*, 1987a). This indicates that the cardiac effects observed *in vivo* during intravenous administration are not simply baroreceptor-mediated reflexes to the CGRP-induced reduction in blood pressure but are also likely to be due to direct chronotropic and inotropic effects of CGRP. In contrast, the increase in heart rate observed during intravenous infusion of SP is probably an indirect effect mediated by reflex activation due to the fall in blood pressure since SP exerts no effect on cardiac contracility *in vitro* (Franco-Cereceda *et al.*, 1987a).

Capsaicin is an irritant agent present in red pepper that has been a useful tool in investigations of sensory nerves since it selectively activates sensory C-fiber afferents. Acute administration of capsaicin induces excitation of sensory nerves with release of both SP and CGRP, which causes vasodilatation and stimulation of cardiac contractility (see Franco-Cereceda, 1988). When applied to human coronary or cerebral arteries, capsaicin evokes relaxation. The effect is not affected by inhibition of endothelium-derived relaxing factor or by tachyphylaxis to SP. Likewise, the vasodilator response to CGRP is, in contrast to that to SP, not endothelium dependent, nor is it influenced by SP tachyphylaxis. This indicates that the relaxatory effects seen upon activation of sensory nerves by capsaicin in these situations are mediated by CGRP rather than by SP (Mejia *et al.*, 1988; Franco-Cereceda and Rudehill, 1989). A similar mechanism may operate in the heart of guinea pig and rat since capsaicin stimulates atrial contractility, an effect that can be mimicked by CGRP but not by SP (Franco-Cereceda, 1988). Capsaicin does not influence cardiac contractility in man, however (Franco-Cereceda *et al.*, 1987a).

A physiological role of the sensory neuropeptides as mediators in neurogenic inflammation by antidromic activation of sensory nerves has been postulated. Thus, it is known from experimental animals that sensory neuropeptides, such as SP, CGRP, and NKA, are released upon sensory nerve stimulation (see B. Pernow, 1983; Owman, 1988). Furthermore, both SP and CGRP cause potent cutaneous vasodilatation and plasma extravasation (see above). Local capsaicin treatment of the skin of healthy subjects has been demonstrated to produce a flare response, suggesting involvement of sensory nerves (Lundblad *et al.*, 1987). After repeated capsaicin treatment the local reaction disappeared and the flare, but not the wheal reaction to allergen exposure, was markedly reduced. These findings suggest that

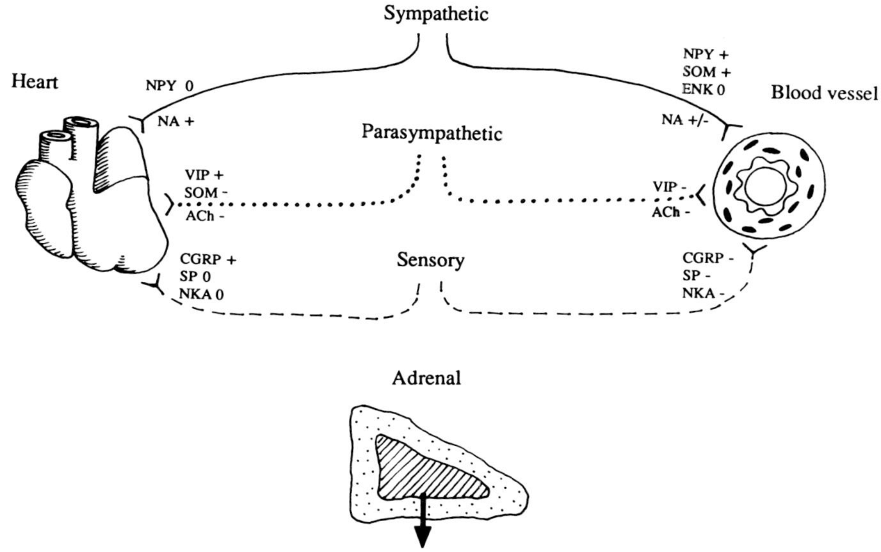

Figure 5 Schematic illustration showing the coexistence of neuropeptides and classical neurotransmitters in different parts of the peripheral nervous system regulating the cardiovascular system in man. Postganglionic sympathetic nerves contain neuropeptide Y (NPY) and noradrenaline (NA) in fibers innervating both the heart and blood vessels. NPY and NA are coreleased upon sympathetic activation. NPY evokes potent vasoconstriction via activation of specific receptors, whereas it exerts little or no effect on the heart. In addition to NPY, enkephalins (ENK) and somatostatin (SOM) are present in sympathetic nerves and in chromaffin cells of the adrenal medulla together with adrenaline (A) and NA. The postganglionic parasympathetic fibers to the heart contain vasoactive intestinal polypeptide (VIP) and SOM, and the fibers innervating blood vessels contain VIP in addition to acetylcholine (ACh). VIP exerts positive inotropic actions on the heart and is a potent vasodilator, whereas SOM exerts negative inotropic and chronotropic effects. In the sensory nerves, calcitonin gene-related peptide (CGRP), substance P (SP), and neurokinin A (NKA) are colocalized and are released from the peripheral nerve endings. CGRP exerts positive inotropic cardiac effects while both SP or NKA are without effects on the human myocardium. All sensory neuropeptides are potent vasodilators. It should be noted that other combinations and additional peptides may be present in these nerves, but experimental data regarding their effects on the human circulation are sparse. + indicates positive chronotropic and/or inotropic effects on the heart and vasoconstrictor effects on blood vessels; − indicates negative chronotropic and/or inotropic effects and vasodilator effects; 0 indicates no effects.

sensory nerve activation caused by histamine from degranulated mast cells evokes release of the neuropeptides and that these peptides are of importance for the cutaneous vasodilatation occurring in neurogenic inflammation upon allergen exposure.

Both SP- and CGRP-like immunoreactivity have been detected in human plasma. The reported levels vary considerably between different studies but are generally in the low picomolar range (B. Pernow, 1983; Tippins, 1986). Circulating SP seems to be degraded in the liver and independent of kidney excretory function (Skrabanek *et al.*, 1976). High plasma levels of SP and especially NPK have been reported in patients with carcinoid tumors, which are known to contain large amounts of SP and other tachykinins (Theodorsson-Norheim *et al.*, 1985). This finding is interesting considering the marked cardiovascular symptoms such as hypotension, tachycardia, and flushing that are associated with carcinoid syndroms. Plasma levels of CGRP increase in patients during hemodialysis and this increase correlates significantly with a reduction in systolic blood pressure (Odar-Cederlöf *et al.*, 1989). Conversely, Edvinsson *et al.* (1989) reported significantly lower levels of CGRP in hypertensive patients and suggested that sensory peptides may be involved in the maintenance of hypertension.

V. Concluding Remarks

The discovery of peptides coexisting with classical neurotransmitters in sympathetic, parasympathetic, and sensory nerves of man, as schematically illustrated in Fig. 5, has opened up new perspectives in the area of cardiovascular regulation. In the sympthetic nerves, NPY, SOM, and enkephalins are present together with noradrenaline. In parasympathetic nerves, VIP, PHM, and to some extent SOM coexist with acetylcholine. Finally, tachykinins and CGRP are present in a subpopulation of sensory nerves. In addition, several different peptides exist in chromaffin cells of the adrenal medulla. The described neuropeptides are, after release, able to evoke a variety of pharmacological actions on both the heart and the peripheral circulation. Many of the peptides have also been demonstrated to exert modulatory effects on transmitter release, indicating prejunctional actions. The potency of the peptide is usually several-fold higher than that of the coexisting classical transmitter. The physiological roles and significance of neuropeptides in the regulation of the circulation and their possible involvement in different cardiovascular disorders are still to a large extent unclear. However, their potent effects and the fact that the plasma levels of some of the neuropeptides are changed both under physiological and pathophysiological conditions indicate that they may be of importance. The exact role of these peptides in the cardiovascular regulation remains to be established with the use of specific receptor antagonists. The development of such agents will undoubtedly be of importance for the understanding of the function of the neuropeptides and possibly also for treatment of cardiovascular disorders.

References

Adrian, T. E., Terenghi, G., Brown, M. J., Allen, J. M., Bacarese-Hamilton, A. J., Polak, J. M., and Bloom, S. R. (1983). Neuropeptide Y in phaeochromocytomas and ganglioneuroblastomas. *Lancet. II*, 540–542.

Ahlborg, G., Lundberg, J., and Sollevi, A. (1989). Metabolism of and circulatory and metabolic effects of NPY in humans. *Clin. Res. 37*, 568A.

Ahlborg, G., Weitzberg, E., Lundberg, J. M., and Sollevi, A. (1992). Splanchnic and renal vasoconstriction and metabolic responses to neuropeptide Y in resting and exercising man. *Acta Physiol. Scand. 145*, 139–149.

Bloom, S. R.(1978). Somatostatin and the gut. *Gastroenterol. 75,* 145–147.
Bloom, S. R., Polak, J. M., and Pearse, A. G. E. (1973). Vasoactive intestinal peptide and watery-diarrhea syndrome. *Lancet. II,* 14–16.
Bosch, J., Kravetz, D., and Rodes, J. (1981). Effects of somatostatin on hepatic and systemic hemodynamics in patients with cirrhosis of the liver: Comparison with vasopressin. *Gastroenterol. 80,* 518–525.
Bosch, J., Kravetz, D., Mastai, R., Navasa, M., Silva, G., Chesta, J., and Rodes, J. (1988). Effects of somatostatin in patients with portal hypertension. *Horm. Res. 29,* 99–102.
Brain, S. D., and Williams, T. J. (1988). Substance P regulates the vasodilator activity of calcitonin gene-related peptide. *Nature 335,* 73–75.
Brain, S. D., Williams, T. J., Tippins, J. R., Morris, H. R., and MacIntyre, I. (1985). Calcitonin gene-related peptide is a potent vasodilator. *Nature 313,* 54–56.
Campbell, G., Gibbins, I. L., Morris, J. L., Furness, J. B., Costa, M., Olivier, J. R., Beardsley, A. M., and Murphy, R. (1982). Somatostatin is contained in and released from cholinergic nerves in the heart of the toad *Bufo marinus. Neurosci. 7,* 2013–2023.
Calquist, M., McDonald, T. J., Go, V. L. W., Bataille, D., Johansson, C., and Mutt, V. (1982). Isolation and amino acid composition of human vasoactive intestinal polypeptide (VIP). *Horm. Metab. Res. 14,* 28–29.
Chernish, S. M., Miller, R. E., Rosenak, B. D., and Brunelle, R. L. (1979). The effect of D-VAL^1, D-TRYP^8, somatostatin on the hematopoetic, cardiovascular systems and gastrointestinal tract. *Gastroenterol. 76,* 1113.
Christophe, J., Waelbroeck, M., Chatelain, P., and Robberecht, P. (1984). Heart receptors for VIP, PHI and secretin are able to activate adenylate cyclase and to mediate inotropic and chronotropic effects. Species variations and physiopathology. *Peptides 5,* 341–353.
Clarke, J. G., Kerwin, R., Larkin, S., Lee, Y., Yacoub, M., Davies, G. J., Hackett, D., Dawbarn, D., Bloom, S. R., and Maseri, A. (1987). Coronary artery infusion of neuropeptide Y in patients with angina pectoris. *Lancet. I,* 1057–1059.
Clement-Jones, V., Corder, R., and Lowry, P. J. (1980a). Isolation of human Met-enkephalin and two groups of putative precursors (2K-Pro-Met-enkephalin) from an adrenal medullary tumor. *Biochem. Biophys. Res. Comm. 95,* 665–673.
Clement-Jones, V., Lowry, P. J., Rees, L. H., and Besser, G. M. (1980b). Met-enkephalin circulates in human plasma. *Nature 283,* 295–297.
Clement-Jones, V., Corder, R., Smith, R., Medbank, S., Lowry, P. J., and Besser, G. M. (1982). Met-enkephalin and related peptides in man. *In* "Regulatory Peptides: From Molecular Biology to Function." (E. Costa and M. Trabucchi, eds.), pp. 379–386. Raven Press, New York.
Costa, M., and Furness, J. B. (1984). Somatostatin is present in a subpopulation of noradrenergic nerve fibers supplying the intestine. *Neurosci. 13,* 911–919.
Crossman, D. C., Larkin, S. W., Fuller, R. W., Davies, G. J., and Maseri, A. (1989). Substance P dilates epicardial coronary arteries and increase coronary blood flow in humans. *Circulation 80,* 475–484.
Dalsgaard, C.-J. (1988). The sensory system. *In* "Handbook of Chemical Neuroanatomy: The Peripheral Nervous System." (A. Björklund, T. Hökfelt, and C. Owman, eds.), Vol. 6, pp. 599–636. Elsevier, Amsterdam.
Day, S. M., Polak, J. M., and Bloom, S. R. (1985). Somatostatin in the human heart and comparison with guinea pig and rat heart. *Br. Heart. J. 53,* 153–157.
Domschke, S., Domschke, W., Bloom, S. R., Mitznegg, P., Mitchell, S. J., Lux, G., and Strunz, U. (1978). Vasoactive intestinal peptide in man: Pharmacokinetics, metabolic and circulatory effects. *Gut. 19,* 1049–1053.
Eckberg, D. L., Rea, R. F., Andersson, O. K., Hedner, T., Pernow, J., Lundberg, J. M., and Wallin, B. G. (1988). Baroreflex modulation of sympathetic activity and sympathetic neurotransmitters in humans. *Acta Physiol. Scand. 133,* 221–231.
Edvinsson, L., Håkanson, R., Steen, S., Sundler, F., Uddman, R., and Wahlestedt, C. (1985). Innerva-

tion of human omental arteries and veins and vasomotor responses to noradrenaline, neuropeptide Y, substance P and vasoactive intestinal peptide. *Regul. Peptides 12,* 67–79.

Edvinsson, L., Ekman, R., Jansen, I., McChulloch, J., and Uddman, R. (1987a). Calcitonin gene-related peptide and cerebral blood vessels: Distribution and vasomotor effects. *J. Cer. Blood Flow Metab. 7,* 720–728.

Edvinsson, L., Ekman, R., Jansen, I., Ottosson, A., and Uddman, R. (1987b). Peptide-containing nerve fibers in human cerebral arteries: Immunohistochemistry, radioimmunoassay, and *in vitro* pharmacology. *Ann. Neurol. 21,* 431–437.

Edvinsson, L., Ekman, R., and Thulin, T. (1989). Circulating perivascular dilatory neuropeptides in hypertension. *J. Hypertension 7* (Suppl. 6), S194–S195.

Edvinsson, L., Ekman, R., and Thulin, T. (1991). Increased plasma levels of neuropeptide Y-like immunoreactivity and catecholamines in severe hypertension remain after treatment to normotension in man. *Regul. Peptides 32,* 279–287.

Eklund, B., Jogestrand, T., and Pernow, B. (1977). Effect of substance P on resistance and capacitance vessels in the human forearm. *In* "Substance P." (U.S. v. Euler and B. Pernow, eds.), pp. 275–286. Raven Press, New York.

Eriksson, L. S., and Wahren, J. (1989). Intravenous and subcutaneous administration of a long-acting somatostatin analogue: Effects on glucose, metabolism and splanchnic haemodynamics in healthy subjects. *Europ. J. Clin. Invest. 19,* 213–219.

Eriksson, L. S., Law, D. H., Sato, Y., and Wahren, J.(1984). Influence of somatostatin on splanchnic haemodynamics in patients with liver cirrhosis. *Clin. Physiol. 4,* 5–11.

Fahrenkrug, J. (1989). VIP and autonomic neurotransmission. *Pharmacol. Ther. 41,* 515–534.

Fahrenkrug, J., Holst Pedersen, J., Yamashita, Y., Ottesen, B., Hökfelt, T., and Lundberg, J. M. (1987). Occurrence of VIP and peptide HM in human pancreas and their influence on pancreatic endocrine secretion in man. *Regul. Peptides 18,* 51–61.

Franco-Cereceda, A. (1988). Calcitonin gene-related peptide and tachykinins in relation to local sensory control of cardiac contractility and coronary vascular tone. *Acta Physiol. Scand. 133* (Suppl. 569), 1–63.

Franco-Cereceda, A., and Lundberg, J. M. (1987). Potent effects of neuropeptide Y and calcitonin gene-related peptide on human coronary vascular tone *in vitro. Acta Physiol. Scand. 131,* 159–160.

Franco-Cereceda, A., and Rudehill, A. (1989). Capsaicin-induced vasodilatation of human coronary arteries *in vitro* is mediated by calcitonin gene-related peptide rather than substance P or neurokinin A. *Acta Physiol. Scand. 136,* 575–580.

Franco-Cereceda, A., Lundberg, J. M., and Hökfelt, T. (1986). Somatostatin: An inhibitory parasympathetic transmitter in the human heart? *Europ. J. Pharmacol. 132,* 101–102.

Franco-Cereceda, A., Bengtsson, L., and Lundberg, J. M. (1987a). Inotropic effects of calcitonin gene-related peptide, vasoactive intestinal polypeptide and somatostatin on the human right atrium *in vitro. Europ. J. Pharmacol. 134,* 69–76.

Franco-Cereceda, A., Henke, H., Lundberg, J. M., Petermann, J. B., Hökfelt, T., and Fishcer, J. A. (1987b). Calcitonin gene-related peptide (CGRP) in capsaicin-sensitive substance P-immunoreactive sensory neurons in animals and man: Distribution and release by capsaicin. *Peptides 8,* 399–410.

Franco-Cereceda, A., Gennari, C., Nami, R., Agnusdei, D., Pernow, J., Lundberg, J. M., and Fischer, J. A. (1987c). Cardiovascular effects of calcitonin gene-related peptides I and II in man. *Circulation Res. 60,* 393–397.

Frase, L. L., Gaffney, F. A., Lane, L. D., Buckey, J. C., Said, S. I., Blomqvist, C. G., and Krejs, G. J. (1987). Cardiovascular effects of vasoactive intestinal peptide in healthy subjects. *Am. J. Cardiol. 60,* 1356–1361.

Ganten, D., Lang, R. E., Archelos, J., and Unger, T. (1984). Peptidergic systems: Effects on blood vessels. *J. Cardiovasc. Pharmacol. 6,* S598–S607.

Gennari, C., and Fischer, J. A. (1985). Cardiovascular action of calcitonin gene-related peptide in humans. *Calcif. Tissue Int. 37,* 581–584.

Greco, A. V., Ghirlanda, G., Barone, C., Bertoli, A., Caputo, S., Uccioli, L., and Manna, R. (1984). Somatostatin in paroxysmal supraventricular and junctional tachycardia. *Br. Med. J. 288,* 28.

Gu, J., Adrian, T. E., Tatemoto, K., Polak, J. M., Allen, J. M., and Bloom, S. R. (1983). Neuropeptide tyrosine (NPY)—a major cardiac neuropeptide. *Lancet. I,* 1008–1010.

Hallemans, R., Naeije, R., Melot, C., Mols, P., and Reding, R. (1981). Systemic and pulmonary haemodynamic effects of somatostatin. *Lancet. I,* 1270.

Hanko, J. H., Törnebrandt, K., Hardebo, J. E., Kåhrström, J., Nobin, A., and Owman, C. (1986). Neuropeptide Y induces and modulates vasoconstriction in intracranial and peripheral vessels of animals and man. *J. Auton. Pharmac. 6,* 117–124.

Hägermark, Ö, Hökfelt, T., and Pernow, B. (1978). Flare and itch induced by substance P in human skin. *J. Invest. Dermatol. 71,* 233–235.

Hou, Z.-Y., Lin, C.-I, Chiu, T.-H., Chiang, B. N., Cheng, K.-K., and Ho, L.-T. (1987). Somatostatin effects in isolated human atrial fibers. *J. Mol. Cell. Cardiol. 19,* 177–185.

Hökfelt, T., Johansson, O., Ljungdahl, Å., Lundberg, J. M., and Schultzberg, M. (1980). Peptidergic neurones. *Nature 284,* 515–521.

Hulting, J., Sollevi, A., Ullman, B., Franco-Cereceda, A., and Lundberg, J. M. (1990). Plasma neuropeptide Y on admission to a coronary care unit: Raised levels in patients with left heart failure. *Cardiovasc. Res. 24,* 102–108.

Jernbeck, J., Edner, M., Dalsgaard, C.-J., and Pernow, B. (1990). The effect of calcitonin gene-related peptide (CGRP) on human forearm blood flow. *Clin. Physiol. 10,* 335–343.

Kaijser, L., Pernow, J., Berglund, B., and Lundberg, J. M. (1990). Neuropeptide Y is released together with noradrenaline from the human heart during exercise. *Clin. Physiol. 10,* 179–188.

Kodama, T., Ito, C., Fujimoto, Y., Ito, Y., Obara, T., and Hirayama, A. (1990). Augmented enkephalin-immunoreactivity in adrenaline-producing phaeochromocytomas. *Virchows Arch. Pathol. Anat. 416,* 417–422.

Kurtz, A., Muff, R., Born, W., Lundberg, J. M., Millberg, B.-I. Gnädinger, M. P., Uehlinger, D. E., Weidamnn, P., Hökfelt, T., and Fischer, J. A. (1988). Calcitonin gene-related peptide is a stimulator of renin secretion. *J. Clin. Invest. 82,* 538–543.

Lang, R. E., Brückner, U. B., Herman, K., Kempf, B., Rascher, W., Sturm, V., Unger, T., and Ganten, D. (1982). Effect of hemorrhagic shock on the concomitant release of endorphin and enkephalin-like peptides from the pituitary and adrenal gland in the dog. *In* "Regulatory Peptides: From Molecular Biology to Function." (E. Costa and M. Trabucchi, eds.), pp. 363–368. Raven Press, New York.

Larsson, O., Dunér-Engström, M., Lundberg, J. M., Fredholm, B. B., and Änggård, A. (1986). Effects of VIP, PHM and substance P on blood vessels and secretory elements of the human submandibular gland. *Regul. Peptides 13,* 319–326.

Linnik, M. D., and Moskowtiz, M. A. (1989). Identification of immunoreactive substance P in human and other mammalian endothelial cells. *Peptides 10,* 957–962.

Lundberg, J. M. (1981). Evidence for coexistence of vasoactive intestinal polypeptide (VIP) and acetylcholine in neurons of cat exocrine glands. *Acta Physiol. Scand. 112* (Suppl. 496), 1–57.

Lundberg, J. M., and Hökfelt, T. (1983). Coexistence of peptides and classical neurotransmitters. *Trends Neurosci. 6,* 325–333.

Lundberg, J. M., Hamberger, B., Schultzberg, M., Hökfelt, T., Granberg, P.-O., Efendic, S., Terenius, L., Goldstein, M., and Luft, R. (1979). Enkephalin- and somatostatin-like immunoreactivity in human adrenal medulla and pheochromocytoma. *Proc. Natl. Acad. Sci. U.S.A. 76,* 4079–4083.

Lundberg, J. M., Terenius, L., Hökfelt, T., and Goldstein, M. (1983). High levels of neuropeptide Y in peripheral noradrenergic neurons in various mammals including man. *Neurosci. Lett. 42,* 167–172.

Lundberg, J. M., Fahrenkrug, J., Hökfelt, T., Martling, C.-R., Larsson, O., Tatemoto, K., and Änggård, A. (1984). Co-existence of peptide HI (PHI) and VIP in nerves regulating blood flow and bronchial smooth muscle tone in various mammals including man. *Peptides 5,* 593–606.

Lundberg, J. M., Martinsson, A., Hemsén, A., Theodorsson-Norheim, E., Svedenhag, J., Ekblom, B., and Hjemdahl, P. (1985a). Co-release of neuropeptide Y and catecholamines during physical exercise in man. *Biochem. Biophys. Res. Comm. 133,* 30–36.

Lundberg, J. M., Torsell, L., Sollevi, A., Pernow, J., Theodorsson-Norheim, E. Änggård, A., and Hamberger, B. (1985b). Neuropeptide Y and sympathetic vascular control in man. *Regul. Peptides 13,* 41–52.

Lundberg, J. M., Hökfelt, T., Hemsén, A., Theodorsson-Norheim, E., Pernow, J., Hamberger, B., and Goldstein, M. (1986a). Neuropeptide Y-like immunoreactivity in adrenaline cells of adrenal medulla and in tumors and plasma of pheochromocytoma patients. *Regul. Peptides 13,* 169–182.

Lundberg, J. M., Hemsén, A., Fried, G., Theodorsson-Norheim, E., and Lagercrantz, H.(1986b). Corelease of neuropeptide Y (NPY)-like immunoreactivity and catecholamines in newborn infants. *Acta Physiol. Scand. 126,* 471–473.

Lundberg, J. M., Pernow, J., Franco-Cereceda, A., and Rudehill, A. (1987). Effects of antihypertensive drugs on sympathetic vascular control in relation to neuropeptide Y. J. *Cardiovasc. Pharmacol. 10* (Suppl. 12), S51–S68.

Lundberg, J. M., Franco-Cereceda, A., Hemsén, A., Lacroix, J. S., and Pernow, J. (1990). Pharmacology of noradrenaline and neuropeptide tyrosine (NPY)-mediated sympathetic cotransmission. *Fundam. Clin. Pharmacol. 4,* 373-391.

Lundblad, L., Lundberg, J. M., Änggård, A., and Zetterström, O. (1987). Capsaicin-sensitive nerves and the cutaneous allergy reaction in man. *Allergy 42,* 20–25.

Maisel, A. S., Scott, N. A., Motulsky, H. J., Michel, M. C., Boublik, J. H., Rivier, J. E., Ziegler, M., Allen, R. S., and Brown, M. R. (1989). Elevation of plasma neuropeptide Y levels in congestive heart failure. *Am. J. Med. 86,* 43–48.

Mantyh, C. R., Gates, T. S., Zimmerman, R. P., Welton, M. L., Passaro, E. P., Vigna, S. R., Maggio, J. E., Kruger, L., and Mantyh, P. W. (1988). Receptor binding sites for substance P, but not substance K or neuromedin K, are expressed in high concentrations by arterioles, venules, and lymph nodules in surgical specimens obtained from patients with ulcerative colitis and Crohn disease. *Proc. Natl. Acad. Sci. U.S.A. 85,* 3235–3239.

McCormack, D. G., Mak, J. C. W., Coupe, M. O., and Barnes, P. J. (1989). Calcitonin gene-related peptide vasodilatation of human pulmonary vessels. *J. Appl. Physiol. 67,* 1265–1270.

McEwan, J., Larkin, S., Davies, G., Chierchia, S., Brown, M., Stevenson, J., MacIntyre, I., and Maseri, A. (1986). Calcitonin gene-related peptide: A potent dilator of human epicardial coronary arteries. *Circulation 74,* 1243–1247.

McEwan, J. R., Benjamin, N., Larkin, S., Fuller, R. W., Dollery, C. T., and MacIntyre, I. (1988). Vasodilatation by calcitonin gene-related peptide and by substance P: A comparison of their effects on resistance and capacitance vessels of human forearms. *Circulation 77,* 1072–1080.

Mejia, J. A., Pernow, J., von Holst, H., Rudehill, A., and Lundberg, J. M. (1988). Effects of neuropeptide Y, calcitonin gene-related peptide, substance P, and capsaicin on cerebral arteries in man and animals. *J. Neurosurg. 69,* 913–918.

Michel, M. C., Wirth, S. C, Zerkowski, H.-R., Maisel, A. S., and Motulsky, H. J. (1989). Lack of inotropic effects of neuropeptide Y in human myocardium. *J. Cardiovasc. Pharmacol. 14,* 919–922.

Morris, M. J., Russell, A. E., Kapoor, V., Cain, M. D., Elliott, J. M., West, M. J., Wing, L. M. H., and Chalmers, J. P. (1986). Increases in plasma neuropeptide Y concentrations during sympathetic activation in man. *J. Auton. Nerv. Syst. 17,* 143–149.

Odar-Cederlöf, I., Theodorsson, E., Eriksson, C.-G., and Kjellstrand, C. M. (1989). Plasma concentrations of calcitonin gene-related peptide increase during haemodialysis: Relation to blood pressure. *J. Int. Med. 226,* 177–182.

Osamura, R. Y., Tsutsumi, Y., Yanaihara, N., Imura, H., and Watanabe, K. (1987). Immunohistochemical studies for multiple peptide-immunoreactivities and co-localization of met-enkephalin-Arg^6-Gly^7-Leu^8, neuropeptide Y and somatostatin in human adrenal medulla and pheochromocytomas. *Peptides 8,* 77–87.

Owman, C. (1988). Autonomic innervation of the cardiovascular system. *In* "Handbook of Chemical Neuroanatomy: The Peripheral Nervous System." (A. Björklund, T. Hökfelt, and C. Owman, eds.), Vol. 6, pp. 327–389. Elsevier, Amsterdam.

Parmer, R., and O'Connor, D. T. (1988). Enkephalins in human phaeochromocytomas: Localization in

immunoreactive, high molecular weight form to the soluble core of chromaffin granules. *J. Hypertension 6,* 187–198.
Pasanisi, F., Sloan, L., and Rubin, P. (1985). Cardiovascular properties of metkephamid, a δ opioid receptor agonist, in man. *Clin. Sci. 68,* 209–213.
Pernow, B. (1983). Substance P. *Pharmacol. Rev. 35,* 85–141.
Pernow, J. (1988). Co-release and functional interactions of neuropeptide Y and noradrenaline in peripheral sympathetic vascular control. *Acta Physiol. Scand. 133* (Suppl. 568), 1–56.
Pernow, J. (1989). Actions of constrictor (NPY and endothelin) and dilator (substance P, CGRP and VIP) peptides on pig splenic and human skeletal muscle arteries: Involvement of the endothelium. *Br. J. Pharmacol. 97,* 983–989.
Pernow, J., and Lundberg, J. M. (1988). Neruopeptide Y induces potent contraction of arterial vascular smooth muscle via an endothelium-independent mechanism. *Acta Physiol. Scand. 134,* 157–158.
Pernow, J., Lundberg, J. M., Kaijser, L., Hjemdahl, P., Theodorsson-Norheim, E., Martinsson, A., and Pernow, B. (1986). Plasma neuropeptide Y-like immunoreactivity and catecholamines during various degrees of sympathetic activation in man. *Clin. Physiol. 6,* 561–578.
Pernow, J., Lundberg, J. M., and Kaijser, L. (1987a). Vasoconstrictor effects *in vivo* and plasma disappearance rate of neuropeptide Y in man. *Life Sci. 40,* 47–54.
Pernow, J., Svenberg, T., and Lundberg, J. M. (1987b). Actions of calcium antagonists on pre- and postjunctional effects of neuropeptide Y on human blood vessels *in vitro. Europ. J. Pharmacol. 136,* 207–218.
Pernow, J., Öhlén, A., Hökfelt, T., Nilsson, O., and Lundberg, J. M. (1987c). Neuropeptide Y: Presence in perivascular noradrenergic neurons and vasoconstrictor effects on skeletal muscle blood vessels in experimental animals and man. *Regul. Peptides. 19,* 313–324.
Pernow, J., Lundberg, J. M., and Kaijser, L. (1988). α-Adrenoceptor influence on plasma levels of neuropeptide Y-like immunoreactivity and catecholamines during rest and sympathoadrenal activation in humans. *J. Cardiovasc. Pharmacol. 12,* 593–599.
Pernow, J., Lundberg, J. M., and Kaijser, L. (1989). Effect of oral administration of nifedipine on neuropeptide Y- and noradrenaline-induced vasoconstriction in the human forearm. *Clin. Physiol. 9,* 67–75.
Peters, W. P., Freidman, P. A., Johnson, M. W., and Mitch, W. E. (1981). Pressor effect of naloxone in septic shock. *Lancet. I,* 529–532.
Pradayrol, L., Jörnvall, H., Mutt, V., and Ribet, A. (1980). N-terminally extended somatostatin: The primary structure of somatostatin-28. *FEBS Lett. 109,* 55–58.
Quirion, R., Regoli, D., Rioux, F., and St-Pierre, S. (1979). An analysis of the negative inotropic action of somatostatin. *Br. J. Pharmacol. 66,* 251–257.
Rechardt, L., Aalto-Setälä, K., Purjeranta, M., Pelto-Huikko, M., and Kyösola, K. (1986). Peptidergic innervation of human atrial myocardium: An electron microscopical and immunocytochemical study. *J. Auton. Nerv. Syst. 17,* 21–32.
Reid, J. L., and Rubin, P. C. (1987). Peptides and central neural regulation of the circulation. *Physiol. Rev. 67,* 725–749.
Rosenthal, J., Escobar-Jiminez, F., and Raptis, S. (1977). Prevention by somatostatin of rise in blood pressure and plasma renin mediated by beta-receptor stimulation. *Clin. Endocrinol. 6,* 455–462.
Rosenthal, J., Raptis, S., Zoupas, C., and Escobar-Jiminez, E. (1978). Inhibition by somatostatin of renin, blood pressure, and cardiac and stroke index in essental hypertension. *Circ. Res. 43* (Suppl. I), 69–76.
Rubin, P., Blaschke, T. F., and Guilleminault, C. (1981). Effect of naloxone, a specific opioid inhibitor, on blood pressure fall during sleep. *Circulation 63,* 117–121.
Rubin, P. C., McLean, K., and Reid, J. L. (1983). Endogenous opioids and baroreflex control in humans. *Hypertension 5,* 535–538.
Rudehill, A., Franco-Cereceda, A., Hemsén, A., Stensdotter, M., Pernow, J., and Lundberg, J. M. (1989). Cigarette smoke-induced elevation of plasma neuropeptide Y levels in man. *Clin. Physiol. 9,* 243–248.

Russell, A. E., Cain, M. D., Kapoor, V., Morris, M. J., and Chalmers, J. P. (1989). Neuropeptide Y-like immunoreactivity of plasma during hypoglycaemia in man. *J. Auton. Nerv. Syst. 26,* 85–88.

Said, S. I., and Mutt, V. (1970). Polypeptide with broad biological activity: Isolation from small intestine. *Science 169,* 1217–1218.

Shawket, S., Hazleman, B., Dickerson, C., and Brown, M. J. (1989). Selective suprasensitivity to calcitonin gene-related peptide in the hands in Raynaud's phenomenon. *Lancet. II,* 1354–1357.

Skrabanek, P., Cannon, D., Kirrane, J., Legge, D., and Powell, D. (1976). Circulating immunoreactive substance P in man. *Ir. J. Med. Sci. 145,* 399–408.

Smitherman, T. C., Popma, J. J., Said, S. I., Krejs, G. J., and Dehmer, G. J. (1989). Coronary hemodynamic effects of intravenous vasoactive intestinal polypeptide. *Am. J. Physiol. 257,* H1254–H1262.

Struthers, A. D., Brown, M. J., MacDonald, D. W. R., Beacham, J. L., Stevenson, J. C., Morris, H. R., and MacIntyre, I. (1986). Human calcitonin gene-related peptide: A potent endogenous vasodilator in man. *Clin. Sci. 70,* 389–393.

Sullivan, S. N., Bloom, S. R., and Polak, J. M. (1978). Enkephalin in peripheral neuroendocrine tumors. *Lancet. I,* 986–987.

Takahashi, K., Mouri, T., Itoi, K., Sone, M., Ohneda, M., Murakami, O., Nozuki, M., Tashibana, Y., and Yoshinaga, K. (1987). Increased plasma immunoreactive neuropeptide Y concentrations in phaeochromocytoma and cronic renal failure. *J. Hypertension 5,* 749–753.

Takahashi, K., Mouri, T., Murakami, O., Itoi, K., Sone, M., Ohneda, M., Nozuki, M., and Yoshinaga, K. (1988). Increases of neuropeptide Y-like immunoreactivity in plasma during insulin-induced hypoglycemia in man. *Peptides 9,* 433–435.

Tatemoto, K. (1982). Neuropeptide Y: Complete amino acid sequence of the brain peptide. *Proc. Natl. Acad. Sci. U.S.A. 79,* 5485–5489.

Theodorsson-Norheim, E., Norheim, I. Öberg, K., Brodin, E., Lundberg, J. M., Tatemoto, K., and Lundgren, P. G. (1985). Neuropeptide K: A major tachykinin in plasma and tumor tissues from carcinoid patients. *Biochem. Biophys. Res. Comm. 131,* 77–83.

Thom, S. M., Hughes, A. D., Goldberg, P., Martin, G., Schachter, M., and Sever, P. S. (1987). The actions of calcitonin gene-related peptide and vasoactive intestinal polypeptide as vasodilators in man *in vivo* and *in vitro*. *Br. J. Clin. Pharmacol. 24,* 139–144.

Thulin, L., Nyberg, B., Tyden, G., and Sonnenfeld, T. (1984). Circulatory effects of VIP in anesthetized man. *Peptides 5,* 319–323.

Tidgren, B. (1989). Renal responses to sympatho-adrenal activation in humans with special reference to renal catecholamine and renin release. Thesis. Karolinska Institutet, Stockholm, Sweden.

Tippins, J. R. (1986). CGRP: A novel neuropeptide from the calcitonin gene is the most potent vasodilator known. *J. Hypertension 4* (Suppl. 5), S102–S105.

Törnebrandt, K., Nobin, A., and Owman, C. (1987). Contractile and dilatory action of neuropeptides on isolated human mesenteric blood vessels. *Peptides 8,* 251–256.

Tseng, C.-J., Robertson, D., Light, R. T., Atkinson, J. R., and Robertson, R. M. (1988). Neuropeptide Y is a vasoconstrictor of human coronary arteries. *Am. J. Med. Sci. 296,* 11–16.

Uddman, R., and Edvinsson, L. (1989). Neuropeptides in the cerebral circulation. *Cerebrovasc. Brain Metab. Rev. 1,* 230–252.

Uddman, R., Edvinsson, L., Jansen, I., Stiernholm, P., Jensen, K., Olesen, J., and Sundler, F. (1986). Peptide-containing nerve fibers in human extracranial tissue: A morphological basis for neuropeptide involvement in extracranial pain? *Pain 27,* 391–399.

Unwin, R. J., Reed, T., Thom, S., Calam, J., and Peart, W. S. (1987). Effects of indomethacin and (±)-propranolol on the cardiovascular and renin responses to vasoactive intestinal polypeptide (VIP) infusion in man. *Br. J. Clin. Pharmacol. 23,* 523–528.

Varndell, I. M., Tapia, F., DeMey, J., Rush, R., Bloom, S., and Polak, J. (1982). Electron immunocytochemical localization of enkephalin-like material in catecholamine-containing cells of the carotid body, the adrenal medulla, and in phaeochromocytomas of animals and other mammals. *J. Histochem. Cytochem. 30,* 682–690.

Viveros, O. H., Diliberto, E. J., Hazum, E., and Chang, K.-J. (1979). Opiate-like materials in adrenal medulla: Evidence for storage and secretion with catecholamines. *Mol. Pharmacol. 16,* 1101–1108.

Wahren, J., and Felig, P. (1976). Influence of somatostatin on carbohydrate disposal and absorption in diabetes mellitus. *Lancet. II,* 1213–1216.

Watson, J. D., Sury, M. R. J., Corder, R., Carson, R., Bouloux, P. M., Lowry, P. J., Besser, G. M., and Hinds, C. J. (1988). Plasma levels of neuropeptide tyrosine Y (NPY) are increased in human sepsis but are unchanged during canine endotoxin shock despite raised catecholamine concentrations. *J. Endocrinol. 116,* 421–426.

Webb, S. C, Krikler, D. M., Hendry, W. G., Adrian, T. E., and Bloom, S. R. (1986). Electrophysiological actions of somatostatin on the atrioventricular junction in sinus rhythm and reentry tachycardia. *Br. Heart J. 56,* 236–241.

Weihe, E., Reinecke, M., Opherk, D., and Forssmann, W. G.(1981). Peptidergic innervation (substance P) in the human heart. *J. Mol. Cell. Cardiol. 13,* 331–333.

Yanase, T., Nawata, H., Kato, K., and Ibayashi, H. (1986). Catecholamines and opioid peptides in human phaeochromocytomas. *Acta Endocrinol. 113,* 378–384.

Yoshimasa, T., Nakao, K., Li, S., Ikeda, Y., Suda, M., Sakamoto, M., and Imura, H. (1983). Plasma methionine-enkephalin and leucine-enkephalin in normal subjects and patients with pheochromocytoma. *J. Clin. Endocrinol. Metab. 57,* 706–712.

Zelis, R., Mansour, J. M., Capone, R. J., Mason, D. T., and Kleckner, R. (1974). The cardiovascular actions of morphine. The peripheral capacitance and resistance vessels in human subjects. *J. Clin. Invest. 54,* 1247–1258.

CHAPTER 17

Perivascular Peptides (NPY, CGRP, and SP) in Hypertension and Congestive Heart Failure

LARS EDVINSSON DAVID ERLINGE THOMAS HEDNER
PAVO HEDNER THOMAS THULIN STIG VALDEMARSSON

I. Introduction

The peripheral vascular system both in animals and man receives innervation by perivascular nerve fibers that broadly may be subdivided into three different categories (Wharton and Gulbenkian, 1985): (1) Sympathetic fibers storing noradrenaline and neuropeptide Y (NPY); (2) parasympathetic fibers containing acetylcholine, vasoactive intestinal peptide, peptide histidine isoleucin/methionine, dynorphin, and sometimes NPY; and (3) fibers that originate in sensory ganglia, being sensitive to capsaicin, store tachykinins (e.g., substance P and neurokinin A) and calcitonin gene-related peptide (Edvinsson and Uddman, 1992). The distribution of these nerve fibers, their transmittor substances, and possible functional roles have been discussed in detail in the preceeding chapters. In the present chapter we will describe current knowledge related to their involvement in two major circulatory disorders in man, arterial hypertension and congestive heart failure. This may provide further insight into the potential regulatory role of perivascular neuropeptides in various cardiovascular diseases.

II. Hypertension

A. Neuropeptide Y and Catecholamines

Although the etiology of primary hypertension still remains unsolved, increased total peripheral vascular resistance represents the hallmark of the established hypertensive disease (Folkow *et al.*, 1970). One early recognized and important regulator of peripheral vascular resistance is the sympathetic nervous system. In the hypertensive state, there is a functional dysregulation of vascular smooth muscle activity that determines the peripheral vascular resistance (Vanhoutte and Lüscher, 1986). In fact, the enhanced activity of sympathetic vasoconstrictor nerves has been shown to be an important factor for the increased tone of peripheral resistance vessels present in the spontaneously hypertensive rat (SHR) (Head, 1989; Kawasaki *et al.*, 1982). Thus, in animals with a hereditary predisposition to develop hypertension there is biochemical as well as functional evidence that peripheral sympathetic nerve activity is inappropriately high (Okamoto *et al.*, 1967; Judy *et al.*, 1976; McCarthy, 1986; Thorén, 1987; Petrovic and Bekk, 1986; Petrovic and Bell, 1985). Furthermore, neonatal destruction of the peripheral sympathetic nervous system retards the development of hypertension and reduces the absolute extent of the blood pressure elevation and the associated cardiovascular hypertrophy during adulthood in SHR (Petrovic and Bell, 1985; Clark, 1977; Liard, 1977; Kline *et al.*, 1978; Winternitz *et al.*, 1980; Provoost, 1984; Weiss *et al.*, 1974). Furthermore, in such models, hypertension can be completely reversed by ganglion blocking agents given early in life (Lee *et al.*, 1987). There are several lines of evidence indicating that an elevated sympathetic drive may represent an etiologic factor in the pathogenesis of the elevated blood pressure in animal models of primary hypertension.

In the sympathetic nervous system, neuropeptide Y (NPY) is colocalized with noradrenaline in some neurons, predominantly those supplying cardiovascular tissues (Lundberg and Hökfelt, 1986). The intraneuronal stores of NPY may be raised by a number of manipulations that are known to increase sympathetic vasomotor drive (Furness *et al.*, 1983). Since the peptide acts as a vasoconstrictor and is released from sympathetic nerve terminals by electrical stimulation (Lundberg and Hökfelt, 1986; Schalling *et al.*, 1988; Allen *et al.*, 1984), NPY is generally considered to be jointly involved in the physiological regulation of cardiovascular tone along with the sympathetic cotransmitters noradrenaline and adenosine triphosphate ATP (Lundberg *et al.*, 1985; Pernow, 1988). Studies of genetic rat models of hypertension have revealed that the number of NPY immunoreactive cells as well as the total number of ganglionic cells are increased in ganglia of young normotensive compared to hypertensive rats (Gurusinghe *et al.*, 1990). However, between the age of 10 and 60 weeks the NPY immunoreactivity and the total cell number fell in the normotensive rat ganglia while it remained constant in ganglia from hypertensive rats. This may suggest a developmental abnormality of NPY in sympathetic neurons in strains of genetically hypertensive rats.

In man a large number of studies in hypertension, reviewed by Goldstein and colleagues (1983), have reported increased plasma noradrenaline levels in patients with primary hypertension. However, just recently has the possibility been examined that neuropeptide Y-like immunoreactivity (NPY-LI) is altered in human hypertension. In a group of patients with moderate hypertension and a diastolic blood pressure of 100 ± 5 mmHg, NPY-LI levels were higher than in age- and sex-matched controls (Solt *et al.*, 1990). A similar increase was observed for basal levels of noradrenaline. Moreover, after an oral glucose tolerance test (2 h following ingestion of a 75-g glucose load) the circulating levels of both noradrenaline and NPY-LI increased in the normotensive as well as in the hypertensive subjects. The

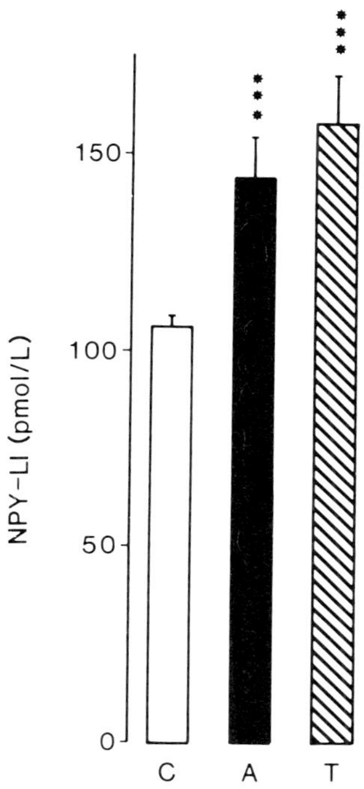

Figure 1 Circulating concentrations of NPY-LI in control (C, *open bar*) ($n = 31$) and in patients with high blood pressure (A, *filled bar*) ($n = 17$). The concentration of NPY-LI was significantly raised acutely (A) and at 2–3 weeks of treatment (T, *hatched bar*) the levels were still raised ($p < 0.001$). Values represent means ± SEM.

degree of increase was similar in both patient groups. Furthermore, the NPY-LI level in fasting subjects has been positively correlated with systolic and diastolic blood pressure (Potter, 1988).

In a group of patients with severe hypertension (diastolic blood pressure 127 ± 7 mmHg), the basal levels of plasma noradrenaline (NA) tended to be raised and NPY-LI was significantly higher than in a control population (Fig. 1) (Edvinsson *et al.*, 1991; Erlinge *et al.*, 1992). After 2–4 weeks of treatment the diastolic blood pressure was reduced to 89 ± 4 mmHg in those with severe hypertension. Despite the lowering of blood pressure, the high levels of NPY-LI remained significantly increased compared to the controls. Furthermore, the finding remained at regular follow-ups during 1 year. NA levels were not elevated in the acute hypertensive group as reported in previous studies but after 2–4 weeks of treatment. It has been shown that NPY-LI is mostly released at intense stimulation of the sympathetic nervous system (Pernow *et al.*, 1986). Thus the domination of NPY-LI in our subjects with severe hypertension might reflect a high degree of activation of the sympathetic nervous system and a depletion of NA depots. NPY-LI and NA did not correlate with each other in the control group or in the untreated hypertensives, but after treatment for 2–4 weeks and 2–12 months they were positively correlated ($p < 0.01$ and $p < 0.05$, respectively). NA was as expected positively correlated ($p < 0.01$) to age, while NPY-LI showed only a tendency ($p = 0.06$) to be positively correlated to age in the normotensive population.

The finding that NPY-LI is increased in a hypertensive population provides fairly strong evidence for an increased level of sympathetic nervous activity among the hypertensives, NPY-LI and NA being coreleased during sympathetic nervous stimulation. Since NPY has many effects at the sympathetic neuroeffector junction, which include direct vasoconstriction, potentiation of cotransmitter, and inhibition of the effect of various vasodilators, such as acetylcholine and substance P (Edvinsson *et al.*, 1987), the raised levels may imply not only an enhanced direct vasoconstriction by NPY but also an activation of the sympathetic cotransmittors NA and ATP.

B. Sensory Nerve-Associated Peptides: Substance P and Calcitonin Gene-Related Peptide

The demonstration of nonadrenergic, noncholinergic vasodilator innervation of the peripheral resistance vessels has added more complexity to the understanding of the neurogenic components involved in the vasomotor regulation. It has been shown that the calcitonin gene-related peptide (CGRP) containing nerves contribute to the vascular resistance in the mesenteric arterial vasculature of the rat (Kawasaki *et al.*, 1988). Studies in SHR (the spontaneously hypertensive rat) have demonstrated that the CGRP containing dilator innervation is markedly decreased during development of hypertension as well as during established hypertension (Kawasaki *et al.*, 1990). It has therefore been suggested that a decreased vasodilator activity of CGRP containing nerves may contribute to the development and maintenance of hypertension.

In addition to CGRP, sensory perivascular nerve fibers around blood vessels also contain substance P as visualized in various parts of the circulation (Gibbins *et al.*, 1985; Lundberg *et al.*, 1985; Uddman *et al.*, 1986). This nerve fiber system originates in sensory ganglia (Uddman *et al.*, 1985) and is sensitive to capsaicin (Wharton *et al.*, 1986). Both substance P and CGRP act as potent vasodilators of the human vasculature *in vitro* (Edvinsson *et al.*, 1985; Edvinsson *et al.*, 1985) as well as *in situ* (McEwan *et al.*, 1988; Thom *et al.*, 1987). In man it has been shown that infusions of both substance P and CGRP induce a decrease in diastolic blood pressure that is accompanied by reflex tachycardia (Fuller *et al.*, 1987; Struthers *et al.*, 1986).

In subjects with severe hypertension (diastolic blood pressure >120 mmHg), the circulating levels of substance P and CGRP have been analyzed before and after normalization of the blood pressure by pharmacological therapy in order to examine whether afferent nerves are involved in the complex mechanisms resulting in elevated vascular resistance in advanced human hypertension (Edvinsson *et al.*, 1989; Edvinsson *et al.*, 1992). In such patients circulating levels of CGRP did not differ

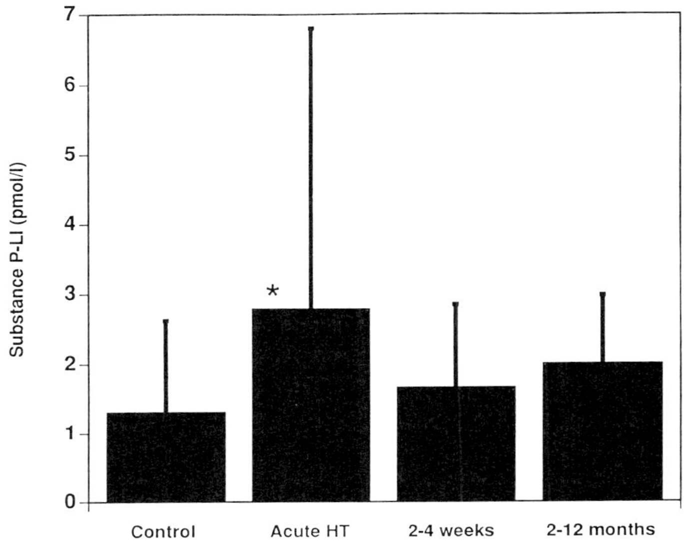

Figure 2 Concentrations of substance P-like immunoreactivity (SP-LI) in normotensive controls, patients with severe hypertension before, after 2–4 weeks, and after 2–12 months of treatment. Values represent mean ± SD; $*p < 0.05$, paired students *t*-test.

from carefully age- and sex-matched controls (Edvinsson *et al.*, 1992). However, a significantly higher concentration of venous substance P was seen in the hypertensive group. Following normalization of blood pressure, the circulating levels of substance P were also normalized (Fig. 2). The observation suggests that in conjunction with high blood pressure, such as in untreated patients with severe hypertension, the perivascular vasodilator peptide substance P is released to counterbalance the increase in vascular smooth muscle tone. This increment in circulating substance P returned essentially to unchanged levels as compared to control during subsequent controls during 1 year. The potent vasodilator CGRP was, however, not elevated before or after the lowering of the blood pressure, but there was a strong correlation between the circulating level of CGRP and diastolic as well as systolic blood pressures in subjects with severe hypertension (Fig 3). This may be taken as further evidence for an association between the blood pressure level and the release of sensory neuropeptides (Edvinsson *et al.*, 1992).

Other clinical studies on plasma CGRP levels have demonstrated raised circulating levels during normal human pregnancy (Stevenson *et al.*, 1986) although it is currently not known whether concentrations are altered during preeclampsia. Furthermore, increased CGRP has also been reported in patients with pheochromocytomas (Takahashi *et al.*, 1988). However, only few authors have reported on CGRP in primary hypertensives as compared to matched normotensive subjects. Tang and colleagues (1989) recently reported that plasma CGRP was significantly

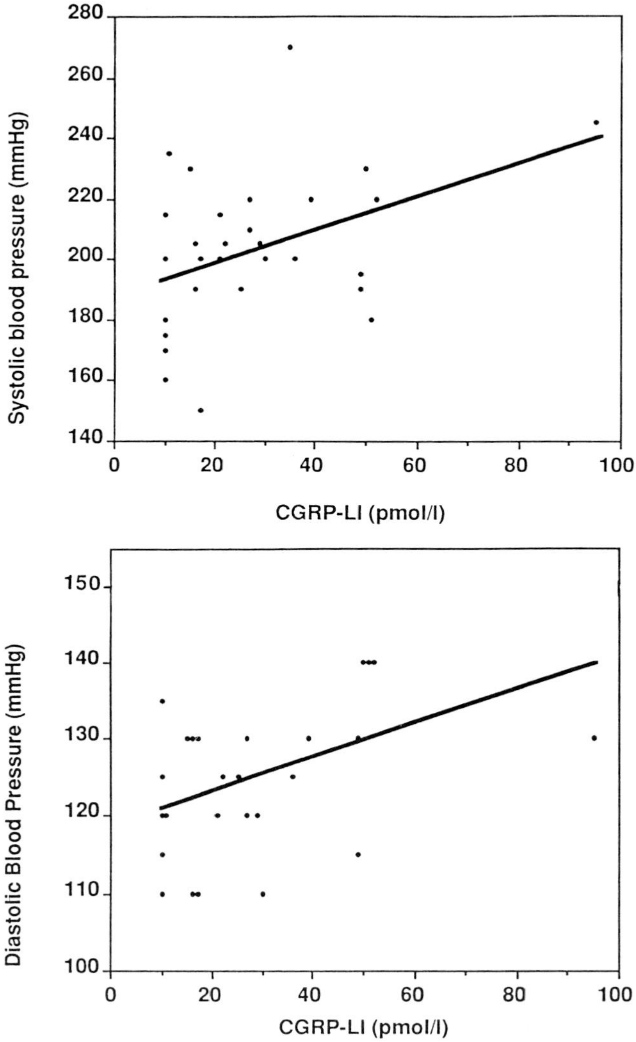

Figure 3 Relationship between levels of calcitonin gene-related peptide-like immunoreactivity (CGRP-LI) and systolic blood pressure, $R = 0.41$, $p < 0.05$ (*top*), and diastolic blood pressure (*bottom*) ($R = 0.42$, $p < 0.05$) in severe hypertensives before treatment. (Linear regression analysis.)

lower in essential hypertensives as compared to a group of subjects with normal blood pressure. This finding was in accordance with an earlier study performed by us using an unmatched control group (Edvinsson *et al.*, 1989). However, after careful matching of the hypertensive patients for age and sex, the difference in circulating CGRP could no longer be demonstrated (Edvinsson *et al.*, 1992). On the other hand, in SHR the issue of altered plasma levels of CGRP is still controversial. Tang *et al.* (1989) reported lower CGRP levels in SHR than in control Wistar-Kyoto (WKY) rats, whereas others (Zaidi and Bevis, 1991) have found increased circulating CGRP levels in the SHR strain.

The reason for the discrepancies in circulating levels of CGRP may not necessarily be due to different methodology, such as sampling, sample preparation, or radioimmunoassay, but may represent a biological difference. Thus, the severity of hypertension, degree of end-organ damage, hormonal balance, duration of the disorder, and treatment of the hypertensive subjects may all be factors that should be analyzed in detail to fully understand the potential role of perivascular peptides in the development and maintenance of hypertension. Due to their role as vasodepressor peptides, CGRP and substance P may be expected to be elevated in hypertensives in order to act as a compensatory mechanism balancing the high blood pressure and the reduced organ perfusion. However, depending on the severity of the disorder and duration of disease, the perivascular nervous system may even be exhausted and hence a lower circulating level could be seen at some stages or in some forms of hypertension. Since the regulating mechanisms for CGRP and substance P are presently not known in detail, further studies are required in order to clarify the potential pathophysiological involvement of sensory neuropeptides in the hypertensive disease.

III. Congestive Heart Failure

A. Neuropeptide Y and Catecholamines

In congestive heart failure (CHF) there are numerous alterations in the hormonal regulation of volume and pressure homeostasis (Dzau, 1988; Caidahl *et al.*, 1988). The reduced pump function is associated with a dysregulation of organ blood flow and an increased efferent sympathetic nerve activity. Thus, high plasma levels of noradrenaline (NA) have been found in patients with congestive heart failure (Thomas and Marks, 1978; Francis, 1984; Caidahl *et al.*, 1988). Plasma concentrations of NA are associated not only to the severity (Caidahl *et al.*, 1988) but also to the prognosis of CHF (Swedberg *et al.*, 1988). For NPY there is yet limited knowledge with regard to CHF, and published studies have all documented increased circulating levels (see below). In patients with left ventricular failure both NPY-LI and NA were significantly higher than in cardiac patients without heart failure (Ullman *et al.*, 1990). In a recent study we compared plasma catecholamine and NPY-LI levels in patients with different degrees of CHF (Edvinsson *et al.* 1990). Circulating levels of NA and adrenaline were increased (Fig. 4) and confirmed previous studies relating to the severity of cardiac failure and the degree of catecholamine release. Interestingly, the NPY-LI levels were significantly raised already in *moderate CHF* grade I–II as classified according to the New York Heart Association (NYHA) (Edvinsson *et al.*, 1990). Furthermore, in *severe* congestive heart failure (NYHA) (III–IV) the levels were not different from patients with moderate

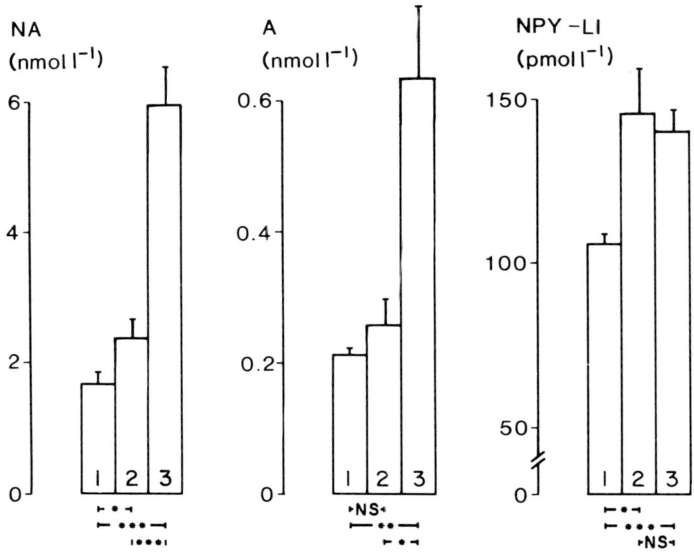

Figure 4 Circulating levels of noradrenaline (NA), adrenaline (A), and neuropeptide Y (NPY-LI) in controls (1), and in patients with congestive heart failure classified according to the New York Heart Association and grouped in moderate (I–II) (2) and severe (III–IV) (3) congestive heart failure. Values given represent means ± SEM. Statistical evaluation using students *t*-test with Bonferroni correction for multiple group comparison; $*p < 0.05$, $**p < 0.01$, $***p < 0.001$; NS, not significant.

degree of heart failure. Thus, in spite of the morphological colocalization of NA and NPY in sympathetic nerve terminals, they do not always seem to be released in parallel and to a similar extent. The observation of an increase in circulating NPY-LI in CHF is in accordance with studies by others (Ullman *et al.*, 1990; Maisel *et al.*, 1989; Hulting *et al.*, 1990).

The plasma levels of catecholamines increased with the degree of heart failure as described earlier (Edvinsson *et al.*, 1990; Thomas and Marks, 1978). Since NPY is not released until sympathetic nerve activity is intense and increased in healthy subjects (Pernow *et al.*, 1986), there may be reasons to believe that CHF is associated with an increased sympathetic activity resulting in enhanced NPY release already in its milder forms (i.e., corresponding to NYHA I–II).

B. Sensory Nerve-Associated Peptides

A number of bioactive circulating perivascular peptides representing vasodilator systems may be activated in order to counterbalance the vasoconstrictive activity induced by sympathetic activation. Thus, in cardiac failure a new neurohormonal balance regulating peripheral organ flow seems to be established between circulating or locally released vasoconstrictive and vasodilating factors. There is an increased release of several hormones that contribute to constriction of arterial as well as capacitance vessels, such as catecholamines, angiotensin II, and vasopressin (Francis, 1988). This enhancement of multiple vasoconstrictive systems is counterbalanced by increased release of atrial natriuretic peptide (ANP) that is generally found in cardiac failure (Burnett *et al.*, 1986; Pettersson *et al.*, 1986; Caidahl *et al.*, 1988) together with an increased synthesis of vasodilator prostaglandins (Dzau *et al.*, 1984). Substance P may participate in regulation of vasodilator tone. The increased plasma level of substance P-like immunoreactivity (LI) found in patients with cardiac failure (Edvinsson *et al.*, 1990) suggests that this peptide may be one of the vasodilators that counterbalance the many vasoconstrictive factors in cardiac failure. In accordance with the increased levels of NPY-LI, release of substance P from the sensory system seems to be already fully activated in milder forms of the disease, without any further increase in patients with more severe CHF, such as NYHA III–IV (Fig. 5). In contrast to substance P, some other peptides with known vasodilator properties, such as CGRP and vasoactive intestinal peptide (VIP), do not appear in the peripheral circulation in increased concentrations (Edvinsson *et al.*, 1990). The potential mechanism for the activation of the substance P- and CGRP-containing sensory nerve terminals is not yet fully known, but animal experiments have demonstrated that sensory ganglia (e.g., the trigeminal) are essential in order to induce a reflex in response to induced vasoconstriction (McCulloch *et al.*, 1986). It should be pointed out that there was a differential release of perivascular peptides in response to induced vasoconstriction (McCulloch *et al.*, 1986) since only the substance P level was increased in congestive heart failure whereas the CGRP-LI levels were unchanged. This indicates an interesting difference compared to, for example, the cerebral vascular sensory fibers: In these vessels there is a selective release of CGRP-LI during a migraine attack (Goadsby *et al.*, 1990) and in patients with a subarachnoid hemorrage (Juul *et al.*, 1990). Interestingly, in these two cerebrovascular disorders there was a significant release of CGRP-LI while the levels of NPY-LI and substance P-LI remained unaltered. The increased plasma level of substance P-LI in CHF suggests that this peptide might act as a vasodilator, counteracting vasoconstrictive factors in CHF. The main part of the circulating level of substance P is generally considered to be derived from perivascular nerves, where it may participate in the regulation of the vascular tone.

C. Treatment with Angiotensin-Converting Enzyme (ACE) Inhibitor

The hemodynamics as well as survival have been found to improve after treatment of mild to severe CHF patients with ACE inhibitors (Consensus, 1987). As can be seen in Fig. 5 (Valdemarsson *et al.*, 1991), several neurohormonal alterations are related to the degree of cardiac dysfunction in CHF patients. In an effort to separate any specific pharmacological effect of ACE inhibition on the neurohormonal balance from the potential effects of an improved circulatory function, we compared CHF patients treated with digoxin and diuretics to a group of patients who were also on continuous ACE inhibition but had a comparable degree of heart failure accord-

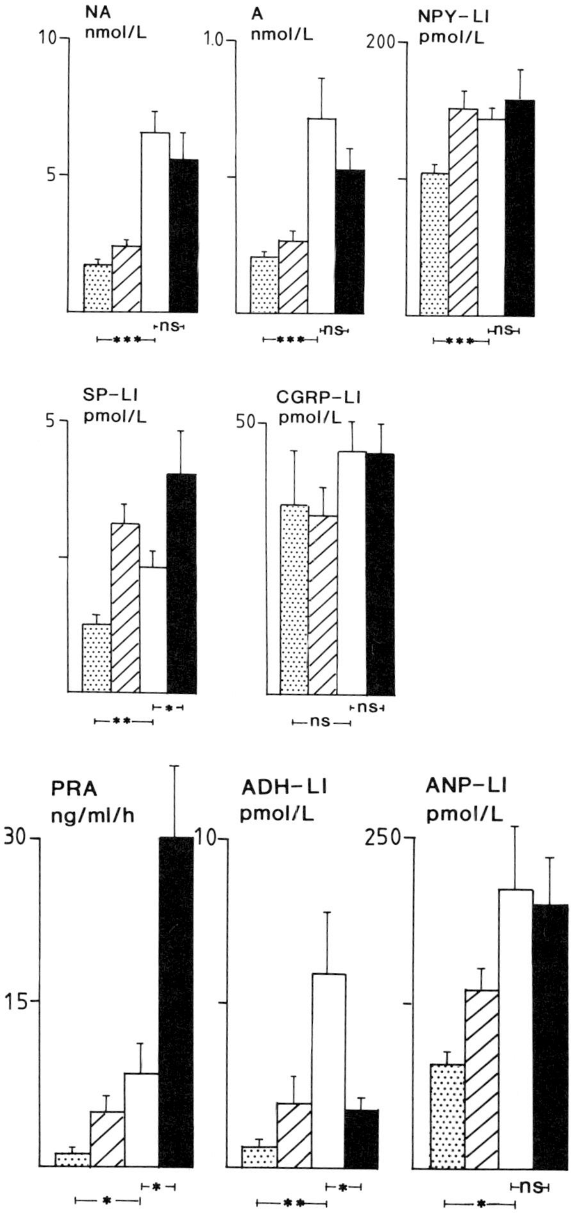

Figure 5 Plasma levels of noradrenaline (NA), adrenaline (A), neuropeptide Y (NPY-LI-like immunoreactivity), substance P (SP-LI), calcitonin gene-related peptide (CGRP-LI), renin activity (PRA), arginine-vasopressin (ADH-LI) and h-Á-atrial natriuretic peptide (ANP-LI) in healthy subjects (n = 31, *dotted bars*), patients with mild CHF (NYHA I-II, n = 10, *hatched bars*), patients with severe CHF (NYHA III–IV, n = 17, *open bars*), and patients with severe CHF on ACE inhibition (NYHA III–IV, n = 15, *closed bars*). Statistical analysis, $*p < 0.05$, $**p < 0.01$, $***p < 0.001$.

ing to NYHA. Our findings corresponded well to the findings in CHF patients presented by others by having increased plasma levels of catecholamines, plasma renin activity, and atrial natriuretic peptide, the latter generally ascribed to backward failure with atrial distention (Swedberg *et al.*, 1990). The increased angiotensin II formation, which results from the activity of renin and ACE, has been suggested to be of importance for the increased vasopressin release found in CHF (Knepel and Meyer, 1980; Riegger and Kochsiek; 1986). Thus, the patient with CHF is subjected to decreased organ blood flow due to vasoconstrictive neurohormones, such as catecholamines and NPY, which are not directly affected by ACE inhibition, in addition to vasoconstriction induced by angiotensin II and vasopressin, which are more closely modified by ACE inhibitors. However, since vasodilating hormones and perivascular peptides are changed in CHF, a new homeostatic balance between vasoconstrictive and vasodilating factors seems to be established in CHF of different severity (Dzau, 1988; Swedberg *et al.*, 1990). Patients classified as NYHA III–IV who were taking ACE inhibitors would perhaps be expected to have increased circulating substance P–LI since ACE is not specific for the conversion of angiotensin I to angiotensin II but also participates in the degradation of several other peptides, among them substance P (Lötvall 1990). This may also be one tentative explanation to the increased circulating substance P-LI concentrations already found in patients with mild forms of cardiac failure taking ACE inhibitors (Valdemarsson *et al.*, 1991).

However, serum concentrations of substance P have been reported to be normal and not changed after ACE inhibition in hypertension (Thysell *et al.*, 1988), which would argue against this hypotheses. Thus, it remains a possibility that an increased vascular tone, as reflected by increased catecholamine levels and plasma renin activity in CHF patients, might be a prerequisite for the release of substance P in CHF (Valdemarsson *et al.*, 1991). Interestingly, we found an inverse correlation between the serum levels of substance P and sodium both in controls and in patients without ACE inhibition, suggesting the possibility of an osmolar- or electrolyte-dependent mechanism taking part in the regulation of the substance P release.

The fact that the ACE is involved also in the degradation of kininogens and tachykinins opens further possible mechanisms for the effect of ACE inhibitors. In *in vitro* systems, ACE has been found to take part in the degradation of substance P to various fragments (Unger *et al.*, 1989). In accordance with this, we found that ACE inhibition was associated with an even higher substance P level in the patient group receiving this treatment as compared to patients with the same degree of CHF but not on ACE inhibition (Valdemarsson *et al.*, 1991). Since substance P has potent vasodilator properties (Håkanson and Sundler, 1985), this finding therefore suggests another way of favorable ACE inhibitor action on counter regulatory mechanisms in severe CHF. Circulating levels of other vasodilator perivascular peptides measured, such as CGRP and VIP, were not increased in CHF (Valdemarsson *et al.*, 1991). Since CGRP and substance P are colocalized in afferent perivascular nerve fibers, this suggests that substance P and CGRP are separately released under these conditions.

IV. Comments

There is a pronounced but variable neurohormonal activation in hypertension and congestive heart failure. In this chapter we have discussed the varying pattern of the perivascular peptides NPY, CGRP, and SP in these circulatory disorders. Knowledge about their potential roles in the treatment and outcome is yet limited but will be analyzed in studies to come. Hopefully studies of the perivascular peptides will provide further understanding of not only the genesis but also the altered neurohormonal, circulatory homeostasis, and compensatory mechanisms being activated in hypertension and congestive heart failure.

Acknowledgments

This work was supported by the Swedish Medical Research Council (project no. 05958 and 08642).

References

Allen, J. M., Bircham, P. M. M., Bloom, S. R., and Edwards, A. V. (1984). Release of neuropeptide Y in response to splanchnic nerve stimulation in the conscious calf. *J. Physiol.* (Lond.) *357,* 401–408.

Burnett, J. C., Jr., Kao, P. C., Hu, D. C., Hesser, P. W., Heublein, D., Ganger, J. P., and Opsgenorth, T. J., Reeder, G. S. (1986). Atrial natriuretic peptide elevation in chronic congestive heart failure. *Science 231,* 1145–1147.

Caidahl, K., Hedner, J., and Hedner, T. (1988). Atrial natriuretic peptide and catecholamines in peripheral blood as indicators of cardiac dysfunction in the general population. *Clin. Sci. 74,* 587–594.

Clark, D. W. J. (1977). Effects of immunosympathectomy on the development of high blood pressure in genetically hypertensive rats. *Circ. Res. 28,* 330–336.

Consensus Trial Study Group. (1987). Effects of enalapril on mortality in severe congestive heart failure: Results of the Cooperative North Scandinavian Enalapril Survival Study (CONSENSUS). *N. Engl. J. Med. 316,* 1429–35.

Crozier, I. G., Nicholls, M. G., Ikram, H., Espiner, E. A., and Yandle, T. G. (1989). Atrial natriuretic peptide levels in congestive heart failure before and during converting enzyme inhibition. *Clin. Exp. Pharmacol. Physiol. 16,* 417–24.

Dzau, V. J. (1988). Contribution of neuroendocrine and local autocrine–paracrine mechanisms to the patophysiology and pharmacology of congestive heart failure. *Am. J. Cardiol. 68* (8), 76E–81E.

Dzau, V. J., Packer, M., Lilly, L. S., Swarts, S. L., Hollenberg, N. K., and Williams, G. H. (1984). Prostaglandins in severe congestive heart failure: Relation to activation of the renin–angiotensin system and hyponatremia. *N. Engl. J. Med. 310,* 347–352.

Edvinsson, L., and Uddman, R. (1992). "Vascular Innervation and Receptor Mechanisms." Academic Press, California.

Edvinsson, L., Ekman, R., Jansen, I., Ottosson, A., and Uddman, R. (1985). Distribution, concentration and effects of neuropeptide Y (NPY), vasoactive intestinal polypeptide (VIP), calcitonin gene-related peptide (CGRP) and substance P (SP) in human cerebral blood vessels. *J. Cereb. Blood Flow Metab. 5* (Suppl. 1), 545–546.

Edvinsson, L., Håkanson, R., Steen, S., Sundler, F., Uddman, R., and Wahlestedt, C. (1985). Innervation of human omental arteries and veins and vasomotor responses to noradrenaline, neuropeptide Y, substance P and vasoactive intestinal peptide. *Regul. Peptides 12,* 67–79.

Edvinsson, L., Håkanson, R., Wahlestedt, C., and Uddman, R. (1987). Effects of neuropeptide Y on the cardiovascular. *TIPS 8,* 231–235.

Edvinsson, L., Ekman, R., and Thulin, T. (1989). Reduced levels of calcitonin gene-related peptide (CGRP) but not substance P during and after treatment of severe hypertension in man. *J. Human Hypertension 3,* 267–270.

Edvinsson, L., Ekman, R., Hedner, P., and Valdemarsson, S. (1990). Congestive heart failure: Involvement of perivascular peptides reflecting activity in sympathetic, parasympathetic and afferent fibres. *Eur. J. Clin. Invest. 20,* 85–89.

Edvinsson, L., Ekman, R., and Thulin, T. (1991). Increased plasma levels of neuropeptide Y-like immunoreactivity and catecholamines in severe hypertension remain after treatment to normotension in man. *Regulat. Pept. 32,* 279–287.

Edvinsson, L., Erlinge, D., Ekman, R., and Thulin, T. (1992). Sensory nerve terminal activity in severe hypertension as reflected by circulating calcitonin gene-related peptide (CGRP) and substance P. *Blood Pressure* (in press).

Erlinge, D., Ekman, R., Thulin, T., and Edvinsson, L. (1992). Neuropeptide Y–like immunoreactivity and hypertension. *J. Hypertension.* (in press).

Folkow, B., Hallbäck, M., Lundgren, Y., and Weiss, L. (1970). Structurally based increase in flow resistance in spontaneously hypertensive rats. *Acta Physiol. Scand. 79,* 373–378.

Francis, G. S. (1984). Plasma catecholamines in patients with congestive heart failure. *Cardiovasc. Rev. Rep. 6,* 444–454.

Francis, G. S. (1988). Neuroendocrine manifestations of congestive heart failure. *Am. J. Cardiol. 62,* 9A–13A.

Francis, G. S., Goldsmith, T. B., Levine, M. T., Olivari, M. T., Cohn, J. N. (1984). The neurohumoral axis in congestive heart failure. *Ann. Intern. Med. 101,* 370–8.

Fuller, R. W., Maxwell, D. L., Dixon, C. M., McGregor, G. P., Barnes, V. F., Bloom, S. R., Barnes, P. J. (1987). Effect of substance P on cardiovascular and respiratory function in normal subjects. *J. Appl. Physiol. 62,* 1473–1479.

Furness, J. B., Costa, M., Emson, P. C., Hakanson, R., Moghimzadeh, E., Sundler, E., Taylor, I. L., and Chance, R. E. (1983). Distribution pathways and reactions to drug treatment of nerves with neuropeptide Y and pancreatic polypeptide-like immunoreactivity in the guinea-pig digestive tract. *Cell. Tissue Res. 234,* 71–92.

Gibbins, I. L., Furness, J. B., Costa, M., MacIntyre, I., Hillyard, C. J., and Girgis, S. (1985). Colocalization of calcitonin gene-related peptide-like immunoreactivity with substance P in cutaneous, vascular and visceral sensory neurones of guinea pigs. *Neurosci. Lett. 57,* 125–129.

Goadsby, P. J., Edvinsson, L., and Ekman, R. (1990). Vasoactive peptide release in the extra cerebral circulation of human during migraine headache. *Ann. Neurol. 28,* 183–187.

Goldstein, D. S., Lake, C. R., Chernow, B., Ziegler, M. G., Coleman, M. D., Taylor, A. A., Mitchell, J. R., Kopin, I. J., Keiser, H. R. (1983). Age-dependence of hypertensive—normotensive differences in plasma norepinephrine. *Hypertension 5,* 100–104.

Gurusinghe, C. J., Harris, P. J., Abbot, D. F., and Bell, C. (1990). Neuropeptide Y in rat sympathetic neurons is altered by genetic hypertension and by age. *Hypertension 16,* 63–71.

Håkanson, R., and Sundler, F. (1985). "Tachykinin Antagonists." Elsevier Science Publ, Amsterdam.

Head, R. J. (1989). Hypernoradrenergic innervation: Its relationship to functional and hyperplastic changes in the vascular of the spontaneously hypertensive rat. *Blood Vessels 26,* 1–20.

Hulting, J., Sollevi, A., Ullman, B., Franco-Cereceda, A., and Lundberg, J. M. (1990). Plasma neuropeptide Y on admission to a coronary care unit. Elevated levels in patients with left heart failure. *Cardiovasc. Res. 24,* 102–108.

Judy, W. V., Watanabe, A. M., Henry, D. P., Besch, H. R., Murphy, W. R., and Hockel, G. M. (1976). Sympathetic nerve activity: Role in regulation of blood pressure in the spontaneously hypertensive rat. *Circ. Res. 38* (suppl. II), II-21–II-29.

Juul, R., Edvinsson, L., Gisvold, S. E., Ekman, R., Brubakk, A. O., and Fredriksen, T. A. (1990). Calcitonin gene-related peptide in subarachnoid hemorrhage in man. Signs of activation of the trigemino-cerebrovascular system? *Br. J. Neurosurg. 4,* 171–180.

Kawasaki, H., Clein, W. H., Jr., and Su, C. (1982). Enhanced presynaptic beta-adrenoceptor-mediated modulation of vascular adrenergic neurotransmission in spontaneously hypertensive rats. *J. Pharmacol. Exp. Ther. 223,* 721–728.

Kawasaki, H., Takasaki, K., Saito, A., and Goto, K. (1988). Calcitonin gene-related peptide as a novel vasodilator neurotransmitter in mesenteric resistance vessels of the rat. *Nature 335,* 164–167.
Kawasaki, H., Saito, A., and Takasaki, K. (1990). Age-related decrease of calcitonin gene-related peptide-containing vasodilator innervation in the mesenteric resistance vessel of the spontaneously hypertensive rat. *Circ. Res. 67,* 733–743.
Kline, R. L., Kelton, P. M., and Mercer, P. F. (1978). Effect of renal denervation on the development of hypertension in spontaneously hypertensive rats. *Can. Physiol. Pharmacol. 56,* 818–822.
Knepel, W., and Meyer, D. K. (1980). Influence of converting enzyme inhibition on the release of vasopressin induced by angiotensin. *Br. J. Pharamacol. 71,* 337–41.
Lee, R. M. K. W., Triggle, C. R., Cheung, D. W. T., and Coughlin, M. D. (1987). Structural and functional consequence of neonatal sympathectomy on the blood vessels on spontaneously hypertensive rats. *Hypertension 10,* 328–338.
Leimbach, W. N., Wallin, B. G., Victor, G. R., Aylward, P. E., Sundlöf, G., and Mark, A. L. (1986). Direct evidence from intraneural recordings of increased central sympathetic outflow in patients with heart failure. *Circulation 73,* 913–919.
Liard, J. F. (1977). Renal denervation delays blood pressure increase in the spontaneously hypertensive rat. *Experientia 33,* 339–340.
Lötvall, J. (1990). Tachykinin- and bradykinin-induced airflow obstruction and airway microvascular leakage. Modulation by neutral endopeptidase and angiotensin converting enzyme. Thesis, University of Gothenburg, pp. 1–76.
Lundberg, J., and Hökfelt, T. (1986). Multiple co-existence of peptides and classical transmitters in peripheral autonomic neurons—functional and pharmacological implications. *Prog. Brain Res. 68,* 241–262.
Lundberg, J. M., Franco-Cereceda, A., Hua, X., Hökfelt, T., and Fischer, J. A. (1985). Co-existence of substance P and calcitonin gene-related peptide-like immunoreactivities in sensory nerves in relation to cardiovascular and bronchoconstrictor effects of capsaicin. *Eur. J. Pharmacol. 108,* 315–319.
Lundberg, J., Martinsson, A., Hemsén, A., Theodorsson-Norheim, E., Svedenhag, J., Ekblom, B., and Hjelmdahl, P. (1985). Co-release of neuropeptide Y and catecholamines during physical exercise in man. *Biochem. Biophys. Res. Commun. 133,* 30–60.
Maisel, A. S., Scott, N. A., Motulsky, H. J., Michel, M. C., Boublik, J. H., Rivier, J. E., Ziegler, M., Allen, R. S., and Brown, M. R. (1989). Evaluation of plasma neuropeptide Y levels in congestive heart failure. *Am. J. Med. 86,* 43–48.
McCarthy, R. Sympathetic nervous system development in hypertensive rats. *J. Hypertens. 4* (suppl. 3), 155–157.
McCulloch, J., Uddman, R., Kingman, T. A., and Edvinsson, L. (1986). Calcitonin gene-related peptide: Functional role in cerebrovascular regulation. *Proc. Natl. Acad. Sci. U.S.A. 83,* 5731–5735.
McEwan, J. R., Benjamin, N., Larkin, S., Fuller, R. W., Dollery, C. T., and MacIntyre, I. (1988). Vasodilation by calcitonin gene-related peptide and by substance P: A comparison of their effects on resistance and capacitance vessels of human forearms. *Circulation 77,* 1072–1080.
Okamoto, K., Nosaka, S., Yamori, Y., and Matsumoto, M. (1967). Participation of neural factor in the pathogenesis of hypertension in the spontaneously hypertensive rat. *Jpn. Heart J. 8,* 168–180.
Pernow, J. (1988). Co-release and functional interactions of neuropeptide Y and noradrenaline in peripheral vascular control. *Acta Physiol. Scand. 133* (suppl), 568.
Pernow, J., Lundberg, J. M., Kaijser, L., Hjemdahl, P., Theodorsson-Norheim, E., Martinsson, A., and Pernow, B. (1986). Plasma neuropeptide Y-like immunoreactivity and catecholamines during various degrees of sympathetic activation in man. *Clin. Physiol. 6,* 561–578.
Petrovic, T., and Bekk, C. (1986). Catecholamines in kidneys of normotensive and genetically hypertensive rats: Effects of salt load. *Hypertension 8,* 122–127.
Petrovic, T., and Bell, C. (1985). Effect of chronic decentralization on the catecholamine content of the superior cervical ganglion of the rat. *Proc. Aust. Physiol. Pharmacol. Soc. 16,* 159P.
Pettersson, A., Hedner, J., Hedner, T., Held, P., Swedberg, K., and Towle, A. C. (1986). Increased levels of ANP in patients with congestive heart failure. *Eur. Heart J. 7,* 693–696.

Potter, E. K. (1988). Neuropeptide Y as an autonomic neurotransmitter. *Pharmacol. Ther. 37,* 251–273.

Provoost, A. P. (1984). Sympathectomy and the development of hypertension, *In* "Experimental and Genetic Models of Hypertension." W. De Jong, ed., pp. 495–517. Amsterdam, Elsevier Science Publishing Co, Inc.

Riegger, G. A., and Kochsiek, K. (1986). Vasopressin, renin and norepihephrine levels before and after captopril administration in patients with congestive heart failure due to idiopathic dilated cardiomyopathi. *Am. J. Cardiol. 58,* 300–303.

Schalling, M., Franco-Cereceda, A., Hökfelt, T., Persson, H., and Lundberg, J. (1988). Increased neuropeptide Y messenger RNA and peptide in sympathetic ganglia after reserpine pretreatment. *Eur. J. Pharmac. 156,* 419–420.

Solt, V. G., Brown, M. R., Keddedy, B., Kolterman, O. G. and Ziegler, M. G. (1990). Elevated Insulin, Norepinephrine, and Neuropeptide Y in Hypertension. *Am. J. Hypertens. 3,* 823–828.

Stevenson, J. C., MacDonald, D. W. R., Warren, R. C., Booker, M. W., and Whitehead, M. I. (1986). Increased concentration of circulation calcitonin gene-related peptide during normal human pregnancy. *Br. Med. J. 293,* 1329–1330.

Struthers, A. D. Brown, M. J., MacDonald, D. W., Beacham, J. L., Stevenson, J. C., Morris, H. R., MacIntyre, I. (1986). Human calcitonin gene-related peptide: A potent endogenous vasodilator in man. *Clin. Sci. 70,* 389–393.

Swedberg, K., Eneroth, P., Kjekshus, J., and Wilhelmsen, L. (1988). For the consensus trial group. Plasma concentrations of hormones affecting cardiovascular function in relation to mortality in severe congestive heart failure. *Circulation 78* (suppl. B), 123–128.

Swedberg, K., Eneroth, P., Kjekshus, J., and Wilhelmsen, L. (1990), Hormones regulating cardiovascular function in patients with severe congestive heart failure and their relation to mortality. *Circulation 82,* 1730–1736.

Takahashi, K., Mouri, T., and Sone, M. (1988). Calcitonin gene-related peptide-like immunoreactivity in pheochromocytoma. *J. Clin. Exper. Med.* (IGAKU NO AYUMI, Japan) *145,* 125–126.

Tang, J., Xu, D., and Yuan, O. X. (1989). Calcitonin gene-related peptide in the pathogenesis and treatment of hypertension. *Chinese med. J. 102,* 897–901.

Thom, S., Hughes, A. D., Goldberg, P., Martin, G., Schachter, M., and Sever, P. S. (1987). The actions of calcitonin gene-related peptide and vasoactive intestinal peptide as vasodilators in man *in vivo* and *in vitro*. *Br. J. Clin. Pharmacol. 24,* 139–144.

Thomas, J. A., and Marks, B. H. (1978). Plasma norepinephrine in congestive heart failure. *Am. J. Cardiol. 41,* 233–243.

Thorén, P. (1987). Efferent renal nerve traffic in the spontaneously hypertensive rat. *Clin. Exp. Hypertens. A9* (suppl. I) 259–280.

Thysell, H., Andersson, K.-E., Andersson, S.-I., and Ekman, R. (1988). Angiotensin-converting enzyme inhibition, cough and the serum concentration of substance P. *Eur. J. Clin. Pharmacol. 34,* 649–50.

Uddman, R., Edvinsson, L., Ekman, R., Kingman, R., and McCulloch, J. (1985). Innervation of the feline cerebral vasculature by nerve fibers containing calcitonin gene-related peptide: trigeminal origin and co-existence with substance P. *Neuros. Lett. 62,* 131–136.

Uddman, R., Edvinsson, L., Ekblad, E., Håkanson, R., and Sundler, F. (1986). Calcitonin gene-related peptide (CGRP): Perivascular distribution and vasodilatory effects. *Regul. Pept. 15,* 1–23.

Ullman, B., Franco-Cereceda, A., Hulting, J., Lundberg, J. M., and Sollevi A. (1990). Elevation of plasma neuropeptide Y-like immunoreactivity and noradrenaline during myocardial ischaemia in man. *J. Int. Med. 228,* 583–589.

Unger, T., Gohlke, P., Ganten, D., and Lang, R. E. (1989). Converting enzyme inhibitors and their effects on the renin-angiotensin system of the blood vessel wall. *J. Cardiovas. Pharmacol. 13* (Supp. 3), 8–16.

Valdemarsson, S., Edvinsson, L., Hedner, P., and Ekman, R. (1990). Hormonal influence on calcitonin gene-related peptide in man: Effects of sex difference and contraceptive pills. *Scand. J. Clin. Lab. Invest. 50,* 385–388.

Valdemarsson, S., Edvinsson, L., Ekman, R., Hedner, P., and Sjöholm, A. (1991). Increased plasma levels of substance P in patients with severe congestive heart failure treated with ACE inhibitors. *J. Int. Med. 230,* 325–331.

Vanhoutte, P. M., and Lüscher, T. F. (1986). Peripheral mechanisms in cardiovascular regulation: Transmitters, receptors and the endothelium, *In* "Handbook of Hypertension." (A. Zanchetti and R. C. Tarazi eds.), Volume 8, pp. 96–123. Pathophysiology of Hypertension. Amsterdam, Elsevier Science Publishing Co, Inc.

Weiss, L., Lundgren, Y., and Folkow, B. (1974). Effects of prolonged treatment with adrenergic β-receptor antagonists on blood pressure, cardiovascular design and reactivity in spontaneously hypertensive rats (SHR). *Acta Physiol. Scand. 91,* 447–457.

Wharton, J. and Gulbenkian, S. (1985). Peptides in the mammalian cardiovascular system. *Experientia 43,* 821–832.

Wharton, J., Gulbenkian, S., and Mulderry, P. K., Ghatei, M. A., McGregor, G. P., Bloom, S. R., Polak, J. M. (1986). Capsaicin induces a depletion of calcitonin gene-related peptide (CGRP)-immunoreactive nerves in the cardiovascular system of the guinea pig and rat. *J. Auton. Nerv. System 16,* 289–309.

Winternitz, S. R., Katholi, R. E., and Oparil, S. (1980). Role of the renal sympathetic nerves in the development and maintenance of hypertension in the spontaneously hypertensive rat. *J. Clin. Invest. 66,* 971–978.

Zaidi, M., and Bevis, P. J. R. (1991). Enhanced circulation levels of neurally derived calcitonin gene-related peptide in spontaneously hypertensive rats. *Cardiovasc. Res. 25,* 125–128.

CHAPTER 18

Perivascular Peptides in the Coronary Circulation: Relation to Pathophysiology

JOHN CLARKE

I. Introduction

Coronary blood flow is regulated in normal individuals to adjust flow to coronary perfusion pressure (i.e., aortic diastolic pressure) and to the demands of the myocardium for oxygen. The smooth muscle layer of coronary arteries has inherent tone that is modulated *in vivo* by autonomic innervation (Vatner *et al.*, 1970) and by vasoactive substances, some produced by vascular endothelial cells (Furchgott and Zawadski, 1980) and some that circulate in the blood. When significant epicardial coronary artery luminal stenosis occurs due to atherosclerosis, coronary blood flow is reduced. In this situation additional reductions of flow caused by physiological constrictor mechanisms may be sufficient to induce transient myocardial ischemia and angina pectoris. Also, the normal regulatory mechanisms may be altered by disease. For example, smooth muscle hyperreactivity is the cause of Prinzmetal's variant angina, and impaired endothelium-dependent coronary vascular smooth muscle relaxation is seen in atherosclerosis. The extent to which the coronary microcirculation is involved in these disease processes is unknown, but there is some evidence that microembolism and microvascular constriction may be a secondary event in some ischemic syndromes (Furchgott, 1983).

The number of vasoactive peptides discovered in cardiovascular tissues has increased in recent years. The physiological and pathophysiological contribution of these peptides is unknown at present. Peptides are associated with endothelial, neural, and muscular elements of the coronary vasculature, as well as with the myocardium and the conducting system. The more recent use of immunocytochemical techniques has made it possible to demonstrate the presence of peptides and transmitter-synthesizing enzymes in cardiovascular nerves, thereby allowing for a further additional distinction to be made between different autonomic nerve types depending not only on the content of "classical" neurotransmitters such as acetylcholine and noradrenaline, but also on their content of other peptide agents. It has been suggested that neuropeptides may be part of a rapidly responding system linked into central mechanisms, whereas blood-borne and endothelial peptides may play a role in medium or longer term control of cardiovascular function. There is an upsurge of interest in smooth muscle proliferation within the heart because, apart from its role in the development of atherosclerotic plaques, proliferation is a major component of restenosis of coronary vessels following angioplasty.

II. Neuropeptide Y

Neuropeptide tyrosine is a member of a family of peptides that have a high degree of sequence homology and include pancreatic polypeptide (PP) and peptide YY (PYY). The neuropeptide Y gene can be regulated at the transcriptional level by at least one substance, nerve growth factor (Sabol and Higuchi, 1990). Two types of binding sites have been described for neuropeptide Y (Wahlestedt *et al.*, 1986; Sheikh *et al.*, 1989). NPY-immunoreactive nerve fibers are the most abundant of all the peptide-containing nerve populations identified to date in the mammalian cardiovascular system. Neuropeptide Y immunoreactive nerve fibers were found throughout mouse and rat heart (Gu *et al.*, 1984), particularly within the sinoatrial and atrioventricular nodes, and the coronary arteries, especially in medium and small coronary vessels. They were also present within the atria, ventricles, papillary muscles, ventricular, and atrial septa in close association with most cardiac muscle cells. The myocardium was more densely innervated than the endocardium or epicardium. Around arteries, NPY-containing nerves form an outer network of nerve bundles containing preterminal axons, which run mainly parallel to the vessel and also a perivascular plexus of fine, mainly varicose fibers and fascicles running around the vessel at the adventitial-medial border (Wharton and Gulbenkian, 1989). The distribution of NPY-immunoreactive nerve fibers parallels noradrenergic, postganglionic, sympathetic neurons (Polak and Bloom, 1984). It is thought that NPY is stored with noradrenaline in large vesicles in the nerve endings whereas noradrenaline alone is stored in small vesicles. So while NPY and noradrenaline coexist in sympathetic cardiovascular terminals, because they are in two distinct storage sites, a differential release of noradrenaline or NPY becomes possible. It has in fact been demonstrated that NPY is preferentially released from nerve endings under high sympathetic stimulation frequency (Allen *et al.*, 1984). Not all noradrenergic nerves contain NPY and conversely not all NPY-immunoreactive nerves are noradrenergic. It has been demonstrated that some intrinsic cardiac neurons may contain NPY immunoreactivity alone (Hassall *et al.*, 1984), suggesting that at least two populations of NPY containing nerves exist in the heart.

In vitro studies have shown that NPY acts directly through its own receptor on vascular smooth muscle by a mechanism dependent on extracellular calcium. Experimental evidence suggests that NPY may also have an indirect action at the neuroeffector junction. NPY has been shown to inhibit presynaptic release of noradrenaline and has also been found to potentiate the postsynaptic action of noradrenaline (Wahlestedt *et al.*, 1986). *In vivo* NPY constricts human forearm resistance vessels through its own receptor, independent of alpha receptors. In this study NPY appears to have no detectable modulating effect on the action of endogenous or infused noradrenaline (Clarke *et al.*, 1991).

NPY has been shown to have displaceable binding sites on human coronary arteries in the smooth muscle layer (tunica media) with an indication of weak binding within the adventitia (Dashwood and Clarke, unpublished observation). No NPY binding occurs to the endothelial layer although it has been demonstrated that NPY induced a concentration-related potentiation of the response to field stimulation that is significantly diminished in the absence of endothelium. (Daly and Hieble 1987). It has been reported that vasoconstriction of human skeletal muscle arteries by NPY is independent of the endothelium (Pernow and Lundberg, 1988).

There is an extensive network of NPY-containing nerves (Allen *et al.*, 1983) and receptors (Nakajima *et al.*, 1986) in the brainstem. Central administration of NPY causes alterations in blood pressure (Chen *et al.*, 1988; Fuxe 1983), suggesting that NPY may be involved in central control of the circulation. In the isolated rabbit heart, NPY has a negative inotropic effect and reduces coronary perfusion (Allen *et al.*, 1983b), whereas positive inotropic and chronotropic effects were observed in the isolated guinea pig atrium. However, in other studies, including studies on human myocardium (Michel *et al.*, 1989), NPY has been found to have no inotropic or chronotropic effects. In the organ bath, NPY is a constrictor of smaller (Franco-Cereceda *et al,*. 1987) but not epicardial (Tippins *et al.*, 1988) human coronary segments. NPY has been shown to constrict the coronary vascular bed of the anesthetized greyhound dog by an action on smaller coronary vessels (Clarke *et al.*, 1987). Intracoronary infusion of NPY in man may transiently induce myocardial ischemia reproducing angina pectoris; the action is on smaller coronary vessels (Clarke *et al.*, 1987b). Decreased levels of NPY are found in areas of experimentally induced myocardial infarction in rats (Han *et al.*, 1989), suggesting that NPY may be continually released from axons following myocardial ischemia, which may further exacerbate the ischemic event. It is possible, therefore, to envisage a situation where high-frequency stimulation of sympathetic nerves, such as those that may occur in man under conditions of emotional stress or physical exertion, could cause selective release of NPY from sympathetic nerve endings in the coronary circulation with subsequent reduction of coronary blood flow leading to myocardial ischemia. Elevated levels of NPY are found in patients with congestive cardiac failure (Maisel *et al.*, 1989).

III. Calcitonin Gene-Related Peptide

The single calcitonin gene transcript has been shown to produce by alternative splicing different mRNA species that encode different polypeptides, either calcitonin or the 37 amino acid peptide calcitonin gene-related peptide (CGRP). Calcitonin is predominantly formed in normal thyroid while in certain neural tissue CGRP is the major product. CGRP-containing nerve fibers are present throughout the cardiovascular system, particularly the carotid bifurcation and in lesser amounts in all regions of the heart, particularly in association with the adventitia of coronary arteries and veins, within the myocardium, epicardium, and papillary muscles, and within the sinoatrial, atrioventricular nodes, and conducting tissue (Mulderry *et al.*, 1985; Wharton *et al.*, 1986; Saito *et al.*, 1986; Lundberg *et al.*, 1985). Relatively few nerve fibers in the human heart and vasculature appear to display CGRP immunoreactivity. CGRP is colocalized with substance P in C-fiber afferents in the heart of several species, including man (Lundberg *et al.*, 1985; Franco-Cereceda *et al.*, 1987b). These nerves also contain related tachykinins from the same gene (Nawa *et al.*, 1983) [i.e., neurokinin A (NKA) and neuropeptide K (NPK)] (Hua *et al.*, 1985).

CGRP binding sites have been reported to occur in both the media and intima of rat coronary arteries and aorta (Sigrist *et al.*, 1986). In human heart the highest concentration of specific receptors for CGRP is in the major coronary arteries, which is consistent with the hypothesis that CGRP is implicated in the control of coronary vasomotor tone (Coupe *et al.*, 1990). Areas of coronary artery with atheroma contain significantly decreased binding sites when compared with normal arteries (Coupe *et al.*, 1990), suggesting a possible alteration in the balance of vasomotor influences in that region in favor of vasoconstriction. No specific binding was seen in myocardium or connective tissue, and CGRP has no reported direct effect on papillary muscle (Sigrist *et al.*, 1986).

Circulating CGRP may control vascular smooth muscle tone (Girgis *et al.*, 1986). CGRP has been identified in normal human and rat plasma and appears to be derived mainly from perivascular nerves (Zaidi *et al.*, 1985). Elevated plasma levels of CGRP have been reported in medullary carcinoma of the thyroid (Edbrooke *et al.*, 1985), a syndrome frequently associated with flushing attacks, and so CGRP may be involved in the pathogenesis of these symptoms. While it has been shown that CGRP is a dilator of bovine coronary artery in the absence of endothelium (Greenberg *et al.*, 1987), other studies suggest that the vasodilatory response may be mediated by an endothelium-dependent mechanism (Brain *et al.*, 1985) through the release of endothelium-derived relaxing factor (EDRF), prostaglandins (Brain *et al.*, 1985), or catecholamines (Fisher *et al.*, 1983; Kubota *et al.*, 1985) and partly via a direct action on the media with the activation of adenylate cyclase (Edvinsson *et al.*, 1985). Whether vasodilation induced by CGRP is dependent on an intact endothelium remains controversial. Recent reports have stated that although the endothelium is not necessary for the relaxation of human pulmonary artery sections (McCormack *et al.*, 1987), or for bovine coronary artery (Greenberg *et al.*, 1987), it is necessary for the actions of CGRP on human coronary arteries (Thom *et al.*, 1987).

CGRP is a potent dose-dependent vasodilator *in vivo* and *in vitro*. The effect is not modified by adrenergic, cholinergic, histaminergic, or neuronal blockade. CGRP has been shown to be a dilator of human epicardial coronary artery segments *in vitro* (Franco-Cereceda and Lundberg, 1987a). *In vivo* in man, CGRP caused local skin dilatation at femtomole doses (Brain *et al.*, 1985). CGRP infusion increased coronary flow in the isolated rabbit heart preparation (Marshall *et al.*, 1986), and vasodilation occurred during intracoronary infusion of CGRP in man

ODEON TCR

(0207) 636 6148 £13.50
(0208) 795 6400 £1.00

Pitch Black 9:35pm

0870 5050007

REPAIR/REQUEST SLIP

DATE..

ROOM NO..

REPAIR/REQUEST..

....

....

....

SIGNED...

(McEwan *et al.*, 1986). In two patients with variant angina, although both the normal artery and the site at which spasm occurred increased in diameter after intracoronary infusion of CGRP, the electrocardiographic changes, pain, and vasospasm induced by ergonovine were not prevented by the prior infusion of CGRP (McEwan *et al.*, 1986). Only at that site at which spasm occurred was there a profound decrease in the coronary artery diameter after ergonovine that resulted in subtotal occlusion of the vessel. In one patient a 1-μg intracoronary bolus of CGRP partially relieved the spasm and its associated features. The inability to prevent focal spasm induced by ergonovine suggests that CGRP is unlikely to be involved pathophysiologically in variant angina.

CGRP has direct inotropic and chronotropic actions on the isolated rat and guinea pig heart. Unlike substance P, CGRP also mimics the nonadrenergic, noncholinergic excitatory response induced by transmural nerve stimulation and capsaicin, suggesting that CGRP may be the neurotransmitter of cardioaccelatory nonadrenergic noncholinergic nerves (Saito *et al.*, 1986). Thus, the heart responds to sensory nerve activation by increasing the outflow of CGRP-LI. It has been suggested that coronary vasodilation induced by hypoxia in guinea pig heart (Wennmalm, 1979) mediated by PGI1 may therefore be due to CGRP release from C-fiber afferents. It has been suggested that CGRP may play a role in vagally induced tachycardia in the dog (Saito *et al.*, 1986), where vagal stimulation in the presence of atropine causes cardioacceleration (Alter *et al.*, 1973) as it has been demonstrated that CGRP colocalizes with substance P in vagal and spinal ganglia and that some nerves containing substance P in the heart are of vagal origin (Lundberg *et al.*, 1983). CGRP has also been shown to have a positive inotropic effect on the isolated human atrium (Franco-Cereceda *et al.*, 1987c).

IV. Tachykinins

The tachykinins are a group of biologically active peptides possessing a common C-terminal amino acid sequence and include substance P, neurokinin A (NKA), and neurokinin B (NKB), the latter two having been recently identified in the mammalian nervous system (Kimura *et al.*, 1983). Substance P occurs in primary sensory neurons that have peripheral branches associated with the cardiovascular system. These immunoreactive nerves are generally sensitive to capsaicin treatment and have been identified in the heart (Furness *et al.*, 1982). They may contain several bioactive peptides, comprising at least two tachykinins (substance P and NKA) and CGRP. The density of the substance P- or CGRP-immunostaining perivascular plexus is about half of that of the noradrenergic network (Dhall *et al.*, 1986). Immunoreactivity for both CGRP and substance P is present in nerve fibers in nerve trunks and around intrinsic neuronal cell bodies in cardiac ganglia, suggesting a possible interaction between afferent nerves and the autonomic innervation of the heart (Wharton *et al.*, 1990). Substance P-containing sensory nerve fibers synapse with postganglionic autonomic neurons in sympathetic ganglia (Matthews and Cuello, 1982; Matthews *et al.*, 1987). Thus substance P may play a role in cardiovascular control by functioning as a sensory modulator of synaptic transmission in autonomic ganglia (Cuello, 1987). This modulation of synaptic autonomic transmission may also occur with neurokinin A (Ogawa *et al.*, 1985). Substance P-like immunoreactivity is present in a small proportion of vascular endothelial cells (Milner *et al.*, 1989).

Three tachykinin receptors have been identified so far, named NK-1, NK-2, and NK-3. Substance P has the highest affinity for NK-1, NKA for NK-2, and NKB for NK-3 (Regoli *et al.*, 1988). Substance P may play a role in neuromodulation of baroreceptor reflex activity within the nucleus tractus solitarius (Reid and Rubin, 1987). Stimulation of spinal cord NK-1 receptors evokes pressor and positive chronotropic responses, resulting from both direct and indirect activation of sympathetic preganglionic neurons (Hassessian *et al.*, 1988; Hassessian and Couture, 1989). Plasma substance P can act on NK-1 receptors located on endothelial cells to produce vasodilatation through the release of an endothelium-derived relaxing factor (Regoli *et al.*, 1988). In contrast to CGRP, substance P does not mimic the actions of capsaicin on the heart and has no direct effect on the contractility of the mammalian heart. NKs can accelerate heart rate by releasing catecholamines through activation of NK-1 and/or NK-2 receptors on noradrenergic fibers and adrenal chromaffin cells. NKs can exert a depressor effect on the cardiovascular system by inducing a vagal reflex, following the activation of NK-3 receptors located on capsaicin-insensitive sensory fibers, or by promoting the release of prostaglandins from mast cells that in turn will stimulate capsaicin-sensitive sensory fibers.

In vivo, while substance P increases large epicardial vessel coronary caliber (Brum *et al.*, 1986b), both dilation and constriction of resistance coronary vessels have been reported (Kulakowski *et al.*, 1983; Losay *et al.*, 1977; Regoli *et al.*, 1977). In anesthetized animals, substance P increased coronary blood flow, but in conscious dogs intravenous substance P failed to increase myocardial perfusion (Brum *et al.*, 1986b; Ezra *et al.*, 1986; Maxwell, 1968; Regoli *et al.*, 1977; Couture and Regoli, 1982). Opposite effects on conductance and resistance coronary vessels have been demonstrated in conscious dogs (Nakamura *et al.*, 1990). In denervated hearts *in vivo,* substance P was shown to be a potent endothelium-dependent dilator of conductance coronary vessels whereas in resistance coronary vessels early transient dilations following bolus intravenous administration of substance P were followed by sustained constriction. Paradoxically epicardial coronary diameter increased in the face of the reductions in coronary blood flow. Endothelial denudation abolished increases in coronary diameter. When substance P was administered into the human coronary circulation, a dose-dependent increase in coronary epicardial caliber and coronary blood flow was seen (Crossman *et al,* 1989). The magnitude of the response was intermediate between that of adenosine and CGRP, suggesting a moderate dilating effect on small vessels. The effect of substance P is very short lived, having a biological half-life of 15 s. Its response exhibits marked tachyphylaxis, therefore making it unsuitable as a possible therapeutic agent. Substance P has also been reported to cause a reflex tachycardia (Delbro, 1987).

V. Vasoactive Intestinal Polypeptide

Vasoactive intestinal polypeptide (VIP) is a 28 amino acid peptide with structural similarities to secretin and glucagon and is a relaxant of vascular and nonvascular smooth muscle with potent vasodilatory effect (Said and Mutt, 1970). It is found in central and peripheral nerves. Another peptide, PHI-27 (peptide with N-terminal histidine and C-terminal isoleucine), has also been isolated (Tatemoto and Mutt, 1980; Tatemoto and Mutt, 1981) and found to have a distribution similar to that of VIP (Bishop *et al.*, 1984; Christofides *et al.*, 1984). The human form of this peptide, PHM (C-terminal methionine), is derived from the same prepro molecule as VIP (Bloom *et al.*, 1983; Itoh *et al.*, 1983). Both peptide sequences are present in postganglionic parasympathetic cardiovascular nerves. In man, the density of the VIP-immunoreactive perivascular nerves appears to be between that of nerves displaying NPY/CPON and CGRP or substance P immunoreactivities, whereas in the guinea pig it is the least dense of all these networks (Dhall *et al.*, 1986). VIP-containing nerves are present in the heart and coronary arteries (Brum *et al.*, 1986a; Della *et al.*, 1983; Rosselin *et al.*, 1982; Weihe and Reinecke, 1981). Relatively few cardiac nerve fibers display VIP-like material. VIP immunoreactivity in mammalian hearts, including human, is reported to occur mainly in nerve fibers associated with the atria, conducting system, and coronary vessels (Brum *et al.*, 1986b). The origin of VIP-containing nerves is uncertain, for while VIP immunoreactive ganglion cells have been described in the dog (Weihe *et al.*, 1984), they were not demonstrated in the human heart (Wharton *et al.*, 1990). VIP receptors, coupled to adenylate cyclase, are present in atrial and ventricular membrane preparations of the dog, monkey, and human heart (Chatelain *et al.*, 1983; Taton *et al.*, 1982). PHI receptors were also identified in the human preparations, VIP and PHI having a similar capacity to stimulate adenylate cyclase activity. VIP immunoreactivity is less prevalent than somatostatin immunoreactivity in man.

VIP has a direct vasodilator action *in vivo* and *in vitro* on systemic vessels from several species. Direct application of VIP relaxes coronary artery strips (Unverferth *et al.*, 1985; Itoh *et al.*, 1990) in the absence and presence of vascular endothelium. It is thought that VIP may mediate neurogenic vasodilatation in a number of vascular beds (Lundberg, 1981). Intravenous infusion of VIP in dogs (Smitherman *et al.*, 1982) and man (Smitherman *et al.*, 1989) is associated with coronary vasodilation. Intracoronary infusion of VIP in dogs increases the diameter of some of the epicardial vessels (Brum *et al.*, 1986b), and intracoronary infusion of VIP in man increases coronary blood flow and decreases coronary vascular resistance (Smitherman *et al.*, 1989) without altering heart rate. Myocardial oxygen uptake or left ventricular dP/dt were not significantly altered during intracoronary infusion, suggesting that VIP has no direct actions on myocardial contractility in man although VIP has been reported to exert positive chronotropic and inotropic effects in other mammalian studies (De Neef *et al.*, 1984). In anesthetized dogs, VIP concentration in the coronary sinus effluent did not increase significantly in response to hypoxia-mediated coronary vasodilation (Smitherman *et al.*, 1982). The mechanism of the coronary vasodilator action of VIP is uncertain. VIP-mediated dilation of pial arteries of cats may be mediated by prostaglandins (Wei *et al.*, 1980), but this may not be a major mechanism for VIP-mediated coronary vasodilatation in humans as, during intravenous VIP administration in man, there was no measurable increase in coronary sinus blood levels of prostaglandin metabolites nor did previous administration of a cyclooxygenase inhibitor reduce the vasodilator response (Smitherman *et al.*, 1989). In addition to VIP, PHI and PHM are also known to have vasoactive properties although they may be less potent than VIP. VIP has been implicated in shock and heart failure (Clark *et al.*, 1983).

VI. Endothelin

Endothelin (ET) is a member of a family of vasoconstrictor peptides derived from vascular endothelium that are encoded by three distinct genes, each present in human, rat, and porcine genome (Inoue *et al.*, 1989). These are named ET-1 (porcine endothelin), ET-2 (human endothelin), and ET-3 (rat endothelin), which have similar but not identical structures. A fourth form, mouse endothelin (VIC or vasoactive intestinal constrictor polypeptide) (Saida *et al.*, 1989), has recently been identified. The sequence of the human ET-1 gene has now been published (Bloch *et al.*, 1989). Only ET-1 appears to be produced by endothelial cells and been shown to be present in the circulating blood (Miauchi *et al.*, 1989; Yanagasawa and Masaki, 1989). ET-2 and ET-1 have a similar pharmacological profile. ET-3 (rat) is markedly different, being a weaker vasoconstrictor but a more potent inhibitor of platelet aggregation than ET-1. VIC is closer to ET-1 than ET-3 in terms of its chemistry. However, it is a weaker vasoconstrictor of porcine coronary artery although a more potent intestinal constrictor than ET-1 (Ishida *et al.*, 1989). The biosynthetic pathway of ET is proposed to be as follows: proET-1 is processed to big ET-1, which in turn is processed to ET-1. In support of this scheme is the fact that the vasoconstrictive potency of big ET-1 is much lower than that of ET-1 (Kashiwabara *et al.*, 1989), indicating that the conversion of big ET-1 to ET-1 appears to be essential for the action of ET-1 *in vivo,* with the conversion from big ET-1 to ET-1 taking place with the help of a putatively named "endothelin-converting enzyme." While endothelin-converting enzyme activity has been found in cultured bovine aortic endothelial cells (Ohnaka *et al.*, 1990), conversion did not appear to be dependent on the presence of an intact endothelium, suggesting that nonendothelial cells, such as vascular smooth muscle cells, may also have a role (Fukuroda *et al.*, 1990). Thus, it appears possible that circulating big ET-1 may be activated when it comes into contact with vascular smooth muscle.

Endothelin has structural homologies to the ion channel toxins of scorpion (Kopeyan *et al.*, 1958), bee (Habermann, 1972), and sarafotoxin, a peptide derived from snake venom (Kloog *et al.*, 1988). These families are linked to the same second-messenger system, raising the possibility that endothelin, like these neurotoxins, acts on membrane ion channels. Indeed, it was initially proposed that endothelin was the endogenous agonist of voltage-dependent calcium channels

(Yanagisawa *et al.*, 1988), but doubt has been cast on this by subsequent studies in some of which endothelin failed to influence specific binding of [^{3}H]-dihydropyridines to voltage-sensitive Ca^{2+} channels (Eglen *et al.*, 1989; Sharif *et al.*, 1989). However, given the large multisubunit structure of the Ca^{2+} channel complex (Catterall *et al.*, 1988), these results do not exclude the possibility that endothelin may interact with some component of the dihydropyridine-sensitive Ca^{2+}-channel complex.

Human endothelial cells have few secretory granules (Leak, 1980.), suggesting that endothelin is produced when required. Biosynthesis of endothelin in cultured endothelial cells is induced by various stimuli, including thrombin, adrenaline, transforming growth factor β (TGFβ) (Kurihara *et al.*, 1989), ionomycin, phorbol esters, and angiotensin II (Emori *et al.*, 1989). Induction may also transiently occur following fluid-mechanical shear stress and hypoxia. The expression of ET-1 mRNA by these agents and also arginine vasopressin, platelet-derived growth factor AA and epidermal growth factor has been shown to be accompanied by synthesis of endothelin peptide (Resink *et al.*, 1990). Many of these inducing agents are present in coronary vessels normally and in increased amounts during ischemic events. That they have the ability to induce production of a potent vasoconstrictor from the coronary endothelium may be particularly relevent to the pathophysiology of ischemic heart disease. Raised plasma concentrations of ET-1 have been observed in patients following myocardial infarction (Miauchi *et al.*, 1989; Salminen *et al.*, 1989), in cardiogenic shock (Cernacek *et al.*, 1989), and subarachnoid hemorrhage with vasospasm (Masaoka *et al.*, 1989). Circulating levels now being reported in health and disease are much lower than threshold levels required to cause constriction of isolated blood vessels (Yanagisawa *et al.*, 1988; Davenport *et al.*, 1988).

ET-1 mRNA is expressed in porcine aortic endothelial cells, all heart chambers, adrenals, and kidney, as well as in human cardiac ventricular tissue, lung, and cultured umbilical endothelial cells (Nunez *et al.*, 1990). Evidence is accumulating to support endothelin production by nonendothelial cells, including neurons of human spinal cord (Giaid *et al.*, 1989), as well as in cultured kidney cells (Kosaka *et al.*, 1989) and canine trachea (Black *et al.*, 1989).

Intracerebroventricular administration of endothelin-1 in rats produced a dose-related increase in blood pressure and heart rate thought to be mediated both sympathetically and through the action of vasopressin (Kawano *et al.*, 1989). In vascular tissue ET-1 may act locally by causing contraction of myocytes in the vascular wall (Yanigasawa *et al.*, 1988; Yanagisawa and Masaki, 1989). Endothelin-1 is a constrictor of human resistance vessels *in vitro*, being 1000-fold more potent than noradrenaline (Watt *et al.*, 1989). The maximum tension developed to endothelin in porcine coronary artery strips is comparable to that in KCl-induced contraction, and occurs at one order of magnitude less than the concentrations required for angiotensin, neuropeptide Y, or vasopressin (Yanigasawa *et al.*, 1988; Kasuya *et al.*, 1989a). The response is slow developing and long lasting (Kasuya *et al.*, 1989b), taking over 2 h to attain maximal contraction, and does not return to the initial level even after repeated washings (Kasuya *et al.*, 1989a). The effect is not modulated by antagonists of norepinephrine, acetylcholine, histamine, and serotonin or by inhibitors of cyclooxygenase or lipoxygenase in pig coronary artery (Kasuya *et al.*, 1989b) or in cultured rat vascular smooth muscle cells (Hirata *et al.*, 1988a). In organ bath studies the response to endothelin has been reported not to be affected by removal of vascular endothelium (Fukuroda *et al.*, 1990; Auguet *et al.*, 1988; Kasuya *et al.*, 1989b), but in isolated perfused rabbit heart the endothelium appears to have an ability to protect against the vasocontrictor action of circulating endothelin in the coronary bed (Stewart and Baffour, 1990). Threshold concentrations of ET-1 potentiate contractions to noradrenaline and serotonin in human internal mammary and left anterior descending coronary artery rings (Yang *et al.*, 1990).

In vivo, intravenous bolus injection of endothelin caused a transient fall followed by a markedly sustained rise in arterial pressure in anesthetized, chemically denervated rats (Yanagisawa *et al.*, 1988) and innervated dogs (Goetz *et al.*, 1988; Miller *et al.*, 1989). The pressor effects of endothelin were significantly buffered by neurohumoral mechanisms while changes in cardiac output were independent of neurohumoral control (Hinojosa-Laborde *et al.*, 1989). ET decreases microvascular blood flow in rabbit skin (Brain *et al.*, 1958). Vasodilation in response to endothelin has been reported in renal, mesenteric (De Nucci *et al.*, 1988), and systemic vascular beds (Wright and Fozard, 1988). The mechanism of the vasodilation is unknown (Yanigasawa *et al.*, 1988). However, it appears that endothelin may induce the production of prostacyclin and/or EDRF from endothelin (De Nucci *et al.*, 1988). Endothelin reduces canine coronary blood flow *in vivo* by acting on smaller coronary vessels, producing long-lasting myocardial ischemia and ultimately terminal ventricular fibrillation (Larkin *et al.*, 1989).

Endothelin is among the most potent positively inotropic agents yet described in mammalian myocardium. High-affinity receptors for endothelin have been identified by several groups in mammalian atria and ventricles (Power *et al.*, 1989; Miyazaki *et al.*, 1989; Hirata *et al.*, 1989a; Gu *et al.*, 1989; Galron *et al.*, 1989; Watanabe *et al.*, 1989). Atrial and ventricular trabeculae from failing *in vitro* human hearts obtained at transplant (Moravec *et al.*, 1989) displayed 74% and 9.9% increases in developed tension, respectively, with endothelin. Endothelin has a positive inotropic effect on isolated guinea pig atria (Ishikawa *et al.*, 1988) and the isolated rat heart (Firth *et al.*, 1990). Like its effect on vascular tissue, endothelin has a characteristic slow onset (4–6 min) and long duration of action of over 8 min, possibly due either to delayed dissociation from sarcolemmal receptors, a prolonged generation of an intracellular signal, or both (Kelly *et al.*, 1990). The effect is abolished by pretreatment with pertussis toxin, thus implicating a guanine nucleotide regulatory (G) protein(s) (Kelly *et al.*, 1990). Since endothelial-derived contracting factors have been reported to be released in response to stretch (Katusic *et al.*, 1987) and anoxia (Rubanyi *et al.*, 1985), both intrinsic to many cardiovascular disease states, the contractility of diseased myocardium may be affected by the release of factors such as endothelin from the endocardium or from the capillary endothelial cells. Endothelin-1 has a potent chronotropic effect in spontaneously beating cultured chick embryo ventricular cells (Concas *et al.*, 1989).

Autoradiographic studies have localized endothelin to the smooth muscle layer of the human coronary vasculature (Chester *et al.*, 1989). Increased endothelin binding is seen in human coronary artery harvested from transplant hearts of patients with myocardial ischemia. Patients with variant angina have elevated plasma and coronary sinus levels of endothelin-1, when compared to controls, even when no episode of vasoconstriction is taking place. No significant difference between coronary sinus plasma levels and peripheral venous plasma levels is observed in these patients (Toyo-oka *et al.*, 1991). During attacks provoked by either acetylcholine or ergonovine, coronary sinus levels fall. Patients who did not respond to provocation demonstrated no change in ET-1 levels during provocation, and ET-1 levels before and after provocation in this group were much lower than those in the spasm provoked cases (Toyo-oka *et al.*, 1991). These findings suggest that an increased plasma level of endothelin may predispose but might not be sufficient in itself to induce coronary spasm. The fall in endothelin levels during attacks may reflect a reduction in coronary blood flow during spasm with a consequent reduction in shear stress, a known inducer of endothelin secretion. Increased concentrations of plasma ET-1 have been reported in peripheral venous blood following cold pressor testing (Fyhrquist *et al.*, 1990), an established stimulus to coronary spasm (Gunther *et al.*, 1981; Gustafsson *et al.*, 1989), suggesting that rapid release of ET-1 may be part of the physiological response to cold exposure and may involve stimulation of the sympathetic nervous system (Green *et al.*, 1965).

In the human forearm in normal subjects, infusion of a fixed dose of endothelin reduces arterial flow by up to 70%. The reduction in blood flow takes 1 h to reach maximum and following cessation of the infusion takes 2 h to return to normal (Clarke *et al.*, 1989). It has been speculated that endothelin, because of its prolonged duration of action, may provide a background tone to the coronary and peripheral vasculature.

VII. Vasopressin

The antidiuretic action of vasopressin released from the anterior pituitary plays an important role in osmoregulation. Vasopressin is also a vasoconstrictor and its release can be induced by hypotension and hypovolemia mediated through the high pressure (carotid sinus) and the low pressure (left atrial) receptors (Baylis, 1987). The pressure responsive release mechanisms are influenced by circulating catecholamines through α- and β-adrenergic pathways. Release of vasopressin may be inhibited by atrial natriuretic peptide (Samson *et al.*, 1987) and stimulated by somatostatin (Brown 1988), both acting through the hypothalamus.

Comparative studies with other vasoconstrictors indicate that cardiac output is decreased by vasopressin but to a lesser extent than by other vasoconstrictors (Heyndrickx *et al.*, 1976; Hof, 1985), suggesting direct effects on the myocardium

(Wilson *et al.,* 1980; Dominguez-Mon *et al.,* 1984), rather than secondary to myocardial ischemia (Zito *et al.,* 1983), may contribute to the decrease in cardiac output seen (Hof 1985). Myocardial depression may also be due to the effects of vasopressin on autonomic reflexes that may affect the contractile function of the heart (Hof, 1986). The effects of vasopressin on the human coronary circulation have recently been examined. Following inhibition of vasopressin using a synthetic analog in patients with coronary atherosclerotic disease, no change in coronary or systemic vascular resistance was found (Glazier *et al.,* 1989). However, there is good clinical evidence that vasopressin constricts the coronary circulation (at least at high plasma levels) as arrythmias, myocardial ischemia, and infarction have been reported in patients receiving an intravenous infusion of vasopressin to treat bleeding esophageal varices (Kelly *et al.,* 1980; Conn, 1982).

VIII. Somatostatin

Somatostatin is present in mammalian sympathetic and sensory nerves (Hokfelt *et al,*. 1980). Studies on guinea pig heart show that it is localized within cardiac nerves, probably postganglionic fibers of the vagus (Day *et al.,* 1985). Somatostatin immunoreactive neuronal cell bodies have been found in the developing (Wharton *et al.,* 1988) and adult human heart, supporting the suggestion that somatostatin-containing nerves represent intrinsic postganglionic parasympathetic neurons (Franco-Cereceda *et al.,* 1986). Somatostatin in human heart is particularly abundant in the right atrium and atrioventricular node where its distribution in nerves is distinct from NPY, CGRP, or substance P immunoreactivity (Wharton *et al.,* 1990). While it is a cotransmitter with acetylcholine in cardiac nerves of lower animals (Campbell *et al.,* 1982; Campbell and Jackson, 1985) and applied somatostatin mediates cardioinhibition in some lower animals (Donald *et al.,* 1990; Diez *et al.,* 1985; Franco-Cereceda *et al.,* 1986; Quirion *et al.,* 1979a), a cotransmission with acetylcholine has not been demonstrated in mammalian heart where the inhibitory innervation is purely cholinergic. However, it has been implicated as the neurotransmitter of adrenergic cardioaccelatory nerves of the guinea pig right atrium (Saito *et al.,* 1986). Somatostatin may act as a modulator of cholinergic transmission since it was recently shown that it can induce acetylcholine release in the canine heart (Wiley *et al.,* 1989). Somatostatin can also act presynaptically in the gut to inhibit the release of acetylcholine from cholinergic nerves (Cohen *et al.,* 1978; Furness and Costa, 1979).

Bolus intravenous infusion in human volunteers increases systolic and diastolic systemic arterial pressure and reduces heart rate (Chermish *et al.,* 1979). When given to patients by intravenous bolus injection, somatostatin can revert supraventricular tachycardia to sinus rhythm with an efficacy similar to verapamil; this effect is probably due to its inhibitory action on cellular calcium uptake (Greco *et al.,* 1984). While verapamil consistently blocks conduction in the atrioventricular node to achieve its effect, somatostatin induces ventricular extrasystoles at the time of reversion (Webb *et al.,* 1986). Atrial contraction is inhibited by somatostatin in humans (Franco-Cereceda *et al.*, 1986) and other mammals (Diez *et al.*, 1985; Quirion *et al.*, 1979a). It has been reported to have both negative (Quirion *et al.*, 1979b) and positive effects on myocardial fiber contractility (Hirai *et al.*, 1989).

IX. Renin–Angiotensin System

Angiotensin II is a circulating vasoconstrictor peptide derived from angiotensin I by the activity of angiotensin-converting enzyme in the lung. Angiotensin I is the product of the action of renin, released by the kidney, on circulating angiotensinogen generated by the liver (Campbell, 1987). Cultured bovine aortic endothelial cells and vascular smooth muscle cells also contain the complete renin angiotensin system (Lilly *et al.*, 1985; Re *et al.*, 1981; Dzau, 1984). When plasma renin levels are elevated, systemic vascular resistance is increased by small vessel vasoconstriction causing hypertension. This action can be reversed by angiotensin-converting enzyme inhibitors. However, angiotensin-converting enzyme inhibition also induces a fall in blood pressure in various types of human hypertension when plasma renin activity is normal or even low (Waeber *et al.*, 1982). This effect is thought to result from inhibition of the local vascular renin angiotensin system.

Angiotensin II facilitates adrenergic function by potentiating the release (Zimmerman, 1972; Hughes *et al.*, 1971) and inhibiting the reuptake (Khairallah, 1972; Campbell *et al.*, 1979) of noradrenaline from sympathetic nerve terminals. Locally generated angiotensin II may potentiate the vasoconstrictor response to sympathetic nerve stimulation (Malik and Nasjletti, 1976). Angiotensin II may also stimulate the release of catecholamines from the adrenal medulla (Feurstein *et al.*, 1977a; Feurstein *et al.*, 1977b) and increase vascular responsiveness to noradrenaline (Struthers *et al.*, 1987; Ichikawa *et al.*, 1978; Day *et al.*, 1976). It has been shown to augment sympathetic vasoconstriction in forearm resistance vessels and veins in man (Benjamin *et al.*, 1989).

In addition to indirect effects on the autonomic nervous system, angiotensin II has direct vasoconstrictor actions (Reid, 1984). Angiotensin II may additionally influence cardiac function by increasing adrenergic input to the heart. By acting centrally it can increase sympathetic nerve activity (Ferrario *et al.*, 1976) and decrease vagal tone to the heart (Potter *et al.*, 1985; Lee *et al.*, 1980). High-affinity membrane angiotensin II receptor binding sites have been reported in several mammalian hearts (Baker *et al.*, 1984; Wright *et al.*, 1983; Mukherjee *et al.*, 1982; Rogers, 1984; Hirakata *et al.*, 1990), including human (Urata *et al.*, 1989). In humans these binding sites have been localized to the myocardium and cardiac adrenergic nerves (Urata *et al.*, 1989). Angiotensin II has positive inotropic and chronotropic effects (Peach, 1977; Goodfriend, 1986). In the isolated rabbit atria (Bonnardeaux *et al.*, 1977) and human atrial trabeculae (Bond *et al.*, 1989), angiotensin II produces an increase in the rate of tension development. Since angiotensin II retains its positive inotropic activity in the denervated heart (Dempsey *et al.*, 1971) and in the presence of β-adrenoceptor blockade (Kobayashi *et al.*, 1978), it is likely that most of the inotropic effect of angiotensin II is direct. It has recently been suggested that angiotensin I may also be positively inotropic; this may be important where high angiotensin I levels are found as occurs during angiotensin-converting enzyme inhibitor therapy (Hirakata *et al.*, 1990).

X. Cholecystokinin

The existence of cholecystokinin (CCK) in nerves is well established (Larsson and Rehfeld, 1979). Colocalization of CCK and neurotensin has been demonstrated in noradrenergic neurons in rat brain (Kawai *et al.*, 1988), but the presence of members of the CCK family in cardiac nerves is not documented. It has been reported that CCK-8 can exert beneficial effects when given to rats experiencing hypovolemia shock, and that these effects are due to the antiopoid actions of CCK-8 (Bertolini *et al.*, 1986). Intravenous injections of the octapeptide in rats consistently produce bradycardia disproportionate to its pressor effects (Marker and Roberts, 1988). The response to CCK-8 is mainly due to a direct action of the peptide on the heart via the classes of cholecystokinin receptors found in brain and exocrine pancreas. The changes in heart rate occurred with no obvious changes in ventricular contractility or perfusion pressure.

XI. Neurotensin

Neurotensin immunoreactive nerve fibers are found in human heart (Reinecke *et al.*, 1981). Neurotensin immunoreactivity appears to be located specifically in preganglionic parasympathetic fibers (Goedert *et al.*, 1984) from where it can be released into the circulation by intravenous bombesin (LeZoche *et al.*, 1979). Within the heart, neurotensin innervation is more pronounced in arteries, arterioles, and metaarterioles than in capillaries, venules, and veins. Neurotensin immunoreactive nerve fibers also occur unrelated to blood vessels in close contact with heart muscle cells. This nonvascular innervation was most pronounced in the atria. The sinoatrial and atrioventricular nodes also exhibited neurotensin immunoreactivity. Neurotensin immunoreactivity is present in higher concentration in the atria than in the ventricles and in higher concentrations on the right side of the heart (Ceconi *et al.*, 1989). Concentrations tend to decrease in the presence of heart failure. Neurotensin is regarded as a vasodilator although it has been reported to cause dose-dependent constriction of rat coronary vessels (Quirion *et al.*, 1979b). Atropine, phenoxybenzamine, or propanolol does not significantly alter the vasodilator effect of neurotensin *in vivo* (Carraway and Leeman, 1973), suggesting a lack of involvement of muscarinic or α- and β-adrenoceptors. However, degranulation of mast cells by pretreatment with compound 48/80 completely abolishes the hypotensive effect of neurotensin, suggesting a possible interaction with histaminergic mechanisms. Triphasic blood pressure responses (Oishi *et al.*, 1981) have also been reported, the first and third phases being through a histamine-mediated process, with the second pressor phase suggested to be mediated by catecholamines released from the adrenal medulla. Systemic administration in rat femoral artery *in vivo* causes a depressor, pressor, and depressor response whereas in the organ bath, neurotensin was without significant effect on isolated rat aortic rings with or without endothelium, suggesting that the hypotension was not due to a direct effect on vascular smooth muscle (Di Paola and Richelson, 1990). In contrast to this, it has also been reported that neurotensin evoked endothelium-dependent relaxations of isolated canine carotid arteries (D'Orleans Juste *et al.*, 1985). No effect of neurotensin was seen on heart rate *in vivo* (Di Paola and Richelson, 1990; Rigel, 1988) whereas previous reports suggest a positive chronotropic effect (Stene-Larsen and Helle, 1979). Application of neurotensin to isolated guinea pig or rat atria and whole hearts elicits a pronounced positive inotropic effect up to 30 times more potent than noradrenaline (Stene-Larsen and Helle, 1979), which is not affected by β-adrenoceptor blockade (Bachelard *et al.*, 1986). On the other hand, no effect on contractility was reported in isolated rabbit hearts and atria (Rosell *et al.*, 1976).

XII. Atrial Natriuretic Peptide

Atrial natriuretic factor (ANF) is present in the atria of the human heart in health but also in the ventricles in patients with heart failure (Edwards *et al.*, 1988). In patients it is released during supraventricular tachycardia and is responsible for the associated polyuria with natriuresis. In the laboratory it can be released by atrial pacing (Ellenbogen *et al.*, 1988) and by stimulation of supraventricular tachycardia (Tsai *et al.*, 1988); the peak plasma level occurs after approximately 30 min of tachycardia and there is a positive correlation with atrial pressure and dimensions. Intravenous bolus infusion in healthy volunteers reduces arterial blood pressure for several hours (Richards *et al.*, 1985; Tikkanen *et al.*, 1985). Intravenous infusion reduces blood pressure in normals when seated but not when supine (Cody *et al.*, 1986). In patients with heart failure the plasma level is elevated; intravenous infusion reduces atrial pressure and systemic vascular resistance and increases the cardiac index. Coronary sinus plasma levels of atrial natriuretic peptide level are from seven- to tenfold that in systemic venous blood (Hirata *et al.*, 1988b).

XIII. Bradykinin

Bradykinin is one of a number of kinins that are generated from kininogens by the action of kallikrein in plasma and rapidly inactivated by kininases (Mullins and Hudgens, 1987). They act as local hormones and therefore plasma levels are low. Bradykinin has direct excitatory effects on cardiac afferent nerves (Uchida and Murao, 1974). Bradykinin is known to influence many important parameters of endothelial cell metabolism and function. It raises intracellular calcium (Colden-Stanfield *et al.*, 1987), causes a transient hyperpolarization (Daut *et al.*, 1989), and evokes the release of EDRF (Cherry *et al.*, 1982). The kallikrein–kinin system is activated and bradykinin is produced in septicemia and endotoxic shock (Aasen *et al.*, 1983). Minute amounts in plasma are sufficient to reduce blood pressure and increase protein permeability (McConn *et al.*, 1983). The hypotension in endotoxic shock can be attenuated by administration of a bradykinin antagonist (Weipert *et al.*, 1988).

Depending on the species and anatomical origin of the blood vessel, application of bradykinin may elicit relaxation, contraction, or no response at all. It has been argued that at least part of the vasodilator response to bradykinin in the coronary vasculature is an indirect effect mediated by release of prostaglandins (PGs). However, in the isolated guinea pig heart, bradykinin released mainly PG12, and the release, but not the simultaneous coronary vasodilatation, could be blocked by pretreatment with indomethacin (Schror *et al.*, 1979). Bradykinin is produced in large amounts during myocardial ischemia (Kimura *et al.*, 1973). Forearm blood flow is increased by brachial arterial infusion in man (De Pasquale and Burch, 1966), and intracoronary infusion dilates coronary arteries (Nakano, 1965). Bradykinin relaxes canine coronary and other canine arteries by endothelium-dependent mechanisms (Cherry *et al.*, 1982). The sensitivity of coronary arteries to bradykinin is considerably higher than other vascular beds (Regoli and Barabe, 1980). In the isolated guinea pig heart, bradykinin-induced changes in coronary resistance were inhibited by oxyhemoglobin and were paralleled by respective changes in the release of nitric oxide and cGMP into the coronary effluent perfusate (Kelm and Schrader, 1990). However, relaxations in cats and rabbit do not seem to be endothelial dependent.

XIV. Transplant Denervation

Studying the denervated transplanted heart allows assessment of the contribution of extrinsic neural control to cardiac function. In health, cardiac tissues receive both efferent sympathetic and parasympathetic nerves and an afferent innervation (Wharton *et al.*, 1988). During heart–lung transplantation, the extrinsic nerves supplying the donor heart are sectioned, resulting in isolation of both efferent and afferent nerves from central influences (Beck *et al.*, 1969; Norvell and Lower, 1973; Moharty *et al.*, 1987). Nerves and ganglion cells are found in tissues from transplanted hearts at postmortem (Milan *et al.*, 1970). Little is known about the nature of these nerves. It appears that in humans, in contrast to experimental animals, functional reinnervation is not a feature of long-term human transplant patients (Horak, 1984). Pronounced changes are observed in the innervation of cardiac tissues after cardiac transplantation. There are fewer nerves in general and a loss of peptide-containing nerves in particular (Wharton *et al.*, 1990). NPY-staining nerves are not found in cardiac allografts, indicating that in man they are of extrinsic origin (Wharton *et al.*, 1990). In comparison with other animals, such as the guinea pig, nerve fibers containing either CGRP or substance P immunoreactivity occur relatively infrequently in the innervated human heart. In cardiac allografts there is an apparent absence of calcitonin gene-related peptide and substance P immunoreactivity (Wharton *et al.*, 1990) consistent with an extrinsic origin for these nerves, presumably representing primary sensory vagal or spinal neurons. Somatostatin-immunostaining fibers, while thought to represent intrinsic postganglionic parasympathetic neurons, are not found in allograft tissue, again suggesting a possible extrinsic origin, but this may equally be explained by the absence of a functional preganglionic innervation results in the downregulation of peptide expression in postganglionic cardiac neurons. The distribution of atrial natriuretic peptide immunoreactivity was comparable in cardiac allografts with that found in the failing innervated hearts (Wharton *et al.*, 1988). This indicates that the capacity of myocardial cells to produce ANP is not dependent on extrinsic autonomic innervation. While basal plasma levels are raised in cardiac transplant patients, the atrial natriuretic peptide secretory and hemodynamic responses to lower body positive pressure are not impaired by extrinsic cardiac denervation (Wilkins *et al.*, 1988).

XV. Growth Factor Activity

The paracrine regulation of smooth muscle cell function and growth by factors either derived from circulating blood platelets or synthesized by endothelial cells is well recognized (Ross, 1986). Smooth muscle cells themselves also secrete a variety of substances that can modulate their function and growth. These compounds include both normal extracellular matrix constituents such as heparin and thrombospondin (Herman and Castellot, 1987; Majack *et al.*, 1985; Scott-Burden and Buhler, 1988), and peptide growth factors such as PDGF-AA (Majesky *et al.*, 1988). Vascular endothelial cell injury plays an important role in inducing atherosclerosis (Ross, 1986). There is much interaction between growth factors. Angiotensin II can increase levels of transcripts for TGF-β and platelet-derived growth factor (PDGF)-A chain, and PDGF-AA homodimer and TGF-β can also increase the levels of both their own and each other's transcripts. ET, like angiotensin II, can also increase PDGF-A chain and TGF-β transcript levels in cultured vascular smooth muscle cells from both rat and human vessels (Resink *et al.*, 1990). In turn, angiotensin II increases expression of ET mRNA in human smooth muscle cells, and human vascular smooth muscle cells secrete ET peptide in response to angiotensin II and arginine vasopressin (Resink *et al.*, 1990).

When adult vascular smooth muscle cells are passaged in culture, they undergo a process of phenotypic modulation from the contractile to the proliferative/secretory form (Chamley-Campbell *et al.*, 1979). This proliferative/secretory form can synthesize and secrete not only extracellular matrix macromolecules but also growth factors such as PDGF-AA and TGF-β (Majesky *et al.*, 1988; Sejersen *et al.*, 1986). Vascular smooth muscle cells *in vivo* are also capable of undergoing reversible phenotypic conversion, a property that may be important in the pathogenesis of hypertension and atherosclerosis (Ross, 1986; Chamley-Campbell *et al.*, 1979). Compounds that have been implicated in both the development and maintenance of the pathological proliferative/secretory smooth muscle phenotype include angiotensin II, PDGF-AA, and TGF-β (Ross, 1986; Majesky *et al.*, 1988; Chamley-Campbell *et al.*, 1979; Sejersen *et al.*, 1986; Powell *et al.*, 1989).

Several different factors have been shown to stimulate angiogenesis and endothelial cell proliferation. These include polypeptide growth factors such as fibroblast growth factor (FGF), endothelial cell growth factor, transforming growth factors α and β, angiogenin, and tumor necrosis factor (Folkman and Klagsbrun, 1987). These factors may be released locally by the endothelium or by cells adjacent to the endothelium, or may circulate in plasma. Peripheral neurons may have trophic effects. In the newt, limb regeneration is prevented by deinnervation (Singer, 1952). Intact sensory innervation is important in wound healing. The sensory peptides neurokinin A and substance P have been shown to induce proliferation of human fibroblasts and rat smooth muscle cells (Nilsson *et al.*, 1985). Vasoactive intestinal polypeptide (VIP) has been shown to inhibit serum-induced proliferation of rat smooth muscle cells and to stimulate proliferation of human keratinocytes (Haegerstrand *et al.*, 1989). Human umbilical vein endothelial cell proliferation was stimulated by CGRP but not by NKA, NPY, or VIP (Haegerstrand *et al.*, 1990). During wound healing, sprouting of sensory nerve fibers in relation to blood vessels toward the wound surface has been described (Hermansson *et al.*, 1987), suggesting a role for these nerves in angiogenesis, including formation of new vessels in ischemia, inflammation, and wound healing (Haegerstrand *et al.*, 1990).

XVI. Conclusion

Studying the role of peptides in disease processes is complex, as peptides may subserve more than one function and may be influenced by hormonal and neurologic factors. Also, animal models of human disease have serious limitations, human coronary tissue is not readily available for *in vitro* studies, and the scope of *in vivo* human studies on patients is limited. While our understanding of peptide actions has improved, to date the only widely used therapeutic agents to result from this knowledge are angiotensin-converting enzyme inhibitors. It has become clear that abnormal endothelial function may exaggerate coronary vasoconstrictor responses to vasoactive agents in disease states (Ludmer *et al.*, 1986; McFadden *et al.*, 1991), and further understanding of endothelial behavior may translate into therapeutic benefits.

References

Aasen, A. O., Smith-Erichsen, N., Amundsen, E. (1983). Plasma kallikrein-kinin system in septicaemia. *Arch. Surg. 118,* 343–345.

Allen, Y. S., Adrian, T. E., Allen, J. M., Tatemoto, K., Crow, T. I., Bloom, S. R., and Polak, J. M. (1983a). Neuropeptide Y distribution in the rat brain. *Science 221,* 877–879.

Allen, J. M., Bircham, P. M. M., Edwards, A. V., Tatemoto, K., Bloom, S. R. (1983b). Neuropeptide Y (NPY) reduces myocardial perfusion and inhibits the force of contraction of the isolated perfused rabbit heart. *Regul. Peptides 6,* 247–253.

Allen, J. M., Bircham, P. M. M., Bloom, S. R., and Edwards, A. V. (1984). Release of neuropeptide Y in response to splanchnic nerve stimulation in the conscious calf. *J. Physiol. 357,* 401–408.

Alter, W. A., Weiss, G. K., and Priola, D. V. (1973). Vagally induced tachycardia in atropinized dogs: Effect of beta-adrenergic blockade. *Eur. J. Pharmacol. 187,* 99–104.

Auguet, M., Delaflotte, S., Chabrier, P., Pirotzky, E., Clostre, F., and Braquet, P. (1988). Endothelin

and Ca^{++} agonist BAY 8644: Different vasoconstrictive properties. *Biochem. Biophys. Res. Comms. 156,* 312–317.
Bachelard, H., St-Pierre, S., and Rioux, F. (1986). The coronary vasodilator effect of neurotensin in the guinea pig isolated heart. *Peptides 7,* 431–435.
Baker, K. M., Campanlie, C. P., Trachte, G. J., and Peach, M. J. (1984). Identification and characterization of the rabbit angiotensin II myocardial receptor. *Circ. Res. 54,* 286–293.
Baylis, P. H. (1987). Osmoregulation and control of vasopressin secretion in healthy humans. *Am. J. Physiol. 253,* R671–R678.
Beck, W., Barnard, C. N., and Schrire, V. (1969). Heart rate after cardiac transplantation. *Circulation 40,* 437–445.
Benjamin, N., Cockcroft, J. R., Collier, J. G., Dollery, C. T., Ritter, J. M., Webb, D. J. (1989). Local inhibition of converting enzyme and vascular responses to angiotensin and bradykinin in the human forearm. *J. Physiol. 412,* 543–555.
Bertolini, A., Guarini, S., Ferrari, W., and Rompianesi, E. (1986). Caerulein and cholecystokinin reverse experimental haemorrhagic shock. *Neuropeptides 8,* 25–31.
Bishop, A. E., Polak, J. M., Yiangou, Y., Christofides, N. D., and Bloom, S. R. (1984). The distribution of PHI and VIP in porcine gut and their co-localisation to a proportion of intrinsic ganglion cells. *Peptides 5,* 255–259.
Black, P. N., Ghatei, M. A., and Takahashi, K. (1989). Formation of endothelin by cultured airway epithelial cells. *FEBS Letts. 255,* 129–132.
Bloch, K. D., Friedrich, P., Lee, M. E., Eddy, R. L., Shows, T. B., and Quertermous, T. (1989). Structural organization and chromosomal assignment of the gene encoding endothelin. *J. Biol. Chem. 86,* 7285–7289.
Bloom, S. R., Christofides, N. D., Delamarter, J., Buell, G., Kawashima, E., Polak, J. M. (1983). Tumour co-production of VIP and PHI explained by single coding gene. *Lancet 2,* 1163–1165.
Bond, M., Moravec, C. S., Stewart, R. W., and Healy, B. P. (1989). Action of angiotensin II on human cardiac muscle *in vitro*. *FASEB J. 3* (part 1), A880.
Bonnardeaux, J. L., Park, W. K., and Regoli, D. (1977). Effects of angiotensins and catecholamines on the transmombrane potential and isometric force of rabbit isolated atria. *Arch. Int. Pharmacodyn. 229,* 83–94.
Brain, S. D., Williams, T. J., Tippins, J. R., Morris, H. R., and MacIntyre, I. (1985). Calcitonin gene-related peptide is a potent vasodilator. *Nature 313,* 54–56.
Brain, S. D., Crossman, D. C., Buckley, T. L., Williams, T. J. (1989). Endothelin-1: Demonstration of potent effects on the microcirculation of humans and other species. *J. Cardiovasc. Pharmacol. 13*(suppl. 5), S147–S149.
Brown, M. R. (1988). Somatostatin-28 effects on central nervous system regulation of vasopressin secretion and blood pressure. *Neuroendocrinology 47,* 556–562.
Brum, J. M., Go, V. L. W., Sufan, Q., Lane, G., Reilly, W., and Bove, A. A. (1986a). Substance P distribution and effects in the canine epicardial coronary arteries. *Regul. Pept. 14,* 41–55.
Brum, J. M., Bove, A. A., Sufan, Q., Reilly, Q., and Go, V. L. W. (1986b). Action and localization of vasoactive intestinal peptide in the coronary circulation: Evidence for nonadrenergic, noncholinergic coronary regulation. *J. Am. Coll. Cardiol. 7,* 406–413.
Campbell, D. J. (1987). Tissue renin-angiotensin system: Sites of angiotensin formation. *J. Cardiovasc. Pharmacol. 10*(suppl. 7), S1–S8.
Campbell, G., and Jackson, F. (1985). Independent co-release of acetylcholine and somatostatin from cardiac vagal neurones in toad. *Neurosci. Lett. 60,* 47–50.
Campbell, G., Gibbins, I. L., Morris, J. L., Furness, J. B., Costa, M., Oliver, J. R., Beardsley, A. M., and Murphy, R. (1982). Somatostatin is contained in and released from cholinergic nerves in the heart of the toad Bufo marinus. *Neuroscience 7,* 2013–2023.
Campbell, W. B., and Jackson, E. K. (1979). Modulation of adrenergic transmission by angiotensins in the perfused rat mesentry. *Am. J. Physiol. 236,* H211–H217.

Carraway, R. E., and Leeman, S. E. (1973). The isolation of a new hypotensive peptide, neurotensin, from bovine hypothalami. *J. Biol. Chem. 248,* 6854–6861.

Catterall, W. A., Seagor, M. J., and Takahashi, M. J. (1988). Molecular properties of dihydropyridine-sensitive calcium channels in skeletal muscle. *J. Biol. Chem. 263,* 3535–3538.

Ceconi, C., Condorelli, E., Quinzanini, M., Rodella, A., Ferrari, R., and Harris, P. (1989). Noradrenaline, atrial natriuretic peptide, bombesin and neurotensin in myocardium and blood or rats in congestive heart failure. *Cardiovasc. Res. 23,* 674–682.

Cernacek, P., and Stewart, D. J. (1989). Immunoreactive endothelin in human plasma: Marked elevations in patients with cardiogenic shock. *Biochem. Biophys. Res. Comm. 161,* 562–567.

Chamley-Campbell, J. H., Campbell, G. R., and Ross, R. (1979). *Physiol. Rev. 59,* 1–55.

Chatelain, P., Robberecht, P., Waelbroeck, M., DeNeef, P., Camus, J. C., Naguyen-Huu, A., Roba, J. and Chrostophe, J. (1983). Topographical distribution of the secretion- and VIP-stimulated adenylate cyclase system in the heart of five animal species. *Pflugers Arch. 397,* 100–105.

Chen, X., Henderson, K., Beinfeld, M. C., and Westfall, T. C. (1988). Alteration in blood pressure of normotensive and hypertensive rats following intrathecal injection of neuropeptide Y. *J. Cardiovasc. Pharmacol. 12,* 473–478.

Chermish, S. M., Miller, R. E., Rosenak, B. D., and Brunelle, R. L. (1979). The effect of D-Val', D-Tryp somatostatin on the hemopoietic, cardiovascular systems and gastrointestinal tract. *Gastroenterology 76,* 113.

Cherry, P. D., Furchgott, R. F., Zawadzki, J. V., and Jothianandan, D. (1982). Role of endothelial cells in relaxation of isolated arteries by bradykinin. *Proc. Natl. Acad. Sci. U.S.A. 79,* 2106–2110.

Chester, A. H., Dashwood, M. R., Clarke, J. G., Larkin, S. W., Tadjkarimi, S., Davies, G. J., Maseri, A., and Yacoub, M. (1989). Influence of endothelin on human coronary arteries and localization of its binding sites. *Am. J. Cardiol. 63,* 1395–1398.

Christofides, N. D., Polak, J. M., and Bloom, S. R. (1984). Studies on the distribution of PHI in mammals. *Peptides 5,* 261–266.

Clark, A. J. L., Adrian, T. E., McMichael, J. B., and Bloom, S. R. (1983). Vasoactive intestinal polypeptide in shock and heart failure. *Lancet 1,* 539.

Clarke, J., Larkin, S., Osinawa, O., Davies, G. J., Taylor, K., and Maseri, A. (1987a). Neuropeptide Y reduces dog coronary blood flow by increased small vessel resistance. *Clin. Sci. 73*(17) (Abs).

Clarke, J. G., Davies, G. J., Kerwin, R., Hackett, D., Larkin, S., Lee, Y., Dawbarn, D., Bloom, S. R., Yacoub, M., and Maseri, A. (1987b). Coronary artery infusion of neuropeptide Y in patients with angina pectoris. *Lancet 1,* 1057–1059.

Clarke, J. G., Benjamin, N., Larkin, S. W., Webb, D. J., Davies, G. J., and Maweri, A. (1989). Endothelin is a potent long-lasting vasoconstrictor in man. *Am. J. Physiol. 257,* H2033–2035.

Clarke, J. G., Benjamin, N., Larkin, S. W., Maseri, A., and Davies, G. J. (1991). Interaction of neuropeptide Y and the sympathetic nervous system in vascular control in man. *Circulation 83,* 774–777.

Cody, R. J., Covit, A. B., Schaer, G. L., Laragh, J. H., Sealey, J. E., and Feldschurh, J. (1986). Sodium and water balance in chronic congestive heart failure. *J. Clin. Invest. 77,* 1441–1452.

Cohen, M. J., Rosing, E., Wiley, K. S., and Slater, I. H. (1978). Somatostatin inhibits adrenergic and cholinergic neurotransmission in smooth muscles. *Life Sci. 23,* 1659–1664.

Colden-Stanfield, M., Schilling, W. P., Ritchie, A. K., Eskin, S. G., Navarro, L. T., and Kunze, D. L. (1987). Bradykinin-induced increases in cytosolic calcium and ionic currents in cultured bovine aortic endothelial cells. *Circ. Res. 61,* 632–640.

Concas, V., Laurent, S., Brisac, A. M., Perret, C., and Safar, M. (1989). Endothelin has potent direct inotropic and chronotropic effects in cultured heart cells. *J. Hypertension 7*(suppl. 6), S96–S97.

Conn, H. D. (1982). A plethora of therapies. *In* "Variceal bleeding." (D. Westaby, B. MacDougall and R. Williams, eds.) pp. 221–252. Pitman Medical, London.

Coupe, M. O., Mak, J. C. W., Yacoub, M., Oldershaw, P. J., and Barnes, P. J. (1990). Autoradiographic mapping of calcitonin gene-related peptide receptors in human and guinea pig hearts. *Circulation 81,* 741–747.

Couture, R., and Regoli, D. (1982). Mini review: smooth muscle pharmacology of substance P. *Pharmacology 24,* 1–25.

Crossman, D. C., Larkin, S., Fuller, R. W., Davies, G. J., and Maseri, A. (1989). Substance P dilates epicardial coronary arteries and increases coronary blood flow in humans. *Circulation 80,* 475–484.

Cuello, A. C. (1987). Peptides as neuromodulators in primary sensory neurons. *Neuropharmacology 26,* 971–979.

Daly, R. N., and Hieble, J. P. (1987). Neuropeptide Y modulates adrenergic neurotransmission by an endothelium dependent mechanism. *Eur. J. Pharmacol. 138,* 445–446.

Daut, J., Dischner, A., and Mehrke, G. (1989). Bradykinin induces a transient hyperpolarization of cultured guinea-pig coronary endothelial cells. *J. Physiol. 410,* 48(abs).

Davenport, A. P., Nunez, D. J., Hall, J. A., Kaumann, A. J., and Brown, M. J. (1988). Autoradiographical localization of binding sites for porcine [^{125}I]endothelin-1 in humans, pigs, and rats: Functional relevance in humans. *J. Cardiovasc. Pharmacol. 13*(suppl. 5), S166–S170.

Day, M. D., and Moore, A. F. (1976). Interaction of angiotensin II with noradrenaline and other spasmogens on rabbit isolated aortic strips. *Arch. Int. Pharmacodyn. 219,* 29–44.

Day, S. M., Gu, J., Polak, J. M., and Bloom, S. R. (1985). Somatostatin in human heart and comparison with guinea pig and rat heart. *Br. Heart J. 53,* 153–157.

Delbro, D. (1987). Substance P induced tachycardia in the anaesthetized rat. *Acta. Physiol. Scand. 130,* 41–45.

Della, N. G., Papka, R. E., Furness, J. B., and Costa, M. (1983). Vasoactive intestinal peptide-like immunoreactivity in nerves associated with the cardiovascular system of guinea pigs. *Neuroscience 9,* 605–619.

Dempsey, P. J., McCallum, Z. T., Kent, K. M., and Cooper, T. (1971). Direct myocardial effects of angiotensin II. *Am. J. Physiol. 220,* 477–481.

De Neef, P., Robberecht, P., Chatelain, P., Waelbroeck, M., and Chrostophe, J. (1984). The *in vitro* chronotropic and inotropic effects of vasoactive intestinal peptide (VIP) on the atria and ventricular papillary muscle from cynomolgus monkey heart. *Reg. Peptides 8,* 237–244.

De Nucci, G. Thomas, R., D'Orleans-Juste, P., Antunes, E., Walder, C., Warner, T. D., and Vane, J. R. (1988). Pressor effects of circulating endothelin are limited by its removal in the pulmonary circulation and by the release of prostacyclin and endothelium-derived relaxing factor. *Proc. Natl. Acad. Sci. U.S.A. 85,* 9797–9800.

De Pasquale, N. P., and Burch, G. E. (1966). Digital vascular responses to intraarterial injections of bradykinin, kallidin and eledoisin in man. *Circulation 34* 211–217.

Dhall, U., Cowen, T., Haven, A. J., and Burnstock, G. (1986). Perivascular noradrenergic and peptide-containing nerves show different patterns of changes during development and ageing in the guinea pig. *J. Auton. Nerv. Syst. 16,* 109–126.

Diez, J., Tamargo, J., and Valenzuela, C. (1985). Negative inotropic effect of somatostatin in guinea pig atrial fibres. *Br. J. Pharmacol. 86,* 547–555.

Di Paola, E. D., and Richelson, E. (1990). Cardiovascular effects of neurotensin and some analogues on rats. *Eur. J. Pharmacol. 175,* 279–283.

Dominguez-Mon, M., Ponce-Hornos, J. E., Gomez, R., Cannata, M., and Taquini, A. C. (1984). Energetic, metabolic and contractile effects of vasopressin in mammalian heart. *Meth. Find Expt. Clin. Pharmac. 6,* 373–378.

Donald, J. A., O'Shea, J. E., and Lillywhite, H. B. (1990). Somatostatin and innervation of the heart of the snake Elaphe obsoleta. *Am. J. Physiol. 258,* R1001–R1007.

D'Orleans Juste, P., Dion, S., Mizrahi, J., and Reegoli, D. (1985). Effects of peptides and non peptides on isolated arterial smooth muscles: Role of endothelium. *Eur. J. Pharmacol. 114,* 9.

Dzau, V. J. (1984). Vascular renin-angiotensin: A possible autocrine or paracrine system in control of vascular function. *J. Cardiovasc. Pharmacol. 6,* S377–S382.

Edbrooke, M. R., Parker, D., McNey, J. M., Riley, J. H., Sorenson, G. D., Pettengill, O. S., and Craig, R. K. (1985). Expression of the human calcitonin/CGRP gene in lung and thyroid carcinoma. *EMBO J. 4,* 715.

Edvinsson, L., Fredholm, B. B., Hamel, E., Hansen, I., and Verrecchia, C. (1985). Perivascular peptides relax cerebral arteries concomitant with stimulation of cyclic adenosine monophosphate accumulation or release of an endothelium derived relaxing factor in the cat. *Neurosci. Lett. 58,* 213–217.

Edwards, B. S., Ackermann, D. M., Lee, M. E., Reeder, G. S., Wold, L. E., and Burnett, J. C. (1988). Identification of atrial natriuretic factor within ventricular tissue in hamster and humans with congestive cardiac failure. *J. Clin. Invest. 81,* 82–86.

Eglen, R. M., Michel, A. D., Sharif, N. A., Swank, S. R., and Whiting, R. L. (1989). The pharmacological properties of the peptide, endothelin. *Br. J. Pharmacol. 97,* 1297–1307.

Ellenbogen, K. A., Mohanty, P. K., Sowers, J. R., Walsh, M., and Thames, M. D. (1988). Atrial natriuretic factor release is enhanced by incremental atrial pacing. *Am. Heart. J. 116,* 489–496.

Emori, Y., Hirata, Y., Ohta, K., Shichiri, M., and Marumo, F. (1989). Secretory mechanisms of immunoreactive endothelin in cultured bovine endothelial cells. *Biochem. Biophys. Res. Comm. 160,* 93–100.

Ezra, D., Laurindo, F. R. M., Eimerl, J., Goldstein, R. E., Peck, C. C., and Feurstein, G. (1986). Tachykinin modulation of coronary blood flow. *Eur. J. Pharmacol. 122,* 135–138.

Ferrario, C. M., McCubbin, J. W., and Berti, G. (1976). Centrally mediated hemodynamic effects of angiotensin. *In* G. Onesti, M. Fernandes, K. E. Kim (eds.), Regulation of blood pressure by the central nervous system. New York, Grune and Stratton Inc. pp. 175–182.

Feurstein, G., Boonyaviroj, P., and Gutman, Y. (1977a). Renin-angiotensin mediation of adrenal catecholamine secretion induced by haemorrhage. *Eur. J. Pharmacol. 44,* 131–142.

Feurstein, G., Boonyaviroj, P., Gutman, Y., Khosla, M. C., and Bumpus, F. M. (1977b). Adrenal catecholamine response to haemorrhage abolished by an angiotensin antagonist. *Eu. J. Pharmacol. 41,* 85–86.

Firth, J. D., Roberts, A. F. C., and Raine, A. E. G. (1990). Effect of endothelin on the function of the isolated perfused working rat heart. *Clin. Sci. 79,* 221–226.

Fisher, L. A., Kakkawa, D. O., Rivier, J. E., Amara, S. B., Evans, R. M., Rosenfeld, M. G., Vale, W. W., and Brown, M. R. (1983). Stimulation of noradrenergic sympathetic outflow by calcitonin gene-related peptide. *Nature 305,* 534–536.

Folkman, J., and Klagsbrun, M. (1987). Angiogenic factors. *Science 235,* 442–447.

Franco-Cereceda, A., Lundberg, J. M., and Hokfelt, T. (1986). Somatostatin: An inhibitory parasympathetic transmitter in the human heart. *Eur. J. Pharmacol. 132,* 101–102.

Franco-Cereceda, A., and Lundberg, J. M. (1987a). Potent effects of neuropeptide Y and calcitonin gene-related peptide on human coronary vascular tone *in vitro. Acta. Physiol. Scand. 131,* 159–160.

Franco-Cereceda, A., Henke, H., Lundberg, J. M., Petermann, J. B., Hokfelt, T., and Fischer, H. A. (1987b). Calcitonin gene-related peptide (CGRP) in capsaicin-sensitive substance P-immunoreactive sensory neurons in animals and in a man: Distribution and release by capsaicin. *Peptides 8,* 399–410.

Franco-Cereceda, A., Bengtsson, L., and Lundberg, J. M. (1987c). Inotropic effects of calcitonin gene-related peptide, vasoactive intestinal peptide and somatostatin on the human right atrium *in vitro. Eur. J. Pharmacol. 134,* 69–76.

Fukuroda, T., Noguchi, K., Tsuchida, S., Nishikibe, M., Ikemoto, F., Okada, K., and Yano, M. (1990). Inhibition of biological actions of big endothelin-1 by phosphoramidon. *Biochem. Biophys. Res. Comm. 172,* 390–395.

Furchgott, R. F. (1983). Role of endothelium in responses of vascular smooth muscle. *Circ. Res. 64,* 557–573.

Furchgott, R. E., and Zawadski, J. V. (1980). The obligatory role of endothelial cells in the relaxation of arterial smooth muscle by acetylcholine. *Nature 288,* 373–376.

Furness, J. B., and Costa, J. (1979). Actions of somatostatin on excitatory and inhibitory nerves in the intestine. *Eur. J. Pharmacol. 56,* 69–74.

Furness, J. B., Papka, R. E., Della, N. G., Costa, M., and Eskay, R. L. (1982). Substance-P like immunoreactivity in nerves associated with the vascular system of guinea-pigs. *Neuroscience 7,* 447–459.

Fuxe, K., Agnati, L. F., Harfstrand, A., Zini, I., Tatemoto, K., Pick, E. M., Hokfelt, T., Mutt, V., and Terenius, L. (1983). Central administration of neuropeptide Y induces hypotension, bradypnea and EEG synchronization in the rat. *Acta Physiol. Scand. 118,* 189–192.

Fyhrquist, F., Saijonmaa, O., Metsarinne, K., Tikkanen, K., Rosenlof, K., and Tikkanen, T. (1990). Raised plasma endothelin-1 concentration following cold pressor test. *Biochem. Biophys. Res. Comm. 169,* 217–221.

Galron, R., Kloog, Y., Bdolah, A., and Sokolovsky, M. (1989). Functional endothelin/sarafotoxin receptors in rat heart myocytes: Structure–activity relationships and receptor subtypes. *Biochem. Biophys. Res. Comm. 163,* 936–943.

Giaid, A., Gibson, S. J., Ibrahim, N. B. N., Legon, S., Bloom, S. R., Yanigasawa, M., Masaki, T., Varndell, I. M., and Polak, J. M. (1989). Endothelin-1, an endothelially derived peptide is expressed in neurons of the human spinal cord and dorsal root gangion. *Proc. Natl. Acad. Sci. U.S.A. 86,* 7634–7638.

Girgis, S. I., Macdonald, D. W. R., Stevenson, J. C., Beris, P. J. R., Lynch, C., Wumalawansa, S. T., Self, C. M., Morris, H. R., and MacIntyre, I. (1986). Calcitonin gene-related peptide: Potent vasodilator and major produce of the calcitonin gene. *Lancet. 2,* 14.

Glazier, J. J., Gavras, H., Mills, R. M., Ruocco, N. A., Bresnahan, M., and Ryan, T. J. (1989). Effects of inhibition of arginine vasopresin on systemic and coronary haemodynamics. *J. Am. Coll. Cardiol. 13,* 172A.

Goedert, M., Sturmey, N., Williams, B. J., and Emson, P. C. (1984). The comparative distribution of xeonopsin- and neurotensin-like immunoreactivity in Xenopus Reevis and rat tissues. *Brain Res. 308,* 273–280.

Goetz, K. L., Wang, B. C., Madwed, J. B., Zhu, J. L., and Leadley, R. J. (1988). Cardiovascular, renal, and endocrine responses to intravenous endothelin in conscious dogs. *Am. J. Physiol. 255,* R1064–R1068.

Goodfriend, T. L. (1986). Physiological effects of angiotensins on blood vessels and the heart, *In* "Handbook of Hypertension." (A. Zanchetti and R. C. Tarazi eds.), vol. 8, pp. 398–420 Elsevier Science Publishing Co., Inc. New York.

Greco, A. V., Ghirlanda, G., Barone, C., Bertoli, A., Caputo, S., Uccioli, L., and Manna, R. (1984). Somatostatin in paroxysmal supraventricular and junctional tachycardia. *Br. Med. J. 288,* 28.

Green, M. A., Boltax, A. J., Lustig, G. A., and Rogow, E. (1965). *Am. J. Cardiol. 16,* 54–60.

Greenberg, B., Rhoden, K., and Barnes, P. (1987). Calcitonin gene-related peptide (CGRP) is a potent non-endothelium-dependent inhibitor of coronary vasomotor tone. *Br. J. Pharmac. 92,* 789–794.

Gu, J., Polak, J. M., Allen, J. M., Huang, W. M., Sheppard, M. N., Bloom, S. R., and Tatemoto, K. (1984). High concentrations of a novel peptide, neuropeptide Y, in the innervation of mouse and rat heart. *J. Histochem. Biochem. 32*(5) 467–472.

Gu, X. H., Casley, D. J., and Nayler, W. G. (1989). Characterization of [^{125}I]endothelin-1 binding sites in rat cardiac membrane fragments. *J. Cardiovasc. Pharmacol. 13,* S171–173.

Gunther, S., Green, L., Muller, J. E., Mudge, G., and Grossman, W. (1981). Prevention by nifedipine abnormal coronary vasoconstriction in patients with coronary artery disease. *Circulation 63,* 849–854.

Gustafsson, R., Manning, F., Kazzam, E., Waldenstrom, A., and Hallgren, R. (1989). Cold-induced reversible myocardial ischaemia in systemic sclerosis. *Lancet. 2,* 475–479.

Habermann, E. (1972). *Science 177,* 314–322.

Haegerstrand, A., Jonzon, B., Dalsgaard, C. J., and Nilsson, J. (1989). *Proc. Natl. Acad. Sci. 86,* 5993–5996.

Haegerstrand, A., Dalsgaard, C. J., Jonzon, B., Larsson, O., and Nilsson, J. (1990). Calcitonin gene-related peptide stimulates proliferation of human endothelial cells. *Proc. Natl. Acad. Sci. U.S.A. 87,*3299–3303.

Han, C., Wang, X., Fiscus, R. R., Gu, L., and McDonald, J. K. (1989). Changes in cardiac neuropeptide Y after experimental myocardial infarction in rat. *Neurosci. Lett. 104,* 141–146.

Hassall, C. J. S., and Burnstock, G. (1984). Neuropeptide Y-like immunoreactivity in cultured intrinsic neurones of the heart. *Neurosci. Lett. 52,* 111–115.

Hassessian, H., and Couture, R. (1989). Cardiovascular responses induced by intrathecal substance P in the conscious freely moving rat. *J. Cardiovasc. Pharmacol. 13,* 594–602.

Hassessian, H., Drapeau, G., and Couture, R. (1988). Spinal action of neurokinins producing cardiovascular responses in the conscious freely moving rat: evidence for a NK-1 receptor mechanism. *Nauym-Schmiedeberg's Arch. Pharmacol. 338,* 649–654.

Herman, I. M., and Castellot, J. J. (1987). Regulation of vascular smooth muscle cell growth by endothelial-synthesized extracellular matrices. *Arteriosclerosis 7,* 463–469.

Hermansson, A., Lindblom, U., Dalsgaard, C. J., and Bjorklund, H. (1987). *Neurosci. Lett. 74,* 377–382.

Heyndrickx, G. R., Boettcher, D. H., and Vatner, S. F. (1976). Effects of angiotensin, vasopressin and methoxamine on cardiac function and blood flow distribution in conscious dogs. *Am. J. Physiol. 231,* 1579–1587.

Hinojosa-Laborde, C., Osborn, J. W., and Cowley, A. W. (1989). Hemodynamic effects of endothelin in conscious rats. *Am. J. Physiol. 256,* H1742–H1746.

Hirai, S., Hasegawa, J., and Mashiba, H. (1989). Positive inotropic effect of somatostatin in guinea pig ventricular muscles. *J. Mol. Cell. Cardiol. 21,* 607–616.

Hirakata, H., Fouad-Tarazi, R. M., Bumpus, F. M., Khosla, M., Healy, B., Husain, A., Urata, H., and Kumagai, H. (1990). Angiotensins and the failing heart: Enhanced positive inotropic response to angiotensin I in cardiomyopathic hamster heart in the presence of captopril. *Circ. Res. 66,* 891–899.

Hirata, Y., Yoshimi, H., Takata, S., Watanabe, T. X., Kumagai, S., and Nakajima, K., Sakakibara, S. (1988a). Cellular mechanism of action by a novel vasoconstrictor endothelin in cultured rat vascular smooth muscle cells. *Biochem. Biophys. Res. Commun. 154,* 868–875.

Hirata, Y., Ishii, M., Matsuoka, H., Sugimoto, T., Iizuka, M., Uchida, Y., Serizawa, T., Sato, H., Kohmoto, O., Mochizuki, T., Sugimoto, T., Miyata, A., Kangawa, K., and Matsuo, H. (1988b). Plasma concentrations of alpha-human atrial natriuretic polypeptide and cyclic GMP in patients with heart disease. *Am. Heart. J. 113,* 1463–1469.

Hirata, Y., Fukuda, Y., Yoshimi, H., Emori, T., Shichiri, M., and Marumo, F. (1989a). Specific receptors for endothelin in cultured rat cardiocytes. *Biochem. Biophys. Res. Commun. 160,* 1438–1444.

Hirata, Y., Takagi, Y., Fukuda, Y., and Marumo, F. (1989b). *Atherosclerosis 78,* 225–228.

Hof, R. P. (1985). Modification of vasopressin- and angiotensin II-induced changes by calcium antagonists in the peripheral circulation of anaesthetized rabbits. *Br. J. Pharmacol. 85,* 75–87.

Hof, R. P. (1986). Vasopressin induced myocardial depression in neurally mediated and not due to impaired coronary blood flow. *Br. J. Pharmacol. 87,* 611–618.

Hokfelt, T., Johansson, O., Ljungdahl, A., Lundberg, J. M., and Schultzberg, M. (1980). Peptidergic neurones. *Nature 284,* 515–521.

Horak, A. R. (1984). Physiology and pharmacology of the transplanted heart. *In* "Heart transplantation." (D. K. C. Cooper and R. P. Lanza eds.), pp. 147–156. MTP Press, Lancaster, UK.

Hua, X. Y., Theodorsson-Norheim, E., Brodin, E., Lundberg, J. M., and Hokfelt, T. (1986). Multiple tachykinins (neurokinin A, neuropeptide K and substance P) in capsaicin-sensitive sensory neurons in the guinea pig. *Reg. Peptides 13,* 1–9.

Hughes, J., and Roth, R. H. (1971). Evidence that angiotensin enhances transmitter release during sympathetic nerve stimulation. *Br. J. Pharmacol. 41,* 239–255.

Ichikawa, S., Johnson, J. A., Fowler, W. L., Payne, C. G., Kurz, K., and Keitzer, W. R. (1978). Pressor responses to norepinephrine in rabbits with 3-day and 30-day renal artery stenosis. *Circ. Res. 43,* 437–446.

Inoue, A., Yanagisawa, M., Kimura, S., Kasuya, Y., Miyauchi, T., Goto, K., and Masaki, T. (1989). The human endothelin family: Three structurally and pharmacologically distinct isopeptides predicted by three separate genes. *Proc. Natl. Acad. Sci. U.S.A. 86,* 2863–2867.

Ishida, N., Tsujioka, K., Tomoi, M., Saida, K., and Mitsui, Y. (1989). Differential activities of two distinct endothelin family peptides on ileum and coronary artery. *FEBS Letts. 247,* 337–340.

Ishikawa, T., Yanagisawa, M., Kimura, S., Goto, K., and Masaki, T. (1988). Positive inotropic action of novel vasoconstrictor peptide endothelin on guinea pig atria. *Am. J. Physiol. 255,* H970–973.

Itoh, N., Obata, K., Yanaihara, N., and Okamoto, H. (1983). Human preprovasoactive intestinal polypeptide contains a novel PHI-27-like peptide, PHM-27. *Nature 304,* 547–549.

Itoh, H., Lederis, K. P., and Rorstad, O. P. (1990). Relaxation of isolated bovine coronary arteries by vasoactive intestinal peptide. *Eur. J. Pharmacol. 181,* 199–205.

Kashiwabara, T., Inagaki, Y., Ohta, H., Iwamatsu, A., Nomizu, M., Morita, A., Nishikori, K. (1989). Putative precursors of endothelin have less vasoconstrictor activity *in vitro* but a potent pressor effect *in vivo. FEBS Lett. 247,* 73–76.

Kasuya, Y., Ishikawa, T., Yanagisawa, M., Kimura, S., Goto, K., and Masaki, T. (1989a). Mechanism of contraction to endothelin in isolated porcine coronary artery. *Am. J. Physiol. 257,* H1828–H1835.

Kasuya, Y., Takuwa, Y., Yanagisawa, M., Kimura, S., Goto, K., and Masaki, T. (1989b). Endothelin-1 induces vasoconstriction through two functionally distinct pathways in porcine coronary artery: Contribution of phosphoinositide turnover. *Biochem. Biophys. Res. Comm. 161,* 1049–1055.

Katusic, Z. S., Shepherd, J. T., and Vanhoutte, P. M. (1987). Endothelium-dependent contraction to stretch in canine basilar artery. *Am. J. Physiol. 252,* H671–H673.

Kawai, Y., Takagi, H., and Tohyama, M. (1988). Co-localization of neurotensin- and cholecystokinin-like immunoreactivities in catecholamine neurons in the rat dorsomedial medulla. *Neuroscience 24,* 227–236.

Kawano, Y., Yoshida, K., Yoshimi, H., Kuramochi, M., and Omae, T. (1989). The cardiovascular effect of intracerebroventricular endothelin in rats. *J. Hypertension 7*(suppl. 6), S22–S23.

Kelly, K. J., Stang, J. M., and Makhjian, H. S. (1980). Vasopressin provocation of ventricular dysrhythmia. *Ann. Int. Med. 92,* 205–206.

Kelly, R. A., Eid, H., Kramer, B. K., O'Neill, M., Liang, B. T., Reers, M., and Smith, T. W. (1990). Endothelin enhances the contractile responsiveness of adult rat ventricular myocytes to calcium by a pertussis toxin-sensitive pathway. *J. Clin. Invest. 86,* 1164–1171.

Kelm, M., and Schrader, J. (1990). Control of coronary vascular tone by nitric oxide. *Circ. Res. 66,* 1561–1575.

Khairallah, P. A. (1972). Action of angiotensin on adrenergic nerve endings: Inhibition of norepinephrine uptake. *Fed. Proc. 31,* 1351–1357.

Kifor, I., and Dzau, V. J. (1986). Endothelial renin-angiotensin pathway: Evidence for intracellular synthesis and secretion of angiotensins. *Circ. Res. 60,* 422–428.

Kimura, E., Hashimoto, K., Furukawa, S., and Hayakawa, H. (1973). Changes in bradykinin level in coronary sinus blood after experimental occlusion of a coronary artery. *Am. Heart. J. 85,* 635–647.

Kimura, S., Okada, M., Sugita, Y., Kanazawa, I., and Munekatat, E. (1983). Novel neuropeptides, neurokinin å and β isolated from porcine spinal cord. *Proc. Jpn. Acad. 59B,* 101–104.

Kloog, Y., Ambar, I., Sokolovsky, M., Kochva, E., Wollberg, Z., and Bdolah, A. (1988). Sarafotoxin, a novel vasoconstrictor peptide: Phosphoinositide hydrolysis in rat heart and brain. *Science 242,* 268–270.

Kobayashi, M., Furukawa, Y., and Chiba, S. (1978). Positive chronotropic and inotropic effects of angiotensin II in the dog heart. *Eur. J. Pharmacol. 50,* 17–25.

Kopeyan, C., Martinez, G., and Rochat, H. (1985). *FEBS Lett. 181,* 211–217.

Kosaka, T., Suzuki, N., Matsumoto, H., Itoh, Y., Yasuhara, T., Onda, H., and Fujino, M. (1989). *FEBS Letts. 249,* 42–46.

Kubota, M., Moseley, J. M., Butera, L., Dusting, G. J., MacDonald, P. S., Martin, T. J.. (1985). Calcitonin gene related peptide stimulates cyclic AMP formation in rat aortic smooth muscle cells. *Biochem. Biophys. Res. Commun. 132,* 86–94.

Kulakowski, E. C., Lampson, W. G., Schaffer, S. W., and Lovenberg, W. (1983). Action of substance P on the working rat heart. *Biochem. Pharmacol. 32,* 1097–1100.

Kurihara, H., Yoshizumi, M., Sugiyama, T., Takaku, F., Yanigasawa, M., Masaki, T., Hamaoki, M., Kato, H., Yasaki, Y. (1989). Transforming growth factor-β stimulates the expression of endothelin mRNA by vascular endothelial cells. *Biochem. Biophys. Res. Comm. 159,* 1435–1440.

Larkin, S., Clarke, J. G., Keogh, B., Taylor, K., Davies, G. J., and Maseri, A. (1989). Endothelin causes transmural myocardial ischaemia in the anaesthetised greyhound model. *Am. J. Cardiol. 64,* 956–958.

Larsson, O. I., and Rehfeld, J. F. (1979). Localization and molecular heterogeneity of cholecystokinin in the central and peripheral nervous system. *Brain Res. 165,* 201–218.

Leak, L. V. (1980). *In* "Electron Microscopy in Human Medicine." (J. V. Johannessen, ed.), vol 5, pp. 87–154, Mc Graw Hill, New York.

Lee, W. B., Ismay, M. J., and Lumbers, E. R. (1980). Method by which angiotensin II affects the heart rate in the conscious sheep. *Circ. Res. 47,* 286–292.

LeZoche, E., Ghatei, M. A., and Carlei, F. (1979). Gut hormone responses to bombesin in man. *Gastroenterology 76,* 1185.

Lilly, L. S., Pratt, R. E., Alexander, R. W., Larson, D. M., Ellison, K. E., Gimbrone, M. A., and Dzau, V. J. (1985). Renin expression by vascular endothelial cells in culture. *Circ. Res. 57,* 312–318.

Losay, J., Mroz, E. A., Tregear, G. W., Leeman, S. E., and Gamble, W. J. (1977). Action of substance P on the coronary blood flow in the isolated dog heart. *In* "Substance P." pp. 287–293. (U. S. von Euler and B. Pernow, eds.), Raven, New York.

Ludmer, P. L., Selwyn, A. P., Shook, T. L., Wayne, R. R., Mudge, G. H., Alexander, R. W., Ganz, P. (1986). Paradoxical vasoconstriction induced by acetylcholine in atherosclerotic coronary arteries. *N. Engl. J. Med. 315,* 1046–1051.

Lundberg, J. (1981). Evidence for the existence of vasoactive intestinal polypeptide (VIP) and acetylcholine neurons in cat exocrine glands. Morphological, anatomical and functional studies. *Acta Physiol. Scand. 496* (Suppl.) 1–57.

Lundberg, J. M., Brodin, E., and Saria, A. (1983). Effects and distribution of vagal capsaicin-sensitive substance P neurons with special reference to the trachea and lungs. *Acta Physiol. Scand. 119,* 243–252.

Lundberg, J. M., Franco-Cercerceda, A., Hua, X., Hokfelt, T., and Fischer, J. A. (1985). Co-existence of substance P and calcitonin gene-related peptide-like immunoreactivities in sensory nerves in relation to cardiovascular and bronchoconstrictor effects of capsaicin. *Eur. J. Pharmacol. 108,* 315–319.

Maisel, A. S., Scott, N. A., Motulsky, N. J., Michel, M. C., Boublik, J. H., Rivier, J. E., Ziegler, M., Allen, R. S., and Brown, M. R. (1989). Elevation of plasma neuropeptide Y levels in congestive heart failure. *Am. J. Med. 86,* 43–48.

Majack, M. A., Cook, S. C., and Bornstein, P. (1985). Platelet derived growth factor and heparin-like glycosaminoglycans regulates thrombospondin synthesis in the matrix by smooth muscle cells. *J. Cell. Biol. 101,* 1059–1070.

Majesky, M. W., Benditt, E. P., and Schwartz, S. M. (1988). *Proc. Natl. Acad. Sci. U.S.A. 85,* 1524–1528.

Malik, K. U., and Nasjletti, A. (1976). Facilitation of adrenergic transmission by local generated angiotensin II in rat mesenteric arteries. *Circ. Res. 38,* 26–30.

Marker, J. D., and Roberts, M. L. (1988). Chronotropic actions of cholecystokinin octapeptide on the rat heart. *Reg. Pept. 20,* 251–259.

Marshall, I., Al-Kazawini, S. J., Roberts, P., Shepperson, N. B., Adams, M., and Craig, R. K. (1986). Cardiovascular effects of human and rat CGRP compared in the rat and other species. *Eur. J. Pharmacol. 123,* 207–216.

Masaoka, H., Suzuki, R., Hirata, Y., Emori, T., Marumo, E., and Hirakawa, K. (1989). Raised plasma endothelin in aneursymal subarachnoid haemorrhage. *Lancet. 2,* 1402.

Matthews, M. R., and Cuello, A. C. (1982). Substance P-immunoreactive peripheral branches of sensory neurones innervate guinea-pig sympathetic neurones. *Proc. Natl. Acad. Sci. U.S.A. 79,* 1668–1672.

Matthews, M. R., Connaughton, M., and Cuello, A. C. (1987). Ultrastructure and distribution of substance P-immunoreactive sensory collaterals in the guinea-pig prevertebral sympathetic ganglia. *J. Comp. Neurol. 258,* 28–51.

Maxwell, G. M. (1968). Actions of substance P on the general, pulmonary and coronary haemodynamics and metabolism of intact dogs. *Br. J. Pharmacol. Chemother. 32,* 514–522.

McConn, R., Wasserman, F., and Haberland, G. (1983). The kallikrein-kinin system in the acutely ill.

(a) Changes in plasma kininogen in acutely ill patients. (b) The efficacy of pulmonary clearance of bradykinin. *Adv. Med. Biol. 1566,* 1019–1035.

McCormack, D. G., Clarke, B. C., and Barnes, P. J. (1987). Characterisation of adenosine receptors in human pulmonary artery. *J. Physiol.* (Lond) *396,* 98.

McEwan, J. R., Larkin, S., Davies, G., Chierchia, S., Brown, M., Stevenson, J., MacIntyre, I., and Maseri, A. (1986). Calcitonin gene-related peptide: A potent dilator of human epicardial coronary arteries. *Circulation 74,* 1243.

McFadden, E., Clarke, J. G., Davies, G. J., Haider, A. W., Kaski, J. C., and Maseri, A. (1991). Human coronary vasomotor responses to 5-HT. *N. Engl. J. Med. 324,* 648–654.

Miauchi, T., Yanagisawa, M., Tomizawa, T., Sugishita, Y., Suzuki, N., Fujino, M., Ajisaka, R., Goto, K., and Masaki, T. (1989). Increased plasma concentration of ET-1 and big ET-1 in acute myocardial infarction. *Lancet. 2,* 53–54.

Michel, M. C., Wirth, S. C., Zerkowski, H. R., Maisel, A. S., and Motulsky, H. J. (1989). Lack of inotropic effects of neuropeptide Y in human mycardium. *J. Cardiovasc. Pharmacol. 14,* 919–922.

Milan, J. D., Shipkey, F. H., Lind, C. J., Nora, J. J., Leachman, R. D., Rochelle, D. G., Bolldwell, R. D., Hallman, G. L., and Cooley, D. A. (1970). Morphological findings in human cardiac allografts. *Circulation 41,* 519–535.

Miller, W. L., Redfield, M. M., and Burnett, J. C. (1989). Integrated cardiac, renal, and endocrine actions of endothelin. *J. Clin. Invest. 83,* 317–320.

Milner, P., Ralevic, V., Hopwood, A. M., Feher, E., Lincoln, J., Kirkpatrick, K. A., and Burnstock, G. (1989). Ultrastructural localization of substance P and choline acetyltransferase in endothelial cells of rat coronary artery and release of substance P and acetylcholine during hypoxia. *Experientia 45,* 121–125.

Miyazaki, H., Kondoh, M., Watanabe, H., Hayashi, T., Murakami, K., Takahashi, M., Yanagisawa, M., Kimura, S., Goto, K., and Masaki, T. (1989). Identification of the endothelin-1 receptor in the chick heart. *J. Cardiovasc. Pharmacol. 13,* S155–S156.

Moharty, P. K., Thames, M. D., Arrowhead, J. A., Sowers, J. R., McNamara, C., and Szentpetery, S. (1987). Impairment of cardiopulmonary baroreflex after cardiac transplantation in humans. *Circulation 75,* 914–921.

Moravec, C. S., Reynolds, E. E., Stewart, R. W., and Bond, M. (1989). Endothelin is a positive inotropic agent in human and rat heart *in vitro. Biochem. Biophys. Res. Comm. 159*(1), 14–18.

Mukherjee, A., Kulkarni, P. V., Haghani, Z., and Sutko, J. L. (1982). Identification and characterization of angiotensin II receptors in cardiac sarcolemma. *Biochem. Biophys. Res. Comm. 105,* 575–581.

Mulderry, P. K., Ghatei, M. A., Rodrigo, J., Allen, J. M., Rosenfeld, M. G., Polak, J. M., and Bloom, S. R. (1985). Calcitonin gene-related peptide in cardiovascular tissues of the rat. *Neuroscience 14,* 947–954.

Mullins, R. J., and Hudgens, R. W. (1987). Increased skin lymph protein clearance after a 6 hr arterial bradykinin infusion. *Am. J. Physiol. 253,* H1462–H1469.

Nakajima, T., Yashima, Y., and Nakamura, K. (1986). Quantitative autoradiographic localization of neuropeptide Y receptors on the rat lower brainstem. *Brain Res. 380,* 144–150.

Nakamura, Y., Parent, R., and Lavallee, M. (1990). Opposite effects of substance P on conductance and resistance vessels in conscious dogs. *Am. J. Physiol. 258,* H565–H573.

Nakano, J. (1965). Effects of synthetic bradykinin on the cardiovascular system. *Arch. Int. Pharmacodyn. Ther. 157,* 1–13.

Nawa, H., Hirose, T., Takashima, H., Inayama, S., and Nakanishi, S. (1983). Nucleotide sequences of cloned cDNAs for two types of bovine brain substance P precursors. *Nature 306,* 32–37.

Nilsson, J., von Eulr, A., and Dalsgaard, C. J. (1985). *Nature 315,* 61–63.

Norvell, J. E., and Lower, R. R. (1973). Degeneration and regeneration of the nerves of the heart after transplantation. *Transplantation 15,* 337–344.

Nunez, D. J. R., Brown, M. J., Davenport, A. P., Neylon, C. B., Schofield, J. P., Wyse, R. K. (1990). Endothelin-1 mRNA is widely expressed in porcine and human tissues. *J. Clin. Invest. 85,* 1537–1541.

Ogawa, T., Kanazawa, I., and Kimura, S. (1985). Regional distribution of substance P, neurokinin alpha and neurokinin beta in rat spinal cord, nerve roots and dorsal root ganglia, and the effects of dorsal root section or spinal transection. *Brain Res. 359,* 152–157.

Oishi, M., Inagaki, C., Fujiwara, M., Takaori, S., Yajima, H., and Akazawa, Y. (1981). Possible mechanisms of the triphasic effects of neurotensin on the rat blood pressure. *Jap. J. Pharmacol. 31,* 1043.

Ohnaka, K., Takayanagi, R., Yamauchi, T., Okazaki, H., Ohashi, M., Umeda, F., and Nawata, H. (1990). Identification and characterization of endothelin converting activity in cultured bovine aortic endothelial cells. *Biochem. Biophys. Res. Comm. 168,* 1128–1136.

Peach, M. J. (1977). Renin-angiotensin system: Biochemistry and mechanism of action. *Physiol. Rev. 57,* 313–370.

Pernow, J., and Lundberg, J. M. (1988). Neuropeptide Y induces potent contraction of arterial smooth muscle via an endothelium-independent mechanism. *Acta. Physiol. Scand. 134,* 157–158.

Polak, J. M. and Bloom, S. R. (1984). Regulatory Peptides—The distribution of two newly discovered peptides: PHI and NPY. *Peptides 5*(1) 79–89.

Potter, E. K., and Reid, I. A. (1985). Intravertebral angiotensin II inhibits cardiac vagal efferent activity in dogs. *Neuroendocrinology 40,* 493–496.

Powell, J. S., CLozel, J. P., Muller, R. K. M., Kuhn, M., Hefti, F., Hosang, M., and Baumgartner, H. R. (1989). Inhibitors of angiotensin converting enzyme prevent myointimal proliferation after vascular injury. *Science 245,* 186–188.

Power, R. F., Wharton, J., Zhao, Y., Bloom, S. R., and Polak, J. M. (1989). Autoradiographic localization of endothelin-1 binding sites in the cardiovascular and respiratory systems. *J. Cardiovasc. Pharmacol. 13,* S50–S56.

Quirion, R., Regoli, D., Rioux, F., and St-Pierre, S. (1979a). An analysis of the negative inotropic action of somatostatin. *Br. J. Pharmacol. 66,* 251–257.

Quirion, R., Rioux, R., Regoli, D., St. and Pierre, S. (1979b). Neurotensin-induced coronary vessel constriction in perfused rat hearts. *Eur. J. Pharmacol. 55,* 221–223.

Re, R., Fallon, J. T., Dzau, V. J., Quay, S., and Haber, E. (1981). Renin synthesis by canine aortic smooth muscle cells in culture. *Life Sci. 30,* 99–106.

Regoli, D., and Barabe, J. (1980). Pharmacology of bradykinin and related kinins. *Pharmacol. Rev. 32,* 1–46.

Regoli, D., Barabe, J., and Theriault, B. (1977). Does indomethacin antagonize the effects of peptides and other agents on the coronary circulation of rabbit isolated hearts? *Can. J. Physiol. Pharmacol. 55,* 307–310.

Regoli, D., Drapeau, G., Dion, S., and Couture, R. (1988). New selective agonists for neurokinin receptors: Pharmacological tools for receptor characterization. *Trends in Pharmacol. Sci. 9,* 290–295.

Reid, I. A. (1984). Endocrine regulation of body fluid balance. *In* "Edema." (N. C. Staub and A. E. Taylor eds), pp. 353–369. Raven Press, New York.

Reid, J. L., and Rubin, P. C. (1987). Peptides and central neural regulation of the circulation. *Physiol. Rev. 67,* 725–749.

Reinecke, M., Weihe, E., Carraway, R. E., and Forssman, W. G. (1981). Neurotensin-immunoreactive nerve fibres in the heart: An immunohistochemical and radioimmunological study. *Neuroscience Lett. 7* (Suppl. 7), S264.

Resink, T. J., Hahn, A. W. A., Scott-Burden, T., Powell, J., Weber, E., and Buhler, F. R. (1990). Inducible endothelin mRNA expression and peptide secretion in cultured human vascular smooth muscle cells. *Biochem. Biophys. Res. Comm. 168,* 1303–1310.

Richards, A. M., Nicholls, M. G., Ikram, H., Webster, M. W. I., Yandle, T. G., and Espiner, E. A. (1985). Renal haemodynamic and hormonal effects of human atrial natriuretic peptide in healthy volunteers. *Lancet. i,* 545–548.

Rigel, D. F. (1988). Effects of neuropeptides on heart rate in dogs: Comparison of VIP, PHI, NPY, CGRP and NT. *Am. J. Physiol. 255,* H311–H317.

Rogers, T. B. (1984). High affinity angiotensin II receptors in myocardial sarcolemmal membranes. *J. Biol. Chem. 259,* 8106–8114.

Rosell, S., Burcher, E., Chang, D., and Folkers, K. (1976). Cardiovascular and metabolic actions of neurotensin and (Gln^4)-neurotensin. *Acta. Physiol. Scand. 98,* 484–491.

Ross, R. (1986). The pathogenesis of atherosclerosis—an update. *N. Engl. J. Med. 314,* 488–500.

Rosselin, G., Maletti, M., Besson, J., and Rostene, W. (1982). A new neuroregulator: The vasoactive intestinal peptide or VIP. *Cell. Endocrinol. 27,* 243–262.

Rubanyi, G. M., and Vanhoutte, P. M. (1985). Hypoxia releases a vasoconstrictor substance from the canine vascular endothelium. *J. Physiol.* (Lond) *364,* 45–56.

Sabol, S. L., and Higuchi, H. (1990). Transcriptional regulation of the neuropeptide Y gene by nerve growth factor, antagonism by glucocorticoids and potentiation by adenosine 3,5-monophosphate and phorbol esters. *Mol. Endocrinol. 4*(3), 384–392.

Said, S. I., and Mutt, V. (1970). Polypeptide with broad biological activity: Isolation from small intestine. *Science 169,* 1217–1218.

Saida, K., Mitsui, Y., and Ishida, N. (1989). A novel peptide, vasoactive intestinal contractor of a new (endothelin) peptide family. *J. Biol. Chem. 264,* 14613–14616.

Saito, A., Kimura, S., and Goto, K. (1986). Calcitonin gene-related peptide as a potential neurotransmitter in guinea pig right atrium. *Am. J. Physiol. 250,* H693–H698.

Salminen, K., Tikkanen, I., Saijonmaa, O., Nieminen, M., Fyhrquist, F., and Frick, M. H. (1989). Modulation of coronary tone in acute myocardial infarction by endothelin. *Lancet. 2,* 747.

Samson, W. K., Aguila, M. C., Martinovic, J., Antunes-Rodrigues, J., and Norris, M. (1987). Hypothalamic action of atrial natriuretic factor to inhibit vasopressin secretion. *Peptides 8,* 449–454.

Schror, K., Metz, U., and Krebs, R. (1979). The bradykinin-induced coronary vasodilatation. Evidence for an additional prostacyclin-independent mechanism. *Nauyn-Schmiedeberg's Arch. Pharmacol. 307,* 213–221.

Scott-Burden, T., and Buhler, F. R. (1988). Regulation of smooth muscle proliferative phenotype by heparinoid-matrix interaction. *Trends Pharmacol. Sci. 9,* 94–98.

Sejersen, T., Betsholtz, C., Sjolund, M., Heldin, C. H., Westermark, B., and Thyberg, J. (1986). Rat skeletal myoblasts and arterial smooth muscle cells express the gene for the A chain but not the gene for the B-chain (c-sis) of platelet-derived growth factor (PDGF) and produce a PDGF-like protein. *Proc. Natl. Acad. Sci. U.S.A. 83,* 6844–6846.

Sharif, N. A., Michel, A. D., and Whiting, R. L. (1989). Human endothelin stimulates phosphoinositide turnover in murine fibrosarcoma cells (HSDM1C1). *Med. Sci. Res. 17,* 55–56.

Sheikh, S. P., O'Hare, M. M. T., Tortora, O., and Schwartz, T. W. (1989). Binding of monoiodinated neuropeptide Y to hippocampal membranes and human neuroblastoma cells lines. *J. Biol. Chem. 264,* 6648–6654.

Sigrist, S., Franco-Cereceda, A., Muff, R., Henke, H., Lundberg, J. M., and Fischer, J. A. (1986). Specific receptor and cardiovascular effects of calcitonin gene-related peptide. *Endocrinology 119,* 381–389.

Singer, M. (1952). *Q. Rev. Biol. 27,* 169–200.

Smitherman, T. C., Sakio, H., Geumei, M., Yoshida, T., Oyamada, M., and Said, S. I. (1982). Coronary vasodilator action of VIP. *In* "Vasoactive intestinal peptide." (S. I. Said ed.), pp. 169–176. Raven, New York.

Smitherman, T. C., Popma, J. J., Said, S. I., Krejs, G. J., and Dehmer, G. J. (1989). Coronary haemodynamic effects of intravenous vasoactive intestinal peptide in humans. *Am. J. Physiol. 257,* H1254–H1262.

Stene-Larsen, G., and Helle, K. B. (1979). Inotropic and chronotropc effects of neurotensin in the rat atrium and of physalaemin in the auricles of *R. esculenta. Comp. Biochem. Physiol. 64c,* 279.

Struthers, A. D., Pai, S., Seidelin, P. H., Coutie, W. J. R., and Morton, J. J. (1987). Evidence in humans for a postsynaptic interaction between noradrenaline and angiotensin II with regard to systolic but not diastolic blood pressure. *J. Hypertens. 5,* 671–676.

Stewart, D. J., and Baffour, R. (1990). Functional state of the endothelium determines the response to endothelin in the coronary circulation. *Cardiovasc. Res. 24,* 7–12.

Tatemoto, K., and Mutt, V. (1980). Isolation of two novel candidate hormones using a chemical method for finding naturally occuring polypeptides. *Nature 285,* 417–418.

Tatemoto, K., and Mutt, V. (1981). Isolation and characterisation of the intestinal peptide porcine PHI (PHI-27), a new member of the glucagon-secretion family. *Proc. Natl. Acad. Sci. U.S.A. 78,* 6603–6607.

Taton, G., Chatelain, P., Delhaye, M., Camus, J. C., De Neef, P., Waelbroeck, M., Tatemoto, K., Robberecht, P., and Christophe, J. (1982). Vasoactive intestinal peptide (VIP) and peptide having N-terminal isoleucine amide (PHI) stimulate adenylate cyclase activity in human heart membranes. *Peptides 3,* 897–900.

Thom, S., Hughes, A. D., Goldberg, P., Martin, G., Schactyer, M., and Sever, P. S. (1987). The actions of calcitonin gene-related peptide and vasoactive intestinal peptide as vasodilators in man *in vivo* and *in vitro*. *Br. J. Clin. Pharmacol. 24,* 139–144.

Tikkanen, I. K., Metsarrine, K., Fyhrquist, F., and Leidenius, R. (1985). Plasma atrial natriuretic peptide in cardiac disease and during infusion in healthy volunteers. *Lancet. i,* 66–69.

Tippins, J. R., Clarke, J., Larkin, S., Yacoub, M., and Maseri, A. (1988). Investigation of the action of neuropeptide Y in the isolated human coronary artery. *Br. J. Pharmacol. 94,* 244.

Toyo-oka, T., Aizawa, T., Suzuki, N., Hirata, T., Miyauchi, T., Shin, W. S., Yanagisawa, M., Masaki, T., and Sugimoto, T. (1991). Increased plasma levels of endothelin-1 and coronary spasm induction in patients with vasospastic angina pectoris. *Circulation 83,* 476–483.

Tsai, R.-C., Yamaji, T., Ishibashi, M., Takaku, F., Pang, S.-C., Yeh, S.-J., Lee, Y.-S., Hung, J.-S., and Delon, W. (1988). Atrial natriuretic peptide during supraventricular tachycardia and relation to haemodynamic changes and renal function. *Am. J. Cardiol. 61,* 1260–1264.

Uchida, U., and Murao, S. (1974). Bradykinin-induced excitation of afferent cardiac sympathetic nerve fibres. *Jpn. Heart J. 15,* 84–91.

Unverferth, D. V., O'Dorisio, T. M., Muir, W. W., White, J., Miller, M. M., Hamlin, R. L., and Magorien, R. D. (1985). Effect of vasoactive intestinal polypeptide on the canine cardiovascular system. *J. Lab. Clin. Med. 106,* 542–550.

Urata, H., Healy, B., Stewart, R. W., Bumpus, F. M., and Husain, A. (1989). Angiotensin II receptors in normal and failing human hearts. *J. Clin. Endocrinol. Metab. 69,* 54–66.

Vatner, S. F., Franklin, D., Van Citters, R. L., and Braunwald, E. (1970). Effects of carotid sinus nerve stimulation on the coronary circulation of the conscious dog. *Circ. Res. 27,* 11–21.

Waeber, B., Gavras, I., Brunner, H. R., Cook, C. A., Charocops, F., and Gavras, H. P. (1982). Prediction of sustained antihypertensive efficacy of chronic captopril therapy: Relationships to immediate blood pressure response and control of plasma renin activity. *Am. Heart. J. 103,* 384–390.

Wahlestedt, C., Yanihara, N., and Hakanson, R. (1986). Evidence for different pre- and post-junctional receptors for NPY and related peptides. *Regul. Pept. 13,* 307–318.

Watanabe, H., Miyazaki, H., Kondoh, M., Masuda, Y., Kimura, S., Yanagisawa, M., Masaki, T., and Murakami, K. (1989). Two distinct types of endothelin receptors are present on chick cardiac membranes. *Biochem. Biophys. Res. Commun. 161,* 1252–1259.

Watt, P. A., Baker, A. R., and Thurston, H. (1989). Vasoconstrictor actions of endothelin-1 in human resistance vessels. *J. Hypertension 7*(suppl. 6), 134–135.

Webb, S. C., Krikler, D. M., Hendry, W. G., Adrian, T. E., and Bloom, S. R. (1986). Electrophysiological actions of somatostatin on the atrioventricular junction in sinus rhythm and re-entry tachycardia. *Br. Heart. J. 56,* 236–241.

Wei, E. P., Kontos, H. A., and Said, S. I. (1980). Mechanisms of action of vasoactive intestinal polypeptide on cerebral arteries. *Am. J. Physiol. 239,* H765–H768.

Weihe, E., and Reinecke, M. (1981). Peptidergic innervation of the mammalian sinus nodes: Vasoactive intestinal polypeptide, neurotensin, substance P. *Neurosci. Lett. 26,* 283–288.

Weihe, E., Reinecke, M., and Forsmann, W. G. (1984). Distribution of vasoactive intestinal polypeptide-like immunoreactivity in the mammalian heart. *Cell. Tiss. Res. 236,* 527–540.

Weipert, J., Hoffman, H., Siebeck, M., and Whalley, E. T. (1988). Attenuation of arterial blood pressure fall in endotoxin shock in the rat using the competitive bradykinin antagonist Lys-Lys-[Hyp2, Thi5,8,DPhe7]-Bk(B4148). *Br. J. Pharmacol. 94,* 282–284.

Wennmalm, A. (1979). Prostacyclin-dependent coronary vasodilation in rabbit and guinea-pig hearts. *Acta Physiol. Scand. 106,* 47–52.

Wharton, J., and Gulbenkian, S. (1989). Peptides in the mammalian cardiovascular system. *Experientia 56,* 292–316.

Wharton, J., Gulbenkian, S., Mulderry, P. K., Ghatei, M. A., McGregor, M. P., Bloom, S. R., and Polak, J. M. (1986). Capsaicin induces a depletion of calcitonin gene-related peptide (CGRP)-immunoreactive nerves in the cardiovascular system of the guinea-pig. *J. Autonom. Nerv. Sys. 16,* 289.

Wharton, J., Gulbenkian, S., Merighi, A., Kuhn, D. M., Jahn, R., Taylor, K. M., and Polak, J. M. (1988). Immunohistochemical and ultrastructural localization of peptide-containing nerves and myocardial cells in the human atrial appendage. *Cell Tiss. Res. 254,* 155–166.

Wharton, J., Polak, J. M., Gordon, L., Banner, N. R., Springall, D. R., Rose, M., Khagani, A., Wallwork, J., and Yacoub, M. H. (1990). Immunohistochemical demonstration of human cardiac innervation before and after transplantation. *Circ. Res. 66,* 900–912.

Wiley, J. W., Uccioli, L., Owyang, C., and Yamada, T. (1989). Somatostatin stimulates scetylcholine release in the canine heart. *Am. J. Physiol. 257,* H483–H487.

Wilkins, M. R., Gammage, M. D., Lewis, H. M., Tan, L. B., and Weisser, P. L. (1988). Effect of lower body positive pressure on blood pressure plasma atrial natriuretic factor concentration and sodium and water excretion in healthy volunteers and cardiac recipients. *Cardiovasc. Res. 22,* 231–235.

Wilson, M. F., Brackett, D. J., Archer, L. T., and Hinshaw, L. B. (1980). Mechanism of impaired cardiac function by vasopressin. *Ann. Surg. 191,* 494–500.

Wright, G. B., Alexander, R. W., Ekstein, L. S., and Gimbrone, M. J. (1983). Characterization of rabbit ventricular myocardial receptor for angiotensin II. *Mol. Pharmacol. 24,* 213–332.

Wright, C. E., and Fozard, J. R. (1988). Regional vasodilation is a prominant feature of the haemodynamic response to endothelin in anaesthetized, spontaneously hypertensive rats. *Eur. J. Pharmacol. 155,* 201–203.

Yanagisawa, M., and Masaki, T. (1989). Molecular biology and biochemistry of the endothelins. *Tr. Pharm. Sci. 10,* 374–378.

Yanagisawa, M., Kurihara, H., Kimura, S., Tomobe, Y., Kobayashi, M., Mitsui, Y., Yazaki, Y., Goto, K., and Masaki, T. (1988). A novel potent vasoconstrictor peptide produced by vascular endothelial cells. *Nature 332,* 411–415.

Yang, Z., Richard, V., von Segesser, L., Bauer, E., Stulz, P., Turina, M., and Luscher, T. F. (1990). Threshold concentrations of endothelin-1 potentiate contractions to norepinephrine and serotonin in human arteries. *Circulation 82,* 188–195.

Zaidi, M., Bevis, P. J. R., Girgis, S. I., Lynch, C., Stevenson, J. C., and MacIntyre, I. (1985). Circulating CGRP comes from the perivascular nerves. *Eur. J. Pharmacol. 117,* 283.

Zimmerman, B. G. (1972). Actions of angiotensin on adrenergic nerve endings: Inhibition of norepinephrine uptake. *Fed. Proc. 37,* 199–202.

Zito, R. A., Diez, A. R., and Groszmann, R. J. (1983). Comparative effect of nitroglycerin and nitroprusside on vasopressin-induced cardiac dysfunction in the dog. *J. Cardiovasc. Pharmacol. 5,* 586–591.

CHAPTER 19

Perivascular Peptides in Subarachnoid Hemorrhage

ROAR JUUL   LARS EDVINSSON

I. Introduction

The cerebral circulation is richly supplied with perivascular nerve fibers (Edvinsson, 1985). These fibers contain acetylcholine, amines, and neuropeptides. The rich plexus of sympathetic fibers stores noradrenaline and neuropeptide Y (NPY). Parasympathetic fibers contain acetylcholinesterase, vasoactive intestinal peptide (VIP), peptide histidine isoleucine amide, and NPY. Finally, sensory afferent fibers contain substance P (SP), neurokinin A, and calcitonin gene-related peptide (CGRP). Denervations and axonal transport techniques have been used to study the origin of these fibers and their projections from cranial ganglia to cerebral vessels in rats, cats, and recently in monkeys (Liu-Chen *et al.*, 1984; Arbab *et al.*, 1986; Uddman *et al.*, 1992; Suzuki *et al.*, 1991).

In the present review we will describe the perivascular nerve supply to the cerebral circulation. Since there are some specie differences, we will mainly describe the innervation pattern in monkey, being the closest to man. Furthermore, we will review the present knowledge on animal experiments and perivascular peptides and finally provide the latest data on peptide release and infusion trials with synthetic CGRP in subarachnoid hemorrhage (SAH) patients. The available evidence suggests an important role for NPY and CGRP in the pathophysiology after a subarachnoid hemorrhage.

II. Peptidergic Innervation of the Monkey Middle Cerebral Artery

The spectral characteristics of True Blue (TB) make it suited for use as a retrograde neuronal tracer in combination with immunofluorescence (Sawchenko and Swanson, 1981; Skagerberg *et al.*, 1985). Due to the restricted localization of the applied TB, it is to be expected that only a portion of the nerve cell bodies that project to the middle cerebral artery will be labeled. After the application a blue fluorescence appeared in nerve cell bodies of the superior cervical and trigeminal ganglia 2 weeks later in monkey (Uddman *et al.*, 1993; Hardebo *et al.*, 1991). Occasionally small clusters of labeled nerve cell bodies could be seen in TB-labeled ganglia, but usually only single nerve cell bodies were visible (Table 1). In the ipsilateral superior cervical ganglion a very rich supply of TB-containing nerve cell bodies were seen. In the contralateral ganglion numerous TB-labeled nerve cell bodies were seen. A rich supply of TB-labeled nerve cell bodies was seen in the ipsilateral trigeminal ganglion. The contralateral trigeminal ganglion contained a moderate number of TB-labeled nerve cell bodies. The ipsilateral sphenopalatine ganglion contained only a few TB-labeled nerve cell bodies and none contralaterally. The findings are basically in agreement with previous studies on the sympathetic and sensory innervation but different for the parasympathetic innervation as performed on the cat middle cerebral (Liu-Chen *et al.*, 1984; Mayberg *et al.*, 1984; Tsai *et al.*, 1985) or the basilar artery (Walters *et al.*, 1986). These studies showed that the superior cervical, sphenopalatine, otic, and trigeminal ganglia contribute to the innervation of the vessels, albeit mainly ipsilaterally. Arbab *et al.* (1986) observed that after application of wheat germ agglutinin conjugated to horseradish peroxidase (WGA-HRP) to the basilar artery of rat, nerve cell bodies were seen bilaterally in the superior cervical and stellate ganglia. Furthermore, observations have been made that TB application around the rat middle cerebral artery resulted in labeling of nerve cell bodies in both the ipsilateral and contralateral superior cervical ganglion (Edvinsson *et al.*, 1989). This is supported by a study in the monkey (Uddman *et al.*, 1992). Noradrenaline and NPY are known to be costored in nerve cell bodies in sympathetic ganglia (Ekblad *et al.*, 1984; Edvinsson *et al.*, 1987). It was found both in monkey and rat that every second TB-labeled nerve cell body in the superior cervical ganglion was NPY-immunoreactive (Edvinsson *et al.*, 1989; Uddman *et al.*, 1993). Information on the origin of the parasympathetic innervation of the brain circulation has, despite decades of research, been elusive (Edvinsson *et al.*, 1972). In previous rat studies neither O'Connor and Van der Kooy (1986) nor Arbab *et al.*

Table 1 Relative Amount of Labeled Nerve Cell Bodies in the Cranial and Cervical Ganglia following Application of True Blue to the Middle Cerebral Artery of the Squirrel Monkey[a]

Ganglion	Ipsilateral	Contralateral
Superior cervical	***	**
Trigeminal	***	*
Sphenopalatine	*	0

Source: Uddman *et al.*, 1993.

[a]The relative amount of labeled nerve cell bodies was assessed in serially sectioned ganglia and graded as follows: *, a few nerve cell bodies; **, a moderate number of nerve cell bodies; ***, a rich supply of nerve cell bodies.

(1986) reported any labeling of the sphenopalatine ganglion. In rat studies a moderate supply of TB-positive nerve cell bodies was seen in the sphenopalatine and otic ganglia following TB application to the middle cerebral artery (Edvinsson *et al.*, 1989), which confirms results obtained with horseradish peroxidase tracing in the cat (Walters *et al.*, 1986). This offers evidence for the origin of at least a considerable proportion of cerebrovascular VIP fibers in the sphenopalatine and otic ganglia (in rat). In the monkey positive staining in only a few nerve cell bodies of the sphenopalatine ganglion was found (Uddman *et al.*, 1993). This sparse supply from the sphenopalatine ganglion is remarkably different from rat and cat tracing. It suggests another source of the VIP supply to the cerebral circulation in monkey. Hardebo *et al.* (1991) found that parasympathetic fibers in monkey originated in the ipsilateral sphenopalatine and internal carotid ganglia, and, according to the number of TB-labeled cells, these two ganglia contribute to innervation to similar extents.

The trigeminal ganglion has long been considered to be the major source of the afferent input to the cerebral circulation (Ray and Wolff, 1940). This is in agreement with tracing studies (Arbab *et al.*, 1986; O'Connor and Van der Kooy, 1986) where labeling was seen in ipsilateral trigeminal cell bodies. However, a bilateral supply has been indicated both in monkey (Uddman *et al.*, 1993) and rat (Edvinsson *et al.*, 1989). In addition, in rat but not in monkey a moderate supply of labeled nerve cell bodies was seen in the dorsal root ganglion at the C_2 level (Edvinsson *et al.*, 1989). This adds further to the complexity of putatative neurogenic mechanisms in the cerebral circulation. In previous studies cerebral vessels have been shown to contain SP- (Edvinsson *et al.*, 1981) and CGRP-immunoreactive fibers (Uddman *et al.*, 1985) that disappear after trigeminal ganglion lesioning (Liu-Chen *et al.*, 1983; Uddman *et al*,. 1985).

The TB-labeled nerve cell bodies in the trigeminal and dorsal root ganglia have shown positive immunostaining for CGRP (Edvinsson *et al.*, 1989; Uddman *et al.*, 1992). In a few instances the labeled nerve cell bodies exhibited SP immunoreactivity. This is in agreement with a similar study by O'Connor and Van der Kooy (1986), establishing the origin of the SP/CGRP fibers.

III. Perivascular Nerves in Subarachnoid Hemorrhage (SAH)

The importance of neurogenic mechanisms in the ethiology and development of chronic cerebral vasospasm after SAH has been debated. Early studies aimed to give support to a neurogenic involvement were based largely on the effects of adrenoceptor agonists and antagonists. At most they offered some support, however, when examined more rigorously, no dramatic improvement was seen.

A. Aminergic Involvement

Vasospasm is primarily associated with areas that have a rich perivascular innervation since it occurs preferrably in the circle of Willis and its branches of proximal location. This part of the cerebral circulation has a rich perivascular nerve supply. It is well known that cerebral arteries exposed to subarachnoid blood lose their catecholamine histofluorescence (Fraser *et al.*, 1970; Peerless and Yassargil, 1971; Svendgaard *et al.*, 1977; Edvinsson *et al.*, 1982; Yoshioka *et al.*, 1984; Delgado *et al.*, 1985). This adrenergic "denervation" lasts for at least 3 weeks post-SAH as seen in animal studies. Comparison of the time course of angiographically visible vasospasm with the duration of decreased fluorescence has revealed a similar time course in the rabbit (Edvinsson *et al.*, 1982). In another study of the monkey it was observed that the fluorescence was reduced even when the vasospasm had disappeared angiographically (Hara *et al.*, 1986). The difference may be due to the amount of blood administered and the severity of the damage to the sympathetic nerve terminals. The adrenergic dysfunction following SAH has been ascertained with quantitative methods measuring noradrenaline content, dopamine-B-hydroxylase activity, and ^{3}H-noradrenaline uptake in the affected cerebral arteries (Svendgaard *et al.*, 1977; Lobato *et al.*, 1980; Edvinsson *et al.*, 1982). The nature of the lesion may in part be explained by the hypoxic condition of the cerebrospinal fluid after SAH, explaining the fall of these functional parameters in adrenergic nerves.

Vasospasm might be initiated by noradrenaline release from the nerve terminals located near the vascular smooth muscle. With electron microscopy, it has been observed that small dense core vesicles in the sympathetic nerve terminals around cat cerebral arteries gradually disappeared after SAH (Endo and Suzuki, 1975). When endogenous noradrenaline is released from perivascular stores by SAH, supramaximal contraction of the smooth muscle will occur since reuptake mechanisms of released amine transmitter, which is the major inactivation mechanism in the adrenergic neuroeffector area, is impaired after SAH. When uptake mechanisms are impaired by cocaine administration in cats, a threefold increase in the potency of noradrenaline-induced contraction of pial arteries occurred (Edvinsson *et al.*, 1975). This finding, coupled with evidence of an altered receptor binding (Tsukahara *et al.*, 1985; Tsukahara *et al.*, 1986), supports the concept that the vasospasm phase may involve supersensitivity to amines, possibly affecting both contractile and relaxant processes. Direct proof has come from experiments on cats and rabbits following the exposure of cerebral arteries to subarachnoid blood (Svendgaard *et al,*. 1977; Lobato *et al.*, 1980). The vessels became supersensitive to noradrenaline and 5-hydroxytryptamine (5-HT) within 24 h and reached the maximum after 3 days. The magnitude of the shift in EC50 values (concentration of agonist giving half-maximum contraction) was reported to be two- to threefold while for 5-hydroxytryptamine the maximum contraction was also enhanced (Svendgaard *et al.*, 1977). However, studies of other species, such as dog and monkey, have failed to demonstrate an increased sensitivity of cerebral arteries to noradrenaline and 5-hydroxytryptamine (Toda *et al.*, 1977; Krueger *et al.*, 1985). This discrepancy may reflect species differences or variations in the methodology used for the SAH induction. However, in each model in which angiographic evidence of vasospasm is available, spastic arteries have shown lower activity. Superior cervical ganglionectomy performed before SAH induction did not influence angiographic vasospasm in rats or dogs (Nagai *et al.*, 1975; Svendgaard *et al.*, 1985). In an *in vivo* study of the cat, topical application of noradrenaline, adrenaline, 5-hydroxytryptamine, fresh blood, or incubated blood to the basilar arteries did not reveal any difference between normal and adrenergically denervated animals (Duff *et al.*, 1986).

Concentrations of noradrenaline have been measured in serum and cerebrospinal fluid (CSF) in patients with SAH. In patients who developed vasospasm, this vasospasm was preceded by an increase of noradrenaline concentration in plasma (Benedict and Loach, 1978). The noradrenaline content of CSF from patients with vasospasm was higher (0.246 ± 0.049 ng/ml) than that of patients without vasospasm (0.075 ± 0.001 ng/ml); this is, however, not high enough to provide marked vasoconstriction of major cerebral vessels (the levels of noradrenaline are in the order of 10^{-9} M) (Shigeno, 1982).

Overall, the various studies using different animal models show wide variation that are both inconclusive and contradictory. Therapeutic intervention with adrenoceptor blockers or sympathomimetic agents has not proven to be dramatically beneficial (Wilkins, 1980). Thus, it still remains unclear whether the aminergic involvement is a portion of the causative mechanism or an epiphenomenon for vasospasm that occurs following SAH.

B. Peptidergic Involvement

SAH may affect various perivascular nerve fibers. Ultrastructural analysis has revealed that SAH affects the structural integrity of different perivascular neural elements, including both clear and dense core vesicles, myelinated as well as nonmyelinated axons (Duff *et al.*,1986).

1. *NPY*. Experimental studies have shown that SAH induces reductions in the NPY-LI content (Jacowski *et al.*, 1989), but this was not confirmed by a study by Edvinsson *et al.* (1991). However, NPY-induced contractions were significantly weaker (lower Emax) in SAH as compared to control rats, indicating dynamic changes in cerebral artery reactivity despite any changes in the sympathetic nerve networks. Plasma NPY-LI has also been shown to be elevated in connection with increased intracranial pressure in the pig (Rudehill *et al.*, 1987).

2. *VIP* immunostaining is also reduced in experimental SAH in some studies (Hara *et al.*, 1986; Uemura *et al.*, 1986; Nozaki *et al.*, 1990), unchanged in one (Edvinsson *et al.*, 1991) but with increased relaxant response in SAH as compared to controls, also indicating dynamic changes.

3. *Substance P* immunostaining has also been shown to be reduced after SAH (Linnik *et al.*, 1989; Edvinsson *et al.*, 1990), however, no significant difference in maximum response or sensitivity to SP could be observed between arteries from SAH and control animals (Edvinsson *et al.*, 1991a).

4. *CGRP*. Quantitative measurements have shown reduced levels of CGRP-LI (-like immunoreactivity) in vessels exposed to SAH, both in rats (Edvinsson *et al.*, 1990) and in man (Edvinsson *et al.*, 1991a). Although unaltered sensitivity to

CGRP was found in SAH, the maximun relaxation was significantly enhanced by SAH compared to controls (Edvinsson *et al.*, 1990). Another study has shown suppressed CGRP immunohistochemistry after SAH, from day 3, and more marked during days 7–14, with normalization at 42 days after SAH (Nozaki *et al.*, 1989).

An interesting approach has been published utilizing lesion experiments of the trigeminal ganglion (McCulloch *et al.*, 1986). In cats in which the trigeminal ganglion was lesioned 2 weeks beforehand, the vasoconstrictor responses to pH increase, prostaglandin F_2 alfa, noradrenaline, $BaCl_2$, and arterial blood were similar to control in magnitude but significantly prolonged in time. This indicates that the trigemino–cerebrovascular system, which includes substance P, neurokinin A, and CGRP as neurotransmitter candidates, could be involved in the restoration of normal cerebrovascular diameter in conditions of pronounced vasoconstriction. CGRP is the more likely candidate as the "antivasoconstrictor" substance in the trigemino–cerebrovascular system. In man all three peptides give strong relaxations, with CGRP being the most potent peptide, while in guinea pig substance P is the more potent peptide. Thus, depending on the species, different peptides may have the key role.

C. Central Neurogenic Components

There is some evidence that the central nervous system may be involved in the etiology of vasospasm. Following craniotomy for intra- or suprasellar tumors, vasospasm is observed from time to time (Simeone and Peerless, 1975). It is also known that head trauma with or without SAH may be accompanied by vasospasm (Wilkins and Odom, 1970; Suwanwela and Suwanwela, 1972; Macpherson and Graham, 1978). Experimental support for a role of a central neurogenic component in vasospasm following SAH has recently appeared. In rats the stereotaxic destruction of ascending catecholamine projections of the median eminence has been found to prevent vasospasm induced by administration of autologous blood intracisternally (Svendgaard *et al.*, 1985; Svendgaard *et al.*, 1986). Although its mechanism is unproven, it seems possible that central or neurohumoral mechanisms may play a role in the development of vasospasm (Wilson and Feild, 1974; Wilkins, 1980).

IV. Studies in Human SAH

A series of studies has recently been carried out in man in order to examine a putative peptidergic involvement in SAH. Plasma samples from the cranial venous outflow have been used to spot cranial peptidergic changes in 20 SAH patients (Juul *et al.*, 1990). Furthermore, peptide level alterations were compared with Doppler ultrasound velocity recordings from both middle cerebral arteries (MCA) and internal carotid arteries (ICA) (Aaslid *et al.*, 1982) as a measure of vasomotor state. We used the hemodynamic index (V mean MCA/V mean ipsilateral ICA) (Lindegaard *et al.*, 1986, 1988). This index is a reliable indicator of cerebral vasospasm. Blood samples from the external jugular vein (EJV) were taken postoperatively on days 1,

2, 3, 5, 7, and 9, with Doppler recordings being done immediately after removal of the blood samples. The blood was processed, frozen, and dispatched to the laboratory on dry ice for radioimmunoassay. Circulating NPY-LI was measured by the method described by Widerløv *et al.* (1988). CGRP-LI by Grunditz *et al.* (1986), SP-LI by Brodin *et al.* (1985), and VIP-LI by the technique described by Helm *et al.* (1985).

V. Neuropeptide Y-LI

When comparing mean hemodynamic index and mean NPY-LI levels in each of the 20 patients, a correlation of $r = 0.75$, $p = 0.0001$ was found (Fig. 1). Higher NPY-LI levels were found in samples taken when MCA velocities were above 120 cm/s than those taken when MCA velocities were below 120 cm/s, 129 ± 9 pmol/l (SEM), and 113 ± 5 pmol/l, respectively, $p = 0.06$.

The highest NPY-LI levels (253 pmol/l) were found in a patient with an unclipable basilar aneurysm. The subject rebled and died 2 days after the operation. The blood sample was taken after the rebleeding, and Doppler recordings indicated vasoconstriction (MCA velocity 138 cm/s, hemodynamic index 13,8). The pulsatility index [(peak systolic velocity − diastolic velocity) / mean velocity, 1.99] indicated high intracranial pressure. High NPY-LI levels (184 pmol/l) were also found in two other subjects who had large cerebral hemodynamic changes and the highest velocities. NPY-LI levels over 130 pmol/l were found in 11 of the 20 patients on single measurements, compared to values in blood from the EJV in controls (116 ± 3 pmol/l, $n = 14$). There seemed to be a relation between velocities and NPY-LI levels in seven patients, while low NPY-LI levels were found in nine patients with corresponding low velocities.

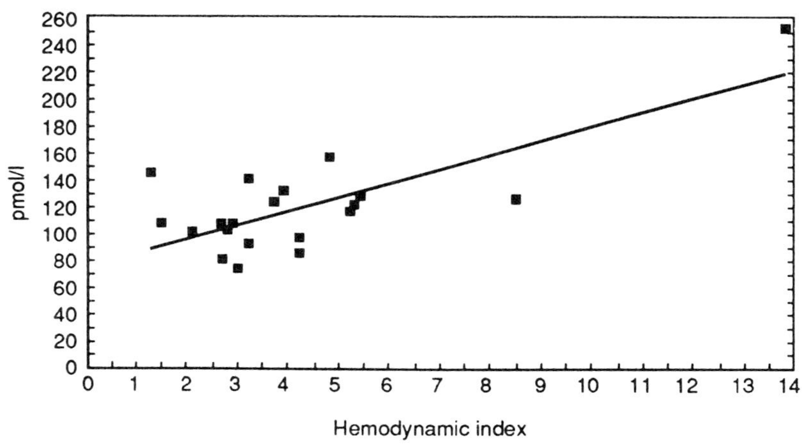

Figure 1 The relationship between mean hemodynamic index and mean NPY-LI in 20 patients is shown, indicating a correlation factor of $r = 0.75$, $p < 0.05$.

VI. Calcitonin Gene-Related Peptide-LI

The highest CGRP-LI levels (63 pmol/l) were found in a patient who developed a infarction on the day after operation. The second highest CGRP-LI levels were found in the patient with the highest velocities/index values. In the patient who rebled from a basilar aneurysm, and with high NPY-LI and index values, CGRP-LI was very low, below 10 pmol/l. CGRP-LI levels above 30 pmol/l were found in 14 of the 20 patients on single measurements. In 13 healthy volunteers CGRP-LI levels in the EJV were 28 ± 3 pmol/l.

When using transcranial Doppler recordings, it is natural to pay particular attention to patients with lateralized lesions because of a good correlation between MCA velocities and angiographic spasm (Lindegaard *et al,* 1988). These results are shown in Fig. 2, comparing the hemodynamic index with CGRP-LI levels in 10 patients with MCA aneurysms, indicating a correlation factor of 0.65, $p = 0.0002$.

VII. CSF Peptides in SAH

Suzuki *et al.* (1989) have published increased NPY-LI levels in the ventricular or basal cisternal CSF in human SAH. We investigated NPY-LI, CGRP-LI, and VIP-LI in CSF in 14 SAH patients. The neuropeptides were quantified using the radioimmunoassay techniques referred to above and correlated to the clinical and hemodynamic course. In total, the CSF NPY-LI levels (mean 67 ± 7 pmol/l) and VIP-LI levels (mean 4.6 ± 0.3 pmol/l) were lower in the SAH group as compared to

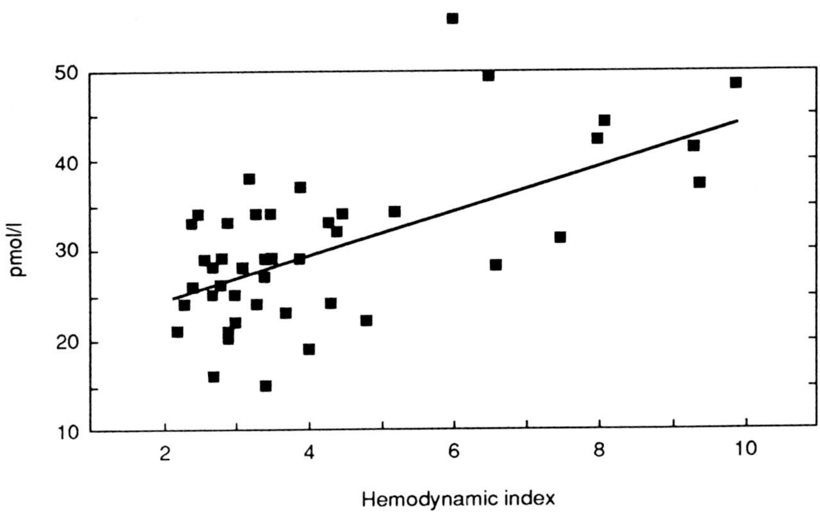

Figure 2 The relationship between the hemodynamic index and CGRP-LI levels in 10 patients with MCA aneurysms is shown, indicating a correlation factor of $r = 0.65$, $p < 0.05$.

lumbar CSF of 10 healthy controls (NPY-LI; mean 140 ± pmol/l, VIP-LI; mean 13 ± 2 pmol/l; $p < 0.05$). In individual patients with marked vasoconstriction, increased CGRP-LI levels (up to 14 pmol/l) and NPY-LI (up to 232 pmol/l) were observed. In conclusion, CSF peptidergic release and alterations could indicate perivascular neural involvement in the pathophysiology of SAH. Hamlyn *et al.* (1991) investigated CGRP-LI and SP-LI in CSF from SAH patients and found decreased CGRP and SP levels and unchanged endothelin levels. Earlier, Masaoka *et al.* (1989) and Sato *et al.* (1990), on the other hand, had investigated plasma and CSF endothelin levels in SAH and found increased levels.

VIII. Infusion Studies

Johnston *et al.* (1990) have published an improvement in the Glascow Coma Scale (GCS) after synthetic CGRP administration to 15 patients. Experimental data show varying results, and a study by Taylor *et al.*, (1991) on the effect of CGRP infusions in anaesthetized rats showed no effect on cerebral blood flow (CBF) until a threshold dose was reached, then the CBF fell. This could indicate different effects in the normal brain as compared to the injured brain. It is also in accordance with infusion results in healthy volunteers, where a small fall in CBF was noted together with increases in heart rate and cardiac output (Stanley *et al.*, 1991).

To evaluate the effect on the injured brain, we performed a study after ethical approval from the Trondheim University on synthetic CGRP administration to SAH patients who either developed neurological deficits (a 2-point drop in GCS) or MCA velocities exceeding 120 cm/s in the postoperative course. Our goal was to evaluate the hemodynamic changes with transcranial and carotid Doppler before, during, and after infusion. We performed eight infusions of CGRP in five patients (infusion rate 0.06 μg/min). Pulse, blood pressure, GCS, and MCA and ICA velocity were monitored before, during, and after infusion. Infusion time varied from 8 to 52 h. In four infusions we also monitored cardiac output before and during infusion. We observed peripheral vasodilation (facial redness) during five infusions.

Collecting our data for all eight infusions, we observed a significant drop in the hemodynamic index during infusion on the "spasm" side (4.3 ± 0.5) compared to before and after infusion (5.8 ± 0.4, $p < 0.05$) as shown in Fig 3. No change was observed on the unaffected side. We observed a significant increase in heart rate. No significant changes were observed in blood pressure, pulsatility index, or GCS during infusion. Cardiac output measurements during four infusions using ultrasound showed a mean increase of 1.9 l/min. A European multicenter study on CGRP infusion versus standard best management in patients with a 2-point drop in GCS has recently been terminated. Of 99 patients enrolled in the study, 66% showed a good outcome in the CGRP group, while in the standard best management group 59% had a good outcome. This difference was not statistically significant (European CGRP in SAH Study Group, 1991). However, there were serious problems with hypotension in the group who received CGRP.

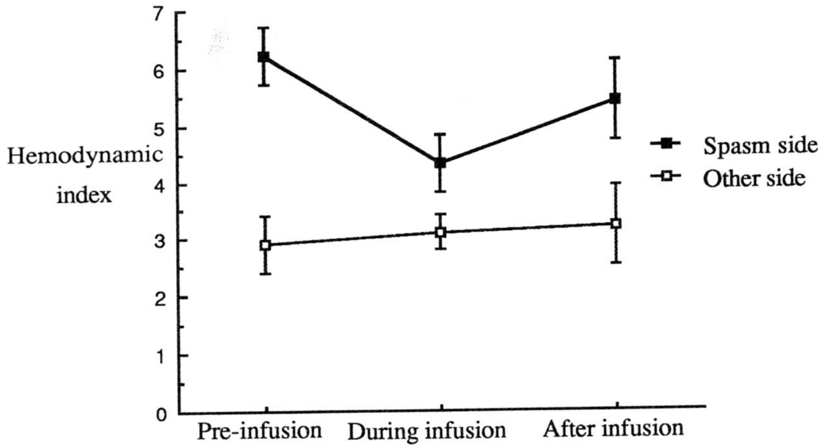

Figure 3 The hemodynamic index before, during, and after infusion is shown on the "spasm" side and on the other side. A significant decrease was found in the hemodynamic index during infusion (4.3 ± 0.5, SEM) compared to pre- and postinfusion (5.8 ± 0.4), $p < 0.05$. Unaltered index values were observed on the other side.

IX. Comments

Plasma and CSF measurements of neuropeptide levels in SAH showed increased levels in a subpopulation of patients with hemodynamical changes monitored with Doppler ultrasound. The increased levels could be explained with activation of the trigemino–cerebrovascular system in SAH as a defense system serving to protect the brain against strong vasoconstriction (McCulloch *et al.*, 1986; Edvinsson *et al.*, 1987). This is also supported by the pilot infusion studies on synthetic CGRP.

Many pathophysiological factors are involved in SAH; recent research has indicated aminergic, peptidergic, and central neurogenic involvement (McCulloch *et al.*, 1986; Delgado *et al.*, 1987; Edvinsson, 1987; Chehrazi *et al.*, 1989). Vascular narrowing, reduced cerebral blood flow, hydrocephalus, blood pressure changes, and intracranial pressure changes are all contributing to the delayed ischemic deficits (Mendelow, 1988). The importance of vasospasm as a clinical problem has diminished in later years, probably due to early operation and the use of calcium antagonists (Allen *et al.*, 1983; Auer, 1984; Pickard *et al.*, 1989). The clinical studies on NPY-LI support the basic research concept that NPY (as a marker for the sympathetic system) may have a role in cerebrovascular regulation and also in the development of vasoconstriction after subarachnoid hemorrhage. The experimental and clinical studies on CGRP-LI indicate that the trigemino–cerebrovascular pathway might be involved in a protective way in the brain circulation, by secreting stored CGRP and tachykinins that will dilate cerebral vessels to alleviate the vasoconstriction induced by the SAH. CGRP seems to be more important than the tachykinins even though direct stimulation of the trigeminal ganglion in cat and man will result in increased levels of CGRP-LI and SP-LI in the EJV (Goadsby *et al.*, 1988). The infusion study results suggest that modulation of spastic vessels is possible, however, no clinical benefits could be observed in the European multicenter CGRP infusion trial. Other administration routes (e.g., intracisternal application) could be an alternative. We may in the future see an increased role for neuropeptides, not only in understanding the pathophysiology but also by therapeutically modulating the vasculature in a subgroup of patients.

Acknowledgments

The work presented was supported by the Swedish MRC (05958), BAYER A. G., and The Søderberg Foundation.

References

Aaslid, R., Markwalder, T. M., and Nornes, H. (1982). Noninvasive transcranial Doppler ultrasound recording of flow velocity in basal cerebral arteries. *J. Neurosurg. 57,* 769–774.

Abel, P. W., and Han, C. (1989). Effects of neuropeptide Y on contraction, relaxation, and membrane potential of rabbit cerebral arteries. *J. Cardiovasc. Pharmacol. 13,* 52–63.

Allen, G. S., Hyo, S. A., Preziosi, T. J., Battye, R., Boone, S. C., Chou, S. N., Kelly, D. L., Weir, B. K., Crabbe, R. A., Lavik, P. J., Rosenbloom, S. B., Dorsey, F. C., Ingram, C. R., Mellits, D. E., Bertsch, L. A., Bolsvert, P. J., Hundley, M. B., Johnson, R. K., Strom, J. A., and Transou, C. R. (1983). Cerebral arterial spasm—a controlled trial of nimodipine in patients with subarachnoid hemorrhage. *N. Engl. J. Med. 308,* 619–624.

Allen, J. M., Todd, N., and Crockard, H. A., Shon, F., Yeats, J. C., and Bloom, S. R. (1984). Presence of neuropeptide Y in human circle of Willis and its possible role in cerebral vasospasm. *Lancet. 2,* 550–552, 1984.

Arbab, M. A. R., Wiklund, L., and Svendgaard, N.-A. (1986). Origin and distribution of cerebral vascular innervation from superior cervical, trigeminal and spinal ganglia investigated with retrograde and anterograde WGA-HRP tracing in the rat. *Neurosci. 19,* 695–708.

Auer, L. M. (1984). Acute operation and preventive nimodipine improve outcome in patients with ruptured cerebral aneurysms. *Neurosurgery 15,* 57–66.

Benedict, C. R., and Loach, A. B. (1978). Sympathetic nervous system activity in patients with subarachnoid hemorrhage. *Stroke 9,* 237–244.

Brodin, E., Lindefors, N., Theodorsson-Norheim, E., Peterson, L.-L., Bartfai, T., Øgren, S.-O., and Rosell, S. (1985). Tachykinins in rat central nervous system: Distribution, molecular forms, release and effects of chronic treatment with antidepressant drugs. *In* "Tachykinin antagonists." (R. Håkanson, and F. Sundler eds.), pp. 15–27. Elsevier Science Publishers, Amsterdam.

Chehrazi, B. B., Giri, S., and Joy, R. M. (1989). Prostaglandins and vasoactive amines in cerebral vasospasm after aneurysmal subarachnoid hemorrhage. *Stroke 20,* 217–224.

Delgado, T. J., Brismar, J., and Svendgaard, N.-A. (1985). Subarachnoid hemorrhage in the rat; Angiography and fluorescence microscopy of the major cerebral arteries. *Stroke 16,* 595–602.

Delgado, T. G., Arbab, M. A.-R., and Rosengrene, E., and Svendgaard, N. A. (1987). The effect of neo-natal 6 hydroxy dopamine treatment on experimental vasospasm following a subarachnoid haemorrhage in the rat. *J. Cereb. Blood Flow Metabol. 7,* 289–294.

Duff, T. A., Scott, G., and Feilbach, J. A. (1986). Ultrastructural evidence of arterial denervation following experimental subarachnoid hemorrhage. *J. Neurosurg. 64,* 292–297.

Edvinsson, L. (1985). Functional role of perivascular peptides in the control of cerebral circulation. *Trends in Neuroscience 3,* 126–131.

Edvinsson, L. (1987). Neurogenic control of cerebral circulation. *In* "Cerebral Ischemia and Hemorheology." pp. 13–23. (A. Hartman and W. Kushinsky, eds.), Springer-Verlag, Heidelberg, Berlin.

Edvinsson, L., Nielsen, K., Owman, C., and Sporrong, R. (1972). Cholinergic mechanisms in pial vessels. Histochemistry, electron microscopy and pharmacology. *Z. Zellforsch 134,* 311–325.

Edvinsson, L., Aubineau, P., Owman, C., Sercombe, R., and Seylaz, J. (1975). Sympathetic innervation of cerebral arteries: Prejunctional supersensitivity to norepinephrine after sympathectomy or cocaine treatment. *Stroke 6,* 525–530.

Edvinsson, L., McCulloch, J., and Uddman, R. (1981). Substance P: Immunohistochemical localization and effect upon cat pial arteries *in vitro* and *in situ*. *J. Physiol. 325,* 251–258.

Edvinsson, L., Egund, N., Owman, C., Sahlin, C., and Svendgaard, N.-A. (1982). Reduced noradrenaline uptake and retention in cerebrovascular nerves associated with angiographically visible vasoconstriction following experimental subarachnoid hemorrhage in rabbits. *Brain Res. Bull. 9,* 799–805.

Edvinsson, L., Copeland, J. R., Emson, P. C., McCulloch, J., and Uddman, R. (1987). Nerve fibers containing neuropeptide Y in the cerebrovascular bed: Immunocytochemistry, radioimmunoassay and vasomotor effects. *J. Cerebr. Blood Flow Metab. 7,* 45–57.

Edvinsson, L., McCulloch, J., Kingman, T. A., and Uddman, R. (1987). Functional significance of the trigeminocerebrovascular innervation: Involvement in cerebrovascular disorders. *In* "Advances in Headache Research." (R. Clifford, ed.), pp. 87–93. John Libbey and Co Ltd, London.

Edvinsson, L., Hara, H., and Uddman, R. (1989). Retrograde tracing of nerve fibers to the rat middle cerebral artery with True Blue, colocalization with different peptides. *J. Cerebr. Blood Flow Metab. 9,* 212–228.

Edvinsson, L., Delgado, T., Ekman, R., Jansen, I., Svendgaard, N-Aa., and Uddman, R. (1990). Involvement of perivascular sensory fibers in the pathophysiology of cerebral vasospasm following subarachnoid hemorrhage. *J. Cerebr. Blood Flow Metab. 10,* 602–607.

Edvinsson, L., Ekman, R., Jansen, I., McCulloch, J., Mortensen, A., and Uddman, R. (1991a). Reduced levels of calcitonin gene-related peptide-like immunoreactivity in human brain vessels after subarachnoid haemorrhage. *Neuroscience Lett. 121,* 151–154.

Edvinsson, L., Alafaci, C., Delgado, T., Ekman, R., Jansen, I., Svendgaard, N. A., and Uddman, R. (1991). Neuropeptide Y and vasoactive intestinal peptide in experimental subarachnoid hemorrhage: Immunocytochemistry, radioimmunoassay and pharmacology. *Acta Neurol. Scand. 83,* 103–109.

Ekblad, E., Edvinsson, L., Wahlestedt, O., Uddman, R., Hakanson, R., and Sundler, R. (1984). Neuropeptide Y coexists and co-operates with noradrenaline in perivascular nerve fibers. *Regul. Pept. 8,* 225–235.

Endo, S., and Suzuki, J. (1975). Experimental cerebral vasospasm after subarachnoid hemorrhage. Participation of adrenergic nerves in cerebral vessel wall. *Stroke, 10,* 703–711.

European CGRP in Subarachnoid Haemorrhage Study Group (1992). The Effect of Calcitonin gene-related peptide on outcome after aneurysmal subarachnoid haemorrhage. Lancet *339,* 831–834.

Fraser, R. A. R., Stein, B. M., Barrett, R. E., and Pool, J. L. (1970). Noradrenergic mediation of experimental cerebrovascular spasm. *Stroke 1,* 356–362.

Goadsby, P. J., Edvinsson, L., and Ekman, R. (1988). Release of vasoactive peptides in the extracerebral circulation of humans and the cat during activation of the trigeminovascular system. *Ann. Neurol. 23,* 193–196.

Grunditz, T., Ekman, R., and Håkanson, R. (1986). Calcitonin gene-related peptide in thyroid nerve fibers and C cells. Effects on thyroid hormone secretion and response to hypercalcemia. *Endocrinology 119,* 2313–2324.

Hamlyn, P. J., Milner, P., Bullock, R., and Burnstock, G. (1991). Substance P and calcitonin gene-related peptide are decreased in CSF following subarachnoid haemorrhage in man: Endothelin is unchanged. *J. Cereb. Blood Flow Met. 11* (Suppl. 2), 241.

Hara, H., and Edvinsson, L. (1988). Perivascular innervation of the cerebral circulation: Involvement in the pathophysiology of subarachnoid hemorrhage. *Rev. Neurosurg. 10,* 171–179.

Hara, H., Hamill, G. S., and Jacobowitz, D. M. (1985). Origin of cholinergic nerves to the rat major cerebral arteries: Coexistence with vasoactive intestinal polypeptide. *Brain Res. 14,* 179–188.

Hara, H., Nosko, M., and Weir, B. (1986). Cerebral perivascular nerves in subarachnoid hemorrhage. A histochemical and immunological study. *J. Neurosurg. 65,* 531–539.

Hardebo, J. E., Arbab, M., Suzuki, N., and Svendgaard, N. A. (1991). Pathways of parasympathetic and sensory cerebrovascular nerves in monkeys. *Stroke 22,* 331–342.

Helm, G., Ekman, R., Rydhstrom, H, Sjoberg, N. O., and Walles, B. (1985). Changes in oviductal VIP content induced by sex steroids and inhibitory effect of VIP on spontaneous oviductal contractility. *Acta Physiol. Scand. 125,* 219–224.

Jackowski, A., Crockard, A., Burnstock, G., and Lincoln, J. (1989). Alterations in serotonin and neuropeptide content of cerebrovascular sympathetic nerves following experimental subarachnoid hemorrhage. *J. Cereb. Blood Flow Metabol. 9,* 271–279.

Jansen, I., Ekman, R., Juul, R., and Edvinsson, L. (1991). Involvement of the sensory neuropeptides calcitonin gene-related peptide and substance P in subarachnoid haemorrhage in man and rat. *J. Cereb. Blood Flow Met. 11* (Suppl. 2), 243.

Johnston, F. G., Bell, B. A., and Miller, J. D., Haliburn, C., O'Shaughnessy, D., Riddell, A. J., and O'Laoire, S. A. (1990). The effect of calcitonin gene related peptide on post-operative neurological deficits following subarachnoid haemorrhage. *Lancet. 335,* 869–872.

Juul, R., Edvinsson, L., and Gisvold, S. E., Ekman, R., Brubakk, A. O., and Fredriksen, T. A. (1990a). Calcitonin gene-related peptide-LI in subarachnoid hemorrhage in man. Signs of activation of the trigemino–cerebrovascular system? *Br. J. Neurosurg. 4,* 171–180.

Juul, R., Edvinsson, L., and Fredriksen, T. A., Ekman, R., Brubakk, A. O., and Gisvold, S. E. (1990b). Changes in the levels of Neuropeptide Y-LI in the external jugular vein in connection with vasoconstriction following subarachnoid hemorrhage in man. *Acta Neurochir. 107:* 75–81.

Krueger, C., Weir, B., Nosko, M., Cook, D., and Norris, S. (1985). Nimodipine and chronic vasospasm in monkeys: Part 2. Pharmacological studies of vessels in spasm. *Neurosurgery 16,* 137–140.

Larsson, L-I., Edvinsson, L., and Fahrenkrug, J., Håkanson, R., Owman, C., Schaffalitzky de Muckadell, O. B., and Sundler, F. (1976). Immunohistochemical localization of a vasodilatory polypeptide (VIP) in cerebrovascular nerves. *Brain Res. 113,* 400–404.

Lee, T. J.-F. (1987). Evidence for and against VIP as a transmitter for vasodilation in cerebral blood vessels. *In* "Peptidergic Mechanisms in the Cerebral Circulation." (L. Edvinsson and J. McCulloch eds.), pp. 64–74. Ellis Horwood, Chichester, U.K.

Lindegaard, K.-F., Grolimund, P., Aaslid, R., and Nornes, H. (1986). Evaluation of cerebral AVMs using transcranial Doppler ultrasound. *J. Neurosurg. 65,* 335–344.

Lindegaard, K.-F., Nornes, H., Bakke, S. J., Sorteberg, W., and Nakstad, P. (1988). Cerebral vasospasm after subarachnoid haemorrhage investigated by means of transcranial Doppler ultrasound. *Acta Neurochirurgica.* (Suppl. 42), 81–84.

Linnik, M. D., Sakas, D. E., Uhl, G. R., and Moskowitz, M. A. (1989). Subarachnoid blood and headache: Altered trigeminal tachykinin gene expression. *Ann. Neurol. 25,* 179–84.

Liu-Chen, L.-Y., Mayberg, M. R., and Moskowitz, M. A. (1983). Immunohistochemical evidence for a substance P-containing trigeminovascular pathway to pial arteries in cat. *Brain Res. 268,* 162–166.

Liu-Chen, L-Y., Gillespie, S. A., Norregaard, T. V., and Moskowitz, M. A. (1984). Colocalization of retrogradely transported wheat germ agglutin and the putative neurotransmitter substance P within trigeminal ganglion cells projecting to cat middle cerebral artery. *J. Comp. Neurol. 225,* 187–192.

Lobato, R. D., Marin, J., Salaices, M., Burgos, J., Rivilla, F., and Garcia, A. G. (1980a). Effect of experimental subarachnoid hemorrhage on the adrenergic innervation of cerebral arteries. *J. Neurosurg. 53,* 477–479.

Lobato, R. D., Marin, J., Salaices, M., Rivilla, F., and Burgos, J. (1980b). Cerebrovascular reactivity to noradrenaline and serotonin following experimental subarachnoid hemorrhage. *J. Neurosurg. 53,* 480–485.

MacPherson, P., and Graham, D. I. (1978). Correlation between angiographic findings and the ischemia of head injury. *J. Neurol. Neurosurg. Psychiatry 41,* 122–127.

Masaoka, H., Suzuki, R., and Hirata, Y., Emori, T., Marumo, F., and Hirakawa, K. (1989). Raised plasma endothelin in aneurysmal subarachnoid haemorrhage. *Lancet* 1402.

Mayberg, M. R., Zervas, N. T., and Moskowitz, M. A. (1984). Trigeminal projection to supratentorial pial and dural blood vessels in cats demonstrated by horse-radish peroxidase histochemistry. *J. Comp. Neurol. 223,* 46–56.

McCulloch, J., and Edvinsson, L. (1980). Cerebral circulatory and metabolic effects of vasoactive intestinal polypeptide. *Am. J. Physiol. 238,* 449–456.

McCulloch, J., Uddman, R., Kingman, T. A., and Edvinsson, L. (1986). Calcitonin gene-related peptide: Functional role in cerebrovascular regulation. *Proc. Natl. Acad. Sci. U.S.A. 83,* 5731–5735.

Mejia, J. A., Pernow, J., von Holst, H., Rudehill, A., and Lundberg, J. M. (1988). Effects of neuropeptide Y, calcitonin gene-related peptide, substance P, and capsaicin on cerebral arteries in man and animals. *J. Neurosurg. 69,* 913–918.

Mendelow, A. D. (1988). Pathophysiology of delayed ischemic dysfunction after subarachnoid haemorrhage: Experimental and Clinical Data *Acta Neurochirurgica. 88* (Suppl. 45), 7–10.

Moskowitz, M. A., Wei, E. P., Saito, K., and Kontos, H. A. (1988). Trigeminalectomy modifies arteriolar responses to hypertension.

Nagai, H., Noda, S., and Mabe, H. (1975). Experimental cerebral vasospasm. Part 2: Effects of vasoactive drugs and sympathectomy on early and late spasm. *J. Neurosurg. 42,* 420–428.

Nozaki, K., Kikuchi, H., and Mizuno, N. (1989). Changes of calcitonin gene-related peptide-like immunoreactivity in cerebrovascular nerve fibers in the dog after experimentally produced subarachnoid hemorrhage. *Neuroscience Lett. 102,* 27–32.

Nozaki, K., Okamoto, S., Uemura, Y., Kikuchi, H., and Mizuno, N. (1990). Erythrocytes suppress calcitonin gene-related peptide- and vasoactive intestinal polypeptide-like immunoreactivities in cerebrovascular nerve fibers after subarachnoid hemorrhage. *Neuroscience Lett. 110,* 22–27.

O'Connor, T. P., and Van der Kooy, D. (1986). Pattern of intracranial and extracranial projections of trigeminal ganglion cells. *J. Neuroscience 6,* 2200–2207.

Peerless, S. J., and Yassargil, M. G. (1971). Adrenergic innervation of the cerebral blood vessels in the rabbit. *J. Neurosurg. 35,* 148–158.

Pickard, J. D., Murray, G. D., and Illingworth, R., *et al.* (1989). Effect of oral nimodipine on cerebral infarction and outcome after subarachnoid haemorrhage: British aneurysm nimodipine trial. *BMJ 298,* 123–128.

Ray, B. S., and Wolff, H. G. (1940). Experimental studies on headache. *Arch. Surg. 41,* 813–856.

Rosenfeld, M. G., Mermod, J. J., and Amara, S. G., *et al.* (1983). Production of a novel neuropeptide encoded by the calcitonin gene via tissue-specific RNA processing. *Nature 304,* 129–135.

Rudehill, A., Lundberg, J. M., Sollevi, A., and Hjemdahl, P. (1987). Elevations of neuropeptide Y-like immunoreactivity and catecholamines in plasma on increased intracranial pressure in the pig. *Acta Anaesthesiol. Scand. 31,* 132–138.

Sato, S., Ishihara, N., Yunoki, K., Nakamigawa, T., Oohira, T., Mitani, S., Ajimi, Y., and Suzuki, H. (1990). "Increased endothelin concentrations in CSF from the patients with SAH." ICCV Tokyo, Abstract book pp. 77.

Sawchenko, P. E., and Swanson, L. W. (1981). A method for tracing biochemically defined pathways in the central nervous system using combined fluorescence retrograde transport and immunohistochemical techniques. *Brain Res. 210,* 31–51.

Shigeno, T. (1982). Norepinephrine in cerebrospinal fluid of patients with cerebral vasospasm. *J. Neurosurg. 56,* 344–349.

Simeone, F. A., and Peerless, S. J. (1975). Prolonged cerebral vasospasm without subarachnoid hemorrhage. *In* "Subarachnoid Hemorrhage and Cerebrovascular Spasm." (R. R. Smith and J. T. Robertson eds.), pp. 206–225. Thomas, Springfield.

Skagerberg, G., Bjørklund, A., and Lindvall, O. (1985). Further studies on the use of the fluorescent retrograde tracer True Blue in combination with monoamine histochemistry. *J. Neurosci. Meth. 14,* 25–40.

Stanley, J., Martin, J., Richards, H., Lovick, A., Barron, M., and Pickard, J. (1991). Effects of calcitonin gene-related peptide on cerebral blood flow in healthy volunteers. *J. Cereb. Blood Flow Met. 11* (Suppl. 2), 273.

Suwanwela, C., and Suwanwela, N. (1973). Intracranial arterial narrowing and spasm in acute head injury. *J. Neurosurg. 36,* 314–323.

Suzuki, H., Sato, S., and Suzuki, Y., Ohtake, R., Hashigami, Y., Namba, J., Ishihara, N., and Shimoda, S. (1989). Pituitary protein 7B2-like immunoreactivity in cerebrospinal fluid: Comparison with other neuropeptides. *J. Lab. Clin. Med. 113,* 743–748.

Suzuki, N., Hardebo, J. E., Kåhrstrøm, J., and Owman, C. (1990). Selective electrical stimulation of postganglionic cerebrovascular parasympathetic nerve fibers originating from the sphenopalatine ganglion enhances cortical blood flow in the rat. *J. Cereb. Blood Flow Met. 10,* 383–391.

Suzuki, Y., Sato, S., Suzuki, H., Namba, J., Ohtake, R., Hashigami, Y., Suga, S., Ishihara, N., and Shimoda, S.-I. (1989). Increased Neuropeptide Y concentrations in cerebrospinal fluid from patients with aneurysmal subarachnoid hemorrhage. *Stroke 20,* 1680--1684.

Svendgaard, N. A., Edvinsson, L., Olin, T., Owman, C., and Sahlin, C. (1977a). On the pathophysiology of cerebral vasospasm: Transmitter changes in perivascular sympathetic nerves, and increased pial artery sensitivity to norepinephrine and serotonin. *In* "Neurogenic Control of Brain Circulation." (C. Owman and L. Edvinsson, eds.), pp. 143–152. Pergamon Press, London.

Svendgaard, N. A., Edvinsson, L., Owman, C., and Sahlin, C. (1977b). Increased sensitivity of the basilar artery to norepinephrine and 5-hydroxytryptamine following experimental subarachnoid hemorrhage. *Surg. Neurol. 8,* 191–195.

Svendgaard, N. A., Brismar, J., Delgado, T. J., and Rosengren, E. (1985). Subarachnoid hemorrhage in the rat: Effect on the development of vasospasm of selective lesions of the catecholamine systems in the lower brain stem. *Stroke 16,* 602–608.

Svendgaard, N.-A. Brismar, J., Delgado, T. J., and Diemer, N. H. (1985). The effect on the development of cerebral vasospasm in the rat of lesioning of the peripheral and central catecholamine systems. *Neurol. Res. 7,* 30–34.

Svendgaard, N.-A., Delgado, T. J., and Brun, A. (1986). Effect of selective lesions in the hypothalamic-pituitary region on the development of cerebral vasospasm following an experimental subarachnoid hemorrhage in the rat. *J. Cereb. Blood Flow Metab. 6,* 650–657.

Taylor, W. A. S., Sydserff, S. G., Bell, B. A., and Ney, U. M. (1991). The effect of calcitonin gene-related peptide in subarachnoid haemorrhage and CBF dose response curve in anaesthetised rats. *J. Cereb. Blood Flow Met. 11* (Suppl. 2), 242.

Toda, N., Ozaki, T., and Ohta, T. (1977). Cerebrovascular sensitivity to vasoconstricting agents induced by subarachnoid hemorrhage and vasospasm in dogs. *J. Neurosurg. 46,* 296–303.

Tsai, S.-H., Lin, S.-Z., Wang, S.-D., Liu, S.-C., and Shih, C.-J. (1985). Retrograde localization of the innervation of the middle cerebral artery with horseradish peroxidase in cats. *Neurosurgery 16,* 463–467.

Tsukahara, T., Taniguchi, T., Fujiwara, M., Handa, H., and Nishikawa, M. (1985). Alterations in alpha adrenergic receptors in human cerebral arteries after subarachnoid hemorrhage. *Stoke 16,* 53–58.

Tsukahara, T., Taniguchi, T., Shimohara, S., Fujiwaea, M., and Handa, H. (1986). Characterization of beta adrenergic receptors in human cerebral arteries and alteration of the receptors after subarachnoid hemorrhage. *Stoke 17,* 202–207.

Uddman, R., Edvinsson, L., Ekman, R., McCulloch, J., and Kingman, T. A. (1985a). Innervation of the feline cerebral vasculature by nerve fibres containing calcitonin gene-related peptide: Trigeminal origin and co-existence with substance P. *Neurosci. Lett. 62,* 131–136.

Uddman, R., Edvinsson, L., Ekman, R., Kingman, T., and McCulloch, J. (1985b). Innervation of the feline cerebral vasculature by nerve fibers containing calcitonin gene-related peptides: Trigeminal origin and co-existence with substance P. *Neurosci. Lett. 62,* 131–136.

Uddman, R., Edvinsson, L., and Svendgaard, N.-A. (1993). Origin of fibers to the middle cerebral artery of the monkey. (To be submitted.)

Uemura, Y., Sugimoto, T., Okamoto, S., Handa, H., and Mizuno, N. (1986). Changes of vasoactive intestinal polypeptide-like immunoreactivity in cerebrovascular nerve fibers after subarachnoid hemorrhage: An experimental study in the dog. *Neurosci. Lett. 71,* 137–141.

Walters, B. B., Gillespie, S. A., and Moskowitz, M. A. (1986). Cerebrovascular projections from the sphenopalatine and otic ganglia to the middle cerebral artery of the cat. *Stroke 17,* 488–494.

Widerløv, E., Heilig, M., Wahlstedt, C., and Ekman, R. (1988). Possible relationship between neuropeptide Y (NPY) and major depression—evidence from human and animal studies. *Nord. Psykiatr. Tidskr. 42,* 131–137.

Wilkins, R. H. (1975). Hypothalamic dysfunction and intracranial arterial spasm. *Surg. Neurol. 4,* 472–480.

Wilkins, R. H. (1980). Attempted preventation or treatment of intracranial arterial spasm: A survey. *In* "Cerebral Arterial Spasm." (R. H. Wilkins ed.), pp. 542–555. Williams and Wilkins, Baltimore.

Wilkins, R. H., and Odom, G. L. (1970). Intracranial arterial spasm associated with craniocerebral trauma. *J. Neurosurg. 32,* 626–633.

Wilson, J. L., and Feild, J. R. (1974). The production of intracranial vascular spasm by hypothalamic extract. *J. Neurosurg. 40,* 473–479.

Yoshihoka, J., Clower, B. R., and Smith, R. R. (1984). The angiopathy of subarachnoid hemorrhage. I. Role of vessel wall catecholamines. *Stroke 15,* 288–294.

CHAPTER 20

Vasoactive Peptides in Migraine and Cluster Headache

PETER J. GOADSBY

Despite considerable advances in the neurosciences over the last three decades, until comparatively recently remarkably little new light has been shed upon the very distressing and common condition of migraine. Recorded in clinical writings since the antiquities (Lance, 1982) and in its form with aura well and widely recognized by both the medical profession and the lay public, the condition defies understanding. It is perhaps a reflection of the enigma that even now argument continues as to whether the condition is primarily neural (Lance *et al.*, 1983) or vascular (Wolff, 1963). This chapter will highlight the contribution of studies of neuropeptides that are now linking our basic understanding of cerebrovascular physiology to the pathogenesis of migraine.

I. Definition

The studies analyzed in this chapter will be divided according to the operational definition of the Classification Committee of the International Headache Society. Migraine is thus defined as headache lasting usually from 4 to 72 h that would be often unilateral, have a pulsating quality, be of moderate or severe intensity, and be aggravated by activity. It is episodic and may often be associated with nausea, vomiting, or photophobia. There should be no neurological abnormality of relevance on either clinical examination or investigation (Olesen, 1988). In clinical practice two varieties of migraine attacks are commonly distinguished: those associated with an aura of neurological symptoms (previously called classical migraine) and a periodic headache without aura (common migraine) (Lance, 1982).

II. Neuroanatomy and Neurochemistry

With the advances in the understanding of the neurobiology of cranial pain processing that have taken place over the last 2 to 3 years, it is essential to review some important aspects of the innervation of the cranial vessels. The data of most importance to the understanding of headache mechanisms is that pertaining to the extrinsic (to the brain) neural innervation of the large cerebral and extracerebral vessels, including the pain sensitive intracranial venous sinuses. This in part stems from the strong clinical impression that the vessels are in some way involved in the disease (Wolff, 1963; Lance, 1982).

Broadly speaking, the innervation of the cerebral circulation is divided into intrinsic and extrinsic neural systems. The intrinsic systems are those that commence within and do not exit the central nervous system, innervating intraparenchymal vessels. Many such systems have been described and are under active investigation (Goadsby and Lance, 1988). Their peptide content is not completely determined and they fall outside the scope of this text. The extrinsic system is, however, almost entirely what this book has characterized. It consists of neurons whose pathways ultimately commence in the central nervous system and then exit synapsing in extraparenchymal ganglia to innervate target vessels. There are three systems that have been well described, the sympathetic system (Section IV), the parasympathetic system (Section V), and the trigeminal system (Section VI). Each is defined anatomically and marked by neuropeptides that give clues as to their possible normal function and role in pathology.

The sympathetic system arises in hypothalamic neurons passing to the intermediolateral cell column of the spinal cord and synapsing before proceeding out to the superior cervical ganglion. Here they again synapse and give rise to the fibers that innervate the vessels. This system is marked by the transmitters or modulator substances noradrenaline, neuropeptide Y (NPY), and peptide YY. It is fundamentally a vasoconstrictor pathway (Olesen and Edvinsson, 1988). The parasympathetic system arises from cell bodies in the superior salivatory nucleus (Spencer *et al.*, 1990), passing out with fibers of the facial nerve (seventh cranial nerve) (Chorobski and Penfield, 1932) and synapsing in the sphenopalatine and otic ganglia (Walters *et al.*, 1986). In some species there are additional microganglia on the internal carotid artery (Gibbins *et al.*, 1984). This system is marked by the neurotransmitters acetylcholine, vasoactive intestinal polypeptide (VIP), and peptide histidine isoleucine [methionine; PHI(M)]. This system is physiologically a vasodilator system. Last, there is a sensory innervation of the cranial vessels and dura mater from the trigeminal system. The cell bodies are bipolar and located in the trigeminal ganglion. They make functional first order connections with neurons in the trigeminal nucleus caudalis and a group of cells at the C_2 level of the spinal cord in the dorsolateral white matter (Goadsby and Zagami, 1991). This system is marked by the peptides calcitonin gene-related peptide (CGRP), substance P (SP), and neurokinin A. It is also an essentially vasodilator system as well as its primary sensory function.

III. Animal Studies

Before turning to work directly carried out in man during headache, the data from animal studies will be considered.

A. Trigeminal Ganglion Stimulation

It is clear, albeit intuitively so, that the trigeminal system must play a key role in migraine. It provides the only known pain-sensitive innervation to the cranial vasculature by virtue of the central convergence of trigeminal and upper cervical pain inputs. This takes place at the second order neurons and has been discussed (Section II). Studies have then examined what takes place when these pathways are activated.

1. Cerebral Blood Flow

Although having no apparent tonic effect on resting cerebral blood flow or glucose utilization (Edvinsson *et al.*, 1986), trigeminal ganglion stimulation increases both bulk extracerebral (Goadsby *et al.*, 1986) and regional cerebral blood flow, specifically in the frontal and parietal cortex (Goadsby and Duckworth, 1987). The effect upon extracerebral blood flow is mediated through strong reflex connections with the facial nerve dilator system that relays some 80% of the dilatation while the remaining 20% is mediated antidromically (Lambert *et al.*, 1984). The cerebral effects are largely mediated by the facial nerve dilator system (Goadsby and Duckworth, 1987) through a classical autonomic nicotinic ganglion (Goadsby, 1991) employing vasoactive intestinal polypeptide as the final transmitter in the system (Goadsby and Shelley, 1990).

2. Neurogenic Inflammation

In a series of experiments Moskowitz's group has looked at changes in vascular permeability and ultrastructural changes in the dura mater of the rat associated with trigeminal ganglion stimulation. It has been demonstrated that trigeminal ganglion stimulation increases vascular permeability in the dura (Markowitz *et al.*, 1987) and that there is coincident mast cell degranulation (Dimitriadou *et al.*, 1990). Both these effects are blocked by pretreatment with the potent antimigraine drugs dihydroergotamine and sumatriptan (Buzzi and Moskowitz, 1990).

3. Peptides

Trigeminal ganglion stimulation results in local (cranial) release of both CGRP and SP in the cat, both acting as markers of trigeminal activation (Goadsby *et al.*, 1988). In a similar study in rat, CGRP has been found to be elevated in the superior sagittal sinus and this increase is attenuated by sumatriptan (Moskowitz, personal communication). Studies of trigeminal ganglion stimulation are hampered by the fact that the ganglion contains cell bodies of not only nociceptive neurons but also proprioceptive, light touch, and others. In short, it has all the sensory modalities of the head and is thus nonspecific in respect of pain. We have therefore turned to the superior sagittal sinus. This structure is pain sensitive in man (McNaughton, 1966) and is predominantly innervated by C fibers. It is clear from electrophysiological studies that most of the neurons excited by sinus stimulation are either nociceptive specific or wide dynamic range (including a nociceptive component) (Zagami and Lambert, 1990). Sagittal sinus stimulation increases regional cerebral blood flow more than does trigeminal ganglion stimulation (Lambert *et al.*, 1988) and leads to the local release of both CGRP and VIP (Zagami *et al.*, 1990). Notably there is no change in either SP or NPY.

IV. Human Data

These data will be considered under the three major headache categories that have been studied, migraine with and without aura and cluster headache. It is important to note, by way of linking the basic science to the clinical data, some observations of direct trigeminal ganglion stimulation in man. Patients being treated for trigeminal neuralgia by thermocoagulation of the ganglion were noted to flush ipsilateral to the side of stimulation. As this flush occurred, blood samples were taken from the external jugular vein and compared with immediately precoagulation samples. There was a marked increase of both CGRP and SP in the patients who flushed compared to the control precoagulation levels (Goadsby *et al.*, 1988). It then appears that much of the animal work that has examined the trigeminovascular system is corroborated by these human studies.

A. *Migraine with Aura (Classical Migraine)*

Early studies of this condition were confined to either peripheral venous blood or cerebrospinal fluid. In peripheral blood, VIP and SP levels are not altered during headache (Blegvad *et al.*, 1986). Furthermore, the often described premonitory symptoms of altered bladder habit, particularly increased urination (Lance, 1982), have led to the suggestion that these may reflect hypothalamic disturbance and therefore be a manifestation of an essentially central nervous system disorder (Lance *et al.*, 1983). Vasopressin levels are unchanged during headache (Blegvad *et al.*, 1986). In an attempt to look more directly at brain function, cerebrospinal fluid levels of various substances have been determined. Morphine-like substances have been reported to be reduced during migraine (Anselmi *et al.*, 1980) although no change in β-endorphin can be detected (Jensen *et al.*, 1985). Interestingly, plasma and platelet met-enkephalin levels are moderately increased during headache (Mosnaim *et al.*, 1985; Ferrari *et al.*, 1987). Considering the contrasting results of these studies and the results of trigeminal ganglion stimulation in man, the levels of various neuropeptides were reexamined in man during migraine. Blood samples were drawn from the cubital fossa and the external jugular vein during headache. Using markers for the sympathetic (NPY), parasympathetic (VIP), and trigeminovascular systems (CGRP, SP), the patients were assessed. There were no changes in the peripheral blood in any of the peptides studied or in NPY, VIP, or SP in the external jugular blood. There were, however, marked increases in CGRP during migraine in man from control levels of 40 pmol/l to 91 ± 11 pmol/l during headache (Goadsby *et al.*, 1990). In two patients with some symptoms similar to cluster headache patients there were increases in VIP, but this was not significant for

Table 1 Peptide Levels during Migraine[a]

	Aura	No aura
Trigeminal		
Substance P	N	N
CGRP	↑	↑
NKA	?	?
Sympathetic		
Neuropeptide Y	N	N
Peptide YY	?	?
Parasympathetic		
VIP	N	N
PHI(M)	?	?
Opioids		
β-Endorphin	N	N
Met-enkephalin	↑	↑
Others		
ACTH	N	N
Vasopressin	N	N

[a]*Abbreviations:* ACTH, adrenocorticotrophin; CGRP, calcitonin gene-related peptide; N indicates the levels were within the normal range for the population studied; NKA, neurokinin A; PHI(M), peptide histidine isoleucine [methionine, PHI(M)]; VIP, vasoactive intestinal polypeptide; ?, data is not available.

the whole group. We have confirmed these observations in a further series of patients who had blood sampling during headache and then after administration of sumatriptan. Remarkably, the CGRP levels return to normal after administration of sumatriptan and successful amelioration of the headache (Goadsby and Edvinsson, 1991). The results of these studies are summarized in Table 1.

B. Migraine without Aura (Common Migraine)

A similar picture emerges for migraine without aura as has been seen for migraine with aura. Plasma levels of SP and VIP are normal both during an attack and interictally (Blegvad *et al.*, 1986). Again as with migraine with aura plasma, vasopressin levels are also unchanged during headache (Blegvad *et al.*, 1986). Conflicting data has emerged concerning the opioids, in two studies met-enkephalin and β-endorphin were shown to be marginally elevated (Mosnaim *et al.*, 1985; Ferrari *et al.*, 1987) while others have found no change specifically in β-endorphin (Jensen *et al.*, 1985). Studies of the cerebrospinal fluid have been limited to demonstrating that there is no change in either SP or somatostatin during headache (Vecchiet *et al.*, 1987).

Again, we have studied patients with migraine without aura during headache by sampling blood from the cubital fossa and external jugular vein (Table 1). As was the case with migraine with aura, substantial increases in CGRP levels were found

Table 2 Peptide Levels during Cluster Headache[a]

	Cluster headache		
	Plasma	CSF	Saliva
Trigeminal			
Substance P	N	N	↑
CGRP	↑	?	N
Sympathetic			
Neuropeptide Y	?	?	?
Parasympathetic			
VIP	N	N	↑
Opioids			
β-Endorphin	↑	N	?
Met-enkephalin	↑	↑	?
Others			
Somatostatin	N	N	?

[a]*Abbreviations:* CGRP, calcitonin gene-related peptide; N indicates the levels were within the normal range for the population studied; VIP, vasoactive intestinal polypeptide; ?, data is not available.

only in the jugular blood while no changes in substance P, VIP, or NPY were seen either in jugular or peripheral blood (Goadsby *et al.*, 1990). Also more recently Nicolodi has used the relatively novel method of salivary sampling and examined peptide levels in association with migraine without aura. They have seen increases in both SP and CGRP along with a reduction in VIP during headache (Nicolodi and Bianco, 1990). The differences in the changes seen with these techniques probably represent the different field of sampling. While the cerebral circulation has a preferentially CGRP-rich trigeminal innervation (O'Connor and van der Kooy, 1988), the innervation of the salivary glands is not similarly characterized. The changes in VIP suggest that some reflex parasympathetically mediated event is also taking place, but this requires further characterization.

C. Cluster Headache

Although this would seem to be an ideal condition to examine in that it is a well-described clear-cut clinical syndrome in which a number of attacks can be sampled, studies of neuropeptide release have not been particularly helpful. Plasma levels of SP and somatostatin show only very modest change (Sicuteri *et al.*, 1985) although met-enkephalin and β-endorphin levels show a consistent elevation during headache (Appenzeller *et al.*, 1981; Hardebo *et al.*, 1985). Cerebrospinal fluid changes have been similar with only changes in met-enkephalin being noted reliably (Table 2) (Hardebo and Ekman, 1987). Interestingly, both substance P and VIP levels are elevated in the saliva during headache. This may well indicate activation of the central and peripheral arms of the trigeminovascular reflex (Goadsby and Lance,

1988), but further studies are needed to clarify the relationship of salivary peptide changes to the problem of headache.

V. Conclusions

The study of neuropeptide levels in migraine and cluster headache is now providing the link between the clinic and research work that is so crucial if the basic pathophysiology of the problem is to be determined. In both migraine with and without aura there are marked changes in cranial levels of CGRP indicating activation of the trigeminal system. These levels are rendered normal by the new and highly effective antimigraine agent sumatriptan in association with the subsiding of the headache. Similarly, CGRP is released with trigeminal activation in animal studies and at least for the rat studies the changes are also inhibited by sumatriptan. These data tell us that the models employed are at least usefully close to the pathways that are activated in migraine. Future work will explore the neural innervation of the cerebral and extracerebral vessels that, along with pharmacological studies of the transmitters involved, will permit both a better understanding of the pathogenesis of migraine and better treatment for its sufferers.

Note Added in Proof

Most recently we have observed marked increases in CGRP in cluster headache patients and a reduction of these levels with oxygen therapy if headache subsides.

Acknowledgments

The work of the author included in this chapter has been supported by the National Health and Medical Research Council of Australia and by grants from Warren and Cheryl Anderson, the J. A. Perini Family Trust, the Basser Trust, and the Australian Brain Foundation. PJG is a Wellcome Senior Research Fellow.

References

Anselmi, B., Baldi, E., Casacci, F., and Salmon, S. (1980). Endogenous opioids in cerebrospinal fluid and blood in idiopathic headache sufferers. *Headache 20,* 294–299.

Appenzeller, O., Atkinson, R. A., and Standefer, J. C. (1981). Serum β-endorphin in cluster headache and common migraine. *In* "Progress in Migraine Research." (F. C. Rose and K. Zilkha, eds.), vol. 1, pp. 106–109. Pitman, Bath.

Blegvad, N., Jensen, K., Fahrenkrug, J., Schaffalitzky de Muckadell, O. B., and Olesen, J. (1986). Plasma-VIP and substance-P during migraine-attack. *Cephalalgia 5* (suppl. 2), 352–353.

Buzzi, M. G., and Moskowitz, M. A. (1990). The antimigraine drug, sumatriptan (GR43175), selectively blocks neurogenic plasma extravasation from blood vessels in dura mater. *Bri. J. Pharmacol. 99,* 202–206.

Chorobski, J., and Penfield, W. (1932). Cerebral vasodilator nerves and their pathway from the medulla oblongata. *Arch. Neurol. Psych. 28,* 1257–1289.

Dimitriadou, V., Buzzi, M. G., Lambracht-Hall, M., Moskowitz, M. A., and Theoharides, T. C. (1990). *In vivo* and *in vitro* ultrastructural evidence for stimulation of dural mast cells by neuropeptides. *Proc. Soc. Neurosci. 16,* 161.

Edvinsson, L., McCulloch, J., Kingman, T. A., and Uddman, R. (1986). On the functional role of the trigemino–cerebrovascular system in the regulation of cerebral circulation. *In* "Neural Regulation of the Cerebral Circulation." (C. H. Owman and J. E. Hardebo, eds.), pp. 407–418. Elsevier Science Publishers B. V., Stockholm.

Ferrari, M. D., Frolich, M., Odink, J., Tapparelli, C., Portielje, J. E. A., and Bruyn, G. W. (1987). Methionine-enkephalin and serotonin in migraine and tension headache. *In* "Advances in Headache Research." (F. C. Rose, ed.), pp. 227–234. John Libbey, London.

Gibbins, I. L., Brayden, J. E., and Bevan, J. A. (1984). Perivascular nerves with immunoreactivity to vasoactive intestinal polypeptide in cephalic arteries of the cat: Distribution, possible origins and functional implications. *Neuroscience 13,* 1327–1346.

Goadsby, P. J. (1991). Characteristics of facial nerve elicited cerebral vasodilatation determined with laser Doppler flowmetry. *Am. J. Physiol. 260,* R255–R262.

Goadsby, P. J., and Duckworth, J. W. (1987). Effect of stimulation of trigeminal ganglion on regional cerebral blood flow in cats. *Am. J. Physiol. 253,* R270–R274.

Goadsby, P. J., and Edvinsson, L. (1991). Sumatriptan reverses the changes in calcitonin gene-related peptide seen in the headache phase of migraine. *Cephalalgia 11,* 3–4.

Goadsby, P. J., and Lance, J. W. (1988). Brainstem effects on intra- and extracerebral circulations. Relation to migraine and cluster headache. *In* "Basic mechanisms of Headache." (J. Olesen and L. Edvinsson, eds.), pp. 413–427. Elsevier Science Publishers, Amsterdam.

Goadsby, P. J., and Shelley, S. (1990). High frequency stimulation of the facial nerve results in local cortical release of vasoactive intestinal polypeptide in the anesthetised cat. *Neurosci. Lett. 112,* 282–289.

Goadsby, P. J., and Zagami, A. S. (1991). Stimulation of the superior sagittal sinus increases metabolic activity and blood flow in certain regions of the brainstem and upper cervical spinal cord of the cat. *Brain 114,* 1001–1011.

Goadsby, P. J., Lambert, G. A., and Lance, J. W. (1986). Stimulation of the trigeminal ganglion increases flow in the extracerebral but not the cerebral circulation of the monkey. *Brain Res. 381,* 63–67.

Goadsby, P. J., Edvinsson, L., and Ekman, R. (1988). Release of vasoactive peptides in the extracerebral circulation of man and the cat during activation of the trigeminovascular system. *Ann. Neurol. 23,* 193–196.

Goadsby, P. J., Edvinsson, L., and Ekman, R. (1990). Vasoactive peptide release in the extracerebral circulation of humans during migraine headache. *Ann. Neurol. 28,* 183–187.

Hardebo, J. E., and Ekman, R. (1987). Substance P and opioids in cluster headache. *In* "Trends in Cluster Headache." (F. Sicuteri, L. Vecchiet, and M. Fanciullacci, eds.), pp. 145–158. Exerpta Medica, Amsterdam.

Hardebo, J. E., Ekman, R., Eriksson, M., Holgersson, S., and Rydberg, B. (1985). CSF opioid levels in cluster headache. Effect of acupuncture. *In* "Migraine: Clinical and Research Advance." (F. C. Rose, ed.), pp. 79–85. Karger, Basel.

Jensen, K., Blegvad, N., Fenger, M., Jordal, R., and Olesen, J. (1985). β-Endorphin and ACTH inplasma during attacks of common and classic migraine. *Cephalalgia 5,* 177–182.

Lambert, G. A., Bogduk, N., Goadsby, P. J., Duckworth, J. W., and Lance, J. W. (1984). Decreased carotid arterial resistance in cats in response to trigeminal stimulation. *J. Neurosurg. 61,* 307–315.

Lambert, G. A., Goadsby, P. J., Zagami, A. S., and Duckworth, J. W. (1988). Comparative effects of stimulation of the trigeminal ganglion and the superior sagittal sinus on cerebral blood flow and evoked potentials in the cat. *Brain Res. 453,* 143–149.

Lance, J. W. (1982). "Mechanism and management of headache (4th edition)." Butterworth Scientific, London.

Lance, J. W., Lambert, G. A., Goadsby, P. J., and Duckworth, J. W. (1983). Brainstem influences on cephalic circulation: Experimental data from cat and monkey of relevance to the mechanism of migraine. *Headache 23,* 258–265.

Markowitz, S., Saito, K., and Moskowitz, M. A. (1987). Neurogenically mediated leakage of plasma proteins occurs from blood vessels in dura mater but not brain. *J. Neurosci. 7,* 4129–4136.

McNaughton, F. L. (1966). The innervation of the intracranial blood vessels and the dural sinuses. *In* "The Circulation of the Brain and Spinal Cord." (S. Cobb, A. M. Frantz, W. Penfield, and H. A. Riley, eds.), pp. 178–200. Hafner Publishing Co. Inc, New York.

Mosnaim, A. D., Wolf, M. E., Chevesich, J., Callaghan, O. H., and Diamond, S. (1985). Plasma methionine enkephalin levels. A biological marker for migraine? *Headache 25,* 259–261.

Nicolodi, M., and Del Bianco, E. (1990). Sensory neuropeptides (substance P, calcitonin gene-related peptide) and vasoactive intestinal polypeptide in human saliva: Their pattern in migraine and cluster headache. *Cephalalgia 10,* 39–50.

O'Connor, T. P., and van der Kooy, D. (1988). Enrichment of a vasoactive neuropeptide (calcitonin gene related peptide) in trigeminal sensory projection to the intracranial arteries. *J. Neurosci. 8,* 2468–2476.

Olesen, J. (1988). Classification and diagnostic criteria for headache disorders, cranial neuralgias and facial pain. *Cephalalgia 8* (suppl. 7), 1–96.

Olesen, J., and Edvinsson, L. (1988). "Basic Mechanisms of Headache." Elsevier Science Publishers, BV, Amsterdam.

Sicuteri, F., Fanciullacci, M., Geppetti, P., Renzi, D., Caleri, D., and Spillantini, M. G. (1985). Substance P mechanisms in cluster headache: Evaluation in plasma and cerebrospinal fluid. *Cephalalgia 5,* 143–149.

Spencer, S. E., Sawyer, W. B., Wada, H., Platt, K. B., and Loewy, A. D. (1990). CNS projections to the pterygopalatine parasympathetic preganglionic neurons in the rat: A retrograde transneuronal viral cell body labeling study. *Brain Res. 534,* 149–169.

Vecchiet, L., Geppetti, P., Marchionni, A., Spillantini, M. G., Fanciullacci, M., and Sicuteri, F. (1987). Cerebrospinal fluid (methionin^5)-enkephalin, substance P and somatostatin-like immunoreactivities in painful and pain-less human diseases. *In* "Trends in Cluster Headache." (F. Sicuteri, L. Vecchiet, and M. Fanciullacci, eds.), pp. 135–143. Exerpta Medica, Amsterdam.

Walters, D. W., Gillespie, S. A., and Moskowitz, M. (1986). Cerebrovascular projections from the sphenopalatine and otic ganglia to the middle cerebral artery of the cat. *Stroke 17,* 488–494.

Wolff, H. G. (1963). "Headache and Other Head Pain." Oxford University Press, New York.

Zagami, A. S., and Lambert, G. A. (1990). Stimulation of cranial vessels excites nociceptive neurones in several thalamic nuclei of the cat. *Experl. Brain Res. 81,* 552–566.

Zagami, A. S., Goadsby, P. J., and Edvinsson, L. (1990). Stimulation of the superior sagittal sinus in the cat causes release of vasoactive peptides. *Neuropeptides 16,* 69–75.

CHAPTER 21

Perivascular Peptides in the Respiratory Tract

J. G. WIDDICOMBE

I. Introduction

The respiratory tract is anatomically very complex, varying from the rigid bony cavity of the nose with its convoluted system of turbinates, through the pharynx and larynx where striated muscle profoundly affects caliber, to the tracheobronchial tree with distensible walls subject to the contraction of airway smooth muscle. Different parts of the respiratory tract have quite different functions. It is therefore not surprising that the vascular beds, for example those in the nose, the trachea, and the bronchi, show considerable structural and functional differences. Nonetheless, the general controls of these vascular beds in terms of neurotransmitters, including peptides, are very similar, even if the functional sequelae diverge.

Little is known of the control of the vasculature of the pharynx and larynx, which therefore will not be described in detail in this chapter, although neuropeptide-containing nerves have been identified at these sites (Albegger *et al.*, 1991). However, the neural controls of the nasal, tracheal, and bronchial vasculatures all have three components: a parasympathetic vasodilator system, a sympathetic vasoconstrictor system, and a sensory nervous system mediating axon reflexes and "neurogenic inflammation" (Figs. 1 and 2). The neuropeptides in these three components are likewise qualitatively similar for the nose and lower airways. However, species differences may exist. The histology of peptide-containing nerves has been reviewed (Uddman and Sundler, 1986, 1987; Uddman *et al.*, 1987; Lundberg *et al.*, 1987; Springall *et al.*, 1988; Lundberg, 1990), as have their vascular physiology and pharmacology (L. A. Laitinen *et al.*, 1987a,b,c; Salonen *et al.*, 1990; Widdicombe, 1990a).

With regard to the clinical implications of peptidergic control of the airway vasculature, this has for long been established for the nose where both allergic rhinitis ("hay fever") and "vasomotor rhinitis" involve considerable vasodilatation and vascular congestion, leading to nasal blockage. For the lower airways it is now agreed that asthma is primarily a condition of mucosal inflammation, and vascular

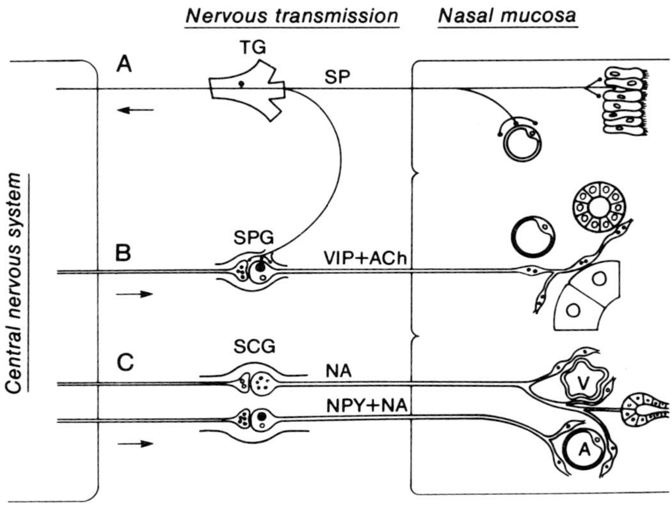

Figure 1 Schematic illustration of the principle peptidergic innervation of the nasal mucosa in relation to "classical" autonomic pathways. (A) Sensory nerve fibers from the trigeminal ganglion containing substance P–like immunoreactivity (SP) with nerve endings around sphenopalatine ganglion cells and around blood vessels, as well as under or within the epithelium. (B) Preganglionic cholinergic fibers synapse in the sphenopalatine ganglion and activate nicotinic receptors on postganglionic cholinergic (ACh) neurons that also contain VIP. (C) Some postganglionic sympathetic adrenergic neurons (NA) innervating arteries (A) also contain neuropeptide Y (NPY) immunoreactivity. The venous sinusoids (V) principally have adrenergic (NA) innervation. Regarding coexistence of peptides, see text. (SCG, superior sympathetic ganglion.) (From Uddman *et al.,* 1987.)

changes in this condition are certainly powerful and may well be basic; they lead to hyperemia and possibly edema with thickening of the mucosa, and with exudate of plasma proteins into the airway lumen.

II. Vascular Anatomy

The nasal vasculature is similar to that of the lower respiratory tract in its general arrangement, namely a copious subepithelial plexus of capillaries with a deeper submucosal supply of arteries, venules, and veins (Widdicombe, 1990b). However, there are several important differences.

First, the nasal mucosa has a system of arteriovenous anastomoses (AVAs) that can be demonstrated both anatomically and physiologically (Anggard, 1974; Baraniuk *et al.,* 1991). These are especially prominent in the anterior respiratory nasal mucosa but are less frequent in the posterior olfactory part. They are absent from the trachea, at least of the sheep (Corfield *et al.,* 1992), and have not been found in the bronchi.

Second, the nasal mucosa, especially in the respiratory region over the septum

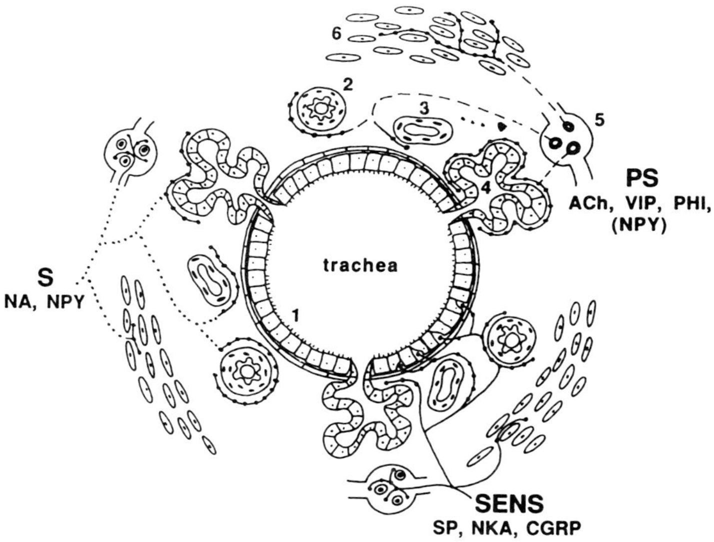

Figure 2 Schematic drawing summarizing the sensory (SENS), parasympathetic (PS), and sympathetic (S) innervation of the trachea. Sensory nerves containing SP-, NKA-, and CGRP-LI are seen close to and within the respiratory epithelium (1), around arteries (2) and veins (3), and close to exocrine glands (4) and the local parasympathetic ganglion cells (5). SP-, NKA-, and CGRP-IR nerves also occur within the tracheobronchial smooth muscle layer (6). Parasympathetic fibers that contain ACh and VIP/PHI surround both the exocrine glands and the blood vessels in the trachea. ACh-, VIP/PHI-, and NPY-IR fibers are also present within the tracheobronchial smooth muscle layer. Sympathetic NA nerves occur around blood vessels and close to local parasympathetic ganglion cells and surround the exocrine glands in the trachea. NPY- and DBH-IR nerves are observed close to the blood vessels, and sparse NA innervation is also seen within the tracheobronchial smooth muscle layer. Sympathetic motor neurons (---), parasympathetic motor neurons (...), and sensory neurons (_____). (From Matran, 1991.)

and turbinates, has an exceptionally well-developed system of capacitance vessels or sinusoids (Cauna, 1982; Eccles, 1982). These have much smooth muscle in their walls, and distension of the capacitance vessels is the main primary cause of nasal blockage during congestion. The lower airways also have capacitance vessels, but the system is less well developed than in the nose and shows considerable species variation (Hughes, 1965; Hill *et al.*, 1989; A. Laitinen *et al.*, 1989).

In the sheep and rabbit the capacitance system is extensive, and its engorgement can almost double the thickness of the mucosa for the trachea (Corfield *et al.*, 1991). In the bronchi, capacitance-like vessels are also seen, especially at points of branching of the larger airways. The system becomes less conspicuous in the smaller airways. In the dog and cat the network is less well developed, with man being intermediate. Another striking difference is that the capacitance vessels in rabbit and man have much smooth muscle in their walls although this is absent in vessels of the sheep. The functional results of these differences are not clear.

A third difference between the vasculatures of the nose and of the lower airways is that the capillaries in the subepithelial plexus of the nose are fenestrated (Watenabe and Watenabe, 1980; Baraniuk, 1991), whereas in most species those in the lower airways are nonfenestrated (McDonald, 1990). One exception applies to the capillaries underlying neuroepithelial bodies, the paracrine structures found in the bronchial epithelium of some species (Sorokin and Hoyt, 1990); these seem usually to be fenestrated. One would suppose that plasma transudation occurs far more readily in the nose compared with the lower airways.

The vasculatures of the trachea and bronchial tree also show striking morphological differences. That of the trachea drains into the systemic veins, whereas most of the bronchial blood enters the pulmonary circulation (Magno, 1990). Some, usually said to be about 25%, drains to the azygos and hemiazygos veins on the systemic side. As a result the bronchial circulation probably flows longitudinally down the bronchi through mainly parallel vessels. The difference between the vasculatures has important hemodynamic consequences. That of the trachea will behave like any other systemic vascular bed, but the bronchial vasculature will be profoundly influenced by events in the pulmonary circulation. Indeed it can be shown that under some circumstances backflow from the pulmonary to the bronchial vessels is considerable (Deffebach, 1990).

Apart from these structural differences, the nasal vasculature has an important functional role not seen in the lower airways; this is to mediate countercurrent mechanisms both in terms of conditioning of inspired and expired air in the nose and in terms of controlling cerebral blood temperature (Cole, 1982). The role of neuropeptides in these unique features of the nasal vasculature has not been studied.

Both in the nose and in the tracheobronchial tree, nerves, usually containing agranular and dense-cored vesicles, can be seen under the electron microscope near or close to the vascular smooth muscle (Fig. 3) (L. A. Laitinen *et al.*, 1987a).

III. Parasympathetic Neuropeptides

Vasoactive intestinal polypeptide (VIP) is the most studied parasympathetic motor neuropeptide, although peptide histidine isoleucine (PHI) and its human variant peptide histidine methionine (PHM) may coexist or be more important in some species. Their actions seem to be similar, and they all come from a common molecular precursor. They coexist in parasympathetic postganglionic nerves with acetylcholine (ACh), the classical parasympathetic neurotransmitter.

A. *Histology*

1. The Nose

Immunostaining methods show that VIP and PHI occur in many nerve fibers in the nasal mucosa, and study of sequential sections suggests that the two peptides coexist. They seem to be especially prominent in the respiratory (anterior) mucosa, although this may merely be due to the fact that the vasculature and submucosal

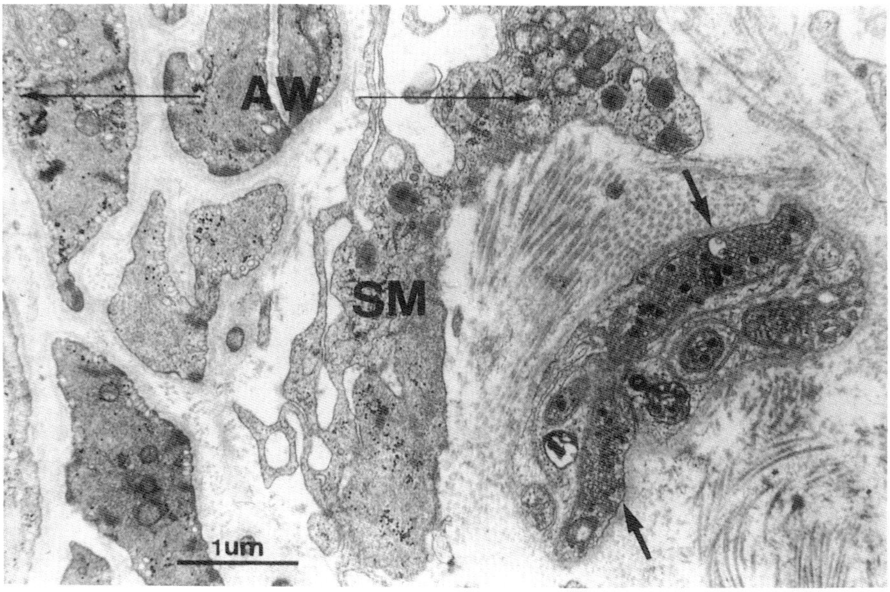

Figure 3 Transmission electron micrograph of canine airway mucosa. Vascular lumen is to the left of the picture. Axon profiles (N_1 and N_2) are seen close to smooth muscle cells (SM) of the arteriolar wall (AW). N_1 contains small granular vesicles (30–40 nm diameter) and may represent an adrenergic nerve. N_2 contains many small agranular vesicles (30–40 nm diameter) but also several large granular vesicles (60–90 nm diameter). The nature of the transmitter in the small agranular and large granular vesicles remains open in conventional electron microscopy, but this type of nerve could represent a peptidergic nerve. Bar, 1 μm. (From L. A. Laitinen *et al.*, 1987a.)

glands are better developed there. Fibers containing these neuropeptides are found around small blood vessels, glands, and beneath the epithelium, including in man (Uddman and Sundler, 1986, 1987; Uddman *et al.*, 1987; Albegger *et al.*, 1991). There are also numerous nerves containing VIP/PHI in the sphenopalatine ganglion (parasympathetic motor to the nose) (Uddman and Sundler, 1987), suggesting the possibility of local reflex modulation at this site. VIP seems to be present in nerve cells that also stain for acetylcholinesterase, supporting the idea that VIP and ACh are colocalized in the same neurons.

Galanin can also be found by immunohistological methods in the nasal mucosa, with the same neural distribution of that of VIP or ACh (Cheung *et al.*, 1985). However, the evidence for galanin as a parasympathetic motor neuropeptide is far less secure than that for VIP, and its actions have been less studied. It may also occur in sensory nerves.

2. The Trachea and Bronchi

The general arrangement of VIP/PHI-containing nerves in the trachea is very similar to that in the nose, namely with a clear association with blood vessels and

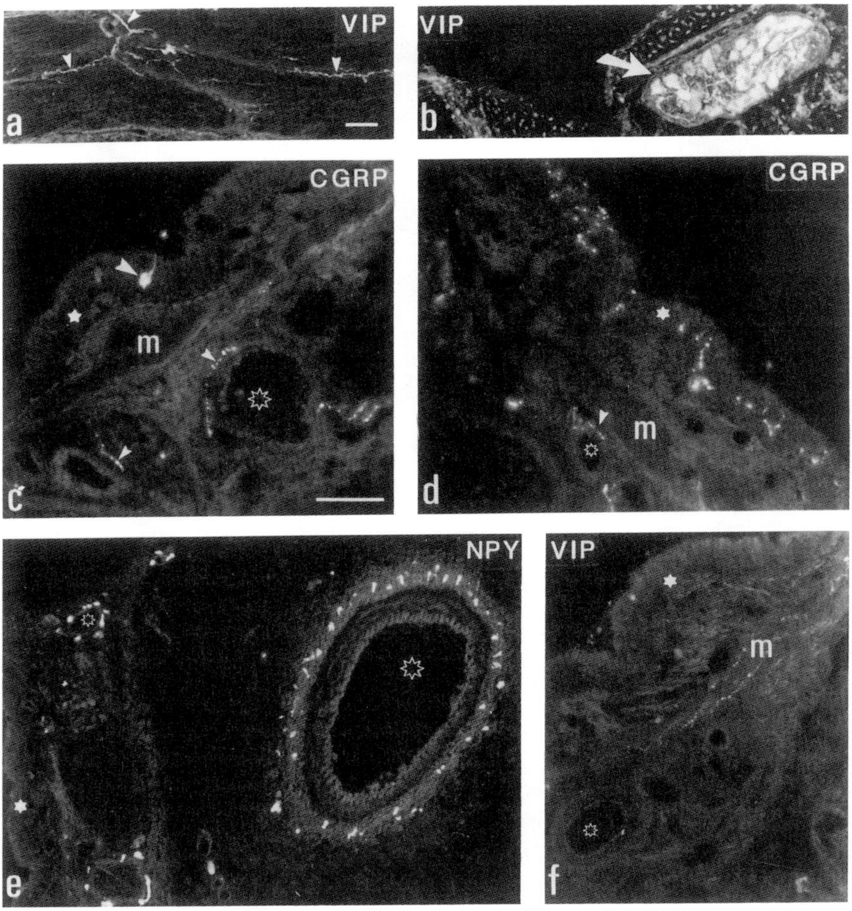

Figure 4 Immunofluorescence micrographs of (a) trachea, (b) a stem bronchus, and (c–f) peripheral bronchi of the pig after incubation with antisera to VIP (a,b,c), CGRP (c,d), or NPY (e). (a) Arrowheads, varicose nerve fibers within the tracheal smooth muscle layer. (b) Large arrow, local ganglion containing many VIP-positive cell bodies. (c,d) Filled star, lining epithelium with an endocrine-like cell (large arrowhead) or intraepithelial nerve fibers. (d) Small arrowheads, perivascular nerve fibers. Open stars are placed in the lumen of thin-walled blood vessels (presumably small veins); m, bronchial smooth muscle. (e) Filled star, bronchial epithelium; small open star, NPY-innervated bronchial blood vessel; large open star, lumen of a pulmonary artery. (f) Filled star, bronchial epithelium; open star, lumen of small bronchial artery; m, bronchial smooth muscle displaying VIP-positive fibers. Bars in (a) and (c), 50 μm; (a), (b), and (c–f) have the same magnification. (From Martling *et al.*, 1990.)

submucosal glands as well as with airway smooth muscle (Fig. 4) (L. A. Laitinen, 1988). Since the two peptides show a similar distribution, they may well be colocalized in the vagal motor nerves. VIP-containing nerves and cells are also found in ganglia in the tracheobronchial mucosa (Uddman *et al.*, 1978) (Fig. 4). However, at least for the pig and probably the cat, a striking observation is that few VIP/PHI-containing nerves are found in the bronchi and smaller airways, their distribution being mainly restricted to the trachea (Martling *et al.*, 1990). This observation correlates with functional studies (see later). In smaller mammals, VIP-containing nerves are more conspicuous around submucosal glands and less frequent in vascular smooth muscle (Springall *et al.*, 1988).

Immunohistological methods show that galanin has a similar distribution to VIP/PHI in many species (Cheung *et al.*, 1985; Springall *et al.*, 1988). The highest concentration is in the main bronchi, and it is found around blood vessels, submucosal glands, and in the airway smooth muscle. If, as has been suggested, galanin coexists with VIP/PHI, then the virtual absence of the latter from the blood vessels of the bronchi and smaller airways of the pig has not been described for galanin, but possibly experiments directly to test this point have not been carried out.

B. Pharmacology

1. The Nose

For the perfused nose of the dog, pig, and cat, VIP is a powerful vasodilator (Lung *et al.*, 1984; Anggard *et al.*, 1979; Uddman *et al.*, 1980; Stjarne, 1991) (Fig. 5). PHI, PHM, and galanin do not seem to have been studied in this way. The methods used do not allow determination of which blood vessels are acted on by VIP, although it increases mucosal volume slightly in the pig (Stjarne, 1991) (Fig. 5); in view of the large decrease of total vascular resistance, it seems likely that at least the arterioles, the main site of vascular resistance, are affected. Measurements of nasal airflow resistance during the vasodilatation due to VIP suggest that the latter is accompanied by very little mucosal congestion or thickening, at least in the dog, since airflow resistance only slightly increases (Lung *et al.*, 1984) (Fig. 6). This is in contrast to the effect of other vasodilators, such as bradykinin and histamine. It is possible that VIP opens AVAs, thereby increasing nasal blood flow, but without distending the nasal capacitance vessels.

Stimulation of parasympathetic nerves to the nose of the cat causes a strong vasodilatation that is atropine-resistant (Anggard *et al.*, 1979). This result, together with the simultaneous release of VIP into the venous outflow of the nose (Lundberg *et al.*, 1981), suggests that VIP is the effective vasodilator from parasympathetic nerves under physiological conditions.

2. The Trachea and Bronchi

Both VIP and PHI are powerful dilators of the tracheal vasculature, at least in the dog and pig (L. A. Laitinen *et al.*, 1987b; Salonen *et al.*, 1988; Matran, 1991;

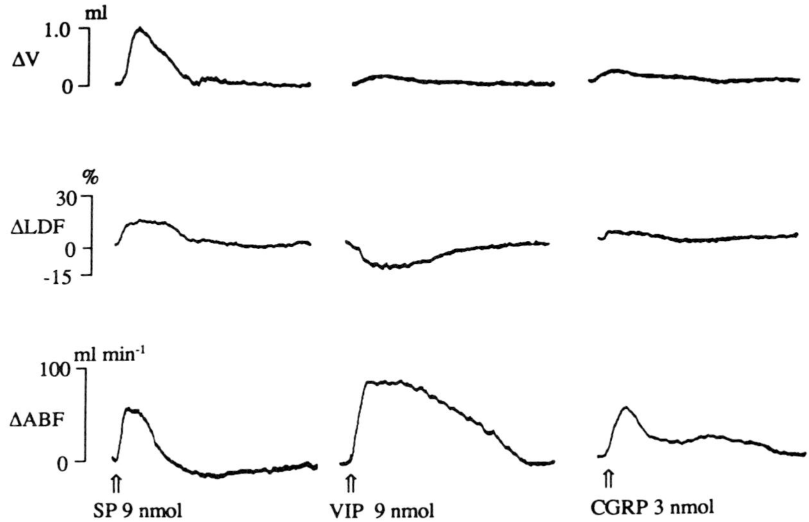

Figure 5 Typical recordings of changes (Δ) in nasal mucosal volume (V), laser-Doppler signal (LDF), and nasal arterial blood flow (ABF) upon infusion of substance P (SP), vasoactive intestinal polypeptide (VIP), and calcitonin gene-related peptide (CGRP) in a pig. All agents caused increased blood flow with various changes in mucosal volume. (From Stjarne, 1991.)

Matran *et al.,* 1989b) (Fig. 7). In the dog, VIP is 10 times more potent than is PHI. The bronchial vasculature also dilates in response to exogenous VIP but, at least in the pig, far less than does that of the trachea (Matran, 1991; Matran *et al.,* 1989a).

Vagal nerve stimulation dilates both the tracheal and bronchial vasculatures of experimental animals (L. A. Laitinen *et al.,* 1987b; Martling *et al.,* 1985; Matran *et al.,* 1989a,c). If the cholinergic component is abolished by atropine, the residual dilatations, due presumably to peptides, differ for the trachea and bronchi. The former seems to be mediated mainly by VIP-like peptides, whereas the latter is abolished by capsaicin desensitization and is presumably due to sensory neuropeptides released antidromically (Matran *et al.,* 1989c). The atropine-sensitive and insensitive components are stimulus frequency dependent, the results being consistent with the view that high-frequency impulses release predominantly VIP-like peptides whereas low-frequency impulses release a preponderance of ACh.

The action of galanin on tracheobronchial vasculature has not been greatly studied. However, at least for the dog it seems to be completely inactive in concentrations that, for other neuropeptides, can cause a very considerable vascular response (Salonen *et al.,* 1988).

IV. Sympathetic Neuropeptides

Neuropeptide-Y (NPY) is the cotransmitter with noradrenaline (NA) in many sympathetic postganglionic fiber systems (Uddman and Sundler, 1986, 1987; Lundberg, 1990; Lundberg *et al.,* 1989). Although in some vascular beds it is only a weak

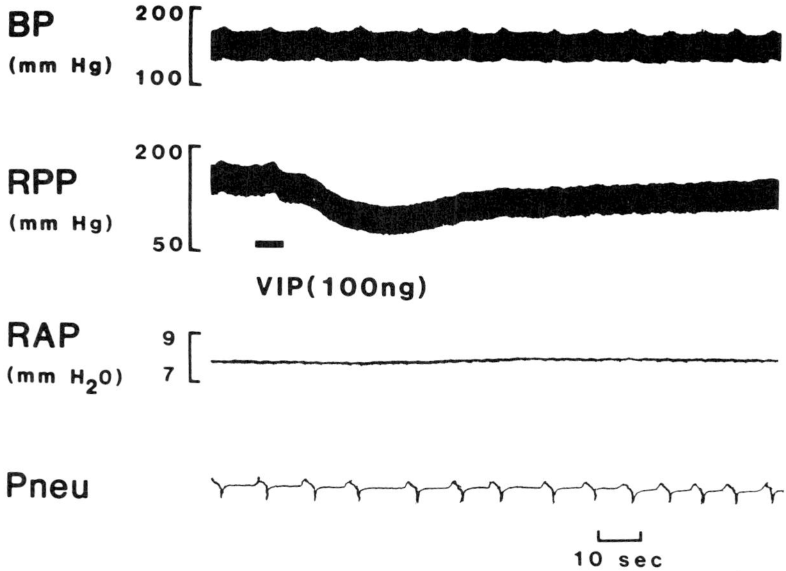

Figure 6 Effects on nasal arterial perfusion pressure in a dog after intraarterial injection of VIP (100 ng injected at signal bar). Traces from above down: systemic arterial blood pressure (BP); right nasal arterial perfusion pressure (RPP); right nasal airway pressure (RAP); tracheal airflow (Pneu). Note the vasodilatation without increase in airways resistance. (From Lung *et al.*, 1984.)

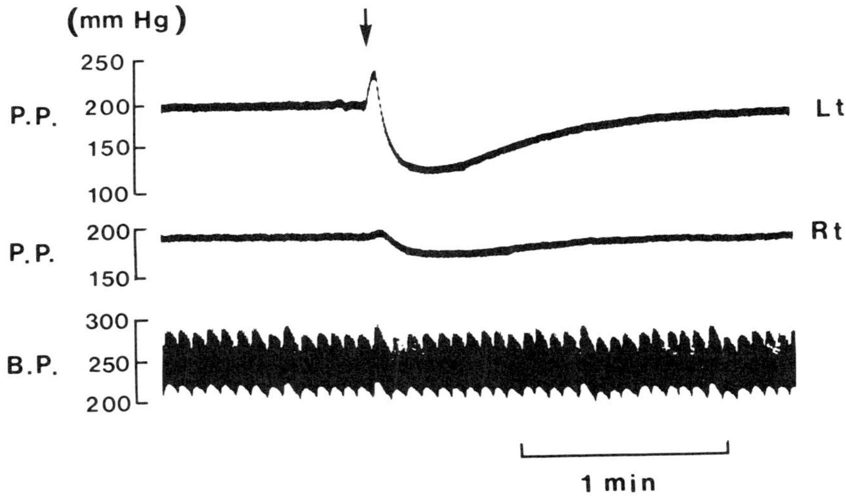

Figure 7 Responses of right and left tracheal vascular beds to close-arterial injection of VIP (60 nmol, at arrow) on the left side. Traces from above down: left arterial perfusion pressure (PP); right arterial perfusion pressure; systemic arterial blood pressure (BP). The transient rise in left arterial perfusion pressure immediately after the injection is a pressure artifact due to the injection. Subsequently, both perfusion pressures fall, with a large effect on the ipsilateral side (43% decrease) and a smaller response on the contralateral side (12% decrease). There are small changes in systemic arterial blood pressure. (From L. A. Laitinen *et al.*, 1987b.)

vasoconstrictor, it can strongly enhance and prolong the vasoconstriction produced by NA.

A. Histology

1. The Nose

Nerves containing NPY, identified by immunohistochemistry, are found in the nasal mucosa, especially around blood vessels, including resistance vessels, capacitance vessels, and AVAs (Baraniuk *et al.*, 1990, 1991; Uddman *et al.*, 1987; Lacroix, 1989) (Fig. 8). The pattern has been mapped out, especially in pig and man. In other tissues NPY coexists with NA in dense-cored vesicles, and there is no reason to doubt that this applies also to the nose. Surgical (sympathectomy) or chemical (with 6-hydroxydopamine or reserpine) interventions lead to histological

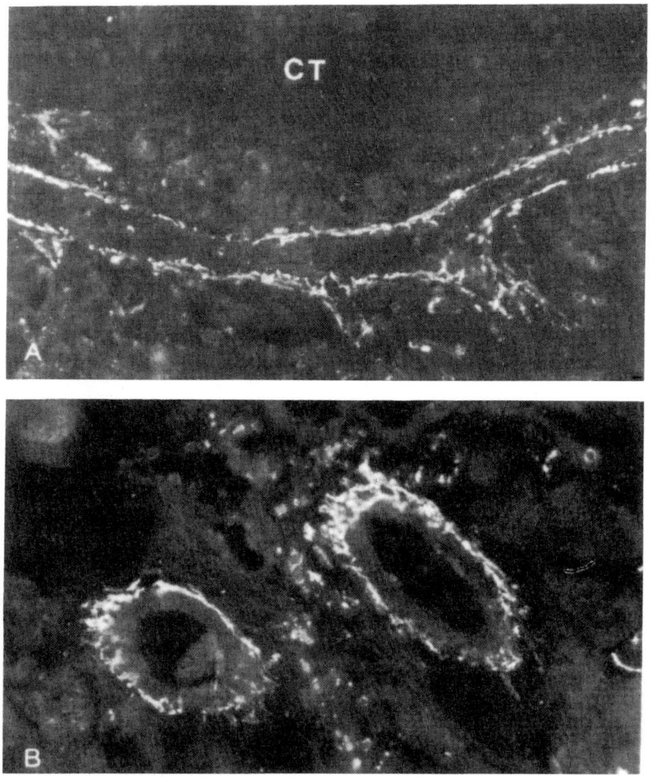

Figure 8 Cat nasal mucosa. Submucosal blood vessels contain a dense supply of neuropeptide Y fibers. (A) Longitudinally cut blood vessel. CT, cartilage. (B) Cross-sectioned blood vessels. Magnification ×150. (From Uddman and Sundler, 1986.)

disappearance of both NA and NPY. In general NPY-containing fibers are most frequent in the walls of arterioles and AVAs, with the venules and capacitance vessels containing far fewer. Receptor-binding studies in man show a predominance of NPY-receptors at the same former vascular sites (Baraniuk *et al.*, 1991). Nerves and receptors are more frequent in the respiratory mucosa than the olfactory, again possibly because the vascular bed is more dense in the former site.

2. The Trachea and Bronchi

Immunohistochemistry identifies nerves containing NPY in the mucosa of the lower respiratory tract (Lundberg *et al.*, 1989; Martling *et al.*, 1990) (Fig. 4). The fibers are particularly conspicuous around blood vessels, especially arterioles, and are also found in airway smooth muscle but are scanty in the region of submucosal glands and the airway epithelium. There are pronounced species differences, the NPY-containing fibers being frequent in the guinea pig but rather infrequent in man and pig (Springall *et al.*, 1988). Comparison of the trachea and the smaller bronchi suggests that the former contains 15 times more fibers than the latter, at least in the pig (Martling *et al.*, 1990). As with the nose and other vascular sites, NPY coexists with NA. Around glands and in the smooth muscle the rather sparse fibers contain not only these two transmitters but some VIP/PHI. Sympathetic ganglia, especially those associated with vasomotor nerve fibers, contain many cell bodies and nerves with NPY (Fig. 9), and the vagal ganglia also contain a little of the peptide (Uddman and Sundler, 1987; Springall *et al.*, 1987). For the tracheobronchial tree, surgical (sympathectomy) and chemical (with 6-hydroxydopamine or reserpine) interventions lead to histological disappearance of the NA and NPY.

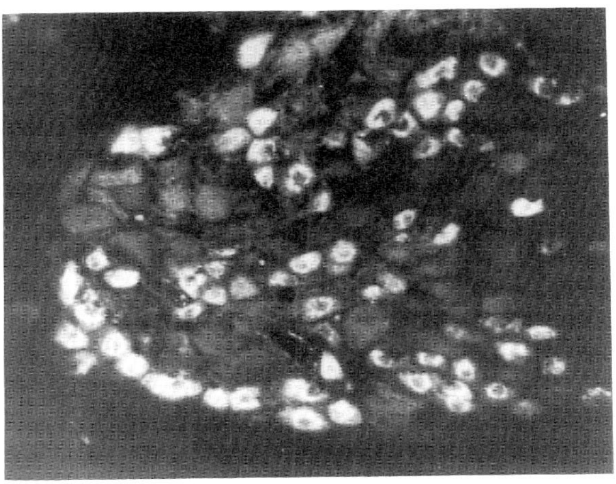

Figure 9 Nasal mucosa of rat. Superior cervical ganglion. Nerve cell bodies displaying NPY-immunoreactivity. Magnification ×225. (From Uddman *et al.*, 1987.)

B. Pharmacology

1. The Nose

Exogenous NPY increases nasal vascular resistance and decreases nasal mucosal volume, indicating decongestion of the nasal mucosa (Salonen *et al.*, 1988; Lacroix, 1989). The vasoconstriction due to NPY has a longer latency and lasts far longer than that due to NA (Fig. 10), and is not influenced by α-adrenoceptor antagonists. Stimulation of sympathetic nerves to the nose causes a pronounced vasoconstriction and decrease in airflow resistance (Lacroix *et al.*, 1988a); these responses are decreased but not eliminated by α-adrenoceptor antagonists (Lacroix *et al.*, 1988b). Measurements of NPY output in the venous effluent of the nose show that this is enhanced by sympathetic nerve stimulation and not prevented (indeed increased) by α-adrenoceptor antagonists (Lundblad *et al.*, 1987; Lacroix *et al.*, 1988b). All of these results are consistent with the view that NPY is a functional nasal "vasoconstrictor," in cooperation with NA. NA release is greater at low-frequency stimulation, and NPY release is greater at high frequencies (Lundblad *et al.*, 1987; Lacroix *et al.*, 1988a). The results are consistent with a colocalization of the two transmitters, with frequency-dependent release of them, and with regulation of their release by prejunctional α-adrenoceptor mechanisms. NPY certainly empties capacitance vessels, and the histological evidence suggests that this may be by

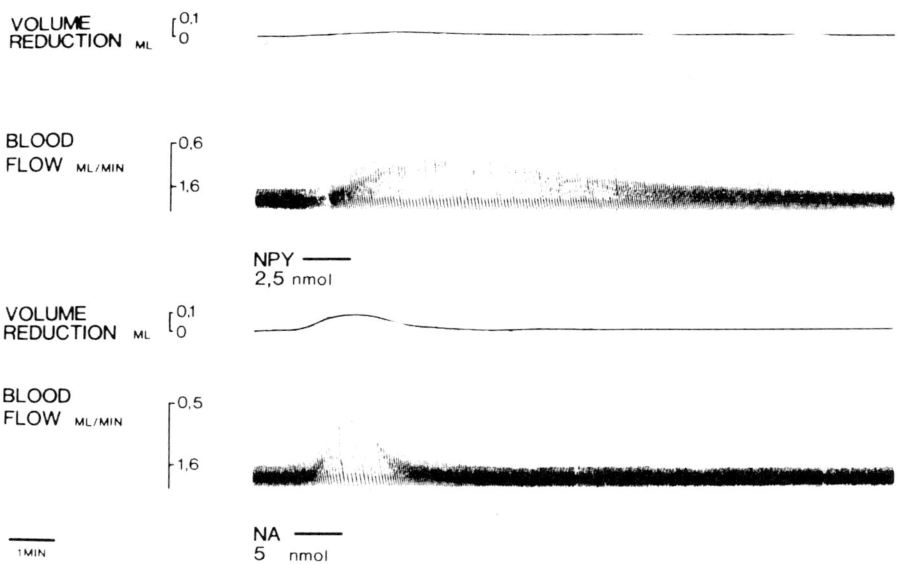

Figure 10 Effects of intraarterial infusions of NPY (2.5 nmol) and NA (5 nmol) on cat mucosal blood flow (ml/min) and volume of the nasal cavity (volume reduction in ml). NPY induces a slowly developing long-lasting vasoconstriction compared to NA. On the other hand, NA induces a more pronounced volume reduction relative to NPY. (From Lundblad *et al.*, 1987.)

contracting the muscle of AVAs. However, the precise vascular sites of action have not been fully worked out.

2. The Trachea and Bronchi

Exogenous NPY increases vascular resistance in the trachea and the bronchi (Salonen *et al.*, 1988; Martling *et al.*, 1990; Matran *et al.*, 1992). Stimulation of the sympathetic nerves to the lower airways causes vasoconstriction and tissue depletion of NA and NPY (Matran, 1991). As for the nose, the two mediator effects depend on stimulus frequency; the response to NA is preferentially at low frequencies whereas that to NPY is greater at high frequencies.

The results for trachea and bronchi compared to nose are qualitatively very similar. However, in addition, the effect of NPY in constricting AVAs has no counterpart for the lower airways where these vessels are absent; the extent to which NPY in the nose constricts capacitance vessels, as distinct from preventing their passive distension, has not been worked out and we do not know if NPY has an action on the rather less well-developed capacitance system of the tracheobronchial tree.

V. Sensory Neuropeptides

It is well established that in many tissues, including the nose and the tracheobronchial tree, there are sensory receptors with branches close to blood vessels and containing one or more neuropeptides (Uddman *et al.*, 1987; Lundberg, 1990). These can be released on activation of the sensory receptors, giving rise to both local axon reflexes ("neurogenic inflammation") (McDonald, 1987) and possibly also central nervous reflexes. For the nose the sensory receptors have not been mapped out well but are probably nonmyelinated (C-fiber) endings with terminals ramifying in the mucosa and epithelium. Their central reflex actions may include sneeze, vasodilatation, and mucus secretion. The neuropeptides they contain include in most species substance P (SP), neurokinin-A (NKA), and calcitonin gene-related peptide (CGRP), and possibly sometimes neurokinin-B (NKB) and galanin (Fig. 11). For the tracheobronchial tree there are probably at least two types of sensory receptor that contain neuropeptides: the nonmyelinated (C-fiber) endings that are probably structurally and functionally very similar to those in the nose; and in addition the rapidly adapting or irritant receptors that have nonmyelinated terminals within the epithelium and deeper parts of the mucosa, and thin myelinated fibers in the vagus nerves. The central reflex actions of the airway C-fiber receptors include rapid shallow breathing and hypotension due mainly to vasodilatation, whereas those from the rapidly adapting receptors include cough.

The sensory receptors of all three groups are activated by irritant gases and aerosols, and by many inflammatory mediators. Their axon reflex responses include local vasodilatation and mucus secretion, and possibly changes in epithelial function.

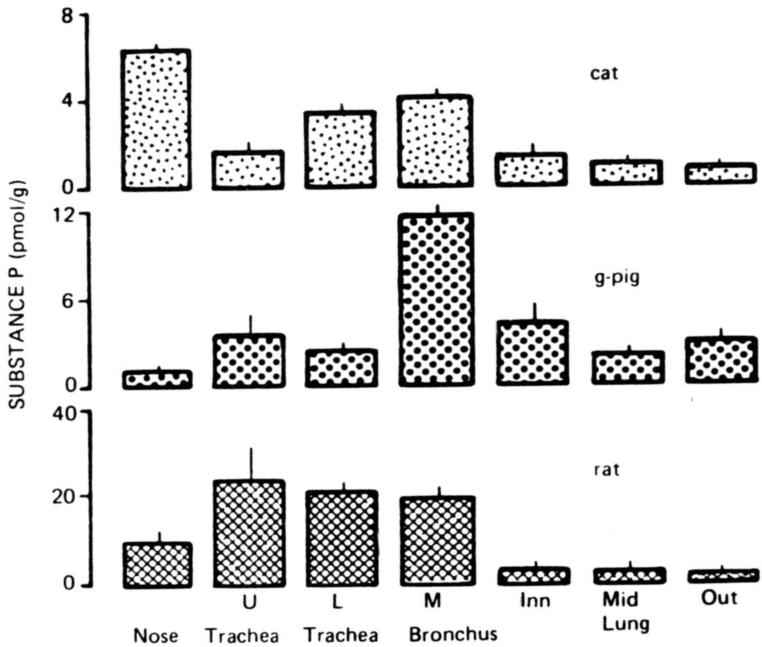

Figure 11 Levels of substance P immunoreactivity in extracts of respiratory tract of cat, guinea pig, and rat. (From Springall *et al.*, 1988.)

A. Histology

1. The Nose

Immunohistochemistry shows a supply of SP-containing fibers in the nasal mucosa, especially around blood vessels and submucosal glands, but also below and within the respiratory epithelium (Uddman and Sundler, 1986; Uddman *et al.*, 1987; Springall *et al.*, 1988; Lundberg, 1990; Albegger *et al.*, 1991). There are species differences, the cat apparently having more fibers than guinea pig and man, although most species seem to have the peptide. Many SP-containing nerve cell bodies and fibers are seen in the trigeminal ganglion (Lundblad, 1984), supporting the idea that sensory nervous receptors may initiate local reflexes with modulation via the ganglia. SP does not seem to have been localized to specific components of the vascular bed, however. Three other neuropeptides, CGRP, NKA, and galanin, have also been identified in nerves in the nasal mucosa, including in man (Baraniuk *et al*,. 1990). These seem to have a distribution very similar to that of SP. CGRP has an especially rich supply (Fig. 12). Finally, gastrin-releasing peptide (GRP), cholecystokinin and somatostatin are found in presumed sensory nerves in the nasal mucosa (Uddman *et al*, 1984; Springall *et al.*, 1988; Albegger *et al.*, 1991). There is less evidence for NKB in the nose.

The presence of four or five sensory neuropeptides in the nose is consistent with their colocation in nerves in other viscera including the tracheobronchial tree. They can be depleted and made to disappear from the nerves by administration of large doses of capsaicin, the extract of red peppers which is now widely used to deplete C-fiber and other sensory receptors of their neuropeptides.

2. The Tracheobronchial Tree

In general, the distribution of sensory neuropeptides in the tracheobronchial tree reflects that in the nose. However, more extensive studies have been done in the

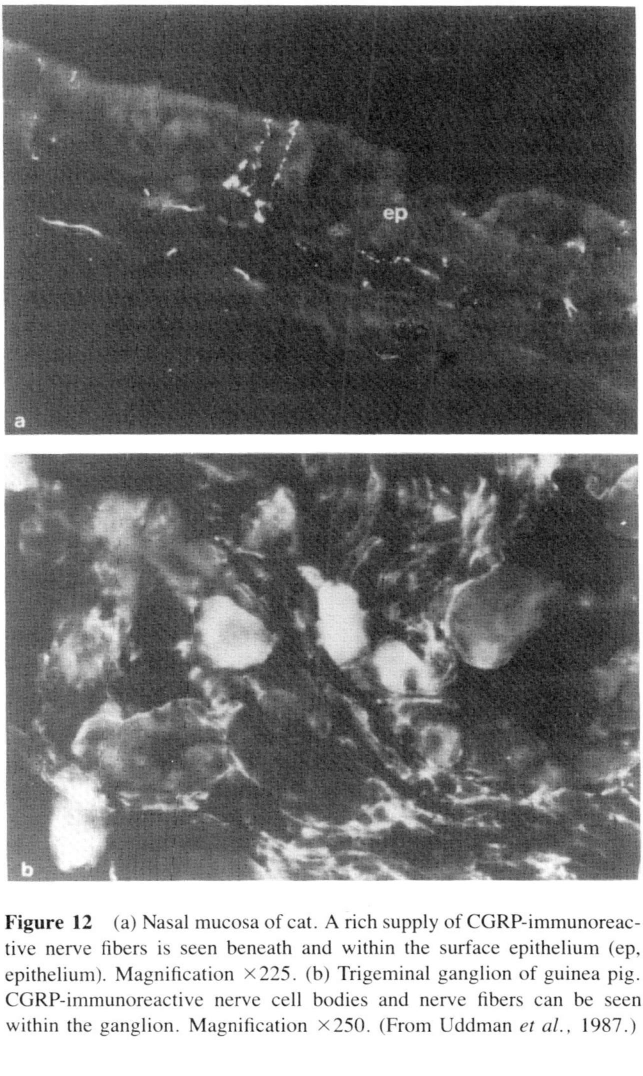

Figure 12 (a) Nasal mucosa of cat. A rich supply of CGRP-immunoreactive nerve fibers is seen beneath and within the surface epithelium (ep, epithelium). Magnification ×225. (b) Trigeminal ganglion of guinea pig. CGRP-immunoreactive nerve cell bodies and nerve fibers can be seen within the ganglion. Magnification ×250. (From Uddman *et al.*, 1987.)

former site. SP and NKA are found in nonmyelinated branches of sensory nerve receptors in the mucosa of all species studied, including man, particularly close to blood vessels, glands, and under or in the epithelium (Uddman *et al.,* 1987; Lundberg *et al.,* 1988a,b; Martling *et al.,* 1990). The smooth muscle seems less well supplied. In view of the fact that SP and NKA have two common molecular precursors, it is perhaps not surprising that they seem to be colocalized. They are also found in the airway ganglia, the nodose (vagal) ganglia (Martling *et al.,* 1990), and in sympathetic ganglia and dorsal roots (Springall *et al.,* 1987). The last observation is important but somewhat surprising since it is usually assumed that the predominant sensory pathway to the lower airways is vagal rather than sympathetic. The SP/NKA supply to the airways shows some species differences, the nerves being far more frequent in rat and guinea pig compared, for example, with cats and man (Springall *et al.,* 1988).

CGRP is also found in the tracheobronchial mucosa (Martling *et al.,* 1990) (Fig. 4). Some workers describe its distribution as identical to that of SP/NKA, whereas others give differences. Differences might not be unexpected in view of the different molecular precursors. In the case of the airway epithelium, CGRP is strongly represented in the neuroepithelial bodies (Martling *et al.,* 1988; McBride *et al.,* 1990; Shimosegawa and Said, 1991), groups of paracrine cells found in some species. However, there is little CGRP in human epithelium. In the rat, CGRP is found relatively more in the vagal jugular ganglion, and SP is found relatively more in the vagal nodose ganglion. In the cat, CGRP is present in the stellate (sympathetic) ganglia (Lindh *et al.,* 1987). For the dorsal root ganglia, differences in the colocalization of SP and CGRP have been described (Springall *et al.,* 1988). Therefore, at least in the airway autonomic ganglia and in the epithelium, there are divergences between localization of CGRP and SP/NKA.

Other neuropeptides with potential action on vascular beds have been described for the tracheobronchial mucosa; these are GRP, eledoisin-like peptide, and pituitary adenylate cyclase-activating peptide (PACAP) (Lundberg *et al,.* 1988b; Uddman *et al.,* 1991). They have not been mapped out well and there is little evidence on their potential vascular function.

B. Pharmacology

1. The Nose

All of the neuropeptides described in the last section cause vasodilatation in the nasal mucosa. This observation is established by direct perfusion of the nasal vascular bed of large animals, such as dog and pig (Lundberg *et al.,* 1984; Matran *et al.,* 1989a; Stjarne, 1991) (Fig. 5), and is obtained from indirect evidence in smaller animals, such as the guinea pig and rat (Lundberg and Saria, 1982; Lundblad, 1984). In these latter species the sensory neuropeptides cause extravasation of Evans blue-labeled albumin into the mucosal interstitium, a result consistent with vasodilation and/or opening of intercellular gaps between the endothelial cells of the end-capillary venules. The response can be mimicked by antidromic stimulation of sensory nerves running to the nasal mucosa, which can be shown also to release CGRP-like material into the venous effluent blood. Which blood vessels are chiefly affected by the sensory neuropeptides is not clear. In the dog, nasal vasodilatation due to SP is associated with little increase in nasal airflow resistance, suggesting that the capacitance vessels are not greatly dilated (Lung *et al.*, 1984). One would suppose that the fenestrated capillaries under the epithelium, and possibly their associated postcapillary venules, would be the most likely site for albumin extravasation, but it does not follow that this is the primary site of the change in vascular resistance.

If the sensory nerves are either destroyed by large doses of capsaicin, or allowed to degenerate after nerve section, the neuropeptides are still effective in causing vasodilatation and/or albumin transudation, but irritation by agents applied locally, such as cigarette smoke, no longer sets up neurogenic inflammation (Lundblad *et al.*, 1985). These results are consistent with nasal sensory nerves mediating axon reflex neurogenic inflammation; they tell us little about the sites of action of the neuropeptides. It is difficult to conduct similar experiments in man, so one cannot say to what extent the results on experimental animals may be relevant to human pathophysiology; however, capsaicin desensitization has been attempted in man and may destroy sensory nerves and block neurogenic inflammation (Wolfe, 1988; Stjarne *et al.*, 1992).

2. The Tracheobronchial Tree

Measurements of blood flow or perfusion of the trachea or bronchi of large mammals show that the above-mentioned sensory neuropeptides, applied exogenously, cause vasodilatation (L. A. Laitinen *et al.*, 1987a,b; Salonen *et al.*, 1988; Martling *et al.*, 1990). Different orders of potency have been identified. For example, for the dog trachea the order is NKA > CGRP > SP (L. A. Laitinen *et al.*, 1987a). For the dog bronchial artery *in vitro* the order is different: SP > NKA > CGRP (McCormack *et al.*, 1989). These ways of determining potency may have little relevance *in vivo* since the concentrations of the peptides on airway vascular smooth muscle are not known.

Tracheal and bronchial vasodilatations can be evoked by antidromic stimulation of the parasympathetic nerves to the airways (L. A. Laitinen *et al.*, 1987c; Martling *et al.*, 1985; Matran *et al.*, 1989b) (Fig. 13). In the presence either of atropine to block cholinergic effects and/or of hexamethonium to prevent ganglionic transmission in motor nerves, there is a residual vasodilatation presumably due to antidromic release of sensory neuropeptides. In the pig this is the dominant vasodilator mechanism of the bronchi where, as mentioned above, parasympathetic motor nerves with release of VIP seem to be rather inactive.

Similarly, sensory neuropeptide release can be promoted by locally applied capsaicin and prevented by previous administration of large doses of capsaicin that either destroy sensory nerves or deplete them of their neuropeptides (Lundberg *et al.*, 1985, 1988b). Irritation of the airway mucosa with agents such as cigarette smoke, which presumably act on the sensory terminals in the epithelium, causes

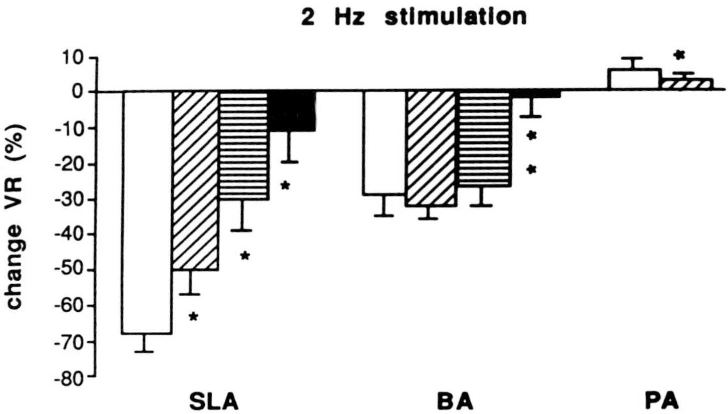

Figure 13 Changes in resistance (VR) in the vascular beds supplied by the superior laryngeal (SLA), bronchial (BA), and pulmonary (PA) arteries caused by vagal nerve stimulation at 2 Hz over 2 min, in controls (*open bars*) and after successive pretreatment with atropine (*diagonal bars*), chlorisondamine (*horizontal bars*), and capsaicin (*solid bars*). Data are given as means ±SEM and expressed as percentage of basal conditions during a control period immediately preceeding the stimulations. $^{*}p < 0.05$, $^{**}p < 0.001$ with the Quade test. (From Matran, 1991.)

vasodilatation and plasma transudation into the interstitium, suggesting the setting up of a local axon reflex, a component of neurogenic inflammation (McDonald, 1987). This response is abolished by capsaicin degeneration of sensory nerves.

VI. Pathophysiological Considerations

It is apparent that the nose, trachea, and bronchi have more common features than differences with regard to the peptidergic innervation of the vascular beds. This is to be expected. However, if more precise delineation of the sites of action of the peptides could be made, the particular anatomical distinctions of the three vascular beds must point to different quantitative effects of the peptides. In this respect we know little about which dilator peptides act on the AVAs of the nose although clearly NPY is a powerful constrictor. The control of mucosal blood flow and blood volume, in resistance and capacitance vessels respectively, has been to some extent worked out in the nose but hardly studied at all in the lower airways. The central reflex interactions at the three sites involving vascular peptides have been little studied, although one might expect interesting results here in view of the different patterns of reflexes originating in the various areas.

It is of interest that both the nose and the tracheobronchial tree have a dual-motor control of their airway vasculature, involving neuropeptides, from the parasympathetic and sympathetic nervous systems. The interaction of these two motor systems and even their relative strength in different physiological and pathological conditions has not been determined.

A. *Local Reflexes*

The presence of regulatory peptides in fibers and cell bodies in the ganglia supplying the airway mucosa points to a modulatory effect here. The peptides could modulate neurotransmission through the ganglia. Insofar as the sensory peptides are in afferent nerves, this points to local reflexes (i.e., when sensory nerves are activated, they not only cause axon and central nervous reflexes but, by connections to motor ganglia, modify mucosal function controlled by parasympathetic or sympathetic pathways). There are many pointers to the existence of this modulation (Barnes, 1986a,b; Coburn, 1989) that is potentially of considerable importance. However, neither for the nose nor for the tracheobronchial tree have the neural pathways or their physiological importance been worked out.

B. *Multiple-Transmitter Release*

Cotransmission to the vascular beds is well established in all airway sites as described earlier (Lundberg, 1990). It can be shown both for the nose and the trachea by stimulating appropriate parasympathetic nerves to achieve a multiple response, first blocking cholinergic mechanisms with atropine, and then blocking other motor pathway effects with a ganglionic blocking agent. The residual vasodilatation is presumably due to release of sensory neuropeptides. The results with this type of analysis by several groups are consistent with the existence of cholinergic motor peptidergic and motor sensory mechanisms interacting to cause dilatation. For the sympathetic nervous system a similar conclusion arises with regard to NPY and NA, as cotransmitters, but sensory antidromic activation by this pathway has not been described. This is of especial interest in view of the evidence that sensory fibers in the lower airways and lungs conduct through the sympathetic trunk to the dorsal roots, as shown by immunohistochemical staining for appropriate peptides (Springall *et al.*, 1987).

C. *Plasma Transudation*

There have been many studies showing that the sensory neuropeptides, usually administered exogenously or released by antidromic stimulation of parasympathetic nerves, cause plasma protein transudation, epithelial structural changes, and protein exudation into the lumen of the airways, both for nose and lower airways. Similar results do not seem to have been established for the parasympathetic motor vasodilator peptides, such as VIP/PHI/PHM. For the sensory neuropeptides an important distinction is that CGRP does not seem to cause transudation, whereas SP and NKA do (Barnes, 1986a). And yet all three are powerful vasodilators. This may suggest that the sensory neuropeptides have different actions on the sites of transudation, presumably the interendothelial junctions of postcapillary venules, even if their relaxant actions on the smooth muscle of the resistance vessels are generally similar. However, for most experiments the application of the sensory peptides has been quantitatively nonphysiological. We know very little about the quantities and concentrations of different neuropeptides released, for example, on sensory nerve activation, and therefore it is difficult to say to what extent the rather artificial experiments described apply to the intact body.

D. Cell Interactions

The above description implies that the vasoactive peptides exert their influences rather independently. In practice they may well act indirectly or induce secondary effects on vascular beds. For example, they may cause release of capillary endothelial mediators, and they profoundly affect epithelial structure and function and presumably may cause the release of epithelial mediators (McDonald, 1990). In addition, they may have an association with mast cells (Alving *et al.*, 1991); SP is known to act on mast cell membranes, and mast cell mediators can act on the sensory nerve fibers that release their neuropeptides. All these interactions may be important in the end result on the airway vascular beds.

E. Airway Pathology

In conditions such as allergic rhinitis and asthma it is well established that a major pathophysiological change in the mucosa is vascular dilatation, possibly with transudation and exudation of plasma proteins (Town and Holgate, 1991). There is a growing belief that axon reflex release of sensory neuropeptides causing neurogenic inflammation may be one important component of this mucosal pathology (Barnes, 1986b) although it has been difficult to establish the importance of this mechanism in man compared with the guinea pig. In addition, there is good evidence that in both allergic rhinitis and asthma the sensory nerves that may mediate axon reflexes also set up central nervous reflexes, as indicated by sneezing and a hyperreactive cough reflex (Widdicombe, 1988). These reflexes in turn would activate autonomic motor pathways to the mucosa with release both of classical transmitters and of the appropriate neuropeptides. The whole field of the role of vascular beds and of related neuropeptides in airway pathophysiology is vast and can only be touched on in this chapter.

VII. Conclusions

The dominant physiological neural control of the nasal and tracheobronchial vascular beds is probably sympathetic and mediated by a combination of NA and NPY. However, this control can be modulated by parasympathetic dilator influences, mediated by ACh and VIP/PHI/PHM. The interplay of these motor controls, and their actions on particular parts of the vascular bed, has not been studied. In pathological conditions activation of sensory nerve receptors will cause a release of neurotransmitters along axon reflex pathways, which have potentially profound effects on the mucosal vascular beds. At the same time the nerves will set up central reflexes that will add to the vasodilatation.

References

Albegger, K., Hauser-Kronberger, C. E., Saria, A., Graf, A.-H., Bernatzky, G., and Hacker, G. W. (1991). Regulatory peptides and general neuroendocrine markers in the human nasal mucosa, the soft palate and the larynx. *Acta Otorhinolaryngol. 111,* 373–378.

Alving, K., Sundstrom, C., Matran, R., Panula, P., Hokfelt, T., and Lundberg, J. M. (1991). Association between histamine-containing mast cells and sensory nerves in the skin and airways of control and capsaicin-treated pigs. *Cell Tiss. Res. 264, 529–538.*

Anggard, A. (1974). Capillary and shunt blood flow in the nasal mucosa of the cat. *Acta Otolaryngol. 78,* 418–422.

Anggard, A., Lundberg, J. M., Hokfelt, T., Nilsson, G., Fahrenkrug, J., and Said, S. (1979). Innervation of the cat nasal mucosa with special reference to relation between peptidergic and cholinergic neurons. *Acta Physiol. Scand. 473* (Suppl.), 50.

Baraniuk, J. N. (1991). Neural control of human nasal secretion. *Pulm. Pharmacol. 4,* 20–31.

Baraniuk, J. N., Lundgren, J. D., Goff, J., Mullol, J., Castellino, S., Merida, M., Shelhamer, J. H., and Kaliner, M. (1990). Calcitonin gene-related peptide (CGRP) in human nasal mucosa. *Am. J. Physiol. 258,* L81–L88.

Baraniuk, J. N., Lundgren, J. D., Okayama, M., Goff, J., Mullol, J., Merida, M., Shelhamer, J. H., and Kaliner, M. A. (1991). Substance P and neurokinin A in human nasal mucosa. *Am. J. Respir. Cell Mol. Biol. 4,* 228–236.

Barnes, P. J. (1986a). Neuropeptides in the airways: Functional significance. *In* "Asthma: Clinical Pharmacology and Therapeutic Progress." (A. B. Kay, ed.), pp. 58–72. Blackwell, Oxford.

Barnes, P. J. (1986b). Asthma as an axon reflex. *Lancet i,* 242–245.

Cauna, N. (1982). Blood and nerve supply to the nasal lining. *In* "The Nose: Upper Airway Physiology and the Atmospheric Environment." (D. F. Proctor and I. B. Andersen, eds.), pp. 45–69. Elsevier Biomedical Press, Amsterdam.

Cheung, A., Polak, J. M., Bauer, F. E., Cadieux, A., Christofides, N. D., Springall, D. R., and Bloom, S. R. (1985). Distribution of galanin immunoreactivity in the respiratory tract of pig, guinea-pig, rat and dog. *Thorax 40,* 889–896.

Coburn, R. F. (1989). Integration of neural inputs in peripheral airway ganglia. *In* "Airway Smooth Muscle in Health and Disease." (R. F. Coburn, ed.), pp. 17–34. Plenum Press, New York.

Cole, P. (1982). Modification of inspired air. *In* "The Nose: Upper Airway Physiology and the Atmospheric Environment." (D. F. Proctor and I. B. Andersen, eds.), pp. 351–376. Elsevier Biomedical Press, Amsterdam.

Corfield, D. R., Hanafi, Z., Webber, S. E., and Widdicombe, J. G. (1991). Changes in mucosal thickness and blood flow in sheep. *J. Appl. Physiol. 71,* 1282–1288.

Corfield, D. R., Webber, S. E., and Widdicombe, J. G. (1992). Distribution of blood flow in the perfused tracheas of sheep: A search for arteriovenous anastomoses. *J. Appl. Physiol.* (in press).

Deffebach, M. E. (1990). Lung mechanical effects on the bronchial circulation. *Eur. Respir. J. 3* (Suppl. 12), 586s–590s.

Eccles, R. (1982). Neurological and pharmacological considerations. *In* "The Nose: Upper Airway Physiology and the Atmospheric Environment." (D. F. Proctor and I. B. Andersen, eds.), pp. 191–214. Elsevier Biomedical Press, Amsterdam.

Hill, P., Goulding, D., Webber, S. E., and Widdicombe, J. G. (1989). Blood sinuses in the submucosa of the large airways of the sheep. *J. Anat. 162,* 235–247.

Hughes, T. (1965). Microcirculation of the tracheonronchial tree. *Nature* (London) *206,* 425–426.

Lacroix, J. S. (1989). Adrenergic and non-adrenergic mechanisms in sympathetic vascular control of the nasal mucosa. *Acta Physiol. Scand. 136* (Suppl.), 1–63.

Lacroix, J. S., Stjarne, P., Anggard, A., and Lundberg, J. M. (1988a). Sympathetic vascular control of the pig nasal mucosa (1): Increased resistance and capacitance vessel responses upon stimulation with irregular bursts compared to continuous impulses. *Acta Physiol. Scand. 132,* 83–90.

Lacroix, J. S., Stjarne, P., Anggard, A., and Lundberg, J. M. (1988b). Sympathetic vascular control of the pig nasal mucosa (2): Reserpine-resistant, nonadrenergic nervous responses in relation to neuropeptide Y and ATP. *Acta Physiol. Scand. 133,* 183–197.

Laitinen, A., Laitinen, L. A., Moss, R., and Widdicombe, J. G. (1989). Organisation and structure of the tracheal and bronchial blood vessels in the dog. *J. Anat. 165,* 133–140.

Laitinen, L. A. (1988). Detailed analysis of neural elements in human airways. *In* "The Airways: Neural Control in Health and Disease." (M. A. Kaliner and P. J. Barnes, eds.), pp. 35–56. Marcel Dekker, New York.

Laitinen, L. A., Laitinen, A., and Widdicombe, J. G. (1987a). Effects of inflammatory and other mediators on airway vascular beds. *Am. Rev. Respir. Dis. 135* (Suppl.), S67–S70.

Laitinen, L. A., Laitinen, M. A., and Widdicombe, J. G. (1987b). Dose-related effects of pharmacological mediators on tracheal vascular resistance in dogs. *Br. J. Pharmacol. 92,* 703–709.

Laitinen, L. A., Laitinen, M. V. A., and Widdicombe, J. G. (1987c). Parasympathetic nervous control of tracheal vascular resistance in the dog. *J. Physiol.* (London) *385,* 135–146.

Lindh, B., Lundberg, J. M., Hokfelt, T., Elfvin, L-G., Fahrenkrug, J., and Fisher, J. (1987). Coexistence of CGRP- and VIP-like immunoreactivities in a population of neurons in the cat stellate ganglion. *Acta Physiol. Scand. 131,* 475–476.

Lundberg, J. M. (1990). Peptide and classical transmitter mechanisms in the autonomic nervous system. *Arch. Int. Pharmacodyn. 303,* 9–19.

Lundberg, J. M., and Saria, A. (1982). Capsaicin-sensitive vagal neurons are involved in control of vascular permeability in rat trachea. *Acta Physiol. Scand. 115,* 521–523.

Lundberg, J. M., Anggard, A., Emson, T., Fahrenkrug, J., and Hokfelt, T. (1981). Vasoactive intestinal polypeptide and cholinergic mechanisms in cat nasal mucosa. Studies on choline acetyltransferase and release of vasoactive intestinal polypeptide. *Proc. Natl. Acad. Sci. U.S.A. 78,* 5255–5259.

Lundberg, J. M., Brodin, E., Hua, X.-Y., and Saria, A. (1984). Vascular permeability changes and smooth muscle contraction in relation to capsaicin-sensitive substance P afference in the guinea-pig. *Acta Physiol. Scand. 120,* 217–227.

Lundberg, J. M., Franco-Cereceda, A., Huang, X.-Y., Hokfelt, T., and Fischer, J. A. (1985). Coexistence of substance P and calcitonin gene-related peptide-like immunoreactivities in sensory nerves in relation to cardiovascular and bronchoconstrictor effects of capsaicin. *Eur. J. Pharmacol. 108,* 315–319.

Lundberg, J. M., Lundblad, L., and Martling, C. R. (1987). Coexistence of multiple peptides and classical neurotransmitters in airway neurons: Functional and pathophysiological aspects. *Am. Rev. Respir. Dis. 136* (Suppl.), S1–S8.

Lundberg, J. M., Martling, C-R., and Hokfelt, T. (1988a). Airways, oral cavity and salivary glands. Classical transmitters and peptides in sensory and autonomic motor neurons. *In* "Handbook of Chemical Neuroanatomy." (A. Bjorklund, T. Hokfelt, and C. Owman, eds.), Vol. 6: The peripheral nervous system, pp. 391–444. Elsevier, Amsterdam.

Lundberg, J. M., Lundblad, L., Anggard, A., Martling, C.-R., Theodorsson-Norheim, E., Stjarne, P., Hokfelt, T. G., and Saria, A. (1988b). Bioactive peptides in capsaicin-sensitive C-fiber afferents of the airways. *In* "The Airways: Neural Control in Health and Disease." (M. A. Kaliner and P. J. Barnes, eds.), pp. 417–445. Marcel Dekker, New York.

Lundberg, J. M., Pernow, J., and Lacroix, J. S. (1989). Neuropeptide Y: Sympathetic cotransmitter and modulator? *News Physiol. Sci. 4,* 13–17.

Lundblad, L. (1984). Protective reflexes and vascular beds in the nasal mucosa elicited by activation of capsaicin-sensitive substance P immunoreactive trigeminal neurons. *Acta Physiol. Scand.* (Suppl. 529), 1–45.

Lundblad, L., Saria, A., Lundberg, J. M., and Anggard, A. (1983). Increased vascular permeability in rat nasal mucosa induced by substance P and stimulation of capsaicin-sensitive trigeminal neurons. *Acta Otolaryngol. 96,* 479–484.

Lundblad, L., Brodin, E., Lundberg, J. M., and Anggard, A. (1985). Effects of nasal capsaicin pretreatment and cryosurgery on sneezing reflexes, neurogenic plasma extravasation, sensory and sympathetic neurons. *Acta Otolaryngol.* (Stockholm) *100*, 117–127.

Lundblad, L., Anggard, A., Saria, A., and Lundberg, J. M. (1987). Neuropeptide Y and non-adrenergic sympathetic vascular control of the cat nasal mucosa. *J. Auton. Nerv. Syst. 20*, 189–197.

Lung, M. A., Phipps, R. C., Wang, J. C. C., and Widdicombe, J. G. (1984). Control of nasal vasculature and airflow resistance in the dog. *J. Physiol.* (London) *349*, 535–551.

Magno, M. (1990). Comparative anatomy of the tracheobronchial circulation. *Eur. Respir. J. 3* (Suppl. 12), 557s–563s.

Martling, C.-R., Anggard, A., and Lundberg, J. M. (1985). Non-cholinergic vasodilation in the tracheobronchial tree of the cat induced by vagal nerve stimulation. *Acta Physiol. Scand. 125*, 343–346.

Martling, C.-R., Saria, A., Fischer, J. A., Hokfelt, T., and Lundberg, J. M. (1988). Calcitonin gene-related peptide and the lung: Neuronal coexistence with substance P release by capsaicin and vasodilatory effect. *Regul. Pept. 20*, 125–139.

Martling, C.-R., Matran, R., Alving, K., Lacroix, J. S., and Lundberg, J. M. (1989). Vagal vasodilatory mechanisms in the pig bronchial circulation preferentially involves sensory nerves. *Neurosci. Lett. 96*, 306–311.

Martling, C.-R., Matran, R., Alving, K., Hokfelt, T., and Lundberg, J. M. (1990). Innervation of lower airways and neuropeptide effects on bronchial and vascular tone in the pig. *Cell Tiss. Res. 260*, 223–233.

Matran, R. (1991). Neural control of lower airway vasculature. *Acta Physiol. Scand. 142* (Suppl.), 1–54.

Matran, R., Alving, K., and Lundberg, J. M. (1989a). Cigarette smoke and capsaicin aerosol-induced vasodilatation in the pig nasal mucosa. *Br. J. Pharmacol. 100*, 535–541.

Matran, R., Alving, K., Martling, C.-R., Lacroix, J. S., and Lundberg, J. M. (1989b). Vagally mediated vasodilatation by motor and sensory nerves in the tracheal and bronchial circulation of the pig. *Acta Physiol. Scand. 135*, 29–37.

Matran, R., Alving, K., Martling, C.-R., Lacroix, J. S., and Lundberg, J. M. (1989c). Effects of neuropeptides and capsaicin on tracheobronchial blood flow of the pig. *Acta Physiol. Scand. 135*, 335–342.

Matran, R., Franco-Cereceda, A., Alving, K., and Lundberg, J. M. (1992). Sympathetic vascular control of the laryngo-tracheal, bronchial and pulmonary circulations in the pig: Evidence for non-adrenergic mechanisms involving neuropeptide Y. *J. Appl. Physiol.* (in press).

McBride, J. T., Springall, D. R., Winter, R. J. D., and Polak, J. M. (1990). Quantitative immunocytochemistry shows calcitonin gene-related peptide-like immunoreactivity in lung neuroendocrine cells is increased by chronic hypoxia in the rat. *Am. J. Respir. Cell Mol. Biol. 3*, 587–593.

McCormack, D. G., Salonen, R. O., and Barnes, P. J. (1989). Effect of sensory neuropeptides on canine bronchial and pulmonary vessels *in vitro*. *Life Sci. 45*, 2405–2412.

McDonald, D. M. (1987). Neurogenic inflammation in the respiratory tract: Actions of sensory nerve mediators on blood vessels and epithelium of the airway mucosa. *Am. Rev. Respir. Dis. 136* (Suppl.), S65–S72.

McDonald, D. M. (1990). The ultrastructure and permeability of tracheobronchial blood vessels in health and disease. *Eur. Respir. J. 3* (Suppl. 12), 572s–585s.

Salonen, R. O., Webber, S. E., and Widdicombe, J. G. (1988). Effects of neuropeptides and capsaicin on the canine tracheal vasculature *in vivo*. *Br. J. Pharmacol. 95*, 1262–1270.

Salonen, R. O., Webber, S. E., and Widdicombe, J. G. (1990). Effects of neurotransmitters on tracheobronchial blood flow. *Eur. Respir. J. 3* (Suppl. 12), 630s–637s.

Shimosegawa, T., and Said, S. (1991). Pulmonary calcitonin gene-related peptide immunoreactivity: Nerve-endocrine cell interrelationships. *Am. J. Respir. Cell Mol. Biol. 4*, 126–134.

Sorokin, S. P., and Hoyt, R. F., Jr. (1990). On the supposed function of neuroepithelial bodies in adult mammalian lungs. *News Physiol. Sci. 5*, 89–95.

Springall, D. R., Cadieux, A., Oliveira, H., Su, H., Royston, D., and Polak, J. M. (1987). Retrograde tracing shows that CGRP-immunoreactive nerves of rat trachea and lung originate from vagal and dorsal root ganglia. *J. Auton. Nerv. Syst. 20,* 155–166.

Springall, D. R., Bloom, S. R., and Polak, J. M. (1988). Distribution, nature, and origin of peptide-containing nerves in mammalian airways. *In* "The Airways: Neural Control in Health and Disease." (M. A. Kaliner and P. J. Barnes, eds.), pp. 299–342. Marcel Dekker, New York.

Stjarne, P. (1991). Sensory and motor reflex control of nasal mucosal blood flow and secretion: Clinical implications in non-allergic nasal hyperreactivity. *Acta Physiol. Scand. 142* (Suppl.), 1–64.

Stjarne, P., Lundblad, L., Lundberg, J. M., and Anggard, A. (1992). Local capsaicin treatment of the nasal mucosa reduces symptoms in patients with nonallergic nasal hyperreactivity. *Am. J. Rhinol.* (in press).

Town, G. I., and Holgate, S. T. (1991). The role of inflammation in airways disease. *In* "Mediators of Pulmonary Inflammation." (M. A. Bray and W. H. Anderson, eds.), pp. 35–79. Marcel Dekker, New York.

Uddman, E., and Sundler F. (1986). Innervation of the upper airways. *Clinic Chest Med. 7,* 210–214.

Uddman, E., and Sundler, F. (1987). Neuropeptides in the airways: A review. *Am. Rev. Respir. Dis. 136* (Suppl.), S3–S8.

Uddman, R., Alumets, J., Densert, O., Hakanson, R., and Sundler, F. (1978). Occurrence and distribution of VIP nerves in the nasal mucosa and tracheobronchial wall. *Acta Orolaryngol. 86,* 443–448.

Uddman, R., Malm, L., and Sundler, F. (1980). Effects of vasoactive intestinal polypeptide on resistance and capacitance vessels in the nasal mucosa. *Acta Otolaryngol. 90,* 304–308.

Uddman, R., Moghimzadeh, E., and Sundler, F. (1984). Occurrence and distrubution of GRP-immunoreactive nerve fibres in the respiratory tract. *Arch. Otorhinolaryngol. 239,* 145–151.

Uddman, R., Anggard, A., and Widdicombe, J. G. (1987). Nerves and neurotransmitters in the nose. *In* "Allergic and Vasomotor Rhinitis: Pathophysiological Aspects." (N. Mygind and U. Pipkorn, eds.), pp. 50–62. Munksgaard, Copenhagen.

Uddman, R., Lutz, A., Arimura, A., and Sundler, F. (1991). Pituitary adenylate cyclase-activating peptide (PACAP), a new vasoactive intestinal polypeptide (VIP)-like peptide in the respiratory tract. *Cell Tiss. Res. 265,* 197–201.

Watanabe, A., and Watanabe, I. (1980). The ultrastructural characteristics of the capillary walls in human nasal mucosa. *Rhinology 18,* 183–195.

Widdicombe, J. G. (1988). Vagal reflexes in the airways. *In* "The Airways: Neural Control in Health and Disease." (M. A. Kaliner and P. J. Barnes, eds.), pp. 187–202. Marcel Dekker, New York.

Widdicombe, J. G. (1990a). The NANC system and airway vasculature. *Arch. Int. Pharmacodyn. 303,* 83–99.

Widdicombe, J. G. (1990b). Comparison between the vascular beds of upper and lower airways. *Eur. Respir. J. 3* (Suppl. 12), 564s–571s.

Wolfe, G. (1988). Neue aspekte zur pathogenese und therapie der hyperflektorischen rhinopathie. *Laryngol. Rhinol. Otol. 67,* 438–445.

CHAPTER 22

Gastrointestinal Circulation in Man and Animals: Regulation by Perivascular Peptides and Classical Neurotransmitters

OVE LUNDGREN

Most major functions of the body are found in the gastrointestinal wall. It contains a muscular layer that ensures the delivery of the luminal contents along the alimentary canal. A variety of substances is transported across the intestinal epithelium by both passive and active mechanisms. Furthermore, the digestive tube is provided with a primitive brain, the enteric nervous system. The vascular beds of the gastrointestinal tract are well adjusted to the needs of the alimentary canal and they represent one of the major vascular circuits of the body. In resting man more than 20% of the cardiac output is distributed to the alimentary canal.

The vascular beds of the gastrointestinal tract contain several parallel coupled vascular circuits that supply the different wall layers. Each vascular circuit consists, in turn, of consecutive vascular sections, such as pre- and postcapillary resistance vessels, exchange vessels (capillaries), and capacitance vessels. Several control mechanisms, including local chemical, hormonal, and nervous, adapt the gastrointestinal blood flow to the local needs and to the overall cardiovascular control. The present review deals with the nervous control of these vascular circuits. When studies performed on humans are available, they are described in some detail.

It was believed for a long time that each neuron contained and released only one transmitter (the so-called Dale's law). Since the mid-1970s morphological and experimental evidence has accumulated strongly suggesting that neurons both in the central and peripheral nervous systems may contain two or more transmitters and/or transmitter candidates. The functional implications of this coexistence of transmitters have been studied *in vitro* and *in vivo* in other vascular beds as discussed in Chapters 7, 9, 10, and 15. Although the knowledge concerning the coexistence in intestinal nerves is considerable (see, e.g., Furness and Costa, 1987), our understanding of its functional implications is limited. Some possible consequences of colocalization are described here in connection with neuropeptide Y (NPY), which coexists with noradrenalin in many adrenergic fibers controlling vascular smooth muscles (Lundberg *et al.*, 1983).

I. The Sympathetic Control of the Gastrointestinal Circulation

The distribution of nerve fibers containing noradrenaline has been mapped out using the Falck–Hillarp technique. The morphology of the vascular innervation revealed by this technique is consistent among mammals. The innervation is located to the outer layer of vascular smooth muscles, where the adrenergic fibers form a so-called ground plexus. Hence the adrenergic nerves, also containing NPY, primarily control only one part of the vascular smooth muscles, the remaining parts of the vascular smooth muscles being influenced indirectly via low-resistance intercellular nexa. This arrangement is described in detail in Chapters 2 and 10.

A. *Intestines*

The most thorough analysis of the sympathetic control of the gastrointestinal circulation has been performed on the small intestine. I will therefore first discuss these studies and then comment on the other vascular beds of the alimentary canal.

1. Hemodynamics

Stimulation of the sympathetic fibers to the small intestine of humans and animals evokes a vasoconstrictor response that has several very characteristic features (Folkow *et al.*, 1964a,b). For obvious ethical reasons it has been most thoroughly investigated in animals, but the studies performed on humans (Hultén *et al.*, 1977) strongly suggest that the sympathetic vascular response is similar in humans and animals. A marked reduction of total intestinal blood flow is seen upon commencing an electrical stimulation of the regional sympathetic fibers (Fig. 1). When rate of stimulation exceeds 2 Hz, blood flow again increases within a couple of minutes despite a continued electrical stimulation. This observation was first reported by Folkow *et al.* (1964a), and the response of the resistance vessels was named "autoregulatory escape from vasoconstrictor fiber influence." Blood flow reaches within a couple of minutes a steady state level. Even at high rates of stimulation of the sympathetic fibers (8–10 Hz), flow resistance in the intestinal vascular bed is only doubled.

Observations have been made on human intestinal blood flow during sympathetic nerve stimulation using an inert gas elimination method (Hultén *et al.*, 1977). The investigations were performed during abdominal surgery and the technique implied the intraarterial injection of a radioactive noble gas, ^{85}Kr, the elimination of which was followed with a scintillation detector (recording radiation from the whole small intestine) and with a Geiger–Müller tube placed at the serosal surface (recording β-radiation only from the intestinal muscle layer). From these two recordings total intestinal blood flow, mucosal blood flow, and muscle layer blood flow can be estimated (for details regarding the technique, consult Hultén *et al.*, 1976). These experiments clearly indicate that the response pattern observed on animals and described earlier is the same as the one seen in humans. For example, when injecting the krypton solution 1 min after starting an electrical stimulation of the regional

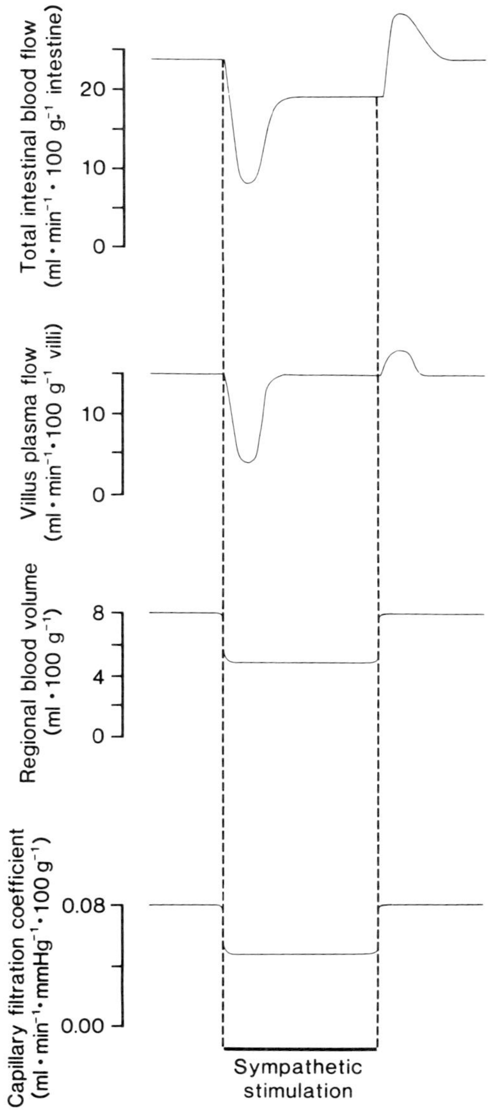

Figure 1 The effect of stimulation of the regional sympathetic fibers (signal) on total intestinal blood flow, villus plasma flow, intestinal blood volume, and intestinal capillary filtration coefficient. The schematic figure is based on studies performed by Folkow *et al.* (1964a,b) and Svanvik (1973). For details, see text.

sympathetic fibers to the small intestine, the elimination curve of the radioactive tracer exhibited an appearance consistent with the presence of an autoregulatory escape from the vasoconstrictor fiber influence also in the human gut (Hultén *et al.*, 1977).

In animal experiments it has been possible to follow also the response of the intestinal veins and the capillary filtration coefficient (Folkow *et al.*, 1964a,b). The latter variable indirectly reflects the number of capillaries patent to flow. The veins are constricted upon sympathetic activation, expelling about 40% of the regional blood volume at most (Fig. 1). In adult humans this corresponds to a mobilization of 50–170 ml from the human small intestine or to 250–500 ml from all organs supplied by the superior and inferior mesenteric arteries (Hultén *et al.*, 1977). Concomitantly, the number of perfused capillaries decreases drastically, whereas mean capillary hydrostatic pressure remains more or less unaltered. At the highest rates of stimulation the number of capillaries patent to flow is reduced to about half of control. In contrast to the reactions of the resistance vessels no "escape" of the veins or the filtration coefficient is seen during the stimulation of the sympathetic fibers to the small intestine (Fig. 1).

An analysis of the blood flow in the different wall layers of the small intestine during sympathetic stimulation has revealed a distinct pattern. Initially a strong vasoconstriction occurs in all wall layers of the gut, including the villi (Fig. 1). During the steady-state phase of vasoconstriction average blood flow in the muscle and mucosal-submucosal layers decreases to the same relative extent as total intestinal blood flow. This has been shown both in animals and in man (see Table 1).

In animals it is possible to record villus and crypt blood flows separately. Using an indicator dilution method, villus plasma flow was estimated repeatedly during a sympathetic vasoconstriction (Svanvik, 1973). These studies show that villus plasma flow is unchanged as compared to control during the "steady-state phase" of vasoconstriction (Fig. 1). This occurs in the face of the clear reduction of total intestinal blood flow observed during the stimulation of the regional vasoconstrictor

Table 1 Blood Flow in the Feline and Human Small and Large Bowel during Control Conditions and during Electrical Stimulation of the Regional Sympathetic Nerves at 8 Hz[a]

	Cat			Homo		
	Control	Stimulation	*n*	Control	Stimulation	*n*
Small intestine						
Total blood flow	33 ± 4	22 ± 28	8	31 ± 11	15 ± 4	6
Mucosa-submucosa	43 ± 7	29 ± 3	8	40 ± 16	19 ± 5	6
Muscularis	17 ± 2	11 ± 2	8	16 ± 5	9 ± 3	6
Colon						
Total blood flow	25 ± 3	18 ± 2	16	24 ± 9	16 ± 6	4
Mucosa-submucosa	35 ± 4	26 ± 3	16	28 ± 10	22 ± 9	4
Muscularis	11 ± 1	7 ± 1	16	19 ± 5	7 ± 3	4

[a]Blood flow is expressed in ml min^{-1} 100 g^{-1} tissue. Mean ± SE.

fibers at 8 Hz. Taken together, these observations suggest that crypt blood flow is markedly diminished upon vasoconstrictor nerve activation. In line with this proposal, Sjövall *et al.* (1983) showed that mucosal blood flow not involved in absorption ("nonabsorptive site blood flow") was decreased to half of control by nerve stimulation at 8 Hz.

The mechanisms explaining the rather drastic changes of intestinal blood flow ("autoregulatory escape") occurring in the early phase of an activation of the sympathetic vasoconstrictor fibers are not known, but several mechanisms have been proposed. One obvious explanation is that an accumulation of vasodilating metabolites occurring during the initial pronounced vasoconstriction counteracts the vasoconstrictor influence from the neurotransmitter(s) released from the nerve terminals. Eventually a steady state is reached between the two conflicting influences on the intestinal resistance vessels. This hypothesis is substantiated by the observation that a vasodilatation occurs upon cessation of the electrical stimulation of the sympathetic fibers.

In vitro studies suggest that the autoregulatory escape may be an inherent property of vascular smooth muscle (Fara and Ross, 1972). According to this hypothesis, the nervous influence on the vasculature is fading not because of any accumulation of metabolites but rather due to a failure of the smooth muscles to respond to the released neurotransmitter(s). The vascular escape is akin to the phenomenon of adaptation. An involvement of β-adrenergic has also been proposed (Ross, 1971). Finally, Remak *et al.* (1989) proposed that capsaicin-sensitive nerves, presumably releasing peptidergic neurotransmitters, are involved in the escape from vasoconstrictor fiber influence.

The vascular response to sympathetic nerve stimulation in the colon has been much less investigated than in the small intestine. The results available on cats and humans suggest that the vascular reactions are largely similar to those in the small intestine. Thus, an autoregulatory escape phenomenon is observed and the capillary filtration coefficient is markedly lowered during the steady-state phase of vasoconstriction. A nervous vasoconstriction is seen in both the mucosa and muscle layer in humans as well as cats. In humans the vasoconstriction seems to be more pronounced in the muscle layer (Table 1), leading to a certain redistribution of flow toward the mucosa.

2. Neurotransmitter Pharmacology

The nerves innervating the intestinal blood vessels contain not only noradrenaline but also two other putative neurotransmitters, NPY and adenosine triphosphate (ATP). The relative contribution of these three neurotransmitters to the nervous vasoconstriction has been investigated in animals. In an extensive series of experiments *in vitro* on isolated mesenteric vessels (outer diameter about 200 μm) from rats performed by Sjöblom *et al.* (1990; for a review see Sjöblom-Widfeldt, 1990), it was shown that adrenergic blockade markedly diminished the vasoconstriction elicited by stimulating the nerves with trains of electrical impulses. Blocking the effects of ATP using αβ-methylene ATP did not influence the maximal force developed by the vascular muscle stimulated by trains of electrical impulses. However, the speed of the initial vascular contraction was diminished by the ATP receptor antagonist. Furthermore, contractions elicited by single impulses were markedly attenuated by administration of the purinergic blocking agent, only 20% of the control response remaining after giving the drug.

The influence of NPY was indirectly studied on the isolated mesenteric artery by giving an α_2-receptor blocker, prazosin. At low frequencies of stimulation this drug totally abolished the nervous response while at higher rates of stimulation (16 Hz) about 10% of the response remained. Thus, it seems that if the sympathetic effect is mediated via the release of NPY, it is most important at higher rates of stimulation. Similar findings have been made with regard to the release of other peptides, such as VIP, from the salivary glands (Lundberg, 1981).

The finding that adrenergic influence is the most important for the sympathetic control of the resistance vessels of the small intestine was also shown in cat experiments *in vivo* by Brunsson *et al.* (1979) and by Sjövall (1983). They showed that the α-adrenergic receptor antagonist phentolamine completely abolished the vascular response to sympathetic nerve stimulation in the feline small intestine. However, studies on the colon and on the submandibular salivary gland in cats suggest that the nonadrenergic component of the sympathetic control of the resistance vessels is more important than would be expected from the results obtained on the vessels of the small intestine. After blockade of both the α- and β-adrenergic receptors, Lundberg and Tatemoto (1982) and Hellström *et al.* (1985) demonstrated that up to 40% of the vascular response to sympathetic nerve stimulation remained. These authors proposed that the remaining response was caused by the release of NPY.

3. Functional Implications

One can only speculate on the significance for functions of the neurally induced vasoconstriction. The studies of the flow distribution within the mucosa of the small intestine show that the vasoconstriction occurs almost exclusively in the crypts, whereas villus blood flow remains unchanged as compared to control. Hence, the absorption occurring in the villi is probably not influenced by the reduction of total intestinal blood flow occurring on sympathetic nerve activation. The reduction of crypt blood flow may, on the other hand, inhibit secretory mechanisms in that part. It is, however, less likely that hemodynamic changes explain the diminished tissue-to-lumen flux of water seen on sympathetic nerve activation (Sjövall *et al.*, 1983).

B. Stomach

The available results from animal experiments clearly suggest that the sympathetic nerves influence the gastric circulation in a manner similar to that seen in the intestinal vascular beds. (Jansson *et al.*, 1970). There is a transient vasoconstriction of the resistance vessels ("autoregulatory escape"), a constriction of the precapillary sphincters as reflected by a lowered capillary filtration coefficient, and an expelling of blood from the stomach as a consequence of the constriction of the capacitance vessels. Mean capillary hydrostatic pressure remains at the control level as judged by a constant volume during the sympathetic vasoconstriction. On cessation of stimulation, tissue volume, blood flow, and capillary filtration coefficient return to control levels within 2–3 min. Quantitative differences may exist between the vascular beds of the gastrointestinal tract, but this has not been studied experimentally neither in animals nor in man.

The stomach differs functionally from the intestines in that the mucosa of the stomach is in principle a secretory structure, whereas the intestinal mucosa is mainly absorptive. An ongoing gastric secretion implies that the gastric tissue contains dilating "metabolites" that tend to counteract the constrictor influence of the neurotransmitter(s) release from the sympathetic nerve endings. It is, however, possible to demonstrate a sympathetic vasoconstriction also during acid secretion evoked by pentagastrin (Reed and Sanders, 1971).

II. The Parasympathetic Control of the Gastrointestinal Vascular Beds

According to the classical view, acetylcholine is the neurotransmitter of the parasympathetic part of autonomic nervous system (ANS). In the alimentary canal this has been demonstrated convincingly with regard to the parasympathetic control of motility. There is, however, also experimental evidence for a parasympathetic noncholinergic control of motility and this also applies to other gastrointestinal functions, such as epithelial transport and blood flow.

As described in Chapter 1, the last two decades have witnessed a revolution of our views of the neurotransmitter pharmacology of the ANS. Several peptides have been isolated chemically and demonstrated to exist in the neurons of ANS in most cases with immunohistochemical techniques. With the exception of NPY (contained in sympathetic nerves), most of these neuropeptides are either contained in neurons controlled by the extrinsic parasympathetic nerve fibers and/or are confined to neurons that are part of intrinsic nerves reflexes of the enteric nervous system (ENS).

Table 2 summarizes the histochemical findings of the peptide neurotransmitters found in the vascular innervation of the gastrointestinal tract of rats, guinea pigs, and humans. The table is based on report by Keast *et al.* (1985). It is apparent from Table 2 that the vascular innervation of the three species studied is rather similar. Vasoactive intestinal polypeptide (VIP)-, substance P (SP)-, and NPY-containing fibers are found in all species, whereas the vascular innervation of the other peptides is scarce or totally absent. It should be pointed out that it is not possible to differentiate between afferent and efferent innervation only from a histochemical study of the distribution of a particular type of nerve. The role of NPY in the vascular control was discussed earlier. The possible roles of VIP and SP will be discussed below.

Our knowledge of the nervous parasympathetic control of the gastrointestinal vascular beds is exclusively based on animal experiments. Hence, the observations reported below were in most cases made on anesthetized animals.

Table 2 Vascular Innervation of Different Peptidergic Neurons in the Gastrointestinal Tract[a]

	ENK	NPY	SOM	SP	VIP
Rat					
Stomach	0	+	0	+	+
Small intestine	0	+	0	+	+
Large intestine	0	+	+	+	+
Guinea pig					
Stomach	(+)	+	0	+	+
Small intestine	+	+	0	+	+
Large intestine	0	+	(+)	+	+
Man					
Stomach	+	+	0	+	+
Small intestine	0	+	+	+	+
Large intestine	0	+	0	+	+

Source: from Keast *et al.* (1985).
[a]*Abbreviations:* ENK, enkefalins; NPY, neuropeptide Y; SOM, somatostatin; SP, substance P. Innervation density was graded according to a four grading system (0, +, ++, +++).

A. *Intestines*

During the first part of this century it was usually assumed that most vascular beds in the mammalian organism were innervated by both vasoconstrictor and vasodilator fibers, considered to be sympathetic and parasympathetic, respectively. In the small intestine this general view was not substantiated in the thorough investigations on cats by Kewenter (1965), who was unable to demonstrate any vasodilator fibers in the vagi, although the characteristics of nervous stimulation he used were varied widely. On the other hand, Neild *et al.* (1990) reported that the submucosal arteries of the guinea pig were supplied with cholinergic vasodilator fibers. However, it is not known to what extent these nerves are activated by stimulating the preganglionic vagal fibers.

Experimental evidence exists for a parasympathetic control of the colonic vasculature. Morphologically, the autonomic innervation of the colon is more complex than that of the small intestine. The parasympathetic supply to the colon consists of two main nerves, the vagal and the pelvic. The distribution areas of the two nerves meet at the transverse portion of the colon.

1. Hemodynamics

Electrical stimulation of the vagal fibers to the colon does not elicit any flow changes apart from those caused by an increase in colonic motility. (Hultén *et al.*, 1969a). Stimulating the pelvic nerves, on the other hand, induces marked changes in colonic hemodynamics. A fairly detailed analysis of this has been performed in the cat (Hultén *et al.*, 1969a,b). Experiments on this species suggest that two vasodilator mechanisms are involved in the pelvic nerve response in the colon: an initial, atropine-resistant hyperemia and an atropine-sensitive flow increase, which may be secondary to an augmented secretion from the crypts. However, it is not possible to entirely exclude the possibility of a direct cholinergic innervation of the colonic blood vessels. Acetylcholine is known to dilate the intestinal circulation.

Pelvic nerve stimulation also influences the venous capacitance vessels. At the maximal rate of stimulation (8 Hz) the increase of tissue volume was 2–2.5 ml/100g, corresponding to a 25% increase in regional blood volume. This capacitance response is probably in part due to the increased transmural pressure in the veins that accompanies vasodilatation and in part due to a direct nervous relaxation of the capacitance vessels.

The intramural flow distribution during nervous vasodilatation has been studied in the cat and man with local microinjections of inert tracers (Hultén *et al.*, 1969c). These studies demonstrated that the initial marked vasodilatation is accompanied by an increase of blood flow exclusively to the colonic mucosa, particularly the superficial layers. Blood flow in the mucosa increases from a resting value 25–40 to 150–200 ml min^{-1} 100 g^{-1} of mucosal tissue at a total colonic blood flow of 60 ml min^{-1} 100 g^{-1} of colonic tissue when the pelvic nerves are stimulated at 4 or 8 Hz.

2. Neurotransmitter Pharmacology

The initial marked colonic vasodilatation is not influenced by atropine (Hultén *et al.*, 1969a) but abolished by a ganglionic blocking agent, implying that preganglionic nerves are cholinergic and that a noncholinergic transmitter is released at the vascular smooth muscle cells. Several observations suggest that the latter effect is mediated by the release of VIP. This hypothesis is substantiated by the demonstration of VIP in nerve fibers in close contact with the colonic vasculature. VIP is a potent vasodilator of the colonic resistance vessels (Eklund *et al.*, 1979). Furthermore, activating the pelvic nerves reflexly or by electrical stimulation produces the release of VIP into the venous effluent from the colon (Fahrenkrug *et al.*, 1978). The release of VIP and the vasodilatation can be abolished by giving apamin, a peptide originally isolated from bee venom (Jodal *et al.*, 1983; Sjöqvist *et al.*, 1983). This effect is probably explained by apamin being an agonist for a presynaptic receptor controlling VIP release. Finally, Blank *et al.* (1990) showed that a VIP antagonist markedly attenuated the vascular response to pelvic nerve stimulation.

Tachykinins are found in the vascular innervation of the colon and they are known to relax vascular smooth muscles (Jodal and Lundgren, 1989; Hellström *et al.* 1991). The evidence that SP or some other tachykinin may be involved in the vascular response to pelvic nerve stimulation is, however, conflicting. Andersson *et al.* (1983) failed to demonstrate any release of SP into blood when stimulating the pelvic nerves. Moreover, Sjöqvist *et al.* (1984) could not influence the colonic nervous vasodilatation with a SP receptor blocker. On the other hand, Hellström *et al.* (1991) reported a marked release of neurokinin A (NKA)-like and SP-like immunoreactivity into the colonic venous effluent. The diverging results of Andersson *et al.* (1983) and Hellström *et al.* (1991) may be ascribed to differences in the immunochemical methods used for determining the peptide concentrations. The question of the possible involvement of tachykinins in the vascular effect of pelvic nerve stimulation probably can not be resolved until specific receptor antagonists for SP and/or NKA become available.

3. Functional Implications

It is apparent from the description above that pelvic nerve stimulation elicits a typical response pattern characterized by a colonic hyperemia and a net secretion of mucus. Furthermore, an atropine-resistant colonic contraction is also induced by the nervous stimulation. The pelvic nerve effects can be reflexly elicited. Mechanical stimulation of rectal and sigmoid mucosa induces the full pelvic response pattern described previously. This response may therefore be involved in the act of defecation, a suggestion supported by observations of a transient colonic vasodilatation during defecation in dogs. Furthermore, Grace *et al.* (1951) in studies of colostomized patients observed that "feelings of anger, resentment and hostility or of anxiety and apprehension were found to be associated with hyperfunction of the colon." These emotional effects are probably mediated via the pelvic nerves. The structures of the central nervous system evoking these reactions are not known.

B. Stomach

There are histological observations to suggest the presence of a peptidergic innervation of the gastric vessels. Table 2 indicates that substance P, VIP, and NPY may function as neurotransmitters at the gastric vasculature. NPY is probably contained in adrenergic nerve fibers, but knowledge of the functions of the other neurotransmitters is limited.

1. Hemodynamics

Electrical stimulations of the vagal fibers to the stomach have revealed, at least in the cat, that the vagal nerves may, from a functional and electrophysiological point of view, be divided into two groups of fibers. Stimulating the vagal low-threshold fibers induces no change in either gastric blood flow or gastric secretion, while stimulating the high-threshold fibers increases both variables. Blood flow augmentation during vagal stimulations is always accompanied by a relaxation of the stomach. The administration of atropine reduces the blood flow increase during vagal stimulation to a large extent and completely abolishes the secretory response (Martinson, 1965).

The increase in gastric blood flow elicited by stimulation of the high-threshold vagal fibers probably occurs mainly in the mucosa. At the highest rate of stimulation (10 Hz) mucosal blood flow approximately doubled. Similar results were reported by Swan and Jacobson (1967) in experiments in which vagal fibers were activated reflexly by insulin hypoglycemia in unanesthetized dogs. Increases of mucosal blood flow and stomach secretion occurred but were abolished by atropine. In my laboratory, studies of the distribution of blood flow in the feline stomach were made before and during electrical vagal stimulation. Stimulating the high-threshold fibers at 8 Hz increases blood flow exclusively in the mucosa, approximately 95% of total blood flow being distributed to this part of the gastric wall (Ivarsson and Lundgren, unpublished observations).

With respect to the vagal control of the consecutive vascular sections, the response of the resistance vessels has already been discussed. The effect on the capacitance vessels is not known. Along with the increase in blood flow, the capillary surface area available for exchange also increases, as suggested by an augmented capillary filtration coefficient. These changes, however, may not only reflect alterations in capillary surface area but also changes in capillary pore size. Observations on the stomach indicate that stimulation of the vagal relaxative fibers may indeed alter the size of the capillary pores (Jansson *et al.*, 1970). This effect may be mediated via peptidergic neurons.

2. Neurotransmitter Pharmacology

The analysis of the neurotransmitter pharmacology in the stomach is complicated by the fact that vagal stimulation evokes an increased acid secretion. Secretion *per se* may evoke an increased blood flow due to the metabolic demands of this process. In early studies by Guth and Smith (1975) and Martinson (1965) a very quick vasodilatation was observed upon vagal stimulation (within 10 s), which argues in favor of direct nervous vascular control. On the other hand, Martinson reported that only 10–30% of the control response to vagal stimulation was observed after atropine was administered in a dose that completely abolished the gastric secretory response.

More recently Guth and collaborators (Thiefin *et al.*, 1990) have addressed this question in the rat stomach with new pharmacological tools. Giving omeprazol, a specific inhibitor of acid secretion, completely abolished acid secretion whereas the nervously evoked increase of gastric mucosal blood flow was reduced only by 50%. A major part of this response (about 75%) was abolished by atropine. The remaining 25% was apparently mediated via nerves sensitive to capsaicin, suggesting that these nerve fibers are afferent. The possible physiological importance of these fibers is discussed below.

To conclude, the interesting study by Thiefin *et al.* (1990) indicates that the vasodilatation evoked by vagal nerve stimulation is in part secondary to the increased gastric secretion and in part explained by a direct effect of an innervation of the gastric vasculature. The vascular innervation is cholinergic as well as noncholinergic.

3. Functional Implications

The response pattern (vasodilatation, secretion, gastric relaxation) elicited by the activation of the high-threshold vagal fibers is probably involved in the receptive relaxation originally described by Cannon and Lieb and also in the cephalic phase of gastric secretion. Furthermore, Wolf and Wolff (1947), studying a human subject, noted that persistent feelings of anxiety are associated with increased gastric blood flow and secretion, presumably caused by an increased rate of firing in the vagal nerves. The mechanisms of the central nervous system underlying these reactions are not known.

III. Intramural Nervous Control of the Gastrointestinal Vascular Beds

The gastrointestinal tract is provided with an extensive nervous system of its own, the enteric nervous system (ENS). It represents a major part of the autonomic nervous system containing almost as many neurons as the spinal cord. The number of efferent fibers (sympathetic and parasympathetic) controlling ENS is small. It has been calculated that there is only one efferent fiber per 300 enteric nerve cells. It is therefore not surprising to find that ENS can function independently. Several reflexes confined to the intestinal wall have been described. The most well known is the peristaltic reflex (for reviews of the ENS, consult Furness and Costa, 1987 or Lundgren *et al.* 1989). Enteric nervous reflexes controlling blood vessels are described below.

A. *Intestines*

1. Types of Reflexes

It is well known that blood flow in the small intestine is increased postprandially. Most investigations have shown that splanchnic blood flow approximately doubles after a meal (for a review see Jodal and Lundgren, 1989). From a teleological point of view it seems feasible that the enteric nerves may mediate, in part, such a change of the circulation, which aims at adjusting blood flow to the increasing metabolic demands of the tissue.

There are observations to suggest that this is, indeed, the case, and two types of ENS blood control have been described. Biber *et al.* (1971, 1974) showed that mechanical stimulation of the intestinal mucosa of the cat with, for example, a plastic tube, approximately doubles intestinal blood flow. This response is evoked via the activation of a reflex confined to enteric nervous system since the hyperemia can be blocked totally by tetrodotoxin (TTX), at low concentrations a specific blocker of sodium gates in excitable tissues. The presence of a nervous vasodilator reflex is also indicated by the observation that transmural electrical field stimulation augments intestinal blood flow via a mechanism that can be blocked by TTX (Biber *et al.*, 1973). The reflex vasodilatation is not influenced by acute or chronic "extrinsic" denervation of the intestine (i.e., cutting all nerves surrounding the superior mesenteric artery). The absence of any effect of chronic denervation indicates that this vascular reflex is not arranged as classical axon reflex.

Evidence also exists for a second type of enteric reflex controlling intestinal blood flow. It is probably arranged as an axon reflex. Rozsa *et al.* (1984, 1986) reported that giving capsaicin, a drug that activates thin afferent fibers (C-fibers), intraarterially, or in the lumen, evoked an intestinal vasodilatation in dogs or rats. It is blocked by intraluminal lidocaine but not by hexamethonium. Furthermore, the capsaicin-sensitive afferents are stimulated by luminal solutions containing bile (10%) and oleic acid (40 mM), inferring that this reflex is stimulated by chemicals in the luminal contents.

2. Neurotransmitter Pharmacology

Pharmacological analyses of the reflex originally described by Biber *et al.* (1971) suggest that the neurotransmitter at the effector cell (vascular smooth muscles) is noncholinergic, nonadrenergic. Several observations indicate that the neurotransmitter is VIP. The most convincing evidence in favor of this proposal is that giving apamin, a peptide in bee venom, blocks the flow increase and VIP release seen upon mechanical or electrical field stimulation (Jodal *et al.*, 1983; Sjöqvist *et al.*, 1983). As pointed out earlier, apamin probably acts via a presynaptic VIP receptor controlling VIP release.

5-Hydroxytryptamine (5-HT; serotonin) is also in some way involved in the vascular reflex described by Biber *et al.* (1971) since 5-HT receptor blockers and 5-HT tachyphylaxis inhibit the reflex hyperemia (Biber *et al.*, 1974). Two sites of action seem possible. 5-HT may be a neurotransmitter in the reflex arch, or it may be released from enterochromaffin cells upon mechanical stimulation. In the latter case, Biber and coworkers proposed that the released 5-HT activates dendrites situated just underneath the enterochromaffin cell. These two possible arrangements of the reflex are illustrated in Fig. 2. Evidence for the presence of a similar reflex in the rat gut and in the cat colon has also been reported.

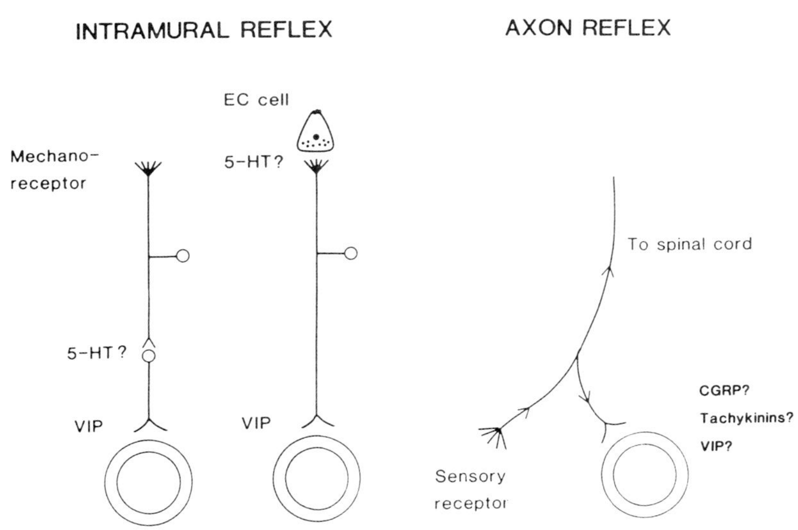

Figure 2 The intrinsic nervous control of intestinal blood vessels schematically illustrated. The left part of the figure illustrates two possible arrangements of the reflex vasodilatation elicited by mechanical stimulation of the intestinal mucosa. The right part of the figure illustrates an axon reflex that evokes an intestinal vasodilatation. This reflex can be activated by luminal chemical stimuli. CGRP, calcitonin gene-related polypeptide; VIP, vasoactive intestinal polypeptide.

The transmitter(s) involved in the vascular axon reflex described by Rozsa *et al.* (1989, 1985, 1986) is not known. Rozsa and collaborators have made several proposals, namely VIP and tachykinins. Calcitonin gene-related peptide (CGRP) is also contained in mesenteric vessels and may be involved in the axon reflex (Kawasaki *et al.*, 1988). The latter peptide has also been demonstrated in mesenteric resistance vessels.

3. Functional Implications

The physiological function of the nervous vascular reflex described by Biber *et al.* (1971, 1973, 1974) is probably to increase blood flow at the site of a food bolus in the intestinal lumen. It also seems possible that the vasodilatation accompanying every migrating myoelectric complex in the conscious dog (Fioramonti and Bueno, 1984) is mediated via the nervous vasodilator reflex reported by Biber *et al.* It is likely that the reflex vasodilatation described by Rozsa *et al.* has a protective function for the intestinal mucosa.

B. Stomach

The study by Thiefin *et al.* (1990) mentioned earlier clearly suggests that the vagal nerves of the rat contain nerve fibers that when stimulated increase blood flow. Part of this response apparently mediated via nerves sensitive to capsaicin treatment. Since this compound selectively destroys thin afferent fibers, Thiefin *et al.* (1990) proposed that part of the vagal vascular control was organized as an axon reflex. The neurotransmitter involved in the reflex is unknown. The axon reflex may have a protective role as suggested by, for example, the study by Holzer *et al.* (1990). They showed that the mucosal damage evoked in the rat stomach by exposing the mucosa to 25% ethanol was significantly attenuated by placing capsaicin in the gastric lumen. This effect was ascribed to the blood flow increase caused by capsaicin.

Acknowledgment

The research performed in my laboratory and reported here was supported by a grant from the Swedish Medical Research Council (2855).

References

Andersson, P.-O., Bloom, S. R., Edwards, A. V., Järhult, J., and Mellander, S. (1983). Neural vasodilator control in the rectum of the cat and its possible mediation by vasoactive intestinal polypeptide. *J. Physiol.* (London) *344*, 49–67.

Biber, B., Lundgren, O., and Svanvik, J. (1971). Studies on the intestinal vasodilatation observed after mechanical stimulation of the mucosa of the gut. *Acta Physiol. Scand. 82*,177–190.

Biber, B., Fara, J., and Lundgren, O. (1973). Intestinal vasodilatation in response to transmural electrical field stimulation. *Acta Physiol. Scand. 87*, 277–282.

Biber, B., Fara, J., and Lundgren, O. (1974). A pharmacological study of intestinal vasodilator mechanisms in the cat. *Acta Physiol. Scand. 90*, 673–683.

Blank, M. A., Kimura, K., Fuortes, M., and Jaffe, B. M. (1990). VIP antagonist {N-Ac-Tyr^1,D-Phe^2}-GRF-(1-29)-NH_2: An inhibitor of vasodilation in the feline colon. *Am. J. Physiol. 259*, G252–G257.

Brunsson, I., Eklund, S., Jodal, M., Lundgren, O., and Sjövall, H. (1979). The effect of vasodilatation and sympathetic nerve activation on net water absorption in the cat's small intestine. *Acta Physiol. Scand. 106*, 61–68.

Eklund, S., Jodal, M., Lundgren, O., and Sjöqvist, A. (1979). Effects of vasoactive intestinal polypeptide on blood flow, motility and fluid transport in the gastrointestinal tract of the cat. *Acta Physiol. Scand. 105,* 461–468.

Fahrenkrug, J., Haglund, U., Jodal, M., Lundgren, O., Olbe, L., and Schaffalitzky de Muckadell, O. B. (1978). Nervous release of vasoactive intestinal polypeptide in the gastrointestinal tract of cats. Possible physiological implications. *J. Physiol.* (London) *284,* 291–305.

Fara, J. W., and Ross, G. (1972). Escape from drug-induced constriction of isolated arterial segments from various vascular beds. *Blood Vessels 9,* 27–33.

Fioramonti, J., and Bueno, L. (1984). Relation between intestinal motility and mesenteric blood flow in the conscious dog. *Am. J. Physiol. 246* (Gastrointest. Liver Physiol.) G108–G113.

Folkow, B., Lewis, D. H., Lundgren, O., Mellander, S., and Wallentin, I. (1964a). The effect of graded vasoconstrictor fibre stimulation on the intestinal resistance and capacitance vessels. *Acta Physiol. Scand. 61,* 445–457.

Folkow, B., Lewis, D. H., Lundgren, O., Mellander, S., and Wallentin, I. (1964b). The effect of the sympathetic vasoconstrictor fibres on the distribution of capillary blood flow in the intestine. *Acta Physiol. Scand. 61,* 458–466.

Furness, J. B., and Costa, M. (1987). "The Eneric Nervous System." Churchill Livingstone, New York.

Grace, W. J., Wolf, S., Wolff, H. G. (1951). "The Human Colon." Hoeber, New York.

Guth, P. H., and Smith, E. (1975a). Neural control of gastric mucosal blood flow in the rat. *Gastroenterology 69,* 935–940.

Guth, P. H., and Smith, E. (1975b). Escape from vasoconstriction in the gastric microcirculation. *Am. J. Physiol. 228,* 1893–1895.

Hellström, P. M., Olerup, O., and Tatemoto, K. (1985). Neuropeptide Y may mediate effects of sympathetic nerve stimulation on colonic motility and blood flow in the cat. *Acta Physiol. Scand. 124,* 613–624.

Hellström, P. M., Söder, O., and Theodorsson, E. (1991). Occurrence, release, and effects of multiple tachykinins in cat colonic tissues and nerves. *Gastroenterology 100,* 431–440.

Holzer, P., Pabst, M. A., Lippe, I. Th., Peskar, B. M., Peskar, B. A., Livingston, E. H., and Guth, P. H. (1990). Afferent nerve-mediated protection against deep mucosal damage in the rat stomach. *Gastroenterology 98,* 838–848.

Hultén, L., Jodal, M., and Lundgren, O. (1969a). Extrinsic nervous control of colonic blood flow. *Acta Physiol. Scand. 335* (Suppl.), 34–49.

Hultén, L., Jodal, M., and Lundgren, O. (1969b). Local and nervous control of the consecutive vascular sections of the colon. *Acta Physiol. Scand.* Suppl. 335, 51–64.

Hultén, L., Jodal, M., and Lundgren, O. (1969c). Nervous control of blood flow in the parallel-coupled vascular sections of the colon. *Acta Physiol. Scand.* Suppl. 335, 65–76.

Hultén, L., Jodal, M., Lindhagen, J., and Lundgren, O. (1976). Colonic blood flow in cat and man as analyzed by an inert gas washout technique. *Gastroenterology 70,* 36–44.

Hultén, L., Lindhagen, J., and Lundgren, O. (1977). Sympathetic nervous control of intramural blood flow in the feline and human intestines. *Gastroenterology 72,* 44–48.

Jansson, G., Lundgren, O., and Martinson, J. (1970). Neurohormonal control of gastric blood flow. *Gastroenterology 58,* 425–429.

Jodal, M., and Lundgren, O. (1989). Neurohormonal control of gastrointestinal blood flow. *In* "Handbook of Physiology. The Gastrointestinal System I." Chapter 46, pp 1667–1711.

Jodal, M., Lundgren, O., and Sjöqvist, A. (1983). The effect of apamin on non-adrenergic, non-cholinergic vasodilator mechanisms in the intestines of the cat. *J. Physiol.* (London) *338,* 207–219.

Kawasaki, H., Takasaki, K., Saito, A., and Goto, K. (1988). Calcitonin gene-related peptide acts as a novel vasodilator neurotransmitter in mesenteric resistance vessels of the rat. *Nature* (London) *335,* 164–167.

Keast, J. R., Furness, J. B., and Costa, M. (1985). Distribution of certain peptide-containing nerve fibres and endocrine cells in the gastrointestinal mucosa in five mammalian species. *J. Comp. Neurol. 236,* 403–422.

Kewenter, J. (1965). The vagal control of the jejunal and ileal motility and blood flow. *Acta Physiol. Scand. Suppl. 251,* 1–68.

Lundberg, J. M. (1981). Evidence for coexistence of vasoactive intestinal polypeptide (VIP) and acetylcholine in neurons of cat exocrine glands. Morphological, biochemical and functional studies. *Acta Physiol. Scand.* Suppl. 496.

Lundberg, J. M., and Tatemoto, K. (1982). Vascular effects of the peptides PYY and PHI: Comparison with APP and VIP. *Eur. J. Pharmacol. 83,* 143–146.

Lundberg, J. M., and Tatemoto, T. (1982). Pancreatic polypeptide family (APP, BPP, NPY and PYY) in relation to sympathetic vasoconstriction resistant to α-adrenoreceptor blockade. *Acta Physiol. Scand. 116,* 393–402.

Lundberg, J. M., Terenius, L., Hökfelt, T., and Goldstein, M. (1983). High levels of neuropeptide Y in peripheral noradrenergic neurons in various mammals including man. *Neurosci. Lett. 42,* 167–172.

Lundgren, O., Svanvik, J., and Jivegård, L. (1989). Enteric nervous system. I. Physiology and pathophysiology of the intestinal tract. *Dig. Dis. and Sci. 34* (2), 264–283.

Martinson, J. (1965). The effect of graded vagal stimulation on gastric motility, secretion and blood flow in the cat. *Acta Physiol. Scand. 65,* 300–309.

Neild, T. O., Shen, K.-Z., and Surprenant, A. (1990). Vasodilatation of arterioles by acetylcholine released from single neurones in the guinea-pig submucosal plexus. *J. Physiol. 420,* 247–265.

Reed, J. D., and Sanders, D. J. (1971). Splanchnic nerve inhibition of gastric acid secretion and mucosal blood flow in anaesthetized cats. *J. Physiol.* (London). *219,* 555–570.

Remak, G., Hottenstein, O. D., and Jacobson, E. D. (1989). Peptidergic nerves are involved in norepinephrine-induced vasoconstriction, but not in escape from norepinephrine. *Gastroenterology 96,* A412.

Ross, G. (1971). Escape of mesenteric vessels from adrenergic and nonadrenergic vasoconstriction. *Am. J. Physiol. 221,* 1217–1222.

Rozsa, Z., Jansco, G., and Varro, V. (1984). Possible involvement of capsaicin sensitive sensory nerves in the regulation of intestinal blood flow in the dog. *Naunyn-Schmiedeberg's Arch Pharmacol. 326,* 353–356.

Rozsa, Z., Varro, V., and Jansco, G. (1985). Use of immunoblockade to study the involvement of peptidergic afferent nerves in the intestinal vasodilatory response to capsaicin in the dog. *Eur. J. Pharmacol. 115,* 59–64.

Rozsa, Z., Sharkey, K. A., and Varro, V. (1986). Evidence for a role of capsaicin-sensitive mucosal afferent nerves in the regulation of mesenteric blood flow in the dog. *Gastroenterology 90,* 906–910.

Sjöblom-Widfeldt, N. (1990). Neuro-muscular transmission in blood vessels: Phasic and tonic components. An *in-vitro* study of mesenteric arteries of the rat. *Acta Physiol Scand. 138* (Suppl.), 587.

Sjöblom-Widfeldt, N., Gustafsson, H., and Nilsson, H. (1990). Transmitter characteristics of small mesenteric arteries from the rat. *Acta Physiol. Scand. 138,* 203–212.

Sjöqvist, A., Fahrenkrug, J., Jodal, M., and Lundgren, O. (1983). Effect of apamin on release of vasoactive intestinal polypeptide (VIP) from the cat intestines. *Acta Physiol. Scand. 119,* 69–76.

Sjöqvist, A., Hellström, P. M., Jodal, M., and Lundgren, O. (1984). Neurotransmitters involved in the colonic contraction and vasodilatation elicited by activation of the pelvic nerves in the cat. *Gastroenterology 86,* 1481–1487.

Sjövall, H., Redfors, S., Hallbäck, D.-A., Eklund, S., Jodal, M., and Lundgren, O. (1983). The effect of splanchnic nerve stimulation on blood flow distribution, villous tissue osmolality and fluid and electrolyte transport in the small intestine of the cat. *Acta Physiol. Scand. 117,* 359–365.

Svanvik, J. (1973). Mucosal hemodynamics in the small intestine of the cat during regional sympathetic vasoconstrictor activation. *Acta Physiol. Scand. 89,* 19–29.

Swan, K. G., and Jacobson, E. D. (1967). Gastric blood flow and secretion in conscious dogs. *Am. J. Physiol. 212,* 891–896.

Thiefin, G., Raybould, H. E., Leung, F. W., Taché, Y., and Guth, P. H. (1990). Capsaicin-sensitive afferent fibers contribute to gastric mucosal blood flow response to electrical vagal stimulation. *Am. J. Physiol. 259* (*Gastrointest. Liver Physiol. 22*), G1037–G1043.

Wolf, S., and Wolff, H. G. (1947). "Human Gastric Function (2nd ed.)." Oxford Univ. Press., New York.

Index

Note: Page numbers printed in boldface type refer to tables or figures